MACHINE ELEMENTS IN MECHANICAL DESIGN

Third Edition

Robert L. Mott, P.E.
University of Dayton

Prentice Hall
Upper Saddle River, New Jersey ▪ Columbus, Ohio

Library of Congress Cataloging-in-Publication Data

Mott, Robert L.

  Machine elements in mechanical design / Robert L.

Mott.—3rd ed.

    p. cm.

  Includes index.

  ISBN 0-13-841446-7 (hardcover)

  1. Machine design.   2. Mechanical movements.

I. Title

TJ230.M68   1999

621.8′15—dc21   98-34439

                                             CIP

Cover art: © Westlight, Inc.

Editor: Stephen Helba

Production Editor: Patricia S. Kelly

Design Coordinator: Karrie M. Converse

Text Designer: Rebecca Bobb

Cover Designer: Dan Eckel

Production Manager: Deidra M. Schwartz

Marketing Manager: Frank Mortimer, Jr.

This book was set in Times Roman and Helvetica by Carlisle Communications, Ltd., and was printed and bound by Courier/Westford, Inc. The cover was printed by Phoenix Color Corp.

Printed in the United States of America

10 9 8 7 6 5 4 3 2

ISBN: 0-13-841446-7

PRENTICE-HALL INTERNATIONAL (UK) LIMITED, LONDON

PRENTICE-HALL OF AUSTRALIA PTY. LIMITED, SYDNEY

PRENTICE-HALL CANADA INC., TORONTO

PRENTICE-HALL HISPANOAMERICANA, S.A., MEXICO

PRENTICE-HALL OF INDIA PRIVATE LIMITED, NEW DELHI

PRENTICE-HALL OF JAPAN, INC., TOKYO

PEARSON EDUCATION ASIA PTE. LTD., SINGAPORE

EDITORA PRENTICE-HALL DO BRASIL, LTDA., RIO DE JANEIRO

To my wife, Marge
our children, Lynné, Robert, Jr., and Stephen
and my Mother and Father

Preface

The objective of this book is to provide the concepts, procedures, data, and decision analysis techniques necessary to design machine elements commonly found in mechanical devices and systems. Students completing a course of study using this book should be able to execute original designs for machine elements and integrate the elements into a system composed of several elements.

This process requires a consideration of the performance requirements of an individual element and of the interfaces between elements as they work together to form a system. For example, a gear must be designed to transmit power at a given speed. The design must specify the number of teeth, pitch, tooth form, face width, pitch diameter, material, and method of heat treatment. But the gear design also affects, and is affected by, the mating gear, the shaft carrying the gear, and the environment in which it is to operate. Furthermore, the shaft must be supported by bearings, which must be contained in a housing. Thus, the designer should keep the complete system in mind while designing each individual element. This book will help the student approach design problems in this way.

This text is designed for those interested in practical mechanical design. The emphasis is on the use of readily available materials and processes and appropriate design approaches to achieve a safe, efficient design. It is assumed that the person using the book will be the designer, that is, the person responsible for determining the configuration of a machine or a part of a machine. Where practical, all design equations, data, and procedures needed to make design decisions are specified.

It is expected that students using this book will have a good background in statics, strength of materials, college algebra, and trigonometry. Helpful, but not required, would be knowledge of kinematics, industrial mechanisms, dynamics, materials, and manufacturing processes.

Among the important features of this book are the following:

1. It is designed to be used at the undergraduate level in a first course in machine design.

2. The large list of topics allows the instructor some choice in the design of the course. The format is also appropriate for a two-course sequence and as a reference for mechanical design project courses.

3. Students should be able to extend their efforts into topics not covered in classroom instruction because explanations of principles are straightforward and include many example problems.

4. The practical presentation of the material leads to feasible design decisions and is useful to practicing designers.

5. The text advocates and demonstrates use of computer spreadsheets in cases requiring long, laborious solution procedures. Using spreadsheets allows the designer to make decisions at several points within the problem while the computer performs all computations; see the chapters on spur gears (Chapter 9), springs (Chapter 19), columns (Chapter 6), shafts (Chapter 12), and shrink fits (Chapter 13).

6. References to other books, standards, and technical papers assist the instructor in presenting alternate approaches or extending the depth of treatment.

7. In addition to the emphasis on original design of machine elements, much of the discussion covers commercially available machine elements and devices, since many design projects require an optimum combination of new, uniquely designed parts and purchased components.

8. For some topics the focus is on aiding the designer in selecting commercially available components, such as rolling contact bearings, flexible couplings, ball screws, electric motors, belt drives, chain drives, clutches, and brakes.

9. Computations and problem solutions use both the International System of Units (SI) and the U.S. Customary System (inch-pound-second) approximately equally. The basic reference for the usage of SI units is Standard E380, *Metric Practice Guide,* issued by the American Society for Testing and Materials (ASTM).

10. Extensive Appendices are included along with detailed tables in many chapters to help the reader to make real design decisions, using only this text.

Features of the Third Edition

While retaining the practical approach of the first and second editions, this new edition includes several changes: new features have been added; data and techniques have been updated; the presentation has been refined; and the chapters have been reorganized into three parts as described below:

■ Each chapter starts with **The Big Picture** (a new feature) to engage the students in a discussion of the many contexts in which the principles in that chapter are used in real, practical design. It draws on the students' own experiences and builds their knowledge of the mechanical design field. It is based on the learning theories that students learn better when they can relate new concepts to past experiences and when they consider the *whole* before tackling the details. It is recommended that instructors use **The Big Picture** to encourage dialogue among the students.

■ Each chapter retains the use of the scenario called **You Are the Designer,** in which a specific design assignment is presented that is relevant to the material in that chapter. This feature, initiated in the second edition, focuses the students' thinking from **The Big Picture** to the details of the chapter.

■ Each chapter also retains the section called **Objectives of This Chapter,** which describes the competencies that the students will acquire as they complete the chapter.

■ The book has been divided into three parts to help students relate chapters to each other and help them envision a cohesive, integrated view of mechanical design.

■ Part I, Chapters 1–6, helps students gain an understanding of design philosophies and build on earlier-learned principles of strength of materials, materials science, and manufacturing processes. An extensive coverage of combined stresses and design for different types of loading provides important extensions from the traditional coverage of strength of materials courses, necessary to virtually all following chapters.

■ Part II, Chapters 7–15, encompasses the design of machine elements that are typically part of a mechanical power transmission: belt drives and chain drives; gears; keys, couplings, and seals; shafts; and rolling contact bearings. But, more importantly, the chapters are tied together to help students see the interrelationships among the individual elements.

- **Chapter 15, Completion of the Design of the Power Transmission,** takes the student through the complete design of a gear-type speed reducer, from conceptualization, decision analysis, and general layout, to design and analysis of the components, tolerances and fits, applied to critical elements, and detail drawings and specifications of the housing and the major components of the reducer. This is a completely new chapter for the third edition.

- Part III, Chapters 16–22, presents methods of analysis and design of several important machine elements that were not pertinent to the design of the power transmission as presented in Part II. These chapters can be covered in any order or can be used as reference material for general design projects. Covered here are plain surface bearings; power screws and ball screws; fasteners; springs; machine frames, bolted connections, and welded joints; electric motors; and clutches and brakes.

- Chapter 23 briefly outlines 30 design projects that instructors can select to give students practice in the application of their mechanical design skills.

My appreciation is extended to all who provided helpful suggestions for the changes to be made from the second edition. I thank the editorial and production staffs of Prentice Hall Publishing Company, those who provided illustrations, and the many users of the book with whom I have had discussions. Special appreciation goes to my colleagues at the University of Dayton, Professors Philip Doepker and David Myszka, who have used the first and second editions of this book and who have offered constructive suggestions for improvement. I also thank those who reviewed the manuscript: Rich Englund, Pennsylvania State University, Erie-Behrend Campus; and Harvey Hoy, Milwaukee School of Engineering. I especially thank my students—past and present—for their encouragement and their positive response to this book.

Contents

Contents

PART I

Principles of Design and Stress Analysis

OBJECTIVES AND CONTENT OF PART I

As you complete the first six chapters of this book, you will gain an understanding of design philosophies, and you will build on earlier-learned principles of strength of materials, materials science, and manufacturing processes. The competencies gained from these chapters are useful throughout the book and in general machine design or product design projects.

Chapter 1: The Nature of Mechanical Design helps you see the big picture of the process of mechanical design. Several examples are shown from different industry sectors: consumer products, manufacturing systems, construction equipment, agricultural equipment, transportation equipment, ships, and space systems. The responsibilities of designers are discussed, along with an illustration of the iterative nature of the design process. Units and conversions complete the chapter.

Chapter 2: Materials in Mechanical Design emphasizes the design properties of materials. Much of this chapter is probably review for you, but it is presented here to emphasize the importance of material selection to the design process and to explain the data for materials presented in the Appendices.

Chapter 3: Stress and Deformation Analysis is a review of the basic principles of stress and deflection analysis. It is essential that you understand the basic concepts summarized here before proceeding with later material. Reviewed are direct tensile, compressive, and shearing stresses; bending stresses; and torsional shear stresses.

Chapter 4: Combined Stresses and Mohr's Circle is important because many general design problems and the design of machine elements covered in later chapters of the book involve combined stresses. You may have covered these topics in a course in strength of materials.

Chapter 5: Design for Different Types of Loading is an in-depth discussion of design factors, fatigue, and many of the details of stress analysis as used in this book.

Chapter 6: Columns discusses the long, slender, axially loaded members that tend to fail by buckling rather than by exceeding the yield, ultimate, or shear stress of the material. Special design and analysis methods are reviewed here.

1

The Nature of Mechanical Design

The Nature of Mechanical Design

Discussion Map

- [ ] To design mechanical components and devices, you must be competent in the design of individual elements that comprise the system.
- [ ] But you must also be able to integrate several components and devices into a coordinated, robust system that meets your customer's needs.

Discover

Think, now, about the many fields in which you can use mechanical design:

What are some of the products of those fields?

What kinds of materials are used in the products?

What are some of the unique features of the products?

How were the components made?

How were the parts of the products assembled?

Consider consumer products, construction equipment, agricultural machinery, manufacturing systems, and transportation systems on the land, in the air, in space, and on and under water.

In this book, you will find the tools to learn the principles of *Machine Elements in Mechanical Design.*

Design of machine elements is an integral part of the larger and more general field of mechanical design. Designers and design engineers create devices or systems to satisfy specific needs. Mechanical devices typically involve moving parts that transmit power and accomplish specific patterns of motion. Mechanical systems are composed of several mechanical devices.

Therefore, to design mechanical devices and systems, you must be competent in the design of individual machine elements that comprise the system. But you must also be able to integrate several components and devices into a coordinated, robust system that meets your customer's needs. From this logic comes the name of this book, *Machine Elements in Mechanical Design.*

Think about the many fields in which you can use mechanical design. Discuss these fields with your instructor and with your colleagues who are studying with you. Talk with people who are doing mechanical design in local industries. Try to visit their companies if possible, or meet designers and design engineers at meetings of professional societies. Consider the following fields where mechanical products are designed and produced.

- *Consumer products:* Household appliances (can openers, food processors, mixers, toasters, vacuum cleaners, clothes washers), lawn mowers, chain saws, power tools, garage door openers, air conditioning systems, and many others. See Figures 1–1 and 1–2 for a few examples of commercially available products.

- *Manufacturing systems:* Material handling devices, conveyors, cranes, transfer devices, industrial robots, machine tools, automated assembly systems, special-purpose processing systems, forklift trucks, and packaging equipment. See Figures 1–3, 1–4, and 1–5.

- *Construction equipment:* Tractors with front-end loaders or backhoes, mobile cranes, power shovels, earthmovers, graders, dump trucks, road pavers, concrete mixers, powered nailers and staplers, compressors, and many others. See Figures 1–5 and 1–6.

- *Agricultural equipment:* Tractors, harvesters (for corn, wheat, tomatoes, cotton, fruit, and many other crops), rakes, hay balers, plows, disc harrows, cultivators, and conveyors. See Figures 1–6, 1–7, and 1–8.

- *Transportation equipment:* (a) Automobiles, trucks, and buses which include hundreds of mechanical devices such as suspension components (springs, shock absorbers, and struts); door and window operators; windshield wiper mechanisms; steering systems; hood and trunk latches and hinges; clutch and braking systems; transmissions; drive shafts; seat adjusters; and numerous parts of the engine systems. (b) Aircraft which include retractable landing gear, flap and rudder actuators, cargo handling devices, seat reclining mechanisms, dozens of latches, structural components, and door operators. See Figures 1–9 and 1–10.

- *Ships:* Winches to haul up the anchor, cargo-handling cranes, rotating radar antennas, rudder steering gear, drive gearing and drive shafts, and the numerous sensors and controls for operating on-board systems.

- *Space systems:* Satellite systems, the space shuttle, and launch systems which contain numerous mechanical systems such as devices to deploy antennas, hatches, docking systems, robotic arms, vibration control devices, devices to secure cargo, positioning devices for instruments, actuators for thrusters, and propulsion systems.

How many examples of mechanical devices and systems can you add to these lists?

What are some of the unique features of the products in these fields?

What kinds of mechanisms are included?

What kinds of materials are used in the products?

How were the components made?

How were the parts assembled into the complete products?

In this book, you will find the tools to learn the principles of *Machine Elements in Mechanical Design.* In the introduction to each chapter, we include a brief scenario called **You Are the Designer.** The purpose of these scenarios is to stimulate your thinking about the material presented in the chapter and to show examples of realistic situations in which you may apply it.

 You Are the Designer

Consider, now, that you are the designer responsible for the design of a new consumer product, such as the band saw for a home workshop shown in Figure 1–1. What kind of technical preparation would you need to complete the design? What steps would you follow? What information would you need? How would you show, by calculation, that the design is safe and that the product will perform its desired function?

The general answers to these questions are presented in this chapter. As you complete the study of this book, you will learn about many design techniques that will aid in your design of a wide variety of machine elements. You will also learn how to integrate several machine elements into a mechanical system by considering the relationships between and among elements.

**1–1
OBJECTIVES OF
THIS CHAPTER**

After completing this chapter, you will be able to:

1. Recognize examples of mechanical systems in which the application of the principles discussed in this book is necessary to complete their design.

2. List what design skills are required to perform competent mechanical design.

3. Describe the importance of integrating individual machine elements into a more comprehensive mechanical system.

4. Write statements of *functions* and *design requirements* for mechanical devices.

FIGURE 1–1 Drill-powered band saw [Courtesy of Black & Decker (U.S.) Inc.]

(*a*) Bandsaw

(*b*) Hand drill powering band saw

(*c*) Parts of a hand drill

FIGURE 1–2 Chain saw (Copyright McCulloch Corporation, Los Angeles, CA)

5. Establish a set of criteria for evaluating proposed designs.

6. Work with appropriate units in mechanical design calculations both in the U.S. Customary Unit System and in SI metric units.

7. Distinguish between *force* and *mass,* and express them properly in both unit systems.

8. Present design calculations in a professional, neat, and orderly manner that can be understood and evaluated by others knowledgeable in the field of machine design.

FIGURE 1–3 Chain conveyor (Richards-Wilcox Manufacturing Company, Aurora, IL)

(*a*) Power chain conveyor

(*b*) Drive system

(c) Conveyor chain

FIGURE 1–4 Automated parts handling device (Pickomatic Systems, Inc., Sterling Heights, MI)

FIGURE 1–5 Industrial crane (Air Technical Industries, Mentor, OH)

FIGURE 1–7 Tractor pulling an implement (J. I. Case, Racine, WI)

FIGURE 1–6 Tractor with a front-end-loader attachment (J. I. Case, Racine, WI)

FIGURE 1–8 Cutaway of a tractor (J. I. Case, Racine, WI)

(*a*) Photograph of installed mechanism

(*b*) Cabin door drive mechanism

FIGURE 1–9   Aircraft door drive mechanism (McDonnell Douglas Corporation, Long Beach, CA)

(*a*) Photograph of installed assembly.

(*b*) Drawing showing landing gear in stowed and deployed positions.

FIGURE 1–10   Aircraft landing gear assembly (McDonnell Douglas Corporation, Long Beach, CA)

**1–2
THE
MECHANICAL
DESIGN
PROCESS**

Mechanical design is not pursued in isolation from other processes and functions that occur in an enterprise that produces a product. Quite the contrary: It is one part of a very comprehensive set of actions that must happen in concert.

In this discussion, the term *product* refers to any device or system that is designed and produced for use by a customer. The customer is often considered to be the person who ultimately buys the product. But in some cases, the customer may be another person in your own company who may use the device you design. It could be a manufacturing department that needs a piece of production machinery. It could be a product testing center that needs a novel performance-evaluation system for some key component of the product. You are advised to think in very broad terms about who the customer is for your mechanical designs.

The chief objective of an enterprise that produces a product is to satisfy the customer. It is essential that you know your customer's desires before beginning a product design.

Many methods are used to determine what the customer wants. One popular method, called *quality function deployment* or *QFD,* seeks (1) to identify all of the features and performance factors that customers desire and (2) to assess the relative importance of these factors. The result of the QFD process is a detailed set of functions and design requirements for the product. (See Reference 7.)

But you do not satisfy customer expectations simply by having a good design for the product. It is important to consider how the design process fits with all of the functions that must happen to deliver a satisfactory product to the customer and to service the product throughout its life cycle. In fact, it is important to consider how the product will be disposed of after it has served its useful life. The total of all such functions that affect the product is sometimes called the *product realization process* or *PRP.* Some of the factors included in PRP are as follows:

- Marketing functions to assess customer requirements
- Research to determine the available technology that can reasonably be used in the product
- Availability of materials and components that can be incorporated into the product
- Product design and development
- Performance testing
- Documentation of the design
- Vendor relationships and purchasing functions
- Consideration of global sourcing of materials and global marketing
- Work-force skills
- Physical plant and facilities available
- Capability of manufacturing systems
- Production planning and control of production systems
- Production support systems and personnel
- Quality systems requirements
- Operation and maintenance of the physical plant
- Distribution systems to get products to the customer
- Sales operations and time schedules
- Cost targets and other competitive issues
- Customer service requirements
- Environmental concerns
- Legal requirements
- Availability of financial capital

Can you add to this list?

You should be able to see that the design of a product is but one part of a very comprehensive process. In this book, we will focus more carefully on the design process itself, but the producibility of your designs must always be considered. This simultaneous consideration of product design and manufacturing process design is often called *concurrent engineering.* Note that this process is a subset of the larger list given above for the product realization process. Other terms used for similar design processes are *integrated product and process design (IPPD)* or *integrated product design (IPD).* (See References 3, 6, 9, 10, and 11.)

1–3
SKILLS NEEDED
IN MECHANICAL
DESIGN

Designers use a wide range of skills and knowledge in their daily work. You should already have completed courses in several supporting technologies, including the following:

Technical drawing and computer-aided design (CAD)

Properties of materials

Manufacturing processes

Statics, dynamics, and strength of materials

Kinematics and mechanisms

Building on these skills, and using this book, you should now be ready to learn design of machine elements and mechanical design.

Designers also draw on knowledge of fluid mechanics, thermodynamics, heat transfer, fluid power, electrical controls, and chemical processing. This book will not discuss these specialized fields; its focus is on mechanical hardware.

1–4
FUNCTIONS
AND DESIGN
REQUIREMENTS

The ultimate objective of mechanical design, of course, is to produce a useful device that is safe, efficient, economical, and practical to manufacture. When beginning the design of a machine or an individual machine element, you should define the *functions* and the *design requirements* for the device clearly and completely.

Statements of *functions* tell what the device is supposed to do. They are often somewhat general, but they should always employ action phrases such as *to transmit, to support,* or *to lift.*

After the functions are defined, a set of *design requirements* is prepared. In contrast to the more general statements of function, design requirements must be detailed and specific, giving quantitative data wherever possible.

Example of Statements of Function and Design Requirements

Consider that you are the designer of a speed reducer that is part of the power transmission train for a small tractor. The tractor's engine operates at a fairly high speed, while the drive for the wheels must rotate more slowly and transmit a higher torque than is available at the output of the engine.

To begin the design process, let us list the functions of the speed reducer. What is it supposed to do? Some answers to this question are as follows:

1. To receive power from the tractor's engine through a rotating shaft.
2. To transmit the power through machine elements that reduce the rotational speed to a desired value.
3. To deliver the power at the lower speed to an output shaft that ultimately drives the wheels of the tractor.

Now the design requirements should be stated. The following list is hypothetical, but if you were on the design team for the tractor, you would be able to identify such requirements from your own experience and ingenuity and/or by consultation with fellow designers, marketing staff, manufacturing engineers, service personnel, suppliers, and customers. The concept of *concurrent engineering* requires that personnel from all of these functions be involved from the earliest stages of design.

1. The reducer must transmit 15.0 hp.

2. The input is from a two-cylinder gasoline engine with a rotational speed of 2 000 rpm.

3. The output delivers the power at a rotational speed in the range of 290 to 295 rpm.

4. A mechanical efficiency of greater than 95% is desirable.

5. The minimum output torque capacity of the reducer should be 3 050 pound-inches (lb·in).

6. The reducer output is connected to the drive shaft for the wheels of a farm tractor. Moderate shock will be encountered.

7. The input and output shafts must be in-line.

8. The reducer is to be fastened to a rigid steel frame of the tractor.

9. Small size is desirable. The reducer must fit in a space no larger than 20 in × 20 in, with a maximum height of 24 in.

10. The tractor is expected to operate 8 hours (h) per day, 5 days per week, with a design life of 10 years.

11. The reducer must be protected from the weather and must be capable of operating anywhere in the United States at temperatures ranging from 0 to 130°F.

12. Flexible couplings will be used on the input and output shafts to prohibit axial and bending loads from being transmitted to the reducer.

13. The production quantity is 10 000 units per year.

14. A moderate cost is critical to successful marketing.

Careful preparation of function statements and design requirements will ensure that the design effort is focused on the desired results. Much time and money can be wasted on designs that, although technically sound, do not meet design requirements. Design requirements should include everything that is needed, but at the same time they should offer ample opportunity for innovation.

1–5 CRITERIA FOR EVALUATING MACHINE DESIGN DECISIONS

Design approaches may have to be adjusted to be compatible with certain industries and markets. For example, devices to be incorporated in aircraft should have low weight, whereas a part for a machine in a manufacturing plant normally does not have weight restrictions. Such considerations can have dramatic effects on the design process.

In approaching a design, the designer should establish criteria that will guide the decision-making processes inherent in any project. Because each design problem has many alternative solutions, each can be evaluated in terms of the list of criteria. There may not be a single best design, but designers should work toward an *optimum* design. That is, the design should maximize benefits and minimize disadvantages.

The following are general criteria for machine design:

Safety

Performance (the degree to which the design meets or exceeds the design objectives)

Reliability (a high probability that the device will reach or exceed its design life)

Ease of manufacture

Ease of service or replacement of components

Ease of operation

Low initial cost

Low operating and maintenance costs

Small size and low weight

Low noise and vibration; smooth operation

Use of readily available materials and purchased components

Prudent use of both uniquely designed parts and commercially available components

Appearance that is attractive and appropriate to the application

You or your coworkers can supply additional criteria and determine the relative importance of the criteria applied in each design. Certainly safety is of paramount concern, and the designer is legally liable if someone is injured because of a design error. You must take into account the reasonably foreseeable uses of the device and ensure the safety of those operating it or those who may be close by. Adequate performance is a primary criterion. The importance of the remaining criteria varies from design to design.

1–6 EXAMPLE OF THE INTEGRATION OF MACHINE ELEMENTS INTO A MECHANICAL DESIGN

Mechanical design is the process of designing and/or selecting mechanical components and putting them together to accomplish a desired function. Of course, machine elements must be compatible, must fit well together, and must perform safely and efficiently. The designer must consider not only the performance of the element being designed at a given time but also the elements with which it must interface.

To illustrate how the design of machine elements must be integrated with a larger mechanical design, let us consider the design of a speed reducer for the small tractor discussed in Section 1–4. Suppose that, to accomplish the speed reduction, you decide to design a double-reduction, spur gear speed reducer. You specify four gears, three shafts, six bearings, and a housing to hold the individual elements in proper relation to each other, as shown in Figure 1–11. You should recognize that you have already made many design decisions by rendering such a sketch. First, you chose *spur gears* rather than helical gears, a worm and wormgear, or bevel gears. In fact, other types of speed reduction devices—belt drives, chain drives, or many others—could be appropriate.

Input shaft
2 000 rpm

Output shaft
292 rpm

End view of first pair of gears

End view of second pair of gears

FIGURE 1–11 Conceptual design for a speed reducer

The arrangement of the gears, the placement of the bearings so that they straddle the gears, and the general configuration of the housing are also design decisions. The design process cannot rationally proceed until these kinds of decisions are made. Notice that the sketch of Figure 1–11 is where *integration* of the elements into a whole design begins. When the overall design is conceptualized, the design of the individual machine elements in the speed reducer can proceed. As each element is discussed, scan the relevant chapters of the book. Part II of this book, including Chapters 7–15, provides details for the elements of the reducer. In the chapters on gears, for example, you can see how some manufacturers have designed gear-type speed reducers (see Chapter 9).

Gears

For the gear pairs, you must specify the number of teeth in each gear, the pitch (size) of the teeth, the pitch diameters, the face width, and the material and its heat treatment. These specifications depend on considerations of strength and wear of the gear teeth and the motion requirements (kinematics). You must also recognize that the gears must be mounted on shafts in a manner that ensures proper location of the gears, adequate torque transmitting capability from the gears to the shafts (as through keys), and safe shaft design.

Shafts

Having designed the gear pairs, you next consider the shaft design (Chapter 12). The shaft is loaded in bending and torsion because of the forces acting at the gear teeth. Thus, its design must consider strength and rigidity, and it must permit the mounting of the gears and bearings. Shafts of varying diameters may be used to provide shoulders against which to seat the gears and bearings. There may be keyseats cut into the shaft (Chapter 11). The input and output shafts will extend beyond the housing to permit coupling with the engine and the drive axle. The type of coupling must be considered, as it can have a dramatic effect on the shaft stress analysis (Chapter 11). Seals on the input and output shafts protect internal components (Chapter 11).

Bearings

Design of the bearings (Chapter 14) is next. If rolling contact bearings are to be used, you will probably select commercially available bearings from a manufacturer's catalog, rather than design a unique one. You must first determine the magnitude of the loads on each bearing from the shaft analysis and the gear designs. The rotational speed and reasonable design life of the bearings and their compatibility with the shaft on which they are to be mounted must also be considered. For example, on the basis of the shaft analysis, you could specify the minimum allowable diameter at each bearing seat location to ensure safe stress levels. The bearing selected to support a particular part of the shaft, then, must have a bore (inside diameter) no smaller than the safe diameter of the shaft. Of course, the bearing should not be grossly larger than necessary. When a specific bearing is selected, the diameter of the shaft at the bearing seat location and allowable tolerances must be specified, according to the bearing manufacturer's recommendations, to achieve proper operation and life expectancy of the bearing.

Keys

Now the keys (Chapter 11) and the keyseats can be designed. The diameter of the shaft at the key determines the key's basic size (width and height). The torque that must be transmitted is used in strength calculations to specify key length and material. Once the working components are designed, the housing design can begin.

Housing

The housing design process must be both creative and practical. What provisions should be made to mount the bearings accurately and to transmit the bearing loads safely through the case to the structure on which the speed reducer is mounted? How will the various elements be assembled into the housing? How will the gears and bearings be lubricated? What housing material should be used? Should the housing be a casting, a weldment, or an assembly of machined parts?

The design process as outlined here implies that the design can progress in sequence: from the gears to the shafts, to the bearings, to the keys and couplings, and finally to the housing. It would be rare, however, to follow this logical path only once for a given design. Usually the designer must go back many times to adjust the design of certain components affected by changes in other components. This process, called *iteration,* continues until an acceptable overall design is achieved. Frequently prototypes are developed and tested during iteration.

Chapter 15 shows how all of the machine elements are finally integrated into a unit.

1–7
COMPUTATIONAL
AIDS IN THIS
BOOK

Because of the usual need for several iterations, and because many of the design procedures require long, complex calculations, spreadsheets, mathematical analysis software, computer programs, or programmable calculators are often useful in performing the design analysis. *Interactive* spreadsheets or programs allow you, the designer, to make design decisions during the design process. In this way, many trials can be made in a short time, and the effects of changing various parameters can be investigated. Spreadsheets are used most frequently as examples in this book for computer-aided design and analysis calculations.

1–8
DESIGN
CALCULATIONS

As you study this book and as you progress in your career as a designer, you will make many design calculations. It is important to record the calculations neatly, completely, and in an orderly fashion. You may have to explain to others how you approached the design, which data you used, and which assumptions and judgments you made. In some cases, someone else will actually check your work when you are not there to comment on it or to answer questions. Also, an accurate record of your design calculations is often useful if changes in design are likely. In all of these situations, you are going to be asked to communicate your design to someone else in written and graphic form.

To prepare a careful design record, you will usually take the following steps:

1. Identify the machine element being designed and the nature of the design calculation.
2. Draw a sketch of the element, showing all features that affect performance or stress analysis.
3. Show in a sketch the forces acting on the element (the free-body diagram), and provide other drawings to clarify the actual physical situation.
4. Identify the kind of analysis to be performed, such as stress due to bending, deflection of a beam, buckling of a column, and so on.

   5. List all given data and assumptions.

   6. Write the formulas to be used in symbol form, and clearly indicate the values and units of the variables involved. If a formula is not well known to a potential reader of your work, give the source. The reader may want to refer to it to evaluate the appropriateness of the formula.

   7. Solve each formula for the desired variable.

   8. Insert data, check units, and perform computations.

   9. Judge the reasonableness of the result.

   10. If the result is not reasonable, change the design decisions and recompute. Perhaps a different geometry or material would be more appropriate.

   11. When a reasonable, satisfactory result has been achieved, specify the final values for all important design parameters, using standard sizes, convenient dimensions, readily available materials, and so on.

   Figure 1–12 shows a sample design calculation. A beam is to be designed to span a 60-in pit to support a large gear weighing 2 050 pounds (lb). The design assumes that a rectangular shape is to be used for the cross section of the beam. Other practical shapes could have been used. The objective is to compute the required dimensions of the cross section, considering both stress and deflection. A material for the beam is also chosen. Refer to Chapter 3 for a review of stress due to bending.

1–9 PREFERRED BASIC SIZES, SCREW THREADS, AND STANDARD SHAPES

One responsibility of a designer is to specify the final dimensions for load-carrying members. After completing the analyses for stress and deformation (strain), the designer will know the minimum acceptable values for dimensions that will ensure that the member will meet performance requirements. The designer then typically specifies the final dimensions to be standard or convenient values that will facilitate the purchase of materials and the manufacture of the parts. This section presents some guides to aid in these decisions and specifications.

Preferred Basic Sizes

Table A2–1 lists preferred basic sizes for fractional-inch, decimal-inch, and metric sizes.[1] You should choose one of these preferred sizes as the final part of your design. An example is at the end of the sample design calculation shown in Figure 1–12. You may, of course, specify another size if there is a sound functional reason.

American Standard Screw Threads

Threaded fasteners and machine elements having threaded connections are manufactured according to standard dimensions to ensure interchangeability of parts and to permit convenient manufacture with standard machines and tooling. Table A2–2 gives the dimensions of American Standard Unified threads. Sizes smaller than 1/4 in are given numbers from

---

[1]Throughout this book, some references to tables and figures have the letter *A* included in their numbers; these tables and figures are in the Appendices in the back of the book. For example, Table A2–1 is the first table in Appendix 2; Figure A15–4 is the fourth figure in Appendix 15. These tables and figures are clearly identified in their captions in the Appendices.

R. L. MOTT

DESIGN OF A BAR TO SUPPORT A GEAR IN A SOAKING PIT

BAR IS TO BE 60 IN LONG BETWEEN SUPPORTS
GEAR WEIGHT 2050 LB
HANGERS TO BE 24 IN APART

BAR IS A BEAM IN BENDING

① $\sigma \approx M/S$

ASSUME A RECTANGULAR SHAPE

S = SECTION MODULUS
$S = th^2/6$
LET $h \approx 3t$
THEN $S = t(3t)^2/6 = 9t^3/6$
$S = 1.5\,t^3$
② REQUIRED $t = \sqrt[3]{S/1.5}$

TRY AISI 1040 HR STEEL BAR
$S_y = 42000$ PSI (YIELD STRENGTH)

LET $\sigma = \sigma_d = S_y/N$ = DESIGN STRESS
N = DESIGN FACTOR
LET N = 2 (DEAD LOAD)
$\sigma_d = 42000/2 = 21000$ PSI

THEN FROM ① : $S = M/\sigma_d$ = REQUIRED SECTION MODULUS

$S = \dfrac{18450 \text{ LB·IN}}{21000 \text{ LB/IN}^2} = 0.879 \text{ IN}^3$

FROM ②
$t = \sqrt[3]{S/1.5} = \sqrt[3]{0.879 \text{ IN}^3/1.5} = 0.837 \text{ IN}$
THEN $h = 3t = 3(0.837 \text{ IN}) = 2.51 \text{ IN}$

SUPPLIER HAS $3/4 \times 2\,3/4$ AVAILABLE   [$h/t = 2.75/0.75 = 3.67$ ok]
CHECK $S = th^2/6 = (0.75 \text{ IN})(2.75 \text{ IN})^2/6 = 0.945 \text{ IN}^3 > 0.837 \text{ IN}^3$ ok
$\sigma = M/S = 18450 \text{ LB· IN}/0.945 \text{ IN}^3 = \underline{19500 \text{ PSI}}$
$N = S_y/\sigma = 42000 \text{ PSI}/19500 \text{ PSI} = \underline{2.15}$ ok

CHECK DEFLECTION AT CENTER: $y = \dfrac{Wa}{24\,EI}\ (3\ell^2 - 4a^2)$ $\left(\dfrac{\text{REF}}{\substack{\text{MACHINERY'S HANDBOOK} \\ \text{25th ED., P. 234, CASE 4}}}\right)$

$y = \dfrac{(1025)(18)[3(60)^2 - 4(18)^2]}{24(30 \times 10^6)(1.30)} = 0.187 \text{ IN}$ $I = th^3/12 = \dfrac{(0.75)(2.75)^3}{12} = 1.30 \text{ IN}^4$
ACCEPTABLE

SPECIFY: $3/4 \times 2\,3/4$ RECTANGULAR STEEL BAR. AISI 1040 HR

LOADS

w   w
1025 LB   1025 LB
18   24   18

1025 LB   1025 LB

SHEAR (LB)
1025
0
−1025
18450

BENDING MOMENT (LB–IN)
0

FIGURE 1–12 Sample design calculation

0 to 12, while fractional-inch sizes are specified for 1/4 in and larger sizes. Two series are listed: UNC is the designation for coarse threads, and UNF designates fine threads. Standard designations are as follows:

6–32 UNC   (number size 6, 32 threads per inch, coarse thread)

12–28 UNF   (number size 12, 28 threads per inch, fine thread)

$\frac{1}{2}$–13 UNC   (fractional size 1/2 in, 13 threads per inch, coarse thread)

$1\frac{1}{2}$–12 UNF   (fractional size $1\frac{1}{2}$ in, 12 threads per inch, fine thread)

Given in the tables are the basic major diameter (D), the number of threads per inch (n), and the tensile stress area (A_t), found from

**Tensile Stress
Area for Threads**

$$A_t = 0.7854 \left(D - \frac{0.9743}{n} \right)^2 \qquad \textbf{(1–1)}$$

When a threaded member is subjected to direct tension, the tensile stress area is used to compute the average tensile stress. It corresponds to the smallest area that would be produced by a transverse cut across the threaded rod. For convenience, some standards use the root area or the gross area and adjust the value of the allowable stress.

Metric Screw Threads

Table A2–3 gives similar dimensions for metric threads. Standard metric thread designations are of the form

$$M10 \times 1.5$$

where M stands for metric
> The following number is the basic major diameter in mm
> The last number is the pitch between adjacent threads in mm

Thus, the designation above would denote a metric thread with a basic major diameter of 10.0 mm and a pitch of 1.5 mm. Note that pitch = $1/n$.

Steel Structural Shapes

Steel manufacturers provide a large array of standard structural shapes that are efficient in the use of material and that are convenient for specification and installation into building structures or machine frames. Included, as shown in Table 1–1, are standard angles (L-shapes), channels (C-shapes), wide-flange beams (W-shapes), American Standard beams (S-shapes), structural tubing, and pipe. Note that the W-shapes and the S-shapes are often referred to in general conversation as "I-beams" because the shape of the cross section looks like the capital letter I.

Appendix 16 gives geometric properties of selected steel structural shapes that cover a fairly wide range of sizes. Note that many more sizes are available as presented in Reference 2. The tables in Appendix 16 give data for the area of the cross section (A), the weight per foot of length, the location of the centroid of the cross section, the moment of inertia (I), the section modulus (S), and the radius of gyration (r). The values of I and S are important in the analysis and design of beams. For column analysis, I and r are needed.

Materials used for structural shapes are typically called *structural steels,* and their characteristics and properties are described more fully in Chapter 2. Refer to Appendix 7 for typical strength data. Rolled structural shapes (W, S, C, and L) are most readily available in ASTM A36, ASTM A572, and ASTM A588. Hollow structural sections (round, square, and rectangular) are most readily available in ASTM A500. Other grades can be used in certain types and sizes.

Steel Angles (L-Shapes)

Table A16–1 shows sketches of the typical shapes of steel angles having equal or unequal leg lengths. Called *L-shapes* because of the appearance of the cross section, angles are often

TABLE 1–1 Designations for steel and aluminum shapes

Name of shape	Shape	Symbol	Example designation and Appendix table
Angle		L	$L4 \times 3 \times \frac{1}{2}$ Table A16–1
Channel		C	$C15 \times 50$ Table A16–2
Wide-flange beam		W	$W14 \times 43$ Table A16–3
American Standard beam		S	$S10 \times 35$ Table A16–4
Structural tubing—square			$4 \times 4 \times \frac{1}{4}$ Table A16–5
Structural tubing—rectangular			$6 \times 4 \times \frac{1}{4}$ Table A16–5
Pipe			4-inch standard weight 4-inch Schedule 40 Table A16–6
Aluminum Association channel		C	$C4 \times 1.738$ Table A17–1
Aluminum Association I-beam		I	$I8 \times 6.181$ Table A17–2

used as tension members of trusses and towers, framing members for machine structures, lintels over windows and doors in construction, stiffeners for large plates used in housings and beams, brackets, and ledge-type supports for equipment. Some refer to these shapes as "angle iron." The standard designation takes the following form, using one example size:

$$L4 \times 3 \times \tfrac{1}{2}$$

where L refers to the L-shape
     4 is the length of the longer leg
     3 is the length of the shorter leg
     $\tfrac{1}{2}$ is the thickness of the legs
     Dimensions are in inches

American Standard Channels (C-Shapes)

See Table A16–2 for the appearance of channels and their geometric properties. Channels are used in applications similar to those described for angles. The flat web and the two flanges provide a generally stiffer shape than angles.

     The sketch at the top of the table shows that channels have tapered flanges and webs with constant thickness. The slope of the flange taper is approximately 2 inches in 12 inches, and this makes it difficult to attach other members to the flanges. Special tapered washers are available to facilitate fastening. Note the designation of the *x*- and *y*-axes in the sketch, defined with the web of the channel vertical which gives it the characteristic C-shape. This is most important when using channels as beams or columns. The *x*-axis is located on the horizontal axis of symmetry, while the dimension *x,* given in the table, locates the *y*-axis relative to the back of the web. The centroid is at the intersection of the *x*- and *y*-axes.

     The form of the standard designation for channels is

$$C15 \times 50$$

where C indicates that it is a standard C-shape
     15 is the nominal (and actual) depth in inches with the web vertical
     50 is the weight per unit length in lb/ft

Wide-Flange Shapes (W-Shapes)

Refer to Table A16–3, which illustrates the most common shape used for beams. W-shapes have relatively thin webs and somewhat thicker, flat flanges with constant thickness. Most of the area of the cross section is in the flanges, farthest away from the horizontal centroidal axis (*x*-axis), thus making the moment of inertia very high for a given amount of material. Note that the properties of moment of inertia and section modulus are very much higher with respect to the *x*-axis than they are for the *y*-axis. Therefore, W-shapes are typically used in the orientation shown in the sketch in Table A16-3. Also, these shapes are best when used in pure bending without twisting because they are quite flexible in torsion.

     The standard designation for W-shapes carries much information. Consider the following example:

$$W14 \times 43$$

where W indicates that it is a W-shape
     14 is the nominal depth in inches
     43 is the weight per unit length in lb/ft

The term depth is the standard designation for the vertical height of the cross section when placed in the orientation shown in Table A16–3. Note from the data in the table that the actual depth is often different from the nominal depth. For the W14 3 43, the actual depth is 13.66 in.

American Standard Beams (S-Shapes)

Table A16–4 shows the properties for S-shapes. Much of the discussion given for W-shapes applies to S-shapes as well. Note that, again, the weight per foot of length is included in the designation such as the S10 $\times$ 35 which weighs 35 lb/ft. For most, but not all, of the S-shapes, the actual depth is the same as the nominal depth. The flanges of the S-shapes are tapered at a slope of approximately 2 inches in 12 inches similar to the flanges of the C-shapes. The x- and y-axes are defined as shown with the web vertical.

Often wide-flange shapes (W-shapes) are preferred over S-shapes because of their relatively wide flanges, the constant thickness of the flanges, and the generally higher section properties for a given weight and depth.

Structural Tubing (Square and Rectangular)

See Table A16–5 for the appearance and properties for steel structural tubing. These shapes are usually formed from flat sheet and welded along the length. The section properties account for the corner radii. Note the sketches showing the x- and y-axes. The standard designation takes the form

$$6 \times 4 \times \tfrac{1}{4}$$

where 6 is the depth of the longer side in inches
  4 is the width of the shorter side in inches
  $\tfrac{1}{4}$ is the wall thickness in inches

Square tubing and rectangular tubing are very useful in machine structures because they provide good section properties for members loaded as beams in bending and for torsional loading (twisting) because of the closed cross section. The flat sides often facilitate fastening of members together or the attachment of equipment to the structural members. Some frames are welded into an integral unit that functions as a stiff space-frame. Square tubing makes an efficient section for columns.

Pipe

Hollow circular sections, commonly called *pipe,* are very efficient for use as beams, torsion members, and columns. The placement of the material uniformly away from the center of the pipe enhances the moment of inertia for a given amount of material and gives the pipe uniform properties with respect to all axes through the center of the cross section. The closed cross-sectional shape gives it high strength and stiffness in torsion as well as in bending.

Table A16–6 gives the properties for American National Standard Schedule 40 welded and seamless wrought steel pipe. This type of pipe is often used to transport water and other fluids, but it also performs well in structural applications. Note that the actual inside and outside diameters are somewhat different from the nominal size, except for the very large sizes. Construction pipe is often called *Standard Weight Pipe,* and it has the same dimensions as the Schedule 40 pipe for sizes from 1/2 in to 10 in. Other "schedules" and "weights" of pipe are available with larger and smaller wall thicknesses.

Other hollow circular sections are commonly available that are referred to as *tubing*. These sections are available in carbon steel, stainless steel, aluminum, copper, brass, titanium, and other materials. See References 1, 2, 5, and 8 for a variety of types and sizes of pipe and tubing.

Aluminum Association Standard Channels and I-Beams

Tables A17–1 and A17–2 give the dimensions and section properties of channels and I-beams developed by the Aluminum Association (see Reference 1). These are extruded shapes having uniform thicknesses of the webs and flanges with generous radii where they meet. The proportions of these sections are somewhat different from those of the rolled steel sections described earlier. The extruded form offers advantages in the efficient use of material and in the joining of members. This book will use the following forms for the designation of aluminum sections:

$$C4 \times 1.738 \quad \text{or} \quad I8 \times 6.181$$

where C or I indicates the basic section shape
      4 or 8 indicates the depth of the shape when in the orientation shown
      1.738 or 6.181 indicates the weight per unit length in lb/ft

1–10
UNIT SYSTEMS

We will perform computations in this book by using either the U.S. Customary Unit System (inch-pound-second) or the International System (SI). Table 1–2 lists the typical units used in the study of machine design. *SI,* the abbreviation for "Le Système International d'Unités," is the standard for metric units throughout the world. (See Reference 4.) For convenience, the term *SI units* will be used instead of *metric units.*

Prefixes applied to the basic units indicate order of magnitude. Only those prefixes listed in Table 1–3, which differ by a factor of 1 000, should be used in technical calculations. The final result for a quantity should be reported as a number between 0.1 and 10 000, times some multiple of 1 000. Then the unit with the appropriate prefix should be specified. Table 1–4 lists examples of proper SI notation.

Sometimes you have to convert a unit from one system to another. Appendix 18 provides tables of conversion factors. Also, you should be familiar with the typical order of magnitude of the quantities encountered in machine design so that you can judge the reasonableness of design calculations (see Table 1–5 for several examples).

Example Problem 1–1

Express the diameter of a shaft in millimeters if it is measured to be 2.755 in.

Solution

Table A18 gives the conversion factor for length to be 1.00 in = 25.4 mm. Then

$$\text{Diameter} = 2.755 \text{ in} \frac{25.4 \text{ mm}}{1.00 \text{ in}} = 69.98 \text{ mm}$$

Example Problem 1–2

An electric motor is rotating at 1 750 revolutions per minute (rpm). Express the speed in radians per second (rad/s).

Solution

A series of conversions is required.

$$\text{Rotational speed} = \frac{1\ 750 \text{ rev}}{\text{min}} \frac{2\pi \text{ rad}}{\text{rev}} \frac{1 \text{ min}}{60 \text{ s}} = 183.3 \text{ rad/s}$$

TABLE 1–2 Typical units used in machine design

Quantity	U.S. Customary unit	SI unit
Length or distance	inch (in)	meter (m)
	foot (ft)	millimeter (mm)
Area	square inch (in^2)	square meter (m^2) or square millimeter (mm^2)
Force	pound (lb)	newton (N)
	kip (K) (1 000 lb)	(1 N = 1 kg·m/s^2)
Mass	slug (lb·s^2/ft)	kilogram (kg)
Time	second (s)	second (s)
Angle	degree (°)	radian (rad) or degree (°)
Temperature	degrees Fahrenheit (°F)	degrees Celsius (°C)
Torque or moment	pound-inch (lb·in) or	newton-meter (N·m)
	pound-foot (lb·ft)	
Energy or work	pound-inch (lb·in)	joule (J)
		(1 J = 1 N·m)
Power	horsepower (hp)	watt (W) or kilowatts (kW)
	(1 hp = 550 lb·ft/s)	(1 W = 1 J/s = 1 N·m/s)
Stress, pressure, or	pounds per square inch	pascal (Pa) (1 Pa = 1 N/m^2)
modulus of elasticity	(lb/in^2, or psi)	kilopascal (kPa) (1 kPa = 10^3 Pa)
	kips per square inch	megapascal (MPa) (1 MPa = 10^6 Pa)
	(K/in^2, or Ksi)	gigapascal (GPa) (1 GPa = 10^9 Pa)
Section modulus	inches cubed (in^3)	meters cubed (m^3) or millimeters cubed (mm^3)
Moment of inertia	inches to the fourth	meters to the fourth power (m^4) or
	power (in^4)	millimeters to the fourth power (mm^4)
Rotational speed	revolutions per min (rpm)	radians per second (rad/s)

TABLE 1–3 Prefixes used with SI units

Prefix	SI symbol	Factor
micro-	μ	10^{-6} = 0.000 001
milli-	m	10^{-3} = 0.001
kilo-	k	10^3 = 1 000
mega-	M	10^6 = 1 000 000
giga-	G	10^9 = 1 000 000 000

TABLE 1–4 Quantities expressed in SI units

Computed result	Reported Result
0.001 65 m	1.65 × 10^{-3} m, or 1.65 mm
32 540 N	32.54 × 10^3 N, or 32.54 kN
1.583 × 10^5 W	158.3 × 10^3 W, or 158.3 kW;
	or 0.158 3 × 10^6 W; or 0.158 3 MW
2.07 × 10^{11} Pa	207 × 10^9 Pa, or 207 GPa

TABLE 1–5 Typical order of magnitude for commonly encountered quantities

Quantity	U.S. Customary unit	SI unit
Dimensions of a wood standard 2 × 4	1.50 in × 3.50 in	38 mm × 89 mm
Moment of inertia of a 2 × 4 (3.50-in side vertical)	5.36 in^4	2.23 × 10^6 mm^4, or 2.23 × 10^{-6} m^4
Section modulus of a 2 × 4 (3.50-in side vertical)	3.06 in^3	5.02 × 10^4 mm^3, or 5.02 × 10^{-5} m^3
Force required to lift 1.0 gal of gasoline	6.01 lb	26.7 N
Density of water	1.94 slugs/ft^3	1 000 kg/m^3, or 1.0 Mg/m^3
Compressed air pressure in a factory	100 psi	690 kPa
Yield point of AISI 1040 hot-rolled steel	42 000 psi, or 42 Ksi	290 MPa
Modulus of elasticity of steel	30 000 000 psi, or 30 × 10^6 psi	207 GPa

**1–11
DISTINCTION
AMONG WEIGHT,
FORCE, AND
MASS**
Distinction must be made among the terms *force, mass,* and *weight. Mass* is the quantity of matter in a body. A *force* is a push or pull applied to a body that results in a change in the body's motion or in some deformation of the body. Clearly these are two different physical phenomena, but the distinction is not always understood. The units for force and mass used in this text are listed in Table 1–2.

The term *weight,* as used in this book, refers to the amount of *force* required to support a body against the influence of gravity. Thus, in response to "What is the weight of 75 kg of steel?" we would use the relationship between force and mass from physics:

⇨ **Weight/Mass
Relationship**

$$F = ma \quad \text{or} \quad w = mg$$

where F = force
$\quad m$ = mass
$\quad a$ = acceleration
$\quad w$ = weight
$\quad g$ = acceleration due to gravity

We will use

$$g = 32.2 \text{ ft/s}^2 \quad \text{or} \quad g = 9.81 \text{ m/s}^2$$

Then, to compute the weight,

$$w = mg = 75 \text{ kg}(9.81 \text{ m/s}^2)$$
$$w = 736 \text{ kg·m/s}^2 = 736 \text{ N}$$

Remember that, as shown in Table 1–2, the newton (N) is equivalent to 1.0 kg·m/s^2. In fact, the newton is defined as the force required to give a mass of 1.0 kg an acceleration of 1.0 m/s^2. In our example, then, we would say that the 75-kg mass of steel has a weight of 736 N.

REFERENCES

1. Aluminum Association. *Aluminum Standards and Data.* Washington, DC: Aluminum Association, 1997.

2. American Institute of Steel Construction. *Manual of Steel Construction,* 9th ed. Chicago: American Institute of Steel Construction, 1989.

3. American Society of Mechanical Engineers. *Integrating the Product Realization Process (PRP) into the Undergraduate Curriculum.* New York: American Society of Mechanical Engineers, 1995.

4. American Society for Testing and Materials. *IEEE/ASTM SI-10 Standard for Use of the International System of Units (SI): The Modern Metric System.* West Conshohocken, PA: American Society for Testing and Materials, 1997.

5. Avallone, Eugene A., and Theodore Baumeister III, eds. *Marks' Standard Handbook for Mechanical Engineers.* 10th ed. New York: McGraw-Hill, 1996.

6. Ertas, Atila, and Jesse C. Jones. *The Engineering Design Process.* New York: John Wiley & Sons, 1993. Discussion of the design process from definition of design objectives through product certification and manufacture.

7. Hauser, J., and D. Clausing. "The House of Quality." *Harvard Business Review* (May–June 1988): 63–73. Discusses Quality Function Deployment.

8. Mott, Robert L. *Applied Fluid Mechanics.* 4th ed. Upper Saddle River, NJ: Prentice-Hall, 1994.

9. National Research Council. *Improving Engineering Design: Designing for Competitive Advantage.* Washington, DC: National Academy Press, 1991. Describes the Product Realization Process (PRP).

10. Pugh, Stuart. *Total Design: Integrated Methods for Successful Product Engineering.* Reading, MA: Addison-Wesley, 1991.

11. Ullman, David G. *The Mechanical Design Process.* 2d ed. New York: McGraw-Hill, 1997.

PROBLEMS

Functions and Design Requirements

For the devices described in Problems 1–14, write a set of functions and design requirements in a similar manner to those in Section 1–4. You or your instructor may add more specific information to the general descriptions given.

1. The hood latch for an automobile

2. A hydraulic jack used for car repair

3. A portable crane to be used in small garages and homes

4. A machine to crush soft-drink or beer cans

5. An automatic transfer device for a production line

6. A device to raise a 55-gallon (gal) drum of bulk materials and dump the contents into a hopper

7. A paper feed device for a copier

8. A conveyor to elevate and load gravel into a truck

9. A crane to lift building materials from the ground to the top of a building during construction

10. A machine to insert toothpaste tubes into cartons

11. A machine to insert 24 cartons of toothpaste into a shipping container

12. A gripper for a robot to grasp a spare tire assembly and insert it into the trunk of an automobile on an assembly line

13. A table for positioning a weldment in relation to a robotic welder

14. A garage door opener

Units and Conversions

For Problems 15–28, perform the indicated conversion of units. (Refer to Appendix 18 for conversion factors.) Express the results with the appropriate prefix as illustrated in Tables 1–3 and 1–4.

15. Convert a shaft diameter of 1.75 in to mm.

16. Convert the length of a conveyor from 46 ft to meters.

17. Convert the torque developed by a motor of 12 550 lb·in to N·m.

18. A wide-flange steel-beam shape, W12 × 14, has a cross-sectional area of 4.12 in^2. Convert the area to mm^2.

19. The W12 × 14 beam shape has a section modulus of 14.8 in^3. Convert it to mm^3.

20. The W12 × 14 beam shape has a moment of inertia of 88.0 in^4. Convert it to mm^4.

21. What standard steel equal leg angle would have a cross-sectional area closest to (but greater than) 750 mm^2? See Table A16–1.

22. An electric motor is rated at 7.5 hp. What is its rating in watts (W)?

23. A vendor lists the ultimate tensile strength of a steel to be 127 000 psi. Compute the strength in MPa.

24. Compute the weight of a steel shaft, 35.0 mm in diameter and 675 mm long. (See Appendix 3 for the density of steel.)

25. A torsional spring requires a torque of 180 lb·in to rotate it 35°. Convert the torque to N·m and the rotation to radians. If the *scale of the spring* is defined as the applied torque per unit of angular rotation, compute the spring scale in both unit systems.

26. To compute the energy used by a motor, multiply the power that it draws by the time of operation. Consider a motor that draws 12.5 hp for 16 h/day, five days per week. Compute the energy used by the motor for one year. Express the result in ft·lb and W·h.

27. One unit used for fluid viscosity in Chapter 16 of this book is the *reyn*, defined as 1.0 lb·s/in^2. If a lubricating oil has a viscosity of 3.75 reyn, convert the viscosity to the standard units in the U.S. Customary System (lb·s/ft^2) and in the SI (N·s/m^2).

28. The life of a bearing supporting a rotating shaft is expressed in number of revolutions. Compute the life of a bearing that rotates 1 750 rpm continuously for 24 h/day for five years.

2

Materials in Mechanical Design

Materials in Mechanical Design

Discussion Map

☐ You must understand the behavior of materials to make good design decisions and to communicate with suppliers and manufacturing staff.

Discover

Examine consumer products, industrial machinery, automobiles, and construction machinery.

What materials are used for the various parts?

Why do you think those materials were specified?

How were they processed?

What material properties were important to the decisions to use particular materials?

Examine the Appendices tables, and refer to them later as you read about specific materials.

This chapter summarizes the design properties of a variety of materials. The Appendices include data for many examples of these materials in many conditions.

Satisfactory performance of machine components and systems depends greatly on the materials that the designer specifies. As a designer, you must understand how materials behave, what properties of the material affect the performance of the parts, and how you should interpret the large amounts of data available on material properties. Your ability to effectively communicate your specifications for materials with suppliers, purchasing agents, metallurgists, manufacturing process personnel, heat treatment personnel, plastics molders, machinists, and quality assurance specialists often has a strong influence on the success of a design.

You should do some exploration on what kinds of materials are used in consumer products, industrial machinery, automobiles, construction machinery, and other devices and systems that you come into contact with each day. Try to make some judgments about why each material was specified for a particular application. Where do you see steel being used? Contrast that usage with where aluminum or other nonferrous materials are used. How are the products produced? Can you find different parts that are machined, cast, forged, roll-formed, and welded? Why do you think those processes were specified for those particular products?

You should also document several applications for plastics and describe the different forms that are available and that have been made by different manufacturing processes. Which are made by plastic molding processes, vacuum forming, blow molding, and others? Can you identify parts made from composite materials that have a significant amount of high-strength fibers embedded in a plastic matrix? Check out sporting goods and parts of cars, trucks, and airplanes.

From the products that you found from the exploration outlined above, try to identify the basic properties of the materials that were important to the designers: strength, rigidity (stiffness), weight (density), corrosion resistance, appearance, machinability, weldability, ease of forming, cost, and others.

This chapter focuses on the use of material property data in design decisions, rather than on the metallurgy or chemistry of the materials. One of the uses of the information in this chapter is as a glossary of terms that you can use throughout the book; important terms are given in *italic* type. Also, there are numerous references to parts of the Appendices where tables of data for material properties are given. Go there now and see what kinds of data are there. Then you can study the tables in more depth as you read the text. Note that many of the problems that you will solve in this book and the design projects that you complete will use data from these tables.

Now apply some of what you have gained from **The Big Picture** exploration to a specific design situation as outlined in **You Are the Designer,** which follows.

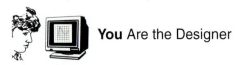

You Are the Designer

You are part of a team responsible for the design of a conveyor system in a plant that packages pharmaceuticals such as aspirin and cold tablets. As you design the frame of the conveyor, what must you consider in the selection of the materials from which it will be made? What materials are available to satisfy the functional requirements? What are their properties, and how do these properties affect your design decisions?

**2–1
OBJECTIVES OF
THIS CHAPTER**

After completing this chapter, you will be able to:

1. State the types of material properties that are important to the design of mechanical devices and systems.

2. Define the following terms: *tensile strength, yield strength, proportional limit, elastic limit, modulus of elasticity in tension, ductility and percent elongation, shear strength, Poisson's ratio, modulus of elasticity in shear, hardness, machinability, impact strength, density, coefficient of thermal expansion, thermal conductivity,* and *electrical resistivity.*

3. Describe the nature of *carbon and alloy steels,* the number-designation system for steels, and the effect of several kinds of alloying elements on the properties of steels.

4. Describe the manner of designating the condition and heat treatment of steels, including *hot rolling, cold drawing, annealing, normalizing, through-hardening, tempering,* and *case hardening by flame hardening, induction hardening,* and *carburizing.*

5. Describe *stainless steels* and recognize many of the types that are commercially available.

6. Describe *structural steels* and recognize many of their designations and uses.

7. Describe *cast irons* and several kinds of *gray iron, ductile iron,* and *malleable iron.*

8. Describe *powdered metals* and their properties and uses.

9. Describe several types of *tool steels* and *carbides* and their typical uses.

10. Describe *aluminum alloys* and their conditions, such as *strain hardening* and *heat treatment.*

11. Describe the nature and typical properties of *zinc, titanium,* and *bronze.*

12. Describe several types of *plastics,* both *thermosetting* and *thermoplastic,* and their typical properties and uses.

13. Describe several kinds of *composite materials* and their typical properties and uses.

**2–2
PROPERTIES
OF METALS**

Machine elements are very often made from one of the metals or metal alloys such as steel, aluminum, or bronze. This section describes the important properties of metals as they affect mechanical design.

Strength, elastic, and ductility properties for metals and other types of materials are usually determined from a *tensile test* in which a sample of the material, typically in the form of a round or flat bar, is clamped between jaws and pulled slowly until it breaks in tension. The magnitude of the force on the bar and the corresponding change in length (strain) are monitored and recorded continuously during the test. Because the stress in the bar is equal to the applied force divided by the area, stress is proportional to the applied force. The data from such tensile tests are often shown on *stress-strain diagrams,* such as those shown in Figures 2–1 and 2–2. In the following paragraphs, several strength, elastic, and ductility properties of metals are defined.

Tensile Strength, s_u

The peak of the stress-strain curve is considered the *ultimate tensile strength* (s_u), sometimes called the *ultimate strength* or simply the *tensile strength.* At this point during the test, the highest *apparent stress* on a test bar of the material is measured. As shown in Figures 2–1 and 2–2, the curve appears to drop off after the peak. However, notice that the instrumentation used to create the diagrams is actually plotting *load versus deflection* rather than *true stress versus strain.* The apparent stress is computed by dividing the load by the original cross-sectional area of the test bar. After the peak of the curve is reached, there is a pronounced decrease in the bar's diameter, referred to as *necking down.* Thus, the load acts over a smaller area, and the *actual stress* continues to increase until failure. It is very

FIGURE 2–1

Typical stress-strain diagram for steel

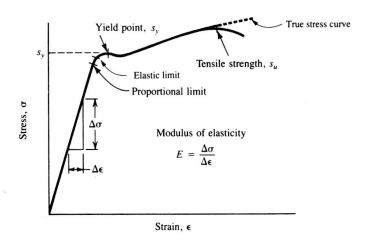

FIGURE 2–2

Typical stress-strain diagram for aluminum and other metals having no yield point

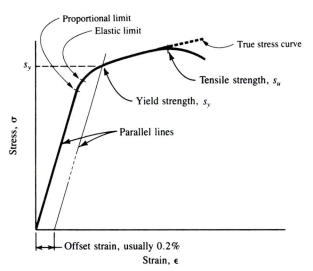

difficult to follow the reduction in diameter during the necking-down process, so it has become customary to use the peak of the curve as the tensile strength, although it is a more conservative value.

Yield Strength, s_y

That portion of the stress-strain diagram where there is a large increase in strain with little or no increase in stress is called the *yield strength* (s_y). This property indicates that the material has, in fact, yielded or elongated plastically, permanently, and to a large degree. If the point of yielding is quite noticeable, as it is in Figure 2–1, the property is called the *yield point* rather than the yield strength.

Figure 2–2 shows the stress-strain diagram form that is typical of a nonferrous metal such as aluminum or titanium or of certain high-strength steels. Notice that there is no pronounced yield point, but the material has actually yielded at or near the stress level indicated as s_y. That point is determined by the *offset method,* in which a line is drawn parallel to the straight-line portion of the curve and is offset to the right by a set amount, usually 0.20% strain (0.002 in/in). The intersection of this line and the stress-strain curve defines the material's yield strength. In this book, the term *yield strength* will be used for s_y, regardless of whether the material exhibits a true yield point or whether the offset method is used.

Proportional Limit

That point on the stress-strain curve where it deviates from a straight line is called the *proportional limit.* That is, at or above that stress value, stress is no longer proportional to strain. Below the proportional limit, Hooke's law applies: Stress is proportional to strain. In mechanical design, materials are rarely used at stresses above the proportional limit.

Elastic Limit

At some point, called the *elastic limit,* a material experiences some amount of plastic strain and thus will not return to its original shape after release of the load. Below that level, the material behaves completely elastically. The proportional limit and the elastic limit lie quite close to the yield strength. Because they are difficult to determine, they are rarely reported.

Modulus of Elasticity in Tension, E

For the part of the stress-strain diagram that is straight, stress is proportional to strain, and the value of E is the constant of proportionality. That is,

⇨ **Modulus of Elasticity in Tension**

$$E = \frac{\text{stress}}{\text{strain}} = \frac{\sigma}{\epsilon}$$

(2–1)

This is the slope of the straight-line portion of the diagram. The modulus of elasticity indicates the stiffness of the material, or its resistance to deformation.

Ductility and Percent Elongation

Ductility is the degree to which a material will deform before ultimate fracture. The opposite of ductility is *brittleness.* When ductile materials are used in machine members, impending failure is detected easily, and sudden failure is unlikely. Also, ductile materials normally resist the repeated loads on machine elements better than brittle materials.

The usual measure of ductility is the percent elongation of the material after fracture in a standard tensile test. Figure 2–3 shows a typical standard tensile specimen before and after the test. Before the test, gage marks are placed on the bar, usually 2.00 in apart. Then, after the bar is broken, the two parts are fitted back together, and the final length between the gage marks is measured. The percent elongation is the difference between the final length and the original length divided by the original length, converted to a percentage. That is,

⇨ **Percent Elongation**

$$\frac{L_f - L_o}{L_o} \times 100\% = \text{percent elongation} \qquad (2\text{–}2)$$

The percent elongation is assumed to be based on a gage length of 2.00 in unless some other gage length is specifically indicated. Theoretically, a material is considered ductile if its percent elongation is greater than 5%. For practical reasons, it is advisable to use a material with a value of 12% or higher for machine members subject to repeated loads or shock or impact.

Percent reduction in area is another indication of ductility. To find this value, compare the original cross-sectional area with the final area at the break for the tensile test specimen.

Shear Strength, s_{ys} and s_{us}

Both the yield strength and the ultimate strength in shear (s_{ys} and s_{us}, respectively) are important properties of materials. Unfortunately, these values are seldom reported. We will use the following estimates:

⇨ **Estimates for s_{ys} and s_{us}**

$$s_{ys} = s_y/2 = 0.50s_y = \text{yield strength in shear} \qquad (2\text{–}3)$$
$$s_{us} = 0.75s_u = \text{ultimate strength in shear} \qquad (2\text{–}4)$$

Poisson's Ratio, ν

When a material is subjected to a tensile strain, there is a simultaneous shortening of the cross-sectional dimensions perpendicular to the direction of the tensile strain. The ratio of the shortening strain to the tensile strain is called *Poisson's ratio,* usually denoted by ν, the Greek letter nu. (The Greek letter mu, μ, is sometimes used for this ratio.) Poisson's ratio is illustrated in Figure 2–4. Typical ranges of values for Poisson's ratio are 0.25–0.27 for cast iron, 0.27–0.30 for steel, and 0.30–0.33 for aluminum and titanium.

FIGURE 2–3
Measurement of percent elongation

FIGURE 2–4
Illustration of Poisson's ratio for an element in tension

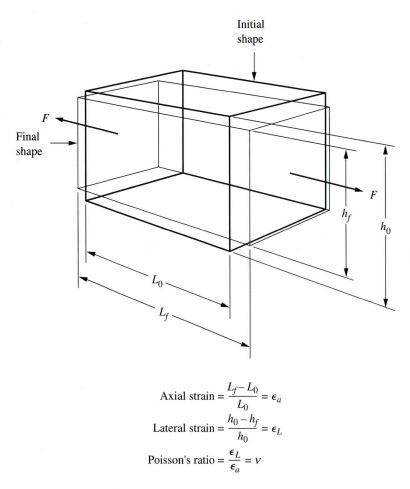

$$\text{Axial strain} = \frac{L_f - L_0}{L_0} = \epsilon_a$$

$$\text{Lateral strain} = \frac{h_0 - h_f}{h_0} = \epsilon_L$$

$$\text{Poisson's ratio} = \frac{\epsilon_L}{\epsilon_a} = v$$

Modulus of Elasticity in Shear, *G*

The *modulus of elasticity in shear* (*G*) is the ratio of shearing stress to shearing strain. This property indicates a material's stiffness under shear loading, that is, the resistance to shear deformation. There is a simple relationship between *E, G,* and Poisson's ratio:

 Modulus of Elasticity in Shear

$$G = \frac{E}{2(1 + v)} \tag{2–5}$$

This equation is valid within the elastic range of the material.

Hardness

The resistance of a material to indentation by a penetrator is an indication of its *hardness.* Several types of devices, procedures, and penetrators measure hardness; the Brinell hardness tester and the Rockwell hardness tester are most frequently used for machine elements. For steels, the Brinell hardness tester employs a hardened steel ball 10 mm in diameter as the penetrator under a load of 3 000-kg force. The load causes a permanent indentation in the test material, and the diameter of the indentation is related to the Brinell hardness number, which is abbreviated BHN or HB. The actual quantity being measured is the load divided by the contact area of the indentation. For steels, the value of HB ranges from approximately 100 for an annealed, low-carbon steel to over 700 for high-strength, high-alloy steels in the as-quenched condition. In the high ranges, above HB 500, the penetrator is sometimes made of tungsten carbide rather than steel. For softer metals, a 500-kg load is used.

The Rockwell hardness tester uses a hardened steel ball with a 1/16-in diameter under a load of 100-kg force for softer metals, and the resulting hardness is listed as Rockwell B, R_B, or HRB. For harder metals, such as heat-treated alloy steels, the Rockwell C scale is used. A load of 150-kg force is placed on a diamond penetrator (a *brale* penetrator) made in a sphero-conical shape. Rockwell C hardness is sometimes referred to as R_C or HRC. Many other Rockwell scales are used.

The Brinell and Rockwell methods are based on different parameters and lead to quite different numbers. However, since they both measure hardness, there is a correlation between them, as noted in Appendix 19. It is also important to note that, especially for highly hardenable alloy steels, there is a nearly linear relationship between the Brinell hardness number and the tensile strength of the steel, according to the equation

Approximate Relationship between Hardness and Strength for Steel ⇨

$$0.50(HB) = \text{approximate tensile strength (Ksi)} \qquad \textbf{(2–6)}$$

This relationship is shown in Figure 2–5.

To compare the hardness scales with the tensile strength, consider Table 2–1. Note that there is some overlap between the HRB and HRC scales. Normally, HRB is used for the softer metals and ranges from approximately 60 to 100, whereas HRC is used for harder metals and ranges from 20 to 65. Using HRB numbers above 100 or HRC numbers below 20 is not recommended. Those shown in Table 2–1 are for comparison purposes only.

Hardness in a steel indicates wear resistance as well as strength. Wear resistance will be discussed in later chapters, particularly with regard to gear teeth.

Machinability

Machinability is related to the ease with which a material can be machined to a good surface finish with reasonable tool life. Production rates are directly affected by machinability. It is difficult to define measurable properties related to machinability, so machinability is usually reported in comparative terms, relating the performance of a given material with some standard.

Impact Strength

Parts subjected to suddenly applied loads, shock, or impact need high values of impact strength. A measure of toughness, *impact strength* is measured by subjecting a notched specimen to a rapidly moving striker having a relatively high mass. The specimen absorbs energy from the striker as fracture occurs, and the amount of energy absorbed is reported as the impact strength. The two most widely used testing methods for impact strength are the Charpy and the Izod tests.

Fatigue Strength or Endurance Strength

Parts subjected to repeated applications of loads or to stress conditions that vary with time over several thousands or millions of cycles fail because of the phenomenon of *fatigue*. Materials are tested under controlled cyclic loading to determine their ability to resist such

FIGURE 2–5
Hardness conversions

TABLE 2–1 Comparison of hardness scales with tensile strength

Material and condition	Hardness			Tensile strength	
	HB	HRB	HRC	Ksi	MPa
1020 annealed	121	70		60	414
1040 hot-rolled	144	79		72	496
4140 annealed	197	93	13	95	655
4140 OQT 1000	341	109	37	168	1 160
4140 OQT 700	461		49	231	1 590

repeated loads. The resulting data are reported as the *fatigue strength,* also called the *endurance strength* of the material. (See Chapter 5.)

Physical Properties

Here we will discuss density, coefficient of thermal expansion, thermal conductivity, and electrical resistivity.

Density.   *Density* is defined as the mass per unit volume of a material. Its usual units are kg/m^3 in the SI and lb/in^3 in the U.S. Customary Unit System, where the pound unit is taken to be pounds-mass. The Greek letter rho (ρ) is the symbol for density.

   In some applications, the term *specific weight* or *weight density* is used to indicate the weight per unit volume of a material. Typical units are N/m^3 in the SI and lb/in^3 in the U.S. Customary Unit System, where the pound is taken to be pounds-force. The Greek letter gamma (γ) is the symbol for specific weight.

Coefficient of Thermal Expansion.   The *coefficient of thermal expansion* is a measure of the change in length of a material subjected to a change in temperature. It is defined by the relation

⟡ **Coefficient of
Thermal Expansion**

$$\alpha = \frac{\text{change in length}}{L_o(\Delta T)} = \frac{\text{strain}}{(\Delta T)} = \frac{\epsilon}{(\Delta T)} \qquad (2\text{–}7)$$

where L_o = original length

ΔT = change in temperature

Virtually all metals and plastics expand with increasing temperature, but different materials expand at different rates. For machines and structures containing parts of more than one material, the different rates can have a significant effect on the performance of the assembly and on the stresses produced.

Thermal Conductivity.   *Thermal conductivity* is the property of a material that indicates its ability to transfer heat. Where machine elements operate in hot environments or where significant internal heat is generated, the ability of the elements or of the machine's housing to transfer heat away can affect machine performance. For example, wormgear speed reducers typically generate frictional heat due to the rubbing contact between the worm and the wormgear teeth. If not adequately transferred, heat causes the lubricant to lose its effectiveness, allowing rapid gear-tooth wear.

Electrical Resistivity.   For machine elements that conduct electricity while carrying loads, the electrical resistivity of the material is as important as its strength. *Electrical resistivity* is a measure of the resistance offered by a given thickness of a material; it is measured in ohm-centimeters (Ω·cm). *Electrical conductivity,* a measure of the capacity of a material to conduct electric current, is sometimes used instead of resistivity. It is often reported as a percentage of the conductivity of a reference material, usually the International Annealed Copper Standard.

**2–3
CLASSIFICATION
OF METALS AND
ALLOYS**

Various industry associations take responsibility for setting standards for the classification of metals and alloys. Each has its own numbering system, convenient to the particular metal covered by the standard. But this leads to confusion at times when there is overlap between two or more standards and when widely different schemes are used to denote the metals. Order has been brought to the classification of metals by the use of the Unified Numbering Systems (UNS) as defined in the Standard E 527-83 (Reapproved 1991), ***Standard Practice for Numbering Metals and Alloys (UNS),*** by the American Society for Testing and Materials, or ASTM. (See Reference 3.) Besides listing materials under the control of ASTM itself, the UNS coordinates designations of the following:

The Aluminum Association (AA)

The American Iron and Steel Institute (AISI)

The Copper Development Association (CDA)

The Society of Automotive Engineers (SAE)

The primary series of numbers within UNS are listed in Table 2–2, along with the organization having responsibility for assigning numbers within each series.

Many alloys within the UNS retain the familiar numbers from the systems used for many years by the various associations as a *part* of the UNS number. Examples are shown in Section 2–5 for carbon and alloy steel. Also, there is still widespread use of the former designations. For these reasons, this book will use the four-digit designation system of the American Iron and Steel Institute (AISI) as described in Section 2–5 for most machine

TABLE 2–2   Unified numbering system (UNS)

Number series	Types of metals and alloys	Responsible organization
Nonferrous metals and alloys		
A00001–A99999	Aluminum and aluminum alloys	AA
C00001–C99999	Copper and copper alloys	CDA
E00001–E99999	Rare earth metals and alloys	ASTM
L00001–L99999	Low-melting metals and alloys	ASTM
M00001–M99999	Miscellaneous nonferrous metals and alloys	ASTM
N00001–N99999	Nickel and nickel alloys	SAE
P00001–P99999	Precious metals and alloys	ASTM
R00001–R99999	Reactive and refractory metals and alloys	SAE
Z00001–Z99999	Zinc and zinc alloys	ASTM
Ferrous metals and alloys		
D00001–D99999	Steels; mechanical properties specified	SAE
F00001–F99999	Cast irons and cast steels	ASTM
G00001–G99999	Carbon and alloy steels (includes former SAE carbon and alloy steels)	AISI
H00001–H99999	H-steels; specified hardenability	AISI
J00001–J99999	Cast steels (except tool steels)	ASTM
K00001–K99999	Miscellaneous steels and ferrous alloys	ASTM
S00001–S99999	Heat- and corrosion-resistant (stainless) steels	ASTM
T00001–T99999	Tool steels	AISI

steels. Many of the designations of the Society of Automotive Engineers (SAE) use the same four numbers. We will also use the designation systems of the American Society for Testing and Materials (ASTM) when referring to structural steels and cast irons.

2–4
VARIABILITY OF
MATERIAL
PROPERTIES
DATA

Tables of data such as those shown in Appendices 3 through 13 normally report single values for the strength, the modulus of elasticity (stiffness), or the percent elongation (ductility) of a particular material at a particular condition created by heat treatment or by the manner in which it was formed. It is important for you to understand the limitations of such data in making design decisions. You should seek information about the bases for the reported data.

Some tables of data report *guaranteed minimum values* for tensile strength, yield strength, and other values. This might be the case when you are using data obtained from a particular supplier. With such data, you should feel confident that the material that actually goes into your product has at least the reported strength. The supplier should be able to provide actual test data and statistical analyses used to determine the reported minimum strengths. Alternatively, you could arrange to have the actual materials to be used in a project tested to determine their minimum strength values. Such tests are costly, but they may be justified in critical designs.

Other tables of data report *typical values* for material properties. Thus, most batches of material (greater than 50%) delivered will have the stated values or greater. However, about 50% will have lower values, and this fact will affect your confidence in specifying a particular material and heat treatment if strength is critical. In such cases, you are advised to use higher than average design factors in your calculations of allowable (design) strength. (See Chapter 5.)

FIGURE 2–6
Normal statistical
distribution of
material strength

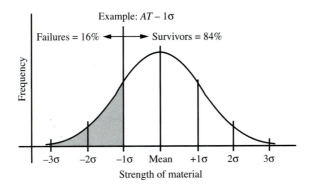

Assuming normal distribution of strength:

Stress level	% Surviving
Mean	50%
-1σ	84%
-2σ	97%
-3σ	99.8%
-4σ	99.99%

Using the *guaranteed minimum values* for strength in design decisions would be the safest approach. However, it is very conservative because most of the material actually delivered would have strengths significantly greater than the listed values.

One way to make the design more favorable is to acquire data for the statistical distribution of strength values taken for many samples. Then applications of probability theories can be used to specify suitable conditions for the material with a reasonable degree of confidence that the parts will perform according to specifications. Figure 2–6 illustrates some of the basic concepts of statistical distribution. The variation of strength over the entire population of samples is often assumed to have a normal distribution around some mean or average value. If you used a strength value that is one standard deviation (1σ) below the mean, 84% of the products would survive. At two standard deviations, greater than 97% would survive; at three standard deviations, over 99.8%; and at four standard deviations, over 99.99%.

As the designer, you must carefully judge the reliability of the data that you use. Ultimately, you should evaluate the reliability of the final product by considering the actual variations in material properties, the manufacturing considerations that may affect performance, and the interactions of various components with each other. There is more discussion on this point in Chapter 5.

**2–5
CARBON AND
ALLOY STEEL**

Steel is possibly the most widely used material for machine elements because of its properties of high strength, high stiffness, durability, and relative ease of fabrication. Many types of steels are available. This section will discuss the methods used for designating steels and will describe the most frequently used types.

The term *steel* refers to an alloy of iron, carbon, manganese, and one or more other significant elements. Carbon has a very strong effect on the strength, hardness, and ductility of any steel alloy. The other elements affect hardenability, toughness, corrosion resistance, machinability, and strength retention at high temperatures. The primary alloying elements present in the various alloy steels are sulfur, phosphorus, silicon, nickel, chromium, molybdenum, and vanadium.

Designation Systems

The AISI uses a four-digit designation system for carbon and alloy steel as shown in Figure 2–7. The first two digits indicate the specific alloy group that identifies the primary alloying elements other than carbon in the steel. (See Table 2–3.) The last two digits indicate the amount of carbon in the steel as described next.

FIGURE 2–7   Steel designation system

TABLE 2–3   Alloy groups in the AISI numbering system

10xx	Plain carbon steel: No significant alloying element except carbon and manganese; less than 1.0% manganese. Also called *nonresulfurized.*
11xx	Free-cutting steel: Resulfurized. Sulfur content (typically 0.10%) improves machinability.
12xx	Free-cutting steel: Resulfurized and rephosphorized. Presence of increased sulfur and phosphorus improves machinability and surface finish.
12Lxx	Free-cutting steel: Lead added to 12xx steel further improves machinability.
13xx	Manganese steel: Nonresulfurized. Presence of approximately 1.75% manganese increases hardenability.
15xx	Carbon steel: Nonresulfurized; greater than 1.0% manganese.
23xx	Nickel steel: Nominally 3.5% nickel.
25xx	Nickel steel: Nominally 5.0% nickel.
31xx	Nickel-chromium steel: Nominally 1.25% Ni; 0.65% Cr.
33xx	Nickel-chromium steel: Nominally 3.5% Ni; 1.5% Cr.
40xx	Molybdenum steel: 0.25% Mo.
41xx	Chromium-molybdenum steel: 0.95% Cr; 0.2% Mo.
43xx	Nickel-chromium-molybdenum steel: 1.8% Ni; 0.5% or 0.8% Cr; 0.25% Mo.
44xx	Molybdenum steel: 0.5% Mo.
46xx	Nickel-molybdenum steel: 1.8% Ni; 0.25% Mo.
48xx	Nickel-molybdenum steel: 3.5% Ni; 0.25% Mo.
5xxx	Chromium steel: 0.4% Cr.
51xx	Chromium steel: Nominally 0.8% Cr.
51100	Chromium steel: Nominally 1.0% Cr; bearing steel, 1.0% C.
52100	Chromium steel: Nominally 1.45% Cr; bearing steel, 1.0% C.
61xx	Chromium-vanadium steel: 0.50%–1.10% Cr; 0.15% V.
86xx	Nickel-chromium-molybdenum steel: 0.55% Ni; 0.5% Cr; 0.20% Mo.
87xx	Nickel-chromium-molybdenum steel: 0.55% Ni; 0.5% Cr; 0.25% Mo.
92xx	Silicon steel: 2.0% silicon.
93xx	Nickel-chromium-molybdenum steel: 3.25% Ni; 1.2% Cr; 0.12% Mo.

Importance of Carbon

Although most steel alloys contain less than 1.0% carbon, it is included in the designation because of its effect on the properties of steel. As Figure 2–7 illustrates, the last two digits indicate carbon content in hundredths of a percent. For example, when the last two digits are 20, the alloy includes approximately 0.20% carbon. Some variation is allowed. The carbon content in a steel with *20 points of carbon* ranges from 0.18% to 0.23%.

As carbon content increases, strength and hardness also increase under the same conditions of processing and heat treatment. Since ductility decreases with increasing carbon content, selecting a suitable steel involves some compromise between strength and ductility.

As a rough classification scheme, a *low-carbon steel* is one having fewer than 30 points of carbon (0.30%). These steels have relatively low strength but good formability. In machine element applications where high strength is not required, low-carbon steels are frequently specified. If wear is a potential problem, low-carbon steels can be carburized (as discussed in Section 2–5) to increase the carbon content in the very outer surface of the part and to improve the combination of properties.

Medium-carbon steels contain 30 to 50 points of carbon (0.30%–0.50%). Most machine elements having moderate to high strength requirements with fairly good ductility and moderate hardness requirements come from this group.

High-carbon steels have 50 to 95 points of carbon (0.50%–0.95%). The high carbon content provides better wear properties suitable for applications requiring durable cutting edges and for applications where surfaces are subjected to constant abrasion. Tools, knives, chisels, and many agricultural implement components are among these uses.

A *bearing steel* nominally contains 1.0% carbon. Common grades are 50100, 51100, and 52100; the usual four-digit designation is replaced by five digits, indicating 100 points of carbon.

Alloy Groups

As indicated in Table 2–3, sulfur, phosphorus, and lead improve the machinability of steels and are added in significant amounts to the 11xx, 12xx, and 12Lxx grades. These grades are used for screw machine parts requiring high production rates where the resulting parts are not subjected to high stresses or wear conditions. In the other alloys, these elements are controlled to a very low level because of their adverse effects, such as increased brittleness.

Nickel improves the toughness, hardenability, and corrosion resistance of steel and is included in most of the alloy steels. Chromium improves hardenability, wear and abrasion resistance, and strength at elevated temperatures. In high concentrations, chromium provides significant corrosion resistance, as discussed in the section on stainless steels. Molybdenum also improves hardenability and high-temperature strength.

The steel selected for a particular application must be economical and must provide optimum properties of strength, ductility, toughness, machinability, and formability. Frequently, metallurgists, manufacturing engineers, and heat treatment specialists are consulted. (See also References 5, 13, 15, and 20.)

Table 2–4 lists some common steels used for machine parts, with typical applications listed for the alloys. You should benefit from the decisions of experienced designers when specifying materials.

TABLE 2–4   Uses of some steels

UNS number	AISI number	Applications
G10150	1015	Formed sheet-metal parts; machined parts (may be carburized)
G10300	1030	General-purpose, bar-shaped parts, levers, links, keys
G10450	1045	Shafts, gears
G10800	1080	Springs; agricultural equipment parts subjected to abrasion (rake teeth, disks, plowshares, mower teeth)
G11120	1112	Screw machine parts
G12144	12L14	Parts requiring good machinability
G41400	4140	Gears, shafts, forgings
G43400	4340	Gears, shafts, parts requiring good through-hardening
G46400	4640	Gears, shafts, cams
G51500	5150	Heavy-duty shafts, springs, gears
G51601	51B60	Shafts, springs, gears with improved hardenability
G52986	E52100	Bearing races, balls, rollers (bearing steel)
G61500	6150	Gears, forgings, shafts, springs
G86500	8650	Gears, shafts
G92600	9260	Springs

Examples of the Relationships between AISI and UNS Numbering Systems

Given in Table 2–4 are both the AISI and the UNS designations for the listed steels. Notice that for most carbon and alloy steels, the four-digit AISI number becomes the first four digits of the UNS number. The final digit in the UNS number is typically zero.

There are some exceptions, however. For high-carbon-bearing steels made in an electric furnace, such as AISI E52100, the UNS designation is G52986. Leaded steels have extra lead to improve machinability, and they have the letter L added between the second and third digits of the AISI number such as AISI 12L14, which becomes UNS G12144. Extra boron is added to some special alloys to improve hardenability. For example, alloy AISI 5160 is a chromium steel that carries the UNS designation G51600. But a similar alloy with added boron is AISI 51B60, and it carries the UNS designation G51601.

**2–6
CONDITIONS
FOR STEELS
AND HEAT
TREATMENT**

The final properties of steels are dramatically affected by the way the steels are produced. Some processes involve mechanical working, such as rolling to a particular shape or drawing through dies. In machine design, many bar-shaped parts, shafts, wire, and structural members are produced in these ways. But most machine parts, particularly those carrying heavy loads, are heat-treated to produce high strength with acceptable toughness and ductility.

Carbon steel bar and sheet forms are usually delivered in the *as-rolled condition;* that is, they are rolled at an elevated temperature that eases the rolling process. The rolling can also be done cold to improve strength and surface finish. Cold-drawn bar and wire have the highest strength of the worked forms, along with a very good surface finish. However, when a material is designated to be *as-rolled,* it should be assumed that it was hot-rolled.

Heat Treating

Heat treating is any process in which steel is subjected to elevated temperatures to modify its properties. Of the several processes available, those most used for machine steels are annealing, normalizing, through-hardening (quench and temper), and case hardening. (See References 4 and 14.)

FIGURE 2–8 Heat treatments for steel

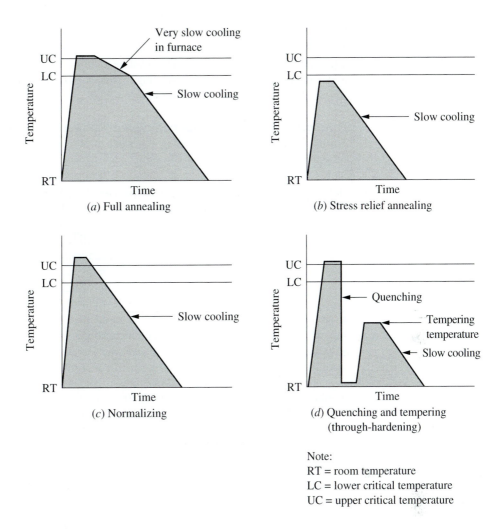

(a) Full annealing

(b) Stress relief annealing

(c) Normalizing

(d) Quenching and tempering (through-hardening)

Note:
RT = room temperature
LC = lower critical temperature
UC = upper critical temperature

Figure 2–8 shows the temperature–time cycles for these heat treatment processes. The symbol RT indicates normal room temperature, and LC refers to the lower critical temperature at which the transformation of ferrite to austenite begins during the heating of the steel. At the upper critical temperature (UC), the transformation is complete. These temperatures vary with the composition of the steel. For most medium-carbon (0.30%–0.50% carbon) steels, UC is approximately 1 500°F (822°C). References giving detailed heat treatment process data should be consulted.

Annealing. *Full annealing* [Figure 2–8(a)] is performed by heating the steel above the upper critical temperature and holding it until the composition is uniform. Then the steel is cooled very slowly in the furnace to below the lower critical temperature. Slow cooling to room temperature outside the furnace completes the process. This treatment produces a soft, low-strength form of the material, free of significant internal stresses. Parts are frequently cold-formed or machined in the annealed condition.

Stress relief annealing [Figure 2–8(b)] is often used following welding, machining, or cold forming to relieve residual stresses and thereby minimize subsequent distortion. The steel is heated to approximately 1 000°F to 1 200°F (540°C–650°C), held to achieve uniformity, and then slowly cooled in still air to room temperature.

Normalizing.   *Normalizing* [Figure 2–8(c)] is performed in a similar manner to annealing, but at a higher temperature, above the transformation range where austenite is formed, approximately 1 600°F (870°C). The result is a uniform internal structure in the steel and somewhat higher strength than annealing produces. Machinability and toughness are usually improved over the as-rolled condition.

Through-hardening and Quenching and Tempering.   *Through-hardening* [Figure 2–8(d)] is accomplished by heating the steel to above the transformation range where austenite forms and then rapidly cooling it in a *quenching* medium. The rapid cooling causes the formation of martensite, the hard, strong form of steel. The degree to which martensite forms depends on the alloy's composition. An alloy containing a minimum of 80% of its structure in the martensite form over the entire cross section has *high hardenability*. This is an important property to look for when selecting a steel requiring high strength and hardness. The common quenching media are water, brine, and special mineral oils. The selection of a quenching medium depends on the rate at which cooling should proceed. Most machine steels use either oil or water quenching.

Tempering is usually performed immediately after quenching and involves reheating the steel to a temperature of 400°F to 1 300°F (200°C–700°C) and then slowly cooling it in air back to room temperature. This process modifies the steel's properties: Tensile strength and yield strength decrease with increasing tempering temperature, whereas ductility improves, as indicated by an increase in the percent elongation. Thus, the designer can tailor the properties of the steel to meet specific requirements. Furthermore, the steel in its as-quenched condition has high internal stresses and is usually quite brittle. Machine parts should normally be tempered at 700°F (370°C) or higher after quenching.

To illustrate the effects of tempering on the properties of steels, several charts in Appendix 4 show graphs of strength versus tempering temperature. Included in these charts are tensile strength, yield point, percent elongation, percent reduction of area, and hardness number HB, all plotted in relation to tempering temperature. Note the difference in the shape of the curves and the absolute values of the strength and hardness when comparing the plain carbon AISI 1040 steel with the alloy steel AISI 4340. Although both have the same nominal carbon content, the alloy steel reaches a much higher strength and hardness. Note also the as-quenched hardness in the upper right part of the heading of the charts; it indicates the degree to which a given alloy can be hardened. When the case-hardening processes (described next) are used, the as-quenched hardness becomes very important.

Appendix 3 lists the range of properties that can be expected for several grades of carbon and alloy steels. The alloys are listed with their AISI numbers and conditions. For the heat-treated conditions, the designation reads, for example, AISI 4340 OQT 1000, which indicates that the alloy was oil-quenched and tempered at 1 000°F. Expressing the properties at the 400°F and 1 300°F tempering temperatures indicates the end-points of the possible range of properties that can be expected for that alloy. To specify a strength between these limits, you could refer to graphs such as those shown in Appendix 4, or you could determine the required heat treatment process from a specialist. For the purposes of material specification in this book, a rough interpolation between given values will be satisfactory. As noted before, you should seek more specific data for critical designs.

Case Hardening.   In many cases, the bulk of the part requires only moderate strength although the surface must have a very high hardness. In gear teeth, for example, high surface hardness is necessary to resist wear as the mating teeth come into contact several million times during the expected life of the gears. At each contact, a high stress develops at the surface of the teeth. For applications such as this, *case hardening* is used; the surface (or *case*) of the part is given a high hardness to a depth of perhaps 0.010 to 0.040 in

(0.25–1.00 mm), although the interior of the part (the *core*) is affected only slightly, if at all. The advantage of surface hardening is that as the surface receives the required wear-resisting hardness, the core of the part remains in a more ductile form, resistant to impact and fatigue. The processes used most often for case hardening are flame hardening, induction hardening, carburizing, nitriding, cyaniding, and carbo-nitriding. (See Reference 16.)

Figure 2–9 shows a drawing of a typical case-hardened gear-tooth section, clearly showing the hard case surrounding the softer, more ductile core. Case hardening is used in applications requiring high wear and abrasion resistance in normal service (gear teeth, crane wheels, wire-rope sheaves, and heavy-duty shafts).

The most commonly used processes for case hardening are described in the following list.

1. ***Flame hardening and induction hardening:*** The processes of flame hardening and induction hardening involve the rapid heating of the surface of the part for a limited time so that a small, controlled depth of the material reaches the transformation range. Upon immediate quenching, only that part above the transformation range produces the high level of martensite required for high hardness.

   Flame hardening uses a concentrated flame impinging on a localized area for a controlled amount of time to heat the part, followed by quenching in a bath or by a stream of water or oil. *Induction hardening* is a process in which the part is surrounded by a coil through which high-frequency electric current is passed. Because of the electrical conductivity of the steel, current is *induced* primarily near the surface of the part. The resistance of the material to the flow of current results in a heating effect. Controlling the electrical power and the frequency of the induction system, and the time of exposure, determines the depth to which the material reaches the transformation temperature. Rapid quenching after heating hardens the surface. (See Reference 21.)

   Note that for flame or induction hardening to be effective, the material must have a good hardenability. Usually the goal of case hardening is to produce a case hardness in the range of Rockwell C hardness HRC 55 to 60 (Brinell hardness approximately HB 550 to 650). Therefore, the material must be capable of being hardened to the desired level. Carbon and alloy steels with fewer than 30 points of carbon typically cannot meet this requirement. Thus, the alloy steels with 40 points or more of carbon are the usual types given flame- or induction-hardening treatments.

2. ***Carburizing, nitriding, cyaniding, and carbo-nitriding:*** The remaining case-hardening processes—carburizing, nitriding, cyaniding, and carbo-nitriding—actually alter the composition of the surface of the material by exposing it to carbon-bearing gases, liquids, or solids at high temperatures that produce carbon and diffuse it into the surface of the part. The concentration and the depth of penetration of carbon depend on the nature of the carbon-bearing material and the time of

FIGURE 2–9
Typical case-hardened
gear-tooth section

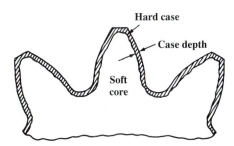

exposure. Nitriding and cyaniding typically result in very hard, thin cases that are good for general wear resistance. Where high load-carrying capability in addition to wear resistance is required, as with gear teeth, carburizing is preferred because of the thicker case.

Several steels are produced as carburizing grades. Among these are 1015, 1020, 1022, 1117, 1118, 4118, 4320, 4620, 4820, and 8620. Appendix 5 lists the expected properties of these carburized steels. Note when evaluating a material for use that the core properties determine its ability to withstand prevailing stresses, and the case hardness indicates its wear resistance. Carburizing, properly done, will virtually always produce a case hardness from HRC 55 to 64 (Rockwell C hardness) or from HB 550 to 700 (Brinell hardness).

Carburizing has several variations that allow the designer to tailor the properties to meet specific requirements. The exposure to the carbon atmosphere takes place at a temperature of approximately 1 700°F (920°C) and usually takes 8 h. Immediate quenching achieves the highest strength, although the case is somewhat brittle. Normally, a part is allowed to cool slowly after carburizing. It is then reheated to approximately 1 500°F (815°C) and then quenched. A tempering at the relatively low temperature of either 300°F or 450°F (150°C or 230°C) follows, to relieve stresses induced by quenching. As shown in Appendix 5, the higher tempering temperature lowers the core strength and the case hardness by a small amount, but in general it improves the part's toughness. The process just described is *single quenching and tempering.*

When a part is quenched in oil and tempered at 450°F, for example, the condition is *case hardening by carburizing, SOQT 450.* Reheating after the first quench and quenching again further refines the case and core properties; this process is *case hardening by carburizing, DOQT 450.* These conditions are listed in Appendix 5.

2–7
STAINLESS
STEELS

The term *stainless steel* characterizes the high level of corrosion resistance offered by alloys in this group. To be classified as a stainless steel, the alloy must have a chromium content of at least 10%. Most have 12% to 18% chromium. (See Reference 6.)

The AISI designates most stainless steels by its 200, 300, and 400 series. As mentioned previously (Section 2–3), another designation system is the Unified Numbering System (UNS) developed by the SAE and the ASTM. Appendix 6 lists the properties of several grades, giving both designations.

The three main groups of stainless steels are austenitic, ferritic, and martensitic. *Austenitic* stainless steels fall into the AISI 200 and 300 series. They are general-purpose grades with moderate strength. Most are not heat-treatable, and their final properties are determined by the amount of working, with the resulting temper referred to as 1/4 hard, 1/2 hard, 3/4 hard, and full hard. These alloys are nonmagnetic and are typically used in food processing equipment.

Ferritic stainless steels belong to the AISI 400 series, designated as 405, 409, 430, 446, and so on. They are magnetic and perform well at elevated temperatures, from 1 300°F to 1 900°F (700°C–1 040°C), depending on the alloy. They are not heat-treatable, but they can be cold-worked to improve properties. Typical applications include heat exchanger tubing, petroleum refining equipment, automotive trim, furnace parts, and chemical equipment.

Martensitic stainless steels are also members of the AISI 400 series, including 403, 410, 414, 416, 420, 431, and 440 types. They are magnetic, can be heat-treated, and have higher strength than the 200 and 300 series, while retaining good toughness. Typical uses include turbine engine parts, cutlery, scissors, pump parts, valve parts, surgical instruments, aircraft fittings, and marine hardware.

There are many other grades of stainless steels, many of which are proprietary to particular manufacturers. A group used for high-strength applications in aerospace, marine, and vehicular applications is of the precipitation-hardening type. They develop very high strengths with heat treatments at relatively low temperatures, from 900°F to 1 150°F (480°C–620°C). This characteristic helps to minimize distortion during treatment. Some examples are 17-4PH, 15-5PH, 17-7PH, PH15-7Mo, and AMS362 stainless steels.

**2–8
STRUCTURAL
STEEL**

Most structural steels are designated by ASTM numbers established by the American Society for Testing and Materials. The most common grade is ASTM A36, which has a minimum yield point of 36 000 psi (248 MPa) and is very ductile. It is basically a low-carbon, hot-rolled steel available in sheet, plate, bar, and structural shapes such as wide-flange beams, American Standard beams, channels, and angles. The geometric properties of some of each of these sections are listed in Appendix 16.

Other standard structural steels are ASTM A283, A284, and A285, which have yield points in the range of 21 000 to 33 000 psi, produced mostly in sheet and plate form.

Many higher-strength grades of structural steel are available for use in construction, vehicular, and machine applications. They provide yield points in the range from 42 000 to 100 000 psi (290–700 MPa). Some of these grades, referred to as *high-strength, low-alloy (HSLA) steels,* are ASTM A242, A440, A441, A514, A572, A588, and A607.

Appendix 7 lists the properties of several structural steels.

**2–9
TOOL STEELS**

The term *tool steels* refers to a group of steels typically used for cutting tools, punches, dies, shearing blades, chisels, and similar uses. Whereas most of these uses are related to the field of manufacturing engineering, they are also pertinent to machine design where the ability to maintain a keen edge under abrasive conditions is required. Also, some tool steels have rather high shock resistance which may be desirable in machine components such as parts for mechanical clutches, pawls, blades, guides for moving materials, and clamps.

The numerous varieties of tool steel materials have been classified into seven general types as shown in Table 2–5. (See Reference 7 for a more extensive discussion of tool steels.)

**2–10
CAST IRON**

Large gears, machine structures, brackets, linkage parts, and other important machine parts are made from cast iron. The several types of grades available span wide ranges of strength, ductility, machinability, wear resistance, and cost. These features are attractive in many applications. The three most commonly used types of cast iron are gray iron, ductile iron, and malleable iron. Appendix 8 lists the properties of several cast irons. (See also Reference 10.)

Gray iron is available in grades having tensile strengths ranging from 20 000 to 60 000 psi (138–414 MPa). Its ultimate compressive strength is much higher, three to five times as high as the tensile strength. One disadvantage of gray iron is that it is brittle and therefore should not be used in applications where impact loading is likely. But it has excellent wear resistance, is relatively easy to machine, has good vibration damping ability, and can be surface-hardened. Applications include engine blocks, gears, brake parts, and machine bases. The gray irons are rated by the ASTM specification A48-83 in classes 20, 25, 30, 40, 50, and 60, where the number refers to the minimum tensile strength in kips/in^2(Ksi). For example, class 40 gray iron has a minimum tensile strength of 40 Ksi or 40 000 psi (276 MPa). Because it is brittle, gray iron does not exhibit the property of yield strength.

TABLE 2–5 Examples of tool steel types

General type	Type symbol	Specific types — Major alloying elements	Examples — AISI No.	Examples — UNS No.	Typical uses (and other common alloys)
High-speed	M	Molybdenum	M2	T11302	General-purpose tool steels for cutting tools and dies for forging, extrusion, bending, drawing, and piercing (M1, M3, M4–M7, M30, M34, M36, M41–M47)
			M10	T11310	
			M42	T11342	
	T	Tungsten	T1	T12001	Similar to uses for M-types (T2, T4, T5, T6, T8)
			T15	T12015	
Hot-worked	H	Chromium	H10	T20810	Cold-heading dies, shearing knives, aircraft parts, low-temperature extrusion and die-casting dies (H1–H19)
		Tungsten	H21	T20821	Higher-temperature dies, hot shearing knives (H20–H39)
		Molybdenum	H42	T20842	Applications that tend to produce high wear (H40–H59)
Cold-worked	D	High-carbon, high-chromium	D2	T30402	Stamping dies, punches, gages (D3–D5, D7)
	A	Medium-alloy, air-hardening	A2	T30102	Punches, thread-rolling dies, die-casting dies (A3–A10)
	O	Oil-hardening	O1	T31501	Taps, reamers, broaches, gages, jigs and fixtures, bushings, machine tool arbors, tool shanks (O2, O6, O7)
Shock-resisting	S		S1	T41901	Chisels, pneumatic tools, heavy-duty punches, machine parts subject to shocks (S2, S4–S7)
Molded steels	P		P2	T51602	Plastic molding dies, zinc die-casting dies (P3–P6, P20, P21)
Special-purpose	L	Low-alloy types	L2	T61202	Tooling and machine parts requiring high toughness (L3, L6)
	F	Carbon-tungsten types	F1	T60601	Similar to L-types but with higher abrasion resistance (F2)
Water-hardened	W		W1	T72301	General-purpose tool and die uses, vise and chuck jaws, hand tools, jigs and fixtures, punches (W2, W5)

Ductile irons have higher strengths than the gray irons and, as the name implies, are more ductile. However, their ductility is still much lower than that of typical steels. A three-part grade designation is used for ductile iron in the ASTM A536-84 specification. The first number refers to the tensile strength in Ksi, the second is the yield strength in Ksi, and the third is the approximate percent elongation. For example, the grade 80-55-06 has a tensile strength of 80 Ksi (552 MPa), a yield strength of 55 Ksi (379 MPa), and a 6% elongation in 2.00 in. Higher-strength cast parts, such as crankshafts and gears, are made from ductile iron.

Malleable iron is a group of heat-treatable cast irons with moderate to high strength, high modulus of elasticity (stiffness), good machinability, and good wear resistance. The five-digit designation roughly indicates the yield strength and the expected percent elongation of the iron. For example, Grade 40010 has a yield strength of 40 Ksi (276 MPa) and a 10% elongation. The strength properties listed in Appendix 8 are for the non-heat-treated condition. Higher strengths would result from heat treating. See ASTM specifications A 47-84 (Reapproved 1989) and A 220-88.

2–11 POWDERED METALS

Making parts with intricate shapes by powder metallurgy can sometimes eliminate the need for extensive machining. Metal powders are available in many formulations whose properties approach those of the wrought form of the metal. The processing involves preparing a preform by compacting the powder in a die under high pressure. Sintering at a high temperature to fuse the powder into a uniform mass is the next step. Re-pressing is sometimes done to improve properties or dimensional accuracy of the part. Typical parts made by the powder metallurgy (PM) process are gears, gear segments, cams, eccentrics, and various machine parts having oddly shaped holes or projections. Dimensional tolerances of 0.001 to 0.005 in (0.025–0.125 mm) are typical.

One disadvantage of PM parts is that they are usually brittle and should not be used in applications where high-impact loading is expected. Another important application is in sintered bearings, which are made to a relatively low density with consequent high porosity. The bearing is impregnated with a lubricant that may be sufficient for the life of the part. This type of material is discussed further in Chapter 16.

Manufacturers of metal powders have many proprietary formulations and grades. However, the Metal Powder Industries Federation (MPIF) is promoting standardization of materials. Figure 2–10 shows photographs of some powder metal parts. (See Reference 4.)

FIGURE 2–10
Examples of powder metal components (Imperial Clevite Inc., Powder Metal Products Division, Salem, IN.)

2–12
ALUMINUM

Aluminum is widely used for structural and mechanical applications. Chief among its attractive properties are light weight, good corrosion resistance, relative ease of forming and machining, and pleasing appearance. Its density is approximately one-third that of steel. However, its strength is somewhat lower, also. (See References 1, 2, and 9.) Table 2–6 lists the commonly used alloy groups.

The standard designations for aluminum alloys listed by the Aluminum Association use a four-digit system. The first digit indicates the alloy type according to the major alloying element. The second digit, if it is other than zero, indicates modifications of another alloy or limits placed on impurities in the alloy. The presence of impurities is particularly important for electrical conductors. Within each group there are several specific alloys, indicated by the last two digits in the designation.

Table 2–7 lists several common alloys, along with the forms in which they are typically produced and some of their major applications. The table also lists several of the fifty or more available alloys that span the range of typical applications. The table should aid you in selecting a suitable alloy for a particular application.

The mechanical properties of the aluminum alloys are highly dependent on their condition. For this reason, the specification of an alloy is incomplete without a reference to its *temper.* The following list describes the usual tempers given to aluminum alloys. Note that some alloys respond to heat treating, and others are processed by strain hardening. *Strain hardening* is controlled cold working of the alloy, in which increased working increases hardness and strength while reducing ductility. Commonly available tempers are described next.

TABLE 2–6 Aluminum alloy groups

Alloy designations (by major alloying element)
1xxx 99.00% or greater aluminum content
2xxx Copper
3xxx Manganese
4xxx Silicon
5xxx Magnesium
6xxx Magnesium and silicon
7xxx Zinc

TABLE 2–7 Common aluminum alloys and their uses

Alloy	Applications	Forms
1060	Chemical equipment and tanks	Sheet, plate, tube
1350	Electrical conductors	Sheet, plate, tube, rod, bar, wire, pipe, shapes
2014	Aircraft structures and vehicle frames	Sheet, plate, tube, rod, bar, wire, shapes, forgings
2024	Aircraft structures, wheels, machine parts	Sheet, plate, tube, rod, bar, wire, shapes, rivets
2219	Parts subjected to high temperatures (to 600°F)	Sheet, plate, tube, rod, bar, shapes, forgings
3003	Chemical equipment, tanks, cooking utensils, architectural parts	Sheet, plate, tube, rod, bar, wire, shapes, pipe, rivets, forgings
5052	Hydraulic tubes, appliances, sheet-metal fabrications	Sheet, plate, tube, rod, bar, wire, rivets
6061	Structures, vehicle frames and parts, marine uses	All forms
6063	Furniture, architectural hardware	Tube, pipe, extruded shapes
7001	High-strength structures	Tube, extruded shapes
7075	Aircraft and heavy-duty structures	All forms except pipe

F (as-fabricated): No special control of properties is provided. Actual limits are unknown. This temper should be accepted only when the part can be thoroughly tested prior to service.

O (annealed): A thermal treatment that results in the softest and lowest strength condition. Sometimes specified to obtain the most workable form of the alloy. The resulting part can be heat-treated for improved properties if it is made from alloys in the 2xxx, 4xxx, 6xxx, or 7xxx series. Also, the working itself may provide some improvement in properties similar to that produced by strain hardening for alloys in the 1xxx, 3xxx, and 5xxx series.

H (strain-hardened): A process of cold working under controlled conditions that produces improved, predictable properties for alloys in the 1xxx, 3xxx, and 5xxx groups. The greater the amount of cold work, the higher the strength and hardness, although the ductility is decreased. The *H* designation is followed by two or more digits (usually 12, 14, 16, or 18) that indicate progressively higher strength. However, several other designations are used.

T (heat-treated): A series of controlled heating and cooling processes applied to alloys in the 2xxx, 4xxx, 6xxx, and 7xxx groups. The letter *T* is followed by one or more numbers to indicate specific processes. The more common designations for mechanical and structural products are T4 and T6.

Property data for aluminum alloys are included in Appendix 9. Because these data are typical values, not guaranteed values, the supplier should be consulted for data at the time of purchase.

For mechanical design applications, alloy 6061 is one of the most versatile types. Note that it is available in virtually all forms, has good strength and corrosion resistance, and is heat-treatable to obtain a wide variety of properties. It also has good weldability. In its softer forms, it is easily formed and worked. Then, if higher strength is required, it can be heat-treated after forming. However, it has low machinability.

2–13
ZINC ALLOYS

Zinc is the fourth most commonly used metal in the world. Much of it is in the form of zinc galvanizing used as a corrosion inhibitor for steels, but very large quantities of zinc alloys are used in castings and for bearing materials. Figure 2–11 shows examples of cast zinc parts. (See Reference 17.)

High production quantities are made using zinc pressure die casting which results in very smooth surfaces and excellent dimensional accuracy. A variety of coating processes can be used to produce desirable finish appearance and to inhibit corrosion. Although the as-cast parts have inherently good corrosion resistance, the performance in some environments can be enhanced with chromate or phosphate treatments or anodizing. Painting and chrome plating are also used to produce a wide variety of attractive surface finishes.

In addition to die casting, zinc products are often made by permanent mold casting, graphite permanent mold casting, sand casting, and shell-mold casting. Other, less frequently used, processes are investment casting, low-pressure permanent mold casting, centrifugal casting, continuous casting, and rubber-mold casting. Plaster-mold casting is often used for prototyping. Continuous casting is used to produce standard shapes (rod, bar, tube, and slabs). Prototypes or finished products can then be machined from these shapes.

Zinc alloys typically contain aluminum and a small amount of magnesium. Some alloys include copper or nickel. The performance of the final products can be very sensitive to small amounts of other elements, and maximum limits are placed on the content of iron, lead, cadmium, and tin in some alloys.

FIGURE 2–11 Cast zinc parts (INTERZINC, Washington, D.C.)

The most widely used zinc casting alloy is called *alloy No. 3,* sometimes referred to as *Zamak 3.* It has 4% aluminum and 0.035% magnesium. Another is called *Zamak 5,* and it also contains 4% aluminum with 0.055% magnesium and 1.0% copper. A group of alloys having higher aluminum content are the ZA-alloys, with ZA-8, ZA-12, and ZA-27 being the most popular. Appendix 10 gives a summary of the composition and the typical properties of these alloys. As with most cast materials, some variations are to be expected with the size of the cast sections, the thermal treatment of the casting, the operating temperature of the product, and the quality assurance during the casting process.

2–14 TITANIUM

The applications of titanium include aerospace structures and components, chemical tanks and processing equipment, fluids-handling devices, and marine hardware. Titanium has very good corrosion resistance and a high strength-to-weight ratio. Its stiffness and density are between those of steel and aluminum; its modulus of elasticity is approximately 16×10^6 psi (110 GPa), and its density 0.160 lb/in^3 (4.429 kg/m^3). Typical yield strengths range from 25 to 175 Ksi (172–1 210 MPa). Disadvantages of titanium include relatively high cost and difficult machining.

The classification of titanium alloys usually falls into four types: commercially pure alpha titanium, alpha alloys, alpha-beta alloys, and beta alloys. Appendix 11 shows the properties of some of these grades. The term *alpha* refers to the hexagonal, close-packed, metallurgical structure that forms at low temperatures, and *beta* refers to the high-temperature, body-centered, cubic structure.

The grades of commercially pure titanium indicate the approximate expected yield strength of the material. For example, Ti-50A has an expected yield strength of 50 000 psi (345 MPa). As a class, these alloys exhibit only moderate strength but good ductility.

One popular grade of alpha alloy is titanium alloyed with 0.20% palladium (Pd), called *Ti-0.2Pd*. Its properties are listed in Appendix 11 for one heat-treat condition. Some alpha alloys have improved high-temperature strength and weldability.

Generally speaking, the alpha-beta alloys and the beta alloys are stronger forms of titanium. They are heat-treatable for close control of their properties. Since several alloys are available, a designer can tailor the properties to meet special needs for formability, machinability, forgeability, corrosion resistance, high-temperature strength, weldability, and creep resistance, as well as basic room-temperature strength and ductility. Alloy Ti-6Al-4V contains 6% aluminum and 4% vanadium and is used in a variety of aerospace applications.

2–15
COPPER, BRASS, AND BRONZE

Copper is widely used in its nearly pure form for electrical and plumbing applications because of its high electrical conductivity and good corrosion resistance. It is rarely used for machine parts because of its relatively low strength compared with that of its alloys, *brass* and *bronze*. (See Reference 4.)

Brass is a family of alloys of copper and zinc, with the content of zinc ranging from about 5% to 40%. Brass is used in marine applications because of its resistance to corrosion in salt water. Many brass alloys also have excellent machinability and are used as connectors, fittings, and other parts made on screw machines. *Yellow brass* contains about 30% or more of zinc and often contains a significant amount of lead to improve machinability. *Red brass* contains 5% to 15% zinc. Some alloys also contain tin, lead, nickel, or aluminum.

Bronze is a class of alloys of copper with several different elements, one of which is usually tin. They are useful in gears, bearings, and other applications where good strength and high wear resistance are desirable.

Wrought bronze alloys are available in four types:

Phosphor bronze: Copper-tin-phosphorus alloy

Leaded phosphor bronze: Copper-tin-lead-phosphorus alloy

Aluminum bronze: Copper-aluminum alloy

Silicon bronze: Copper-silicon alloy

Cast bronze alloys have four main types:

Tin bronze: Copper-tin alloy

Leaded tin bronze: Copper-tin-lead alloy

Nickel tin bronze: Copper-tin-nickel alloy

Aluminum bronze: Copper-aluminum alloy

The cast alloy called *manganese bronze* is actually a high-strength form of brass because it contains zinc, the characteristic alloying element of the brass family. Manganese bronze contains copper, zinc, tin, and manganese.

In the UNS, copper alloys are designated by the letter *C,* followed by a five-digit number. Numbers from 10000 to 79900 refer to wrought alloys; 80000 to 99900 refer to casting alloys. See Appendix 12 for typical properties.

2–18
COMPOSITE
MATERIALS

Composite materials are composed of two or more different materials that act together to produce properties that are different from, and generally superior to, those of the individual components. Typical composites include a polymeric resin matrix material with fibrous reinforcing material dispersed within it. Some advanced composites have a metal matrix. (See References 11 and 18.)

Designers can tailor the properties of composite materials to meet the specific needs of a particular application by the selection of each of several variables that determine the performance of the final product. Among the factors under the designer's control are the following:

1. Matrix resin or metal
2. Type of reinforcing fibers
3. Amount of fiber contained in the composite
4. Orientation of the fibers
5. Number of individual layers used
6. Overall thickness of the material
7. Orientation of the layers relative to each other
8. Combination of two or more types of composites or other materials into a composite structure

Typically, the filler is a strong, stiff material, whereas the matrix has a relatively low density. When the two materials bond together, much of the load-carrying ability of the composite is produced by the filler material. The matrix serves to hold the filler in a favorable orientation relative to the manner of loading and to distribute the loads to the filler. The result is a somewhat optimized composite that has high strength and high stiffness with low weight. Table 2–9 lists some of the composites formed by combinations of resins and fibers and their general characteristics and uses.

A virtually unlimited variety of composite materials can be produced by combining different matrix materials with different fillers in different forms and in different orientations. Some typical materials are listed below.

Matrix Materials

The following are among the more frequently used matrix materials:

- Thermoplastic polymers: Polyethylene, nylon, polypropylene, polystyrene, polyamides
- Thermosetting polymers: Polyester, epoxy, phenolic polyimide

TABLE 2–9   Examples of composite materials and their uses

Type of composite	Typical applications
Glass/epoxy	Automotive and aircraft parts, tanks, sporting goods, printed wiring boards
Boron/epoxy	Aircraft structures and stabilizers, sporting goods
Graphite/epoxy	Aircraft and spacecraft structures, sporting goods, agricultural equipment, material handling devices, medical devices
Aramid/epoxy	Filament-wound pressure vessels, aerospace structures and equipment, protective clothing, automotive components
Glass/polyester	Sheet-molding compound (SMC), body panels for trucks and cars, large housings

- Ceramics and glass
- Carbon and graphite
- Metals: Aluminum, magnesium, titanium

Forms of Filler Materials

Many forms of filler materials are used:

- Continuous fiber strand consisting of many individual filaments bound together
- Chopped strands in short lengths (0.75 to 50 mm or 0.03 to 2.00 in)
- Chopped strands randomly spread in the form of a mat
- Roving: A group of parallel strands
- Woven fabric made from roving or strands
- Metal filaments or wires
- Solid or hollow microspheres
- Metal, glass, or mica flakes
- Single-crystal whiskers of materials such as graphite, silicon carbide, and copper

Types of Filler Materials

Fillers, also called *fibers,* come in many types based on both organic and inorganic materials. The following are some of the more popular fillers:

- Glass fibers in five different types:
  A-glass: Good chemical resistance because it contains alkalis such as sodium oxide
  C-glass: Special formulations for even higher chemical resistance than A-glass
  E-glass: Widely used glass with good electrical insulating ability and good strength
  S-glass: High-strength, high-temperature glass
  D-glass: Better electrical properties than E-glass
- Quartz fibers and high-silica glass: Good properties at high temperatures up to 2 000°F (1 095°C)
- Carbon fibers made from PAN-base carbon (PAN is polyacrylonitrile): Approximately 95% carbon with very high modulus of elasticity
- Graphite fibers: Greater than 99% carbon and an even higher modulus of elasticity than carbon; the stiffest fibers typically used in composites
- Boron coated onto tungsten fibers: Good strength and a higher modulus of elasticity than glass
- Silicon carbide coated onto tungsten fibers: Strength and stiffness similar to those of boron/tungsten, but with higher temperature capability
- Aramid fibers: A member of the polyamide family of polymers; higher strength and stiffness with lower density as compared with glass; very flexible. (Aramid fibers produced by the DuPont Company carry the name *Kevlar.*™)

Processing of Composites

One method that is frequently used to produce composite products is first to place layers of sheet-formed fabrics on a form having the desired shape and then to impregnate the fabric with wet resin. Each layer of fabric can be adjusted in its orientation to produce special properties of the finished article. After the lay-up and resin impregnation are completed, the entire system is subjected to heat and pressure while a curing agent reacts with the base resin to produce cross-linking that binds all of the elements into a three-dimensional, unified structure. The polymer binds to the fibers and holds them in their preferred position and orientation during use.

An alternative method of fabricating composite products starts with a process of preimpregnating the fibers with the resin material to produce strands, tape, braids, or sheets. The resulting form, called a *prepreg,* can then be stacked into layers or wound onto a form to produce the desired shape and thickness. The final step is the curing cycle as described for the wet process.

Polyester-based composites are often produced as *sheet-molding compounds (SMC)* in which preimpregnated fabric sheets are placed into a mold and shaped and cured simultaneously under heat and pressure. Large body panels for automotive applications can be produced in this manner.

Pultrusion is a process in which the fiber reinforcement is coated with resin as it is pulled through a heated die to produce a continuous form in the desired shape. This process is used to produce rod, tubing, structural shapes (I-beams, channels, angles, and so on), tees, and hat sections used as stiffeners in aircraft structures.

Filament winding is used to make pipe, pressure vessels, rocket motor cases, instrument enclosures, and odd-shaped containers. The continuous filament can be placed in a variety of patterns, including helical, axial, and circumferential, to produce desired strength and stiffness characteristics.

Advantages of Composites

Designers typically seek to produce products that are safe, strong, stiff, lightweight, and highly tolerant of the environment in which the product will operate. Composites often excel in meeting these objectives when compared with alternative materials such as metals, wood, and unfilled plastics. Two parameters that are used to compare materials are *specific strength* and *specific modulus,* defined as follows:

> **Specific strength** *is the ratio of the tensile strength of a material to its specific weight.*

> **Specific modulus** *is the ratio of the modulus of elasticity of a material to its specific weight.*

Because the modulus of elasticity is a measure of the stiffness of a material, the specific modulus is sometimes called *specific stiffness.*

Although obviously not a length, both of these quantities have the *unit* of length, derived from the ratio of the units for strength or modulus of elasticity and the units for specific weight. In the U.S. Customary System, the units for tensile strength and modulus of elasticity are lb/in^2, whereas specific weight (weight per unit volume) is in lb/in^3. Thus, the unit for specific strength or specific modulus is inches. In the SI, strength and modulus are expressed in N/m^2 (pascals), whereas specific weight is in N/m^3. Then the unit for specific strength or specific modulus is meters.

Table 2–10 gives comparisons of the specific strength and specific stiffness of selected composite materials with certain steel, aluminum, and titanium alloys. Figure 2–12 shows a comparison of these materials using bar charts. Figure 2–13 is a plot of these data with specific strength on the vertical axis and specific modulus on the horizontal axis. When weight is critical, the ideal material will lie in the upper-right part of this chart. Note that

TABLE 2–10 Comparison of specific strength and specific modulus for selected materials

Material	Tensile strength, s_u (Ksi)	Specific weight, γ (lb/in^3)	Specific strength (in)	Specific modulus (in)
Metals				
Steel ($E = 30 \times 10^6$ psi)				
AISI 1020 HR	55	0.283	0.194×10^6	1.06×10^8
AISI 5160 OQT 700	263	0.283	0.929×10^6	1.06×10^8
Aluminum ($E = 10.0 \times 10^6$ psi)				
6061-T6	45	0.098	0.459×10^6	1.02×10^8
7075-T6	83	0.101	0.822×10^6	0.99×10^8
Titanium ($E = 16.5 \times 10^6$ psi)				
Ti-6Al-4V, quenched and aged at 1000°F	160	0.160	1.00×10^6	1.03×10^8
Composites				
Glass/epoxy composite ($E = 4.0 \times 10^6$ psi)				
34% fiber content	114	0.061	1.87×10^6	0.66×10^8
Aramid/epoxy composite ($E = 11.0 \times 10^6$ psi)				
60% fiber content	200	0.050	4.0×10^6	2.20×10^8
Boron/epoxy composite ($E = 30.0 \times 10^6$ psi)				
60% fiber content	270	0.075	3.60×10^6	4.00×10^8
Graphite/epoxy composite ($E = 19.7 \times 10^6$ psi)				
62% fiber content	278	0.057	4.86×10^6	3.45×10^8
Graphite/epoxy composite ($E = 48 \times 10^6$ psi)				
Ultrahigh modulus	160	0.058	2.76×10^6	8.28×10^8

data in these charts and figures are for composites having the filler materials aligned in the most favorable direction to withstand the applied loads.

Advantages of composites can be summarized as follows:

1. Specific strengths for composite materials can range as high as five times those of high-strength steel alloys.

2. Specific modulus values for composite materials can be as high as eight times those for steel, aluminum, or titanium alloys.

3. Composite materials typically perform better than steel or aluminum in applications where cyclic loads can lead to the potential for fatigue failure.

4. Where impact loads and vibrations are expected, composites can be specially formulated with materials that provide high toughness and a high level of damping.

5. Some composites have much higher wear resistance than metals.

6. Careful selection of the matrix and filler materials can provide superior corrosion resistance.

7. Dimensional changes due to changes in temperature are typically much less for composites than for metals.

8. Because composite materials have properties that are highly directional, designers can tailor the placement of reinforcing fibers in directions that provide the required strength and stiffness under the specific loading conditions to be encountered.

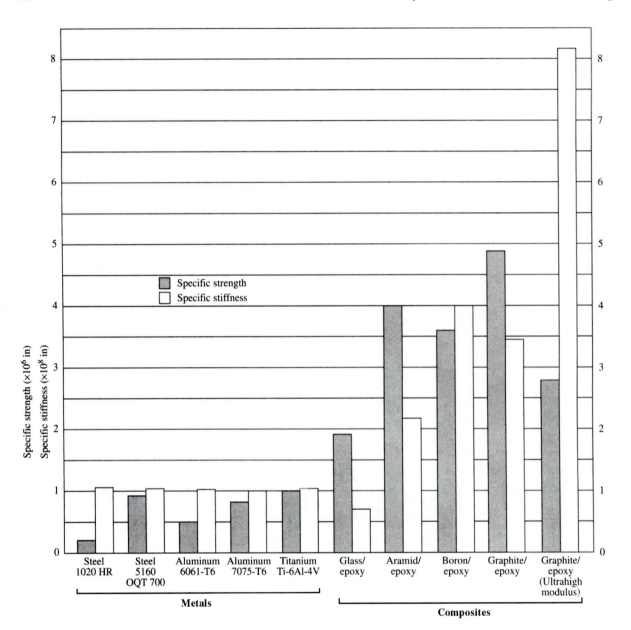

FIGURE 2–12   Comparison of specific strength and specific stiffness for selected metals and composites

9. Composite structures can often be made in complex shapes in one piece, thus reducing the number of parts in a product and the number of fastening operations required. The elimination of joints typically improves the reliability of such structures as well.

10. Composite structures are typically made in their final form directly or in a near-net shape, thus reducing the number of secondary operations required.

Limitations of Composites

Designers must balance many properties of materials in their designs while simultaneously considering manufacturing operations, costs, safety, life, and service of the product. Listed next are some of the major concerns when using composites.

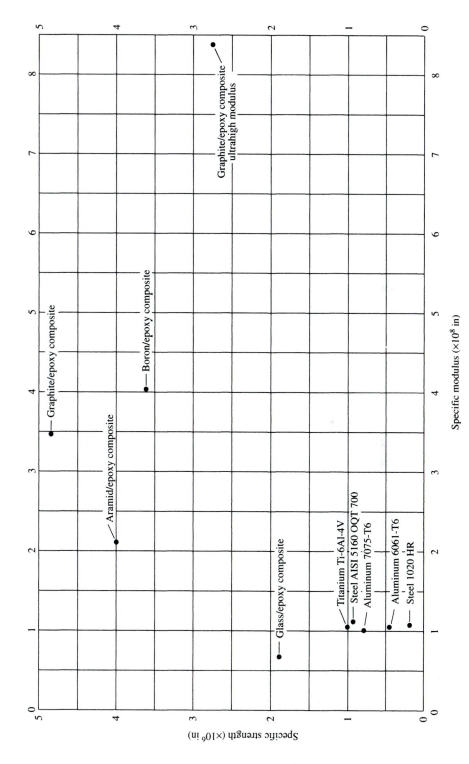

FIGURE 2–13 Specific strength versus specific modulus for selected metals and composites

1. Material costs for composites are typically higher than for many alternative materials.

2. Fabrication techniques are quite different from those used to shape metals. New manufacturing equipment may be required, along with additional training for production operators.

3. The performance of products made from some composite production techniques is subject to a wider range of variability than the performance of products made from most metal fabrication techniques.

4. The operating temperature limits for composites having a polymeric matrix are typically 500°F (260°C). (But ceramic or metal matrix composites can be used at higher temperatures such as those found in engines.)

5. The properties of composite materials are not isotropic: Properties vary dramatically with the direction of the applied loads. Designers must account for these variations to ensure safety and satisfactory operation under all expected types of loading.

6. At this time, many designers lack understanding of the behavior of composite materials and the details of predicting failure modes. Whereas major advancements have been made in certain industries such as the aerospace and recreational equipment fields, there is a need for more general understanding about designing with composite materials.

7. The analysis of composite structures requires detailed knowledge of more properties of the materials than would be required for metals.

8. Inspection and testing of composite structures are typically more complicated and less precise than for metal structures. Special nondestructive techniques may be required to ensure that there are no major voids in the final product that could seriously weaken the structure. Testing of the complete structure may be required rather than testing of simply a sample of the material because of the interaction of different parts on each other and because of the directionality of the material properties.

9. Repair and maintenance of composite structures are serious concerns. Some of the initial production techniques require special environments of temperature and pressure that may be difficult to reproduce in the field when damage repair is required. Bonding of a repaired area to the parent structure may also be difficult.

Laminated Composite Construction

Many structures made from composite materials are made from several layers of the basic material containing both the matrix and the reinforcing fibers. The manner in which the layers are oriented relative to one another affects the final properties of the completed structure.

As an illustration, consider that each layer is made from a set of parallel strands of the reinforcing filler material, such as E-glass fibers, embedded in the resin matrix, such as polyester. As mentioned previously, in this form, the material is sometimes called a *prepreg,* indicating that the filler has been preimpregnated with the matrix prior to the forming of the structure and the curing of the assembly. To produce the maximum strength and stiffness in a particular direction, several layers or plies of the prepreg could be laid on top of one another with all of the fibers aligned in the direction of the expected tensile load. This is called a *unidirectional laminate.* After curing, the laminate would have a very high strength and stiffness when loaded in the direction of the strands, called the *longitudinal* direction. However, the resulting product would have a very low strength and stiffness in the direction perpendicular to the fiber direction, called the *transverse* direction. If any off-axis loads are encountered, the part may fail or deform significantly. Table 2–11 gives sample data for a unidirectional laminated, carbon/epoxy composite.

FIGURE 2–14 Multilayer, laminated, composite construction designed to produce quasi-isotropic properties

TABLE 2–11 Examples of the effect of laminate construction on strength and stiffness

| | Tensile strength | | | | Modulus of elasticity | | | |
| | Longitudinal | | Transverse | | Longitudinal | | Transverse | |
Laminate type	Ksi	MPa	Ksi	MPa	10^6 psi	GPa	10^6 psi	GPa
Unidirectional	200	1 380	5	34	21	145	1.6	11
Quasi-isotropic	80	552	80	552	8	55	8	55

To overcome the lack of off-axis strength and stiffness, laminated structures should be made with a variety of orientations of the layers. One popular arrangement is shown in Figure 2–14. Naming the longitudinal direction of the surface layer the *0° ply,* this structure is referred to as

$$0°, 90°, +45°, -45°, -45°, +45°, 90°, 0°$$

The symmetry and the balance of this type of layering technique result in more nearly uniform properties in two directions. The term *quasi-isotropic* is sometimes used to describe such a structure. Note that the properties perpendicular to the faces of the layered structure (through the thickness) are still quite low because fibers do not extend in that direction. Also, the strength and the stiffness in the primary directions are somewhat lower than if the plies were aligned in the same direction. Table 2–11 also shows sample data for a quasi-isotropic laminate compared with one having unidirectional fibers in the same matrix.

Predicting Composite Properties

The following discussion summarizes some of the important variables needed to define the properties of a composite. The subscript c refers to the composite, m refers to the matrix,

FIGURE 2–15   Stress versus strain for fiber and matrix materials

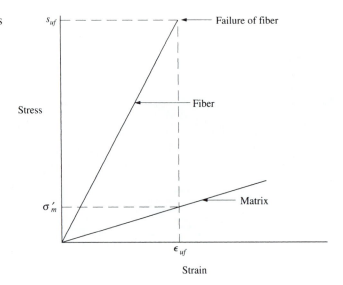

and f refers to the fibers. The strength and the stiffness of a composite material depend on the elastic properties of the fiber and matrix components. But another parameter is the relative volume of the composite composed of fibers, V_f, and that composed of the matrix material, V_m. That is,

$$V_f = \text{volume fraction of fiber in the composite}$$
$$V_m = \text{volume fraction of matrix in the composite}$$

Note that for a unit volume, $V_f + V_m = 1$; thus, $V_m = 1 - V_f$.

We will use an ideal case to illustrate the way in which the strength and the stiffness of a composite can be predicted. Consider a composite with unidirectional, continuous fibers aligned in the direction of the applied load. The fibers are typically much stronger and stiffer than the matrix material. Furthermore, the matrix will be able to undergo a larger strain before fracture than the fibers can. Figure 2–15 shows these phenomena on a plot of stress versus strain for the fibers and the matrix. We will use the following notation for key parameters from Figure 2–15:

$$s_{uf} = \text{ultimate strength of fiber}$$
$$\epsilon_{uf} = \text{strain in the fiber corresponding to its ultimate strength}$$
$$\sigma'_m = \text{stress in the matrix at the same strain as } \epsilon_{uf}$$

The ultimate strength of the composite, s_{uc}, is at some intermediate value between s_{uf} and σ'_m, depending on the volume fraction of fiber and matrix in the composite. That is,

**Rule of Mixtures for
Ultimate Strength**

$$s_{uc} = s_{uf}V_f + \sigma'_m V_m \qquad (2\text{–}8)$$

At any lower level of stress, the relationship among the overall stress in the composite, the stress in the fibers, and the stress in the matrix follows a similar pattern:

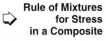
**Rule of Mixtures
for Stress
in a Composite**

$$\sigma_c = \sigma_f V_f + \sigma_m V_m \qquad (2\text{–}9)$$

Figure 2–16 illustrates this relationship on a stress-strain diagram.

FIGURE 2–16
Relationship among stresses and strains for a composite and its fiber and matrix materials

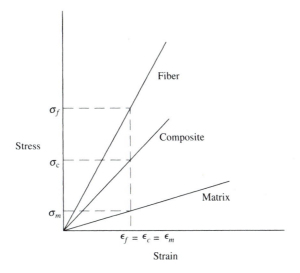

Both sides of Equation (2–9) can be divided by the strain at which these stresses occur. Since for each material, $\sigma/\epsilon = E$, the modulus of elasticity for the composite can be shown as

Rule of Mixtures for Modulus of Elasticity

$$E_c = E_f V_f + E_m V_m \qquad (2\text{–}10)$$

The density of a composite can be computed in a similar fashion:

Rule of Mixtures for Density of a Composite

$$\rho_c = \rho_f V_f + \rho_m V_m \qquad (2\text{–}11)$$

As mentioned previously (Section 2–2), density is defined as *mass per unit volume.* A related property, specific weight, is defined as *weight per unit volume* and is denoted by the symbol γ (the Greek letter gamma). The relationship between density and specific weight is simply $\gamma = \rho g$, where g is the acceleration due to gravity. Multiplying each term in Equation (2–11) by g gives the formula for the specific weight of a composite:

Rule of Mixtures for Specific Weight of a Composite

$$\gamma_c = \gamma_f V_f + \gamma_m V_m \qquad (2\text{–}12)$$

The forms of Equations (2–8) through (2–12) are examples of *the rules of mixtures.*

Table 2–12 lists example values for the properties of some matrix and filler materials. Remember that wide variations can occur in such properties, depending on the exact formulation and the condition of the materials.

Example Problem 2–1   Compute the expected properties of ultimate tensile strength, modulus of elasticity, and specific weight of a composite made from unidirectional strands of carbon-PAN fibers in an epoxy matrix. The volume fraction of fibers is 30%. Use data from Table 2–12.

Solution   Objective   Compute the expected values of s_{uc}, E_c, and γ_c for the composite.

Given   Matrix-epoxy: $s_{um} = 18$ Ksi; $E_m = 0.56 \times 10^6$ psi; $\gamma_m = 0.047$ lb/in³.

Fiber-carbon-PAN: $s_{uf} = 470$ Ksi; $E_f = 33.5 \times 10^6$ psi; $\gamma_f = 0.064$ lb/in³.

Volume fraction of fiber: $V_f = 0.30$, and $V_m = 1.0 - 0.30 = 0.70$.

TABLE 2–12   Example properties of matrix and filler materials

	Tensile strength		Tensile modulus		Specific weight	
	Ksi	MPa	10^6 psi	GPa	lb/in^3	kN/m^3
Matrix materials:						
Polyester	10	69	0.40	2.76	0.047	12.7
Epoxy	18	124	0.56	3.86	0.047	12.7
Aluminum	45	310	10.0	69	0.100	27.1
Titanium	170	1 170	16.5	114	0.160	43.4
Filler materials:						
S-glass	600	4 140	12.5	86.2	0.09	24.4
Carbon-PAN	470	3 240	33.5	231	0.064	17.4
Carbon-PAN (high-strength)	820	5 650	40	276	0.065	17.7
Carbon (high-modulus)	325	2 200	100	690	0.078	21.2
Aramid	500	3 450	19.0	131	0.052	14.1

Analysis and Results   The ultimate tensile strength, s_{uc}, is computed from Equation (2–8):

$$s_{uc} = s_{uf}V_f + \sigma'_m V_m$$

To find σ'_m, we first find the strain at which the fibers would fail at s_{uf}. Assume that the fibers are linearly elastic to failure. Then

$$\epsilon_f = s_{uf}/E_f = (470 \times 10^3 \text{ psi})/(33.5 \times 10^6 \text{ psi}) = 0.014$$

At this same strain, the stress in the matrix is

$$\sigma'_m = E_m\epsilon = (0.56 \times 10^6 \text{ psi})(0.014) = 7\ 840 \text{ psi}$$

Then, in Equation (2–8),

$$s_{uc} = (470\ 000 \text{ psi})(0.30) + (7\ 840 \text{ psi})(0.70) = 146\ 500 \text{ psi}$$

The modulus of elasticity computed from Equation (2–10):

$$E_c = E_fV_f + E_mV_m = (33.5 \times 10^6)(0.30) + (0.56 \times 10^6)(0.70)$$
$$E_c = 10.4 \times 10^6 \text{ psi}$$

The specific weight is computed from Equation (2–12):

$$\gamma_c = \gamma_fV_f + \gamma_mV_m = (0.064)(0.30) + (0.047)(0.70) = 0.052 \text{ lb/in}^3$$

Summary of Results
$$s_{uc} = 146\ 500 \text{ psi}$$
$$E_c = 10.4 \times 10^6 \text{ psi}$$
$$\gamma_c = 0.052 \text{ lb/in}^3$$

Comment   Note that the resulting properties for the composite are intermediate between those for the fibers and the matrix.

Design Guidelines for Members Made from Composites

The most important difference between designing with metals and designing with composites is that metals are typically taken to be homogeneous with isotropic strength and stiffness properties, whereas composites are decidedly *not* homogeneous or isotropic.

The failure modes of composite materials are complex. Tensile failure when the load is in-line with continuous fibers occurs when the individual fibers break. If the composite is made with shorter, chopped fibers, failure occurs when the fibers are pulled free from the matrix. Tensile failure when the load is perpendicular to continuous fibers occurs when the matrix itself fractures. If the fibers are in a woven form, or if a mat having shorter, randomly oriented fibers is used, other failure modes, such as fiber breakage or pullout, prevail. Such composites would have more nearly equal properties in any direction, or, as shown in Figure 2–13, multilayer laminate construction can be used.

Thus, an important design guideline to produce optimum strength is as follows:

Align the fibers with the direction of the load.

Another important failure mode is *interlaminar shear,* in which the plies of a multilayer composite separate under the action of shearing forces. The following is another design guideline:

Avoid shear loading, if possible.

Connections to composite materials are sometimes difficult to accomplish and provide places where fractures or fatigue failure could initiate. The manner of forming composites often allows the integration of several components into one part. Brackets, ribs, flanges, and the like, can be molded in along with the basic form of the part. The design guideline, then, is the following:

Combine several components into an integral structure.

When high panel stiffness is desired to resist flexure, as in beams or in broad panels such as floors, the designer can take advantage of the fact that the most effective material is near the outside surfaces of the panel or beam shape. Placing the high-strength fibers on these outer layers while filling the core of the shape with a light, yet rigid, material produces an efficient design in terms of weight for a given strength and stiffness. Figure 2–17 illustrates some examples of such designs. Another design guideline follows:

Use light core material covered with strong composite layers.

Because most composites use a polymeric material for the matrix, the temperatures that they can withstand are limited. Both strength and stiffness decrease as temperature increases. The polyimides provide better high-temperature properties [up to 600°F (318°C)] than most other polymer matrix materials. Epoxies are typically limited to from 250°F to 350°F (122°C–178°C). Any application above room temperature should be checked with material suppliers. The following is a design guideline:

Avoid high temperatures.

As described earlier in this section, many different fabrication techniques are used for composite materials. The shape can dictate a part's manufacturing technique. This is a good reason to implement the principles of concurrent engineering and adopt another design guideline:

Involve manufacturing considerations early in the design.

FIGURE 2–17
Laminated panels with
lightweight cores

Composite outer skin

Foam core

Composite inner skin

(*a*) Curved panel with foam core and composite skins

Glass/epoxy layers
with decorative surface

Honeycomb core

Glass/epoxy layer

(*b*) Flat panel with honeycomb core and composite skins

REFERENCES

1. Aluminum Association. *Aluminum Standards and Data.* Washington, DC: The Aluminum Association, 1997.

2. American Society for Testing and Materials. *Metals and Alloys in the Unified Numbering System.* 7th ed. West Conshohocken, PA: American Society for Testing and Materials. Publication DS 56F, 1996. Jointly developed by ASTM and the Society of Automotive Engineers (SAE).

3. American Society for Testing and Materials. *Standard Practice for Numbering Metals and Alloys (UNS).* West Conshohocken, PA: American Society for Testing and Materials. Standard E527-83 (1991), 1997.

4. ASM International. *ASM Handbook.* Vol. 1, *Properties and Selection: Iron, Steels, and High-Performance Alloys* (1990). Vol. 2, *Properties and Selection: Nonferrous Alloys and Special-Purpose Materials* (1991). Vol. 3, *Alloy Phase Diagrams* (1992). Vol. 4, *Heat Treating* (1991). Vol. 7, *Powder Metallurgy* (1984). Vol. 20, *Materials Selection and Design* (1997). Materials Park, OH: ASM International.

5. ASM International. *ASM Specialty Handbook: Carbon and Alloy Steels.* Edited by J. R. Davis. Materials Park, OH: ASM International, 1996.

6. ASM International. *ASM Specialty Handbook: Stainless Steels.* Edited by J. R. Davis. Materials Park, OH: ASM International, 1994.

7. ASM International. *ASM Specialty Handbook: Tool Materials.* Edited by J. R. Davis. Materials Park, OH: ASM International, 1995.

8. ASM International. *ASM Specialty Handbook: Heat-Resistant Materials.* Edited by J. R. Davis. Materials Park, OH: ASM International, 1997.

9. ASM International. *ASM Specialty Handbook: Aluminum and Aluminum Alloys.* Edited by J. R. Davis. Materials Park, OH: ASM International, 1993.

10. ASM International. *ASM Specialty Handbook: Cast Irons.* Edited by J. R. Davis. Materials Park, OH: ASM International, 1996.

11. ASM International. *ASM Engineered Materials Handbook: Composites.* Materials Park, OH: ASM International, 1987.

12. ASM International. *ASM Engineered Materials Handbook: Engineering Plastics.* Materials Park, OH: ASM International, 1988.

13. Bethlehem Steel Corporation. *Modern Steels and Their Properties.* Bethlehem, PA: Bethlehem Steel Corporation, 1980.

14. Brooks, Charlie R. *Principles of the Heat Treatment of Plain Carbon and Low Alloy Steels.* Materials Park, OH: ASM International, 1996.

15. Budinski, Kenneth G. *Engineering Materials: Properties and Selection.* 5th ed. Upper Saddle River, NJ: Prentice-Hall, 1996.

16. Budinski, Kenneth G. *Surface Engineering for Wear Resistance.* Upper Saddle River, NJ: Prentice-Hall, 1988.

17. INTERZINC. *Zinc Casting: A Systems Approach.* Algonac, MI: INTERZINC.

18. Jang, Bor Z. *Advanced Polymer Composites: Principles and Applications.* Materials Park, OH: ASM International, 1994.

19. Muccio, E. A. *Plastic Part Technology.* Materials Park, OH: ASM International, 1991.

20. Penton Publishing. *Machine Design Magazine* vol. 69. Cleveland, OH: Penton Publishing, 1997.

21. Zinn, S., and S. L. Semiatin. *Elements of Induction Heating: Design, Control, and Applications.* Materials Park, OH: ASM International, 1984.

PROBLEMS

1. Define *ultimate tensile strength*.

2. Define *yield point*.

3. Define *yield strength* and tell how it is measured.

4. What types of materials would have a yield point?

5. What is the difference between proportional limit and elastic limit?

6. Define *Hooke's law*.

7. What property of a material is a measure of its stiffness?

8. What property of a material is a measure of its ductility?

9. If a material is reported to have a percent elongation in a 2.00-in gage length of 2%, is it ductile?

10. Define *Poisson's ratio*.

11. If a material has a tensile modulus of elasticity of 114 GPa and a Poisson's ratio of 0.33, what is its modulus of elasticity in shear?

12. A material is reported to have a Brinell hardness of 525. What is its approximate hardness on the Rockwell C scale?

13. A steel is reported to have a Brinell hardness of 450. What is its approximate tensile strength?

   For Problems 14–17, describe what is wrong with each statement.

14. "After annealing, the steel bracket had a Brinell hardness of 750."

15. "The hardness of that steel shaft is HRB 120."

16. "The hardness of that bronze casting is HRC 12."

17. "Based on the fact that this aluminum plate has a hardness of HB 150, its approximate tensile strength is 75 Ksi."

18. Name two tests used to measure impact strength.

19. What are the principal constituents in steels?

20. What are the principal alloying elements in AISI 4340 steel?

21. How much carbon is in AISI 4340 steel?

22. What is the typical carbon content of a low-carbon steel? Of a medium-carbon steel? Of a high-carbon steel?

23. How much carbon does a bearing steel typically contain?

24. What is the main difference between AISI 1213 steel and AISI 12L13 steel?

25. Name four materials that are commonly used for shafts.

26. Name four materials that are typically used for gears.

27. Describe the properties desirable for the auger blades of a post hole digger, and suggest a suitable material.

28. Appendix 3 lists AISI 5160 OQT 1000. Describe the basic composition of this material, how it was processed, and its properties in relation to other steels listed in that table.

29. If a shovel blade is made from AISI 1040 steel, would you recommend flame hardening to give its edge a surface hardness of HRC 40? Explain.

30. Describe the differences between through-hardening and carburizing.

31. Describe the process of induction hardening.

32. Name 10 steels used for carburizing. What is their approximate carbon content prior to carburizing?

33. What types of stainless steels are nonmagnetic?

34. What is the principal alloying element that gives a stainless steel corrosion resistance?

35. Of what material is a typical wide-flange beam made?

36. With regard to structural steels, what does the term *HSLA* mean? What strengths are available in HSLA steel?

37. Name three types of cast iron.

38. Describe the following cast iron materials according to type, tensile strength, yield strength, ductility, and stiffness:

   ASTM A48-83, Grade 30

   ASTM A536-84, Grade 100-70-03

   ASTM A47-84, Grade 35018

   ASTM A220-88, Grade 70003

39. Describe the process of making parts from powdered metals.

40. What properties are typical for parts made from Zamak 3 zinc casting alloy?

41. What are the typical uses for Group D tool steels?

42. What does the suffix *O* in aluminum 6061-O represent?

43. What does the suffix *H* in aluminum 3003-H14 represent?

44. What does the suffix *T* in aluminum 6061-T6 represent?

45. Name the aluminum alloy and condition that has the highest strength of those listed in Appendix 9.

46. Which is one of the most versatile aluminum alloys for mechanical and structural uses?

47. Name three typical uses for titanium alloys.

48. What is the principal constituent of bronze?

49. Describe the bronze having the UNS designation C86200.

50. Name two typical uses for bronze in machine design.

51. Describe the difference between thermosetting plastics and thermoplastics.

52. Suggest a suitable plastic material for each of the following uses:
(a) Gears
(b) Football helmets
(c) Transparent shield
(d) Structural housing
(e) Pipe
(f) Wheels
(g) Electrical switch-gear, structural part

53. Name eight factors over which the designer has control when specifying a composite material.

54. Define the term *composite.*

55. Name four base resins often used for composite materials.

56. Name four types of reinforcement fibers used for composite materials.

57. Name three types of composite materials used for sporting equipment, such as tennis rackets, golf clubs, and skis.

58. Name three types of composite materials used for aircraft and aerospace structures.

59. What base resin and reinforcement are typically used for sheet-molding compound (SMC)?

60. For what applications are sheet-molding compounds used?

61. Describe six forms in which reinforcing fibers are produced.

62. Describe *wet processing* of composite materials.

63. Describe *preimpregnated materials.*

64. Describe the production processing of sheet-molding compounds.

65. Describe *pultrusion,* and list four shapes produced by this process.

66. Describe *filament winding* and four types of products made by this process.

67. Define the term *specific strength* as it is applied to structural materials.

68. Define the term *specific stiffness* as it is applied to structural materials.

69. Discuss the advantages of composite materials relative to metals with regard to specific strength and specific stiffness.

70. Compare the specific strength of AISI 1020 hot-rolled steel with that of AISI 5160 OQT 700 steel, the two aluminum alloys 6061-T6 and 7075-T6, and titanium Ti-6Al-4V.

71. Compare the specific stiffness of AISI 1020 hot-rolled steel with that of AISI 5160 OQT 700 steel, the two aluminum alloys 6061-T6 and 7075-T6, and titanium Ti-6Al-4V.

72. Compare the specific strengths of each of the five composite materials shown in Figure 2–12 with that of AISI 1020 hot-rolled steel.

73. Compare the specific stiffness of each of the five composite materials shown in Figure 2–12 with that of AISI 1020 hot-rolled steel.

74. Describe the general construction of a composite material identified as [0/+30/−30/90].

75. List and discuss six design guidelines for the application of composite materials.

76. Why is it desirable to form a composite material in layers or plies with the angle of orientation of the different plies in different directions?

77. Why is it desirable to form a composite structural element with relatively thin skins of the stronger composite material over a core of light foam?

78. Describe why concurrent engineering and early manufacturing involvement are important when you are designing parts made from composite materials.

3

Stress and Deformation Analysis

Stress and Deformation Analysis

Discussion Map

- As a designer you are responsible for ensuring the safety of the components and the systems that you design.
- You must apply your prior knowledge of the principles of strength of materials.

Discover

How could consumer products and machines fail?
Describe some product failures that you have seen.

This chapter presents a brief review of the fundamentals of stress analysis. It will help you design products that do not fail, and it will prepare you for other topics later in this book.

A designer is responsible for ensuring the safety of the components and systems that he or she designs. Many factors affect safety, but one of the most critical aspects of design safety is that the level of stress to which a machine component is subjected must be safe under reasonably foreseeable conditions. This principle implies, of course, that nothing actually breaks. Safety may also be compromised if components are permitted to deflect excessively, even though nothing breaks.

You have already studied the principles of strength of materials to learn the fundamentals of stress analysis. Thus, at this point, you should be competent to analyze load-carrying members for stress and deflection due to direct tensile and compressive loads, direct shear, torsional shear, and bending.

Think, now, about consumer products and machines with which you are familiar, and try to explain how they *could fail.* Of course, we do not expect them to fail, because most such products are well designed. But some do fail. Can you recall any? How did they fail? What were the operating conditions when they failed? What was the material of the components that failed? Can you visualize and describe the kinds of loads that were placed on the components that failed? Were they subjected to bending, tension, compression, shear, or torsion? Could there have been more than one type of stress acting at the same time? Are there evidences of accidental overloads? Should such loads have been anticipated by the designer? Could the failure be due to the manufacture of the product rather than its design?

Talk about product and machine failures with your associates and your instructor. Consider parts of your car, home appliances, lawn maintenance equipment, or equipment where you have worked. If possible, bring failed components to the meetings with your associates, and discuss the components and their failure.

Most of this book emphasizes developing special methods to analyze and design machine elements. But these methods are all based on the fundamentals of stress analysis, and it is assumed that you have completed a course in strength of materials. This chapter presents a brief review of the fundamentals. (See References 1, 3, 4, and 6.)

 You Are the Designer

In Chapter 2, you were the designer of a conveyor for a pharmaceutical company. Continuing that scenario, consider that you are in the process of designing the frame of the conveyor. What do you need to know to complete the design? What loads are exerted on the frame? How and where are they applied? What type of structure are you going to propose? What will be the shape of the members that comprise the structure? What materials will they be made from? After answers to these questions are known, you must complete a very important part of the design process, the *stress analysis,* to determine the final dimensions of the frame members that will ensure that the frame is safe and is able to provide a sturdy, rigid support for the functional components of the conveyor.

**3–1
OBJECTIVES OF
THIS CHAPTER**

After completing this chapter, you will:

1. Have reviewed the principles of stress and deformation analysis for several kinds of stresses, including the following:

   Direct tension and compression

   Direct shear

   Torsional shear for both circular and noncircular sections

   Vertical shearing stresses in beams

   Bending

2. Be able to interpret the nature of the stress at a point by drawing the *stress element* at any point in a load-carrying member for a variety of types of loads.

3. Have reviewed the importance of the *flexural center* of a beam cross section with regard to the alignment of loads on beams.

4. Have reviewed beam-deflection formulas.

5. Be able to analyze beam-loading patterns that produce abrupt changes in the magnitude of the bending moment in the beam.

6. Be able to use the principle of superposition to analyze machine elements that are subjected to loading patterns that produce combined stresses.

7. Be able to properly apply stress concentration factors in stress analyses.

**3–2
PHILOSOPHY OF
A SAFE DESIGN**

In this book, every design approach will ensure that the stress level is below yield in ductile materials, automatically ensuring that the part will not break under a static load. For brittle materials, we will ensure that the stress levels are well below the ultimate tensile strength. We will also analyze deflection where it is critical to safety or performance of a part.

Two other failure modes that apply to machine members are fatigue and wear. *Fatigue* is the response of a part subjected to repeated loads (see Chapter 5). *Wear* is discussed within the chapters devoted to the machine elements, such as gears, bearings, and chains, for which it is a major concern.

**3–3
REPRESENTING
STRESSES ON
A STRESS
ELEMENT**

To visualize the state of stress at a point within a load-carrying member, it is helpful to consider a small element, called the *stress element,* as shown in Figure 3–1. In most cases it is sufficient to use a two-dimensional element that shows the stress condition in a given plane such as *x-y, x-z,* or *y-z.* Three different types of stresses are shown as examples:

1. Direct tensile stress acting in the *x*-direction.

2. Direct compressive stress acting in the *y*-direction.

3. Shear stress in the *x-y* plane.

The *x-* and *y*-directions must be related to a specific orientation of the load-carrying member. Note that the stress elements must be shown to be in equilibrium under the action of the applied stresses. There can be no unbalanced forces or moments.

Normal stresses, either tension or compression, are shown as vectors acting perpendicular to opposite faces of the element. Tensile stresses tend to pull on the element, while compressive stresses tend to crush it.

FIGURE 3–1 Stress elements for three types of stresses

(*a*) Direct tension (*b*) Direct compression (*c*) Pure shear

Shear stresses are created by direct shear, vertical shear in beams, or torsion. In each case, the action on an element subjected to shear is a tendency to *cut* the element by exerting a stress downward on one face while simultaneously exerting a stress upward on the opposite, parallel face. This action is that of a simple pair of shears or scissors. But note that if only one pair of shear stresses acts on a stress element, it will not be in equilibrium. Rather, it will tend to spin because the pair of shear stresses forms a couple. To produce equilibrium, a second pair of shear stresses on the other two faces of the element must exist, acting in a direction that opposes the first pair.

In summary, shear stresses on an element will always be shown as two pairs of equal stresses acting on (parallel to) the four sides of the element. Figure 3–1(c) shows an example.

Sign Convention for Shear Stresses

This book adopts the following convention:

Positive shear stresses tend to rotate the element in a clockwise direction.

Negative shear stresses tend to rotate the element in a counterclockwise direction.

A double subscript notation is used to denote shear stresses in a plane. For example, in Figure 3–1(c), drawn for the *x-y* plane, the pair of shear stresses, τ_{xy}, indicates a shear stress acting on the element face that is perpendicular to the *x*-axis and parallel to the *y*-axis. Then τ_{yx} acts on the face that is perpendicular to the *y*-axis and parallel to the *x*-axis. In this example, τ_{xy} is positive and τ_{yx} is negative.

**3–4
DIRECT
STRESSES:
TENSION AND
COMPRESSION**

Stress can be defined as the internal resistance offered by a unit area of a material to an externally applied load. *Normal stresses* (σ) are either *tensile* (positive) or *compressive* (negative).

For a load-carrying member in which the external load is uniformly distributed across the cross-sectional area of the member, the magnitude of the stress can be calculated from the direct stress formula:

⇨ **Direct Tensile or Compressive Stress**

$$\sigma = \text{force/area} = F/A \tag{3–1}$$

The units for stress are always expressed as *force per unit area*. (See Table 1–2.) In the U.S. Customary Unit System, the units for stress are generally expressed in lb/in^2 or psi. For convenience, stress is sometimes expressed as kips/in^2 (Ksi). In the SI, the unit for force is the newton (N). Thus, the standard unit for stress is N/m^2, called the *pascal* (Pa). Since the pascal is a very small unit, the typical stress level in machine elements, particularly in metallic parts, is in the *megapascal* (10^6 pascal) range. It is also good to remember that

$$1.0 \text{ MPa} = 1.0 \text{ N/mm}^2$$

This relationship is useful since the cross-sectional area of typical machine parts is expressed in mm².

Example Problem 3–1   A tensile force of 9 500 N is applied to a 12-mm-diameter round bar, as shown in Figure 3–2. Compute the direct tensile stress in the bar.

FIGURE 3–2 Tensile stress in a round bar

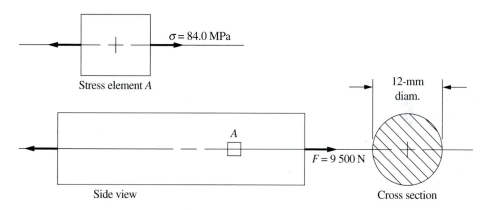

Solution   Objective   Compute the tensile stress in the round bar.

Given   Force = F = 9 500 N; diameter = D = 12 mm.

Analysis   Use the direct tensile stress formula, Equation (3–1): $\sigma = F/A$. Compute the cross-sectional area from $A = \pi D^2/4$.

Results $A = \pi D^2/4 = \pi(12 \text{ mm})^2/4 = 113 \text{ mm}^2$

$\sigma = F/A = (9\ 500 \text{ N})/(113 \text{ mm}^2) = 84.0 \text{ N/mm}^2 = 84.0 \text{ MPa}$

Comment   This level of stress will be uniform on all parts of the cross section of the bar.

The conditions on the use of Equation (3–1) are as follows:

1. The load-carrying member must be straight.
2. The line of action of the load must pass through the centroid of the cross section of the member.
3. The member must be of uniform cross section near where the stress is being computed.
4. The material must be homogeneous and isotropic.
5. In the case of compression members, the member must be short to prevent buckling. The conditions under which buckling is expected are discussed in Chapter 6.

3–5
DEFORMATION UNDER DIRECT AXIAL LOADING

The following formula computes the stretch due to a direct axial tensile load or the shortening due to a direct axial compressive load:

⇨ **Deformation Due to Direct Axial Load**

$$\delta = FL/EA \tag{3–2}$$

where δ = total deformation of the member carrying the axial load
  F = direct axial load
  L = original total length of the member
  E = modulus of elasticity of the material
  A = cross-sectional area of the member

Noting that $\sigma = F/A$, we can also compute the deformation from

$$\delta = \sigma L/E \tag{3–3}$$

Example Problem 3–2

For the round bar subjected to the tensile load shown in Figure 3–2, compute the total deformation if the original length of the bar is 3 600 mm. The bar is made from a steel having a modulus of elasticity of 207 GPa.

Solution   Objective   Compute the deformation of the bar.

Given   Force = F = 9 500 N; diameter = D = 12 mm.
Length = L = 3 600 mm; E = 207 GPa.

Analysis   From Example Problem 3–1, we found that $\sigma = 84.0$ MPa. Use Equation (3–3).

Results $$\delta = \frac{\sigma L}{E} = \frac{(84.0 \times 10^6 \text{ N/m}^2)(3\ 600 \text{ mm})}{(207 \times 10^9 \text{ N/m}^2)} = 1.46 \text{ mm}$$

3–6
DIRECT SHEAR STRESS

Direct shear stress occurs when the applied force tends to cut through the member as scissors or shears do or when a punch and a die are used to punch a slug of material from a sheet. Another important example of direct shear in machine design is the tendency for a key to be sheared off at the section between the shaft and the hub of a machine element when transmitting torque. Figure 3–3 shows the action.

The method of computing direct shear stress is similar to that used for computing direct tensile stress because the applied force is assumed to be uniformly distributed across the cross section of the part that is resisting the force. But that kind of stress is *shear stress* rather than *normal stress*. The symbol used for shear stress is the Greek letter tau (τ). The formula for direct shear stress can thus be written

⇨ **Direct Shear Stress**

$$\tau = \text{shearing force/area in shear} = F/A_s \tag{3–4}$$

This stress is more properly called the *average shearing stress*, but we will make the simplifying assumption that the stress is uniformly distributed across the shear area.

FIGURE 3–3 Direct
shear on a key

(*a*) Shaft/sheave arrangement

Key ½ × ½ × 1¾

Hub

Shear
plane

F

R

D

Shaft

(*b*) Enlarged view of hub/shaft/key

b = 0.50 in

Shear area = $A_s = bL$ = (0.50 in) (1.75 in) = 0.875 in²

L = 1.75 in

(*c*)

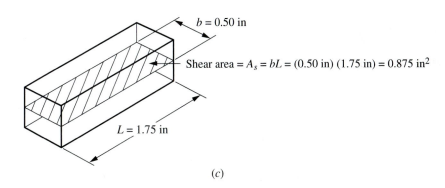

Example Problem 3–3   Figure 3–3 shows a shaft carrying two sheaves that are keyed to the shaft. Part (b) shows that a force F is transmitted from the shaft to the hub of the sheave through a square key. The shaft is 2.25 inches in diameter and transmits a torque of 14 063 lb·in. The key has a square cross section, 0.50 in on a side, and a length of 1.75 in. Compute the force on the key and the shear stress caused by this force.

Solution   Objective   Compute the force on the key and the shear stress.

Given   Layout of shaft, key, and hub shown in Figure 3–3.

Torque = T = 14 063 lb·in; key dimensions = 0.5 × 0.5 × 1.75 in.

Shaft diameter = D = 2.25 in; radius = $R = D/2$ = 1.125 in.

Analysis   Torque T = force $F \times$ radius R. Then $F = T/R$.

Use equation (3–4) to compute shearing stress: $\tau = F/A_s$.
Shear area is the cross section of the key at the interface between the shaft and the hub:
$A_s = bL$.

Results $F = T/R = (14\,063\text{ lb·in})/(1.125\text{ in}) = 12\,500\text{ lb}$

$A_s = bL = (0.50\text{ in})(1.75\text{ in}) = 0.875\text{ in}^2$

$\tau = F/A = (12\,500\text{ lb})/(0.875\text{ in}^2) = 14\,300\text{ lb/in}^2$

Comment   This level of shearing stress will be uniform on all parts of the cross section of the key.

3–7 RELATIONSHIP AMONG TORQUE, POWER, AND ROTATIONAL SPEED

⇨ **Power–Torque–Speed Relationship**

The relationship among the power (P), the rotational speed (n), and the torque (T) in a shaft is described by the equation

$$T = P/n \qquad (3\text{–}5)$$

In SI units, power is expressed in the unit of *watt* (W) or its equivalent, *newton meter per second* (N·m/s), and the rotational speed is in *radians per second* (rad/s).

Example Problem 3–4   Compute the amount of torque in a shaft transmitting 750 W of power while rotating at 183 rad/s. (*Note:* This is equivalent to the output of a 1.0-hp, 4-pole electric motor, operating at its rated speed of 1 750 rpm. See Chapter 21.)

Solution   Objective   Compute the torque T in the shaft.

Given   Power = P = 750 W = 750 N·m/s.
Rotational speed = n = 183 rad/s.

Analysis   Use Equation (3–5).

Results $T = P/n = (750\text{ N·m/s})/(183\text{ rad/s})$
$T = 4.10\text{ N·m/rad} = 4.10\text{ N·m}$

Comments   In such calculations, the unit of N·m/rad is dimensionally correct, and some advocate its use. Most, however, consider the radian to be dimensionless, and thus torque is expressed in N·m or other familiar units of force times distance.

In the U.S. Customary Unit System, power is typically expressed as *horsepower,* equal to 550 ft·lb/s. The typical unit for rotational speed is rpm, or revolutions per minute. But the most convenient unit for torque is the pound-inch (lb·in). Considering all of these quantities and making the necessary conversions of units, we use the following formula to compute the torque (in lb·in) in a shaft carrying a certain power P (in hp) while rotating at a speed of n rpm.

**P–T–n
Relationship for U.S.
Customary Units**

$$T = 63\ 000P/n \tag{3–6}$$

The resulting torque will be in pound-inches. You should verify the value of the constant, 63 000.

Example Problem 3–5   Compute the torque on a shaft transmitting 1.0 hp while rotating at 1 750 rpm. Note that these conditions are approximately the same as those for which the torque was computed in Example Problem 3–4 using SI units.

Solution   Objective   Compute the torque in the shaft.

Given $P = 1.0$ hp; $n = 1\ 750$ rpm.

Analysis   Use Equation (3–6).

Results $T = 63\ 000P/n = [63\ 000(1.0)]/1\ 750 = 36.0$ lb·in

---

**3–8
TORSIONAL
SHEAR STRESS**   When a *torque,* or twisting moment, is applied to a member, it tends to deform by twisting, causing a rotation of one part of the member relative to another. Such twisting causes a shear stress in the member. For a small element of the member, the nature of the stress is the same as that experienced under direct shear stress. However, in *torsional shear,* the distribution of stress is *not* uniform.

The most frequent case of torsional shear in machine design is that of a round circular shaft transmitting power. Chapter 12 covers shaft design.

Torsional Shear Stress Formula

When subjected to a torque, the outer surface of a solid round shaft experiences the greatest shearing strain and therefore the largest torsional shear stress. See Figure 3–4. The value of the maximum torsional shear stress is found from

**Maximum Torsional
Shear Stress in a
Circular Shaft**

$$\tau_{max} = Tc/J \tag{3–7}$$

where c = radius of the shaft
J = polar moment of inertia

See Appendix 1 for formulas for J.

Example Problem 3–6   Compute the maximum torsional shear stress in a shaft having a diameter of 10 mm when it carries a torque of 4.10 N·m.

Solution   Objective   Compute the torsional shear stress in the shaft.

Given   Torque $= T = 4.10$ N·m; shaft diameter $= D = 10$ mm.

                  $c =$ radius of the shaft $= D/2 = 5.0$ mm.

Analysis   Use Equation (3–7) to compute the torsional shear stress: $\tau_{max} = Tc/J$.
            J is the polar moment of inertia for the shaft: $J = \pi D^4/32$ (see Appendix 1).

Results $J = \pi D^4/32 = [(\pi)(10 \text{ mm})^4]/32 = 982 \text{ mm}^4$

$$\tau_{max} = \frac{(4.10 \text{ N·m})(5.0 \text{ mm})}{982 \text{ mm}^4} \frac{10^3 \text{ mm}}{\text{m}} = 20.9 \text{ N/mm}^2 = 20.9 \text{ MPa}$$

Comment   The maximum torsional shear stress occurs at the outside surface of the shaft around its entire circumference.

If it is desired to compute the torsional shear stress at some point inside the shaft, the more general formula is used:

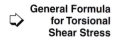
General Formula for Torsional Shear Stress

$$\tau = Ty/J \qquad (3\text{–}8)$$

where $y =$ radial distance from the center of the shaft to the point of interest

Figure 3–4 shows graphically that this equation is based on the linear variation of the torsional shear stress from zero at the center of the shaft to the maximum value at the outer surface.

Equations (3–7) and (3–8) apply also to hollow shafts (Figure 3–5 shows the distribution of shear stress). Again note that the maximum shear stress occurs at the outer surface. Also note that the entire cross section carries a relatively high stress level. As a result, the hollow shaft is more efficient. Notice that the material near the center of the solid shaft is not highly stressed.

For design, it is convenient to define the *polar section modulus, Z_p*:

Polar Section Modulus

$$Z_p = J/c \qquad (3\text{–}9)$$

FIGURE 3–4 Stress distribution in a solid shaft

FIGURE 3–5 Stress distribution in a hollow shaft

Then the equation for the maximum torsional shear stress is

$$\tau_{max} = T/Z_p \tag{3–10}$$

Formulas for the polar section modulus are also given in Appendix 1.

3–9 TORSIONAL DEFORMATION

When a shaft is subjected to a torque, it undergoes a twisting in which one cross section is rotated relative to other cross sections in the shaft. The angle of twist is computed from

⇨ **Torsional Deformation**

$$\theta = TL/GJ \tag{3–11}$$

where θ = angle of twist (radians)

L = length of the shaft over which the angle of twist is being computed

G = modulus of elasticity of the shaft material in *shear*

Example Problem 3–7   Compute the angle of twist of a 10-mm-diameter shaft carrying 4.10 N·m of torque if it is 250 mm long and made of steel with G = 80 GPa. Express the result in both radians and degrees.

Solution   **Objective**   Compute the angle of twist in the shaft.

Given   Torque = T = 4.10 N·m; length = L = 250 mm.

Shaft diameter = D = 10 mm; G = 80 GPa.

Analysis   Use Equation (3–11). For consistency, let $T = 4.10 \times 10^3$ N·mm and $G = 80 \times 10^3$ N/mm^2. From Example Problem 3–6, $J = 982$ mm^4.

Results $$\theta = \frac{TL}{GJ} = \frac{(4.10 \times 10^3 \text{ N·mm})(250 \text{ mm})}{(80 \times 10^3 \text{ N/mm}^2)(982 \text{ mm}^4)} = 0.013 \text{ rad}$$

Using π rad = 180°,

$$\theta = (0.013 \text{ rad})(180 \text{ deg}/\pi \text{ rad}) = 0.75 \text{ deg}$$

Comment   Over the length of 250 mm, the shaft twists 0.75 deg.

3–10 TORSION IN MEMBERS HAVING NONCIRCULAR CROSS SECTIONS

The behavior of members having noncircular cross sections when subjected to torsion is radically different from that for members having circular cross sections. However, the factors of most use in machine design are the maximum stress and the total angle of twist for such members. The formulas for these factors can be expressed in similar forms to the formulas used for members of circular cross section (solid and hollow round shafts).

The following two formulas can be used:

⇨ **Torsional Shear Stress**

$$\tau_{max} = T/Q \tag{3–12}$$

⇨ **Deflection for Noncircular Sections**

$$\theta = TL/GK \qquad\qquad (3\text{--}13)$$

Note that Equations (3–12) and (3–13) are similar to Equations (3–10) and (3–11), with the substitution of Q for Z_p and K for J. Refer to Figure 3–6 for the methods of determining the values for K and Q for several types of cross sections useful in machine design. These values are appropriate only if the ends of the member are free to deform. If either end is fixed, as by welding to a solid structure, the resulting stress and angular twist are quite different. (See References 2, 4, and 6.)

Example Problem 3–8   A 2.00-in-diameter shaft carrying a chain sprocket has one end milled in the form of a square to permit the use of a hand crank. The square is 1.75 in on a side. Compute the maximum shear stress on the square part of the shaft when a torque of 15 000 lb·in is applied.

FIGURE 3–6
Methods for determining values for K and Q for several types of cross sections

Cross-sectional shape	K = for use in $\theta = TL/GK$ Q = for use in $\tau = T/Q$	Black dot (●) denotes location of τ_{max}
Square	$K = 0.141a^4$ $Q = 0.208a^3$	τ_{max} at midpoint of each side
Rectangle	$K = bh^3\left[\frac{1}{3} - 0.21\frac{h}{b}\left(1 - \frac{(h/b)^4}{12}\right)\right]$ $Q = \dfrac{bh^2}{[3 + 1.8(h/b)]}$	(Approximate; within ≈ 5%) τ_{max} at midpoint of long sides
Triangle (equilateral)	$K = 0.0217a^4$ $Q = 0.050a^3$	

Shaft with One Flat

$K = C_1 r^4$
$Q = C_2 r^3$

h/r	0	0.2	0.4	0.6	0.8	1.0
C_1	0.30	0.51	0.78	1.06	1.37	1.57
C_2	0.35	0.51	0.70	0.92	1.18	1.57

Shaft with Two Flats

$K = C_3 r^4$
$Q = C_4 r^3$

h/r	0.5	0.6	0.7	0.8	0.9	1.0
C_3	0.44	0.67	0.93	1.19	1.39	1.57
C_4	0.47	0.60	0.81	1.02	1.25	1.57

Hollow Rectangle

t (uniform)

$$K = \frac{2t(a-t)^2(b-t)^2}{(a+b-2t)}$$

$$Q = 2t(a-t)(b-t)$$

Gives average stress; good approximation of maximum stress if t is small

Inner corners should have generous fillets

Split Tube
Mean radius (r)

$$K = 2\pi r t^3/3$$

$$Q = \frac{4\pi^2 r^2 t^2}{(6\pi r + 1.8t)}$$

t must be small

t (uniform)

Also, if the length of the square part is 8.00 in, compute the angle of twist over this part. The shaft material is steel with $G = 11.5 \times 10^6$ psi.

Solution   Objective   Compute the maximum shear stress and the angle of twist in the shaft.

Given   Torque $= T = 15\,000$ lb·in; length $= L = 8.00$ in.

The shaft is square; thus, $a = 1.75$ in.

$G = 11.5 \times 10^6$ psi.

Analysis   Figure 3–6 shows the methods for calculating the values for Q and K for use in Equations (3–12) and (3–13).

Results $Q = 0.208a^3 = (0.208)(1.75 \text{ in})^3 = 1.115 \text{ in}^3$

$K = 0.141a^4 = (0.141)(1.75 \text{ in})^4 = 1.322 \text{ in}^4$

Now the stress and the deflection can be computed.

$$\tau_{max} = \frac{T}{Q} = \frac{15\,000 \text{ lb·in}}{(1.115 \text{ in}^3)} = 13\,460 \text{ psi}$$

$$\theta = \frac{TL}{GK} = \frac{(15\,000 \text{ lb·in})(8.00 \text{ in})}{(11.5 \times 10^6 \text{ lb/in}^2)(1.322 \text{ in}^4)} = 0.0079 \text{ rad}$$

Convert the angle of twist to degrees:

$$\theta = (0.0079 \text{ rad})(180 \text{ deg}/\pi \text{ rad}) = 0.452 \text{ deg}$$

Comments   Over the length of 8.00 in, the square part of the shaft twists 0.452 deg. The maximum shear stress is 13 460 psi, and it occurs at the midpoint of each side as shown in Figure 3–6.

**3–11
TORSION IN
CLOSED, THIN-
WALLED TUBES**

A general approach for closed, thin-walled tubes of virtually any shape uses Equations (3–12) and (3–13) with special methods of evaluating K and Q. Figure 3–7 shows such a tube having a constant wall thickness. The values of K and Q are

$$K = 4A^2t/U \qquad\qquad \textbf{(3–14)}$$
$$Q = 2tA \qquad\qquad\quad\ \textbf{(3–15)}$$

where A = area enclosed by the median boundary (indicated by the dashed line in Figure 3–7)

t = wall thickness (which must be uniform and thin)

U = length of the median boundary

FIGURE 3–7
Closed, thin-walled
tube with a constant
wall thickness

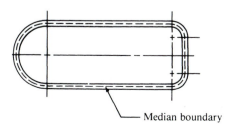

Median boundary

The shear stress computed by this approach is the *average stress* in the tube wall. However, if the wall thickness t is small (a thin wall), the stress is nearly uniform throughout the wall, and this approach will yield a close approximation of the maximum stress. For the analysis of tubular sections having nonuniform wall thickness, see References 2, 4, and 7.

　　To design a member to resist torsion only, or torsion and bending combined, it is advisable to select hollow tubes, either round or rectangular, or some other closed shape. They possess good efficiency both in bending and in torsion.

**3–12
OPEN TUBES
AND A
COMPARISON
WITH CLOSED
TUBES**

The term *open tube* refers to a shape that appears to be tubular but is not completely closed. For example, some tubing is manufactured by starting with a thin, flat strip of steel that is roll-formed into the desired shape (circular, rectangular, square, and so on). Then the seam is welded along the entire length of the tube. It is interesting to compare the properties of the cross section of such a tube before and after it is welded. The following example problem illustrates the comparison for a particular size of circular tubing.

Example Problem 3–9

Figure 3–8 shows a tube before [Part (b)] and after [Part (a)] the seam is welded. Compare the stiffness and the strength of each shape.

Solution　　Objective

Compare the torsional stiffness and the strength of the closed tube of Figure 3–8(a) with those of the open-seam (split) tube shown in Figure 3–8(b).

Given

The tube shapes are shown in Figure 3–8. Both have the same length, diameter, and wall thickness, and both are made from the same material.

Analysis

Equation (3–13) gives the angle of twist for a noncircular member and shows that the angle is inversely proportional to the value of K. Similarly, Equation (3–11) shows that the angle of twist for a hollow circular tube is inversely proportional to the polar moment of inertia J. All other terms in the two equations are the same for each design. Therefore, the ratio of θ_{open} to θ_{closed} is equal to the ratio J/K. From Appendix 1, we find

$$J = \pi(D^4 - d^4)/32$$

From Figure 3–6, we find

$$K = 2\pi rt^3/3$$

　　Using similar logic, Equations (3–12) and (3–8) show that the maximum torsional shear stress is inversely proportional to Q and Z_p for the open and closed tubes, respectively.

FIGURE 3–8
Comparison of closed and open tubes

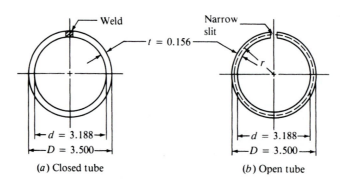

(a) Closed tube   (b) Open tube

Then we can compare the strengths of the two forms by computing the ratio Z_p/Q. By Equation (3–9), we find that

$$Z_p = J/c = J/(D/2)$$

The equation for Q for the split tube is listed in Figure 3–6.

Results   We make the comparison of torsional stiffness by computing the ratio J/K.

For the closed, hollow tube,

$$J = \pi(D^4 - d^4)/32$$

$$J = \pi(3.500^4 - 3.188^4)/32 = 4.592 \ \text{in}^4$$

For the open tube before the slit is welded, from Figure 3–6,

$$K = 2\pi r t^3/3$$

$$K = [(2)(\pi)(1.672)(0.156)^3]/3 = 0.013\ 3 \ \text{in}^4$$

$$\text{Ratio} = J/K = 4.592/0.013\ 3 = 345$$

Then we make the comparison of the strengths of the two forms by computing the ratio Z_p/Q. The value of J has already been computed to be 4.592 in⁴. Then

$$Z_p = J/c = J/(D/2) = (4.592 \ \text{in}^4)/[(3.500 \ \text{in})/2] = 2.624 \ \text{in}^3$$

For the open tube,

$$Q = \frac{4\pi^2 r^2 t^2}{(6\pi r + 1.8t)} = \frac{4\pi^2(1.672 \ \text{in})^2(0.156 \ \text{in})^2}{[6\pi(1.672 \ \text{in}) + 1.8(0.156 \ \text{in})]} = 0.084\ 5 \ \text{in}^3$$

Then the strength comparison is

$$\text{Ratio} = Z_p/Q = 2.624/0.084\ 5 = 31.1$$

Comments   Thus, for a given applied torque, the slit tube would twist 345 times as much as the closed tube. The stress in the slit tube would be 31.1 times higher than in the closed tube. Also note that if the material for the tube is thin, it will likely buckle at a relatively low stress level, and the tube will collapse suddenly. This comparison shows the dramatic superiority of the closed form of a hollow section to an open form. A similar comparison could be made for shapes other than circular.

3–13
VERTICAL
SHEARING
STRESS

A beam carrying loads transverse to its axis will experience shearing forces, denoted by V. In the analysis of beams, it is usual to compute the variation in shearing force across the entire length of the beam and to draw the *shearing force diagram*. Then the resulting vertical shearing stress can be computed from

**Vertical Shearing
Stress in Beams**

$$\tau = VQ/It \tag{3–16}$$

where I = rectangular moment of inertia of the cross section of the beam
     t = thickness of the section at the place where the shearing stress is to be computed

For most section shapes, the maximum vertical shearing stress occurs at the neutral axis. Specifically, if the thickness is not less at a place away from the neutral axis, then it is assured that the maximum vertical shearing stress occurs at the neutral axis.

The term Q in Equation (3–16), called the *statical moment* of the section, is the moment of the area above where the shearing stress is to be calculated. That is,

⟳ τ$_{max}$ **for Rectangle** $$\tau_{max} = 3V/2A \quad \text{(exact)} \qquad\qquad \textbf{(3–18)}$$

where A = total cross-sectional area of the beam

⟳ τ$_{max}$ **for Circle** $$\tau_{max} = 4V/3A \quad \text{(exact)} \qquad\qquad \textbf{(3–19)}$$

⟳ τ$_{max}$ **for I-Shape** $$\tau_{max} \simeq V/th \quad \text{(approximate: about 15\% low)} \qquad \textbf{(3–20)}$$

where t = web thickness
h = height of the web (for example, a wide-flange beam)

⟳ **τ$_{max}$ for**
Thin-walled Tube $$\tau_{max} \simeq 2V/A \quad \text{(approximate: a little high)} \qquad \textbf{(3–21)}$$

In all of these cases, the maximum shearing stress occurs at the neutral axis.

Example Problem 3–11   Compute the maximum shearing stress in the beam described in Example Problem 3–10 using the special shearing stress formula for a rectangular section.

Solution   Objective   Compute the maximum shearing stress τ in the beam in Figure 3–10.

Given   The data are the same as stated in Example Problem 3–10 and as shown in Figure 3–10.

Analysis   Use Equation (3–18) to compute $\tau = 3V/2A$. For the rectangle, $A = th$.

Results $$\tau_{max} = \frac{3V}{2A} = \frac{3(1\ 000\ \text{lb})}{2[(2.0\ \text{in})(8.0\ \text{in})]} = 93.8\ \text{psi}$$

Comment   This result is the same as that obtained for Example Problem 3–10, as expected.

3–15
STRESS DUE TO
BENDING

A *beam* is a member that carries loads transverse to its axis. Such loads produce bending moments in the beam, which result in the development of bending stresses. Bending stresses are *normal stresses,* that is, either tensile or compressive. The maximum bending stress in a beam cross section will occur in the part farthest from the neutral axis of the section. At that point, the *flexure formula* gives the stress:

⟳ **Flexure Formula**
for Maximum
Bending Stress

$$\sigma = Mc/I \qquad\qquad \textbf{(3–22)}$$

where M = magnitude of the bending moment at the section
I = moment of inertia of the cross section with respect to its neutral axis
c = distance from the neutral axis to the outermost fiber of the beam cross section

The magnitude of the bending stress varies linearly within the cross section from a value of zero at the neutral axis, to the maximum tensile stress on one side of the neutral axis, and to the maximum compressive stress on the other side. Figure 3–12 shows a typical stress distribution in a beam cross section.

Note that *positive bending* occurs when the deflected shape of the beam is concave upward, resulting in compression on the upper part of the cross section and tension on the lower part. Such a beam could be said to "hold water." Conversely, *negative bending* causes the beam to be concave downward.

FIGURE 3–12
Typical bending stress
distribution in a beam
cross section

(*a*) Beam loading

(*b*) Shear and bending
moment diagrams

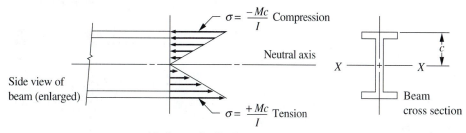

(*c*) Stress distribution on beam section

(*d*) Stress element in compression
in top part of beam

(*e*) Stress element in tension
in bottom part of beam

The flexure formula was developed subject to the following conditions:

1. The beam must be in pure bending. Shearing stresses must be zero or negligible. No axial loads are present.
2. The beam must not twist or be subjected to a torsional load.
3. The material of the beam must obey Hooke's law.
4. The modulus of elasticity of the material must be the same in both tension and compression.
5. The beam is initially straight and has a constant cross section.
6. Any plane cross section of the beam remains plane during bending.
7. No part of the beam shape fails because of local buckling or wrinkling.

If condition 1 is not strictly met, you can continue the analysis by using the method of combined stresses presented in Chapter 4. In most practical beams, which are long relative to their height, shear stresses are sufficiently small as to be negligible. Furthermore, the maximum bending stress occurs at the outermost fibers of the beam section, where the shear stress is in fact zero. A beam with varying cross section, which would violate condition 5, can be analyzed by the use of stress concentration factors discussed later in this chapter.

For design, it is convenient to define the term *section modulus, S,* as

$$S = I/c \tag{3–23}$$

The flexure formula then becomes

▷ **Flexure Formula**
$$\sigma = M/S \qquad\qquad \textbf{(3–24)}$$

Since I and c are geometrical properties of the cross section of the beam, S is also. Then, in design, it is usual to define a design stress, σ_d, and, with the bending moment known, solve for S:

▷ **Required Section Modulus**
$$S = M/\sigma_d \qquad\qquad \textbf{(3–25)}$$

This results in the required value of the section modulus. From it, the required dimensions of the beam cross section can be determined.

Example Problem 3–12   For the beam shown in Figure 3–12, the load F due to the pipe is 12 000 lb. The distances are a = 4 ft and b = 6 ft. Determine the required section modulus for the beam to limit the stress due to bending to 21 600 psi, the recommended design stress for a typical structural steel in static bending.

Solution   Objective   Compute the required section modulus S for the beam in Figure 3–12.

Given   The layout and the loading pattern are shown in Figure 3–10.

Lengths: Overall length = L = 10 ft; a = 4 ft; b = 6 ft.

Load = F = 12 000 lb.

Design stress = σ_d = 21 600 psi.

Analysis   Use Equation (3–25) to compute the required section modulus S. Compute the maximum bending moment that occurs at the point of application of the load using the formula shown in Part (b) of Figure 3–12.

Results
$$M_{max} = R_1 a = \frac{Fba}{a+b} = \frac{(12\ 000\ \text{lb})(6\ \text{ft})(4\ \text{ft})}{(6\ \text{ft} + 4\ \text{ft})} = 28\ 800\ \text{lb·ft}$$

$$S = \frac{M}{\sigma_d} = \frac{28\ 800\ \text{lb·ft}}{21\ 600\ \text{lb/in}^2}\ \frac{12\ \text{in}}{\text{ft}} = 16.0\ \text{in}^3$$

Comments   We could select a beam section having at least this value for the section modulus. An American Standard beam shape, S8 × 23, has S = 16.2 in^3, as found in Table A16–4. A lighter, but taller, section that could be used is the W12 × 16 wide-flange shape having a section modulus of 17.1 in^3.

**3–16
FLEXURAL
CENTER FOR
BEAMS**   A beam section must be loaded in a way that ensures symmetrical bending; that is, there must be no tendency for the section to twist under the load. Figure 3–13 shows several shapes that are typically used for beams having a vertical axis of symmetry. If the line of action of the loads on such sections passes through the axis of symmetry, then there is no tendency for the section to twist, and the flexure formula applies.

When there is no vertical axis of symmetry, as with the sections shown in Figure 3–14, care must be exercised in placement of the loads. If the line of action of the loads were shown

FIGURE 3–13
Symmetrical sections.
A load applied through
the axis of symmetry
results in pure bending
in the beam.

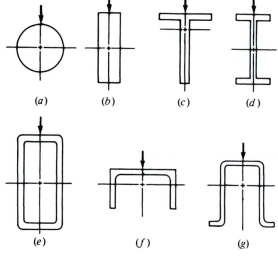

(a)   (b)   (c)   (d)

(e)   (f)   (g)

FIGURE 3–14
Nonsymmetrical
sections. A load applied
as at F_1 would cause
twisting; loads applied
as at F_2 through the
flexural center Q would
cause pure bending.

(a)   (b)   (c)

as F_1 in the figure, the beam would twist and bend, so the flexure formula would not give accurate results for the stress in the section. For such sections, the load must be placed in line with the *flexural center,* sometimes called the *shear center.* Figure 3–14 shows the approximate location of the flexural center for these shapes (indicated by the symbol Q). Applying the load in line with Q, as shown with the forces labeled F_2, would result in pure bending. A table of formulas for the location of the flexural center is available (see Reference 7).

3–17
BEAM
DEFLECTIONS

The bending loads applied to a beam cause it to deflect in a direction perpendicular to its axis. A beam that was originally straight will deform to a slightly curved shape. In most cases, the critical factor is either the maximum deflection of the beam or its deflection at specific locations.

Consider the double-reduction speed reducer shown in Figure 3–15. The four gears (A, B, C, and D) are mounted on three shafts, each of which is supported by two bearings. The action of the gears in transmitting power creates a set of forces that in turn act on the shafts to cause bending. One component of the total force on the gear teeth acts in a direction that tends to separate the two gears. Thus, gear A is forced upward while gear B is forced downward. For good gear performance, the net deflection of one gear to the other should not exceed 0.005 in (0.13 mm) for medium-sized industrial gearing.

To evaluate the design, there are many methods of computing shaft deflections. We will review briefly those methods using deflection formulas, superposition, and a general analytical approach.

A set of formulas for computing the deflection of beams at any point or at selected points is useful in many practical problems. Appendix 14 includes several cases.

FIGURE 3–15 Shaft deflection analysis for a double-reduction speed reducer

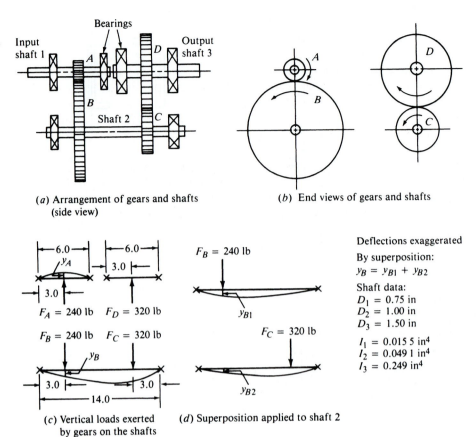

(*a*) Arrangement of gears and shafts
(side view)

(*b*) End views of gears and shafts

$F_B = 240$ lb

Deflections exaggerated

By superposition:
$y_B = y_{B1} + y_{B2}$

Shaft data:
$D_1 = 0.75$ in
$D_2 = 1.00$ in
$D_3 = 1.50$ in

$I_1 = 0.015\ 5$ in^4
$I_2 = 0.049\ 1$ in^4
$I_3 = 0.249$ in^4

$F_A = 240$ lb $F_D = 320$ lb

$F_B = 240$ lb $F_C = 320$ lb

$F_C = 320$ lb

(*c*) Vertical loads exerted
by gears on the shafts

(*d*) Superposition applied to shaft 2

For many additional cases, superposition is useful if the actual loading can be divided into parts that can be computed by available formulas. The deflection for each loading is computed separately, and then the individual deflections are summed at the points of interest.

Many commercially available computer software programs allow the modeling of beams having rather complex loading patterns and varying geometry. The results include reaction forces, shearing force and bending moment diagrams, and deflections at any point. It is important that you understand the principles of beam deflection, studied in strength of materials and reviewed here, so that you can apply such programs accurately and interpret the results carefully.

Example Problem 3–13   For the two gears, *A* and *B*, in Figure 3–15, compute the relative deflection between them in the plane of the paper that is due to the forces shown in Part (c). These *separating forces*, or *normal forces*, are discussed in Chapters 9 and 10. It is customary to consider the loads at the gears and the reactions at the bearings to be concentrated. The shafts carrying the gears are steel and have uniform diameters as listed in the figure.

Solution   Objective   Compute the relative deflection between gears *A* and *B* in Figure 3–15.

Given   The layout and loading pattern are shown in Figure 3–15. The separating force between gears *A* and *B* is 240 lb. Gear *A* pushes downward on gear *B*, and the reaction force of gear *B* pushes upward on gear *A*. Shaft 1 has a diameter of 0.75 in and a moment of inertia of

0.0155 in^4. Shaft 2 has a diameter of 1.00 in and a moment of inertia of 0.0491 in^4. Both shafts are steel. Use $E = 30 \times 10^6$ psi.

Analysis Use the deflection formulas from Appendix 14 to compute the upward deflection of shaft 1 at gear A and the downward deflection of shaft 2 at gear B. The sum of the two deflections is the total deflection of gear A with respect to gear B.

Case (a) from Table A14–1 applies to shaft 1 because there is a single concentrated force acting at the midpoint of the shaft between the supporting bearings. We will call that deflection y_A.

Shaft 2 is a simply supported beam carrying two nonsymmetrical loads. No single formula from Appendix 14 matches that loading pattern. But we can use superposition to compute the deflection of the shaft at gear B by considering the two forces separately as shown in Part (d) of Figure 3–15. Case (b) from Table A14–1 is used for each load.

We first compute the deflection at B due only to the 240-lb force, calling it y_{B1}. Then we compute the deflection at B due to the 320-lb force, calling it y_{B2}. The total deflection at B is $y_B = y_{B1} + y_{B2}$.

Results The deflection of shaft 1 at gear A is

$$y_a = \frac{F_A L_1^3}{48EI} = \frac{(240)(6.0)^3}{48(30 \times 10^6)(0.015\,5)} = 0.002\,3 \text{ in}$$

The deflection of shaft 2 at B due only to the 240-lb force is

$$y_{B1} = -\frac{F_B a^2 b^2}{3EI_2 L_2} = -\frac{(240)(3.0)^2(11.0)^2}{3(30 \times 10^6)(0.049\,1)(14)} = -0.004\,2 \text{ in}$$

The deflection of shaft 2 at B due only to the 320-lb force at C is

$$y_{B2} = -\frac{F_C bx}{6EI_2 L_2}(L_2^2 - b^2 - x^2)$$

$$y_{B2} = -\frac{(320)(3.0)(3.0)}{6(30 \times 10^6)(0.049\,1)(14)}[(14)^2 - (3.0)^2 - (3.0)^2]$$

$$y_{B2} = -0.004\,1 \text{ in}$$

Then the total deflection at gear B is

$$y_B = y_{B1} + y_{B2} = -0.004\,2 - 0.004\,1 = -0.008\,3 \text{ in}$$

Because shaft 1 deflects upward and shaft 2 deflects downward, the total *relative deflection* is the sum of y_A and y_B:

$$y_{\text{total}} = y_A + y_B = 0.002\,3 + 0.008\,3 = 0.010\,6 \text{ in}$$

Comment This deflection is very large for this application. How could the deflection be reduced?

**3–18
EQUATIONS FOR
DEFLECTED
BEAM SHAPE**

The general principles relating the deflection of a beam to the loading on the beam and its manner of support are presented here. The result will be a set of relationships among the load, the vertical shearing force, the bending moment, the slope of the deflected beam shape, and the actual deflection curve for the beam. Figure 3–16 shows diagrams for these five factors, with θ as the slope and y indicating deflection of the beam from its initial straight position. The product of modulus of elasticity and the moment of inertia, *EI*, for the beam is a measure of its stiffness or resistance to bending deflection. It is convenient to combine *EI* with the slope and deflection values to maintain a proper relationship, as discussed next.

FIGURE 3–17
Bending moment in a
bell crank

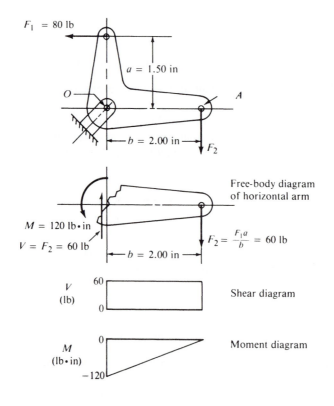

Below the drawing of the complete crank, we have drawn a sketch of the horizontal part, isolating it from the vertical part. The internal force and moment at the cut section are shown. The externally applied downward force F_2 is reacted by the upward reaction at the pin. Also, because F_2 causes a moment with respect to the section at the pin, an internal reaction moment exists, where

$$M = F_2 \cdot b = (60 \text{ lb})(2.00 \text{ in}) = 120 \text{ lb·in}$$

The shear and moment diagrams can then be shown in the conventional manner. The result looks much like a cantilever that is built into a rigid support. The difference here is that the reaction moment at the section through the pin is developed in the vertical arm of the crank.

Comments   Note that the shape of the moment diagram for the horizontal part shows that the maximum moment occurs at the section through the pin and that the moment decreases linearly as we move out toward point A. As a result, the shape of the crank is optimized, having its largest cross section (section modulus) at the section of highest bending moment. You could complete the design of the crank using the techniques reviewed in Section 3–15.

Example Problem 3–15   Figure 3–18 represents a print head for a typewriter or a computer printer. The force F moves the print head toward the left against the ribbon, imprinting the character on the paper that is backed up by the platen. Draw the free-body diagram for the horizontal portion of the print head, along with the shearing force and bending moment diagrams.

FIGURE 3–18

Bending moment on a
print head

Solution **Objective** Draw the free-body diagram of the horizontal part of the print head in Figure 3–18. Draw
the shearing force and bending moment diagrams for that part.

Given The layout from Figure 3–18.

Analysis The horizontal force of 35 lb acting to the left is reacted by an equal 35-lb horizontal force
produced by the platen pushing back to the right on the print head. The guides provide simple supports in the vertical direction. The applied force also produces a moment at its base
where it joins the horizontal part of the print head.

We create the free-body diagram for the horizontal part by breaking it at its right end
and replacing the removed part with the internal force and moment acting at the break. The
shearing force and bending moment diagrams can then be drawn.

Results The free-body diagram for the horizontal portion is shown below the complete sketch. Note
that at the right end (section *D*) of the print head, the vertical arm has been removed and replaced with the internal horizontal force of 35.0 N and a moment of 875 N·mm caused by
the 35.0-N force acting 25 mm above it. Also note that the 25-mm-moment arm for the force
is taken from the line of action of the force *to the neutral axis of the horizontal part.* The
35.0-N reaction of the platen on the print head tends to place the head in compression over
the entire length. The rotational tendency of the moment is reacted by the couple created by
R_1 and R_2 acting 45 mm apart at *B* and *C*.

Below the free-body diagram is the vertical shearing force diagram in which a constant shear of 19.4 N occurs only between the two supports.

The bending moment diagram can be derived from either the left end or the right
end. If we choose to start at the left end at *A*, there is no shearing force from *A* to *B*, and
therefore there is no change in bending moment. From *B* to *C*, the positive shear causes
an increase in bending moment from zero to 875 N·mm. Because there is no shear from
C to *D*, there is no change in bending moment, and the value remains at 875 N·mm. The
counterclockwise-directed concentrated moment at *D* causes the moment diagram to
drop abruptly, closing the diagram.

Example Problem 3–16   Figure 3–19 shows a crank in which it is necessary to visualize the three-dimensional arrangement. The 60-lb downward force tends to rotate the shaft *ABC* around the *x*-axis. The reaction torque acts only at the end of the shaft outboard of the bearing support at *A*. Bearings *A* and *C* provide simple supports. Draw the complete free-body diagram for the shaft *ABC*, along with the shearing force and bending moment diagrams.

Solution   Objective   Draw the free-body diagram of the shaft *ABC* in Figure 3–19. Draw the shearing force and bending moment diagrams for that part.

Given   The layout from Figure 3–19.

Analysis   The analysis will take the following steps:

1. We must determine the magnitude of the torque in the shaft between the left end and point *B* where the crank arm is attached.
2. We will analyze the connection of the crank at point *B* to determine the force and moment transferred to the shaft *ABC* by the crank.
3. We will compute the vertical reactions at supports *A* and *C*.
4. We will draw the shearing force and bending moment diagrams considering the concentrated moment applied at point *B*, along with the familiar relationships between shearing force and bending moments.

Results   The free-body diagram is shown as viewed looking into the *x-z* plane. Note that the free body must be in equilibrium in all force and moment directions. Considering first the torque (rotating moment) about the *x*-axis, note that the crank force of 60 lb acts 5.0 in from the axis. The torque, then, is

FIGURE 3–19
Bending moment on a shaft carrying a crank

$$T = (60 \text{ lb})(5.0 \text{ in}) = 300 \text{ lb·in}$$

This level of torque acts from the left end of the shaft to section B, where the crank is attached to the shaft.

Now the loading at B should be described. One way to do so is to visualize that the crank itself is separated from the shaft and is replaced with a force and moment caused by the crank. First, the downward force of 60 lb pulls down at B. Also, because the 60-lb applied force acts 3.0 in to the left of B, it causes a concentrated moment *in the x-z plane* of 180 lb·in to be applied at B.

Both the downward force and the moment at B affect the magnitude and direction of the reaction forces at A and C. First, summing moments about A,

$$(60 \text{ lb})(6.0 \text{ in}) - 180 \text{ lb·in} - R_C(10.0 \text{ in}) = 0$$

$$R_C = [(360 - 180) \text{ lb·in}]/(10.0 \text{ in}) = 18.0 \text{ lb upward}$$

Now, summing moments about C,

$$(60 \text{ lb})(4.0 \text{ in}) + 180 \text{ lb·in} - R_A(10.0 \text{ in}) = 0$$

$$R_A = [(240 + 180) \text{ lb·in}]/(10.0 \text{ in}) = 42.0 \text{ lb upward}$$

Now the shear and bending moment diagrams can be completed. The moment starts at zero at the simple support at A, rises to 252 lb·in at B under the influence of the 42-lb shear force, then drops by 180 lb·in due to the counterclockwise concentrated moment at B, and finally returns to zero at the simple support at C.

Comments   In summary, shaft ABC carries a torque of 300 lb·in from point B to its left end. The maximum bending moment of 252 lb·in occurs at point B where the crank is attached. The bending moment then suddenly drops to 72 lb·in under the influence of the concentrated moment of 180 lb·in applied by the crank.

**3–20
COMBINED
NORMAL
STRESSES:
SUPERPOSITION
PRINCIPLE**

When the same cross section of a load-carrying member is subjected to both a direct tensile or compressive stress and a stress due to bending, the resulting normal stress can be computed by the method of superposition. The formula is

$$\sigma = \pm Mc/I \pm F/A \qquad (3\text{–}26)$$

where tensile stresses are positive and compressive stresses are negative

Figure 3–20 shows a beam subjected to a load applied downward and to the right through a bracket below the beam. Resolving the load into horizontal and vertical components shows that its effect can be broken into three parts:

1. The vertical component tends to place the beam in bending with tension on the top and compression on the bottom.

2. The horizontal component, because it acts away from the neutral axis of the beam, causes bending with tension on the bottom and compression on the top.

3. The horizontal component causes direct tensile stress across the entire cross section.

We can proceed with the stress analysis by considering each of these cases individually and then combining the stress at any point.

$$M_1 = F_y a$$
$$M_2 = F_x e$$
$$\text{Axial force} = F_x$$

(a) Loaded beam

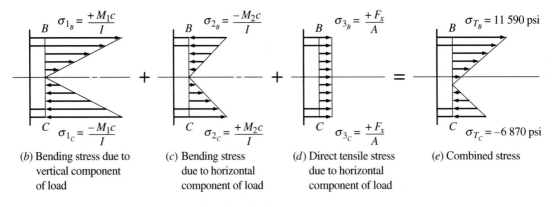

(b) Bending stress due to vertical component of load

(c) Bending stress due to horizontal component of load

(d) Direct tensile stress due to horizontal component of load

(e) Combined stress

FIGURE 3–20   Beam subjected to combined normal stresses

Example Problem 3–17   The cantilever beam in Figure 3–20 is a steel American Standard beam, S6 × 12.5 (section modulus = S = 7.37 in^3; area = A = 3.67 in^2). The force F is 10 000 lb and acts at an angle of 30° below the horizontal, as shown. Use a = 24 in and e = 6.0 in. Compute the normal stress at B and C, where the beam joins the rigid column.

Solution   **Objective**   Compute the normal stress at points B and C for the beam shown in Figure 3–20.

Given   The layout from Figure 3–20. Force = F = 10 000 lb; angle = θ = 30°.

The beam shape: S6 × 12.5; length = a = 24 in.

Section modulus = S = 7.37 in^3; area = A = 3.67 in^2.

Eccentricity of load = e = 6.0 in. Note that this dimension is the vertical distance from the point of application of the load on the bracket to the neutral axis of the beam.

Analysis   The analysis will take the following steps:

1. It is convenient to resolve the applied force that acts at an angle into horizontal and vertical components.

2. The vertical component acts as a concentrated force bending the beam downward as a cantilever.

3. The horizontal component has two effects: It places the beam in tension, and it also tends to bend it upward because of the eccentricity of its line of action.

4. We will determine the parts of the total stress due to each component of the load by using the principles of direct tensile stress and bending stress. These calcula-

tions will be done both at point B on the top of the beam and at point C on the bottom.

5. We will algebraically sum the components of the stress at each point to determine the combined stress.

Results First, resolve the force into horizontal and vertical components:

$$F_x = F \cos(30°) = 10\,000 \cos(30°) = 8\,660 \text{ lb}$$
$$F_y = F \sin(30°) = 10\,000 \sin(30°) = 5\,000 \text{ lb}$$

Now, compute the bending moments. The bending moment due to F_y is

$$M_1 = F_y a = (5\,000)(24) = 120\,000 \text{ lb·in}$$

The bending moment due to F_x is

$$M_2 = F_x e = (8\,660)(6.0) = 51\,960 \text{ lb·in}$$

Next compute the individual stress components. To compute the bending stress due to M_1, let $\sigma = M/S$ because the section is symmetrical.

$$\sigma_1 = \pm M_1/S = \pm(120\,000 \text{ lb·in})/(7.37 \text{ in}^3) = \pm 16\,280 \text{ psi}$$

This stress is tensile at B and compressive at C [Figure 3–20(b)].

The bending stress due to M_2 is

$$\sigma_2 = \pm M_2/S = \pm(51\,960 \text{ lb·in})/(7.37 \text{ in}^3) = \pm 7\,050 \text{ psi}$$

This stress is compressive at B and tensile at C [Figure 3–20(c)].

The direct tensile stress due to F_x is

$$\sigma_3 = +F_x/A = (+8\,660 \text{ lb})/(3.67 \text{ in}^2) = +2\,360 \text{ psi}$$

This stress is uniform across the entire cross section [Figure 3–20(d)].

Now combine the stresses at B:

$$\sigma_{T_B} = +\sigma_1 - \sigma_2 + \sigma_3 = 16\,280 - 7\,050 + 2\,360$$
$$\sigma_{T_B} = +11\,590 \text{ psi} \quad \text{(tensile)}$$

Now combine the stresses at C:

$$\sigma_{T_C} = -\sigma_1 + \sigma_2 + \sigma_3 = -16\,280 + 7\,050 + 2\,360$$
$$\sigma_{T_C} = -6\,870 \text{ psi} \quad \text{(compressive)}$$

Comments The resulting combined stress is shown in Figure 3–20(e). The maximum combined stress is 11 590-psi tension on the top surface of the beam at B where it attaches to the column.

3–21
STRESS
CONCENTRATIONS

The formulas reviewed earlier for computing simple stresses due to direct tensile and compressive forces, bending moments, and torsional moments are applicable under certain conditions. One condition is that the geometry of the member is uniform throughout the section of interest.

In many typical machine design situations, inherent geometric discontinuities are necessary for the parts to perform their desired functions. For example, as shown in Figure 12–2 in Chapter 12, shafts carrying gears, chain sprockets, or belt sheaves usually have several diameters that create a series of shoulders that seat the power transmission members and support bearings. Grooves in the shaft allow the installation of retaining rings. Keyseats milled into the shaft enable keys to drive the elements. Similarly, tension members in linkages may be designed with retaining ring grooves, radial holes for pins, screw threads, or reduced sections.

Any of these geometric discontinuities will cause the actual maximum stress in the part to be higher than the simple formulas predict. Defining *stress concentration factors* as the factors by which the actual maximum stress exceeds the nominal stress, σ_{nom} or τ_{nom}, predicted from the simple equations allows the designer to analyze these situations. The symbol for these factors is K_t. In general, the K_t factors are used as follows:

$$\sigma_{\text{max}} = K_t \sigma_{\text{nom}} \quad \text{or} \quad \tau_{\text{max}} = K_t \tau_{\text{nom}} \tag{3–27}$$

depending on the kind of stress produced for the particular loading. The value of K_t depends on the shape of the discontinuity, the specific geometry, and the type of stress. Appendix 15 includes several charts for stress concentration factors. (See Reference 5.) Note that the charts indicate the method of computing the nominal stress. Usually, we compute the nominal stress by using the net section in the vicinity of the discontinuity. For example, for a flat plate with a hole in it subjected to a tensile force, the nominal stress is computed as the force divided by the minimum cross-sectional area through the location of the hole.

But there are other cases in which the gross area is used in calculating the nominal stress. For example, we analyze keyseats by applying the stress concentration factor to the computed stress in the full-diameter portion of the shaft.

Figure 3–21 shows an experimental device that demonstrates the phenomenon of stress concentrations. A model of a beam having several different cross-sectional heights is made from a special plastic that reacts to the presence of varying stresses at different points. When the model is viewed through a polarizing filter, several black *fringes* appear. Where there are many closely spaced fringes, the stress is changing rapidly. We can compute the actual magnitude of stress by knowing the optical characteristics of the plastic.

The beam in Figure 3–21 is simply supported near each end and is loaded vertically at its middle. The largest stress occurs to the left of the middle where the height of the cross section is reduced. Note that the fringes are very close together in the vicinity of the fillet connecting the smaller section to the larger part where the load is applied. This indicates that the highest stress occurs in the fillet. Figure A15–2 gives data for the values of the stress concentration factor, K_t. The nominal stress, σ_{nom}, is computed from the classic flexure formula, and the section modulus is based on the smaller cross section near the fillet. These formulas are listed near the stress concentration chart.

An interesting observation can be made from Figure A15–3 showing the stress concentration factors for a flat plate with a central hole. Curves A and B refer to tension loading, while curve C is for bending. The nominal stress for each case is computed on the basis of the net cross section accounting for the material removed by the hole. Curve C

FIGURE 3–21
Illustration of stress concentrations (Source: Measurements Group, Inc., Raleigh, North Carolina, U.S.A.)

indicates that the stress concentration factor is taken to be 1.0 for smaller holes, with the ratio of the hole diameter to the width of the plate <0.50.

Figure A15–4 covers the case of a round shaft that has a circular hole completely through it. The three curves cover tension, bending, and torsional loading, and each is based on the stress in the gross section, that is, the geometry of the shaft without the hole. Therefore, the value of K_t includes the effects of both the material removal and the discontinuity created by the presence of the hole. The values of K_t are relatively high, even for the smaller holes, indicating that you should be cautious when applying shafts with holes to ensure that the local stresses are small.

The following are guidelines on the use of stress concentration factors:

1. The worst case occurs for those areas in tension.

2. Always use stress concentration factors in analyzing members under fatigue loading because fatigue cracks usually initiate near points of high local tensile stress.

3. Stress concentrations can be ignored for static loading of ductile materials because if the local maximum stress exceeds the yield strength of the material, the load is redistributed. The resulting member is actually stronger after the local yielding occurs.

4. The stress concentration factors in Appendix 15 are empirical values based only on the geometry of the member and the manner of loading.

5. Use stress concentration factors when analyzing brittle materials under either static or fatigue loading. Because the material does not yield, the stress redistribution described in item 3 cannot occur.

6. Even scratches, nicks, corrosion, excessive surface roughness, and plating can cause stress concentrations. Chapter 5 discusses the care essential to manufacturing, handling, and assembling components subjected to fatigue loading.

Example Problem 3–18   Compute the maximum stress in a round bar subjected to an axial tensile force of 9 800 N. The geometry is shown in Figure 3–22.

Solution   Objective   Compute the maximum stress in the stepped bar shown in Figure 3–22.

Given   The layout from Figure 3–22. Force = F = 9 800 N.
The shaft has two diameters joined by a fillet with a radius of 1.5 mm.
Larger diameter = D = 12 mm; smaller diameter = d = 10 mm.

Analysis   The presence of the change in diameter at the step causes a stress concentration to occur. The general situation is a round bar subjected to an axial tensile load. We will use the top graph of Figure A15–1 to determine the stress concentration factor. That value is used in Equation (3–27) to determine the maximum stress.

FIGURE 3–22
Stepped round bar subjected to axial tensile force

Results   Figure A15–1 indicates that the nominal stress is computed for the smaller of the two diameters of the bar. The stress concentration factor depends on the ratio of the two diameters and the ratio of the fillet radius to the smaller diameter.

$$D/d = 12 \text{ mm}/10 \text{ mm} = 1.20$$

$$r/d = 1.5 \text{ mm}/10 \text{ mm} = 0.15$$

From these values, we can find that $K_t = 1.60$. The stress is

$$A = \frac{\pi d^2}{4} = \frac{(\pi)(10 \text{ mm}^2)}{4} = 78.5 \text{ mm}^2$$

$$\sigma = \frac{K_t F}{A} = \frac{1.60(9\,800 \text{ N})}{78.5 \text{ mm}^2} = 199.6 \text{ N/mm}^2 = 199.6 \text{ MPa}$$

Comments   The maximum tensile stress of 199.6 MPa occurs in the fillet near the smaller diameter. This value is 1.60 times higher than the nominal stress that occurs in the 10-mm-diameter shaft. To the left of the shoulder, the stress reduces dramatically as the effect of the stress concentration diminishes and because the area is larger.

3–22 NOTCH SENSITIVITY AND STRENGTH REDUCTION FACTOR

The amount by which a load-carrying member is weakened by the presence of a stress concentration (notch), considering both the material and the sharpness of the notch, is defined as

K_f = fatigue strength reduction factor

$$K_f = \frac{\text{endurance limit of a notch-free specimen}}{\text{endurance limit of a notched specimen}}$$

This factor could be determined by actual test. However, it is typically found by combining the stress concentration factor, K_t, defined in the previous section, and a material factor called the *notch sensitivity, q*. We define

$$q = (K_f - 1)/(K_t - 1) \qquad\qquad \textbf{(3–28)}$$

When q is known, K_f can be computed from

$$K_f = 1 + q(K_t - 1) \qquad\qquad \textbf{(3–29)}$$

Values of q range from zero to 1.0, and therefore K_f varies from 1.0 to K_t. Under repeated bending loads, very ductile steels typically exhibit values of q from 0.5 to 0.7. High-strength steels with hardness approximately HB 400 ($s_u \cong 200$ Ksi or 1 400 MPa) have values of q from 0.90 to 0.95. (See Reference 2 for further discussion of values of q.)

Because reliable values of q are difficult to obtain, the problems in this book will assume that $q = 1.0$ and $K_f = K_t$, the safest, most conservative value.

REFERENCES

1. Blake, Alexander. *Practical Stress Analysis in Engineering Design.* New York: Marcel Dekker, 1982.

2. Boresi, A. P., O. M. Sidebottom, F. B. Seely, and J. O. Smith. *Advanced Mechanics of Materials.* 4th ed. New York: Halsted Press, 1985.

3. Mott, R. L. *Applied Strength of Materials.* 3d ed. Upper Saddle River, NJ: Prentice-Hall, 1996.

4. Muvdi, B. B., and J. W. McNabb. *Engineering Mechanics of Materials.* 2d ed. New York: Macmillan, 1984.

5. Pilkey, Walter D. *Peterson's Stress Concentration Factors.* 2d ed. New York: John Wiley & Sons, 1997.

6. Popov, E. P. *Engineering Mechanics of Solids.* Upper Saddle River, NJ: Prentice-Hall, 1990.

7. Young, W. C. *Roark's Formulas for Stress and Strain.* 6th ed. New York: McGraw-Hill, 1989.

PROBLEMS

Direct Tension and Compression

1. A tensile member in a machine structure is subjected to a steady load of 4.50 kN. It has a length of 750 mm and is made from a steel tube having an outside diameter of 18 mm and an inside diameter of 12 mm. Compute the tensile stress in the tube and the axial deformation.

2. Compute the stress in a round bar having a diameter of 10.0 mm and subjected to a direct tensile force of 3 500 N.

3. Compute the stress in a rectangular bar having cross-sectional dimensions of 10.0 mm by 30.0 mm when a direct tensile force of 20.0 kN is applied.

4. A link in a packaging machine mechanism has a square cross section 0.40 in on a side. It is subjected to a tensile force of 860 lb. Compute the stress in the link.

5. Two circular rods support the 3 800-lb weight of a space heater in a warehouse. Each rod has a diameter of 0.375 in and carries 1/2 of the total load. Compute the stress in the rods.

6. A tensile load of 5.00 kN is applied to a square bar, 12 mm on a side and having a length of 1.65 m. Compute the stress and the axial deformation in the bar if it is made from (a) AISI 1020 hot-rolled steel, (b) AISI 8650 OQT 1000 steel, (c) ductile iron A536-77 (60-40-18), (d) aluminum 6061-T6, (e) titanium Ti-6Al-4V, (f) rigid PVC plastic, and (g) phenolic plastic.

7. An aluminum rod is made in the form of a hollow square tube, 2.25 in outside, with a wall thickness of 0.120 in. Its length is 16.0 in. What axial compressive force would

cause the tube to shorten by 0.004 in? Compute the resulting compressive stress in the aluminum.

8. Compute the stress in the middle portion of rod *AC* in Figure 3–23 if the vertical force on the boom is 2 500 lb. The rod is rectangular, 1.50 in by 3.50 in.

FIGURE 3–23 (Problems 8, 16, and 56)

9. Compute the forces in the two angled rods in Figure 3–24 for an applied force, $F = 1\,500$ lb, if the angle θ is 45°.

10. If the rods from Problem 9 are circular, determine their required diameter if the load is static and the allowable stress is 18 000 psi.

11. Repeat Problems 9 and 10 if the angle θ is 15°.

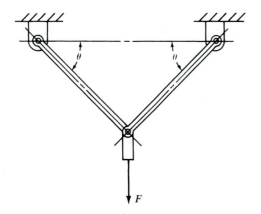

FIGURE 3–24 (Problems 9, 17, and 18)

12. Figure 3–25 shows a small truss spanning between solid supports and suspending a 10.5-kN load. The cross sections for the three main types of truss members are shown. Compute the stresses in all of the members of the truss near their midpoints away from the connections. Consider all joints to be pinned.

13. The truss shown in Figure 3–26 spans a total space of 18.0 ft and carries two concentrated loads on its top chord. The members are made from standard steel angle and channel shapes as indicated in the figure. Consider all

joints to be pinned. Compute the stresses in all members near their midpoints away from the connections.

14. Figure 3–27 shows a short leg for a machine that carries a direct compression load. Compute the compressive stress if the cross section has the shape shown and the applied force is $F = 52\,000$ lb.

15. Consider the short compression member shown in Figure 3–28. Compute the compressive stress if the cross section has the shape shown and the applied load is 640 kN.

Direct Shear Stress

16. Refer to Figure 3–23. Each of the pins at A, B, and C has a diameter of 0.50 in and is loaded in double shear. Compute the shear stress in each pin.

17. Compute the shear stress in the pins connecting the rods shown in Figure 3–24 when a load of $F = 1\,500$ lb is carried. The pins have a diameter of 0.75 in. The angle $\theta = 40°$.

18. Repeat Problem 17, but change the angle to $\theta = 15°$.

19. Refer to Figure 3–3. Compute the shear stress in the key if the shaft transmits a torque of $1\,600$ N·m. The shaft diameter is 60 mm. The key is square with $b = 12$ mm, and it has a length of 45 mm.

20. A punch is attempting to cut a slug having the shape shown in Figure 3–29 from a sheet of aluminum having a thickness of 0.060 in. Compute the shearing stress in the

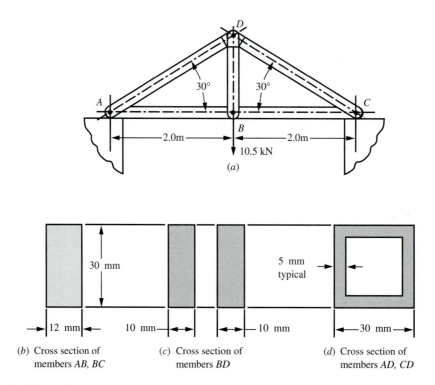

(a)

(b) Cross section of members *AB, BC*

(c) Cross section of members *BD*

(d) Cross section of members *AD, CD*

FIGURE 3–25 (Problem 12)

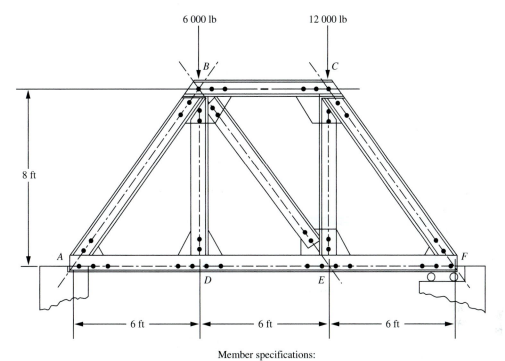

Member specifications:

AD, DE, EF L2 × 2 × 1/8 – doubled ⎯⎯ ⌐L

BD, CE, BE L2 × 2 × 1/8 – single ⎯⎯ ⌐

AB, BC, CF C3 × 4.1 – doubled ⎯⎯ ⊐⊏

FIGURE 3–26   (Problem 13)

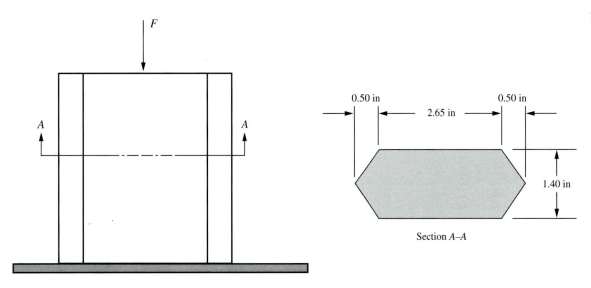

0.50 in   2.65 in   0.50 in

1.40 in

Section A–A

FIGURE 3–27   (Problem 14)

FIGURE 3–28 (Problem 15)

Section A–A

FIGURE 3–29 (Problem 20)

FIGURE 3–30 (Problem 21)

aluminum when a force of 52 000 lb is applied by the punch.

21. Figure 3–30 shows the shape of a slug that is to be cut from a sheet of steel having a thickness of 2.0 mm. If the punch exerts a force of 225 kN, compute the shearing stress in the steel.

Torsion

22. Compute the torsional shear stress in a circular shaft with a diameter of 50 mm that is subjected to a torque of 800 N·m.

23. If the shaft of Problem 22 is 850 mm long and is made of steel, compute the angle of twist of one end in relation to the other.

24. Compute the torsional shear stress due to a torque of 88.0 lb·in in a circular shaft having a 0.40-in diameter.

25. Compute the torsional shear stress in a solid circular shaft having a diameter of 1.25 in that is transmitting 110 hp at a speed of 560 rpm.

26. Compute the torsional shear stress in a hollow shaft with an outside diameter of 40 mm and an inside diameter of 30 mm when transmitting 28 kilowatts (kW) of power at a speed of 45 rad/s.

27. Compute the angle of twist for the hollow shaft of Problem 26 over a length of 400 mm. The shaft is steel.

Noncircular Members in Torsion

28. A square steel bar, 25 mm on a side and 650 mm long, is subjected to a torque of 230 N·m. Compute the shear stress and the angle of twist for the bar.

29. A 3.00-in-diameter steel bar has a flat milled on one side, as shown in Figure 3–31. If the shaft is 44.0 in long and carries a torque of 10 600 lb·in, compute the stress and the angle of twist.

30. A commercial steel supplier lists rectangular steel tubing having outside dimensions of 4.00 by 2.00 in and a wall thickness of 0.109 in. Compute the maximum torque that

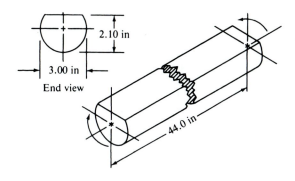

FIGURE 3–31   (Problem 29)

can be applied to such a tube if the shear stress is to be limited to 6 000 psi. For this torque, compute the angle of twist of the tube over a length of 6.5 ft.

Beams

31. A beam is simply supported and carries the load shown in Figure 3–32. Specify suitable dimensions for the beam if it is steel and the stress is limited to 18 000 psi, for the following shapes:
(a) Square
(b) Rectangle with height three times the width
(c) Rectangle with height one-third the width
(d) Solid circular section
(e) American Standard beam section
(f) American Standard channel with the legs down
(g) Standard steel pipe

FIGURE 3–32   (Problems 31, 32, and 33)

32. For each beam of Problem 31, compute its weight if the steel weighs 0.283 lb/in^3.

33. For each beam of Problem 31, compute the maximum deflection and the deflection at the loads.

34. For the beam loading of Figure 3–33, draw the complete shearing force and bending moment diagrams, and determine the bending moments at points *A, B,* and *C.*

35. For the beam loading of Figure 3–33, design the beam, choosing a shape that will be reasonably efficient and will limit the stress to 100 MPa.

36. Figure 3–34 shows a beam made from 4-in steel pipe. Compute the deflection at points *A* and *B* for two cases: (a) the simple cantilever and (b) the supported cantilever.

FIGURE 3–33   (Problems 34 and 35)

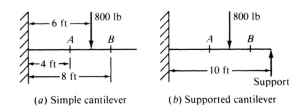

(a) Simple cantilever   (b) Supported cantilever

FIGURE 3–34   (Problem 36)

37. Select an aluminum I-beam shape to carry the load shown in Figure 3–35 with a maximum stress of 12 000 psi. Then compute the deflection at each load.

FIGURE 3–35   (Problem 37)

38. Figure 3–36 represents a wood joist for a platform, carrying a uniformly distributed load of 120 lb/ft and two concentrated loads applied by some machinery. Compute the maximum stress due to bending in the joist and the maximum vertical shear stress.

FIGURE 3–36   (Problem 38)

Beams with Concentrated Bending Moments

For Problems 39 through 50, draw the free-body diagram of only the horizontal beam portion of the given figures. Then draw the complete shear and bending moment diagrams. Where used, the symbol X indicates a simple support capable of exerting a reaction force in any direction but having no moment resistance. For beams having unbalanced axial loads, you may specify which support offers the reaction.

39. Use Figure 3–37.

40. Use Figure 3–38.

41. Use Figure 3–39.

42. Use Figure 3–40.

43. Use Figure 3–41.

44. Use Figure 3–42.

45. Use Figure 3–43.

46. Use Figure 3–44.

47. Use Figure 3–45.

48. Use Figure 3–46.

49. Use Figure 3–47.

50. Use Figure 3–48.

FIGURE 3–41   (Problems 43 and 59)

FIGURE 3–42   (Problem 44)

FIGURE 3–37   (Problems 39 and 57)

FIGURE 3–43   (Problem 45)

FIGURE 3–38   (Problem 40)

FIGURE 3–44   (Problem 46)

FIGURE 3–39   (Problem 41)

FIGURE 3–40   (Problems 42 and 58)

FIGURE 3–45   (Problem 47)

FIGURE 3–46 (Problem 48)

FIGURE 3–47 (Problem 49)

FIGURE 3–48 (Problems 50 and 60)

Combined Normal Stresses

51. Compute the maximum tensile stress in the bracket shown in Figure 3–49.

52. Compute the maximum tensile and compressive stresses in the horizontal beam shown in Figure 3–50.

53. For the lever shown in Figure 3–51(a), compute the stress at section A near the fixed end. Then redesign the lever to the tapered form shown in Part (b) of the figure by adjust-

FIGURE 3–49 (Problem 51)

FIGURE 3–50 (Problem 52)

(a) Original design of lever

(b) Modified design of lever

FIGURE 3–51 (Problem 53)

ing only the height of the cross section at sections B and C so that they have no greater stress than section A.

54. Compute the maximum tensile stress at sections A and B on the crane boom shown in Figure 3–52.

FIGURE 3–52 (Problem 54)

55. Refer to Figure 3–18. Compute the maximum tensile stress in the print head just to the right of the right guide. The head has a rectangular cross section, 5.0 mm high in the plane of the paper and 2.4 mm thick.

56. Refer to Figure 3–23. Compute the maximum tensile and compressive stresses in the member *B–C* if the load *F* is 1 800 lb. The cross section of *B–C* is a 6 × 4 × 1/4 rectangular tube.

57. Refer to Figure 3–37. The vertical member is to be made from steel with a maximum allowable stress of 12 000 psi. Specify the required size of a standard square cross section if sizes are available in increments of 1/16 in.

58. Refer to Figure 3–40. Compute the maximum stress in the horizontal portion of the bar, and tell where it occurs on the cross section. The left support resists the axial force.

59. Refer to Figure 3–41. Compute the maximum stress in the horizontal portion of the bar, and indicate where it occurs on the cross section. The right support resists the unbalanced axial force.

60. Refer to Figure 3–48. Specify a suitable diameter for a solid circular bar to be used for the top horizontal member, which is supported in the bearings. The left bearing resists the axial load. The allowable normal stress is 25 000 psi.

Stress Concentrations

61. Figure 3–53 shows a valve stem from an engine subjected to an axial tensile load applied by the valve spring. For a force of 1.25 kN, compute the maximum stress at the fillet under the shoulder.

Automotive valve assembly

FIGURE 3–53   (Problem 61)

62. The conveyor fixture shown in Figure 3–54 carries three heavy assemblies (1 200 lb each). Compute the maximum stress in the fixture, considering stress concentrations at the fillets and assuming that the load acts axially.

63. For the flat plate in tension in Figure 3–55, compute the stress at each hole, assuming that the holes are sufficiently far apart that their effects do not interact.

For Problems 64 through 68, compute the maximum stress in the member, considering stress concentrations.

FIGURE 3–54   (Problem 62)

FIGURE 3–55   (Problem 63)

64. Use Figure 3–56.

65. Use Figure 3–57.

66. Use Figure 3–58.

67. Use Figure 3–59.

68. Use Figure 3–60.

Problems of a General Nature

69. Figure 3–61 shows a horizontal beam supported by a vertical tension link. The cross sections of both the beam and the link are 20 mm square. All connections use 8.00-mm-diameter cylindrical pins in double shear. Compute the tensile stress in member *A–B*, the stress due to bending in *C–D,* and the shearing stress in the pins *A* and *C.*

FIGURE 3–56   (Problem 64)

FIGURE 3–57   (Problem 65)

FIGURE 3–58   (Problem 66)

FIGURE 3–59   (Problem 67)

FIGURE 3–60   (Problem 68)

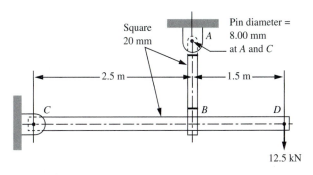

FIGURE 3–61   (Problem 69)

70. Figure 3–62 shows a tapered flat bar that has a uniform thickness of 20 mm. The depth tapers from $h_1 = 40$ mm near the load to $h_2 = 20$ mm at each support. Compute the stress due to bending in the bar at points spaced 40 mm apart from the support to the load. Let the load $P = 5.0$ kN.

71. For the flat bar shown in Figure 3–62, compute the stress in the middle of the bar if a hole of 25-mm diameter is drilled directly under the load on the horizontal centerline. The load is $P = 5.0$ kN. See data in Problem 70.

72. The beam shown in Figure 3–63 is a stepped, flat bar having a constant thickness of 1.20 in. It carries a single concentrated load at C of 1 500 lb. Compare the stresses at the following locations:
    (a) In the vicinity of the load
    (b) At the section through the smaller hole to the right of section C
    (c) At the section through the larger hole to the right of section C
    (d) Near section B where the bar changes height

73. Figure 3–64 shows a stepped, flat bar having a constant thickness of 8.0 mm. It carries three concentrated loads as shown. Let $P = 200$ N, $L_1 = 180$ mm, $L_2 = 80$ mm, and $L_3 = 40$ mm. Compute the maximum stress due to bending, and state where it occurs. The bar is braced against lateral bending and twisting. Note that the dimensions in the figure are not drawn to scale.

74. Figure 3–65 shows a bracket carrying opposing forces of $F = 2 500$ N. Compute the stress in the upper horizontal part through one of the holes as at B. Use $d = 15.0$ mm for the diameter of the holes.

75. Repeat Problem 74, but use a hole diameter of $d = 12.0$ mm.

76. Figure 3–66 shows a lever made from a rectangular bar of steel. Compute the stress due to bending at the fulcrum (20 in from the pivot) and at the section through the bottom hole. The diameter of each hole is 1.25 in.

77. For the lever in Figure 3–66, determine the maximum stress if the attachment point is moved to each of the other two holes.

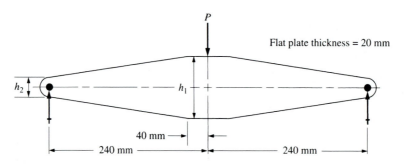

FIGURE 3–62 Tapered flat bar for Problems 70 and 71

FIGURE 3–63 (Problem 72)

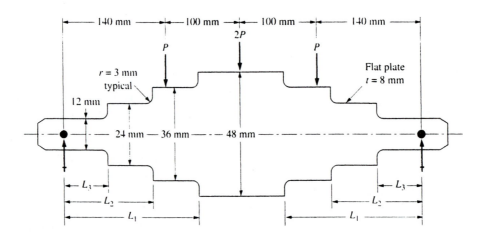

FIGURE 3–64 Stepped flat bar for Problem 73

78. Figure 3–67 shows a shaft that is loaded only in bending. Bearings are located at points *B* and *D* to allow the shaft to rotate. Pulleys at *A*, *C*, and *E* carry cables that support loads from below while allowing the shaft to rotate. Compute the maximum stress due to bending in the shaft considering stress concentrations.

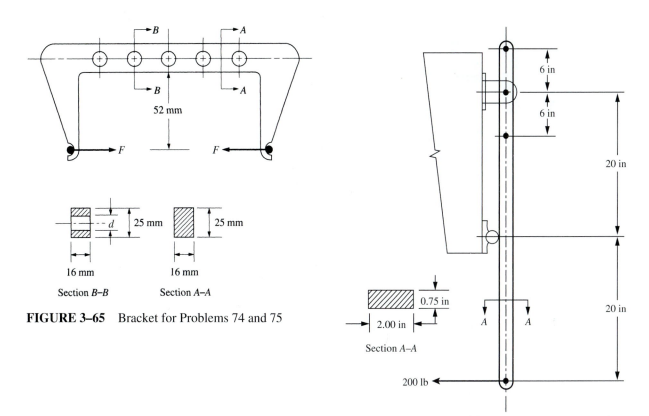

FIGURE 3–65   Bracket for Problems 74 and 75

FIGURE 3–66   Lever for Problems 76 and 77

FIGURE 3–67   Data for Problem 78

4

Combined Stresses
and Mohr's Circle

Combined Stresses and Mohr's Circle

Discussion Map

☐ You must build your ability to analyze more complex parts and loading patterns.

Discover
Find products around you that have complex geometries or loading patterns.
Discuss these products with your colleagues.

This chapter helps you analyze complex objects to determine maximum stresses. We will use *Mohr's circle,* a graphical tool for stress analysis, as an aid in understanding how stresses vary within a load-carrying member.

In Chapter 3, you reviewed the basic principles of stress and deformation analysis, practiced the application of those principles to machine design problems, and solved some problems by superposition when two or more types of loads caused normal, either tensile or compressive, stresses.

But what happens when the loading pattern is more complex?

Many practical machine components experience combinations of normal and shear stresses. Sometimes the pattern of loading or the geometry of the component causes the analysis to be very difficult to solve directly using the methods of basic stress analysis.

Look around you and identify products, parts of structures, or machine components that have a more complex loading or geometry. Perhaps some of those identified in **The Big Picture** for Chapter 3 have this characteristic.

Discuss how the selected items are loaded, where the maximum stresses are likely to occur, and how the loads and the geometry are related. Did the designer tailor the shape of the object to be able to carry the applied loads in an efficient manner? How are the shape and the size of critical parts of the item related to the expected stresses?

When we move on to **Chapter 5: Design for Different Types of Loading,** we will need tools to determine the magnitude and the direction of maximum shear stresses or maximum principal (normal) stresses.

Completing this chapter will help you develop a clear understanding of the distribution of stress in a load-carrying member, and it will help you determine the maximum stresses, either normal or shear, so that you can complete a reliable design or analysis.

Some of the techniques of combining stresses require the application of fairly involved equations. A graphical tool, *Mohr's circle,* can be used as an aid in completion of the analysis. Applied properly, the method is precise and should aid you in understanding how the stresses vary within a complex load-carrying member. It should also help you correctly use commercially available stress analysis software.

 You Are the Designer

Your company is designing a special machine to test a high-strength fabric under prolonged exposure to a static load to determine whether it continues to deform a greater amount with time. The tests will be run at a variety of temperatures requiring a controlled environment around the test specimen. Figure 4–1 shows the general

layout of one proposed design. Two rigid supports are available at the rear of the machine with a 24-in gap between them. The line of action of the load on the test fabric is centered on this gap and 15.0 in out from the middle of the supports. You are asked to design a bracket to hold the upper end of the load frame.

Assume that one of your design concepts uses the arrangement shown in Figure 4–2. Two circular bars are bent 90°. One end of each bar is securely welded to the

FIGURE 4–1 Layout of the load frame supports—top view

24 in

12 in

Rigid support

Rigid support

15 in

Point of application of the load

Element *A*

Support

Support

Circular bar

15 in

Top view

6 in

6 in

Rigid support

Yoke

Rigid support

Yoke

Fabric sample

Support

Yoke

Yoke

Static weight

Static weight

Front view

Side view

FIGURE 4–2 Proposed bracket design

vertical support surface. A flat bar is attached across the outboard end of each bar so that the load is shared evenly by the two bars.

One of your design problems is to determine the maximum stress that exists in the bent bars to ensure that they are safe. What kinds of stress are developed in the bars? Where are the stresses likely to be the greatest? How can the magnitude of the stresses be computed? Note that the part of the bar near its attachment to the support has a combination of stresses exerted on it.

Consider the element on the top surface of the bar, labeled element *A* in Figure 4–2. The moment caused by the force acting at an extension of 6.0 in from the support places element *A* in tension due to the bending action. The torque caused by the force acting 15.0 in out from the axis of the bar at its point of support creates a torsional shear stress on element *A*. Both of these stresses act in the *x-y* plane, subjecting element *A* to a combined normal and shear stress. How do you analyze such a stress condition? How do the tensile and shear stresses act together? What are the maximum normal stress and the maximum shear stress on element *A*, and where do they occur?

You would need such answers to complete the design of the bars. The material in this chapter will enable you to complete the necessary analyses.

4–1
OBJECTIVES OF
THIS CHAPTER

After completing this chapter, you will be able to:

1. Illustrate a variety of combined stresses on stress elements.

2. Analyze a load-carrying member subjected to combined stress to determine the maximum normal stress and the maximum shear stress on any given element.

3. Determine the directions in which the maximum stresses are aligned.

4. Determine the state of stress on an element in any specified direction.

5. Draw the complete Mohr's circle as an aid in completing the analyses for the maximum stresses.

4–2
GENERAL CASE
OF COMBINED
STRESS

To visualize the general case of combined stress, it is helpful to consider a small element of the load-carrying member on which combined normal and shear stresses act. For this discussion we will consider a two-dimensional stress condition, as illustrated in Figure 4–3. The x- and y-axes are aligned with corresponding axes on the member being analyzed.

The normal stresses, σ_x and σ_y, could be due to a direct tensile force or to bending. If the normal stresses were compressive (negative), the vectors would be pointing in the opposite sense, into the stress element.

The shear stress could be due to direct shear, torsional shear, or vertical shear stress. The double-subscript notation helps to orient the direction of shear stresses. For example, τ_{xy} indicates the shear stress acting on the element face that is perpendicular to the x-axis and parallel to the y-axis. A *positive shear stress* is one that tends to rotate the stress element clockwise. In Figure 4–3, τ_{xy} is positive, and τ_{yx} is negative. Their magnitudes must be equal to maintain the element in equilibrium.

It is necessary to determine the magnitudes and the signs of each of these stresses in order to show them properly on the stress element. Example Problem 4–1 which follows the definition of principal stresses illustrates the process.

With the stress element defined, the objectives of the remaining analysis are to determine the maximum normal stress, the maximum shear stress, and the planes on which these stresses occur. The governing formulas follow.

FIGURE 4–3
General two-dimensional stress element

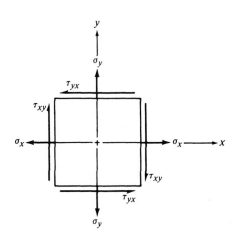

Maximum Normal Stresses: Principal Stresses

The combination of the applied normal and shear stresses that produces the maximum normal stress is called the *maximum principal stress, σ_1*. The magnitude of σ_1 can be computed from the following equation:

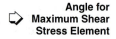

Maximum Principal Stress

$$\sigma_1 = \frac{\sigma_x + \sigma_y}{2} + \sqrt{\left(\frac{\sigma_x - \sigma_y}{2}\right)^2 + \tau_{xy}^2} \qquad (4\text{--}1)$$

The combination of the applied stresses that produces the minimum normal stress is called the *minimum principal stress, σ_2*. Its magnitude can be computed from

Minimum Principal Stress

$$\sigma_2 = \frac{\sigma_x + \sigma_y}{2} - \sqrt{\left(\frac{\sigma_x - \sigma_y}{2}\right)^2 + \tau_{xy}^2} \qquad (4\text{--}2)$$

Particularly in experimental stress analysis, it is important to know the orientation of the principal stresses. The angle of inclination of the planes on which the principal stresses act, called the *principal planes,* can be found from

Angle for Principal Stress Element

$$\phi_\sigma = \tfrac{1}{2} \arctan\left[2\tau_{xy}/(\sigma_x - \sigma_y)\right] \qquad (4\text{--}3)$$

The angle ϕ_σ is measured from the positive x-axis of the original stress element to the maximum principal stress, σ_1. Then the minimum principal stress, σ_2, is on the plane 90° from σ_1.

When the stress element is oriented as discussed so that the principal stresses are acting on it, the shear stress is zero. The resulting stress element is shown in Figure 4–4.

Maximum Shear Stress

On a different orientation of the stress element, the maximum shear stress will occur. Its magnitude can be computed from

Maximum shear stress

$$\tau_{\max} = \sqrt{\left(\frac{\sigma_x - \sigma_y}{2}\right)^2 + \tau_{xy}^2} \qquad (4\text{--}4)$$

The angle of inclination of the element on which the maximum shear stress occurs is computed as follows:

Angle for Maximum Shear Stress Element

$$\phi_\tau = \tfrac{1}{2} \arctan\left[-(\sigma_x - \sigma_y)/2\tau_{xy}\right] \qquad (4\text{--}5)$$

FIGURE 4–4
Principal stress element

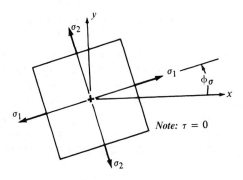

The angle between the principal stress element and the maximum shear stress element is always 45°.

On the maximum shear stress element, there will be normal stresses of equal magnitude acting perpendicular to the planes on which the maximum shear stresses are acting. These normal stresses have the value

▷ **Average Normal Stress**

$$\sigma_{avg} = (\sigma_x + \sigma_y)/2 \qquad (4\text{–}6)$$

Note that this is the *average* of the two applied normal stresses. The resulting maximum shear stress element is shown in Figure 4–5. Note, as stated above, that the angle between the principal stress element and the maximum shear stress element is always 45°.

Summary and General Procedure for Analyzing Combined Stresses

The following list gives a summary of the techniques presented in this section; it also outlines the general procedure for applying the techniques to a given stress analysis problem.

General Procedure for Computing Principal Stresses and Maximum Shear Stresses

1. Decide for which point you want to compute the stresses.

2. Clearly specify the coordinate system for the object and the direction of forces.

3. Compute the stresses on the selected point due to the applied forces, and show the stresses acting on a stress element at the desired point with careful attention to directions. Figure 4–3 is a model for how to show these stresses.

4. Compute the principal stresses on the point and the directions in which they act. Use Equations (4–1), (4–2), and (4–3).

5. Draw the stress element on which the principal stresses act, and show its orientation relative to the original x-axis. It is recommended that the principal stress element be drawn beside the original stress element to illustrate the relationship between them.

6. Compute the maximum shear stress on the element and the orientation of the plane on which it acts. Also, compute the normal stress that acts on the maximum shear stress element. Use Equations (4–4), (4–5), and (4–6).

FIGURE 4–5
Maximum shear stress element

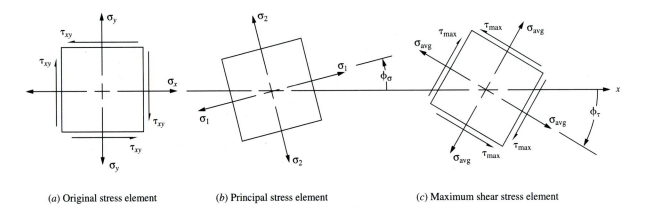

(a) Original stress element   (b) Principal stress element   (c) Maximum shear stress element

FIGURE 4–6   Relationships among original stress element, principal stress element, and maximum shear stress element for a given loading

7. Draw the stress element on which the maximum shear stress acts, and show its orientation to the original x-axis. It is recommended that the maximum shear stress element be drawn beside the maximum principal stress element to illustrate the relationship between them.

8. The resulting set of three stress elements will appear as shown in Figure 4–6.

The following example problem illustrates the use of this procedure.

Example Problem 4–1   The shaft shown in Figure 4–7 is supported by two bearings and carries two chain sprockets. The tensions in the chains exert horizontal forces on the shaft, tending to bend it in the x-z plane. Sprocket B exerts a clockwise torque on the shaft when viewed toward the origin of the coordinate system along the x-axis. Sprocket C exerts an equal but opposite torque on the shaft. For the loading condition shown, determine the principal stresses and the maximum shear stress on element K on the front surface of the shaft (on the positive z-side) just to the right of sprocket B. Follow the general procedure for analyzing combined stresses given in this section.

Solution   **Objective**   Compute the principal stresses and the maximum shear stresses on element K.

Given   Shaft and loading pattern shown in Figure 4–7.

Analysis   Use the general procedure for analyzing combined stresses.

Results   Element K is subjected to bending that produces a tensile stress acting in the x-direction. Also, there is a torsional shear stress acting at K. Figure 4–8 shows the shearing force and bending moment diagrams for the shaft and indicates that the bending moment at K is 1 540 lb·in. The bending stress is therefore

$$\sigma_x = M/S$$
$$S = \pi D^3/32 = [\pi(1.25 \text{ in})^3]/32 = 0.192 \text{ in}^3$$
$$\sigma_x = (1\,540 \text{ lb·in})/(0.192 \text{ in}^3) = 8\,030 \text{ psi}$$

The torsional shear stress acts on element K in a way that causes a downward shear stress on the right side of the element and an upward shear stress on the left side. This action results

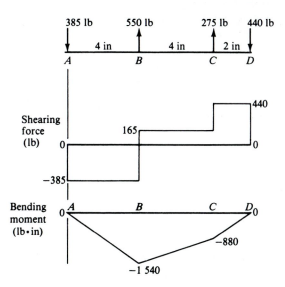

(a) Pictorial view of shaft

(b) Forces acting on shaft at B and C caused by chain drives

(c) Normal view of forces on shaft in x - z plane with reactions at bearings

(d) Enlarged view of element K on front of shaft

FIGURE 4–7 Shaft supported by two bearings and carrying two chain sprockets

FIGURE 4–8
Shearing force and bending moment diagrams for the shaft

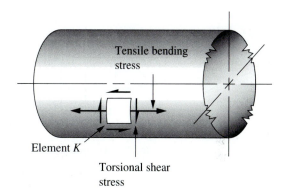

in a tendency to rotate the element in a *clockwise* direction, which is the *positive* direction for shear stresses according to the standard convention. Also, the notation for shear stresses uses double subscripts. For example, τ_{xy} indicates the shear stress acting on the face of an element that is perpendicular to the x-axis and parallel to the y-axis. Thus, for element K,

$$\tau_{xy} = T/Z_p$$
$$Z_p = \pi D^3/16 = \pi(1.25 \text{ in})^3/16 = 0.383 \text{ in}^3$$
$$\tau_{xy} = (1\,100 \text{ lb·in})/0.383 \text{ in}^3 = 2\,870 \text{ psi}$$

FIGURE 4–9
Stresses on element K

FIGURE 4–10
Principal stress element

(a) Original stress element   (b) Principal stress element

The values of the normal stress, σ_x, and the shear stress, τ_{xy}, are shown on the stress element K in Figure 4–9. Note that the stress in the y-direction is zero for this loading. Also, the value of the shear stress, τ_{yx}, must be equal to τ_{xy}, and it must act as shown in order for the element to be in equilibrium.

We can now compute the principal stresses on the element, using Equations (4–1) through (4–3). The maximum principal stress is

$$\sigma_1 = \frac{\sigma_x + \sigma_y}{2} + \sqrt{\left(\frac{\sigma_x - \sigma_y}{2}\right)^2 + \tau_{xy}^2}$$ (4–1)

$$\sigma_1 = (8\,030/2) + \sqrt{(8\,030/2)^2 + (2\,870)^2}$$

$$\sigma_1 = 4\,015 + 4\,935 = 8\,950 \text{ psi}$$

The minimum principal stress is

$$\sigma_2 = \frac{\sigma_x + \sigma_y}{2} - \sqrt{\left(\frac{\sigma_x - \sigma_y}{2}\right)^2 + \tau_{xy}^2}$$ (4–2)

$$\sigma_2 = (8\,030/2) - \sqrt{(8\,030/2)^2 + (2\,870)^2}$$

$$\sigma_2 = 4\,015 - 4\,935 = -920 \text{ psi} \quad \text{(compression)}$$

The direction in which the maximum principal stress acts is

$$\phi_\sigma = \tfrac{1}{2} \arctan[2\tau_{xy}/(\sigma_x - \sigma_y)]$$ (4–3)

$$\phi_\sigma = \tfrac{1}{2} \arctan[(2)(2\,870)/(8\,030)] = 17.8°$$

The principal stresses can be shown on a stress element as illustrated in Figure 4–10. Note that the element is shown in relation to the original element to emphasize the direction of the principal stresses in relation to the original x-axis. The positive sign for ϕ_σ indicates that the principal stress element is rotated *clockwise* from its original position.

(a) Original stress element, K   (b) Principal stress element   (c) Maximum shear stress element

FIGURE 4–11   Relation of maximum shear stress element to the original stress element and the principal stress element

Now the maximum shear stress element can be defined, using Equations (4–4) through (4–6):

$$\tau_{max} = \sqrt{\left(\frac{\sigma_x - \sigma_y}{2}\right)^2 + \tau_{xy}^2}$$ (4–4)

$$\tau_{max} = \sqrt{(8\ 030/2)^2 + (2\ 870)^2}$$

$$\tau_{max} = \pm 4\ 935 \text{ psi}$$

The two pairs of shear stresses, $+\tau_{max}$ and $-\tau_{max}$, are equal in magnitude but opposite in direction.

The orientation of the element on which the maximum shear stress acts is found from Equation (4–5):

$$\phi_\tau = \tfrac{1}{2} \arctan\left[-(\sigma_x - \sigma_y)/2\tau_{xy}\right]$$ (4–5)

$$\phi_\tau = \tfrac{1}{2} \arctan\left(-8\ 030/[(2)(2\ 870)]\right) = -27.2°$$

The negative sign calls for a *counterclockwise* rotation of the element.

There are equal normal stresses acting on the faces of this stress element, which have the value of

$$\sigma_{avg} = (\sigma_x + \sigma_y)/2$$ (4–6)

$$\sigma_{avg} = 8\ 030/2 = 4\ 015 \text{ psi}$$

Comments   Figure 4–11 shows the stress element on which the maximum shear stress acts in relation to the original stress element. Note that the angle between this element and the principal stress element is 45°.

In Example Problem 4–1, all of the values necessary to draw the requested stress elements were obtained from calculations. Although these values are fairly straightforward, *Mohr's circle,* which is discussed next, is somewhat easier to use and offers added insight into the actual stress condition at a point.

**4–3
MOHR'S CIRCLE**

Because of the many terms and signs involved, and the many calculations required in the computation of the principal stresses and the maximum shear stress, there is a rather high probability of error. Using the graphic aid *Mohr's circle* helps to minimize errors and gives a better "feel" for the stress condition at the point of interest.

After Mohr's circle is constructed, it can be used for the following:

1. Finding the maximum and minimum principal stresses and the directions in which they act.

2. Finding the maximum shear stresses and the orientation of the planes on which they act.

3. Finding the value of the normal stresses that act on the planes where the maximum shear stresses act.

4. Finding the values of the normal and shear stresses that act on an element with any orientation.

The data needed to construct Mohr's circle are, of course, the same as those needed to compute the preceding values, because the graphical approach is an exact analogy to the computations.

If the normal and shear stresses that act on any two mutually perpendicular planes of an element are known, the circle can be constructed and any of items 1 through 4 can be found. This method is particularly valuable in experimental stress analysis work because the results obtained from many types of standard strain gage instrumentation techniques give the necessary inputs for the creation of Mohr's circle. When the principal stresses and the maximum shear stress are known, the complete design and analysis can be done, using the various theories of failure discussed in Chapter 5.

Procedure for Constructing Mohr's Circle

Refer to Figures 4–12 and 4–13.

FIGURE 4–12
Partially completed
Mohr's circle,
Steps 1–7

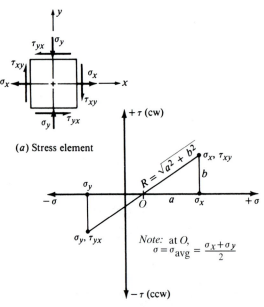

(*a*) Stress element

$$\text{Note: at } O,\ \sigma = \sigma_{\text{avg}} = \frac{\sigma_x + \sigma_y}{2}$$

(*b*) Partially completed
Mohr's circle

1. Perform the stress analysis to determine the magnitudes and directions of the normal and shear stresses acting at the point of interest.

2. Draw the stress element at the point of interest. Normal stresses on any two mutually perpendicular planes are drawn with tensile stresses positive: projecting outward from the element. Compressive stresses are negative: directed inward on the face. Note that the *resultants* of all normal stresses acting in the chosen directions are plotted. Shear stresses are considered to be positive if they tend to rotate the element in a *clockwise* (cw) direction, and negative otherwise.

   Note that on the stress element illustrated, σ_x is positive, σ_y is negative, τ_{xy} is positive, and τ_{yx} is negative. This assignment is arbitrary for the purpose of illustration. In general, any combination of positive and negative values could exist.

3. Set up a rectangular coordinate system in which the positive horizontal axis represents positive (tensile) normal stresses, and the positive vertical axis represents positive (clockwise) shear stresses. Thus, the plane created will be referred to as the σ-τ *plane.*

4. Plot points on the σ-τ plane corresponding to the stresses acting on the faces of the stress element. If the element is drawn in the *x-y* plane, the two points to be plotted are σ_x, τ_{xy} and σ_y, τ_{yx}.

5. Draw the line connecting the two points.

6. The resulting line crosses the σ-axis at the center of Mohr's circle at the *average* of the two applied normal stresses, where

$$\sigma_{avg} = (\sigma_x + \sigma_y)/2$$

The center of Mohr's circle is called O in Figure 4–12.

7. Note in Figure 4–12 that a right triangle has been formed, having the sides *a, b,* and *R,* where

$$R = \sqrt{a^2 + b^2}$$

FIGURE 4–13
Completed Mohr's
circle, Steps 8–14

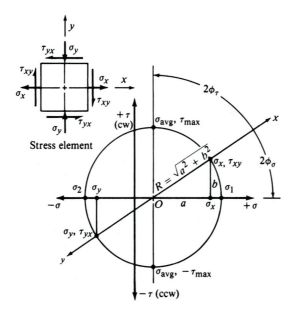

By inspection, we can see that

$$a = (\sigma_x - \sigma_y)/2$$
$$b = \tau_{xy}$$

The point labeled O is at a distance of $\sigma_x - a$ from the origin of the coordinate system. We can now proceed with the construction of the circle.

8. Draw the complete circle with the center at O and a radius of R, as shown in Figure 4–13.

9. The point where the circle crosses the σ-axis at the right gives the value of the maximum principal stress, σ_1. Note that $\sigma_1 = \sigma_{avg} + R$.

10. The point where the circle crosses the σ-axis at the left gives the minimum principal stress, σ_2. Note that $\sigma_2 = \sigma_{avg} - R$.

11. The coordinates of the top of the circle give the maximum shear stress and the average normal stress that act on the element having the maximum shear stress. Note that $\tau_{max} = R$.

   Note: The following steps relate to determining the angles of inclination of the principal stress element and the maximum shear stress element in relation to the original x-axis. It is important to realize that angles on Mohr's circle are actually *double* the true angles. Refer to Figure 4–13; the line from O through the first point plotted, σ_x, τ_{xy}, represents the original x-axis, as noted in the figure. The line from O through the point σ_y, τ_{yx} represents the original y-axis. Of course, on the original element, these axes are 90° apart, not 180°, illustrating the double-angle feature of Mohr's circle. Having made this observation, we can continue with the development of the process.

12. The angle $2\phi_\sigma$ is measured from the x-axis on the circle to the σ-axis. Note that

$$2\phi_\sigma = \arctan(b/a)$$

It is also important to note the direction from the x-axis to the σ-axis (clockwise or counterclockwise). This is necessary for representing the relation of the principal stress element to the original stress element properly.

13. The angle from the x-axis on the circle to the vertical line through τ_{max} gives $2\phi_\tau$. From the geometry of the circle, we can see that

$$2\phi_\tau = 90° - 2\phi_\sigma$$

Again it is important to note the direction from the x-axis to the τ_{max}-axis for use in orienting the maximum shear stress element. You should also note that the σ-axis and the τ_{max}-axis are always 90° apart on the circle and therefore 45° apart on the actual element.

14. The final step in the process of using Mohr's circle is to draw the resulting stress elements in their proper relation to the original element, as shown in Figure 4–14.

   We will now illustrate the construction of Mohr's circle by using the same data as in Example Problem 4–1, in which the principal stresses and the maximum shear stress were computed directly from the equations.

FIGURE 4–14
Display of results from
Mohr's circle

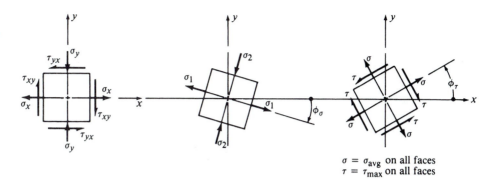

$$\sigma = \sigma_{avg} \text{ on all faces}$$
$$\tau = \tau_{max} \text{ on all faces}$$

(*a*) Original stress element   (*b*) Principal stress element   (*c*) Maximum shear stress element

---

Example Problem 4–2   The shaft shown in Figure 4–7 is supported by two bearings and carries two chain sprockets. The tensions in the chains exert horizontal forces on the shaft, tending to bend it in the
x-z plane. Sprocket *B* exerts a clockwise torque on the shaft when viewed toward the origin of the coordinate system along the *x*-axis. Sprocket *C* exerts an equal but opposite
torque on the shaft. For the loading condition shown, determine the principal stresses and
the maximum shear stress on element *K* on the front surface of the shaft (on the positive
z-side) just to the right of sprocket *B*. Use the procedure for constructing Mohr's circle included in this section.

Solution   Objective   Determine the principal stresses and the maximum shear stresses on element *K*.

Given   Shaft and loading pattern shown in Figure 4–7.

Analysis   Use the procedure for constructing Mohr's circle. Some intermediate results will be taken
from the solution to Example Problem 4–1 and from Figures 4–7, 4–8, and 4–9.

Results   ***Steps 1 and 2.*** The stress analysis for the given loading was completed in Example
Problem 4–1. Figure 4–15 is identical to Figure 4–9 and represents the results of Step
2 of the Mohr's circle procedure.

Steps 3–6. Figure 4–16 shows the results. The first point plotted was

$$\sigma_x = 8\,030 \text{ psi}, \tau_{xy} = 2\,870 \text{ psi}$$

FIGURE 4–15
Stresses on element *K*

FIGURE 4–16
Partially completed
Mohr's circle

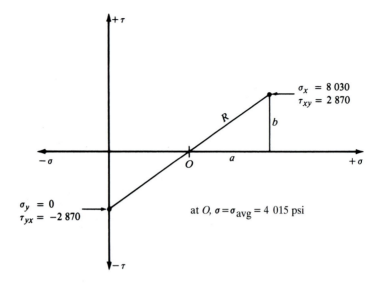

The second point was plotted at

$$\sigma_y = 0 \text{ psi}, \tau_{yx} = -2\,870 \text{ psi}$$

Then a line was drawn between them, crossing the σ-axis at O. The value of the stress at O is

$$\sigma_{avg} = (\sigma_x + \sigma_y)/2 = (8\,030 + 0)/2 = 4\,015 \text{ psi}$$

Step 7. We compute the values for a, b, and R from

$$a = (\sigma_x - \sigma_y)/2 = (8\,030 - 0)/2 = 4\,015 \text{ psi}$$
$$b = \tau_{xy} = 2\,870 \text{ psi}$$
$$R = \sqrt{a^2 + b^2} = \sqrt{(4\,015)^2 + (2\,870)^2} = 4\,935 \text{ psi}$$

Step 8. Figure 4–17 shows the completed Mohr's circle. The circle has its center at O and the radius R. Note that the circle passes through the two points originally plotted. It must do so because the circle represents all possible states of stress on the element K.

Step 9. The maximum principal stress is at the right side of the circle.

$$\sigma_1 = \sigma_{avg} + R$$
$$\sigma_1 = 4\,015 + 4\,935 = 8\,950 \text{ psi}$$

Step 10. The minimum principal stress is at the left side of the circle.

$$\sigma_2 = \sigma_{avg} - R$$
$$\sigma_2 = 4\,015 - 4\,935 = -920 \text{ psi}$$

FIGURE 4–17
Completed Mohr's circle

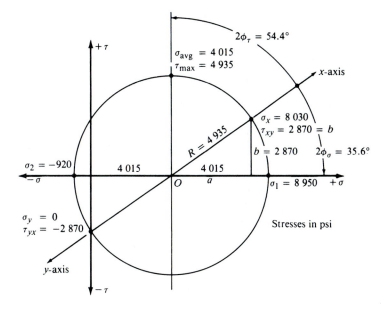

Step 11. At the top of the circle,

$$\sigma = \sigma_{avg} = 4\,015 \text{ psi}$$
$$\tau = \tau_{max} = R = 4\,935 \text{ psi}$$

The value of the normal stress on the element that carries the maximum shear stress is the same as the coordinate of O, the center of the circle.

Step 12. Compute the angle $2\phi_\sigma$ and then ϕ_σ. Use the circle as a guide./

$$2\phi_\sigma = \arctan(b/a) = \arctan(2\,870/4\,015) = 35.6°$$
$$\phi_\sigma = 35.6°/2 = 17.8°$$

Note that ϕ_σ must be measured *clockwise* from the original x-axis to the direction of the line of action of σ_1.

Step 13. Compute the angle $2\phi_\tau$ and then ϕ_τ. From the circle we see that

$$2\phi_\tau = 90° - 2\phi_\sigma = 90° - 35.6° = 54.4°$$

Then

$$\phi_\tau = 54.4°/2 = 27.2°$$

Note that the stress element on which the maximum shear stress acts must be rotated *counterclockwise* from the orientation of the original element.

(a) Original stress element, K, from Figure 4–7

(b) Principal stress element

(c) Maximum shear stress element

$\sigma = \sigma_{avg} = 4\,015$ psi   on all faces
$\tau = \tau_{max} = 4\,935$ psi   on all faces

FIGURE 4–18   Results from Mohr's circle analysis

Step 14. Figure 4–18 shows the required stress elements. They are identical to those shown in Figure 4–11.

**4–4
MOHR'S CIRCLE
PRACTICE
PROBLEMS**

To a person seeing Mohr's circle for the first time, it may seem confusing. But with practice under a variety of combinations of normal and shear stresses, you should be able to execute the 14 steps quickly and accurately.

Table 4–1 gives six sets of data (Example Problems 4–3 through 4–8) for normal and shear stresses in the x-y plane. You are advised to complete the Mohr's circle for each before looking at the solutions in Figures 4–19 through 4–24. From the circle, determine the two principal stresses, the maximum shear stress, and the planes on which these stresses act. Then draw the given stress element, the principal stress element, and the maximum shear stress element, all oriented properly with respect to the x- and y-directions.

TABLE 4–1   Practice problems for Mohr's circle

Example Problem	σ_x	σ_y	τ_{xy}	Fig. No.
4–3	+10.0 Ksi	−4.0 Ksi	+5.0 Ksi	4–19
4–4	+10.0 Ksi	−2.0 Ksi	−4.0 Ksi	4–20
4–5	+4.0 Ksi	−10.0 Ksi	+4.0 Ksi	4–21
4–6	+120 MPa	−30 MPa	+60 MPa	4–22
4–7	−80 MPa	+20 MPa	−50 MPa	4–23
4–8	−80 MPa	+20 MPa	+50 Mpa	4–24

Example Problem 4–3

FIGURE 4–19
Solution for Example
Problem 4–3

Given:
σ_x $= +10.0$ Ksi
σ_y $= -4.0$ Ksi
τ_{xy} $= +5.0$ Ksi (cw)

Results:
σ_1 $= +11.60$ Ksi
σ_2 $= -5.60$ Ksi
ϕ_σ $= 17.8°$ cw
τ_{max} $= 8.60$ Ksi
ϕ_τ $= 27.2°$ ccw
σ_{avg} $= +3.0$ Ksi

(a) Complete Mohr's
circle

(b) Original stress
element

(c) Principal stress
element

(d) Maximum shear
stress element

Example Problem 4–4

FIGURE 4–20
Solution for Example
Problem 4–4

Given:
σ_x $= +10.0$ Ksi
σ_y $= -2.0$ Ksi
τ_{xy} $= -4.0$ Ksi (ccw)

Results:
σ_1 $= +11.21$ Ksi
σ_2 $= -3.21$ Ksi
ϕ_σ $= 16.8°$ ccw
τ_{max} $= 7.21$ Ksi
ϕ_τ $= 28.2°$ cw to $-\tau_{max}$
σ_{avg} $= +4.0$ Ksi

(a) Complete Mohr's circle

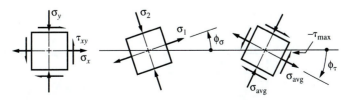

(b) Original stress
element

(c) Principal stress
element

(d) Maximum shear
stress element

Example Problem 4–5

FIGURE 4–21
Solution for Example
Problem 4–5

Given:

$\sigma_x = +4.0$ Ksi
$\sigma_y = -10.0$ Ksi
$\tau_{xy} = +4.0$ Ksi

Results:

$\sigma_1 = +5.06$ Ksi
$\sigma_2 = -11.06$ Ksi
$\phi_\sigma = 14.9°$ cw
$\tau_{max} = 8.06$ Ksi
$\phi_\tau = 30.1°$ ccw
$\sigma_{avg} = -3.0$ Ksi

(a) Complete Mohr's circle

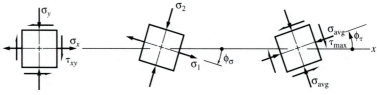

(b) Original stress
element

(c) Principal stress
element

(d) Maximum shear
stress element

Example Problem 4–6

FIGURE 4–22
Solution for Example
Problem 4–6

Given:

$\sigma_x = +120$ MPa
$\sigma_y = -30$ MPa
$\tau_{xy} = +60$ MPa

Results:

$\sigma_1 = +141$ MPa
$\sigma_2 = -51$ MPa
$\phi_\sigma = 19.3°$ cw
$\tau_{max} = 96$ MPa
$\phi_\tau = 25.7°$ ccw
$\sigma_{avg} = +45$ MPa

(a) Complete Mohr's circle

(b) Original stress
element

(c) Principal
stress element

(d) Maximum shear
stress element

Example Problem 4–7

FIGURE 4–23
Solution for Example
Problem 4–7

Given:
σ_x = −80 MPa
σ_y = +20 MPa
τ_{xy} = −50 MPa

Results:
σ_1 = +40.7 MPa
σ_2 = −100.7 MPa
ϕ_σ = 67.5°ccw
τ_{max} = 70.7 MPa
ϕ_τ = 22.5°ccw to $-\tau_{max}$
σ_{avg} = −30 MPa

(a) Complete Mohr's circle

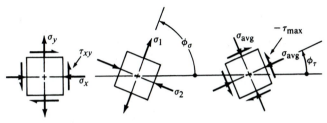

(b) Original stress
element

(c) Principal
stress element

(d) Maximum shear
stress element

Example Problem 4–8

FIGURE 4–24
Solution for Example
Problem 4–8

Given:
σ_x = −80 MPa
σ_y = +20 MPa
τ_{xy} = +50 MPa

Results:
σ_1 = +40.7 MPa
σ_2 = −100.7 MPa
ϕ_σ = 67.5°cw
τ_{max} = 70.7 MPa
ϕ_τ = 22.5°cw
σ_{avg} = −30 MPa

(a) Complete Mohr's circle

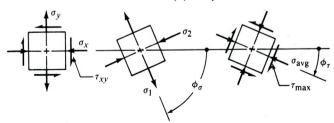

(b) Original stress
element

(c) Principal stress
element

(d) Maximum shear
stress element

4–5
CASE WHEN
BOTH PRINCIPAL
STRESSES HAVE
THE SAME SIGN

Remember that all of the problems presented thus far have been plane stress problems, also called *biaxial stress* problems because stresses are acting in only two directions within one plane. Obviously, real load-carrying members are three-dimensional objects. The assumption here is that if no stress is given for the third direction, it is zero. In most cases, the solutions as given will produce the true maximum shear stress, along with the two principal stresses for the given plane. This will always be true if the two principal stresses have opposite signs, that is, if one is tensile and the other is compressive.

But the true maximum shear stress on the element will not be found if the two principal stresses are of the same sign. In such cases, you must consider the three-dimensional case.

Familiar examples of real products in which two principal stresses have the same sign are various forms of pressure vessels. A hydraulic cylinder with closed ends contains fluids under high pressure that tend to burst the walls of the cylinder. In strength of materials, you learned that the outer surfaces of the walls of such cylinders are subjected to tensile stresses in two directions: (1) tangent to its circumference and (2) axially, parallel to the axis of the cylinder. The stress perpendicular to the wall at the outer surface is zero.

Figure 4–25 shows the stress condition on an element of the surface of the cylinder. The tangential stress, also called *hoop stress,* is aligned with the *x*-direction and is labeled σ_x. The axially directed stress, also called *longitudinal stress,* acts in line with the *y*-direction and is labeled σ_y.

In strength of materials, you learned that if the wall of the cylinder is relatively thin, the maximum hoop stress is

$$\sigma_x = pD/2t$$

where p = internal pressure in the cylinder

D = diameter of the cylinder

t = thickness of the cylinder wall

Also, the longitudinal stress is

$$\sigma_y = pD/4t$$

Both stresses are tensile, and the hoop stress is twice as large as the longitudinal stress.

FIGURE 4–25   Thin-walled cylinder subjected to pressure with its ends closed

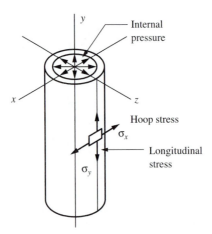

The analysis would be similar for any kind of thin-walled cylindrical vessel carrying an internal pressure. Examples are storage tanks for compressed gases, pipes carrying moving fluids under pressure, and the familiar beverage can that releases internal pressure when the top is popped open.

Let's use the hydraulic cylinder as an example for illustrating the special use of Mohr's circle when both principal stresses have the same sign. Consider that Figure 4–25 shows a cylinder with closed ends carrying an internal pressure of 500 psi. The wall thickness is $t = 0.080$ in, and the diameter of the cylinder is $D = 4.00$ in. The ratio of $D/t = 50$ indicates that the cylinder can be considered thin-walled. Any ratio over 20 is typically considered to be thin-walled.

The computed hoop and longitudinal stresses in the wall are

➪ **Hoop Stress**

$$\sigma_x = \frac{pD}{2t} = \frac{(500 \text{ psi})(4.0 \text{ in})}{(2)(0.080 \text{ in})} = 12\ 500 \text{ psi} \quad \text{(tension)}$$

➪ **Longitudinal Stress**

$$\sigma_y = \frac{pD}{4t} = \frac{(500 \text{ psi})(4.0 \text{ in})}{(4)(0.080 \text{ in})} = 6\ 250 \text{ psi} \quad \text{(tension)}$$

There are no shear stresses applied in the x- and y-directions.

Figure 4–26(a) shows the stress element for the x-y plane, and Part (b) shows the corresponding Mohr's circle. Because there are no applied shear stresses, σ_x and σ_y are the principal stresses for the plane. The circle would predict the maximum shear stress to be equal to the radius of the circle, 3 125 psi.

But notice Part (c) of the figure. We could have chosen the x-z plane for analysis, instead of the x-y plane. The stress in the z-direction is zero because it is perpendicular to the free face of the element. Likewise, there are no shear stresses on this face. The Mohr's circle for this plane is shown in Part (d) of the figure. The maximum shear stress is equal to the radius of the circle, 6 250 psi, or *twice* as much as would be predicted from the x-y plane. This approach should be used any time the two principal stresses in a biaxial stress problem have the same sign.

In summary, on a general three-dimensional stress element, there will be one orientation of the element in which there are no shear stresses acting. The normal stresses on the three perpendicular faces are then the three principal stresses. If we call these stresses σ_1, σ_2, and σ_3, taking care to order them such that $\sigma_1 > \sigma_2 > \sigma_3$, then the maximum shear stress on the element will always be

$$\tau_{\text{max}} = \frac{\sigma_1 - \sigma_3}{2}$$

Figure 4–27 shows the three-dimensional element.

For the cylinder of Figure 4–25, we can conclude that

$$\sigma_1 = \sigma_x = 12\ 500 \text{ psi}$$
$$\sigma_2 = \sigma_y = 6\ 250 \text{ psi}$$
$$\sigma_3 = \sigma_z = 0$$
$$\tau_{\text{max}} = (\sigma_1 - \sigma_3)/2 = (12\ 500 - 0)/2 = 6\ 250 \text{ psi}$$

Figure 4–28 shows two additional examples in which the two principal stresses in the given plane have the same sign. Then the zero stress in the third direction is added to the diagram, and the new Mohr's circle is superimposed on the original one. This serves to illustrate that the maximum shear stress will occur on the Mohr's circle having the largest radius.

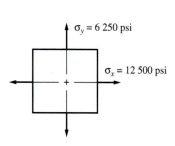

(a) Stress element
in x-y plane

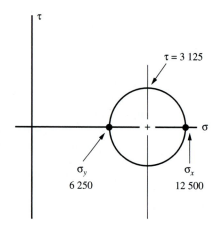

(b) Mohr's circle for x-y plane

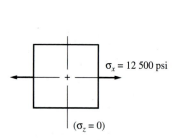

(c) Stress element
in x-z plane

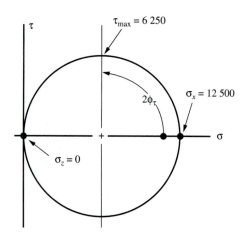

(d) Mohr's circle for x-z plane

FIGURE 4–26   Stress analysis for a thin-walled cylinder

FIGURE 4–27
Three-dimensional
stress element

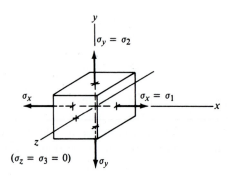

FIGURE 4–28

Mohr's circle for cases in which two principal stresses have the same sign

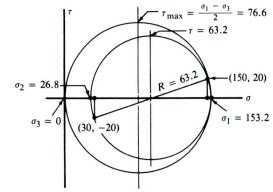

Given:

$\sigma_x = +150$ MPa
$\sigma_y = +30$ MPa
$\sigma_z = 0$
$\tau_{xy} = +20$ MPa

Results:

$\sigma_1 = 153.2$ MPa
$\sigma_2 = 26.8$ MPa
$\sigma_3 = 0$
$\tau_{max} = 76.6$ MPa

$\tau_{max} = \frac{\sigma_1 - \sigma_3}{2} = 76.6$
$\tau = 63.2$
$\sigma_2 = 26.8$
$R = 63.2$
$(150, 20)$
$\sigma_3 = 0$
$(30, -20)$
$\sigma_1 = 153.2$

(a) σ_x and σ_y both positive

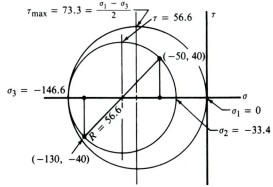

Given:

$\sigma_x = -50$ MPa
$\sigma_y = -130$ MPa
$\sigma_z = 0$
$\tau_{xy} = 40$ MPa

Results:

$\sigma_1 = 0$
$\sigma_2 = -33.4$ MPa
$\sigma_3 = -146.6$ MPa
$\tau_{max} = 73.3$ MPa

$\tau_{max} = 73.3 = \frac{\sigma_1 - \sigma_3}{2}$
$\tau = 56.6$
$(-50, 40)$
$\sigma_3 = -146.6$
$R = 56.6$
$\sigma_1 = 0$
$(-130, -40)$
$\sigma_2 = -33.4$

(b) σ_x and σ_y both negative

4–6 MOHR'S CIRCLE FOR SPECIAL STRESS CONDITIONS

Mohr's circle is used here to demonstrate the relationship among the applied stresses, the principal stresses, and the maximum shear stress for the following special cases:

Pure uniaxial tension

Pure uniaxial compression

Pure torsional shear

Uniaxial tension combined with torsional shear

These are important, frequently encountered stress conditions, and they will be used in later chapters to illustrate failure theories and design methods. These failure theories are based on the values of the principal stresses and the maximum shear stress.

Pure Uniaxial Tension

The stress condition produced in all parts of a standard tensile test specimen is pure uniaxial tension. Figure 4–29 shows the stress element and the corresponding Mohr's circle. Note that the maximum principal stress, σ_1, is equal to the applied stress, σ_x; the minimum principal stress, σ_2, is zero; and the maximum shear stress, τ_{max}, is equal to $\sigma_x/2$.

Pure Uniaxial Compression

Figure 4–30 shows pure uniaxial compression as it would be produced by a standard compression test. Mohr's circle shows that $\sigma_1 = 0$; $\sigma_2 = \sigma_x$ (a negative value); and the magnitude of the maximum shear stress is $\tau_{max} = \sigma_x/2$.

FIGURE 4–29
Mohr's circle for pure
uniaxial tension

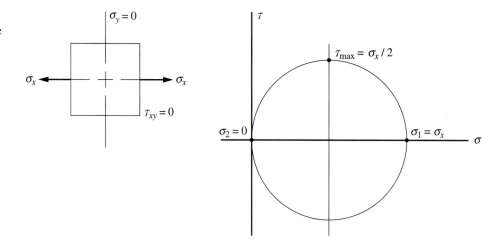

FIGURE 4–30
Mohr's circle for pure
uniaxial compression

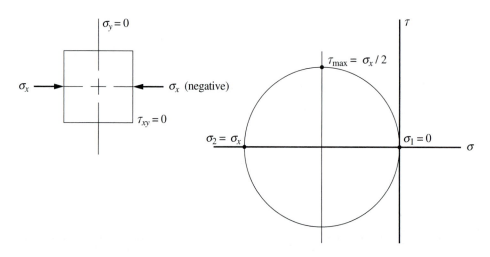

Pure Torsion

Figure 4–31 shows that the Mohr's circle for this special case has its center at the origin of the σ-τ axes and that the radius of the circle is equal to the value of the applied shear stress, τ_{xy}. Therefore, $\tau_{max} = \tau_{xy}$; $\sigma_1 = \tau_{xy}$; and $\sigma_2 = -\tau_{xy}$.

Uniaxial Tension Combined with Torsional Shear

This is an important special case because it describes the stress condition in a rotating shaft carrying bending loads while simultaneously transmitting torque. This is the type of stress condition on which the procedure for designing shafts, presented in Chapter 12, is based. If the applied stresses are called σ_x and τ_{xy}, the Mohr's circle in Figure 4–32 shows that

$$\tau_{max} = R = \text{radius of circle} = \sqrt{(\sigma_x/2)^2 + \tau_{xy}^2} \tag{4–7}$$

$$\sigma_1 = \sigma_x/2 + R = \sigma_x/2 + \sqrt{(\sigma_x/2)^2 + \tau_{xy}^2} \tag{4–8}$$

$$\sigma_2 = \sigma_x/2 - R = \sigma_x/2 - \sqrt{(\sigma_x/2)^2 + \tau_{xy}^2} \tag{4–9}$$

A convenient and useful concept called *equivalent torque* can be developed from Equation (4–7) for the special case of a body subjected to only bending and torsion.

FIGURE 4–31
Mohr's circle for pure
torsional shear

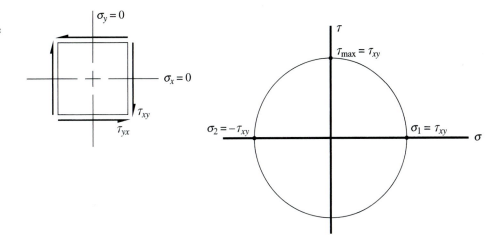

FIGURE 4–32
Mohr's circle for
uniaxial tension
combined with
torsional shear

$$R = \sqrt{(\sigma_x/2)^2 + \tau_{xy}^2} = \tau_{max}$$

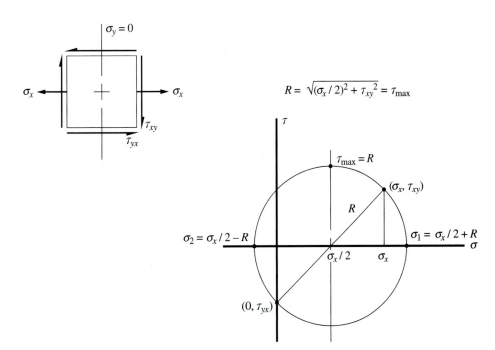

An example is shown in Figure 4–33, where a circular bar is loaded at one end by a downward force and a torsional moment. The force causes bending in the bar with the maximum moment at the point where the bar is attached to the support. The moment causes a tensile stress on the top of the bar in the *x*-direction at the point called *A,* where the magnitude of the stress is

$$\sigma_x = M/S \qquad\qquad (4\text{–}10)$$

where S = section modulus of the round bar

Now the torsional moment causes torsional shear stress in the *x-y* plane at point *A* having a magnitude of

$$\tau_{xy} = T/Z_p \qquad\qquad (4\text{–}11)$$

where Z_p = polar section modulus of the bar

FIGURE 4–33
Circular bar in
bending and torsion

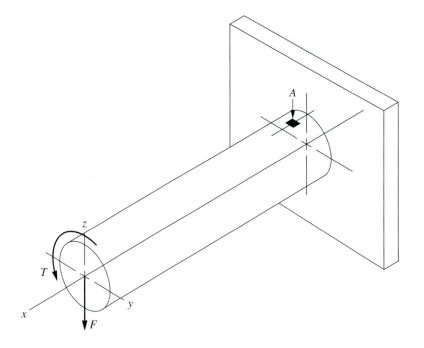

Point A then is subjected to a tensile stress combined with shear, the special case shown in the Mohr's circle of Figure 4–32. The maximum shear stress can be computed from Equation (4–7). If we substitute Equations (4–10) and (4–11) in Equation (4–7), we get

$$\tau_{max} = \sqrt{(M/2S)^2 + (T/Z_p)^2} \tag{4–12}$$

Note from Appendix 1 that $Z_p = 2S$. Equation (4–12) can then be written as

$$\tau_{max} = \frac{\sqrt{M^2 + T^2}}{Z_p} \tag{4–13}$$

It is convenient to define the quantity in the numerator of this equation to be the *equivalent torque*, T_e. Then the equation becomes

$$\tau_{max} = T_e/Z_p \tag{4–14}$$

**4–7
ANALYSIS OF
COMPLEX
LOADING
CONDITIONS**

The examples shown in this chapter involved relatively simple part geometries and loading conditions for which the necessary stress analysis could be performed using familiar methods of statics and strength of materials. If more complex geometries or loading conditions were involved, you may not have been able to complete the required analysis to create the original stress element from which the Mohr's circle was derived.

Consider, for example, a cast wheel for a high-performance racing car. The geometry would likely involve webs or spokes of a unique design connecting the hub to the rim to create a lightweight wheel. The loading would be a complex combination of torsion, bending, and compression generated by the cornering action of the wheel.

One method of analysis of such a load-carrying member would be accomplished by experimental stress analysis using strain gages or photoelastic techniques. The results would identify the stress levels at selected points in certain specified directions that could be used as the input to the construction of the Mohr's circle for critical points in the structure.

Another method of analysis would involve the modeling of the geometry of the wheel as a *finite-element model*. The three-dimensional model would be divided into several hundred small-volume elements. Points of support and restraint would be defined on the model, and then external loads would be applied at appropriate points. The complete data set would be input to a special type of computer analysis program called *finite-element analysis*. The output from the program lists the stresses and the deflection for each of the elements. These data can be plotted on the computer model so that the designer can visualize the stress distribution within the model. Most such programs list the principal stresses and the maximum shear stress for each element, eliminating the need to actually draw the Mohr's circle. A special form of stress, called the *von Mises stress*, is often computed by combining the principal stresses. (See Section 5–7 for a more complete discussion of the von Mises stress and its use.) Several different finite-element analysis programs are commercially available for use on personal computers, on engineering work stations, or on mainframe computers.

PROBLEMS

For the sets of given stresses on an element given in Table 4–2, draw a complete Mohr's circle, find the principal stresses and the maximum shear stress, and draw the principal stress element and the maximum shear stress element. Any stress components not shown are assumed to be zero.

TABLE 4–2 Stresses for Problems 1–30

Problem	σ_x	σ_y	τ_{xy}
1.	20 Ksi	0	10 Ksi
2.	85 Ksi	−40 Ksi	30 Ksi
3.	40 Ksi	−40 Ksi	−30 Ksi
4.	−80 Ksi	−40 Ksi	−30 Ksi
5.	120 Ksi	40 Ksi	20 Ksi
6.	20 Ksi	140 Ksi	20 Ksi
7.	20 Ksi	−40 Ksi	0
8.	120 Ksi	−40 Ksi	100 Ksi
9.	100 MPa	0	80 MPa
10.	250 MPa	−80 MPa	110 MPa
11.	50 MPa	−80 MPa	40 MPa
12.	−150 MPa	−80 MPa	−40 MPa
13.	150 MPa	80 MPa	−40 MPa
14.	50 MPa	180 MPa	40 MPa
15.	250 MPa	−80 MPa	0
16.	50 MPa	−80 MPa	−30 MPa
17.	400 MPa	−300 MPa	200 MPa
18.	−120 MPa	180 MPa	−80 MPa
19.	−30 MPa	20 MPa	40 MPa
20.	220 MPa	−120 MPa	0 MPa
21.	40 Ksi	0 Ksi	0 Ksi
22.	0 Ksi	0 Ksi	40 Ksi
23.	38 Ksi	−25 Ksi	−18 Ksi
24.	55 Ksi	15 Ksi	−40 Ksi
25.	22 Ksi	0 Ksi	6.8 Ksi

Problem	σ_x	σ_y	τ_{xy}
26.	−4 250 psi	3 250 psi	2 800 psi
27.	300 MPa	100 MPa	80 MPa
28.	250 MPa	150 MPa	40 MPa
29.	−840 kPa	−335 kPa	−120 kPa
30.	−325 kPa	−50 kPa	−60 kPa

31. Refer to Figure 3–19 in Chapter 3. For the shaft aligned with the *x*-axis, create a stress element on the bottom of the shaft just to the left of section *B*. Then draw the Mohr's circle for that element.

32. Refer to Figure 3–42 in Chapter 3. For the shaft *ABC*, create a stress element on the bottom of the shaft just to the right of section *B*. The torque applied to the shaft at *B* is resisted at support *C* only. Draw the Mohr's circle for the stress element.

33. Repeat Problem 32 for the shaft in Figure 3–43 in Chapter 3.

34. Refer to Figure 3–44 in Chapter 3. For the shaft *AB*, create a stress element on the bottom of the shaft just to the right of section *A*. The torque applied to the shaft by the crank is resisted at support *B* only. Draw the Mohr's circle for the stress element.

35. A short cylindrical bar having a diameter of 4.00 in is subjected to an axial compressive force of 75 000 lb and a torsional moment of 20 000 lb·in. Draw a stress element on the surface of the bar. Then draw the Mohr's circle for the element.

36. A torsion bar is used as a suspension element for a vehicle. The bar has a diameter of 20 mm. It is subjected to a torsional moment of 450 N·m and an axial tensile force of 3.60 kN. Draw a stress element on the surface of the bar, and then draw the Mohr's circle for the element.

5

Design for Different Types of Loading

Design for Different Types of Loading

Discussion Map

☐ You should be familiar with different types of loading and with different materials that are used in load-carrying members.

☐ You should be able to anticipate how a given part could fail.

Discover

Identify parts that are subjected to static loads.

Identify parts that experience fluctuating loads.

Discuss some failed parts, and try to describe why they failed.

Keep these examples in mind as you study this chapter.

For the concepts considered in this chapter, the big picture is truly huge. In this chapter, you will learn how to do the following:

1. Recognize the manner of loading for a part: Is it *static, repeated and reversed, fluctuating, shock,* or *impact?*

2. Select the appropriate method to analyze the stresses produced.

3. Determine the strength property for the material that is appropriate to the kind of loading and to the kind of material: Is the material a metal or a nonmetal? Is it brittle or ductile? Does it have a defined endurance limit?

4. Specify a suitable design factor, often called a *factor of safety.*

5. Design a wide variety of load-carrying members to be safe under their particular expected loading patterns.

So the big picture is the universe of components that need to be designed. The following paragraphs show by example some of the situations to be studied in this chapter.

An ideal *static load* is one that is applied slowly and is never removed. Some loads that are applied slowly and removed and replaced very infrequently can also be considered to be static. What examples can you think of for products or their components that are subjected to static loads? Consider load-carrying members of structures, parts of furniture pieces, and brackets or support rods holding equipment in your home or in a business or factory. Try to identify specific examples, and describe them to your colleagues. Discuss how the load is applied and which parts of the load-carrying member are subjected to the higher stress levels. Some of the examples that you discovered during **The Big Picture** discussion for Chapter 3 could be used again here.

Fluctuating loads are those that vary during the normal service of the product. They typically are applied for quite a long time so the part experiences many thousands or millions of cycles of stress during its expected life. There are many examples in consumer products around your home, in your car, in commercial buildings, and in manufacturing facilities. Consider virtually anything that has moving parts. Again, try to identify specific examples, and describe them to your colleagues. How does the load fluctuate? Is it applied and then completely removed each cycle? Or is there always some level of mean or average load with an alternating load superimposed on it? Does the load swing from a positive maximum value to a negative minimum value of equal magnitude during each cycle of loading? Consider parts with rotating shafts, such as engines or agricultural, production, and construction machinery.

Consider products that have failed. You may have identified some from **The Big Picture** discussion for Chapter 3. Did they fail the first time they were used? Or did they fail after some fairly long service? Why do you think they were able to operate for some time before failure?

Can you find components that failed suddenly because the material was brittle, such as cast iron, some ceramics, or some plastics? Can you find others that failed only after some considerable deformation? Such failures are called *ductile fractures.*

What were the consequences of the failures that you have found? Was anyone hurt? Was there damage to some other valuable component or property? Or was the failure simply an inconvenience? What was the order of magnitude of cost related to the failure? The answer to some of these questions can help you make rational decisions about design factors to be used in your designs.

It is the designer's responsibility to ensure that a machine part is safe for operation under reasonably foreseeable conditions. This requires that a stress analysis be performed in which the predicted stress levels in the part are compared with the *design stress,* or that level of stress permitted under the operating conditions.

The stress analysis can be performed either analytically or experimentally, depending on the degree of complexity of the part, the knowledge about the loading conditions, and the material properties. The designer must be able to verify that the stress to which a part is subjected is safe.

The manner of computing the design stress depends on the manner of loading and on the type of material. Loading types include the following:

Static

Repeated and reversed

Fluctuating

Shock or impact

Random

Material types are many and varied. Among the metallic materials, the chief classification is between *ductile* and *brittle* materials. Other considerations include the manner of forming the material (casting, forging, rolling, machining, and so on), the type of heat treatment, the surface finish, the physical size, the environment in which it is to operate, and the geometry of the part. Different factors must be considered for plastics, composites, ceramics, wood, and others.

This chapter outlines methods of analyzing load-carrying machine parts to ensure that they are safe. Several different cases are described in which knowledge of the combinations of material types and loading patterns leads to the determination of the appropriate method of analysis. It will then be your job to apply these tools correctly and judiciously as you continue your career.

 You Are the Designer

Recall the task presented at the start of Chapter 4, in which you were the designer of a bracket to hold a fabric sample during a test to determine its long-term stretch characteristics. Figure 4–2 showed a proposed design.

Now you are asked to continue this design exercise by selecting a material from which to make the two bent circular bars that are welded to the rigid support. Also, you must specify a suitable diameter for the bars when a certain load is applied to the test material.

5–1
OBJECTIVES OF THIS CHAPTER

After completing this chapter, you will be able to:

1. Identify various kinds of loading commonly encountered by machine parts, including *static, repeated and reversed, fluctuating, shock or impact,* and *random.*

2. Define the term *stress ratio* and compute its value for the various kinds of loading.

3. Define the concept of *fatigue.*

4. Define the material property of *endurance strength* and determine estimates of its magnitude for different materials.

5. Recognize the factors that affect the magnitude of endurance strength.

6. Define the term *design factor.*

7. Specify a suitable value for the design factor.

8. Define the *maximum normal stress theory of failure.*

9. Define the *maximum shear stress theory of failure.*

10. Define the *distortion energy theory,* also called the *von Mises theory* or the *Mises-Hencky theory.*

11. Describe the *Soderberg criterion* and apply it to the design of parts subjected to fluctuating stresses.

5–2
TYPES OF LOADING AND STRESS RATIO

The primary factors to consider when specifying the type of loading to which a machine part is subjected are the manner of variation of the load and the resulting variation of stress with time. Stress variations are characterized by four key values:

1. Maximum stress, σ_{max}

2. Minimum stress, σ_{min}

3. Mean (average) stress, σ_m

4. Alternating stress, σ_a (*stress amplitude*)

The maximum and minimum stresses are usually computed from known information by stress analysis or finite-element methods, or they are measured using experimental stress analysis techniques. Then the mean and alternating stresses can be computed from

$$\sigma_m = (\sigma_{max} + \sigma_{min})/2 \qquad (5\text{–}1)$$

$$\sigma_a = (\sigma_{max} - \sigma_{min})/2 \qquad (5\text{–}2)$$

The behavior of a material under varying stresses is dependent on the manner of the variation. One method used to characterize the variation is called *stress ratio.* Two types of stress ratios are commonly used, defined as

$$\text{Stress ratio } R = \frac{\text{minimum stress}}{\text{maximum stress}} = \frac{\sigma_{min}}{\sigma_{max}} \qquad (5\text{–}3)$$

$$\text{Stress ratio } A = \frac{\text{alternating stress}}{\text{mean stress}} = \frac{\sigma_a}{\sigma_m}$$

Static Stress

When a part is subjected to a load that is applied slowly, without shock, and is held at a constant value, the resulting stress in the part is called *static stress.* An example is the load on

FIGURE 5–1 Static
stress

Stress ratio $R = 1.0$

a structure due to the dead weight of the building materials. Figure 5–1 shows a diagram of stress versus time for static loading. Because $\sigma_{max} = \sigma_{min}$, the stress ratio for static stress is $R = 1.0$.

Static loading can also be assumed when a load is applied and is removed slowly and then reapplied, if the number of load applications is small, that is, under a few thousand cycles of loading.

Repeated and Reversed Stress

A stress reversal occurs when a given element of a load-carrying member is subjected to a certain level of tensile stress followed by the *same level* of compressive stress. If this stress cycle is repeated many thousands of times, the stress is called *repeated and reversed.*

An important example in machine design is a rotating circular shaft loaded in bending such as that shown in Figure 5–2. In the position shown, an element on the bottom of the shaft experiences tensile stress while an element on the top of the shaft sees a compressive stress of equal magnitude. As the shaft is rotated 180° from the given position, these two elements experience a complete reversal of stress. Now if the shaft continues to rotate, all parts of the shaft that are in bending see repeated, reversed stress. This is a description of the classical loading case of *reversed bending.*

Figure 5–3 shows the diagram of stress versus time for repeated and reversed stress. Because $\sigma_{min} = -\sigma_{max}$, the stress ratio is $R = -1.0$, and the mean stress is zero.

This type of loading is often called *fatigue loading,* and a machine of the type shown in Figure 5–2 is called a *fatigue test device.* Such machines are used to test materials for their ability to resist repeated loads. The material property called *endurance strength* is measured in this manner. More is said later in this chapter about endurance strength. Actually, reversed bending is only a special case of fatigue loading, since any stress that varies with time can lead to fatigue failure of a part.

Fluctuating Stress

When a load-carrying member is subjected to an alternating stress with a nonzero mean, the loading produces *fluctuating stress.* Figure 5–4 shows four diagrams of stress versus time for this type of stress. Differences among the four diagrams occur in whether the various stress levels are positive (tensile) or negative (compressive). Any varying stress with a

FIGURE 5–2
Fatigue test device

FIGURE 5–3
Repeated, reversed
stress

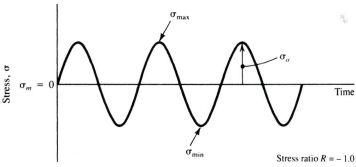

nonzero mean is considered a fluctuating stress. Figure 5–4 also shows the possible ranges of values for the stress ratio R for the given loading patterns.

An example of a machine part subjected to fluctuating stress of the type shown in Figure 5–4(a) is shown in Figure 5–5, in which a reciprocating cam follower feeds spherical balls one at a time from a chute. The follower is held against the cam by a flat spring loaded as a cantilever. When the follower is farthest to the left, the spring is deflected from its free (straight) position by an amount $y_{min} = 3.0$ mm. When the follower is farthest to the right, the spring is deflected to $y_{max} = 8.0$ mm. Then, as the cam continues to rotate, the spring sees the cyclic loading between the minimum and maximum values. A point at the base of the spring on the convex side experiences the varying tensile stresses of the type shown in Figure 5–4(a). Example Problem 5–1 completes the analysis of the stress in the spring at point A.

FIGURE 5–4
Fluctuating stresses

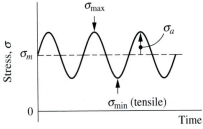

(a) Tensile mean stress —
all stresses tensile
$0 < R < 1.0$

(b) Tensile mean stress —
σ_{max} tensile
σ_{min} compressive
$-1.0 < R < 0$

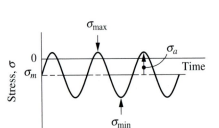

(c) Compressive mean stress —
σ_{max} tensile
σ_{min} compressive
$-\infty < R < -1.0$

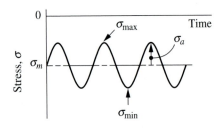

(d) Compressive mean stress —
all stresses compressive
$1.0 < R < \infty$

FIGURE 5–5
Example of cyclic
loading in which the
flat spring is subjected
to fluctuating stress

$y_{max} = 8.0\ mm$
$y_{min} = 3.0\ mm$

(a) Cross section of spring
(enlarged)

(b) Ball feed device

Example Problem 5–1   For the flat steel spring shown in Figure 5–5, compute the maximum stress, the minimum stress, the mean stress, and the alternating stress. Also compute the stress ratio, R. The length L is 65 mm. The dimensions of the spring cross section are $t = 0.80$ mm and $b = 6.0$ mm.

Solution   Objective   Compute the maximum, minimum, mean, and alternating tensile stresses in the flat spring. Compute the stress ratio, R.

Given   Layout shown in Figure 5–5. The spring is steel: $L = 65$ mm.
Spring cross section dimensions: $t = 0.80$ mm and $b = 6.0$ mm.
Maximum deflection of the spring at the follower = 8.0 mm.
Minimum deflection of the spring at the follower = 3.0 mm.

Analysis   Determine the force exerted on the spring by the follower for each level of deflection using the formulas from Table A14–2, Case (a). Compute the bending moment at the base of the spring for each deflection. Then compute the stresses using the bending stress formula, $\sigma = Mc/I$. Use Equations (5–1), (5–2), and (5–3) for the mean and alternating stresses and R.

Results   Case (a) of Table A14–2 gives the following formula for the amount of deflection of a cantilever for a given applied force:

$$y = PL^3/3EI$$

Solve for the force as a function of deflection:

$$P = 3EIy/L^3$$

Appendix 3 gives the modulus of elasticity for steel to be $E = 207$ GPa. The moment of inertia, I, for the spring cross section is found from

$$I = bt^3/12 = (6.00 \text{ mm})(0.80 \text{ mm})^3/12 = 0.256 \text{ mm}^4$$

Then the force on the spring when the deflection y is 3.0 mm is

$$P = \frac{3(207 \times 10^9 \text{ N/m}^2)(0.256 \text{ mm}^4)(3.0 \text{ mm})}{(65 \text{ mm})^3} \frac{(1.0 \text{ m}^2)}{(10^6 \text{ mm}^2)} = 1.74 \text{ N}$$

The bending moment at the support is

$$M = P{\cdot}L = (1.74 \text{ N})(65 \text{ mm}) = 113 \text{ N·mm}$$

The bending stress caused by this moment is

$$\sigma = \frac{Mc}{I} = \frac{(113 \text{ N·mm})(0.40 \text{ mm})}{0.256 \text{ mm}^4} = 176 \text{ N/mm}^2 = 176 \text{ MPa}$$

This is the lowest stress that the spring sees in service, and therefore $\sigma_{\min} = 176$ MPa.
      Because the force on the spring is proportional to the deflection, the force exerted when the deflection is 8.0 mm is

$$P = (1.74 \text{ N})(8.0 \text{ mm})/(3.0 \text{ mm}) = 4.63 \text{ N}$$

The bending moment is

$$M = P{\cdot}L = (4.63 \text{ N})(65 \text{ mm}) = 301 \text{ N·mm}$$

The bending stress is

$$\sigma = \frac{Mc}{I} = \frac{(301 \text{ N·mm})(0.40 \text{ mm})}{0.256 \text{ mm}^4} = 470 \text{ N/mm}^2 = 470 \text{ MPa}$$

This is the maximum stress that the spring sees, and therefore $\sigma_{max} = 470$ MPa.
Now the mean stress can be computed:

$$\sigma_m = (\sigma_{max} + \sigma_{min})/2 = (470 + 176)/2 = 323 \text{ MPa}$$

Finally, the alternating stress is

$$\sigma_a = (\sigma_{max} - \sigma_{min})/2 = (470 - 176)/2 = 147 \text{ MPa}$$

The stress ratio is found using Equation (5–3):

$$\text{Stress ratio } R = \frac{\text{minimum stress}}{\text{maximum stress}} = \frac{\sigma_{min}}{\sigma_{max}} = \frac{176 \text{ MPa}}{470 \text{ MPa}} = 0.37$$

Comments   The sketch of stress versus time shown in Figure 5–4(a) illustrates the form of the fluctu-ating stress on the spring. In Section 5–9, you will see how to design parts subjected to this kind of stress.

A special, frequently encountered case of fluctuating stress is *repeated, one-direction stress,* in which the load is applied and removed many times. As shown in Figure 5–6, the stress varies from zero to a maximum with each cycle. Then, by observation,

$$\sigma_{min} = 0$$
$$\sigma_m = \sigma_a = \sigma_{max}/2$$
$$R = \sigma_{min}/\sigma_{max} = 0$$

Shock or Impact Loading

Loads applied suddenly and rapidly cause shock or impact. Examples include a hammer blow, a weight falling onto a structure, and the action inside a rock crusher. The design of machine members to withstand shock or impact involves an analysis of their energy-absorption capability, a topic not considered in this book. (See References 4, 5, 7, 10, 11, 13, and 14.)

FIGURE 5–6
Repeated, one-direction
stress

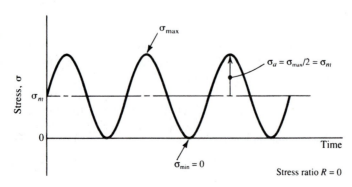

Random Loading

When varying loads are applied that are not regular in their amplitude, the loading is called *random*. Statistical analysis is used to characterize random loading for purposes of design and analysis. This topic is not covered in this book.

**5–3
ENDURANCE
STRENGTH**

The *endurance strength* of a material is its ability to withstand fatigue loads. In general, it is the stress level that a material can survive for a given number of cycles of loading. If the number of cycles is infinite, the stress level is called the *endurance limit*.

Endurance strengths are usually charted on a graph like that shown in Figure 5–7, called an *S-N diagram*. Curves *A, B,* and *D* are representative of a material that does exhibit an endurance limit, such as a plain carbon steel. Curve *C* is typical of most nonferrous metals, such as aluminum, which do not exhibit an endurance limit. For such materials, the number of cycles to failure should be reported for the given endurance strength.

We will use the symbol s_n to denote the basic endurance strength of the material.

**5–4
FACTORS
AFFECTING THE
ENDURANCE
STRENGTH**

Published data for endurance strength are determined by special fatigue testing devices, which typically use a polished specimen subjected to a reversed bending load, similar to that sketched in Figure 5–2. If the actual operating conditions of a part in a machine are different, and they usually are, the fatigue strength must be reduced from the reported value. Some of the factors that decrease the endurance strength are discussed next. (See References 1, 2, 4, 9, and 12.)

This discussion relates to the endurance strength for materials subjected only to tensile normal stresses, that is, tensile stresses resulting from bending actions or axial tension. Fatigue failures are most likely to occur in regions of high tensile stress rather than

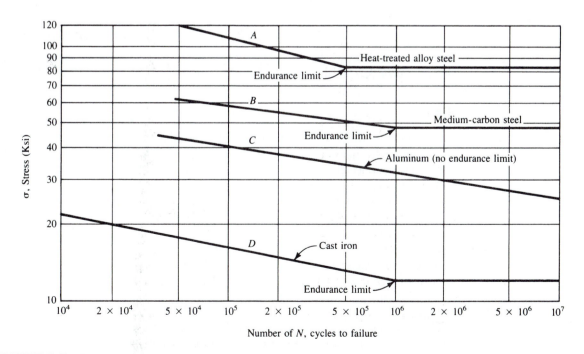

FIGURE 5–7 Representative endurance strengths

compressive stress. Cases involving fluctuating torsional shear stresses are discussed separately in Section 5–9.

Size of the Section

The test specimen is usually 0.30 inch (7.6 mm) in diameter. Larger section sizes exhibit lower strengths, have a less favorable stress distribution, and have less uniformity of properties, particularly with heat-treated parts. Reference 1 includes a suggested method of determining the size factor for rotating shafts up to 10.0 inches (250 mm) in diameter. We will use that method here also because we are basing our analysis on the reversed bending phenomenon. Table 5–1 shows the suggested relationships for determining a *size factor,* C_s, to be applied to the endurance strength to account for the size of the cross section. Figure 5–8 shows plots of the equations from Table 5–1, with some blending of the curves. Reading from the curves should provide acceptable accuracy.

When the component being designed is not circular like a shaft, judgment is required to determine the characteristic dimension to use in the formulas in Table 5–1. For flat, rolled shapes, the thickness should be used. Note that the use of these equations is approximate.

Surface Finish

Any deviation from a polished surface reduces endurance strength. Figure 5–9 shows rough estimates for the endurance strength s_n compared with the ultimate tensile strength of steels for several practical surface conditions. It is critical that parts subjected to fatigue loading be protected from nicks, scratches, and corrosion because they drastically reduce fatigue strength.

Stress Concentrations

Sudden changes in geometry, especially sharp grooves and notches where high stress concentrations occur, are likely places for fatigue failures to occur. Care should be taken in the design and manufacture of cyclically loaded parts to keep stress concentration factors to a low value. We will apply the stress concentration factors, as found from the methods of Section 3–9, to the computed stresses, rather than to the allowable strengths. See Section 5–9.

Flaws

Internal flaws of the material, especially likely in cast parts, are places in which fatigue cracks initiate. Critical parts can be inspected by x-ray techniques for internal flaws. If they are not inspected, a higher-than-average design factor should be specified for cast parts, and a lower endurance strength should be used.

Temperature

Most materials have a lower endurance strength at high temperatures. The reported values are for room temperatures. Operation above 160°F (72°C) will reduce the endurance strength of most ductile materials.

Nonuniform Material Properties

Many materials have different material properties in different directions because of the manner in which the material was processed. Rolled sheet or bar products are typically

FIGURE 5–8 Size factor

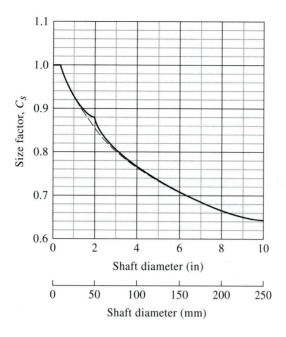

TABLE 5–1 Suggested size factors

U.S. Customary Units		SI Units	
Size range	For D in inches	Size range	For D in mm
$D < 0.40$ in	$C_s = 1.0$	$D < 10$ mm	$C_s = 1.0$
$0.40 < D < 2.0$ in	$C_s = (D/0.3)^{-0.068}$	$10 < D < 50$ mm	$C_s = (D/7.6)^{-0.068}$
$2.0 < D < 10.0$ in	$C_s = D^{-0.19}$	$50 < D < 250$ mm	$C_s = 1.85 D^{-0.19}$

FIGURE 5–9
Endurance strength s_n versus tensile strength for wrought steel for various surface conditions

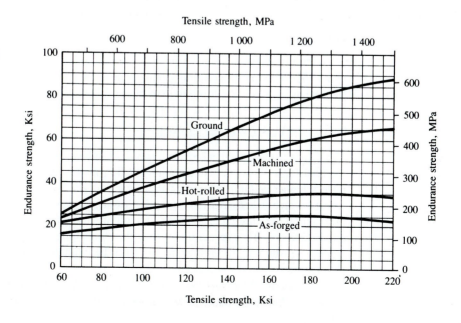

151

stronger in the direction of rolling than they are in the transverse direction. Fatigue tests are likely to have been run on test bars oriented in the stronger direction. Stressing of such materials in the transverse direction may result in lower endurance strength.

Nonuniform properties are also likely to exist in the vicinity of welds because of incomplete weld penetration, slag inclusions, and variations in the geometry of the part at the weld. Also, welding of heat-treated materials may alter the strength of the material because of local annealing near the weld. Some welding processes may result in the production of residual tensile stresses that decrease the effective endurance strength of the material. Annealing or normalizing after welding is often used to relieve these stresses, but the effect of such treatments on the strength of the base material must be considered.

Residual Stresses

Fatigue failures typically initiate at locations of relatively high tensile stress. Any manufacturing process that tends to produce residual tensile stress will decrease the endurance strength of the component. Welding has already been mentioned as a process that may produce residual tensile stress. Grinding and machining, especially with high material removal rates, also cause undesirable residual tensile stresses. Critical areas of cyclically loaded components should be machined or ground in a gentle fashion.

Processes that produce residual *compressive* stresses can prove to be beneficial. Shot blasting and peening are two such methods. *Shot blasting* is performed by directing a high-velocity stream of hardened balls or pellets at the surface to be treated. *Peening* uses a series of hammer blows on the surface. Crankshafts, springs, and other cyclically loaded machine parts can benefit from these methods.

Corrosion and Environmental Factors

Endurance strength data are typically measured with the specimen in air. Operating conditions that expose a component to water, salt solutions, or other corrosive environments can significantly reduce the effective endurance strength. Corrosion may cause harmful local surface roughness and may also alter the internal grain structure and chemistry of the material. Steels exposed to hydrogen are especially affected adversely.

Nitriding

Nitriding is a surface-hardening process for alloy steels in which the material is heated to 950°F (514°C) in a nitrogen atmosphere, typically ammonia gas, followed by slow cooling. Improvement of endurance strength of 50% or more can be achieved with nitriding.

Wrought versus Cast Materials

Metal alloys having similar chemical compositions can be either wrought or cast to produce the final form. Wrought materials are usually rolled or drawn. Wrought materials usually have a higher endurance strength than cast materials of similar composition in regions where no significant stress concentration exists. However, in the vicinity of notches and other discontinuities, the endurance strength of wrought and cast materials is more nearly equal. One possible explanation of this phenomenon is that the cast material is likely to have more isotropic material properties than the wrought material and is less affected by the presence of the stress concentration.

To use the more conservative approach, it is recommended that a factor of 0.8 be applied to the basic endurance strength if a *cast steel* is used. For *cast iron,* a factor of 0.70 is recommended.

Type of Stress

Endurance strength data are obtained from the rotating beam test that produces completely reversed and repeated normal stresses. The maximum stress is produced at the surface of the specimen, and the stress varies linearly to zero at the center of the circular cross section. If the actual loading is different from bending, a factor for the type of loading should be applied to the endurance strength.

Axial Tension

Under pure tension, all of the material—not just the surface—is subjected to the maximum stress. A factor of 0.80 is suggested to be applied to the bending endurance strength to reflect this different behavior.

Effect of Stress Ratio on Endurance Strength

Figure 5–10 shows the general variation of endurance-strength data for a given material when the stress ratio R varies from -1.0 to $+1.0$, covering the range of cases including the following:

- Repeated, reversed stress (Figure 5–3); $R = -1.0$
- Partially reversed fluctuating stress with a tensile mean stress [Figure 5–4(b)]; $-1.0 < R < 0$

FIGURE 5–10
Effect of stress ratio R on endurance strength of a material

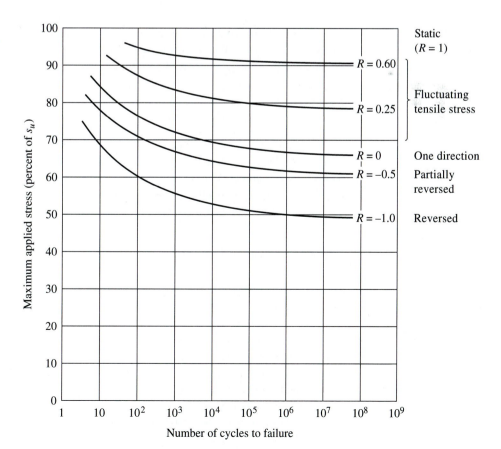

■ Repeated, one-direction tensile stress (Figure 5–6); $R = 0$

■ Fluctuating tensile stress [Figure 5–4(a)]; $0 < R < 1.0$

■ Static stress (Figure 5–1); $R = 1$

Note that Figure 5–10 is only an example, and it should not be used to determine actual data points. If such data are desired for a particular material, specific data for that material must be found either experimentally or in published literature.

The most damaging kind of stress among those listed is the repeated, reversed stress with $R = -1$. (See Reference 2, page 27.) Recall that the rotating shaft in bending as shown in Figure 5–2 is an example of a load-carrying member subjected to a stress ratio $R = -1$.

Fluctuating stresses with a compressive mean stress as shown in Parts (c) and (d) of Figure 5–4 do not significantly affect the endurance strength of the material because fatigue failures tend to originate in regions of tensile stress.

Note that the curves of Figure 5–10 show estimates of the endurance strength, s_n, as a function of the ultimate tensile strength for steel. These data apply to ideal polished specimens and do not include any of the other factors discussed in this section. For example, the curve for $R = -1.0$ (reversed bending) shows that the endurance strength for steel is approximately 0.5 times the ultimate strength ($0.50 \times s_n$) for large numbers of cycles of loading (approximately 10^5 or higher). This is a good general estimate for steels. The chart also shows that types of loads producing R greater than -1.0 but less than 1.0 have less of an effect on the endurance strength. This illustrates that using data from the reversed bending test is the most conservative.

We will not use Figure 5–10 directly for problems in this book because our procedure for estimating the actual endurance strength starts with the use of Figure 5–9 which presents data from reversed bending tests. Therefore, the effect of stress ratio is already included. Section 5–9 of this chapter includes methods of analysis for loading cases in which the fluctuating stress produces a stress ratio different from $R = -1.0$.

Reliability

The data for endurance strength for steel shown in Figure 5–9 represent average values derived from many tests of specimens having the appropriate ultimate strength and surface conditions. Naturally, there is variation among the data points; that is, half are higher and half are lower than the reported values on the given curve. The curve, then, represents a reliability of 50%, indicating that half of the parts would fail. Obviously, it is advisable to design for a higher reliability, say, 90%, 99%, or 99.9%. A factor can be used to estimate a lower endurance strength that can be used for design to produce the higher reliability values. Ideally, a statistical analysis of actual data for the material to be used in the design should be obtained. Reference 8 shows a method for analyzing such data. By making certain assumptions about the form of the distribution of strength data, Reference 8 also reports the values in Table 5–2 as approximate reliability factors, C_R. We will apply these factors to the average endurance strength from Figure 5–9 as shown in the next section.

TABLE 5–2
Approximate reliability factors, C_R

Desired reliability	C_R
0.50	1.0
0.90	0.90
0.99	0.81
0.999	0.75

**5–5
ESTIMATING
ACTUAL
ENDURANCE
STRENGTH, s_n'**

This section presents a method for estimating the actual expected endurance strength that a material will have in service, considering the basic endurance strength of the material modified by the several factors discussed in Section 5–4. Note that many of the factors presented are appropriate only for carbon and alloy steels and that the factors are approximate. You are advised to search for the most accurate data possible for a given design, considering the type of material, the processing conditions, and the environment to which the component is to be subjected. Specialized references, such as that by Boyer (2), should be consulted. Also, it is highly recommended that testing be done on actual components subjected to fatigue loading under conditions as close as possible to the actual service expected.

In problem solutions, we will use the following procedure for estimating actual endurance strengths for steel parts, called s_n':

Procedure for Estimating Actual Endurance Strength

1. Specify the material from which the part is to be made, and determine its ultimate strength from published data or actual tests, considering its condition and heat treatment.

2. Specify the manufacturing process used to produce the final part. The condition of the surface of the part in the most highly stressed area is the most important factor.

3. Use Figure 5-9 to estimate the modified endurance strength s_n by entering at the value of the tensile strength on the horizontal axis and projecting to the appropriate curve representing the expected surface condition. Then project to the vertical axis to read the endurance strength. Note that the bottom and left axes are used for U.S. Customary Units and that the top and right axes are used for SI units. If the tensile strength of the material is greater than 220 Ksi (1 500 MPa), use the value $s_u = 220$ Ksi. Experimental data show that there is little increase in the endurance strength of the material at higher tensile strength levels. (See Reference 2.)

4. Apply a size factor, C_s, from Table 5-1 or Figure 5-8.

5. Apply a material factor, C_m:

$$C_m = 1.0 \text{ for wrought steel}$$
$$C_m = 0.80 \text{ for cast steel}$$
$$C_m = 0.70 \text{ for cast iron}$$

6. Apply a factor for the type of stress, C_{st}:

$$C_{st} = 1.0 \text{ for bending}$$
$$C_{st} = 0.80 \text{ for axial tension}$$

7. Apply a reliability factor, C_R, from Table 5-2. The desired reliability is a design decision based on many factors, such as the criticality of the application, cost, weight, and so forth. For problems in this book, we will use a reliability of 0.99 unless stated otherwise. Theoretically, this would predict one failure in 100 parts. However, we will also be applying a design factor (factor of safety) as explained in the next section. Therefore, the actual expected reliability should be higher. For a reliability of 0.99, use $C_R = 0.81$.

8. Compute the estimated endurance strength, s_n', from

Estimated Actual Endurance Strength

$$s_n' = s_n(C_s)(C_m)(C_{st})(C_R) \tag{5-4}$$

These are the only factors that will be used in this book to estimate the endurance strength. If other factors exist, such as temperature or residual stress, they should be included.

Example Problems for Estimating Endurance Strengths

Example Problem 5–2   Estimate the actual endurance strength of AISI 1050 cold-drawn steel when used in a shaft subjected to rotating bending only. The shaft will be machined to a diameter of approximately 1.75 in.

Solution   Objective   Compute the estimated actual endurance strength of the shaft material.

Given   AISI 1050 cold-drawn steel, machined.
Size of section: $D = 1.75$ in.
Type of stress: Reversed bending.

Analysis   Use Equation (5–4) to estimate s_n'.
Ultimate tensile strength from Appendix 3: $s_u = 100$ Ksi.
Modified endurance strength from Figure 5–9, machined surface: $s_n = 38$ Ksi.
Size factor for $D = 1.75$ in from Figure 5–8: $C_s = 0.88$.
Material factor for wrought steel: $C_m = 1.00$.
Factor for type of stress for reversed bending: $C_{st} = 1.00$.
Reliability factor: This is a design decision because the desired reliability is not stated. We will use a desired reliability of 0.99 for which $C_R = 0.81$.

Results $s_n' = s_n(C_s)\,(C_m)\,(C_{st})\,(C_R) = (38\ \text{Ksi})(0.88)(1.00)(1.00)(0.81) = 27.1$ Ksi

Comments   This is the level of stress that would be expected to produce fatigue failure in a rotating shaft due to the action of reversed bending. It accounts for the basic endurance strength of the wrought AISI 1050 cold-drawn material, the effect of the machined surface, the size of the section, and the desired reliability.

Example Problem 5–3   Estimate the actual endurance strength of cast steel having an ultimate strength of 120 Ksi when used in a connecting rod in a linkage subjected to reversed, repeated, axial load. The rod will be machined to a square cross section, 1.50 in on a side.

Solution   Objective   Compute the estimated actual endurance strength of the rod material.

Given   Cast steel, machined: $s_u = 120$ Ksi.
Size of section: $S = 1.50$ in.
Type of stress: Repeated, reversed axial stress.

Analysis   Use Equation (5–4) to estimate s_n'.
Ultimate tensile strength, given: $s_u = 120$ Ksi.
Modified endurance strength from Figure 5–9, machined surface: $s_n = 44$ Ksi.
Size factor for $S = 1.50$ in from Figure 5–8. $C_s = 0.90$.
Material factor for cast steel: $C_m = 0.80$.
Factor for type of stress for reversed axial stress: $C_{st} = 0.80$.
Reliability factor: This is a design decision because the desired reliability is not stated. We will use a desired reliability of 0.99 for which $C_R = 0.81$.

Results $s_n' = s_n(C_s)\,(C_m)\,(C_{st})\,(C_R) = (44\ \text{Ksi})(0.90)(0.80)(0.80)(0.81) = 20.5$ Ksi

**5–6
DESIGN FOR
DIFFERENT
TYPES OF
LOADING**

It is the designer's responsibility to ensure that a machine part is safe for operation under reasonably foreseeable conditions. This is done by performing a stress analysis, either analytically or experimentally, and comparing the predicted stress with a design stress. The *design stress* is the level of stress that the part will be permitted to see under operating conditions. Some designers prefer the terms *working stress, allowable stress,* or *safe stress.* We will use *design stress* to emphasize the role of the designer in determining the appropriate value to use. (See Reference 3.)

The decision on what design stress to use involves, at least, the consideration of the type of load on the part and the material from which it is made. The types of loads considered here are the following:

Static

Repeated and reversed

Fluctuating

Shock or impact

It is important to consider the ductility of the material, in addition to its strength and stiffness. The modes of failure are dramatically different for brittle materials and ductile materials. (Review the discussion of ductility in Chapter 2.) Most machine parts are made from ductile materials, especially if they are subjected to fatigue loads or to shock or impact. Wrought metals, such as steel, aluminum, and copper, are usually ductile. For fatigue loading, it is advised that the percent elongation of the material be significantly higher than 5%, say, 10% or higher, to ensure against brittle-type fracture. Gray cast iron, some highly heat-treated steels, some plastics, and some parts made from powder metal are brittle.

**5–7
PREDICTIONS OF
FAILURE**

Four different ways of predicting failure are considered in this section:

1. Maximum normal stress theory
2. Maximum shear stress theory
3. Distortion energy theory (also called the *von Mises theory* or the *Mises-Hencky theory*)
4. The Soderberg criterion

Comparing the actual expected stress in a part with the stress that would predict failure allows the determination of a *factor of safety,* or *design factor, N.* These two terms are used interchangeably in this book.

Machine elements can also fail because of excessive deformation or vibration. You should be able to compute deformations due to axial tensile or compressive loads, bending, or torsion from your study of strength of materials. Some of the basic concepts are reviewed in Chapter 3. For more complex shapes or loading patterns, computer-based analysis techniques such as the finite-element method and beam-design software are important tools.

Failure due to deformation is often highly dependent on the use to which the machine is to be put and on more subjective considerations. Will excessive deformation cause two or more members to touch when they should not? Will the desired precision of the machine be compromised? Will the part feel or look too flexible and flimsy? Will the required relative positions of component parts be out of tolerance? Will parts vibrate excessively at the frequencies experienced during operation? Will resonance occur because of vibrations? Will shafts experience a critical speed during operation, resulting in wild oscillations of parts carried by the shaft?

This chapter will not pursue the quantitative analysis of deformation, leaving that to be your responsibility as the design of a machine evolves. Later chapters do address some critical cases such as the interference fit between two mating parts (Chapter 13), the position of the teeth of one gear relative to its mating gear (Chapter 9), the axial clearance between a journal bearing and the shaft rotating within it (Chapter 16), and the deformation of springs (Chapter 19). Also, Section 5–10 suggests some guidelines for limiting deflections due to bending and torsion for general machine parts, for parts expected to exhibit moderate precision such as production machinery, and for high-precision parts such as measuring instruments.

Maximum Normal Stress Theory of Failure

This theory states that a material will fracture when the maximum normal stress (tension or compression) exceeds the ultimate strength of the material. It is used in very limited situations, namely, for brittle materials under pure, uniaxial tension or compression. The standard tension test and compression test are examples of this type of loading. Therefore, the actual use of the part must subject the material to a similar loading for this theory to give a valid prediction of failure.

Maximum Shear Stress Theory of Failure

This theory states that a ductile material begins to yield when the maximum shear stress in a load-carrying part exceeds the maximum shear stress present in a tensile-test specimen when yielding begins. In Section 4–6 it was shown that in this situation, the magnitude of the maximum shear stress is one-half of the applied tensile stress. Then, when yielding begins, the applied stress is by definition equal to s_y. The corresponding maximum shear stress is $s_{sy} = s_y/2$.

The maximum shear stress theory of failure has been shown by experimentation to be slightly conservative for ductile materials subjected to any combination of normal and shear stresses. It is relatively easy to use and is often chosen by designers for parts to be made from ductile materials under static loads or completely reversed normal, shear, or combined stresses. It should not be used for fluctuating stresses.

Distortion Energy Theory

Also sometimes called the *von Mises theory* or the *Mises-Hencky theory,* this theory provides a very accurate prediction of failure of ductile materials under static loads or completely reversed normal, shear, or combined stresses. It requires the definition of the term *von Mises stress,* indicated by the symbol σ'. If the two principal stresses from Mohr's circle are called σ_1 and σ_2, the von Mises stress is

 von Mises Stress

$$\sigma' = \sqrt{\sigma_1^2 + \sigma_2^2 - \sigma_1\sigma_2} \qquad (5-5)$$

Then the distortion energy theory of failure states that failure occurs when the von Mises stress exceeds the yield strength of the material, that is, when $\sigma' > s_y$.

The Soderberg Criterion for Fluctuating Stresses

Recall from Section 5–2 that the term *fluctuating stress* refers to the condition in which the part is subjected to a mean stress greater than zero with an alternating stress superimposed on the mean stress (see Figure 5–4). One method of failure prediction uses the

Soderberg diagram, as shown in Figure 5–11, which is drawn for the case of normal stresses only. Mean stress is plotted along the horizontal axis, and alternating stress is plotted along the vertical axis. Failure when the alternating stress is zero (steady stress) would be predicted as the stress reaches the yield strength of the material. Failure under alternating stress with a zero mean stress (reversed, repeated load) would be predicted as the stress reaches the endurance strength of the material. A line drawn between these two points is called the *Soderberg line.* Then for fluctuating stress, if the point representing the actual combination of mean and alternating stress plots on the diagram below the Soderberg line, it should be safe. Conversely, a combination outside the line would be predicted to cause failure.

Applying a design factor to the Soderberg line produces the design stress line on the diagram. Actual operating stress levels should be kept below this line for safety.

5–8
DESIGN
FACTORS

The term *design factor* is a measure of the relative safety of a load-carrying component. We will use the symbol N to denote the design factor. In most cases, the strength of the material from which the component is to be made is divided by the design factor to determine a *design stress, σ_d,* sometimes called the *allowable stress.* Then the actual stress to which the component is subjected should be less than the design stress. For some kinds of loading, it is more convenient to set up a relationship from which the design factor, N, can be computed from the actual applied stresses and the strength of the material. Still in other cases, particularly for the case of the buckling of columns, as discussed in Chapter 6, the design factor is applied to the *load* on the column rather than the strength of the material.

Section 5–9 presents methods for computing the design stress or design factor for several different kinds of loading and materials.

The designer must determine what a reasonable value for the design factor should be in any given situation. Often the value of the design factor or the design stress is governed by codes established by standards-setting organizations such as the American Society of Mechanical Engineers, the American Gear Manufacturers Association, the U.S. Department of Defense, the Aluminum Association, or the American Institute of Steel Construction. For structures, local or state building codes often prescribe design factors or design stresses. Some companies have adopted their own policies specifying design factors based on past experience with similar conditions.

FIGURE 5–11
Soderberg criterion for fluctuating stresses

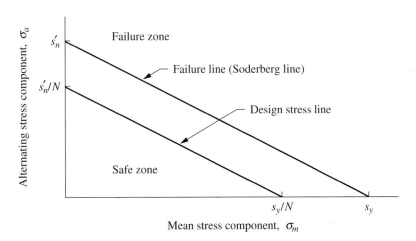

In the absence of other specifications, this book will use the following:

1. *Normal case for structures or machine elements:* Ductile material under conditions of moderate uncertainty with regard to material properties, nature of loading, or adequacy of the stress analysis. Use $N = 3$.

2. *Static structures:* Ductile materials with high confidence in the knowledge of material properties, magnitude of loads, and the adequacy of the stress analysis. Use $N = 2$.

3. *Static structures:* Brittle materials with high confidence in the knowledge of the operating conditions. Use $N = 3$.

4. *Machine elements:* Brittle materials with moderate uncertainty about material properties, loads, or stress analysis. Use $N = 4$ or higher.

5. *Machine elements:* Ductile materials with uncertainty about some *combination* of material properties, loads, or stress analysis, particularly under shock or impact loading. Use $N = 4$ or higher.

5–9 METHODS OF COMPUTING DESIGN FACTOR OR DESIGN STRESS

Nine different cases are considered here with different combinations of types of material (ductile or brittle) and manner of loading. Careful attention should be paid to the strength basis for the different cases. The following symbols are used:

s_u or s_{ut} = ultimate tensile strength
s_{uc} = ultimate compressive strength
s_y = yield strength or yield point
s_{sy} = yield strength in shear
s_n' = endurance strength of material under actual conditions
s_{sn}' = endurance strength in shear under actual conditions

Case A: Brittle Materials under Static Loads

When the actual applied stress, σ, is simple tension or compression in only one direction, use the maximum normal stress theory of failure. Because brittle materials do not yield, you should always apply stress concentration factors when computing the applied stress.

Case A1: Tensile Stress

$$\sigma_d = s_{ut}/N \quad \text{or} \quad N = s_{ut}/\sigma \tag{5–6}$$

Case A2: Compressive Stress

$$\sigma_d = s_{uc}/N \quad \text{or} \quad N = s_{uc}/\sigma \tag{5–7}$$

Case A3: Biaxial Stress.   Use Mohr's circle to determine the principal stresses, σ_1 and σ_2. If both principal stresses are of the same sign, either tensile or compressive, use Case A1 or A2. If they are of different signs, use

$$\frac{1}{N} = \frac{\sigma_1}{s_{ut}} + \frac{\sigma_2}{s_{uc}} \tag{5–8}$$

Any stress concentration factors should be applied to the computed stresses.

Case B: Brittle Materials under Fatigue Loads

No specific recommendation will be given for brittle materials under fatigue loads because it is usually not desirable to use a brittle material in such cases. When it is necessary to do so, testing should be done to ensure safety under actual conditions of service.

Case C: Ductile Materials under Static Loads

Two failure theories will be discussed here. The maximum shear stress theory of failure is the more conservative of the two and is somewhat easier to use. The distortion energy theory is the more accurate predictor of impending failure.

Case C1: Maximum Shear Stress Theory.   Failure occurs when the maximum shear stress exceeds the yield strength of the material in shear, s_{ys}. Determine the maximum shear stress from the Mohr circle. Then

$$N = s_{sy}/\tau_{max} \tag{5–9}$$

We will use $s_{sy} = 0.5\ s_y$. Then

$$N = 0.5\ s_y/\tau_{max} \quad \text{or} \quad \tau_d = 0.5\ s_y/N = \text{design shear stress} \tag{5–10}$$

It can be shown from a Mohr's circle analysis that for the special case of uniaxial tension or compression with no applied shear stresses, Equation (5–10) is the same as

$$N = s_y/\sigma_{max} \quad \text{or} \quad \sigma_d = s_y/N = \text{design stress}$$

Case C2: Distortion Energy Theory.   Failure occurs when the von Mises stress, σ', exceeds the tensile yield strength of the material. To find the von Mises stress, first use Mohr's circle to determine the two principal stresses. Then

$$\sigma' = \sqrt{\sigma_1^2 + \sigma_2^2 - \sigma_1\sigma_2} \tag{5–11}$$

The design factor is then found from

$$N = s_y/\sigma' \tag{5–12}$$

The advantage of this theory, in addition to its being a more accurate predictor of failure, is that it allows the use of the tensile yield strength for analysis.

For static loading on ductile materials, it is not necessary to include a stress concentration factor when geometric discontinuities occur. If the local stress at a point of stress concentration reaches the yield strength of the material, the material will, in fact, yield, causing the stress to be redistributed and bringing the level back below the yield strength.

Case D: Reversed, Repeated Normal Stress

Determine the peak amplitude of the repeated stress (as σ_{max} in Figure 5–3) and the endurance strength for the material, s_n', adjusted for actual conditions as in Section 5–5. Then

$$N = s_n'/\sigma_{max} \quad \text{or} \quad \sigma_d = s_n'/N \tag{5–13}$$

Stress concentration factors must be considered in computing the maximum stress.

Case E: Reversed, Repeated Shear Stress

Again the maximum shear stress theory or the distortion energy theory can be used. First compute the maximum repeated shear stress, τ_{max}, including any stress concentration factor.

Case E1: Maximum Shear Stress Theory

$$s'_{sn} = 0.5s'_n \quad \text{(estimate for endurance strength in shear)}$$
$$N = s'_{sn}/\tau_{max} \quad \text{or} \quad \tau_d = s'_{sn}/N \tag{5–14}$$

Case E2: Distortion Energy Theory

$$s'_{sn} = 0.577\, s'_n$$
$$N = s'_{sn}/\tau_{max} = 0.577 s'_n/\tau_{max} \tag{5–15}$$

The factor 0.577 can be derived from the definition of the von Mises stress for the special case of pure torsion. Drawing the Mohr circle, you will find that $\sigma_1 = +\tau_{max}$, and $\sigma_2 = -\tau_{max}$. Then, from Equation (5–11), $\sigma' = \sqrt{3}\tau_{max} = 1.732\tau_{max}$. Putting this into the basic form of the distortion energy theory, we have

$$N = s'_n/\sigma' = s'_n/(1.732\tau_{max}) = 0.577 s'_n/\tau_{max}$$

as used in Equation (5–15).

Case F: Reversed Combined Stress

Use the Mohr circle to find the maximum shear stress and the two principal stresses by using the maximum values of the applied stresses.

Case F1: Maximum Shear Stress Theory

$$N = 0.5\, s'_n/\tau_{max} \tag{5–16}$$

Case F2: Distortion Energy Theory

$$N = 0.577\, s'_n/\tau_{max} \tag{5–17}$$

Or compute σ' from Equation (5–11) and use

$$N = s'_n/\sigma' \tag{5–18}$$

Case G: Fluctuating Normal Stresses: The Soderberg Criterion

The Soderberg criterion was described in Section 5–7 and illustrated in Figure 5–11. A satisfactory design results if the combination of the mean stress and the alternating stress produces a point in the *safe zone* shown in Figure 5–11. Special attention should be paid to the application of stress concentration factors. Tests have shown that stress concentrations affect only the alternating stress component and not the mean stress. Then you can use the following equation to evaluate the design factor for fluctuating loads:

$$\frac{1}{N} = \frac{\sigma_m}{s_y} + \frac{K_t\sigma_a}{s'_n} \tag{5–19}$$

Case H: Fluctuating Shear Stresses

The preceding development of the Soderberg criterion can also be done for fluctuating shear stresses instead of normal stresses. The design factor equation would then be

$$\frac{1}{N} = \frac{\tau_m}{s_{sy}} + \frac{K_t \tau_a}{s'_{sn}} \tag{5–20}$$

Experimental work indicates that this form of the Soderberg equation is very conservative. If the maximum shear stress theory of failure is used, $s_{sy} = 0.5s_y$, and $s'_{sn} = 0.5s'_n$. For the distortion energy theory, $s_{sy} = 0.577s_y$, and $s'_{sn} = 0.577s'_n$.

Case I: Fluctuating Combined Stresses

The approach presented here is similar to the Soderberg criterion described previously. But in this case, the effect of the combined stresses is first determined by the use of the Mohr circle.

Case I1. For the maximum shear stress theory, draw two Mohr's circles, one for the mean stresses and one for the alternating stresses. From the first circle, determine maximum mean shear stress, $(\tau_m)_{max}$. From the second circle, determine the maximum alternating shear stress, $(\tau_a)_{max}$. Then use these values in the design equation

$$\frac{1}{N} = \frac{(\tau_m)_{max}}{s_{sy}} + \frac{K_t (\tau_a)_{max}}{s'_{sn}} \tag{5–21}$$

where $s_{sy} = 0.5s_y$
$s'_{sn} = 0.5s'_n$

Case I2. For the distortion energy theory, draw two Mohr's circles, one for the mean stresses and one for the alternating stresses. From these circles, determine the maximum and minimum principal stresses. Then compute the von Mises stresses for both the mean and the alternating components from

$$\sigma'_m = \sqrt{\sigma_{1m}^2 + \sigma_{2m}^2 - \sigma_{1m}\sigma_{2m}}$$
$$\sigma'_a = \sqrt{\sigma_{1a}^2 + \sigma_{2a}^2 - \sigma_{1a}\sigma_{2a}}$$

The Soderberg equation then becomes

$$\frac{1}{N} = \frac{\sigma'_m}{s_y} + \frac{K_t \sigma'_a}{s'_n} \tag{5–22}$$

**5–10
GENERAL
DESIGN
PROCEDURE**

The earlier parts of this chapter have provided guidance related to the many factors involved in design of machine elements that must be safe when carrying the applied loads. This section brings these factors together so that you can complete the design. The general design procedure described here is meant to give you a feel for the process. It is not practical to provide a completely general procedure. You will have to adapt it to the specific situations that you encounter.

The procedure is set up assuming that the following factors are known or can be specified or estimated:

- General design requirements: Objectives and limitations on size, shape, weight, desired precision, and so forth.
- Nature of the loads to be carried.
- Types of stresses produced by the loads.
- Type of material from which the element is to be made.
- General description of the manufacturing process to be used, particularly with regard to the surface finish that will be produced.
- Desired reliability.

General Design Procedure

1. Specify the objectives and limitations, if any, of the design, including desired life, size, shape, and appearance.

2. Determine the environment in which the element will be placed, considering such factors as corrosion potential and temperature.

3. Determine the nature and characteristics of the loads to be carried by the element, such as

   Static, dead, slowly applied loads.

   Dynamic, live, varying, repeated loads that may potentially cause fatigue failure.

   Shock or impact loads.

4. Determine the magnitudes for the loads and the operating conditions, such as

   Maximum expected load.

   Minimum expected load.

   Mean and alternating levels for fluctuating loads.

   Frequency of load application and repetition.

   Expected number of cycles of loading.

5. Analyze how loads are to be applied to determine the type of stresses produced, such as

   Direct normal stress, bending stress, direct shear stress, torsional shear stress, or some combination of stresses.

6. Propose the basic geometry for the element, paying particular attention to

   Its ability to carry the applied loads safely.

   Its ability to transmit loads to appropriate support points. Consider the *load paths*.

   The use of efficient shapes according to the nature of the loads and the types of stresses encountered. This applies to the general shape of the element and to each of its cross sections. Achieving efficiency involves optimizing the amount and the type of material involved. In Chapter 20, Section 20–2 gives some suggestions for efficient design of frames and members in bending and torsion.

Providing appropriate attachments to supports and to other elements in the machine or structure.

Providing for the positive location of other components that may be installed on the element being designed. This may call for shoulders, grooves, holes, retaining rings, keys and keyseats, pins, or other means of fastening or holding parts.

7. Propose the method of manufacturing the element with particular attention to the precision required for various features and the surface finish that is desired. Will it be cast, machined, ground, or polished, or produced by some other process? These are design decisions that have important impacts on the performance of the element, its ability to withstand fatigue loading, and the cost to produce it.

8. Specify the material from which the element is to be made, along with its condition. For metals the specific alloy should be specified, and the condition could include such processing factors as hot rolling, cold drawing, and a specific heat treatment. For nonmetals, it is often necessary to consult with vendors to specify the composition and the mechanical and physical properties of the desired material. Consult Chapter 2 and Section 20–2 in Chapter 20 for additional guidance.

9. Determine the expected properties of the selected material, for example

   Ultimate tensile strength, s_u.

   Ultimate compressive strength, s_{uc}, if appropriate.

   Yield strength, s_y.

   Ductility as represented by percent elongation.

   Stiffness as represented by modulus of elasticity, E or G.

10. Specify an appropriate design factor, N, for the stress analysis using the guidelines discussed in Section 5–8.

11. Determine which stress analysis method outlined in Section 5–9 applies to the design being completed.

12. Compute the appropriate design stress for use in the stress analysis. If fatigue loading is involved, the actual expected endurance strength of the material should be computed as outlined in Section 5–5. This requires the consideration of the expected size of the section, the type of material to be used, the nature of the stress, and the desired reliability. Because the size of the section is typically unknown at the start of the design process, an estimate must be made to allow the inclusion of a reasonable size factor, C_s. You should check the estimate at the end of the design process to verify that reasonable values were assumed at this stage of the design.

13. Determine the nature of any stress concentrations that may exist in the design at places where geometry changes occur. Stress analysis should be considered at all such places because of the likelihood of localized high tensile stresses that may produce fatigue failure. If the geometry of the element in these areas is known, determine the appropriate stress concentration factor, K_t. If the geometry is not yet known, it is advisable to estimate the expected magnitude of K_t. The estimate must then be checked at the end of the design process.

14. Complete the required stress analyses at all points where the stress may be high and at changes of cross section to determine the minimum acceptable dimensions for critical areas.

15. Specify suitable, convenient dimensions for all features of the element. Many design decisions are required, such as

The use of preferred basic sizes as listed in Table A2–1.

The size of any part that will be installed on or attached to the element being analyzed. Examples of this are shown in Chapter 12 on shaft design where gears, chain sprockets, bearings, and other elements are to be installed on the shafts. But many machine elements have similar needs to accommodate mating elements.

Elements should not be significantly oversized without good reason in order to achieve an efficient overall design.

Sometimes the manufacturing process to be used has an effect on the dimensions. For example, a company may have a preferred set of cutting tools for use in producing the elements. Casting, rolling, or molding processes often have limitations on the dimensions of certain features such as the thickness of ribs, radii produced by machining or bending, variation in cross section within different parts of the element, and convenient handling of the element during manufacture.

Consideration should be given to the sizes and shapes that are commercially available in the desired material. This could result in significant cost reductions both in material and in processing.

Sizes should be compatible with standard company practices if practical.

16. After completing all necessary stress analyses and proposing the basic sizes for all features, check all assumptions made earlier in the design to ensure that the element is still safe and reasonably efficient. See Steps 7, 12, and 13.

17. Specify suitable tolerances for all dimensions, considering the performance of the element, its fit with mating elements, the capability of the manufacturing process, and cost. Chapter 8 may be consulted. The use of computer-based tolerance-analysis techniques may be appropriate.

18. Check to determine whether some part of the component may deflect excessively. If that is an issue, complete an analysis of the deflection of the element as designed to this point. Sometimes there are known limits for deflection based on the operation of the machine of which the element being designed is a part. In the absence of such limits, the following guidelines may be applied based on the degree of precision desired:

Deflection of a Beam Due to Bending

General machine part:	0.000 5 to 0.003 in/in of beam length
Moderate precision:	0.000 01 to 0.000 5 in/in
High precision:	0.000 001 to 0.000 01 in/in

Deflection (Rotation) Due to Torsion

General machine part:	0.001° to 0.01°/in of length
Moderate precision:	0.000 02° to 0.000 4°/in
High precision:	0.000 001° to 0.000 02°/in

See also Section 20–2 in Chapter 20 for additional suggestions for efficient design. The results of the deflection analysis may cause you to redesign the component. Typically, when high stiffness and precision are required, deflection, rather than strength, will govern the design.

19. Document the final design with drawings and specifications.

20. Maintain a careful record of the design analyses for future reference. Keep in mind that others may have to consult these records whether or not you are still involved in the project.

5–11
DESIGN
EXAMPLES

Example design problems are shown here to give you a feel for the application of the process outlined in Section 5–10. It is not practical to illustrate all possible situations, and you must develop the ability to adapt the design procedure to the specific characteristics of each problem. Also note that there are many possible solutions to any given design problem. The selection of a final solution is the responsibility of you, the designer.

In most design situations, a great deal more information will be available than is given in the problem statements in this book. But, often, you will have to seek out that information. We will make certain assumptions in the examples that allow the design to proceed. In your job, you must ensure that such assumptions are appropriate. The design examples focus on only one or a few of the components of the given systems. In real situations, you must ensure that each design decision is compatible with the totality of the design.

Design Example 5–1

A large electrical transformer is to be suspended from a roof truss of a building. The total weight of the transformer is 32 000 lb. Design the means of support.

Solution Objective Design the means of supporting the transformer.

Given The total load is 32 000 lb. The transformer will be suspended below a roof truss inside a building. The load can be considered to be static. It is assumed that it will be protected from the weather and that temperatures are not expected to be severely cold or hot in the vicinity of the transformer.

Basic Design Decisions Two straight, cylindrical rods will be used to support the transformer, connecting the top of its casing to the bottom chord of the roof truss. The ends of the rod will be threaded to allow them to be secured by nuts or by threading them into tapped holes. This design example will be concerned only with the rods. It is assumed that appropriate attachment points are available to allow the two rods to share the load equally during service. However, it is possible that only one rod will carry the entire load at some point during installation. Therefore, each rod will be designed to carry the full 32 000 lb.

We will use steel for the rods, and because neither weight nor physical size is critical in this application, a plain, medium-carbon steel will be used. We specify AISI 1040 cold-drawn steel. From Appendix 3, we find that it has a yield strength of 71 Ksi and moderately high ductility as represented by its 12% elongation. The rods should be protected from corrosion by appropriate coatings.

The objective of the design analysis that follows is to determine the size of the rod.

Analysis The rods are to be subjected to direct normal tensile stress. Assuming that the threads at the ends of the rods are cut or rolled into the nominal diameter of the rods, the critical place for stress analysis is in the threaded portion.

Use the direct tensile stress formula, Equation (3–1): $\sigma = F/A$. We will first compute the design stress and then compute the required cross-sectional area to maintain the stress in service below that value. Finally, a standard thread will be specified from the data in Chapter 18 on fasteners.

Case C from Section 5–9 applies for computing the design stress because the rod is made from a ductile steel and it carries a static load. The design stress is

$$\sigma_d = s_y/N$$

We specify a design factor of $N = 3$, because it is typical for general machine design and because there is some uncertainty about the actual production and installation procedures that may be used (see Section 5–8). Then

$$\sigma_d = s_y/N = (71\ 000\ \text{psi})/3 = 23\ 667\ \text{psi}$$

Results   In the basic tensile stress equation, $\sigma = F/A$, we know F, and we will let $\sigma = \sigma_d$. Then the required cross-sectional area is

$$A = F/\sigma_d = (32\ 000\ \text{lb})/(23\ 667\ \text{lb/in}^2) = 1.35\ \text{in}^2$$

A standard size thread will now be specified from the data in Chapter 18 on fasteners. You should be familiar with such data from earlier courses. Table A2–2(b) lists the tensile stress area for American Standard threads. A $1\frac{1}{2}$–6 UNC thread ($1\frac{1}{2}$-in-diameter rod with 6 threads per in) has a tensile stress area of 1.405 in^2 which should be satisfactory for this application.

Comments   The final design specifies a $1\frac{1}{2}$-in-diameter rod made from AISI 1040 cold-drawn steel with $1\frac{1}{2}$–6 UNC threads machined on each end to allow the attachment of the rods to the transformer and to the truss.

Design Example 5–2   A part of a conveyor system for a production operation is shown in Figure 5–12. Design the pin that connects the horizontal bar to the fixture. The empty fixture weighs 85 lb. A cast iron engine block weighing 225 lb is hung on the fixture to carry it from one process to another, where it is then removed. It is expected that the system will experience many thousands of cycles of loading and unloading of the engine blocks.

Solution   **Objective**   Design the pin for attaching the fixture to the conveyor system.

Given   The general arrangement is shown in Figure 5–12. The fixture places a shearing load that is alternately 85 lb and 310 lb (85 + 225) on the pin many thousands of times in the expected life of the system.

Basic Design Decisions   It is proposed to make the pin from AISI 1020 cold-drawn steel. Appendix 3 lists $s_y = 51$ Ksi and $s_u = 61$ Ksi. The steel is ductile with 15% elongation. This material is inexpensive, and it is not necessary to achieve a particularly small size for the pin.

The connection of the fixture to the bar is basically a clevis joint with two tabs at the top of the fixture, one on each side of the bar. There will be a close fit between the tabs and the bar to minimize bending action on the pin. Also, the pin will be a fairly close fit in the holes while still allowing rotation of the fixture relative to the bar.

Analysis   Case H from Section 5–9 applies for completing the design analysis because fluctuating shearing stresses are experienced by the pin. Therefore, we will have to determine relationships for the mean and alternating stresses (τ_m and τ_a) in terms of the applied loads and the cross-sectional area of the bar. Note that the pin is in double shear, so two cross sections resist the applied shearing force. In general, $\tau = F/2A$.

FIGURE 5–12 Conveyor system

Now we will use the basic forms of Equations (5–1) and (5–2) to compute the values for the mean and alternating forces on the pin:

$$F_m = (F_{max} + F_{min})/2 = (310 + 85)/2 = 198 \text{ lb}$$
$$F_a = (F_{max} - F_{min})/2 = (310 - 85)/2 = 113 \text{ lb}$$

The stresses will be found from $\tau_m = F_m/2A$ and $\tau_a = F_a/2A$.

The material strength values needed in Equation (5–20) for Case H are

$$s_{sy} = 0.5s_y = 0.5(51 \text{ Ksi}) = 25.5 \text{ Ksi} = 25\,500 \text{ psi}$$
$$s'_{sn} = 0.5s'_n$$

We must find the value of s'_n using the method from Section 5–5. We find from Figure 5–9 that $s_n = 21$ Ksi for the machined pin having a value of $s_u = 61$ Ksi. It is expected that the pin will be fairly small, so we will use $C_s = 1.0$. The material is wrought steel rod, so $C_m = 1.0$. Let's use $C_{st} = 1.0$ to be conservative because there is little information about such factors for shearing stress. A high reliability is desired for this application, so let's use $C_R = 0.75$ to produce a reliability of 0.999. Then

$$s'_n = C_R(s_n) = (0.75)(21 \text{ Ksi}) = 15.75 \text{ Ksi} = 15\,750 \text{ psi}$$

Finally,

$$s'_{sn} = 0.5s'_n = 0.5(15\ 750\ \text{psi}) = 7\ 885\ \text{psi}$$

We can now apply Equation (5–20) from Case H:

$$\frac{1}{N} = \frac{\tau_m}{s_{sy}} + \frac{K_t \tau_a}{s'_{sn}} s'_{sn}$$

Because the pins will be of uniform diameter, $K_t = 1.0$.

Substituting $\tau_m = F_m/2A$ and $\tau_a = F_a/2A$ found earlier gives

$$\frac{1}{N} = \frac{F_m}{2As_{sy}} + \frac{F_a}{2As'_{sn}}$$

Let's use $N = 4$ because mild shock can be expected.

Note that we now know all factors in this equation except the cross-sectional area of the pin, A. We can solve for the required area:

$$A = \frac{N}{2}\left[\frac{F_m}{s_{sy}} + \frac{F_a}{s'_{sn}}\right]$$

Finally, we can compute the minimum allowable pin diameter, D, from $A = \pi D^2/4$ and $D = \sqrt{4A/\pi}$.

Results   The required area is

$$A = \frac{4}{2}\left[\frac{198\ \text{lb}}{25\ 500\ \text{lb/in}^2} + \frac{113\ \text{lb}}{7\ 885\ \text{lb/in}^2}\right] = 0.044\ 2\ \text{in}^2$$

Now the required diameter is

$$D = \sqrt{4A/\pi} = \sqrt{4(0.0442\ \text{in}^2/\pi} = 0.237\ \text{in}$$

Final Design Decisions   The computed value for the minimum required diameter for the pin, 0.237 in, is quite small.
and Comments   Other considerations such as bearing stress and wear at the surfaces that contact the tabs of
the fixture and the bar indicate that a larger diameter would be preferred. Let's specify $D =$
0.50 in for the pin at this location. The pin will be of uniform diameter within the area of
the bar and the tabs. It should extend beyond the tabs, and it could be secured with cotter
pins or retaining rings.

This completes the design of the pin. But the next design example deals with the horizontal bar for this same system. There are pins at the conveyor hangers to support the bar.
They would also have to be designed. However, note that each of these pins carries only
half the load of the pin in the fixture connection. These pins would experience less relative
motion as well, so wear should not be so severe. Thus, let's use pins with $D = 3/8$ in $=$
0.375 in at the ends of the horizontal bar.

Design Example 5–3

A part of a conveyor system for a production operation is shown in Figure 5–12. The complete system will include several hundred hanger assemblies like this one. Design the horizontal bar that extends between two adjacent conveyor hangers and that supports a fixture
at its midpoint. The empty fixture weighs 85 lb. A cast iron engine block weighing 225 lb
is hung on the fixture to carry it from one process to another, where it is then removed. It
is expected that the bar will experience several thousand cycles of loading and unloading
of the engine blocks. Design Example 5–2 considered this same system with the objective
of specifying the diameter of the pins. The pin at the middle of the horizontal bar where the
fixture is hung has been specified to have a diameter of 0.50 in. Those at each end where
the horizontal bar is connected to the conveyor hangers are each 0.375 in.

Solution Objective Design the horizontal bar for the conveyor system.

Given The general arrangement is shown in Figure 5–12. The bar is simply supported at points 24 in apart. A vertical load that is alternately 85 lb and 310 lb (85 + 225) is applied at the middle of the bar through the pin connecting the fixture to the bar. The load will cycle between these two values many thousands of times in the expected life of the bar. The pin at the middle of the bar has a diameter of 0.50 in, while the pins at each end are 0.375 in.

Basic Design Decisions It is proposed to make the bar from steel in the form of a rectangular bar with the long dimension of its cross section vertical. Cylindrical holes will be machined on the neutral axis of the bar at the support points and at its center to receive cylindrical pins that will attach the bar to the conveyor carriers and to the fixture. Figure 5–13 shows the basic design for the bar.

The thickness of the bar, t, should be fairly large to provide a good bearing surface for the pins and to ensure lateral stability of the bar when subjected to the bending stress. A relatively thin bar would tend to buckle along its top surface where the stress is compressive. As a design decision, we will use a thickness of $t = 0.50$ in. The design analysis will determine the required height of the bar, h, assuming that the primary mode of failure is stress due to bending. Other possible modes of failure are discussed in the comments at the end of this example.

An inexpensive steel is desirable because several hundred bars will be made. We specify AISI 1020 hot-rolled steel having a yield strength of $s_y = 30$ Ksi and an ultimate strength of $s_u = 55$ Ksi (Appendix 3).

Analysis Case G from Section 5–9 applies for completing the design analysis because fluctuating normal stress due to bending is experienced by the bar. Equation (5–19) will be used:

$$\frac{1}{N} = \frac{\sigma_m}{s_y} + \frac{K_t \sigma_a}{s_n'}$$

In general, the bending stress in the bar will be computed from the flexure formula:

$$\sigma = M/S$$

where M = bending moment

S = section modulus of the cross section of the bar

Our approach will be to first determine the values for both the mean and the alternating bending moments experienced by the bar at its middle. Then the yield and endurance strength values for the steel will be found. Furthermore, as shown in Figure A15–3, the stress concentration factor for this case can be taken as $K_t = 1.0$ if the ratio of the hole diameter, d, to the height of the bar, h, is less than 0.50. We will make that assumption and check it later. Finally, Equation (5–19) includes the design factor, N. Based on the application conditions, let's use $N = 4$ as advised in item 5 in Section 5–8 because the actual use pattern for this conveyor system in a factory environment is somewhat uncertain.

Bending Moments

Figure 5–13 shows the shearing force and bending moment diagrams for the bar when carrying just the fixture and then both the fixture and the engine block. The maximum bending moment occurs at the middle of the bar where the load is applied. The values are $M_{max} = 1\,860$ lb·in with the engine block on the fixture and $M_{min} = 510$ lb·in for the fixture alone. Now the values for the mean and alternating bending moments are calculated using modified forms of Equations (5–1) and (5–2):

$$M_m = (M_{max} + M_{min})/2 = (1\,860 + 510)/2 = 1\,185 \text{ lb·in}$$
$$M_a = (M_{max} - M_{min})/2 = (1\,860 - 510)/2 = 675 \text{ lb·in}$$

The stresses will be found from $\sigma_m = M_m/S$ and $\sigma_a = M_a/S$.

FIGURE 5–13   Basic design of the horizontal bar and the load, shearing force, and bending moment diagrams

Material Strength Values

The material strength properties required are the yield strength s_y and the estimated actual endurance strength s_n'. We know that $s_y = 30$ Ksi and that the ultimate strength $s_u = 55$ Ksi. We now find s_n' using the method outlined in Section 5–5.

Size factor, C_s: For the rectangular cross section, the properties are most affected by the thickness because the bar is rolled flat. For $t = 0.50$ in, $C_s = 0.97$ is reasonable, taken from Figure 5–8.

Material factor, C_m: Use $C_m = 1.0$ for the wrought, hot-rolled steel.

Stress-type factor, C_{st}: Use $C_{st} = 1.0$ for repeated bending stress.

Reliability factor, C_R: A high reliability is desired. Let's use $C_R = 0.75$ to achieve a reliability of 0.999 as indicated in Table 5–2.

The value of $s_n = 20$ Ksi is found from Figure 5–9 for hot-rolled steel having an ultimate strength of 55 Ksi.

Now, applying Equation (5–4) from Section 5–5, we have

$$s_n' = (C_s)(C_m)(C_{st})(C_R)s_n = (0.97)(1.0)(1.0)(0.75)(20\text{ Ksi}) = 14.6\text{ Ksi}$$

Solution for the Required Section Modulus

At this point, we have specified all factors in Equation (5–19) except the section modulus of the cross section of the bar that is involved in each expression for stress as shown above. We will now solve the equation for the required value of S.

Recall that we showed earlier that $\sigma_m = M_m/S$ and $\sigma_a = M_a/S$. Then

$$\frac{1}{N} = \frac{\sigma_m}{s_y} + \frac{K_t\sigma_a}{s_n'} = \frac{M_m}{Ss_y} + \frac{K_t M_a}{Ss_n'} = \frac{1}{S}\left[\frac{M_m}{s_y} + \frac{K_t M_a}{s_n'}\right]$$

$$S = N\left[\frac{M_m}{s_y} + \frac{K_t M_a}{s_n'}\right] = 4\left[\frac{1\,185\text{ lb·in}}{30\,000\text{ lb/in}^2} + \frac{1.0\,(675\text{ lb·in})}{14\,600\text{ lb/in}^2}\right]$$

$$S = 0.343\text{ in}^3$$

Results   The required section modulus has been found to be $S = 0.343$ in^3. We observed earlier that $S = th^2/6$ for a solid rectangular cross section, and we decided to use this form to find an initial estimate for the required height of the section, h. We have specified $t = 0.50$ in. Then the estimated minimum acceptable value for the height h is

$$h = \sqrt{6S/t} = \sqrt{6(0.343\text{ in}^3)/(0.50\text{ in})} = 2.03\text{ in}$$

The table of preferred basic sizes in the decimal-inch system (Table A2–1) recommends $h = 2.20$ in. We should first check the earlier assumption that the ratio $d/h < 0.50$ at the middle of the bar. The actual ratio is

$$d/h = (0.50\text{ in})/(2.20\text{ in}) = 0.23 \quad (\text{okay})$$

This indicates that our earlier assumption that $K_t = 1.0$ is correct.

We will now compute the actual value for the section modulus of the cross section with the hole in it.

$$S = \frac{t(h^3 - d^3)}{6h} = \frac{(0.50\text{ in})[(2.20\text{ in})^3 - (0.50\text{ in})^3]}{6(2.20\text{ in})} = 0.399\text{ in}^3$$

This value is larger than the minimum required value of 0.343 in^3. Therefore, the size of the cross section is satisfactory with regard to stress due to bending.

Final Design Decisions and Comments   In summary, the following are the design decisions for the horizontal bar of the conveyor hanger shown in Figure 5–13.

1. *Material:* AISI 1020 hot-rolled steel.

2. *Size:* Rectangular cross section. Thickness $t = 0.50$ in; height $h = 2.20$ in.

3. *Overall design:* Figure 5–13 shows the basic features of the bar.

4. *Other considerations:* Remaining to be specified are the tolerances on the dimensions for the bar and the finishing of its surfaces. The potential for corrosion

should be considered and may call for paint, plating, or some other corrosion protection. The size of the cross section can likely be used with the as-received tolerances on thickness and height, but this is somewhat dependent on the design of the fixture that holds the engine block and the conveyor hangers. So the final tolerances will be left open pending later design decisions. The holes in the bar for the pins should be designed to produce a close sliding fit with the pins, and the details of specifying the tolerances on the hole diameters for such a fit are discussed in Chapter 13.

5. *Other possible modes of failure:* The analysis used in this problem assumed that failure would occur due to the bending stresses in the rectangular bar. The dimensions were specified to preclude this from happening. Other possible modes are discussed here:

   a. *Deflection of the bar as an indication of stiffness:* The type of conveyor system described in this problem should not be expected to have extreme rigidity because moderate deflection of members should not impair its operation. However, if the horizontal bar deflects so much that it appears to be rather flexible, it would be deemed unsuitable. This is a subjective judgment. We can use Case (a) in Table A14–2 to compute the deflection.

   $$y = FL^3/48EI$$

   In this design,

   $$F = 310 \text{ lb} = \text{maximum load on the bar}$$
   $$L = 24.0 \text{ in} = \text{distance between supports}$$
   $$E = 30 \times 10^6 \text{ psi} = \text{modulus of elasticity of steel}$$
   $$I = th^3/12 = \text{moment of inertia of the cross section}$$
   $$I = (0.50 \text{ in})(2.20 \text{ in})^3/12 = 0.444 \text{ in}^4$$

   Then

   $$y = \frac{(310 \text{ lb})(24.0 \text{ in})^3}{48(30 \times 10^6 \text{ lb/in}^2)(0.444 \text{ in}^4)} = 0.0067 \text{ in}$$

   This value seems satisfactory. In Section 5–10, some guidelines were given for deflection of machine elements. One stated that bending deflections for general machine parts should be limited to the range of 0.000 5 to 0.003 in/in of beam length. For the bar in this design, the ratio of y/L can be compared to this range:

   $$y/L = (0.006\ 7 \text{ in})/(24.0 \text{ in}) = 0.000\ 3 \text{ in/in of beam length}$$

   Therefore, this deflection is well within the recommended range.

   b. *Buckling of the bar:* When a beam with a tall, thin, rectangular cross section is subjected to bending, it would be possible for the shape to distort due to buckling before the bending stresses would cause failure of the material. This is called *elastic instability,* and a complete discussion is beyond the scope of this book. However, Reference 15 shows a method of computing the critical buckling load for this kind of loading. The pertinent geometrical feature is the ratio of the thickness t of the bar to its height h. It can be shown that the bar as designed will not buckle.

   c. *Bearing stress on the inside surfaces of the holes in the beam:* Pins transfer loads between the bar and the mating elements in the conveyor system. It is possible that the bearing stress at the pin–hole interface could be large, leading to excessive deformation or wear. The analysis is also beyond the scope of this book. The parameters are the size of the pin and its mating hole (the closeness of the fit), the magnitude of applied loads, the hardness of the mating surfaces, the quality of lubrication, and the presence of contaminants.

Design Example 5–4　A bracket is made by welding a rectangular bar to a circular rod as shown in Figure 5–14. Design the bar and the rod to carry a static load of 250 lb.

Solution　Objective　The design process will be divided into two parts:
1. Design the rectangular bar for the bracket.
2. Design the circular rod for the bracket.

Rectangular Bar

Given　The bracket design is shown in Figure 5–14. The rectangular bar carries a load of 250 lb vertically downward at its end. Support is provided by the weld at its left end where the loads are transferred to the circular rod. The bar acts as a cantilever beam, 12 in long. The design task is to specify the material for the bar and the dimensions of its cross section.

Basic Design Decisions　We will use steel for both parts of the bracket because of its relatively high stiffness, the ease of welding, and the wide range of strengths available. Let's specify AISI 1340 annealed steel having $s_y = 63$ Ksi and $s_u = 102$ Ksi (Appendix 3). The steel is highly ductile, with a 26% elongation.

　　　　The objective of the design analysis that follows is to determine the size of the cross section of the rectangular bar. Assuming that the loading and processing conditions are well known, we will use a design factor of $N = 2$ because of the static load.

Analysis and Results　The free-body diagram of the cantilever bar is shown in Figure 5–15, along with the shearing force and bending moment diagrams. This should be a familiar case, leading to the judgment that the maximum tensile stress occurs at the top of the bar near to where it is supported by the circular rod. This point is labeled element A in Figure 5–15. The maximum bending moment there is $M = 3\,000$ lb·in. The stress at A is

$$\sigma_A = M/S$$

where S = section modulus of the cross section of the bar

FIGURE 5–14
Bracket design

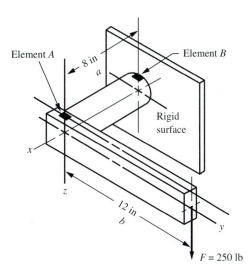

FIGURE 5–15 Free-body diagram of bar

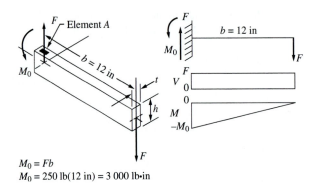

$M_0 = Fb$
$M_0 = 250 \text{ lb}(12 \text{ in}) = 3\,000 \text{ lb·in}$

We will first compute the minimum allowable value for S and then determine the dimensions for the cross section.

Case C1 from Section 5–9 applies because of the static loading. We will first compute the design stress from

$$\sigma_d = s_y/N$$
$$\sigma_d = s_y/N = (63\,000 \text{ psi})/2 = 31\,500 \text{ psi}$$

Now we must ensure that the expected maximum stress $\sigma_A = M/S$ does not exceed the design stress. We can substitute $\sigma_A = \sigma_d$ and solve for S.

$$S = M/\sigma_d = (3\,000 \text{ lb·in})/(31\,500 \text{ lb/in}^2) = 0.095 \text{ in}^3$$

The relationship for S is

$$S = th^2/6$$

As a design decision, let's specify the approximate proportion for the cross-sectional dimensions to be $h = 3t$. Then

$$S = th^2/6 = t(3t)^2/6 = 9t^3/6 = 1.5t^3$$

The required minimum thickness is then

$$t = \sqrt[3]{S/1.5} = \sqrt[3]{(0.095 \text{ in}^3)/1.5} = 0.399 \text{ in}$$

The nominal height of the cross section should be, approximately,

$$h = 3t = 3(0.399 \text{ in}) = 1.20 \text{ in}$$

Final Design Decisions and Comments   In the fractional-inch system, standard sizes are selected to be $t = 3/8 \text{ in} = 0.375 \text{ in}$ and $h = 1\frac{1}{4} \text{ in} = 1.25 \text{ in}$ (see Table A2–1). Note that we chose a slightly smaller value for t but a slightly larger value for h. We must check to see that the resulting value for S is satisfactory.

$$S = th^2/6 = (0.375 \text{ in})(1.25 \text{ in})^2/6 = 0.0977 \text{ in}^3$$

This is larger than the required value of 0.095 in^3, so the design is satisfactory.

Circular Rod

Given   The bracket design is shown in Figure 5–14. The design task is to specify the material for the rod and the diameter of its cross section.

Basic Design Decisions   Let's specify AISI 1340 annealed steel, the same as that used for the rectangular bar. Its properties are steel having $s_y = 63$ Ksi and $s_u = 102$ Ksi.

FIGURE 5–16 Free-body diagram of rod

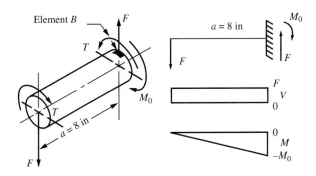

$M_0 = Fa = 250\ \text{lb}(8\ \text{in}) = 2\ 000\ \text{lb·in}$

$T_0 = Fb = 250\ \text{lb}(12\ \text{in}) = 3\ 000\ \text{lb·in}$

Analysis and Results

Figure 5–16 is the free-body diagram for the rod. The rod is loaded at its left end by the reactions at the end of the rectangular bar, namely, a downward force of 250 lb and a moment of 3 000 lb·in. The figure shows that the moment acts as a torque on the circular rod, and the 250-lb force causes bending with a maximum bending moment of 2 000 lb·in at the right end. Reactions are provided by the weld at its right end where the loads are transferred to the support. The rod then is subjected to a combined stress due to torsion and bending. Element *B* on the top of the rod is subjected to the maximum combined stress.

The manner of loading on the circular rod is identical to that analyzed earlier in Section 4–6 in Chapter 4. It was shown that when only bending and torsional shear occur, a procedure called the *equivalent torque method* can be used to complete the analysis. First we define the equivalent torque, T_e:

$$T_e = \sqrt{M^2 + T^2} = \sqrt{(2\ 000)^2 + (3\ 000)^2} = 3\ 606\ \text{lb·in}$$

Then the shear stress in the bar is

$$\tau = T_e/Z_p$$

where Z_p = polar section modulus

For a solid circular rod,

$$Z_p = \pi D^3/16$$

Our approach is to determine the design shear stress and T_e and then solve for Z_p. Case C1 using the maximum shear stress theory of failure can be applied. The design shear stress is

$$\tau_d = 0.50 s_y/N = (0.5)(63\ 000\ \text{psi})/2 = 15\ 750\ \text{psi}$$

We let $\tau = \tau_d$ and solve for Z_p:

$$Z_p = T_e/\tau_d = (3\ 606\ \text{lb·in})/(15\ 750\ \text{lb/in}^2) = 0.229\ \text{in}^3$$

Now that we know Z_p, we can compute the required diameter from

$$D = \sqrt[3]{16Z_p/\pi} = \sqrt[3]{16(0.229\ \text{in}^3)/\pi} = 1.053\ \text{in}$$

This is the minimum acceptable diameter for the rod.

Final Design Decisions and Comments

The circular rod is to be welded to the side of the rectangular bar, and we have specified the height of the bar to be $1\frac{1}{4}$ in. Let's specify the diameter of the circular rod to be machined to 1.10 in. This will allow welding all around its periphery.

REFERENCES

1. American Society of Mechanical Engineers. ANSI Standard B106.1M-1985. *Design of Transmission Shafting.* New York: American Society of Mechanical Engineers, 1985.

2. Boyer, Howard E. *Atlas of Fatigue Curves.* Metals Park, OH: American Society for Metals, 1986.

3. Boyer, H. E., ed. *Metals Handbook No. 10: Failure Analysis and Prevention.* 8th ed. Metals Park, OH: American Society for Metals, 1975.

4. Fuchs, H. O., and R. I. Stephens. *Metal Fatigue in Engineering.* New York: John Wiley & Sons, 1980.

5. Harris, C. M., and C. E. Crede. *Shock and Vibration Handbook.* New York: McGraw-Hill, 1976.

6. Krulick, T. G. "A Road Map for Stress Analysis." *Machine Design Magazine* (August 1981).

7. Juvinall, R. C. *Engineering Considerations of Stress, Strain, and Strength.* New York: McGraw-Hill, 1967.

8. Juvinall, R. C., and K. M. Marshek. *Fundamentals of Machine Component Design.* 2d ed. New York: John Wiley & Sons, 1991.

9. Madayag, A. F. *Metal Fatigue: Theory and Design.* New York: John Wiley & Sons, 1969.

10. Rinehart, J. S., and J. Pearson. *Behavior of Metals under Impulsive Loads.* Metals Park, OH: American Society for Metals, 1954.

11. Shigley, J. E., and C. R. Mischke. *Mechanical Engineering Design.* 5th ed. New York: McGraw-Hill, 1989.

12. Sines, G., and J. L. Waisman, eds. *Metal Fatigue.* New York: McGraw-Hill, 1959.

13. Spotts, M. F., and T. E. Shoup. *Design of Machine Elements.* 7th ed. Upper Saddle River, NJ: Prentice-Hall, 1998.

14. Vigness, Irwin, and W. P. Welch. "Shock and Impact Considerations in Design." *ASME Handbook: Metals Engineering—Design.* Oscar J. Horger, ed. New York: McGraw-Hill, 1965.

15. Young, Warren C. *Roark's Formulas for Stress and Strain.* 6th ed. New York: McGraw-Hill, 1989.

PROBLEMS

Stress Ratio

For each of Problems 1–9, draw a sketch of the variation of stress versus time, and compute the maximum stress, minimum stress, mean stress, alternating stress, and stress ratio, R. For Problems 6–9, analyze the beam at the place where the maximum stress would occur at any time in the cycle.

1. A link in a mechanism is made from a round bar having a diameter of 10.0 mm. It is subjected to a tensile force that varies from 3 500 to 500 N in a cyclical fashion as the mechanism runs.

2. A strut in a space frame has a rectangular cross section of 10.0 mm by 30.0 mm. It sees a load that varies from a tensile force of 20.0 kN to a compressive force of 8.0 kN.

3. A link in a packaging machine mechanism has a square cross section of 0.40 in on a side. It is subjected to a load that varies from a tensile force of 860 lb to a compressive force of 120 lb.

4. A circular rod with a diameter of 3/8 in supports part of a storage shelf in a warehouse. As products are loaded and unloaded, the rod is subjected to a tensile load that varies from 1 800 to 150 lb.

5. A part of a latch for a car door is made from a circular rod having a diameter of 3.0 mm. With each actuation, it sees a tensile force that varies from 780 to 360 lb.

6. A part of the structure for a factory automation system is a beam that spans 30.0 in as shown in Figure 5–17. Loads are applied at two points, each 8.0 in from a support. The left load F_1 = 1 800 lb remains constantly applied, while the right load F_2 = 1 800 lb is applied and removed frequently as the machine cycles.

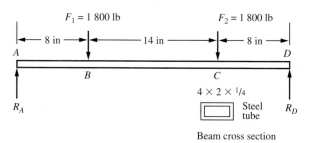

FIGURE 5–17 (Problems 6 and 23)

7. A cantilevered boom is part of an assembly machine and is made from an American Standard steel beam, S4 × 7.7. A tool with a weight of 500 lb moves continuously from the end of the 60-in beam to a point 10 in from the support.

8. A part of a bracket in the seat assembly of a bus is shown in Figure 5–18. The load varies from 1 450 to 140 N as passengers enter and exit the bus.

9. A flat steel strip is used as a spring to maintain a force against part of a cabinet latch in a commercial printer as shown in Figure 5–19. When the cabinet door is open, the

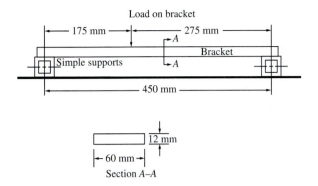

FIGURE 5–18   Seat bracket (Problems 8, 19, and 20)

FIGURE 5–19   Cabinet latch spring (Problems 9 and 22)

spring is deflected by the latch pin by an amount $y_1 = 0.25$ mm. The pin causes the deflection to increase to 0.40 mm when the door is closed.

Endurance Strength

For Problems 10–14, use the method outlined in Section 5–5 to determine the expected actual endurance strength for the material.

10. Compute the estimated actual endurance strength for a 0.75-in-diameter rod made from AISI 1040 cold-drawn steel. It is to be used in the as-rolled condition and subjected to repeated bending stress. A reliability of 99% is desired.

11. Compute the estimated actual endurance strength for AISI 5160 OQT 1300 steel rod with a diameter of 20.0 mm. It is to be machined and subjected to repeated bending stress. A reliability of 99% is desired.

12. Compute the estimated actual endurance strength for AISI 4130 WQT 1300 steel bar with a rectangular cross section of 20.0 mm by 60 mm. It is to be machined and subjected to repeated bending stress. A reliability of 99% is desired.

13. Compute the estimated actual endurance strength for AISI 301 stainless steel rod, 1/2 hard, with a diameter of 0.60 in. It is to be machined and subjected to repeated axial tensile stress. A reliability of 99.9% is desired.

14. Compute the estimated actual endurance strength for ASTM A242 steel bar with a rectangular cross section of 0.375 in by 3.50 in. It is to be machined and subjected to repeated bending stress. A reliability of 99% is desired.

Design and Analysis

15. A link in a mechanism is to be subjected to a tensile force that varies from 3 500 to 500 N in a cyclical fashion as the mechanism runs. It has been decided to use AISI 1040 cold-drawn steel. Complete the design of the link, specifying a suitable cross-sectional shape and dimensions.

16. A circular rod is to support part of a storage shelf in a warehouse. As products are loaded and unloaded, the rod is subjected to a tensile load that varies from 1 800 to 150 lb. Specify a suitable shape, material, and dimensions for the rod.

17. A strut in a space frame sees a load that varies from a tensile force of 20.0 kN to a compressive force of 8.0 kN. Specify a suitable shape, material, and dimensions for the strut.

18. A part of a latch for a car door is to be made from a circular rod. With each actuation, it sees a tensile force that varies from 780 to 360 lb. Small size is important. Complete the design, and specify a suitable shape, material, and dimensions for the rod.

19. A part of a bracket in the seat assembly of a bus is shown in Figure 5–18. The load varies from 1 450 to 140 N as passengers enter and exit the bus. The bracket is made from AISI 1020 hot-rolled steel. Determine the resulting design factor.

20. For the bus seat bracket described in Problem 19 and shown in Figure 5–18, propose an alternate design for the bracket, different from that shown in the figure, to achieve a lighter design with a design factor of approximately 4.0.

21. A cantilevered boom is part of an assembly machine. A tool with a weight of 500 lb moves continuously from the end of the 60-in beam to a point 10 in from the support. Specify a suitable design for the boom, giving the material, the cross-sectional shape, and the dimensions.

22. A flat steel strip is used as a spring to maintain a force against part of a cabinet latch in a commercial printer as shown in Figure 5–19. When the cabinet door is open, the spring is deflected by the latch pin by an amount $y_1 = 0.25$ mm. The pin causes the deflection to increase to 0.40 mm when the door is closed. Specify a suitable material for the spring if it is made to the dimensions shown in the figure.

23. A part of the structure for a factory automation system is a beam that spans 30.0 in as shown in Figure 5–17. Loads are applied at two points, each 8.0 in from a support. The left load $F_1 = 1\,800$ lb remains constantly applied, while the right load $F_2 = 1\,800$ lb is applied and removed frequently as the machine cycles. If the rectangular tube is made from ASTM A500 Grade B steel, is the proposed design satisfactory? Improve the design to achieve a lighter beam.

24. Figure 5–20 shows a hydraulic cylinder that pushes a heavy tool during the outward stroke, placing a compressive load of 400 lb in the piston rod. During the return stroke, the rod pulls on the tool with a force of 1 500 lb. Compute the resulting design factor for the 0.60-in-diameter rod when subjected to this pattern of forces for many cycles. The material is AISI 4130 WQT 1300 steel. If the resulting design factor is much different from 4.0, determine the size of rod that would produce $N = 4.0$.

FIGURE 5–20   (Problem 24)

25. The cast iron cylinder shown in Figure 5–21 carries only an axial compressive load of 75 000 lb. (The torque $T = 0$.) Compute the design factor if it is made from gray cast iron, Grade 40, having a tensile ultimate strength of 40 Ksi and a compressive ultimate strength of 140 Ksi.

26. Repeat Problem 25, except using a tensile load with a magnitude of 12 000 lb.

27. Repeat Problem 25, except using a load that is a combination of an axial compressive load of 75 000 lb and a torsion of 20 000 lb·in.

28. The shaft shown in Figure 5–22 is supported by bearings at each end, which have bores of 20.0 mm. Design the shaft to carry the given load if it is steady and the shaft is stationary. Make the dimension a as large as possible

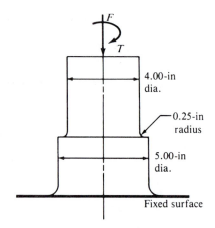

FIGURE 5–21   (Problems 25, 26, and 27)

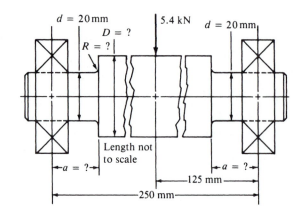

FIGURE 5–22   (Problems 28, 29, and 30)

while keeping the stress safe. Determine the required diameter in the middle portion. The maximum fillet permissible is 2.0 mm. Use AISI 1137 cold-drawn steel. Use a design factor of 3.

29. Repeat Problem 28, except using a rotating shaft.

30. Repeat Problem 28, except using a shaft that is rotating and transmitting a torque of 150 N·m from the left bearing to the middle of the shaft. Also, there is a profile keyseat at the middle under the load.

31. Figure 5–23 shows a proposed design for a seat support. The vertical member is to be a standard pipe (see Table A16–6). Specify a suitable pipe to resist static loads simultaneously in the vertical and horizontal directions, as shown. The tube has properties similar to those of AISI 1020 hot-rolled steel. Use a design factor of 3.

32. A torsion bar is to have a solid circular cross section. It is to carry a fluctuating torque from 30 to 65 N·m. Use AISI 4140 OQT 1000 for the bar, and determine the required diameter for a design factor of 2. Attachments produce a stress concentration of 2.5 near the ends of the bar.

FIGURE 5–23 (Problem 31)

33. Determine the required size for a square bar to be made from AISI 1213 cold-drawn steel. It carries a constant axial tensile load of 1 500 lb and a bending load that varies from zero to a maximum of 800 lb at the center of the 48-in length of the bar. Use a design factor of 3.

34. Repeat Problem 33, but add a constant torsional moment of 1 200 lb·in to the other loads.

In some of the following problems, you are asked to compute the design factor resulting from the design proposed for the given loading. Unless stated otherwise, assume that the element being analyzed has a machined surface. If the design factor is significantly different from $N = 3$, redesign the component to achieve approximately $N = 3$. (See the figures in Chapter 3.)

35. A tensile member in a machine structure is subjected to a steady load of 4.50 kN. It has a length of 750 mm and is made from a steel tube, AISI 1040 hot-rolled, having an outside diameter of 18 mm and an inside diameter of 12 mm. Compute the resulting design factor.

36. A steady tensile load of 5.00 kN is applied to a square bar, 12 mm on a side and having a length of 1.65 m. Compute the stress in the bar and the resulting design factor if it is made from (a) AISI 1020 hot-rolled steel; (b) AISI 8650 OQT 1000 steel; (c) ductile iron A536-84 (60-40-18); (d) aluminum alloy 6061-T6; (e) titanium alloy Ti-6Al-4V, annealed; (f) rigid PVC plastic; and (g) phenolic plastic.

37. An aluminum rod, made from alloy 6061-T6, is made in the form of a hollow square tube, 2.25 in outside with a wall thickness of 0.125 in. Its length is 16.0 in. It carries an axial compressive force of 12 600 lb. Compute the resulting design factor. Assume that the tube does not buckle.

38. Compute the design factor in the middle portion only of the rod AC in Figure 3–23 if the steady vertical force on the boom is 2 500 lb. The rod is rectangular, 1.50 in by 3.50 in, and is made from AISI 1144 cold-drawn steel.

39. Compute the forces in the two angled rods in Figure 3–24 for a steady applied force, $F = 1 500$ lb, if the angle θ is 45°. Then design the middle portion of each rod to be cir-

cular and made from AISI 1040 hot-rolled steel. Specify a suitable diameter.

40. Repeat Problem 39 if the angle θ is 15°.

41. Figure 3–22 shows a portion of a circular bar that is subjected to a repeated and reversed force of 7 500 N. If the bar is made from AISI 4140 OQT 1000, compute the resulting design factor.

42. Compute the torsional shear stress in a circular shaft having a diameter of 50 mm when subjected to a torque of 800 N·m. If the torque is completely reversed and repeated, compute the resulting design factor. The material is AISI 1040 WQT 1000.

43. If the torque in Problem 42 fluctuates from zero to the maximum of 800 N·m, compute the resulting design factor.

44. Compute the torsional shear stress in a circular shaft 0.40 inch in diameter that is due to a steady torque of 88.0 lb·in. Specify a suitable aluminum alloy for the rod.

45. Compute the required diameter for a solid circular shaft if it is transmitting a maximum of 110 hp at a speed of 560 rpm. The torque varies from zero to the maximum. There are no other significant loads on the shaft. Use AISI 4130 WQT 700.

46. Specify a suitable material for a hollow shaft with an outside diameter of 40 mm and an inside diameter of 30 mm when transmitting 28 kilowatts (kW) of steady power at a speed of 45 radians per second (rad/s).

47. Repeat Problem 46 if the power fluctuates from 15 to 28 kW.

48. Figure 5–24 shows part of a support bar for a heavy machine, suspended on springs to soften applied loads. The tensile load on the bar varies from 12 500 lb to a minimum of 7 500 lb. Rapid cycling for many million cycles is expected. The bar is made from AISI 6150 OQT 1300 steel. Compute the design factor for the bar in the vicinity of the hole.

FIGURE 5–24 (Problem 48)

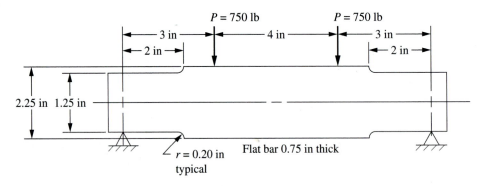

FIGURE 5–25 Screwdriver for Problem 57

FIGURE 5–26 Beam for Problem 58

49. Figure 3–53 shows a valve stem from an engine subjected to an axial tensile load applied by the valve spring. The force varies from 0.80 to 1.25 kN. Compute the resulting design factor at the fillet under the shoulder. The valve is made from AISI 8650 OQT 1300 steel.

50. A conveyor fixture shown in Figure 3–54 carries three heavy assemblies (1 200 lb each). The fixture is machined from AISI 1144 OQT 900 steel. Compute the resulting design factor in the fixture, considering stress concentrations at the fillets and assuming that the load acts axially. The load will vary from zero to the maximum as the conveyor is loaded and unloaded.

51. For the flat plate in tension in Figure 3–55, compute the minimum resulting design factor, assuming that the holes are sufficiently far apart that their effects do not interact. The plate is machined from stainless steel, UNS S17400 in condition H1150. The load is repeated and varies from 4 000 to 6 200 lb.

For Problems 52–56, select a suitable material for the member, considering stress concentrations, for the given loading to produce a design factor of $N = 3$.

52. Use Figure 3–56. The load is steady. The material is to be some grade of gray cast iron, ASTM A47-77.

53. Use Figure 3–57. The load varies from 20.0 to 30.3 kN. The material is to be titanium.

54. Use Figure 3–58. The torque varies from zero to 2 200 lb·in. The material is to be steel.

55. Use Figure 3–59. The bending moment is steady. The material is to be ductile iron, ASTM A48-76.

56. Use Figure 3–60. The bending moment is completely reversed. The material is to be stainless steel.

57. Figure 5–25 shows part of an automatic screwdriver designed to drive several million screws. The maximum torque required to drive a screw is 100 lb·in. Compute the design factor for the proposed design if the part is made from AISI 8740 CD.

58. The beam in Figure 5–26 carries two steady loads, $P = 750$ lb. Evaluate the design factor that would result if the beam were made from class 40 gray cast iron.

59. A tension link is subjected to a repeated, one-direction load of 3 000 lb. Specify a suitable material if the link is to be steel and is to have a diameter of 0.50 in.

60. One member of an automatic transfer device in a factory must withstand a repeated tensile load of 800 lb and must not elongate more than 0.010 inch in its 25.0-in length. Specify a suitable steel material and the dimensions for the rod if it has a square cross section.

61. Figure 5–27 shows two designs for a beam to carry a repeated central load of 600 lb. Which design would have the highest design factor for a given material?

62. Refer to Figure 5–27. By reducing the 8.0-in dimension, redesign the beam in Part (b) of the figure so that it has a design factor equal to or higher than that for the design in Part (a).

63. Refer to Figure 5–27. Redesign the beam in Part (b) of the figure by first increasing the fillet radius to 0.40 in and then by reducing the 8.0-in dimension so that the new design has a design factor equal to or higher than that for the design in Part (a).

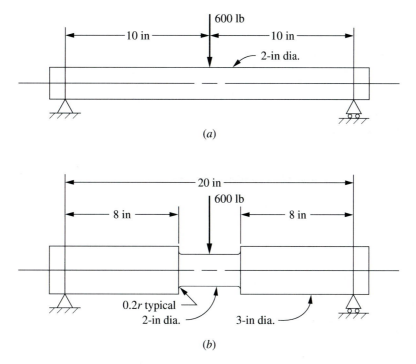

FIGURE 5–27   Beam for Problems 61, 62, and 63

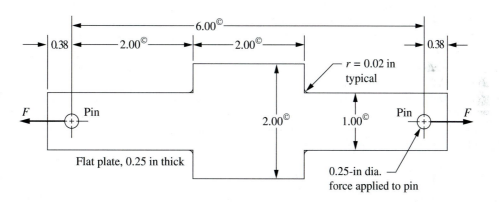

FIGURE 5–28   Beam for Problems 64 and 65

64. The part shown in Figure 5–28 is made from AISI 1040 HR steel. It is to be subjected to a repeated, one-direction force of 5 000 lb applied through two 0.25-in-diameter pins in the holes at each end. Compute the resulting design factor.

65. For the part described in Problem 64, make at least three improvements in the design that will significantly reduce the stress without increasing the weight. The dimensions marked © are critical and cannot be changed. After the redesign, specify a suitable material to achieve a design factor of at least 3.

66. The link shown in Figure 5–29 is subjected to a tensile force that varies from 3.0 to 24.8 kN. Evaluate the design factor if the link is made from AISI 1040 CD steel.

67. The beam shown in Figure 5–30 carries a repeated, reversed load of 400 N applied at section C. Compute the resulting design factor if the beam is made from AISI 1340 OQT 1300.

68. For the beam described in Problem 67, change the steel's tempering temperature to achieve a design factor of at least 3.0.

69. The cantilever shown in Figure 5–31 carries a downward load that varies from 300 to 700 lb. Compute the resulting design factor if the bar is made from AISI 1050 HR steel.

70. For the cantilever described in Problem 69, increase the size of the fillet radius to improve the design factor to at least 3.0 if possible.

0.625-in dia.

F   Pin   Hole   Pin   F

1.50 in   Flat plate thickness = 0.375 in

FIGURE 5–29   Link for Problem 66

FIGURE 5–30   Beam for Problems 67 and 68

FIGURE 5–31   Cantilever for Problems 69, 70, and 71

71. For the cantilever described in Problem 69, specify a suitable material to achieve a design factor of at least 3.0 without changing the geometry of the beam.

72. Figure 5–32 shows a rotating shaft carrying a steady downward load of 100 lb at *C*. Specify a suitable material.

73. The stepped rod shown in Figure 5–33 is subjected to a direct tensile force that varies from 8 500 to 16 000 lb. If the rod is made from AISI 1340 OQT 700 steel, compute the resulting design factor.

74. For the rod described in Problem 73, complete a redesign that will achieve a design factor of at least 3.0. The two diameters cannot be changed.

75. The beam shown in Figure 5–34 carries a repeated, reversed load of 800 lb alternately applied upward and downward. If the beam is made from AISI 1144 OQT 1100, specify the smallest acceptable fillet radius at *A* to ensure a design factor of 3.0.

FIGURE 5–32   Shaft for Problem 72

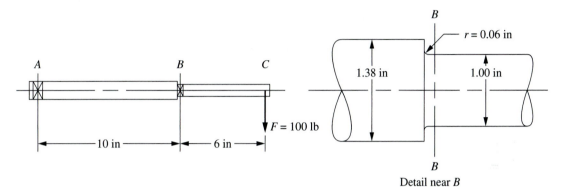

FIGURE 5–33   Rod for Problems 73 and 74

FIGURE 5–34   Beam for Problems 75 and 76

FIGURE 5–35   Link for Problem 77

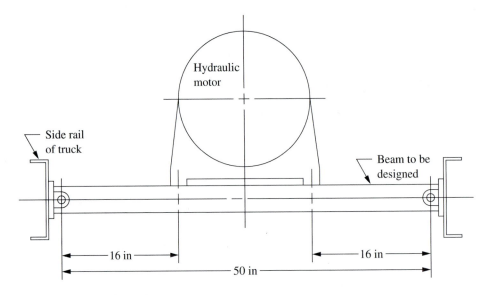

FIGURE 5–36   Beam for Problem 78

76. For the beam described in Problem 75, design the section at *B* to achieve a minimum design factor of 3.0. Specify the shape, dimensions, and fillet radius where the smaller part joins the 2.00-by-2.00-in section.

Design Problems

For each of the following problems, complete the requested design to achieve a minimum design factor of 3.0. Specify the shape, the dimensions, and the material for the part to be designed. Work toward an efficient design that will have a low weight.

77. The link shown in Figure 5–35 carries a load of 3 000 N that is applied and released many times. The link is machined from a square bar, 12.0 mm on a side, from AISI 1144 OQT 1100 steel. The ends must remain 12.0 mm square to facilitate the connection to mating parts. It is desired to reduce the size of the middle part of the link to reduce weight. Complete the design.

78. Complete the design of the beam shown in Figure 5–36 to carry a large hydraulic motor. The beam is attached to the

two side rails of the frame of a truck. Because of the vertical accelerations experienced by the truck, the load on the beam varies from 1 200 lb upward to 5 000 lb downward. One-half of the load is applied to the beam by each foot of the motor.

79. A tensile member in a truss frame is subjected to a load that varies from 0 to 6 500 lb as a traveling crane moves across the frame. Design the tensile member.

80. A hanger for a conveyor system extends outward from two supports as shown in Figure 5–37. The load at the right end varies from 600 to 3 800 lb. Design the hanger.

81. Figure 5–38 shows a yoke suspended beneath a crane beam by two rods. Design the yoke if the loads are applied and released many times.

82. For the system shown in Figure 5–38, design the two vertical rods if the loads are applied and released many times.

83. Design the connections between the rods and the yoke and the crane beam shown in Figure 5–38.

FIGURE 5–37   Hanger for the conveyor system for Problem 80

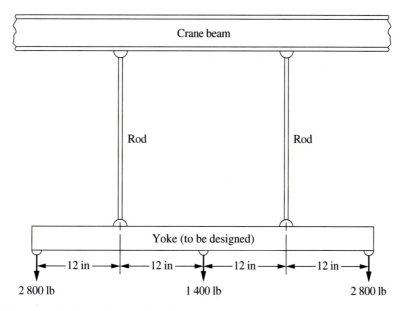

FIGURE 5–38   Yoke and rods for Problems 81, 82, and 83

FIGURE 6–2
Buckling of a thin,
rectangular column.
(a) General
appearance of the
buckled column.
(b) Radius of
gyration for *Y-Y* axis.
(c) Radius of
gyration for *X-X* axis

(a)

Column cross section

Critical buckling axis

For axis *Y-Y*: $r = 0.289t$

(b)

For axis *X-X*: $r = 0.289h$

(c)

prohibits any tendency for the fixed end of the column to rotate. A column end securely welded to a rigid base plate is also a good approximation of a fixed-end column.

The *free end* can be illustrated by the example of a flagpole. The top end of a flagpole is unrestrained and unguided, the worst case for column loading.

The manner of support of both ends of the column affects the *effective length* of the column, defined as

 Effective Length

$$L_e = KL \tag{6–2}$$

where L = actual length of the column between supports

     K = constant dependent on the end fixity, as illustrated in Figure 6–3

The first values given for K are theoretical values based on the shape of the deflected column. The second values take into account the expected fixity of the column ends in real, practical structures. It is particularly difficult to achieve a true fixed-end column because of the lack of complete rigidity of the support or the means of attachment. Therefore, the higher value of K is recommended.

FIGURE 6–3 Values of K for effective length, $L_e = KL$, for different end connections

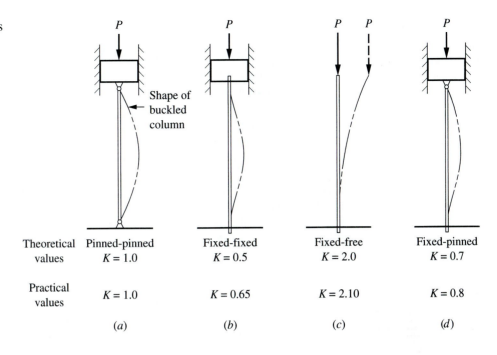

| | Theoretical values | Pinned-pinned $K = 1.0$ | Fixed-fixed $K = 0.5$ | Fixed-free $K = 2.0$ | Fixed-pinned $K = 0.7$ |

| | Practical values | $K = 1.0$ | $K = 0.65$ | $K = 2.10$ | $K = 0.8$ |
| | | (a) | (b) | (c) | (d) |

6–4 SLENDERNESS RATIO

⇨ **Slenderness Ratio**

The *slenderness ratio* is the ratio of the effective length of the column to its least radius of gyration. That is,

$$\text{Slenderness ratio} = L_e/r_{\min} = KL/r_{\min} \qquad \textbf{(6–3)}$$

We will use the slenderness ratio to aid in the selection of the method of performing the analysis of straight, centrally loaded columns.

6–5 TRANSITION SLENDERNESS RATIO

In the following sections, two methods for analyzing straight, centrally loaded columns are presented: (1) the Euler formula for long, slender columns and (2) the J. B. Johnson formula for short columns.

The choice of which method to use depends on the value of the actual slenderness ratio for the column being analyzed in relation to the *transition slenderness ratio,* or *column constant, C_c,* defined as

⇨ **Column Constant**

$$C_c = \sqrt{\frac{2\pi^2 E}{s_y}} \qquad \textbf{(6–4)}$$

where E = modulus of elasticity of the material of the column

s_y = yield strength of the material

The use of the column constant is illustrated in the following procedure for analyzing straight, centrally loaded columns.

Procedure for Analyzing Straight, Centrally Loaded Columns

1. For the given column, compute its actual slenderness ratio.

2. Compute the value of C_c.

3. Compare C_c with KL/r. Because C_c represents the value of the slenderness ratio that separates a long column from a short one, the result of the comparison indicates which type of analysis should be used.

4. If the actual KL/r is greater than C_c, the column is *long*. Use Euler's equation, as described in Section 6–6.

5. If KL/r is less than C_c, the column is *short*. Use the J. B. Johnson formula, described in Section 6–7.

Figure 6–4 is a logical flowchart for this procedure.

The value of the column constant, or transition slenderness ratio, is dependent on the material properties of modulus of elasticity and yield strength. For any given class of material, for example, steel, the modulus of elasticity is nearly constant. Thus, the value of C_c varies inversely as the square root of the yield strength. Figures 6–5 and 6–6 show the resulting values for steel and aluminum, respectively, for the range of yield strengths expected for each material. The figures show that the value of C_c decreases as the yield strength increases. The importance of this observation is discussed in the following section.

6–6
LONG COLUMN ANALYSIS: THE EULER FORMULA

Analysis of a long column employs the Euler formula:

Euler Formula for Long Columns

$$P_{cr} = \frac{\pi^2 EA}{(KL/r)^2} \qquad \text{(6–5)}$$

The equation gives the critical load, P_{cr}, at which the column would begin to buckle.

An alternative form of the Euler formula is often desirable. Note that, from Equation (6–5),

$$P_{cr} = \frac{\pi^2 EA}{(KL/r)^2} = \frac{\pi^2 EA}{(KL)^2/r^2} = \frac{\pi^2 EA r^2}{(KL)^2}$$

But, from the definition of the radius of gyration, r,

$$r = \sqrt{I/A}$$
$$r^2 = I/A$$

Then

Alternative Form of Euler Formula

$$P_{cr} = \frac{\pi^2 EA}{(KL)^2} \frac{I}{A} = \frac{\pi^2 EI}{(KL)^2} \qquad \text{(6–6)}$$

This form of the Euler equation aids in a design problem in which the objective is to specify a size and a shape of a column cross section to carry a certain load. The moment of inertia for the required cross section can be easily determined from Equation (6–6).

Notice that the buckling load is dependent only on the geometry (length and cross section) of the column and the stiffness of the material represented by the modulus of elasticity. The strength of the material is not involved at all. For these reasons, it is often of no benefit to specify a high-strength material in a long column application. A lower-strength material having the same stiffness, E, would perform as well.

FIGURE 6–4
Analysis of a straight, centrally loaded column

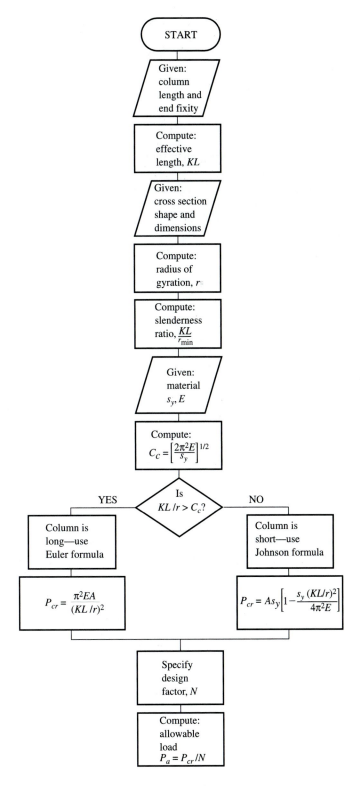

FIGURE 6–5
Transition slenderness ratio C_c vs. yield strength for steel

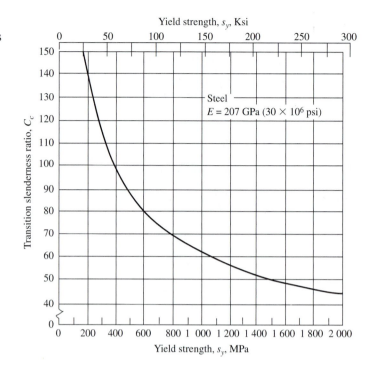

FIGURE 6–6
Transition slenderness ratio C_c vs. yield strength for aluminum

Design Factor and Allowable Load

Because failure is predicted to occur at a limiting load, rather than a stress, the concept of a design factor is applied differently than it is for most other load-carrying members. Rather than applying the design factor to the yield strength or the ultimate strength of the material, we apply it to the critical load, from Equation (6–5) or (6–6). For typical machine design applications, a design factor of 3 is used. For stationary columns with well-known loads and end fixity, a lower factor can be used, such as 2.0. A factor of 1.92 is used in some construction applications. Conversely, for very long columns, where there is some uncertainty about the loads or the end fixity, or where special dangers are presented, larger factors are advised. (See References 1 and 2.)

In summary, the objective of column analysis and design is to ensure that the load applied to a column is safe, well below the critical buckling load. The following definitions of terms must be understood:

$$P_{cr} = \text{critical buckling load}$$
$$P_a = \text{allowable load}$$
$$P = \text{actual applied load}$$
$$N = \text{design factor}$$

Then

 Allowable Load

$$P_a = P_{cr}/N$$

The actual applied load, P, must be less than P_a.

Example Problem 6–1 A column has a solid circular cross section, 1.25 inches in diameter; it has a length of 4.50 ft and is pinned at both ends. If it is made from AISI 1020 cold-drawn steel, what would be a safe column loading?

Solution   Objective   Specify a safe loading for the column.

Given   Solid circular cross section: diameter = d = 1.25 in; length = L = 4.50 ft.
Both ends of the column are pinned.
Material: AISI 1020 cold-drawn steel.

Analysis   Use the procedure in Figure 6–4.

Results   **Step 1.** For the pinned-end column, the end-fixity factor is K = 1.0. The effective length equals the actual length; KL = 4.50 ft = 54.0 in.

Step 2. From Appendix 1, for a solid round section,
$$r = D/4 = 1.25/4 = 0.312\ 5 \text{ in}$$

Step 3. Compute the slenderness ratio:
$$\frac{KL}{r} = \frac{1.0(54)}{0.312\ 5} = 173$$

Step 4. Compute the column constant from Equation (6–4). For AISI 1020 cold-drawn steel, the yield strength is 51 000 psi, and the modulus of elasticity is 30 × 10⁶ psi. Then

$$C_c = \sqrt{\frac{2\pi^2 E}{s_y}} = \sqrt{\frac{2\pi^2 (30 \times 10^6)}{51\ 000}} = 108$$

Step 5. Because KL/r is greater than C_c, the column is long, and Euler's formula should be used. The area is

$$A = \frac{\pi D^2}{4} = \frac{\pi(1.25)^2}{4} = 1.23 \text{ in}^2$$

Then the critical load is

$$P_{cr} = \frac{\pi^2 EA}{(KL/r)^2} = \frac{\pi^2(30 \times 10^6)(1.23)}{(173)^2} = 12\ 200 \text{ lb}$$

At this load, the column should just begin to buckle. A safe load would be a reduced value, found by applying the design factor to the critical load. Let's use $N = 3$ to compute the *allowable load*, $P_a = P_{cr}/N$:

$$P_a = (12\ 200)/3 = 4\ 067 \text{ lb}$$

Comment   The safe load on the column is 4 067 lb.

6–7
SHORT COLUMN
ANALYSIS: THE
J. B. JOHNSON
FORMULA

When the actual slenderness ratio for a column, KL/r, is less than the transition value, C_c, then the column is short, and the J. B. Johnson formula should be used. Use of the Euler formula in this range would predict a critical load greater than it really is.

The J. B. Johnson formula is written as follows:

⇨ **J. B. Johnson**
Formula for Short
Columns

$$P_{cr} = As_y \left[1 - \frac{s_y(KL/r)^2}{4\pi^2 E} \right] \tag{6–7}$$

Figure 6–7 shows a plot of the results of this equation as a function of the slenderness ratio, KL/r. Notice that it becomes tangent to the result of the Euler formula at the transition

FIGURE 6–7
Johnson formula curves

slenderness ratio, the limit of its application. Also, at very low values for the slenderness ratio, the second term of the equation approaches zero, and the critical load approaches the yield load. Curves for three different materials are included in the figure to illustrate the effect of E and s_y on the critical load and the transition slenderness ratio.

The critical load for a short column is affected by the strength of the material in addition to its stiffness, E. As shown in the preceding section, strength is not a factor for a long column when the Euler formula is used.

Example Problem 6–2

Determine the critical load on a steel column having a rectangular cross section, 12 mm by 18 mm, and a length of 280 mm. It is proposed to use AISI 1040 hot-rolled steel. The lower end of the column is inserted into a close-fitting socket and is welded securely. The upper end is pinned (see Figure 6–8).

Solution   Objective

Compute the critical load for the column.

Given

Solid rectangular cross section: $B = 12$ mm; $H = 18$ mm; $L = 280$ mm.
The bottom of column is fixed; the top is pinned (see Figure 6–8).
Material: AISI 1040 hot-rolled steel.

Analysis

Use the procedure in Figure 6–4.

Results

Step 1. Compute the slenderness ratio. The radius of gyration must be computed about the axis that gives the least value. This is the Y-Y axis, for which

$$r = \frac{B}{\sqrt{12}} = \frac{12 \text{ mm}}{\sqrt{12}} = 3.46 \text{ mm}$$

The column has a fixed-pinned end fixity for which $K = 0.8$. Then

$$KL/r = [(0.8)(280)]/3.46 = 64.7$$

FIGURE 6–8   Steel column

Section A–A

(*a*) Column cross section

(*b*) Sketch of column installation

Step 2. Compute the transition slenderness ratio. For the AISI 1040 hot-rolled steel, $E = 207$ GPa and $s_y = 290$ MPa. Then, from Equation (6–4),

$$C_c = \sqrt{\frac{2\pi^2(207 \times 10^9 \text{ Pa})}{290 \times 10^6 \text{ Pa}}} = 119$$

Step 3. Then $KL/r < C_c$; thus the column is short. Use the J. B. Johnson formula to compute the critical load:

$$P_{cr} = As_y\left[1 - \frac{s_y(KL/r)^2}{4\pi^2 E}\right] \tag{6–7}$$

$$P_{cr} = (216 \text{ mm}^2)(290 \text{ N/mm}^2)\left[1 - \frac{(290 \times 10^6 \text{ Pa})(64.7)^2}{4\pi^2(207 \times 10^9 \text{ Pa})}\right]$$

$$P_{cr} = 53.3 \times 10^3 \text{ N} = 53.3 \text{ kN}$$

Comments   This is the critical buckling load. We would have to apply a design factor to determine the allowable load. Specifying $N = 3$ results in $P_a = 17.8$ kN.

**6–8
COLUMN
ANALYSIS
SPREADSHEET**

Completing the process described in Figure 6–4 using a calculator, pencil, and paper is tedious. A spreadsheet automates the calculations after you have entered the pertinent data for the particular column to be analyzed. Figure 6–9 shows the output of a spreadsheet used to solve Example Problem 6–1. The layout of the spreadsheet could be done in many ways, and you are encouraged to develop your own style. The following comments describe the features of the given spreadsheet:

1. At the top of the sheet, instructions to the user are given for entering data and for units. This sheet is for U.S. Customary units only. A different sheet would be used if SI metric data were to be used. (See Figure 6–10 which gives the solution for Example Problem 6–2.)

2. On the left side of the sheet are listed the various data that must be provided by the user to run the calculations. On the right are listed the output values. Formulas for computing L_e, C_c, KL/r, and allowable load are written directly into the cell where the computed values show. The output data for the message "Column is: *long*" and the critical buckling load are produced by *functions* set up within *macros* written in Visual Basic and placed on a separate sheet of the spreadsheet. Figure 6–11 shows the two macros used. The first (*LorS*) carries out the decision process to test whether the column is long or short as indicated by comparison of its slenderness ratio with the column constant. The second (*Pcr*) computes the critical buckling load using either the Euler formula or the J. B. Johnson formula, depending on the result of the *LorS* macro. These functions are called by statements in the cells where "long" and the computed value of the critical buckling load (12 197 lb) are located.

3. Having such a spreadsheet can enable you to analyze several design options quickly. For example, the given problem statement indicated that the ends were pinned, resulting in an end-fixity value of $K = 1$. What would happen if both ends were fixed? Simply changing the value of that one cell to $K = 0.65$ would cause the entire sheet to be recalculated, and the revised value of critical buckling load would be available almost instantly. The result is that $P_{cr} = 28\ 868$ lb, an increase of 2.37 times the original value. With that kind of improvement, you, the designer, might be inclined to change the design to produce fixed ends.

COLUMN ANALYSIS PROGRAM		Data from:	Example Problem 6–1
Refer to Figure 6–4 for analysis logic.			
Enter data for variables in *italics in shaded boxes.*		Use consistent U.S. Customary units.	
Data to Be Entered:		**Computed Values:**	
Length and End Fixity:			
Column length, L= 54 in *End fixity, K=* 1	→	Eq. length, $L_e = KL =$ 54.0 in	
Material Properties:			
Yield strength, s_y= 51,000 psi *Modulus of elasticity, E= 3.00E+07 psi*	→	Column constant, $C_c =$ 107.8	
Cross Section Properties:			
[Note: Enter *r* or compute $r = \text{sqrt}(I/A)$.] [Always enter Area.] [Enter zero for *I* or *r* if not used.]			
Area, A= 1.23 in^2 *Moment of inertia, I=* 0 in^4 **Or** *Radius of gyration, r=* 0.3125 in	→	Slender ratio, $KL/r =$ 172.8	
		Column is: *long*	
		Critical buckling load= *12,197 lb*	
Design Factor:			
Design factor on load, N= 3	→	**Allowable load= *4,066 lb***	

FIGURE 6–9   Spreadsheet for column analysis with data from Example Problem 6–1

COLUMN ANALYSIS PROGRAM	Data from:	Example Problem 6–2
Refer to Figure 6–4 for analysis logic.		
Enter data for variables in *italics in shaded boxes*.	Use consistent SI metric units.	
Data to Be Entered:	**Computed Values:**	
Length and End Fixity:		
Column length, *L*= 280 mm End fixity, *K*= 0.8	→	Eq. length, L_e= KL= 224.0 mm
Material Properties:		
Yield strength, s_y= 290 MPa Modulus of elasticity, *E*= 207 GPa	→	Column constant, C_c= 118.7
Cross Section Properties:		
[Note: Enter *r* or compute *r*= sqrt(*I/A*).] [Always enter Area.] [Enter zero for *I* or *r* if not used.]		
Area, *A*= 216 mm^2 Moment of inertia, *I*= 0 mm^4 **Or** Radius of gyration, *r*= 3.5 mm	→	Slender ratio, KL/r= 64.7
	Column is: ***short***	
Design Factor:	**Critical buckling load**= ***53.32 kN***	
Design factor on load, *N*= 3	→	**Allowable load**= ***17.77 kN***

FIGURE 6–10 Spreadsheet for column analysis with data from Example Problem 6–2

FIGURE 6–11
Macros used in the column analysis spreadsheet

```
' LorS Macro
' Determines if column is long or short.
Function LorS(SR, CC)
    If SR > CC Then
        LorS = "long"
    Else
        LorS = "short"
    End If
End Function

' Critical Load Macro
' Uses Euler formula for long columns
' Uses Johnson formula for short columns
Function Pcr(LorS, SR, E, A, Sy)
Const Pi = 3.1415926
    If LorS = "long" Then
        Pcr = Pi ^ 2 * E * A / SR ^ 2
        ' Euler Equation; Eq. (6-4)
    Else
        Pcr = A * Sy(1 - (Sy * SR ^ 2 / (4 * Pi ^ 2 * E)))
        ' Johnson Equation; Eq. (6-7)
    End If
End Function
```

6–9 EFFICIENT SHAPES FOR COLUMN CROSS SECTIONS

An *efficient shape* is one that provides good performance with a small amount of material. In the case of columns, the shape of the cross section and its dimensions determine the value of the radius of gyration, r. From the definition of the slenderness ratio, KL/r, we can see that as r gets larger, the slenderness ratio gets smaller. In the critical load equations, a smaller slenderness ratio results in a larger critical load, the most desirable situation. Therefore, it is desirable to maximize the radius of gyration to design an efficient column cross section.

Unless end fixity varies with respect to the axes of the cross section, the column will tend to buckle with respect to the axis with the *least* radius of gyration. So a column with equal values for the radius of gyration in any direction is desirable.

Review again the definition of the radius of gyration:

$$r = \sqrt{I/A}$$

This equation indicates that for a given area of material, we should try to maximize the moment of inertia to maximize the radius of gyration. A shape with a high moment of inertia has its area distributed far away from its centroidal axis.

Shapes that have the desirable characteristics described include circular hollow pipes and tubes, square hollow tubing, and fabricated column sections made from structural shapes placed at the outer boundaries of the section. Solid circular sections and solid square sections are also good, although not as efficient as the hollow sections. Figure 6–12(a–d) illustrates some of these shapes. The built-up section in (e) gives a rigid, boxlike section approximating the hollow square tube in larger sizes. In the case of the section in Figure 6–12(f), the angle sections at the corners provide the greatest contribution to the moment of inertia. The lacing bars merely hold the angles in position. The H-column in (g) has an equal depth and width and relatively heavy flanges and web. The moment of inertia with respect to the *y-y* axis is still smaller than for the *x-x* axis, but they are more nearly equal

FIGURE 6–12
Column cross sections

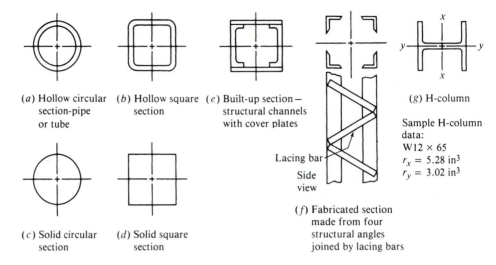

(*a*) Hollow circular   (*b*) Hollow square   (*e*) Built-up section —
section-pipe   section   structural channels
or tube   with cover plates

(*g*) H-column

Sample H-column
data:
W12 × 65
$r_x = 5.28$ in^3
$r_y = 3.02$ in^3

Lacing bar

Side
view

(*c*) Solid circular   (*d*) Solid square
section   section

(*f*) Fabricated section
made from four
structural angles
joined by lacing bars

than for most other I-sections designed to be used as beams with bending in only one direction. Thus, this shape would be more desirable for columns.

6–10
THE DESIGN
OF COLUMNS

In a design situation, the expected load on the column would be known, along with the length required by the application. The designer would then specify the following:

1. The manner of attaching the ends to the structure that affects the end fixity.
2. The general shape of the column cross section (for example, round, square, rectangular, and hollow tube).
3. The material for the column.
4. The design factor, considering the application.
5. The final dimensions for the column.

It may be desirable to propose and analyze several different designs to approach an optimum for the application. A computer program or spreadsheet facilitates the process.

It is assumed that items 1 through 4 are specified by the designer for any given trial. For some simple shapes, such as the solid round or square section, the final dimensions are computed from the appropriate formula: the Euler formula, Equation (6–5) or (6–6); or the J. B. Johnson formula, Equation (6–7). If an algebraic solution is not possible, iteration can be done.

In a design situation, the unknown cross-sectional dimensions make computing the radius of gyration and therefore the slenderness ratio, KL/r, impossible. Without the slenderness ratio, we cannot determine whether the column is long (Euler) or short (Johnson). Thus, the proper formula to use is not known.

We overcome this difficulty by making an assumption that the column is either long or short and proceeding with the corresponding formula. Then, after the dimensions are determined for the cross section, the actual value of KL/r will be computed and compared with C_c. This will show whether or not the correct formula has been used. If so, the computed answer is correct. If not, the alternate formula must be used and the computation repeated to determine new dimensions. Figure 6–13 shows a flowchart for the design logic described here.

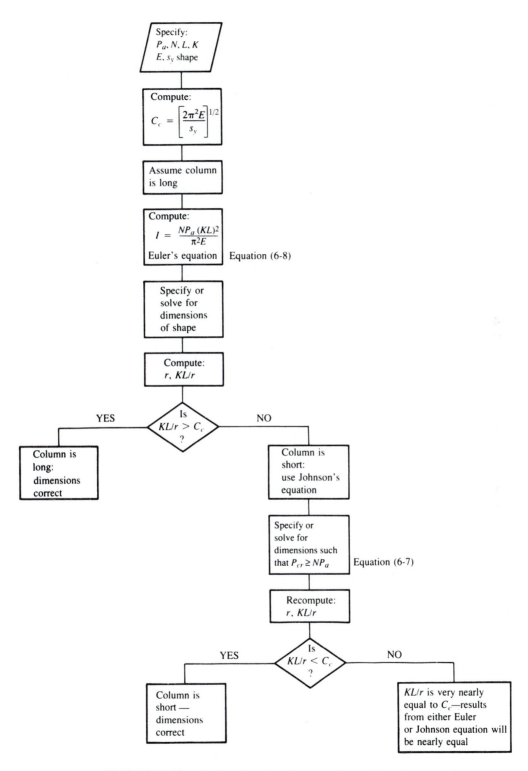

FIGURE 6–13 Design of a straight, centrally loaded column

Design: Assuming a Long Column

Euler's formula is used if the assumption is that the column is long. Equation (6–6) would be the most convenient form because it can be solved for the moment of inertia, I:

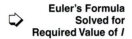

Euler's Formula Solved for Required Value of I

$$I = \frac{P_{cr}(KL)^2}{\pi^2 E} = \frac{NP_a(KL)^2}{\pi^2 E} \qquad (6\text{–}8)$$

where P_a = allowable load, usually set equal to the actual maximum expected load

Having the required value for I, we can determine the dimensions for the shape by additional computations or by scanning tables of data of the properties of commercially available sections.

The solid circular section is one for which it is possible to derive a final equation for the characteristic dimension, the diameter. The moment of inertia is

$$I = \frac{\pi D^4}{64}$$

Substituting this into Equation (6–8) gives

$$I = \frac{\pi D^4}{64} = \frac{NP_a(KL)^2}{\pi^2 E}$$

Solving for D yields

Required Diameter for a Long, Solid Circular Column

$$D = \left[\frac{64 N P_a (KL)^2}{\pi^3 E} \right]^{1/4} \qquad (6\text{–}9)$$

Design: Assuming a Short Column

The J. B. Johnson formula is used to analyze a short column. It is difficult to derive a convenient form for use in design. In the general case, then, trial and error is used.

For some special cases, including the solid circular section, it is possible to solve the Johnson formula for the characteristic dimension, the diameter:

$$P_{cr} = A s_y \left[1 - \frac{s_y (KL/r)^2}{4\pi^2 E} \right] \qquad (6\text{–}7)$$

But

$$A = \pi D^2/4$$
$$r = D/4 \quad \text{(from Appendix 1)}$$
$$P_{cr} = N P_a$$

Then

$$N P_a = \frac{\pi D^2}{4} s_y \left[1 - \frac{s_y (KL)^2}{4\pi^2 E (D/4)^2} \right]$$

$$\frac{4 N P_a}{\pi s_y} = D^2 \left[1 - \frac{s_y (KL)^2 (16)}{4\pi^2 E D^2} \right]$$

Solving for D gives

Required Diameter
for a Short, Solid
Circular Column

$$D = \left[\frac{4NP_a}{\pi s_y} + \frac{4s_y(KL)^2}{\pi^2 E} \right]^{1/2}$$

(6–10)

Example Problem 6–3   Specify a suitable diameter of a solid, round cross section for a machine link if it is to carry 9 800 lb of axial compressive load. The length will be 25 in, and the ends will be pinned. Use a design factor of 3. Use AISI 1020 hot-rolled steel.

Solution   Objective   Specify a suitable diameter for the column.

Given   Solid circular cross section: $L = 25$ in; use $N = 3$.
Both ends are pinned.
Material: AISI 1020 hot-rolled steel.

Analysis   Use the procedure in Figure 6–13. Assume first that the column is long.

Results   From Equation (6–9),

$$D = \left[\frac{64NP_a(KL)^2}{\pi^3 E} \right]^{1/4} = \left[\frac{64(3)(9\ 800)(25)^2}{\pi^3(30 \times 10^6)} \right]^{1/4}$$

$$D = 1.06\ \text{in}$$

The radius of gyration can now be found:

$$r = D/4 = 1.06/4 = 0.265\ \text{in}$$

The slenderness ratio is

$$KL/r = [(1.0)(25)]/0.265 = 94.3$$

For the AISI 1020 hot-rolled steel, $s_y = 30\ 000$ psi. The graph in Figure 6–5 shows C_c to be approximately 138. Thus, the actual KL/r is less than the transition value, and the column must be redesigned as a short column, using Equation (6–10) derived from the Johnson formula:

$$D = \left[\frac{4NP_a}{\pi s_y} + \frac{4s_y(KL)^2}{\pi^2 E} \right]^{1/2}$$

$$D = \left[\frac{4(3)(9\ 800)}{(\pi)(30\ 000)} + \frac{4(30\ 000)(25)^2}{\pi^2(30 \times 10^6)} \right]^{1/2} = 1.23\ \text{in}$$

(6–10)

Checking the slenderness ratio again, we have

$$KL/r = [(1.0)(25)]/(1.23/4) = 81.3$$

Comments   This is still less than the transition value, so our analysis is acceptable. A preferred size of $D = 1.25$ in could be specified.

An alternate method of using spreadsheets to design columns is to use an analysis approach similar to that shown in Figure 6–9 but to use it as a convenient "trial and error" tool. You could compute data by hand, or you could look them up in a table for A, I, and r for any desired cross-sectional shape and dimensions and insert the values into the spreadsheet. Then you could compare the computed allowable load with the required value and choose smaller or larger sections to bring the computed value close to the required value. Many iterations could be completed in a short amount of time. For shapes that allow computing r and A fairly simply, you could add a new section to the spreadsheet to calculate these values. An example is shown in Figure 6–14, where the differently shaded box shows the calculations for the properties of a round cross section. The data are from Example Problem 6–3, and the result shown was arrived at with only four iterations.

6–11 CROOKED COLUMNS

The Euler and Johnson formulas assume that the column is straight and that the load acts in line with the centroid of the cross section of the column. If the column is somewhat crooked, bending occurs in addition to the column action (see Figure 6–15).

The crooked column formula allows an initial crookedness, a, to be considered (see References 6, 7, and 8):

⇨ **Crooked Column Formula**

$$P_a^2 - \frac{1}{N}\left[s_y A + \left(1 + \frac{ac}{r^2}\right)P_{cr}\right]P_a + \frac{s_y A P_{cr}}{N^2} = 0 \qquad \textbf{(6–11)}$$

where $c =$ distance from the neutral axis of the cross section about which bending occurs to its outer edge

P_{cr} is defined to be the critical load found from the *Euler formula.*

Although this formula may become increasingly inaccurate for shorter columns, it is not appropriate to switch to the Johnson formula as it is for straight columns.

The crooked column formula is a quadratic with respect to the allowable load P_a. Evaluating all constant terms in Equation (6–11) produces an equation of the form

$$P_a^2 + C_1 P_a + C_2 = 0$$

Then, from the solution for a quadratic equation,

$$P_a = 0.5[-C_1 - \sqrt{C_1^2 - 4C_2}]$$

The smaller of the two possible solutions is selected.

FIGURE 6–15
Illustration of crooked and eccentric columns

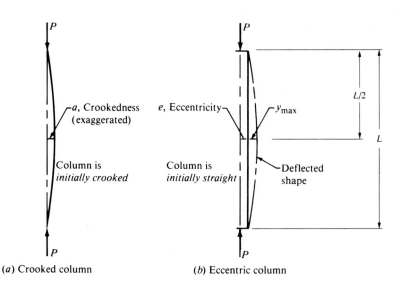

(*a*) Crooked column   (*b*) Eccentric column

CIRCULAR COLUMN ANALYSIS		Data from:	Example Problem 6–3
Refer to Figure 6–4 for analysis logic.			
Enter data for variables in *italics in shaded boxes*.		Use consistent U.S. Customary units.	
Data to Be Entered:		**Computed Values:**	
Length and End Fixity:			
Column length, L= 25 in *End fixity, K=* 1	→	Eq. length, $L_e = KL =$ 25.0 in	
Material Properties:			
Yield strength, s_y= 30 000 psi *Modulus of elasticity, E=* 3.00E+07 psi	→	Column constant, $C_c =$ 140.5	
Cross Section Properties:			
[Note: A and r computed from] [dimensions for circular cross section] [in following section of this spreadsheet.]			
Area, A= 1.188 in^2			
Radius of gyration, r= 0.3075 in	→	Slender ratio, $KL/r =$ 81.3	
Properties for round column:			
Diameter for round column= 1.23 in Area, A= 1.188 in^2 Radius of gyration, r= 0.3075 in		Column is: **short** **Critical buckling load= 29,679 lb**	
Design Factor:			
Design factor on load, N= 3	→	**Allowable load= 9,893 lb**	

FIGURE 6–14   Spreadsheet for column analysis used as a tool to design a column with a round cross section

---

Example Problem 6–4   A column has both ends pinned and has a length of 32 in. It has a circular cross section with a diameter of 0.75 in and an initial crookedness of 0.125 in. The material is AISI 1040 hot-rolled steel. Compute the allowable load for a design factor of 3.

Solution   Objective   Specify the allowable load for the column.

Given   Solid circular cross section: $D = 0.75$ in; $L = 32$ in; use $N = 3$.
Both are ends pinned. Initial crookedness $= a = 0.125$ in.
Material: AISI 1040 hot-rolled steel.

Analysis   Use Equation (6–11). First evaluate C_1 and C_2. Then solve the quadratic equation for P_a.

Results
$$s_y = 42\ 000 \text{ psi}$$
$$A = \pi D^2/4 = (\pi)(0.75)^2/4 = 0.442 \text{ in}^2$$
$$r = D/4 = 0.75/4 = 0.188 \text{ in}$$
$$c = D/2 = 0.75/2 = 0.375 \text{ in}$$
$$KL/r = [(1.0)(32)]/0.188 = 171$$
$$P_{cr} = \frac{\pi^2 EA}{(KL/r)^2} = \frac{\pi^2 (30\ 000\ 000)(0.442)}{(171)^2} = 4\ 476 \text{ lb}$$
$$C_1 = \frac{-1}{N}\left[s_y A + \left(1 + \frac{ac}{r^2}\right)P_{cr}\right] = -9\ 649$$
$$C_2 = \frac{s_y A P_{cr}}{N^2} = 9.232 \times 10^6$$

The quadratic is therefore
$$P_a^2 - 9\ 649 P_a + 9.232 \times 10^6 = 0$$

Comment   From this, $P_a = 1\ 077$ lb is the allowable load.

Figure 6–16 shows the solution of Example Problem 6–4 using a spreadsheet. Whereas its appearance is similar to that of the earlier column analysis spreadsheets, the details follow the calculations needed to solve Equation (6–11). On the lower left, two special data values are needed: (1) the crookedness a and (2) the distance c from the neutral axis for buckling to the outer surface of the cross section. In the middle of the right part are listed some intermediate values used in Equation (6–11): C_1 and C_2 as defined in the solution to Example Problem 6–4. The result, the allowable load, P_a, is at the lower right of the spreadsheet. Above that, for comparison, the computed value of the critical buckling load is given for a straight column of the same design. Note that this solution procedure is most accurate for long columns. If the analysis indicates that the column is *short* rather than *long*, the designer should take note of how short it is by comparing the slenderness ratio, KL/r, with the column constant, C_c. If the column is quite short, the designer should not rely on the accuracy of the result from Equation (6–11).

6–12
ECCENTRICALLY
LOADED
COLUMNS

An *eccentric load* is one that is applied away from the centroidal axis of the cross section of the column, as shown in Figure 6–15(b). Such a load exerts bending in addition to column action that results in the deflected shape shown in the figure. The maximum stress in the deflected column occurs in the outermost fibers of the cross section at the midlength

CROOKED COLUMN ANALYSIS		*Data from:*	Example Problem 6–4
Solves Equation 6–11 for allowable load.			
Enter data for variables in *italics in shaded boxes*.		Use consistent U.S. Customary units.	
Data to Be Entered:		**Computed Values:**	
Length and End Fixity:			
Column length, $L=$ 32 in End fixity, $K=$ 1	$\rightarrow$	Eq. length, $L_e = KL =$	32.0 in
Material Properties:			
Yield strength, $s_y=$ 42,000 psi Modulus of elasticity, $E=$ 3.00E+07 psi	$\rightarrow$	Column constant, $C_c=$	18.7
Cross Section Properties:		Euler buckling load$=$	4,491 lb
[Note: Enter r or compute $r=$ sqrt(I/A).] [Always enter Area.] [Enter zero for I or r if not used.]		C_1 in Eq. (6–11)$=$ $-9,678$ C_2 in Eq. (6–11)$=$ 9.259E+06	
Area, $A=$ 0.442 in^2 Moment of inertia, $I=$ 0 in^4			
Radius of gyration, $r=$ 0.188 in	$\rightarrow$	Slender ratio, $KL/r=$	170.7
Values for Eq. (6–11):		Column is: **long**	
Initial crookedness$= a=$ 0.125 in Neutral axis to outside$= c=$ 0.375 in		**Straight Column**	
		Critical buckling load$=$	**4,491 lb**
Design Factor:		**Crooked Column**	
Design factor on load, $N=$ 3	$\rightarrow$	**Allowable load$=$**	**1,076 lb**

FIGURE 6–16   Spreadsheet for analysis of crooked columns

of the column where the maximum deflection, y_{max}, occurs. Let's denote the stress at this point as $\sigma_{L/2}$. Then, for any applied load, P,

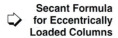
Secant Formula for Eccentrically Loaded Columns

$$\sigma_{L/2} = \frac{P}{A}\left[1 + \frac{ec}{r^2}\sec\left(\frac{KL}{2r}\sqrt{\frac{P}{AE}}\right)\right] \tag{6–12}$$

(See References 3, 4, 5, and 9.) Note that this stress is *not* directly proportional to the load. When evaluating the secant in this formula, note that its argument in the parentheses is in *radians*. Also, because most calculators do not have the secant function, recall that the secant is equal to 1/cosine.

For design purposes, we would like to specify a design factor, N, that can be applied to the *failure load* similar to that defined for straight, centrally loaded columns. However, in this case, failure is predicted when the maximum stress in the column exceeds the yield strength of the material. Let's now define a new term, P_y, to be the load applied to the eccentrically loaded column when the maximum stress is equal to the yield strength. Equation (6–12) then becomes

$$s_y = \frac{P_y}{A}\left[1 + \frac{ec}{r^2}\sec\left(\frac{KL}{2r}\sqrt{\frac{P_y}{AE}}\right)\right]$$

Now, if we define the *allowable load* to be

$$P_a = P_y/N$$

or

$$P_y = NP_a$$

this equation becomes

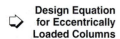
Design Equation for Eccentrically Loaded Columns

$$\text{Required}\quad s_y = \frac{NP_a}{A}\left[1 + \frac{ec}{r^2}\sec\left(\frac{KL}{2r}\sqrt{\frac{NP_a}{AE}}\right)\right] \tag{6–13}$$

This equation cannot be solved for either N or P_a. Therefore, an iterative solution is required, as will be demonstrated in Example Problem 6–6.

Another critical factor may be the amount of deflection of the axis of the column due to the eccentric load:

Maximum Deflection in an Eccentrically Loaded Column

$$y_{max} = e\left[\sec\left(\frac{KL}{2r}\sqrt{\frac{P}{AE}}\right) - 1\right] \tag{6–14}$$

Note that the argument of the secant is the same as that used in Equation (6–12).

Example Problem 6–5   For the column of Example Problem 6–4, compute the maximum stress and deflection if a load of 1 075 lb is applied with an eccentricity of 0.75 in.

Solution   Objective   Compute the stress and the deflection for the eccentrically loaded column.

Given   Data from Example Problem 6–4, but eccentricity $= e = 0.75$ in.
Solid circular cross section: $D = 0.75$ in; $L = 32$ in.
Both ends are pinned; $KL = 32$ in; $r = 0.188$ in; $c = D/2 = 0.375$ in.
Material: AISI 1040 hot-rolled steel; $E = 30 \times 10^6$ psi.

Analysis   Use Equation (6–12) to compute maximum stress. Then use Equation (6–14) to compute maximum deflection.

Results   All terms have been evaluated before. Then the maximum stress is found from Equation (6–12):

$$\sigma_{L/2} = \frac{1\,075}{0.422}\left[1 + \frac{(0.75)(0.375)}{(0.188)^2} \sec\left(\frac{32}{2(0.188)}\sqrt{\frac{1\,075}{(0.442)(30 \times 10^6)}}\right)\right]$$

$$\sigma_{L/2} = 29\,300 \text{ psi}$$

The maximum deflection is found from Equation (6–14):

$$y_{max} = 0.75\left[\sec\left(\frac{32}{2(0.188)}\sqrt{\frac{1\,075}{(0.442)(30 \times 10^6)}}\right) - 1\right] = 0.293 \text{ in}$$

Comments   The maximum stress is 29 300 psi at the midlength of the column. The deflection there is 0.293 in.

Example Problem 6–6   The stress in the column found in Example Problem 6–5 seems high for the AISI 1040 hot-rolled steel. Redesign the column to achieve a design factor of at least 3.

Solution   Objective   Redesign the eccentrically loaded column of Example Problem 6–5 to reduce the stress and achieve a design factor of at least 3.

Given   Data from Example Problems 6–4 and 6–5.

Analysis   Use a larger diameter. Use Equation (6–13) to compute the required strength. Then compare that with the strength of AISI 1040 hot-rolled steel. Iterate until the stress is satisfactory.

Results   Appendix 3 gives the value for the yield strength of AISI 1040 HR to be 42 000 psi. If we choose to retain the same material, the cross-sectional dimensions of the column must be increased to decrease the stress. Equation (6–13) can be used to evaluate a design alternative.
      The objective is to find suitable values for A, c, and r for the cross section such that $P_a = 1\,075$ lb; $N = 3$; $L_e = 32$ in; $e = 0.75$ in; and the value of the entire right side of the equation is less than 42 000 psi. The original design had a circular cross section with a diameter of 0.75 in. Let's try increasing the diameter to $D = 1.00$ in. Then

$$A = \pi D^2/4 = \pi(1.00 \text{ in})^2/4 = 0.785 \text{ in}^2$$

$$r = D/4 = (1.00 \text{ in})/4 = 0.250 \text{ in}$$

$$r^2 = (0.250 \text{ in})^2 = 0.062\,5 \text{ in}^2$$

$$c = D/2 = (1.00 \text{ in})/2 = 0.50 \text{ in}$$

Now let's call the right side of Equation (6–13) s'_y. Then

$$s'_y = \frac{3(1\,075)}{0.785}\left[1 + \frac{(0.75)(0.50)}{(0.062\,5)} \sec\left(\frac{32}{2(0.250)}\sqrt{\frac{(3)(1\,075)}{(0.785)(30 \times 10^6)}}\right)\right]$$

$$s'_y = 37\,740 \text{ psi} = \text{required value of } s_y$$

This is a satisfactory result because it is just slightly less than the value of s_y of 42 000 psi for the steel.

Now we can evaluate the expected maximum deflection with the new design using Equation (6–14):

$$y_{max} = 0.75 \left[\sec \left(\frac{32}{2(0.250)} \sqrt{\frac{1\ 075}{(0.785)(30 \times 10^6)}} \right) - 1 \right]$$

$$y_{max} = 0.076 \text{ in}$$

Comments   The diameter of 1.00 in is satisfactory. The maximum deflection for the column is 0.076 in.

Figure 6–17 shows the solution of the eccentric column problem of Example Problem 6–6 using a spreadsheet to evaluate Equations (6–13) and (6–14). It is a design aid that facilitates the iteration required to determine an acceptable geometry for a column to carry a specified load with a desired design factor. Note that the data are in U.S. Customary units. At the lower left of the spreadsheet, data required for Equations (6–13) and (6–14) are entered by the designer, along with the other data discussed for earlier column analysis spreadsheets. The "**FINAL RESULTS**" at the lower right show the computed value of the required yield strength of the material for the column and compare it with the given value entered by the designer near the upper left. The designer must ensure that the actual value is greater than the computed value. The last part of the right side of the spreadsheet gives the computed maximum deflection of the column that occurs at its midlength.

ECCENTRIC COLUMN ANALYSIS		Data from:	Example Problem 6–6

Solves Equation (6–13) for design stress and Equation (6–14) for maximum deflection.

Enter data for variables in *italics in shaded boxes*.		Use consistent U.S. Customary units.
Data to Be Entered:		**Computed Values:**
Length and End Fixity:		
Column length, $L=$ 32 in End fixity, $K=$ 1	→	Eq. length, $L_e= KL=$ 32.0 in
Material Properties:		
Yield strength, $s_y=$ 42,000 psi Modulus of elasticity, $E=$ 3.00E+07 psi	→	Column constant, $C_c=$ 118.7
Cross Section Properties:		Argument of sec= 0.749 for strength Value of secant= 1.3654

[Note: Enter r or compute $r=$ sqrt(I/A).]
 [Always enter Area.]
 [Enter zero for I or r if not used.]
         Area, $A=$ 0.785 in^2
    Moment of inertia, $I=$ 0 in^4
 OR
   Radius of gyration, $r=$ 0.250 in
		Argument of sec= 0.432 for deflection
		 Value of secant= 1.1014

	→	Slender ratio, $KL/r=$ 128.0

Values for Eqs. (6–13) and (6–14):
     Eccentricity, $e=$ 0.75 in
 Neutral axis to outside, $c=$ 0.5 in
    Allowable load, $P_a=$ 1,075 lb
		Column is: **long**

		FINAL RESULTS
Design Factor: Design factor on load, $N=$ 3		**Req'd yield strength= 37,764 psi** ***Must be less than actual yield strength:*** ***$s_y=$ 42,000 psi***
		Max deflection, $y_{max}=$ 0.076 in

FIGURE 6–17   Spreadsheet for analysis of eccentric columns

You Are the Designer

A plant in Louisiana that produces sugar needs a drive system designed for a machine that chops long pieces of sugar cane into short lengths prior to processing. The machine's drive shaft is to rotate slowly at 30 rpm so that the cane is chopped smoothly and not beaten. The large machine requires a torque of 31 500 lb·in to drive the chopping blades.

Your company is asked to design the drive, and you are given the assignment. What kind of power source should be used? You might consider an electric motor, a gasoline engine, or a hydraulic motor. Most of these run at relatively high speeds, significantly higher than 30 rpm. Therefore, some type of speed reduction is needed. Perhaps you decide to use a drive similar to that shown in Figure 7–1.

Three stages of speed reduction are used. The input sheave of the belt drive rotates at the speed of the motor, while the larger driven sheave rotates at a slower speed and delivers the power to the input of the gear reducer. The larger part of the speed reduction is likely to be accomplished in the gear reducer, with the output shaft rotating slowly and providing a large torque. Remember, as the speed of rotation of a rotating shaft decreases, the torque delivered increases for a given power transmitted. But because there are only a limited number of reducer designs available, the output speed of the reducer will probably not be ideal for the cane chopper input shaft. The chain drive then provides the last stage of reduction.

As the designer, you must decide what type and size of belt drive to use and what the speed ratio between the driving and the driven sheave should be. How is the driving sheave attached to the motor shaft? How is the driven sheave attached to the input shaft of the gear reducer? Where should the motor be mounted in relation to the gear reducer, and what will be the resulting center distance between the two shafts? What speed reduction ratio will the gear reducer provide? What type of gear reducer should be used: helical gears, a worm and worm-gear drive, or bevel gears? How much additional speed reduction must the chain drive provide to deliver the proper speed to the cane-chopper shaft? What size and type of chain should be specified? What is the center distance between the output of the gear reducer and the input to the chopper? Then what length of chain is required? Finally, what motor power is required to drive the entire system at the stated conditions? The information in this chapter will help you answer questions about the design of power transmission systems incorporating belts and chains.

**7–1
OBJECTIVES OF
THIS CHAPTER**

After completing this chapter, you will be able to:

1. Describe the basic features of a belt drive system.

2. Describe several types of belt drives.

3. Specify suitable types and sizes of belts and sheaves to transmit a given level of power at specified speeds for the input and output sheaves.

4. Specify the primary installation variables for belt drives, including center distance and belt length.

5. Describe the basic features of a chain drive system.

6. Describe several types of chain drives.

7. Specify suitable types and sizes of chains and sprockets to transmit a given level of power at specified speeds for the input and output sprockets.

8. Specify the primary installation variables for chain drives, including center distance between the sheaves, chain length, and lubrication requirements.

**7–2
TYPES OF
BELT DRIVES**

A belt is a flexible power transmission element that seats tightly on a set of pulleys or sheaves. Figure 7–2 shows the basic layout. When the belt is used for speed reduction, the typical case, the smaller sheave is mounted on the high-speed shaft, such as the shaft of an electric motor. The larger sheave is mounted on the driven machine. The belt is designed to ride around the two sheaves without slipping.

The belt is installed by placing it around the two sheaves while the center distance between them is reduced. Then the sheaves are moved apart, placing the belt in a rather high initial tension. When the belt is transmitting power, friction causes the belt to grip the driving sheave, increasing the tension in one side, called the "tight side," of the drive. The tensile force in the belt exerts a tangential force on the driven sheave, and thus a torque is applied to the driven shaft. The opposite side of the belt is still under tension, but at a smaller value. Thus, it is called the "slack side."

Many types of belts are available: flat belts; grooved or cogged belts; standard V-belts; double-angle V-belts; and others. See Figure 7–3 for examples. References 2–5 and 8–15 give more examples and technical data.

FIGURE 7–2 Basic belt drive geometry

(a) Wrapped construction

(b) Die cut, cog type

(c) Timing belt

(d) Poly-rib belt

(e) Vee-band

(f) Double angle V-belt

FIGURE 7–3 Examples of V-belt construction (Dayco Corp., Dayton, OH)

FIGURE 7–4 Cross section of V-belt and sheave groove

The *flat belt* is the simplest type, often made from leather or rubber-coated fabric. The sheave surface is also flat and smooth, and the driving force is therefore limited by the pure friction between the belt and the sheave. Some designers prefer flat belts for delicate machinery because the belt *will* slip if the torque tends to rise to a level high enough to damage the machine.

Grooved belts, sometimes called *timing belts* [see Figure 7–3(c)], ride on sheaves having mating grooves into which the cogs on the belt seat. This is a more positive drive, limited only by the tensile strength of the belt and the shear strength of the cogs.

Some cog belts, such as that shown in Figure 7–3(b), are applied to standard V-grooved sheaves. The cogs give the belt greater flexibility and higher efficiency compared with standard belts. They can operate on smaller sheave diameters.

But the most widely used type of belt, particularly in industrial drives and vehicular applications, is the *V-belt drive,* shown in Figures 7–3(a) and 7–4. The V-shape causes the belt to wedge tightly into the groove, increasing friction and allowing high torques to be transmitted before slipping occurs. Most belts have high-strength cords positioned at the pitch diameter of the belt cross section to increase the tensile strength of the belt. The cords, made from natural fibers, synthetic strands, or steel, are embedded in a firm rubber compound to provide the flexibility needed to allow the belt to pass around the sheave. Often an outer fabric cover is added to give the belt good durability.

The selection of commercially available V-belt drives is discussed in the next section.

7–3
V-BELT DRIVES

The typical arrangement of the elements of a V-belt drive is shown in Figure 7–2. The important observations to be derived from this arrangement are summarized here:

1. The pulley, with a circumferential groove carrying the belt, is called a *sheave* (usually pronounced "shiv").

2. The size of a sheave is indicated by its pitch diameter, slightly smaller than the outside diameter of the sheave.

3. The speed ratio between the driving and the driven sheaves is inversely proportional to the ratio of the sheave pitch diameters. This follows from the observation that there is no slipping (under normal loads). Thus, the linear speed of the pitch line of both sheaves is the same as and equal to the belt speed, v_b. Then

$$v_b = R_1\omega_1 = R_2\omega_2 \tag{7–1}$$

But $R_1 = D_1/2$ and $R_2 = D_2/2$. Then

$$v_b = \frac{D_1\omega_1}{2} = \frac{D_2\omega_2}{2} \tag{7–1A}$$

The angular velocity ratio is

$$\frac{\omega_1}{\omega_2} = \frac{D_2}{D_1} \tag{7–2}$$

4. The relationships between pitch length, L, center distance, C, and the sheave diameters are

$$L = 2C + 1.57(D_2 + D_1) + \frac{(D_2 - D_1)^2}{4C} \tag{7–3}$$

$$C = \frac{B + \sqrt{B^2 - 32\,(D_2 - D_1)^2}}{16} \tag{7–4}$$

where $B = 4L - 6.28(D_2 + D_1)$

5. The angle of contact of the belt on each sheave is

$$\theta_1 = 180° - 2 \sin^{-1}\left[\frac{D_2 - D_1}{2C}\right] \tag{7–5}$$

$$\theta_2 = 180° + 2 \sin^{-1}\left[\frac{D_2 - D_1}{2C}\right] \tag{7–6}$$

These angles are important because commercially available belts are rated with an assumed contact angle of 180°. This will occur only if the drive ratio is 1 (no speed change). The angle of contact on the smaller of the two sheaves will always be less than 180°, requiring a lower power rating.

6. The length of the span between the two sheaves, over which the belt is unsupported, is

$$S = \sqrt{C^2 - \left[\frac{D_2 - D_1}{2}\right]^2} \tag{7–7}$$

This is important for two reasons: You can check the proper belt tension by measuring the amount of force required to deflect the belt at the middle of the span by a given amount. Also, the tendency for the belt to vibrate or whip is dependent on this length.

7. The contributors to the stress in the belt are as follows:
   a. The tensile force in the belt, maximum on the tight side of the belt.
   b. The bending of the belt around the sheaves, maximum as the tight side of the belt bends around the smaller sheave.
   c. Centrifugal forces created as the belt moves around the sheaves.

   The maximum total stress occurs where the belt enters the smaller sheave, and the bending stress is a major part. Thus, there are recommended minimum sheave diameters for standard belts. Using smaller sheaves drastically reduces belt life.

8. The design value of the ratio of the tight side tension to the slack side tension is 5.0 for V-belt drives. The actual value may range as high as 10.0.

Standard Belt Cross Sections

Commercially available belts are made to one of the standards shown in Figures 7–5 through 7–8. The alignment between the inch sizes and the metric sizes indicates that the

FIGURE 7–5 Heavy-duty industrial V-belts

FIGURE 7–6 Industrial narrow-section V-belts

FIGURE 7–7 Light-duty, fractional horsepower (FHP) V-belts

paired sizes are actually the same cross section. A "soft conversion" was used to rename the familiar inch sizes with the number for the metric sizes giving the nominal top width in millimeters.

    The nominal value of the included angle between the sides of the V-groove ranges from 30° to 42°. The angle on the belt may be slightly different to achieve a tight fit in the groove. Some belts are designed to "ride out" of the groove somewhat.

FIGURE 7–8
Automotive V-belts

Inch sizes:	0.250	0.315	0.380	0.440	0.500
Metric sizes:	6A	8A	10A	11A	13A

Inch sizes:	11/16	3/4	7/8	1
Metric sizes:	15A	17A	20A	23A

Many automotive applications use synchronous belt drives similar to that called a *timing belt* in Figure 7–3(c) or V-ribbed belts similar to that called a *poly-rib belt* in Figure 7–3(d). The following standards of the Society of Automotive Engineers (SAE) give dimensions and performance standards for automotive belts.

SAE Standard J636c: V-belts and pulleys

SAE Standard J637b: Automotive V-belt drives

SAE Standard J1278: SI (metric) synchronous belts and pulleys

SAE Standard J1313: Automotive synchronous belt drives

SAE Standard J1459: V-ribbed belts and pulleys

**7–4
V-BELT DRIVE
DESIGN**

The factors involved in selection of a V-belt and the driving and driven sheaves and proper installation of the drive are summarized in this section. Abbreviated examples of the data available from suppliers are given for illustration. Catalogs contain extensive data, and step-by-step instructions are given for their use. The basic data required for drive selection are the following:

- The rated power of the driving motor or other prime mover
- The service factor based on the type of driver and driven load
- The center distance
- The power rating for one belt as a function of the size and speed of the smaller sheave
- The belt length
- The size of the driving and driven sheaves
- The correction factor for belt length
- The correction factor for the angle of wrap on the smaller sheave
- The number of belts
- The initial tension on the belt

FIGURE 7–14 Angle of wrap correction factor, C_θ

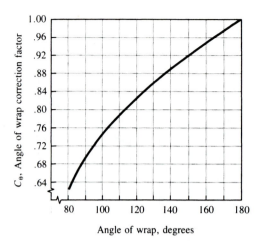

C_θ, Angle of wrap correction factor

Angle of wrap, degrees

FIGURE 7–15 Belt length correction factor, C_L

C_L, Length correction factor

Belt length, in

TABLE 7–2 Standard belt lengths for 3V, 5V, and 8V belts (in)

3V only	3V and 5V	3V, 5V, and 8V	5V and 8V	8V only
25	50	100	150	375
26.5	53	106	160	400
28	56	112	170	425
30	60	118	180	450
31.5	63	125	190	475
33.5	67	132	200	500
35.5	71	140	212	
37.5	75		224	
40	80		236	
42.5	85		250	
45	90		265	
47.5	95		280	
			300	
165			315	
			335	
			355	

Figure 7–15 gives the value of the correction factor, C_L, for belt length. A longer belt is desirable because it reduces the frequency with which a given part of the belt encounters the stress peak as it enters the small sheave. Only certain standard belt lengths are available (Table 7–2).

Example Problem 7–1 illustrates the use of the design data.

Example Problem 7–1   Design a V-belt drive that has the input sheave on the shaft of an electric motor (normal torque) rated at 50.0 hp at 1 160-rpm, full-load speed. The drive is to a bucket elevator in a potash plant that is to be used 16 hours (h) daily at approximately 675 rpm.

Solution   Objective   Design the V-belt drive.

Given   Power transmitted = 50 hp to bucket elevator

Speed of motor = 1 160 rpm; output speed = 675 rpm

Analysis   Use the design data presented in this section. The solution procedure is developed within the Results section of the problem solution.

Results   *Step 1.* Compute the design power. From Table 7–1, for a normal torque electric motor running 16 h daily driving a bucket elevator, the service factor is 1.40. Then the design power is 1.40(50.0 hp) = 70.0 hp.

Step 2. Select the belt section. From Figure 7–9, a 5V belt is recommended for 70.0 hp at 1 160-rpm input speed.

Step 3. Compute the nominal speed ratio:

$$\text{Ratio} = 1\ 160/675 = 1.72$$

Step 4. Compute the driving sheave size that would produce a belt speed of 4 000 ft/min, as a guide to selecting a standard sheave:

$$\text{Belt speed} = v_b = \frac{\pi D_1 n_1}{12} \text{ ft/min}$$

Then the required diameter to give v_b = 4 000 ft/min is

$$D_1 = \frac{12v_b}{\pi n_1} = \frac{12(4\ 000)}{\pi n_1} = \frac{15\ 279}{n_1} = \frac{15\ 279}{1\ 160} = 13.17 \text{ in}$$

Step 5. Select trial sizes for the input sheave, and compute the desired size of the output sheave. Select a standard size for the output sheave, and compute the actual ratio and output speed.

For this problem, the trials are given in Table 7–3 (diameters are in inches).

The two trials in **boldface** in Table 7–3 give only about 1% variation from the desired output speed of 675 rpm, and the speed of a bucket elevator is not critical. Because no space limitations were given, let's choose the larger size.

Step 6. Determine the rated power from Figure 7–10, 7–11, or 7–12.

For the 5V belt that we have selected, Figure 7–11 is appropriate. For a 12.4-in sheave at 1 160 rpm, the basic rated power is 26.4 hp. Multiple belts will be required. The ratio is relatively high, indicating that some added power rating can be used. This value can be estimated from Figure 7–11 or taken directly from Figure 7–13 for the 5V belt. Power added is 1.15 hp. Then the actual rated power is 26.4 + 1.15 = 27.55 hp.

TABLE 7–3

Standard driving sheave size, D_1	Approximate driven sheave size $(1.72D_1)$	Nearest standard sheave, D_2	Actual output speed (rpm)
13.10	22.5	21.1	720
12.4	**21.3**	**21.1**	**682**
11.7	20.1	21.1	643
10.8	18.6	21.1	594
10.2	17.5	15.9	744
9.65	16.6	15.9	704
9.15	**15.7**	**15.9**	**668**
8.9	15.3	14.9	693

Step 7. Specify a trial center distance.

We can use Equation (7–8) to determine a nominal acceptable range for C:

$$D_2 < C < 3(D_2 + D_1)$$
$$21.1 < C < 3(21.1 + 12.4)$$
$$21.1 < C < 100.5 \text{ in}$$

In the interest of conserving space, let's try $C = 24.0$ in.

Step 8. Compute the required belt length from Equation (7–3):

$$L = 2C + 1.57(D_2 + D_1) + \frac{(D_2 - D_1)^2}{4C}$$

$$L = 2(24.0) + 1.57(21.1 + 12.4) + \frac{(21.1 - 12.4)^2}{4(24.0)} = 101.4 \text{ in}$$

Step 9. Select a standard belt length from Table 7–2, and compute the resulting actual center distance from Equation (7–4).

In this problem, the nearest standard length is 100.0 in. Then, from Equation (7–4),

$$B = 4L - 6.28(D_2 + D_1) = 4(100) - 6.28(21.1 + 12.4) = 189.6$$

$$C = \frac{189.6 + \sqrt{(189.6)^2 - 32(21.1 - 12.4)^2}}{16} = 23.30 \text{ in}$$

Step 10. Compute the angle of wrap of the belt on the small sheave from Equation (7–5):

$$\theta_1 = 180° - 2 \sin^{-1}\left[\frac{D_2 - D_1}{2C}\right] = 180° - 2 \sin^{-1}\left[\frac{21.1 - 12.4}{2(23.30)}\right] = 158°$$

Step 11. Determine the correction factors from Figures 7–14 and 7–15. For $\theta = 158°$, $C_\theta = 0.94$. For $L = 100$ in, $C_L = 0.96$.

Step 12. Compute the corrected rated power per belt and the number of belts required to carry the design power:

$$\text{Corrected power} = C_\theta C_L P = (0.94)(0.96)(27.55 \text{ hp}) = 24.86 \text{ hp}$$

$$\text{Number of belts} = 70.0/24.86 = 2.82 \text{ belts} \quad \text{(Use 3 belts.)}$$

Comments **Summary of Design**

Input: Electric motor, 50.0 hp at 1 160 rpm

Service factor: 1.4

Design power: 70.0 hp

Belt: 5V cross section, 100-in length, 3 belts

Sheaves: Driver, 12.4-in pitch diameter, 3 grooves, 5V. Driven, 21.1-in pitch diameter, 3 grooves, 5V

Actual output speed: 682 rpm

Center distance: 23.30 in

Belt Tension

The initial tension given to a belt is critical because it ensures that the belt will not slip under the design load. At rest, the two sides of the belt have the same tension. As power is being transmitted, the tension in the tight side increases while the tension in the slack side decreases. Without the initial tension, the slack side would go totally loose, and the belt would not seat in the groove; thus, it would slip. Manufacturers' catalogs give data for the proper belt-tensioning procedures.

7–5
CHAIN DRIVES

A chain is a power transmission element made as a series of pin-connected links. The design provides for flexibility while enabling the chain to transmit large tensile forces. See References 1, 6, and 7 for more technical information and manufacturers' data.

When transmitting power between rotating shafts, the chain engages mating toothed wheels, called *sprockets.* Figure 7–16 shows a typical chain drive.

The most common type of chain is the *roller chain,* in which the roller on each pin provides exceptionally low friction between the chain and the sprockets. Other types include a variety of extended link designs used mostly in conveyor applications (see Figure 7–17).

Roller chain is classified by its *pitch,* the distance between corresponding parts of adjacent links. The pitch is usually illustrated as the distance between the centers of adjacent pins. Standard roller chain carries a size designation from 40 to 240, as listed in Table 7–4. The digits (other than the final zero) indicate the pitch of the chain in eighths of an inch, as in the table. For example, the no. 100 chain has a pitch of 10/8 or $1\frac{1}{4}$ in. A series of heavy-duty sizes, with the suffix H on the designation (60H–240H), has the same basic dimensions as the standard chain of the same number except for thicker side plates. In addition, there are the smaller and lighter sizes: 25, 35, and 41.

FIGURE 7–16
Roller chain drive
(Rexnord, Inc.,
Milwaukee, WI)

Working strand (tight side)

Driven
sprocket

Driving
sprocket

Slack strand

FIGURE 7–17 Some roller chain styles (Rexnord, Inc., Milwaukee, WI)

(a) Standard roller chain, single strand

(b) Standard roller chain, two-strand (also available with three and four strands)

(c) Heavy series roller chain

(d) Double-pitch drive chain

(e) Double-pitch conveyor chain

TABLE 7–4   Roller chain sizes

Chain number	Pitch (in)	Average tensile strength (lb)
25	1/4	925
35	3/8	2 100
41	1/2	2 000
40	1/2	3 700
50	5/8	6 100
60	3/4	8 500
80	1	14 500
100	$1\frac{1}{4}$	24 000
120	$1\frac{1}{2}$	34 000
140	$1\frac{3}{4}$	46 000
160	2	58 000
180	$2\frac{1}{4}$	80 000
200	$2\frac{1}{2}$	95 000
240	3	130 000

The average tensile strengths of the various chain sizes are also listed in Table 7–4. These data can be used for very low speed drives or for applications in which the function of the chain is to apply a tensile force or to support a load. It is recommended that only 10% of the average tensile strength be used in such applications. For power transmission, the rating of a given chain size as a function of the speed of rotation must be determined, as explained later in this chapter.

A wide variety of attachments are available to facilitate the application of roller chain to conveying or other material handling uses. Usually in the form of extended plates or tabs with holes provided, the attachments make it easy to connect rods, buckets, parts pushers, part support devices, or conveyor slats to the chain. Figure 7–18 shows some attachment styles.

FIGURE 7–18 Chain attachments (Rexnord, Inc., Milwaukee, WI)

(*a*) Slats assembled to attachments to form a flat conveying surface

(*b*) *V* block assembled to attachments to convey round objects of varying diameters

(*c*) Attachments used as spacers to convey and position long objects

Figure 7–19 shows a variety of chain types used especially for conveying and similar applications. Such chain typically has a longer pitch than standard roller chain (usually twice the pitch), and the link plates are heavier. The larger sizes have cast link plates.

7–6 DESIGN OF CHAIN DRIVES

The rating of chain for its power transmission capacity considers three modes of failure: (1) fatigue of the link plates due to the repeated application of the tension in the tight side of the chain, (2) impact of the rollers as they engage the sprocket teeth, and (3) galling between the pins of each link and the bushings on the pins.

The ratings are based on empirical data with a smooth driver and a smooth load (service factor = 1.0) and with a rated life of approximately 15 000 h. The important variables are the pitch of the chain and the size and rotational speed of the smaller sprocket. Lubrication is critical to the satisfactory operation of a chain drive. Manufacturers recommend the type of lubrication method for given combinations of chain size, sprocket size, and speed. Details are discussed later.

Tables 7–5, 7–6, and 7–7 list the rated power for three sizes of standard chain: no. 40 (1/2 in), no. 60 (3/4 in), and no. 80 (1.00 in). These are typical of the types of data available for all chain sizes in manufacturers' catalogs. Notice these features of the data:

1. The ratings are based on the speed of the smaller sprocket.

2. For a given speed, the power capacity increases with the number of teeth on the sprocket. Of course, the larger the number of teeth, the larger the diameter of the sprocket. Note that the use of a chain with a small pitch on a large sprocket produces the quieter drive.

3. For a given sprocket size (a given number of teeth), the power capacity increases with increasing speed up to a point; then it decreases. Fatigue due to the tension

FIGURE 7–19
Conveyor chains
(Rexnord, Inc.,
Milwaukee, WI)

Mill, narrow series
(drive and conveyor sizes)
Offset cast-link chain used primarily in the lumber industry for conveyor applications.

Combination mill
(wide conveyor sizes)
Cast block links and steel sidebar construction for drag conveyor applications.

Heavy-duty drag chain
Cast steel offset block links. Used in ash and clinker conveyors.

Pintle chain
Chain constructed of a series of cast offset links coupled by steel pins or rivets. Suitable for slow-to moderate-speed drive, conveyor and elevator service.

Roller-top transfer
Cast links with top rollers used in several strands to convey material transversely.

Roof-top
Cast roof-shaped links used in several strands on transfer conveyors.

Detachable
Consists of unit links, each with an open-type hook that flexes on the end bar of the adjacent link. Used for slow- to moderate-speed drive and conveyor application.

Drop-forged
Drop-forged inner and outer links coupled by headed pins. Used for trolley, scraper, flight and similar conveyors.

in the chain governs at the low to moderate speeds; impact on the sprockets governs at the higher speeds. Each sprocket size has an absolute upper-limit speed due to the onset of galling between the pins and the bushings of the chain. This explains the abrupt drop in power capacity to zero at the limiting speed.

4. The ratings are for a single strand of chain. Although multiple strands do increase the power capacity, they do not provide a direct multiple of the single-strand capacity. Multiply the capacity in the tables by the following factors.

Two strands: Factor = 1.7

Three strands: Factor = 2.5

Four strands: Factor = 3.3

5. The ratings are for a service factor of 1.0. Specify a service factor for a given application according to Table 7–8.

TABLE 7–5 Horsepower ratings, standard single-strand roller chain, no. 40, 1/2-in pitch

No. of teeth, small spkt.	Revolutions per minute, small sprocket																								
	10	25	50	100	200	300	400	500	700	900	1 000	1 200	1 400	1 600	1 800	2 100	2 400	2 700	3 000	3 500	4 000	5 000	6 000	7 000	8 000
9	0.04	0.10	0.19	0.35	0.65	0.93	1.21	1.48	2.00	2.51	2.75	3.25	3.73	4.12	3.45	2.74	2.24	1.88	1.60	1.27	1.04	0.75	0.57	0.45	0.37
10	0.05	0.11	0.21	0.39	0.73	1.04	1.35	1.65	2.24	2.81	3.09	3.64	4.18	4.71	4.04	3.21	2.63	2.20	1.88	1.49	1.22	0.87	0.66	0.53	0.43
11	0.05	0.12	0.23	0.43	0.80	1.16	1.50	1.83	2.48	3.11	3.42	4.03	4.63	5.22	4.66	3.70	3.03	2.54	2.17	1.72	1.41	1.01	0.77	0.61	0.50
12	0.06	0.14	0.25	0.47	0.88	1.27	1.65	2.01	2.73	3.42	3.76	4.43	5.09	5.74	5.31	4.22	3.45	2.89	2.47	1.96	1.60	1.15	0.87	0.69	0.57
13	0.06	0.15	0.28	0.52	0.96	1.39	1.80	2.20	2.97	3.73	4.10	4.83	5.55	6.26	5.99	4.76	3.89	3.26	2.79	2.21	1.81	1.29	0.98	0.78	0.64
14	0.07	0.16	0.30	0.56	1.04	1.50	1.95	2.38	3.22	4.04	4.44	5.23	6.01	6.78	6.70	5.31	4.35	3.65	3.11	2.47	2.02	1.45	1.10	0.87	0.71
15	0.07	0.17	0.32	0.60	1.12	1.62	2.10	2.56	3.47	4.35	4.78	5.64	6.47	7.30	7.43	5.89	4.82	4.04	3.45	2.74	2.24	1.60	1.22	0.97	0.79
16	0.08	0.19	0.35	0.65	1.20	1.74	2.25	2.75	3.72	4.66	5.13	6.04	6.94	7.83	8.18	6.49	5.31	4.45	3.80	3.02	2.47	1.77	1.34	1.07	0.87
17	0.08	0.20	0.37	0.69	1.29	1.85	2.40	2.93	3.97	4.98	5.48	6.45	7.41	8.36	8.96	7.11	5.82	4.88	4.17	3.31	2.71	1.94	1.47	1.17	0.96
18	0.09	0.21	0.39	0.73	1.37	1.97	2.55	3.12	4.22	5.30	5.82	6.86	7.88	8.89	9.76	7.75	6.34	5.31	4.54	3.60	2.95	2.11	1.60	1.27	0
19	0.09	0.22	0.42	0.78	1.45	2.09	2.71	3.31	4.48	5.62	6.17	7.27	8.36	9.42	10.5	8.40	6.88	5.76	4.92	3.91	3.20	2.29	1.74	1.38	0
20	0.10	0.24	0.44	0.82	1.53	2.21	2.86	3.50	4.73	5.94	6.53	7.69	8.83	9.96	11.1	9.07	7.43	6.22	5.31	4.22	3.45	2.47	1.88	1.49	0
21	0.11	0.25	0.46	0.87	1.62	2.33	3.02	3.69	4.99	6.26	6.88	8.11	9.31	10.5	11.7	9.76	7.99	6.70	5.72	4.54	3.71	2.66	2.02	1.60	0
22	0.11	0.26	0.49	0.91	1.70	2.45	3.17	3.88	5.25	6.58	7.23	8.52	9.79	11.0	12.3	10.5	8.57	7.18	6.13	4.87	3.98	2.85	2.17	1.72	0
23	0.12	0.27	0.51	0.96	1.78	2.57	3.33	4.07	5.51	6.90	7.59	8.94	10.3	11.6	12.9	11.2	9.16	7.68	6.55	5.20	4.26	3.05	2.32	1.84	0
24	0.13	0.29	0.54	1.00	1.87	2.69	3.48	4.26	5.76	7.23	7.95	9.36	10.8	12.1	13.5	11.9	9.76	8.18	6.99	5.54	4.54	3.25	2.47	1.96	
25	0.13	0.30	0.56	1.05	1.95	2.81	3.64	4.45	6.02	7.55	8.30	9.78	11.2	12.7	14.1	12.7	10.4	8.70	7.43	5.89	4.82	3.45	2.63	0	
26	0.14	0.31	0.58	1.09	2.04	2.93	3.80	4.64	6.28	7.88	8.66	10.2	11.7	13.2	14.7	13.5	11.0	9.23	7.88	6.25	5.12	3.66	2.79	0	
28	0.15	0.34	0.63	1.18	2.20	3.18	4.11	5.03	6.81	8.54	9.39	11.1	12.7	14.3	15.9	15.0	12.3	10.3	8.80	6.99	5.72	4.09	3.11	0	
30	0.16	0.37	0.68	1.27	2.38	3.42	4.43	5.42	7.33	9.20	10.1	11.9	13.7	15.4	17.2	16.7	13.6	11.4	9.76	7.75	6.34	4.54	3.45	0	
32	0.17	0.39	0.73	1.36	2.55	3.67	4.75	5.81	7.86	9.86	10.8	12.8	14.7	16.5	18.4	18.4	15.0	12.6	10.8	8.64	6.99	5.00	0		
35	0.19	0.43	0.81	1.50	2.81	4.04	5.24	6.40	8.66	10.9	11.9	14.1	16.2	18.2	20.3	21.0	17.2	14.4	12.3	9.76	7.99	5.72	0		
40	0.22	0.50	0.93	1.74	3.24	4.67	6.05	7.39	10.0	12.5	13.8	16.3	18.7	21.1	23.4	25.7	21.0	17.6	15.0	11.9	9.76	6.99	0		
45	0.25	0.57	1.06	1.97	3.68	5.30	6.87	8.40	11.4	14.2	15.7	18.5	21.2	23.9	26.6	30.5	25.1	21.0	17.9	14.2	11.7	0			
	Type I								Type II											Type III					

SOURCE: Reprinted from *Chains for Power Transmission and Material Handling*, p. 147, by courtesy of Marcel Dekker, Inc.

NOTES: The limiting rpm for each lubrication type is read from the column to the left of the boundary line shown.

TYPE I: Manual or drip lubrication

TYPE II: Bath or disc lubrication

TYPE III: Oil stream lubrication

TABLE 7–6 Horsepower ratings, standard single-strand roller chain, no. 60, 3/4-in pitch

No. of teeth, small spkt.	Revolutions per minute, small sprocket																								
	10	25	50	100	150	200	300	400	500	600	700	800	900	1 000	1 100	1 200	1 400	1 600	1 800	2 000	2 500	3 000	3 500	4 000	4 500
9	0.15	0.33	0.62	1.16	1.67	2.16	3.12	4.04	4.94	5.82	6.68	7.54	8.38	9.21	9.99	8.77	6.96	5.70	4.77	4.08	2.92	2.22	1.76	1.44	1.21
10	0.16	0.37	0.70	1.30	1.87	2.43	3.49	4.53	5.53	6.52	7.49	8.44	9.39	10.3	11.2	10.3	8.15	6.67	5.59	4.77	3.42	2.60	2.06	1.69	1.41
11	0.18	0.41	0.77	1.44	2.07	2.69	3.87	5.02	6.13	7.23	8.30	9.36	10.4	11.4	12.5	11.9	9.41	7.70	6.45	5.51	3.94	3.00	2.38	1.95	1.63
12	0.20	0.45	0.85	1.58	2.28	2.95	4.25	5.51	6.74	7.94	9.12	10.3	11.4	12.6	13.7	13.5	10.7	8.77	7.35	6.28	4.49	3.42	2.71	2.22	1.86
13	0.22	0.50	0.92	1.73	2.49	3.22	4.64	6.01	7.34	8.65	9.94	11.2	12.5	13.7	14.9	15.2	12.1	9.89	8.29	7.08	5.06	3.85	3.06	2.50	0
14	0.24	0.54	1.00	1.87	2.69	3.49	5.02	6.51	7.96	9.37	10.8	12.1	13.5	14.8	16.2	17.0	13.5	11.1	9.26	7.91	5.66	4.31	3.42	2.80	0
15	0.25	0.58	1.08	2.01	2.90	3.76	5.41	7.01	8.57	10.1	11.6	13.1	14.5	16.0	17.4	18.8	15.0	12.3	10.3	8.77	6.28	4.77	3.79	3.10	0
16	0.27	0.62	1.16	2.16	3.11	4.03	5.80	7.52	9.19	10.8	12.4	14.0	15.6	17.1	18.7	20.2	16.5	13.5	11.3	9.66	6.91	5.26	4.17	3.42	0
17	0.29	0.66	1.24	2.31	3.32	4.30	6.20	8.03	9.81	11.6	13.3	15.0	16.7	18.3	19.9	21.6	18.1	14.8	12.4	10.6	7.57	5.76	4.57	3.74	0
18	0.31	0.70	1.31	2.45	3.53	4.58	6.59	8.54	10.4	12.3	14.1	15.9	17.7	19.5	21.2	22.9	19.7	16.1	13.5	11.5	8.25	6.28	4.98	4.08	0
19	0.33	0.75	1.39	2.60	3.74	4.85	6.99	9.05	11.1	13.0	15.0	16.9	18.8	20.6	22.5	24.3	21.4	17.5	14.6	12.5	8.95	6.81	5.40	4.42	0
20	0.35	0.79	1.47	2.75	3.96	5.13	7.38	9.57	11.7	13.8	15.8	17.9	19.8	21.8	23.8	25.7	23.1	18.9	15.8	13.5	9.66	7.35	5.83	0	
21	0.36	0.83	1.55	2.90	4.17	5.40	7.78	10.1	12.3	14.5	16.7	18.8	20.9	23.0	25.1	27.1	24.8	20.3	17.0	14.5	10.4	7.91	6.28	0	
22	0.38	0.87	1.63	3.05	4.39	5.68	8.19	10.6	13.0	15.3	17.5	19.8	22.0	24.2	26.4	28.5	26.6	21.8	18.2	15.6	11.1	8.48	6.73	0	
23	0.40	0.92	1.71	3.19	4.60	5.96	8.59	11.1	13.6	16.0	18.4	20.8	23.1	25.4	27.7	29.9	28.4	23.3	19.5	16.7	11.9	9.07	7.19	0	
24	0.42	0.96	1.79	3.35	4.82	6.24	8.99	11.6	14.2	16.8	19.3	21.7	24.2	26.6	29.0	31.3	30.3	24.8	20.8	17.8	12.7	9.66	7.67	0	
25	0.44	1.00	1.87	3.50	5.04	6.52	9.40	12.2	14.9	17.5	20.1	22.7	25.3	27.8	30.3	32.7	32.2	26.4	22.1	18.9	13.5	10.3	8.15	0	
26	0.46	1.05	1.95	3.65	5.25	6.81	9.80	12.7	15.5	18.3	21.0	23.7	26.4	29.0	31.6	34.1	34.2	28.0	23.4	20.0	14.3	10.9	8.65		
28	0.50	1.13	2.12	3.95	5.69	7.37	10.6	13.8	16.8	19.8	22.8	25.7	28.5	31.4	34.2	37.0	38.2	31.3	26.2	22.4	16.0	12.2	0		
30	0.54	1.22	2.28	4.26	6.13	7.94	11.4	14.8	18.1	21.4	24.5	27.7	30.8	33.8	36.8	39.8	42.4	34.7	29.1	24.8	17.8	13.5	0		
32	0.57	1.31	2.45	4.56	6.57	8.52	12.3	15.9	19.4	22.9	26.3	29.7	33.0	36.3	39.5	42.7	46.7	38.2	32.0	27.3	19.6	14.9	0		
35	0.63	1.44	2.69	5.03	7.24	9.38	13.5	17.5	21.4	25.2	29.0	32.7	36.3	39.9	43.5	47.1	53.4	43.7	36.6	31.3	22.4	17.0	0		
40	0.73	1.67	3.11	5.81	8.37	10.8	15.6	20.2	24.7	29.1	33.5	37.7	42.0	46.1	50.3	54.4	62.5	53.4	44.7	38.2	27.3	0			
45	0.83	1.89	3.53	6.60	9.50	12.3	17.7	23.0	28.1	33.1	38.0	42.0	47.7	52.4	57.1	61.7	70.9	63.7	53.4	45.6	32.6	0			
	Type I									Type II									Type III						

SOURCE: Reprinted from *Chains for Power Transmission and Material Handling*, p. 149, by courtesy of Marcel Dekker, Inc.

NOTES: The limiting rpm for each lubrication type is read from the column to the left of the boundary line shown.

TYPE I: Manual or drip lubrication

TYPE II: Bath or disc lubrication

TYPE III: Oil stream lubrication

TABLE 7-7 Horsepower ratings, standard single-strand roller chain, no. 80, 1-in pitch

No. of teeth, small spkt.	Revolutions per minute, small sprocket																								
	10	25	50	100	150	200	300	400	500	600	700	800	900	1 000	1 100	1 200	1 400	1 600	1 800	2 000	2 200	2 400	2 700	3 000	3 400
9	0.34	0.78	1.45	2.71	3.90	5.05	7.28	9.43	11.5	13.6	15.6	17.6	17.0	14.5	12.6	11.0	8.76	7.17	6.01	5.13	4.45	3.90	3.27	2.79	2.32
10	0.38	0.87	1.63	3.03	4.37	5.66	8.16	10.6	12.9	15.2	17.5	19.7	19.9	17.0	14.7	12.9	10.3	8.40	7.04	6.01	5.21	4.57	3.83	3.27	2.71
11	0.42	0.97	1.80	3.36	4.84	6.28	9.04	11.7	14.3	16.9	19.4	21.9	23.0	19.6	17.0	14.9	11.8	9.69	8.12	6.93	6.01	5.27	4.42	3.77	1.70
12	0.47	1.06	1.98	3.69	5.32	6.89	9.93	12.9	15.7	18.5	21.3	24.0	26.2	22.3	19.4	17.0	13.5	11.0	9.25	7.90	6.85	6.01	5.04	4.30	0
13	0.51	1.16	2.16	4.03	5.80	7.52	10.8	14.0	17.1	20.2	23.2	26.2	29.1	25.2	21.8	19.2	15.2	12.5	10.4	8.91	7.72	6.78	5.68	4.85	0
14	0.55	1.25	2.34	4.36	6.29	8.14	11.7	15.2	18.6	21.9	25.1	28.4	31.5	28.2	24.4	21.4	17.0	13.9	11.7	9.96	8.63	7.57	6.35	5.42	0
15	0.59	1.35	2.52	4.70	6.77	8.77	12.6	16.4	20.0	23.6	27.1	30.6	34.0	31.2	27.1	23.8	18.9	15.4	12.9	11.0	9.57	8.40	7.04	6.01	0
16	0.63	1.45	2.70	5.04	7.26	9.41	13.5	17.6	21.5	25.3	29.0	32.8	36.4	34.4	29.8	26.2	20.8	17.0	14.2	12.2	10.5	9.25	7.76	6.62	0
17	0.68	1.55	2.88	5.38	7.75	10.0	14.5	18.7	22.9	27.0	31.0	35.0	38.9	37.7	32.7	28.7	22.7	18.6	15.6	13.3	11.5	10.1	8.49	7.25	0
18	0.72	1.64	3.07	5.72	8.25	10.7	15.4	19.9	24.4	28.7	33.0	37.2	41.4	41.1	35.6	31.2	24.8	20.3	17.0	14.5	12.6	11.0	9.25	7.90	0
19	0.76	1.74	3.25	6.07	8.74	11.3	16.3	21.1	25.8	30.4	35.0	39.4	43.8	44.5	38.6	33.9	26.9	22.0	18.4	15.7	13.6	12.0	10.0	8.57	0
20	0.81	1.84	3.44	6.41	9.24	12.0	17.2	22.3	27.3	32.2	37.0	41.7	46.3	48.1	41.7	36.6	29.0	23.8	19.9	17.0	14.7	12.9	10.8	0	
21	0.85	1.94	3.62	6.76	9.74	12.6	18.2	23.5	28.8	33.9	39.0	43.9	48.9	51.7	44.8	39.4	31.2	25.6	21.4	18.3	15.9	13.9	11.7	0	
22	0.90	2.04	3.81	7.11	10.2	13.3	19.1	24.8	30.3	35.7	41.0	46.2	51.4	55.5	48.1	42.2	33.5	27.4	23.0	19.6	17.0	14.9	12.5	0	
23	0.94	2.14	4.00	7.46	10.7	13.9	20.1	26.0	31.8	37.4	43.0	48.5	53.9	59.3	51.4	45.1	35.8	29.3	24.6	21.0	18.2	15.9	13.4	0	
24	0.98	2.24	4.19	7.81	11.3	14.6	21.0	27.2	33.2	39.2	45.0	50.8	56.4	62.0	54.8	48.1	38.2	31.2	26.2	22.3	19.4	17.0	14.2	0	
25	1.03	2.34	4.37	8.16	11.8	15.2	21.9	28.4	34.7	40.9	47.0	53.0	59.0	64.8	58.2	51.1	40.6	33.2	27.8	23.8	20.6	18.1	15.1	0	
26	1.07	2.45	4.56	8.52	12.3	15.9	22.9	29.7	36.2	42.7	49.1	55.3	61.5	67.6	61.8	54.2	43.0	35.2	29.5	25.2	21.8	19.2	16.1	0	
28	1.16	2.65	4.94	9.23	13.3	17.2	24.8	32.1	39.3	46.3	53.2	59.9	66.7	73.3	69.0	60.6	48.1	39.4	33.0	28.2	24.4	21.4	0		
30	1.25	2.85	5.33	9.94	14.3	18.5	26.7	34.6	42.3	49.9	57.3	64.6	71.8	78.9	76.6	67.2	53.3	43.6	36.6	31.2	27.1	23.8	0		
32	1.34	3.06	5.71	10.7	15.3	19.9	28.6	37.1	45.4	53.5	61.4	69.2	77.0	84.6	84.3	74.0	58.7	48.1	40.3	34.4	29.8	26.2	0		
35	1.48	3.37	6.29	11.7	16.9	21.9	31.6	40.9	50.0	58.9	67.6	76.3	84.8	93.3	96.5	84.7	67.2	55.0	46.1	39.4	34.1	0			
40	1.71	3.89	7.27	13.6	19.5	25.3	36.4	47.2	57.7	68.0	78.1	88.1	98.0	108	117	103	82.1	67.2	56.3	48.1	20.0	0			
45	1.94	4.42	18.25	15.4	22.2	28.7	41.4	53.6	65.6	77.2	88.7	100	111	122	133	123	98.0	80.2	67.2	54.1	0				

Type I · Type II · Type III

SOURCE: Reprinted from *Chains for Power Transmission and Material Handling*, p. 149, by courtesy of Marcel Dekker, Inc.

NOTES: The limiting rpm for each lubrication type is read from the column to the left of the boundary line shown.

TYPE I: Manual or drip lubrication

TYPE II: Bath or disc lubrication

TYPE III: Oil stream lubrication

Type I (170 to 650 ft/min).   *Manual or drip lubrication:* For manual lubrication, oil is applied with a brush or a spout can, preferably at least once every 8 h of operation. For drip feed lubrication, oil is fed directly onto the link plates of each chain strand.

Type II (650 to 1 500 ft/min).   *Bath or disc lubrication:* The chain cover provides a sump of oil into which the chain dips continuously. Alternatively, a disc or a slinger can be attached to one of the shafts to lift oil to a trough above the lower strand of chain. The trough then delivers a stream of oil to the chain. The chain itself, then, does not need to dip into the oil.

Type III (above 1 500 ft/min).   *Oil stream lubrication:* An oil pump delivers a continuous stream of oil on the lower part of the chain.
   Example Problem 7–2 presents a chain drive design.

Example Problem 7–2   Design a chain drive for a uniformly loaded conveyor to be driven by a gasoline engine through a mechanical drive. The input speed will be 900 rpm, and the desired output speed is 230 to 240 rpm. The conveyor requires 15.0 hp.

Solution   Objective   Design the chain drive.

Given   Power transmitted = 15 hp to a conveyor
Speed of motor = 900 rpm; output speed range = 230 to 240 rpm

Analysis   Use the design data presented in this section. The solution procedure is developed within the Results section of the problem solution.

Results   *Step 1.*   Specify a service factor and compute the design power. From Table 7–7, for moderate shock and a gasoline engine drive through a mechanical drive, $SF = 1.4$. A lower value can be used because of the uniform loading:

$$\text{Design power} = 1.4(15.0) = 21.0 \text{ hp}$$

Step 2.   Compute the desired ratio. Using the middle of the required range of output speeds, we have

$$\text{Ratio} = (900 \text{ rpm})/(235 \text{ rpm}) = 3.83$$

Step 3.   Refer to the tables for power capacity (Tables 7–5, 7–6, and 7–7), and select the chain pitch. For a single strand, the no. 60 chain with $p = 3/4$ in seems best. A 21-tooth sprocket is rated at 20.9 hp at 900 rpm. At this speed, type II lubrication (oil bath) is required.

Step 4.   Compute the required number of teeth on the large sprocket:

$$N_2 = N_1 \times \text{ratio} = 21(3.83) = 80.43$$

Let's use the integer: 80 teeth.

Step 5.   Compute the actual expected output speed:

$$n_2 = n_1(N_1/N_2) = 900 \text{ rpm}(21/80) = 236 \text{ rpm} \quad \text{(Okay!)}$$

Step 6. Compute the pitch diameters of the sprockets:

$$D_1 = \frac{p}{\sin(180°/N_1)} = \frac{0.75}{\sin(180°/21)} = 5.032 \text{ in}$$

$$D_2 = \frac{p}{\sin(180°/N_2)} = \frac{0.75}{\sin(180°/80)} = 19.103 \text{ in}$$

Step 7. Specify the nominal center distance. Let's use the middle of the recommended range, 40 pitches.

Step 8. Compute the required chain length in pitches from Equation (7–9):

$$L = 2C + \frac{N_2 + N_1}{2} + \frac{(N_2 - N_1)^2}{4\pi^2 C} \tag{7–9}$$

$$L = 2(40) + \frac{80 + 21}{2} + \frac{(80 - 21)^2}{4\pi^2(40)} = 132.7 \text{ pitches}$$

Step 9. Specify an integral number of pitches for the chain length, and compute the actual theoretical center distance. Let's use 132 pitches, an even number. Then, from Equation (7–10),

$$C = \frac{1}{4}\left[L - \frac{N_2 + N_1}{2} + \sqrt{\left[L - \frac{N_2 + N_1}{2}\right]^2 - \frac{8(80 - 21)^2}{4\pi^2}}\right] \tag{7–10}$$

$$C = \frac{1}{4}\left[132 - \frac{80 + 21}{2} + \sqrt{\left[132 - \frac{80 + 21}{2}\right]^2 - \frac{8(80 - 21)^2}{4\pi^2}}\right]$$

$$C = 39.638 \text{ pitches} = 39.638(0.75 \text{ in}) = 29.73 \text{ in}$$

Comments   **Summary of Design**

Figure 7–21 shows a sketch of the design to scale.

Pitch: No. 60 chain, 3/4-in pitch

Length: 132 pitches = 132(0.75) = 99.0 in

Center distance: $C = 29.73$ in (maximum)

Sprockets: Single-strand, no. 60, 3/4-in pitch

FIGURE 7–21   Scale
layout of chain drive

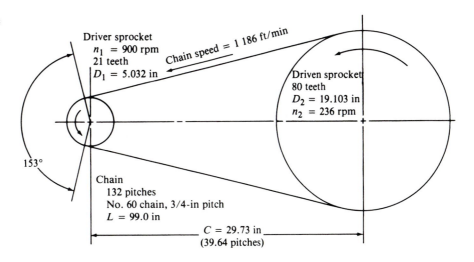

FIGURE 8–1   Pair of spur gears. The pinion drives the gear.

Pinion: 11 teeth = N_P

Pitch circle

Outside diameter

Tooth profile (involute)

n_P

Pitch circle

Pitch circle

Center distance

Pitch diameter

n_G

Gear: 18 teeth = N_G

Spur gears have teeth that are straight and arranged parallel to the axis of the shaft that carries the gear. The curved shape of the faces of the spur gear teeth have a special geometry called an *involute curve,* described later in this chapter. This shape makes it possible for two gears to operate together with smooth, positive transmission of power. Figure 8–1 also shows the side view of spur gear teeth, and the involute curve shape is evident there. The shafts carrying the gears are parallel.

The teeth of *helical gears* are arranged so that they lie at an angle with respect to the axis of the shaft. The angle, called the *helix angle,* can be virtually any angle. Typical helix angles range from approximately 10° to 30°, but angles up to 45° are practical. The helical teeth operate more smoothly than equivalent spur gear teeth, and stresses are lower. Therefore, a smaller helical gear can be designed for a given power-transmitting capacity as compared with spur gears. One disadvantage of helical gears is that an axial force, called a *thrust force,* is generated in addition to the driving force that acts tangent to the basic cylinder on which the teeth are arranged. The designer must consider the thrust force when selecting bearings that will hold the shaft during operation. Shafts carrying helical gears are typically arranged parallel to each other. However, a special design, called *crossed helical gears,* has 45° helix angles, and their shafts operate 90° to each other.

Bevel gears have teeth that are arranged as elements on the surface of a cone. The teeth of straight bevel gears appear to be similar to spur gear teeth, but they are tapered, being wider at the outside and narrower at the top of the cone. Bevel gears typically operate on shafts that are 90° to each other. Indeed, this is often the reason for specifying bevel gears in a drive system. Specially designed bevel gears can operate on shafts that are at some angle other than 90°. When bevel gears are made with teeth that form a helix angle similar to

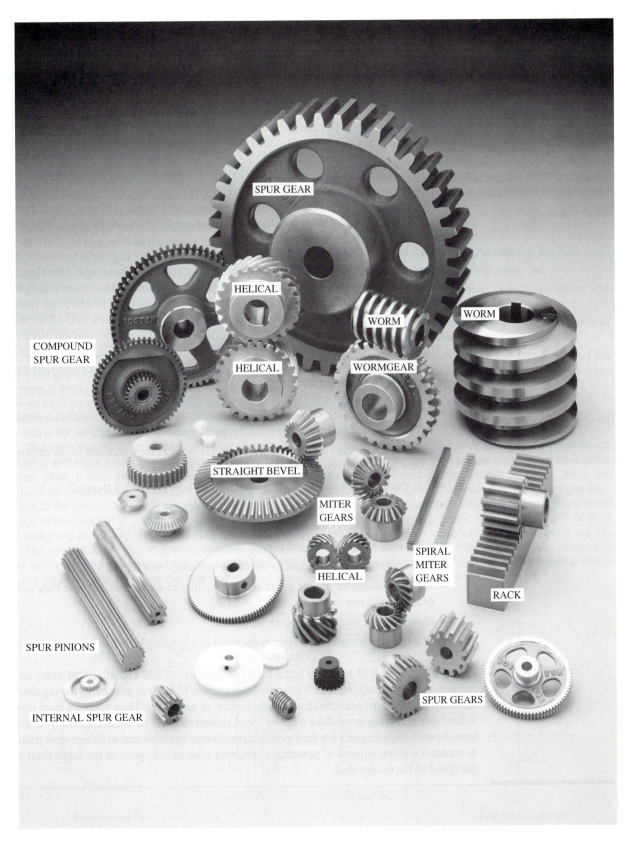

FIGURE 8–2 A variety of gear types (Boston Gear, IMO Industries, Quincy, MA)

TABLE 8–5   Recommended minimum backlash for coarse pitch gears

A. Diametral pitch system (backlash in inches)

P_d	Center distance, C (in)				
	2	4	8	16	32
18	0.005	0.006			
12	0.006	0.007	0.009		
8	0.007	0.008	0.010	0.014	
5		0.010	0.012	0.016	
3		0.014	0.016	0.020	0.028
2			0.021	0.025	0.033
1.25				0.034	0.042

B. Metric module system (backlash in millimeters)

Module, m	Center distance, C (mm)				
	50	100	200	400	800
1.5	0.13	0.16			
2	0.14	0.17	0.22		
3	0.18	0.20	0.25	0.35	
5		0.26	0.31	0.41	
8		0.35	0.40	0.50	0.70
12			0.52	0.62	0.82
18				0.80	1.00

SOURCE: Extracted from AGMA 2002-B88 Standard, *Tooth Thickness Specification and Measurement,* with permission of the publisher, American Gear Manufacturers Association, 1500 King Street, Suite 201, Alexandria, VA 22314.

Pressure Angle

> *The* **pressure angle** *is the angle between the tangent to the pitch circles and the line drawn normal (perpendicular) to the surface of the gear tooth (see Figure 8–12).*

The normal line is sometimes referred to as the *line of action.* When two gear teeth are in mesh and are transmitting power, the force transferred from the driver to the driven gear tooth acts in a direction along the line of action. Also, the actual shape of the gear tooth depends on the pressure angle, as illustrated in Figure 8–13. The teeth in this figure were drawn according to the proportions for a 20-tooth, 5-pitch gear having a pitch diameter of 4.000 in.

All three teeth have the same tooth thickness because, as stated in Equation (8–16), the thickness at the pitch line depends only on the pitch. The difference between the three teeth shown is due to the different pressure angles because the pressure angle determines the size of the base circle. Remember that the base circle is the circle from which the involute is generated. The line of action is always tangent to the base circle. Therefore, the size of the base circle can be found from

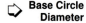
Base Circle Diameter

$$D_b = D \cos \phi \qquad\qquad\qquad (8\text{–}20)$$

FIGURE 8–12
Pressure angle

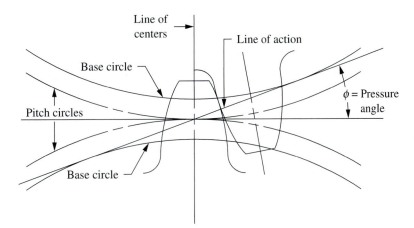

FIGURE 8–13 Full-depth, involute-tooth form for varying pressure angles

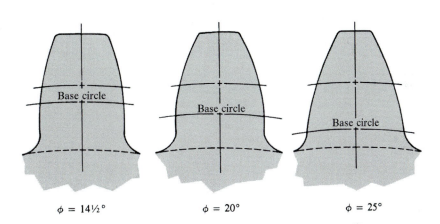

$$\phi = 14\tfrac{1}{2}° \qquad \phi = 20° \qquad \phi = 25°$$

Standard values of the pressure angle are established by gear manufacturers, and the pressure angles of two gears in mesh must be the same. Current standard pressure angles are $14\tfrac{1}{2}°$, $20°$, and $25°$, as illustrated in Figure 8–13. Actually, the $14\tfrac{1}{2}°$ tooth form is considered to be obsolete. Although it is still available, it should be avoided for new designs. The $20°$ tooth form is the most readily available at this time. The advantages and disadvantages of the different values of pressure angle relate to the strength of the teeth, the occurrence of interference, and the magnitude of forces exerted on the shaft. Interference is discussed next. The other points are discussed in a later chapter.

Example Problem 8–1

For the pair of gears shown in Figure 8–1, compute all of the features of the gear teeth described in this section. The gears conform to the standard AGMA form and have a diametral pitch of 12 and a 20° pressure angle.

Solution

Given $P_d = 12$; $N_P = 11$; $N_G = 18$; $\phi = 20°$.

Analysis We use Equations (8–2) through (8–20) and Table 8–4 to compute the features. Note that gears are precision mechanical components. Dimensions are typically produced to at least the nearest thousandth of an inch (0.001 in). Often, for more accurate gears, controlling to the nearest 0.0001 in is important. Also, in the inspection of gear features using metrology techniques, it is important to know the standard dimension to a high degree of precision.

The results for this problem are presented to a minimum of three decimal places or, for small dimensions, to four decimal places. A similar level of precision is expected in practice problems using this book.

Results **Pitch Diameters**

For the pinion,

$$D_P = N_P/P_d = 11/12 = 0.916\ 7 \text{ in}$$

For the gear,

$$D_G = N_G/P_d = 18/12 = 1.500 \text{ in}$$

Circular Pitch

Three different approaches could be used. First, using Equation (8–3), note that either the pinion or the gear data may be used. For the pinion,

$$p = \pi D_P/N_P = \pi(0.916\ 7 \text{ in})/11 = 0.261\ 8 \text{ in}$$

For the gear,

$$p = \pi D_G/N_G = \pi(1.500 \text{ in})/18 = 0.261\ 8 \text{ in}$$

Now, we can also use Equation (8–5):

$$p = \pi/P_d = \pi/12 = 0.261\ 8 \text{ in}$$

Addendum

From Table 8–4,

$$a = 1/P_d = 1/12 = 0.083\ 3 \text{ in}$$

Dedendum

From Table 8–4, note that the 12-pitch gear is considered to be *coarse.* Thus,

$$b = 1.25/P_d = 1.25/12 = 0.104\ 2 \text{ in}$$

Clearance

From Table 8–4,

$$c = 0.25/P_d = 0.25/12 = 0.020\ 8 \text{ in}$$

Outside Diameters

Use of Equation (8–10) is preferred for accuracy. For the pinion,

$$D_{oP} = (N_P + 2)/P_d = (11 + 2)/12 = 1.083\ 3 \text{ in}$$

For the gear,

$$D_{oG} = (N_G + 2)/P_d = (18 + 2)/12 = 1.666\ 7 \text{ in}$$

Root Diameters

We use Equation (8–12). First, for the pinion,

$$D_{RP} = D_P - 2b = 0.916\ 7 \text{ in} - 2(0.104\ 2 \text{ in}) = 0.708\ 3 \text{ in}$$

For the gear,

$$D_{RG} = D_G - 2b = 1.500 \text{ in} - 2(0.104\ 2 \text{ in}) = 1.291\ 7 \text{ in}$$

Whole Depth

Using Equation (8–13), we have

$$h_t = a + b = 0.083\ 3\ \text{in} + 0.104\ 2\ \text{in} = 0.187\ 5\ \text{in}$$

Working Depth

Using Equation (8–14), we have

$$h_k = 2a = 2(0.083\ 3\ \text{in}) = 0.166\ 7\ \text{in}$$

Tooth Thickness

Using Equation (8–16), we have

$$t = \pi/2P_d = \pi/2(12) = 0.130\ 9\ \text{in}$$

Center Distance

Use of Equation (8–18) is preferred:

$$C = (N_G + N_P)/(2P_d) = (18 + 11)/[2(12)] = 1.208\ 3\ \text{in}$$

Base Circle Diameter

Using Equation (8–20), we have

$$D_{bP} = D_P \cos \phi = (0.916\ 7\ \text{in})\cos(20°) = 0.861\ 4\ \text{in}$$

$$D_{bg} = D_G \cos \phi = (1.500\ \text{in})\cos(20°) = 1.409\ 5\ \text{in}$$

**8–5
INTERFERENCE
BETWEEN
MATING SPUR
GEAR TEETH**

For certain combinations of numbers of teeth in a gear pair, there is interference between the tip of the teeth on the pinion and the fillet or root of the teeth on the gear. Obviously this cannot be tolerated because the gears simply will not mesh. The probability that interference will occur is greatest when a small pinion drives a large gear, with the worst case being a small pinion driving a rack. A *rack* is a gear with a straight pitch line; it can be thought of as a gear with an infinite pitch diameter [see Figure 8–4(c)].

It is the designer's responsibility to ensure that interference does not occur in a given application. The surest way to do this is to control the minimum number of teeth in the pinion to the limiting values shown on the left side of Table 8–6. With this number of teeth or a greater number, there will be no interference with a rack or with any other gear. A designer who desires to use fewer than the listed number of teeth can use a graphical layout to test the combination of pinion and gear for interference. Texts on kinematics provide the necessary procedure. The right side of Table 8–6 indicates the maximum number of gear teeth that you can use for a given number of pinion teeth to avoid interference. (See Reference 9.)

Using the information in Table 8–6, we can draw the following conclusions:

1. If a designer wants to be sure that there will not be interference between any two gears when using the $14\frac{1}{2}°$, full-depth, involute system, the pinion of the gear pair must have no fewer than 32 teeth.

2. For the 20°, full-depth, involute system, using no fewer than 18 teeth will ensure that no interference occurs.

TABLE 8–6 Number of pinion teeth to ensure no interference

| For a pinion meshing with a rack | | For a 20°, full-depth pinion meshing with a gear | |
Tooth form	Minimum number of teeth	Number of pinion teeth	Maximum number of gear teeth
14½°, involute, full-depth	32	17	1 309
20°, involute, full-depth	18	16	101
25°, involute, full-depth	12	15	45
		14	26
		13	16

3. For the 25°, full-depth, involute system, using no fewer than 12 teeth will ensure that no interference occurs.

4. If a designer desires to use fewer than 18 teeth in a pinion having 20°, full-depth teeth, there is an upper limit to the number of teeth that can be used on the mating gear without interference. For 17 teeth in the pinion, any number of teeth on the gear can be used up to 1 309, a very high number. Most gear drive systems use no more than about 200 teeth in any gear. But a 17-tooth pinion *would* have interference with a *rack* which is effectively a gear with an infinite number of teeth or an infinite pitch diameter. Similarly, the following requirements apply:

A 16-tooth pinion requires a gear having 101 or fewer teeth, producing a maximum velocity ratio of $N_G/N_P = 101/16 = 6.31$.

A 15-tooth pinion requires a gear having 45 or fewer teeth, producing a maximum velocity ratio of 45/15 = 3.00.

A 14-tooth pinion requires a gear having 26 or fewer teeth, producing a maximum velocity ratio of 26/14 = 1.85.

A 13-tooth pinion requires a gear having 16 or fewer teeth, producing a maximum velocity ratio of 16/13 = 1.23.

As noted earlier, the 14½° system is considered to be obsolete. The data in Table 8–6 indicate one of the main disadvantages with that system: its potential for causing interference.

If a proposed design encounters interference, there are ways to make it work. But caution should be exercised because the tooth form or the alignment of the mating gears is changed, causing the stress and wear analysis to be inaccurate. With this in mind, the designer can provide for undercutting, modification of the addendum on the pinion or the gear, or modification of the center distance:

Undercutting *is the process of cutting away the material at the fillet or root of the gear teeth, thus relieving the interference.*

Figure 8–14 shows the result of undercutting. It should be obvious that this process weakens the tooth; this point is discussed further in the section on stresses in gear teeth.

To alleviate the problem of interference, increase the addendum of the pinion while decreasing the addendum of the gear. The center distance can remain the same as its theo-

FIGURE 8–14
Undercutting of a gear tooth

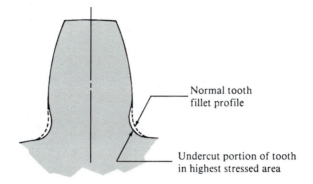

Normal tooth
fillet profile

Undercut portion of tooth
in highest stressed area

retical value for the number of teeth in the pair. But the resulting gears are, of course, non-standard. (See Reference 10.) It is possible to make the pinion of a gear pair larger than standard while keeping the gear standard if the center distance for the pair is enlarged. (See Reference 9.)

**8–6
VELOCITY RATIO
AND GEAR
TRAINS**

A gear train is one or more pairs of gears operating together to transmit power.

Normally there is a speed change from one gear to the next due to the different sizes of the gears in mesh. The fundamental building block of the total speed change ratio in a gear train is the *velocity ratio* between two gears in a single pair.

Velocity Ratio

*The **velocity ratio (VR)** is defined as the ratio of the rotational speed of the input gear to that of the output gear for a single pair of gears.*

To develop the equation for computing the velocity ratio, it is helpful to view the action of two gears in mesh, as shown in Figure 8–15. The action is equivalent to the action of two smooth wheels rolling on each other without slipping, with the diameters of the two wheels equal to the pitch diameters of the two gears. Remember that when two gears are in mesh, their pitch circles are tangent and that, obviously, the gear teeth prohibit any slipping.

As shown in Figure 8–15, without slipping there is no relative motion between the two pitch circles at the pitch point, and therefore the linear velocity of a point on either pitch circle is the same. We will use the symbol v_t for this velocity. The linear velocity of a point that is in rotation at a distance R from its center of rotation and rotating with an angular velocity, ω, is found from

⇨ **Pitch Line
Speed of a Gear**

$$v_t = R\omega \tag{8–21}$$

Using the subscript P for the pinion and G for the gear for two gears in mesh, we have

$$v_t = R_P\omega_P \quad \text{and} \quad v_t = R_G\omega_G$$

FIGURE 8–15 Two gears in mesh

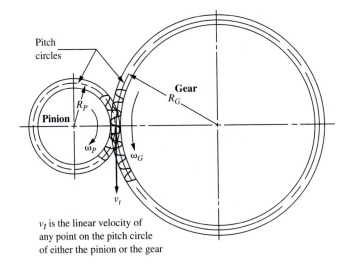

Pitch circles

Pinion

R_P

Gear R_G

ω_P

ω_G

v_t

v_t is the linear velocity of any point on the pitch circle of either the pinion or the gear

This set of equations says that the pitch line speeds of the pinion and the gear are the same. Equating these two and solving for ω_P/ω_G gives our definition for the velocity ratio, *VR:*

$$VR = \omega_P/\omega_G = R_G/R_P$$

In general, it is convenient to express the velocity ratio in terms of the pitch diameters, the rotational speeds, or the numbers of teeth of the two gears in mesh. Remember that

$$R_G = D_G/2$$
$$R_P = D_P/2$$
$$D_G = N_G/P_d$$
$$D_P = N_P/P_d$$
$$n_P = \text{rotational speed of the pinion} \quad \text{(in rpm)}$$
$$n_G = \text{rotational speed of the gear} \quad \text{(in rpm)}$$

The velocity ratio can then be defined in any of the following ways:

▷ **Velocity Ratio for Gear Pair**

$$VR = \frac{\omega_P}{\omega_G} = \frac{n_P}{n_G} = \frac{R_G}{R_P} = \frac{D_G}{D_P} = \frac{N_G}{N_P} = \frac{\text{speed}_P}{\text{speed}_G} = \frac{\text{size}_G}{\text{size}_P} \qquad \textbf{(8–22)}$$

Most gear drives are *speed reducers;* that is, their output speed is lower than their input speed. This results in a velocity ratio greater than 1. If a *speed increaser* is desired, then *VR* is less than 1. Note that not all books and articles use the same definition for velocity ratio. Some define it as the ratio of the output speed to the input speed, the inverse of our definition. It is thought that the use of *VR* greater than 1 for the reducer, that is, the majority of the time, is more convenient.

Train Value

When more than two gears are in mesh, the term **train value (TV)** *refers to the ratio of the input speed (for the first gear in the train) to the output speed (for the last gear in the train). By definition the* **train value** *is the product of the values of VR for each gear pair in the train. In this*

FIGURE 8–16
Double-reduction gear train

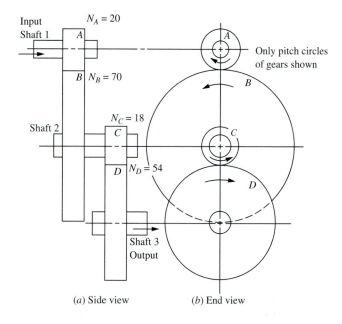

(a) Side view   (b) End view

definition, a **gear pair** *is any set of two gears with a driver and a follower (driven) gear.*

Again, *TV* will be greater than 1 for a reducer and less than 1 for an increaser. For example, consider the gear train sketched in Figure 8–16. The input is through the shaft carrying gear *A*. Gear *A* drives gear *B*. Gear *C* is on the same shaft with gear *B* and rotates at the same speed. Gear *C* drives gear *D*, which is connected to the output shaft. Then gears *A* and *B* constitute the first gear pair, and gears *C* and *D* constitute the second pair. The velocity ratios are

$$VR_1 = n_A/n_B$$
$$VR_2 = n_C/n_D$$

The train value is

$$TV = (VR_1)(VR_2) = \frac{n_A}{n_B}\frac{n_C}{n_D}$$

But, because they are on the same shaft, $n_B = n_C$, and the preceding equation reduces to

$$TV = n_A/n_D$$

This is the input speed divided by the output speed, the basic definition of the train value. This process can be expanded to any number of stages of reduction in a gear train.

Remember that any of the forms for velocity ratio shown in Equation (8–22) can be used for computing the train value. In design, it is often most convenient to express the velocity ratio in terms of the numbers of teeth in each gear because they must be integers. Then, once the diametral pitch or module is defined, the values of the diameters or radii can be determined.

The train value of the double-reduction gear train in Figure 8–16 can be expressed in terms of the numbers of teeth in the four gears as follows:

$$VR_1 = N_B/N_A$$

Note that this is the number of teeth in the *driven gear B* divided by the number of teeth in the *driving gear A*. This is the typical format for velocity ratio. Then VR_2 can be found similarly:

$$VR_2 = N_D/N_C$$

Thus, the train value is

$$TV = (VR_1)(VR_2) = (N_B/N_A)(N_D/N_C)$$

This is usually shown in the form

⟳ **Train Value**

$$TV = \frac{N_B}{N_A}\frac{N_D}{N_C} = \frac{\text{product of number of teeth in the driven gears}}{\text{product of number of teeth in the driving gears}} \quad \textbf{(8–23)}$$

This is the form for train value that we will use most often.

The direction of rotation can be determined by observation, noting that there is a direction reversal for each pair of external gears.

We will use the term positive train value *to refer to one in which the input and output gears rotate in the same direction. Conversely, if they rotate in the opposite direction, the train value will be negative.*

Example Problem 8–2

For the gear train shown in Figure 8–16, if the input shaft rotates at 1 750 rpm clockwise, compute the speed of the output shaft and its direction of rotation.

Solution

We can find the output speed if we can determine the train value:

$$TV = n_A/n_D = \text{input speed/output speed}$$

Then

$$n_D = n_A/TV$$

But

$$TV = (VR_1)(VR_2) = \frac{N_B}{N_A}\frac{N_D}{N_C} = \frac{70}{20}\frac{54}{18} = \frac{3.5}{1}\frac{3.0}{1} = \frac{10.5}{1} = 10.5$$

Now

$$n_D = n_A/TV = (1\ 750\ \text{rpm})/10.5 = 166.7\ \text{rpm}$$

Gear *A* rotates clockwise; gear *B* rotates counterclockwise.
Gear *C* rotates counterclockwise; gear *D* rotates clockwise.

Thus, the train in Figure 8–16 is a positive train.

Example Problem 8–3

Determine the train value for the train shown in Figure 8–17. If the shaft carrying gear *A* rotates at 1 750 rpm clockwise, compute the speed and the direction of the shaft carrying gear *E*.

Solution

Look first at the direction of rotation. Remember that a gear pair is defined as any two gears in mesh (a driver and a follower). There are actually three gear pairs:

FIGURE 8–17
Double-reduction gear
train with an idler. Gear
D is an idler.

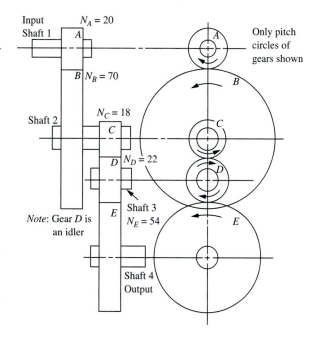

Gear *A* drives gear *B: A* rotates clockwise; *B,* counterclockwise.

Gear *C* drives gear *D: C* rotates counterclockwise; *D,* clockwise.

Gear *D* drives gear *E: D* rotates clockwise; *E,* counterclockwise.

Because gears *A* and *E* rotate in opposite directions, the train value is negative. Now

$$TV = -(VR_1)(VR_2)(VR_3)$$

In terms of the number of teeth,

$$TV = -\frac{N_B}{N_A}\frac{N_D}{N_C}\frac{N_E}{N_D}$$

Note that the number of teeth in gear *D* appears in both the numerator and the denominator and thus can be canceled. The train value then becomes

$$TV = -\frac{N_B}{N_A}\cdot\frac{N_E}{N_C} = -\frac{70}{20}\cdot\frac{54}{18} = -\frac{3.5}{1}\frac{3.0}{1} = -10.5$$

Gear *D* is called an *idler.* As demonstrated here, it has no effect on the magnitude of the train value, but it does cause a direction reversal. The output speed is then found from

$$TV = n_A/n_E$$

$$n_E = n_A/TV = (1\ 750\ \text{rpm})/(-10.5) = -166.7\ \text{rpm}\quad\text{(counterclockwise)}$$

Idler Gear

Example Problem 8–2 introduced the concept of an *idler gear,* defined as follows:

> ***Any gear in a gear train that performs as both a driving gear and a driven gear is called an*** idler gear ***or simply an*** idler.

The main features of an idler are as follows:

1. An idler does not affect the train value of a gear train because, since it is both a driver and a driven gear, its number of teeth appears in both the numerator and the denominator of the train value equation, Equation (8–23). Thus, any pitch diameter size and any number of teeth may be used for the idler.

2. Placing an idler in a gear train causes a direction reversal of the output gear.

3. An idler gear may be used to fill a space between two gears in a gear train when the desired distance between their centers is greater than the center distance for the two gears alone.

Internal Gear

*An **internal gear** is one for which the teeth are machined on the inside of a ring instead of on the outside of a gear blank.*

An internal gear mating with a standard, external pinion is illustrated in Figure 8–2, along with a variety of other kinds of gears.

Figure 8–18 is a sketch of an external pinion driving an internal gear. Note the following:

$$\text{Center distance} = \frac{D_G - D_P}{2}$$

FIGURE 8–18   Internal gear driven by an external pinion

1. The gear rotates in the *same direction* as the pinion. This is different from the case when an external pinion drives an external gear.

2. The center distance is

**Center Distance
Internal Gear**

$$C = D_G/2 - D_P/2 = (D_G - D_P)/2 = (N_G/P_d - N_P/P_d)/2 = (N_G - N_P)/(2P_d) \quad \textbf{(8–24)}$$

The last form is preferred because its factors are all integers for typical gear trains.

3. The descriptions of most other features of internal gears are the same as those for external gears presented earlier. Exceptions for an internal gear are as follows:

The addendum, *a,* is the radial distance from the pitch circle to the inside of a tooth. The inside diameter, D_i, is

$$D_i = D - 2a$$

The root diameter, D_R, is

$$D_R = D + 2b$$

where b = dedendum

Internal gears are used when it is desired to have the same direction of rotation for the input and the output. Also note that less space is taken for an internal gear mating with an external pinion compared with two external gears in mesh.

Velocity of a Rack

Figure 8–19 shows the basic configuration of a *rack-and-pinion* drive. The function of such a drive is to produce a linear motion of the rack from the rotational motion of the driving pinion. The opposite is also true: If the driver produces the linear motion of the rack, it produces a rotational motion of the pinion.

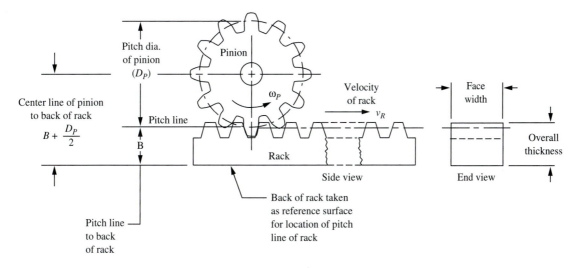

FIGURE 8–19 Rack driven by a pinion

The linear velocity of the rack, v_R, must be the same as the pitch line velocity of the pinion, v_t, as defined by Equation (8–21), repeated here. Recall that ω_P is the angular velocity of the pinion:

$$v_R = v_t = R_P\omega_P = (D_P/2)\omega_P \qquad \textbf{(8–21)}$$

You must carefully consider units when using this equation. The angular velocity ω_P should be expressed in rad/s. Then the units for linear velocity will be in/s if the pitch diameter is in inches. If the pitch diameter is in mm, as in the metric module system, then the units for velocity will be mm/s. These units could be converted to m/s if that is more convenient.

The concept of center distance does not apply directly for a rack-and-pinion set because the center of the rack is at infinity. But it is critical that the pitch circle of the pinion be tangent to the pitch line of the rack as shown in Figure 8–19. The rack will be machined so that there is a specified dimension between the pitch line and a reference surface, typically the back of the rack. This is dimension B in Figure 8–19. Then the location of the center of the pinion can be computed using the relationships shown in the figure.

Example Problem 8–4 Determine the linear velocity of the rack in Figure 8–19 if the driving pinion rotates at 1 750 rpm. The pinion has 24 teeth and a diametral pitch of 6.

Solution We will use Equation (8–21). First the pitch diameter of the pinion is computed using Equation (8–4):

$$D_P = N_P/P_d = 24/6 = 4.000 \text{ in}$$

Now the rotational speed is converted to rad/s:

$$\omega_P = (1\,750 \text{ rev/min})(2\pi \text{ rad/rev})(1 \text{ min/60 s}) = 183.25 \text{ rad/s}$$

Then the pitch line speed of the pinion and the linear velocity of the rack are both equal to

$$v_R = v_t = (D_P/2)\omega_P = (4.000 \text{ in/2})(183.25 \text{ rad/s}) = 366.5 \text{ in/s}$$

8–7 HELICAL GEAR GEOMETRY

Helical and spur gears are distinguished by the orientation of their teeth. On spur gears, the teeth are straight and are aligned with the axis of the gear. On helical gears, the teeth are inclined at an angle with the axis, that angle being called the *helix angle*. If the gear were very wide, it would appear that the teeth wind around the gear blank in a continuous, helical path. However, practical considerations limit the width of the gears so that the teeth normally appear to be merely inclined with respect to the axis. Figure 8–20 shows two examples of commercially available helical gears.

The forms of helical gear teeth are very similar to those discussed for spur gears. The basic task is to account for the effect of the helix angle.

Helix Angle

The helix for a given gear can be either *left-hand* or *right-hand*. The teeth of a right-hand helical gear would appear to lean to the right when the gear is lying on a flat surface. Con-

FIGURE 8-20
Helical gears. These
gears have a 45° helix
angle. (Emerson Power
Transmission
Corporation, Drive and
Component Division,
Ithaca, NY.)

Left
hand
helix

Right
hand
helix

(*a*) Helical gears with parallel shafts   (*b*) Crossed helical gears, shafts
at right angle

versely, the teeth of a left-hand helical gear would lean to the left. In normal installation, helical gears would be mounted on parallel shafts as shown in Figure 8-20(a). To achieve this arrangement, it is required that one gear be of the right-hand design and that the other be left-hand with an equal helix angle. If both gears in mesh are of the same hand, as shown in Figure 8-20(b), the shafts will be at 90° to each other. Such gears are called *crossed helical gears.*

The parallel shaft arrangement for helical gears is preferred because it results in a much higher power-transmitting capacity for a given size of gear than the crossed helical arrangement. In this book, we will assume that the parallel shaft arrangement is being used unless otherwise stated.

Figure 8-21(a) shows the pertinent geometry of helical gear teeth. To simplify the drawing, only the pitch surface of the gear is shown. The pitch surface is the cylinder that passes through the gear teeth at the pitch line. Thus, the diameter of the cylinder is equal to the pitch diameter of the gear. The lines drawn on the pitch surface represent elements of each tooth where the surface would cut into the face of the tooth. These elements are inclined with respect to a line parallel to the axis of the cylinder, and the angle of inclination is the *helix angle,* ψ (the Greek letter *psi*).

The main advantage of helical gears over spur gears is smoother engagement because a given tooth assumes its load gradually instead of suddenly. Contact starts at one end of a tooth near the tip and progresses across the face in a path downward across the pitch line to the lower flank of the tooth, where it leaves engagement. Simultaneously, other teeth are coming into engagement before a given tooth leaves engagement, with the result that a larger average number of teeth are engaged and are sharing the applied loads compared with a spur gear. The lower average load per tooth allows a greater power transmission capacity for a given size of gear, or a smaller gear can be designed to carry the same power.

The main disadvantage of helical gears is that an *axial thrust load* is produced as a natural result of the inclined arrangement of the teeth. The bearings that hold the shaft carrying the helical gear must be capable of reacting against the thrust load.

The helix angle is specified for each given gear design. A balance should be sought to take advantage of the smoother engagement of the gear teeth when the helix angle is high while maintaining a reasonable value of the axial thrust load that increases with increasing helix angle. A typical range of values of helix angles is from 15° to 45°.

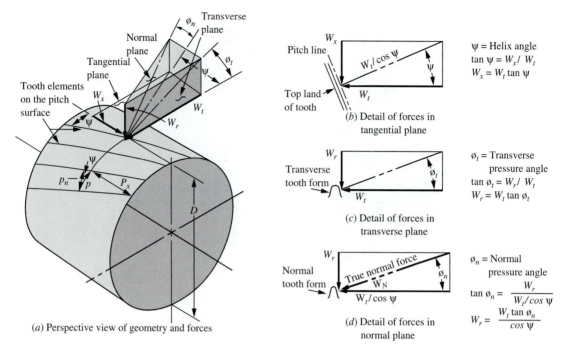

FIGURE 8–21   Helical gear geometry and forces

Pressure Angles, Primary Planes, and Forces for Helical Gears

To completely describe the geometry of helical gear teeth, we need to define two different pressure angles in addition to the helix angle. The two pressure angles are related to the three primary planes that are illustrated in Figure 8–21: (1) the *tangential plane,* (2) the *transverse plane,* and (3) the *normal plane.* Note that these planes contain the three orthogonal components of the true total normal force that is exerted by a tooth of one gear on a tooth of the mating gear. More will be said about the forces later, but it may help you to understand the geometry of the teeth, and the importance of that geometry, to see how it affects the forces.

We refer to the *true normal force* as W_N. It acts normal (perpendicular) to the curved surface of the tooth. In reality, we typically do not use the normal force itself in analyzing the performance of the gear. Instead we use its three orthogonal components:

- The *tangential force* (also called the *transmitted force*), W_t, acts tangential to the pitch surface of the gear and perpendicular to the axis of the shaft carrying the gear. This is the force that actually drives the gear. Stress analysis and pitting resistance are both related to the magnitude of the tangential force. It is similar to W_t used in spur gear design and analysis.

- The *radial force,* W_r, acts toward the center of the gear along a radius and tends to separate the two gears in mesh. It is similar to W_r used in spur gear design and analysis.

- The *axial force,* W_x, acts in the tangential plane parallel to the axis of the shaft carrying the gear. Another name for the axial force is the *thrust.* It tends to push the gear along the shaft. The thrust must be reacted by one of the bearings that carry the shaft, so this force is generally undesirable. Spur gears generate no such force because the teeth are straight and are parallel to the axis of the gear.

The plane containing the tangential force, W_t, and the axial force, W_x, is the *tangential plane* [see Figure 8–21(b)]. It is tangential to the pitch surface of the gear and acts through the pitch point at the middle of the face of the tooth being analyzed.

The plane containing the tangential force, W_t, and the radial force, W_r, is the *transverse plane* [see Figure 8–21(c)]. It is perpendicular to the axis of the gear and acts through the pitch point at the middle of the face of the tooth being analyzed. The *transverse pressure angle*, ϕ_t, is defined in this plane as shown in the figure.

The plane containing the true normal force, W_N, and the radial force, W_r, is the *normal plane* [see Figure 8–21(d)]. The angle between the normal plane and the tangential plane is the helix angle, ψ. Within the normal plane we can see that the angle between the tangential plane and the true normal force, W_N, is the *normal pressure angle*, ϕ_n.

In design of a helical gear, there are three angles of interest: (1) the helix angle, ψ; (2) the *normal pressure angle*, ϕ_n; and (3) the *transverse pressure angle*, ϕ_t. Designers must specify the helix angle and one of the two pressure angles. The other pressure angle can be computed from the following relationship:

$$\tan \phi_n = \tan \phi_t \cos \psi \tag{8–25}$$

For example, one manufacturer's catalog offers standard helical gears with a normal pressure angle of $14\frac{1}{2}°$ and a $45°$ helix angle. Then the transverse pressure angle is found from

$$\tan \phi_n = \tan \phi_t \cos \psi$$
$$\tan \phi_t = \tan \phi_n / \cos \psi = \tan(14.5°)/\cos(45°) = 0.3657$$
$$\phi_t = \tan^{-1}(0.3657) = 20.09°$$

Pitches for Helical Gears

To obtain a clear picture of the geometry of helical gears, you must understand the following five different pitches.

Circular Pitch, *p*.   *Circular pitch* is the distance from a point on one tooth to the corresponding point on the next adjacent tooth, measured at the pitch line in the transverse plane. This is the same definition used for spur gears. Then

▷ Circular Pitch
$$p = \pi D/N \tag{8–26}$$

Normal Circular Pitch, *p_n*.   *Normal circular pitch* is the distance between corresponding points on adjacent teeth measured on the pitch surface in the normal direction. Pitches p and p_n are related by the following equation:

▷ Normal
Circular Pitch
$$p_n = p \cos \psi \tag{8–27}$$

Diametral Pitch, *P_d*.   *Diametral pitch* is the ratio of the number of teeth in the gear to the pitch diameter. This is the same definition as the one for spur gears; it applies in considerations of the form of the teeth in the diametral or transverse plane. Thus, this pitch is sometimes called the *transverse diametral pitch:*

▷ Diametral Pitch
$$P_d = N/D \tag{8–28}$$

Normal Diametral Pitch, *P_{nd}*.   *Normal diametral pitch* is the equivalent diametral pitch in the plane normal to the teeth:

Normal Diametral Pitch

$$P_{nd} = P_d/\cos \psi \qquad (8\text{-}29)$$

It is helpful to remember these relationships:

$$P_d p = \pi \qquad (8\text{-}30)$$
$$P_{nd} p_n = \pi \qquad (8\text{-}31)$$

Axial Pitch, P_x. *Axial pitch* is the distance between corresponding points on adjacent teeth, measured on the pitch surface in the axial direction:

Axial Pitch

$$P_x = p/\tan \psi = \pi/(P_d \tan \psi) \qquad (8\text{-}32)$$

It is necessary to have at least two axial pitches in the face width to have the benefit of full helical action and its smooth transfer of the load from tooth to tooth.

The use of equations (8–25) through (8–29) and Equation (8–32) is now illustrated in the following example problem.

Example Problem 8–5

A helical gear has a transverse diametral pitch of 12, a transverse pressure angle of $14\frac{1}{2}°$, 28 teeth, a face width of 1.25 in, and a helix angle of 30°. Compute circular pitch, normal circular pitch, normal diametral pitch, axial pitch, pitch diameter, and the normal pressure angle. Compute the number of axial pitches in the face width.

Solution

Circular Pitch

Use Equation (8–30):

$$p = \pi/P_d = \pi/12 = 0.262 \text{ in}$$

Normal Circular Pitch

Use Equation (8–27):

$$p_n = p \cos \psi = (0.262)\cos(30) = 0.227 \text{ in}$$

Normal Diametral Pitch

Use Equation (8–29):

$$P_{nd} = P_d/\cos \psi = 12/\cos(30) = 13.856$$

Axial Pitch

Use Equation (8–32):

$$P_x = p/\tan \psi = 0.262/\tan(30) = 0.453 \text{ in}$$

Pitch Diameter

Use Equation (8–28):

$$D = N/P_d = 28/12 = 2.333 \text{ in}$$

Normal Pressure Angle

Use Equation (8–25):

$$\phi_n = \tan^{-1}(\tan\phi_t \cos\psi)$$
$$\phi_n = \tan^{-1}[\tan(14\tfrac{1}{2})\cos(30)] = 12.62°$$

Number of Axial Pitches in the Face Width

$$F/P_x = 1.25/0.453 = 2.76 \text{ pitches}$$

Since this is greater than 2.0, there will be full helical action.

**8–8
BEVEL GEAR
GEOMETRY**

Bevel gears are used to transfer motion between nonparallel shafts, usually at 90° to one another. The four primary styles of bevel gears are straight bevel, spiral bevel, ZEROL® bevel, and hypoid. Figure 8–22 shows the general appearance of these four types of bevel gear sets. The surface on which bevel gear teeth are machined is inherently a part of a cone. The differences occur in the specific shape of the teeth and in the orientation of the pinion relative to the gear. (See References 3, 5, 13, and 14.)

Straight Bevel Gears

The teeth of a straight bevel gear are straight and lie along an element of the conical surface. Lines along the face of the teeth through the pitch circle meet at the apex of the pitch cone. As shown in Figure 8–22(f), the centerlines of both the pinion and the gear also meet at this apex. In the standard configuration, the teeth are tapered toward the center of the cone.

Key dimensions are specified either at the outer end of the teeth or at the mean, mid-face position. The relationships that control some of these dimensions are listed in Table 8–7 for the case when the shafts are at the 90° angle. The pitch cone angles for the pinion and the gear are determined by the ratio of the number of teeth, as shown in the table. Note that their sum is 90°. Also, for a pair of bevel gears having a ratio of unity, each has a pitch cone angle of 45°. Such gears, called *miter gears,* are used simply to change the direction of the shafts in a machine drive without affecting the speed of rotation.

You should understand that many more features need to be specified before the gears can be produced. Furthermore, many successful, commercially available gears are made in some nonstandard form. For example, the addendum of the pinion is often made longer than that of the gear. Some manufacturers modify the slope of the root of the teeth to produce a uniform depth, rather than using the standard, tapered form. Reference 5 gives many more data.

The pressure angle, ϕ, is typically 20°, but 22.5° and 25° are often used to avoid interference. The minimum number of teeth for straight bevel gears is typically 12. More is said about the design of straight bevel gears in Chapter 10.

The mounting of bevel gears is critical if satisfactory performance is to be achieved. Most commercial gears have a defined mounting distance similar to that shown in Figure 8–22(f). It is the distance from some reference surface, typically the back of the hub of the gear, to the apex of the pitch cone. Because the pitch cones of the mating gears have coincident apexes, the mounting distance also locates the axis of the mating gear. If the gear is mounted at a distance smaller than the recommended mounting distance, the teeth will likely bind. If it is mounted at a greater distance, there will be excessive backlash, causing noisy and rough operation.

(a) Straight bevel

(b) Spiral bevel

(c) ZEROL ® bevel

(d) Hypoid

(e) Photograph of straight bevel gear pair

(f) Key dimensions of straight bevel gear pair

FIGURE 8–22   Types of bevel gears [Parts (a) through (d) extracted from ANSI/AGMA 2005-C96, *Design Manual for Bevel Gears,* with the permission of the publisher, the American Gear Manufacturers Association, 1500 King Street, Suite 201, Alexandria, VA 22314. Photograph in (e) courtesy of (Emerson Power Transmission Corporation, Drive and Component Division, Ithaca, N.Y.)

TABLE 8–7   Geometrical features of straight bevel gears

Given   Diametral pitch $= P_d = N_P/d = N_G/D$

where $N_P =$ number of teeth in pinion
$N_G =$ number of teeth in gear

Dimension	Formula
Gear ratio	$m_G = N_G/N_P$
Pitch diameters:	
Pinion	$d = N_P/P_d$
Gear	$D = N_G/P_d$
Pitch cone angles:	
Pinion	$\gamma = \tan^{-1}(N_P/N_G)$ (lowercase Greek *gamma*)
Gear	$\Gamma = \tan^{-1}(N_G/N_P)$ (uppercase Greek *gamma*)
Outer cone distance	$A_o = 0.5D/\sin(\Gamma)$
Face width must be specified.	$F =$
Nominal face width	$F_{\text{nom}} = 0.30A_o$
Maximum face width	$F_{\text{max}} = A_o3$ or $F_{\text{max}} = 10/P_d$ (whichever is less)
Mean cone distance	$A_m = A_o - 0.5F$
	(*Note:* A_m is defined for the gear, also called A_{mG}.)
Mean circular pitch	$p_m = (\pi/P_d)(A_m/A_o)$
Mean working depth	$h = (2.00/P_d)(A_m/A_o)$
Clearance	$c = 0.125h$
Mean whole depth	$h_m = h + c$
Mean addendum factor	$c_1 = 0.210 + 0.290/(m_G)^2$
Gear mean addendum	$a_G = c_1h$
Pinion mean addendum	$a_P = h - a_G$
Gear mean dedendum	$b_G = h_m - a_G$
Pinion mean dedendum	$b_P = h_m - a_P$
Gear dedendum angle	$\delta_G = \tan^{-1}(b_G/A_{mG})$
Pinion dedendum angle	$\delta_P = \tan^{-1}(b_P/A_{mG})$
Gear outer addendum	$a_{oG} = a_G + 0.5F\tan\delta_P$
Pinion outer addendum	$a_{oP} = a_P + 0.5F\tan\delta_G$
Gear outside diameter	$D_o = D + 2a_{oG}\cos\Gamma$
Pinion outside diameter	$d_o = d + 2a_{oP}\cos\gamma$

SOURCE: Extracted from ANSI/AGMA 2005-C96, *Design Manual for Bevel Gears,* with the permission of the publisher, the American Gear Manufacturers Association, 1500 King Street, Suite 201, Alexandria, VA 22314.

Spiral Bevel Gears

The teeth of a spiral bevel gear are curved and sloped with respect to the surface of the pitch cone. Spiral angles, ψ, of 20° to 45° are used, with 35° being typical. Contact starts at one end of the teeth and moves along the tooth to its end. For a given tooth form and number of teeth, more teeth are in contact for spiral bevel gears than for straight bevel gears. The gradual transfer of loads and the greater average number of teeth in contact make spiral bevel gears smoother and allow smaller designs than for typical straight bevel gears. Recall that similar advantages were described for a helical gear relative to a spur gear.

The pressure angle, ϕ, is typically 20° for spiral bevel gears, and the minimum number of teeth is typically 12 to avoid interference. But nonstandard spiral gears allow as few as five teeth in the pinion of high-ratio sets if the tips of the teeth are trimmed to avoid interference. The rather high average number of teeth in contact (high contact ratio) for spiral gears makes this approach acceptable and can result in a very compact design.

Reference 5 gives the relationships for computing the geometric features of spiral bevel gears that are extensions of those given in Table 8–7.

ZEROL® Bevel Gears

The teeth of a ZEROL® bevel gear are curved somewhat as in a spiral bevel gear, but the spiral angle is zero. These gears can be used in the same mountings as straight bevel gears, but they operate more smoothly.

Hypoid Gears

The major difference between hypoid gears and the others just described is that the centerline of the pinion for a set of hypoid gears is offset either above or below the centerline of the gear. The teeth are designed specially for each combination of offset distance and spiral angle of the teeth. A major advantage is the more compact design that results, particularly when applied to vehicle drive trains and machine tools. (See References 5, 13, and 14 for more data.)

The hypoid gear geometry is the most general form, and the others are special cases. The hypoid gear has an offset axis for the pinion, and its curved teeth are cut at a spiral angle. Then the spiral bevel gear is a hypoid gear with a zero offset distance. A ZEROL® bevel gear is a hypoid gear with a zero offset and a zero spiral angle. A straight bevel gear is a hypoid gear with a zero offset, a zero spiral angle, and straight teeth.

Example Problem 8–6 Compute the values for the geometrical features listed in Table 8–7 for a pair of straight bevel gears having a diametral pitch of 8, a 20° pressure angle, 16 teeth in the pinion, and 48 teeth in the gear. The shafts are at 90°.

Solution Given $P_d = 8$; $N_P = 16$; $N_G = 48$.

Computed Values **Gear Ratio**

$$m_G = N_G/N_P = 48/16 = 3.000$$

Pitch Diameter
For the pinion,
$$d = N_P/P_d = 16/8 = 2.000 \text{ in}$$
For the gear,
$$D = N_G/P_d = 48/8 = 6.000 \text{ in}$$

Pitch Cone Angles
For the pinion,
$$\gamma = \tan^{-1}(N_P/N_G) = \tan^{-1}(16/48) = 18.43°$$
For the gear,
$$\Gamma = \tan^{-1}(N_G/N_P) = \tan^{-1}(48/16) = 71.57°$$

Outer Cone Distance

$$A_o = 0.5D/\sin(\Gamma) = 0.5(6.00 \text{ in})/\sin(71.57°) = 3.162 \text{ in}$$

Face Width

The face width must be specified:

$$F = 1.000 \text{ in}$$

Based on the following guidelines:

Nominal face width:

$$F_{\text{nom}} = 0.30A_o = 0.30(3.162 \text{ in}) = 0.949 \text{ in}$$

Maximum face width:

$$F_{\text{max}} = A_o/3 = (3.162 \text{ in})/3 = 1.054 \text{ in}$$

or

$$F_{\text{max}} = 10/P_d = 10/8 = 1.25 \text{ in}$$

Mean Cone Distance

$$A_m = A_{mG} = A_o - 0.5F = 3.162 \text{ in} - 0.5(1.00 \text{ in}) = 2.662 \text{ in}$$

Ratio $A_m/A_o = 2.662/3.162 = 0.842$ (This ratio occurs in several following calculations.)

Mean Circular Pitch

$$p_m = (\pi/P_d)(A_m/A_o) = (\pi/8)(0.842) = 0.331 \text{ in}$$

Mean Working Depth

$$h = (2.00/P_d)(A_m/A_o) = (2.00/8)(0.842) = 0.210 \text{ in}$$

Clearance

$$c = 0.125h = 0.125(0.210 \text{ in}) = 0.026 \text{ in}$$

Mean Whole Depth

$$h_m = h + c = 0.210 \text{ in} + 0.026 \text{ in} = 0.236 \text{ in}$$

Mean Addendum Factor

$$c_1 = 0.210 + 0.290/(m_G)^2 = 0.210 + 0.290/(3.00)^2 = 0.242$$

Gear Mean Addendum

$$a_G = c_1h = (0.242)(0.210 \text{ in}) = 0.051 \text{ in}$$

Pinion Mean Addendum

$$a_P = h - a_G = 0.210 \text{ in} - 0.051 \text{ in} = 0.159 \text{ in}$$

Gear Mean Dedendum

$$b_G = h_m - a_G = 0.236 \text{ in} - 0.051 \text{ in} = 0.185 \text{ in}$$

Pinion Mean Dedendum

$$b_P = h_m - a_P = 0.236 \text{ in} - 0.159 \text{ in} = 0.077 \text{ in}$$

Gear Dedendum Angle

$$\delta_G = \tan^{-1}(b_G/A_{mG}) = \tan^{-1}(0.185/2.662) = 3.98°$$

Pinion Dedendum Angle

$$\delta_P = \tan^{-1}(b_P/A_{mG}) = \tan^{-1}(0.077/2.662) = 1.66°$$

Gear Outer Addendum

$$a_{oG} = a_G + 0.5F \tan \delta_P$$
$$a_{oG} = (0.051 \text{ in}) + (0.5)(1.00 \text{ in}) \tan(1.657°) = 0.065\ 5 \text{ in}$$

Pinion Outer Addendum

$$a_{oP} = a_P + 0.5F \tan \delta_G$$
$$a_{oP} = (0.159 \text{ in}) + (0.5)(1.00 \text{ in}) \tan(3.975°) = 0.193\ 7 \text{ in}$$

Gear Outside Diameter

$$D_o = D + 2a_{oG} \cos \Gamma$$
$$D_o = 6.000 \text{ in} + 2(0.065\ 5 \text{ in}) \cos(71.57°) = 6.041 \text{ in}$$

Pinion Outside Diameter

$$d_o = d + 2a_{oP} \cos \gamma$$
$$d_o = 2.000 \text{ in} + 2(0.193\ 7 \text{ in}) \cos(18.43°) = 2.368 \text{ in}$$

8–9
TYPES OF
WORMGEARING

Wormgearing is used to transmit motion and power between nonintersecting shafts, usually at 90° to each other. The drive consists of a worm on the high-speed shaft which has the general appearance of a power screw thread: a cylindrical, helical thread. The worm drives a wormgear, which has an appearance similar to that of a helical gear. Figure 8–23 shows a typical worm and wormgear set. Sometimes the wormgear is referred to as a *worm wheel* or simply a *wheel* or *gear*. (See Reference 6.)

Several variations of the geometry of wormgear drives are available. The most common one, shown in Figures 8–23 and 8–24, employs a cylindrical worm mating with a wormgear having teeth that are throated, wrapping partially around the worm. This is called a *single-enveloping type* of wormgear drive. The contact between the threads of the worm and wormgear teeth is along a line, and the power transmission capacity is quite good. Many manufacturers offer this type of wormgear set as a stocked item. Installation of the worm is relatively easy because axial alignment is not very critical. However, the wormgear must be carefully aligned radially in order to achieve the benefit of the enveloping action. Figure 8–25 shows a cutaway of a commercial wormgear reducer.

A simpler form of wormgear drive allows a special cylindrical worm to be used with a standard spur gear or helical gear. Neither the worm nor the gear must be aligned with great accuracy, and the center distance is not critical. However, the contact between the

FIGURE 8-23   Worms and wormgears (Emerson Power Transmission Corporation, Drive and Component Division, Ithaca, NY.)

FIGURE 8-24
Single-enveloping
wormgear set

worm threads and the wormgear teeth is theoretically a point, drastically reducing the power transmission capacity of the set. Thus, this type is used mostly for nonprecision positioning applications at low speeds and low power levels.

A third type of wormgear set is the *double-enveloping type* in which the worm is made in an hourglass shape and mates with an enveloping type of wormgear. This results in area contact rather than line or point contact and allows a much smaller system to transmit a given power at a given reduction ratio. However, the worm is more difficult to manufacture, and the alignment of both the worm and the wormgear is very critical.

FIGURE 8–25
Wormgear reducer
(Rockwell
Automation/Reliance
Electric, Rockwell
International
Corporation, Costa
Mesa, CA.)

Cast gray iron housing

Lubrication system

Flange for mounting
a C-face motor

Ball bearing

Worm gear

Tapered roller
bearing

Solid output
shaft with key

Shaft seal

Single reduction
with Relialube system

Bored input shaft mates
with motor shaft

Worm

**8–10
GEOMETRY OF
WORMS AND
WORMGEARS**

Pitches, p and P_d

A basic requirement of the worm and wormgear set is that the *axial pitch* of the worm must be equal to the *circular pitch* of the wormgear in order for them to mesh. Figure 8–24 shows the basic geometric features of a single-enveloping worm and wormgear set. *Axial pitch, P_x,* is defined as the distance from a point on the worm thread to the corresponding point on the next adjacent thread, measured axially on the pitch cylinder. As before, the circular pitch is defined for the wormgear as the distance from a point on a tooth on the pitch circle of the gear to the corresponding point on the next adjacent tooth, measured along the pitch circle. Thus, the circular pitch is an arc distance that can be calculated from

 Circular Pitch

$$p = \pi D_G / N_G \tag{8–33}$$

where D_G = pitch diameter of the gear
 N_G = number of teeth in the gear

Some wormgears are made according to the circular pitch convention. But, as noted with spur gears, commercially available wormgear sets are usually made to a diametral pitch convention with the following pitches readily available: 48, 32, 24, 16, 12, 10, 8, 6, 5, 4, and 3. The diametral pitch is defined for the gear as

Diametral Pitch

$$P_d = N_G / D_G \tag{8–34}$$

The conversion from diametral pitch to circular pitch can be made from the following equation:

$$P_d p = \pi \tag{8–35}$$

FIGURE 8–26 Lead angle

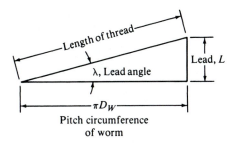

Pitch circumference of worm

Number of Worm Threads, N_W

Worms can have a single thread, as in a typical screw, or multiple threads, usually 2 or 4, but sometimes 3, 5, 6, 8, or more. It is common to refer to the number of threads as N_W and then to treat that number as if it were the number of teeth in the worm. The number of threads in the worm is frequently referred to as the number of *starts;* this is convenient because if you look at the end of a worm, you can count the number of threads that start at the end and wind down the cylindrical worm.

Lead, L

The *lead* of a worm is the axial distance that a point on the worm would move as the worm is rotated one revolution. Lead is related to the axial pitch by

 Lead

$$L = N_W P_x \qquad (8\text{–}36)$$

Lead Angle, λ

The *lead angle* is the angle between the tangent to the worm thread and the line perpendicular to the axis of the worm. To visualize the method of calculating the lead angle, refer to Figure 8–26, which shows a simple triangle that would be formed if one thread of the worm were unwrapped from the pitch cylinder and laid flat on the paper. The length of the hypotenuse is the length of the thread itself. The vertical side is the lead, L. The horizontal side is the circumference of the pitch cylinder, πD_W, where D_W is the pitch diameter of the worm. Then

Lead Angle

$$\tan \lambda = L/\pi D_W \qquad (8\text{–}37)$$

Pitch Line Speed, v_t

As before, the pitch line speed is the linear velocity of a point on the pitch line for the worm or the wormgear. For the worm having a pitch diameter D_W in, rotating at n_W rpm,

Pitch Line Speed for Worm

$$v_{tW} = \frac{\pi D_W n_W}{12} \text{ ft/min}$$

For the wormgear having a pitch diameter D_G in, rotating at n_G rpm,

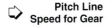

**Pitch Line
Speed for Gear**

$$V_{tG} = \frac{\pi D_G n_G}{12} \text{ ft/min}$$

Note that these two values for pitch line speed are *not* equal.

Velocity Ratio, *VR*

It is most convenient to calculate the velocity ratio of a worm and wormgear set from the ratio of the input rotational speed to the output rotational speed:

**Velocity Ratio for
Worm/Wormgear Set**

$$VR = \frac{\text{speed of worm}}{\text{speed of gear}} = \frac{n_W}{n_G} = \frac{N_G}{N_W} \qquad \text{(8–38)}$$

Example Problem 8–7   A wormgear has 52 teeth and a diametral pitch of 6. It mates with a triple-threaded worm that rotates at 1 750 rpm. The pitch diameter of the worm is 2.000 in. Compute the circular pitch, the axial pitch, the lead, the lead angle, the pitch diameter of the wormgear, the center distance, the velocity ratio, and the rotational speed of the wormgear.

Solution   **Circular Pitch**

$$p = \pi/P_d = \pi/6 = 0.523\ 6 \text{ in}$$

Axial Pitch

$$P_x = p = 0.523\ 6 \text{ in}$$

Lead

$$L = N_W P_x = (3)(0.523\ 6) = 1.570\ 8 \text{ in}$$

Lead Angle

$$\lambda = \tan^{-1}(L/\pi D_W) = \tan^{-1}(1.570\ 8/\pi 2.000)$$
$$\lambda = 14.04°$$

Pitch Diameter

$$D_G = N_G/P_d = 52/6 = 8.667 \text{ in}$$

Center Distance

$$C = (D_W + D_G)/2 = (2.000 + 8.667)/2 = 5.333 \text{ in}$$

Velocity Ratio

$$VR = N_G/N_W = 52/3 = 17.333$$

Gear rpm

$$n_G = n_W/VR = 1\ 750/17.333 = 101 \text{ rpm}$$

Pressure Angle

Most commercially available wormgears are made with pressure angles of $14\frac{1}{2}°$, 20°, 25°, or 30°. The low pressure angles are used with worms having a low lead angle and/or a low diametral pitch. For example, a $14\frac{1}{2}°$ pressure angle may be used for lead angles up to about 17°. For higher lead angles and with higher diametral pitches (smaller teeth), the 20° or 25° pressure angle is used to eliminate interference without excessive undercutting. The 20° pressure angle is the preferred value for lead angles up to 30°. From 30° to 45° of lead angle, the 25° pressure angle is recommended. Either the normal pressure angle, ϕ_n, or the transverse pressure angle, ϕ_t, may be specified. These are related by

 Pressure Angle

$$\tan \phi_n = \tan \phi_t \cos \lambda \qquad \text{(8–39)}$$

Self-locking Wormgear Sets

Self-locking is the condition in which the worm drives the wormgear, but if torque is applied to the gear shaft, the worm does not turn. It is locked! The locking action is produced by the friction force between the worm threads and the wormgear teeth, and this is highly dependent on the lead angle. It is recommended that a lead angle no higher than about 5.0° be used in order to ensure that self-locking will occur. This lead angle usually requires the use of a single-threaded worm. Note that the triple-threaded worm in Example Problem 8–7 has a lead angle of 14.04°. It is *not* likely to be self-locking.

8–11
TYPICAL
GEOMETRY OF
WORMGEAR
SETS

Considerable latitude is permissible in the design of wormgear sets because the worm and wormgear combination is designed as a unit. However, there are some guidelines.

Typical Tooth Dimensions

Table 8–8 shows typical values used for the dimensions of worm threads and gear teeth.

Worm Diameter

The diameter of the worm affects the lead angle, which in turn affects the efficiency of the set. For this reason, small diameters are desirable. But for practical reasons and proper proportion with respect to the wormgear, it is recommended that the worm diameter be approximately $C^{0.875}/2.2$, where C is the center distance between the worm and the wormgear. Variation of about 30% is allowed. (See Reference 6.) Thus, the worm diameter should fall in the range

$$1.6 < \frac{C^{0.875}}{D_W} < 3.0 \qquad \text{(8–40)}$$

But some commercially available wormgear sets fall outside this range, especially in the smaller sizes. Also, those worms designed to have a through-hole bored in them for installation on a shaft are typically larger than you would find from Equation (8–40). Proper proportion and efficient use of material should be the guide. The worm shaft must also be checked for deflection under operating loads. For worms machined integral with the shaft,

TABLE 8–8   Typical tooth dimensions for worms and wormgears

Dimension	Formula
Addendum	$a = 0.318\ 3P_x = 1/P_d$
Whole depth	$h_t = 0.686\ 6P_x = 2.157/P_d$
Working depth	$h_k = 2a = 0.636\ 6P_a = 2/P_d$
Dedendum	$b = h_t - a = 0.368\ 3P_x = 1.157/P_d$
Root diameter of worm	$D_{rW} = D_W - 2b$
Outside diameter of worm	$D_{oW} = D_W + 2a = D_W + h_k$
Root diameter of gear	$D_{rG} = D_G - 2b$
Throat diameter of gear	$D_t = D_G + 2a$

SOURCE: Standard AGMA 341.02-1965 (R1970), *Design of General Industrial Coarse-Pitch Cylindrical Wormgearing,* with the permission of the publisher, American Gear Manufacturers Association, 1500 King Street, Suite 201, Alexandria, VA 22314.

the root of the worm threads determines the minimum shaft diameter. For worms having bored holes, sometimes called *shell worms,* care must be exercised to leave sufficient material between the thread root and the keyway in the bore. Figure 8–27 shows the recommended thickness above the keyway to be a minimum of one-half the whole depth of the threads.

Wormgear Dimensions

We are concerned here with the single-enveloping type of wormgear, as shown in Figure 8–28. Its addendum, dedendum, and depth dimensions are assumed to be the same as those listed in Table 8–8, measured at the throat of the wormgear teeth. The throat is in line with the vertical centerline of the worm. The recommended face width for the wormgear is

Face Width of Wormgear

$$F_G = (D_{oW}^2 - D_W^2)^{1/2} \qquad \text{(8–41)}$$

This corresponds to the length of the line tangent to the pitch circle of the worm and limited by the outside diameter of the worm. Any face width beyond this value would not be effective in resisting stress or wear, but a convenient value slightly greater than the minimum should be used. The outer edges of the wormgear teeth should be chamfered approximately as shown in Figure 8–28.

Another recommendation, which is convenient for initial design, is that the face width of the gear should be approximately 2.0 times the circular pitch. Because we are working in the diametral pitch system, we will use

$$F_G = 2p = 2\pi/P_d \qquad \text{(8–42)}$$

However, since this is only approximate and 2π is approximately 6, we will use

$$F_G = 6/P_d \qquad \text{(8–43)}$$

If the gear web is thinned, a rim thickness at least equal to the whole depth of the teeth should be left.

FIGURE 8–27 Shell worm

FIGURE 8–28
Wormgear details

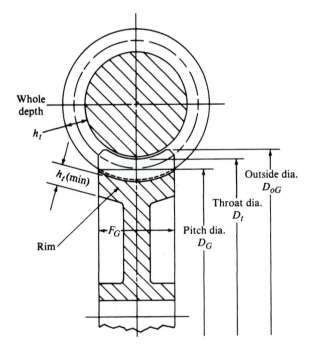

Face Length of the Worm

For maximum load sharing, the worm face length should extend to at least the point where the outside diameter of the worm intersects the throat diameter of the wormgear. This length is

⇨ **Face Length of Worm**

$$F_W = 2[(D_t/2)^2 - (D_G/2 - a)^2]^{1/2} \tag{8–44}$$

Example Problem 8–8   A worm and wormgear set is to be designed to produce a velocity ratio of 40. It has been proposed that the diametral pitch of the wormgear be 8, based on the torque that must be transmitted. (This will be discussed in Chapter 10.) Using the relationships presented in this section, specify the following:

Worm diameter, D_W

Number of threads in the worm, N_W

Number of teeth in the gear, N_G

Actual center distance, C

Face width of the gear, F_G

Face length of the worm, F_W

Minimum thickness of the rim of the gear

Solution Many design decisions need to be made, and multiple solutions could satisfy the requirements. Presented here is one solution, along with comparisons with the various guidelines discussed in this section. This type of analysis precedes the stress analysis and the determination of the power-transmitting capacity of the worm and wormgear drive which is discussed in Chapter 10.

Trial Design

Let's specify a double-threaded worm: $N_W = 2$. Then there must be 80 teeth in the wormgear to achieve a velocity ratio of 40. That is,

$$VR = N_G/N_W = 80/2 = 40$$

With the known diametral pitch, $P_d = 8$, the pitch diameter of the wormgear is

$$D_G = N_G/P_d = 80/8 = 10.000 \text{ in}$$

An estimate for the magnitude of the center distance is approximately $C = 7.50$ in. We know that it will be greater than 5.00 in, the radius of the wormgear. Using Equation (8–40), the recommended minimum size of the worm is

$$D_W = C^{0.875}/3.0 = 1.94 \text{ in}$$

Similarly, the maximum diameter should be

$$D_W = C^{0.875}/1.6 = 3.64 \text{ in}$$

A small worm diameter is desirable. Let's specify $D_W = 2.25$ in. The actual center distance is

$$C = (D_W + D_G)/2 = 6.125 \text{ in}$$

Worm Outside Diameter

$$D_{oW} = D_W + 2a = 2.25 + 2(1/P_d) = 2.25 + 2(1/8) = 2.50 \text{ in}$$

Whole Depth

$$h_t = 2.157/P_d = 2.157/8 = 0.270 \text{ in}$$

Face Width for Gear

Let's use Equation (8–41):

$$F_G = (D_{oW}^2 - D_W^2)^{1/2} = (2.50^2 - 2.25^2) = 1.090 \text{ in}$$

Let's specify $F_G = 1.25$ in.

Addendum

$$a = 1/P_d = 1/8 = 0.125 \text{ in}$$

Throat Diameter of Wormgear

$$D_t = D_G + 2a = 10.000 + 2(0.125) = 10.250 \text{ in}$$

Recommended Minimum Face Length of Worm

$$F_W = 2[(D_t/2)^2 - (D_G/2 - a)^2]^{1/2} = 3.16 \text{ in}$$

Let's specify $F_W = 3.25$ in.

Minimum Thickness of the Rim of the Gear

The rim thickness should be greater than the whole depth:

$$h_t = 0.270 \text{ in}$$

**8–12
TRAIN VALUE
FOR COMPLEX
GEAR TRAINS**

The concept of train value was introduced in Section 8–6 and was applied to gear trains having all spur gears and a modest reduction ratio. Trains of two to five gears having a single speed reduction or a double reduction were included. This section expands on the concept of train value to include a wider variety of gear types, higher reduction ratios, and the opportunity to use several different arrangements of gears.

The basic definition of *train value* as shown in Equation (8–23) will continue to be used. Its general form is repeated here:

$$TV = \frac{\text{product of number of teeth in the driven gears}}{\text{product of number of teeth in the driving gears}} = \frac{\text{input speed}}{\text{output speed}}$$

This is equivalent to saying that the train value is the product of the velocity ratios of the individual gear pairs in the train.

When spur, helical, and bevel gears are used, the specific geometry of the gears and their teeth does not affect the train value. Also, when a worm/wormgear set is a part of the train, its data can be entered into the equation if we recall that the number of *threads* in the worm can be considered equivalent to the number of *teeth* in the driver of that set.

Sketches of the gear trains are valuable to illustrate the arrangement of gears and to enable you to track the flow of power through the train, that is, how the motion is transferred from the input shaft through the entire train to the output shaft. The sketches can be somewhat schematic, but they should show the relative positions of the gears and the shafts that carry the gears. Although sizes need not be drawn to scale, it is helpful to suggest the nominal size of the gears to aid in judging whether a given pair will produce a speed reduction or a speed increase. Showing the pitch diameters of the gears will suffice. Recall that the motion of one gear on its mating gear is kinematically similar to the rolling of smooth cylinders on one another without slipping where the diameters for the cylinders are equal to the pitch diameters of the gears. But be sure to understand that any two gears in mesh must have compatible tooth forms. Only the kinematics of the train are being considered here. Later, when the actual gears, shafts, bearings, and the housing are designed, many other geometric features and operating parameters must be considered.

The sketching technique shown in Figure 8–29 will be used in this section and in the Problems section of this chapter. Gears are given letter designations, and the shafts carrying the gears are numbered. The data for the train shown in the figure should be interpreted as follows:

- Gears *A, B, C,* and *D* are external type, either spur or helical. Their shafts (shafts 1, 2, and 3) are parallel and are indicated only by their centerlines. Shaft 1 is the input to the train. It would typically be connected directly to a drive such as an electric motor.

- Gear *A* drives gear *B* with a speed reduction because gear *A* is smaller than gear *B*. A change in direction of rotation occurs. Gear *C* is on the same shaft as gear *B* and rotates at the same speed and in the same direction. Gear *C* drives gear *D* with a speed reduction and a direction reversal.

- Gears *E* and *F* are bevel gears, straight, or spiral, or some other tooth form. Their shafts (3 and 4) are perpendicular.

- Gear *G* is a worm, and it is driving wormgear *H*. Their shafts (4 and 5) are perpendicular, with shaft 5 directed out of the page. Note that we are seeing the side view of the pitch cylinder of the worm rotating about a vertical axis (shaft 4). We see the end view of wormgear *H*.

- The small pinion *I* is also mounted on shaft 5 and rotates at the same speed and in the same direction as wormgear *H*.

- Pinion *I* drives the internal gear *J* mounted on shaft 6 which is the output gear of the train. Note that shaft 6 rotates in the same direction as shaft 5.

Data for the gears in the train may be provided in a number of ways. If only speed of rotation is to be considered, the number of teeth in each gear will normally be given. If physical size of the train, center distances, bearing placement, housing design, and detailed layout are to be considered, then other geometrical features such as diametral pitch, face

FIGURE 8–29 Gear train for Example Problem 8–9

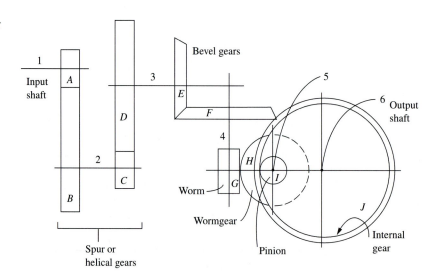

width, gear blank style, hub style, and gear diameters will have to be specified. Most of these features are discussed in later chapters. We will give only the numbers of teeth in most cases in this chapter. But notice that if pitch diameter and diametral pitch of a pinion and gear are given, their numbers of teeth can be determined from

$$P_d = N_P/D_P = N_G/D_G$$
$$N_P = P_d D_P$$
$$N_G = P_d D_G$$

Also, we can use the definition of diametral pitch to show the relationship between the ratio of the numbers of teeth in a pair of gears and the ratio of their diameters:

$$P_d = N_P/D_P = N_G/D_G$$
$$N_G/N_P = D_G/D_P$$

This shows that the diameter ratio could replace the teeth ratio in the train value equation.

Example Problem 8–9   Refer to Figure 8–29. Shaft 1 is a motor shaft and rotates at 1 160 rpm. Compute the rotational speed of the output shaft, shaft 6. Data for the gears are as follows:

$N_A = 18$	$N_B = 34$	$N_C = 20$	$N_D = 62$
$N_E = 30$	$N_F = 60$	$N_G = 2$ (worm threads)	$N_H = 40$
$N_I = 16$	$N_J = 88$		

Solution   We will compute the train value first:

$$TV = \frac{(N_B)(N_D)(N_F)(N_H)(N_J)}{(N_A)(N_C)(N_E)(N_G)(N_I)} = \frac{(34)(62)(60)(40)(88)}{(18)(20)(30)(2)(16)} = 1\ 288.2$$

But $TV = n_1/n_6$. Then

$$n_6 = n_1/TV = (1\ 160 \text{ rpm})/1\ 288.2 = 0.900 \text{ rpm}$$

Shaft 6 rotates 0.900 rpm.

**8–13
DEVISING GEAR
TRAINS**

Now we will show several methods for devising gear trains to produce a desired train value. The result will typically be the specification of the number of teeth in each gear and the general arrangement of the gears relative to each other. The determination of the types of gears will generally not be considered except for how they may affect the direction of rotation or the general alignment of the shafts. Additional details can be specified after completion of the study of the design procedures in later chapters.

A few general principles that were discussed earlier in this chapter are reviewed first:

1. The velocity ratio for any pair of gears can be computed in a variety of ways as indicated in Equation (8–22).

2. The number of teeth in any gear must be an integer.

3. Gears in mesh must have the same tooth form and the same pitch.

4. When external gears mesh, there is a direction reversal of their shafts.

5. When an external pinion meshes with an internal gear, their shafts rotate in the same direction.

6. An idler is a gear that performs as both a driver and a driven gear in the same train. Its size and number of teeth have no effect on the magnitude of the train value, but the direction of rotation is changed.

7. Spur and helical gears operate on parallel shafts.

8. Bevel gears and a worm/wormgear set operate on shafts perpendicular to each other.

9. The number of teeth in the pinion of a gear pair should not be such that it causes interference with the teeth of its mating gear. Refer to Table 8–6.

10. In general, the number of teeth in the gear should not be larger than about 150. This is somewhat arbitrary, but it is typically more desirable to use a double-reduction gear train rather than a very large, single-reduction gear pair.

The several design procedures will now be demonstrated through example problems. It is not practical to outline a completely general procedure because of the many variables in any given design situation. You are advised to study the examples for their general approach which you can adapt to future problems as needed.

Design of a Single Pair of Gears to Produce a Desired Velocity Ratio

Example Problem 8–10   Devise a gear train to reduce the speed of rotation of a drive from an electric motor shaft operating at 3 450 rpm to approximately 650 rpm.

Solution   First we will compute the nominal train value:

$$TV = \text{(input speed)}/\text{(output speed)} = 3\ 450/650 = 5.308$$

If a single pair of gears is used then the train value is equal to the velocity ratio for that pair. That is, $TV = VR = N_G/N_P$.

Let's decide that spur gears having 20°, full-depth, involute teeth are to be used. Then we can refer to Table 8–6 and determine that no fewer than 16 teeth should be used for the pinion in order to avoid interference. We can specify the number of teeth in the pinion and use the velocity ratio to compute the number of teeth in the gear:

$$N_G = (VR)(N_P) = (5.308)(N_P)$$

Some examples are given in Table 8–9.

Conclusion and Comments   The combination of $N_P = 26$ and $N_G = 138$ gives the most ideal result for the output speed. But all of the trial values give output speeds reasonably close to the desired value. Only one is more than 2.0 rpm off the desired value. It remains a design decision as to how close the output speed must be to the stated value of 650 rpm. Note that the input speed is given as 3 450 rpm, the full load speed of an electric motor. But how accurate is that? The actual speed of the input will vary depending on the load on the motor. Therefore, it is not likely that the ratio must be precise.

TABLE 8-9

N_P	Computed $N_G = (5.308)(N_P)$	Nearest integer N_G	Actual VR: $VR = N_G/N_P$	Actual output speed (rpm): $n_G = n_P/VR = n_P(N_P/N_G)$
16	84.92	85	85/16 = 5.31	649.4
17	90.23	90	90/17 = 5.29	651.7
18	95.54	96	96/18 = 5.33	646.9
19	100.85	101	101/19 = 5.32	649.0
20	106.15	106	106/20 = 5.30	650.9
21	111.46	111	111/21 = 5.29	652.7
22	116.77	117	117/22 = 5.32	648.7
23	122.08	122	122/23 = 5.30	650.4
24	127.38	127	127/24 = 5.29	652.0
25	132.69	133	133/25 = 5.32	648.5
26	138.00	138	138/26 = 5.307	650.0 Exact
27	143.31	143	143/27 = 5.30	651.4
28	148.61	149	149/28 = 5.32	648.3
29	153.92	154 **Too large**		

Equal Reduction Ratios for Compound Gear Trains

Example Problem 8-11

Devise a gear train for a machine tool drive. The input is a shaft that rotates at exactly 1 800 rpm. The output speed must be within the range of 31.5 and 32.5 rpm. Use 20°, full-depth, involute teeth and no more than 150 teeth in any gear.

Solution

Permissible Train Values

First let's compute the nominal train value that will produce an output speed of 32.0 rpm at the middle of the allowable range:

$$TV_{nom} = \text{(input speed)}/\text{(nominal output speed)} = 1\ 800/32 = 56.25$$

Similarly, we can compute the minimum and maximum allowable ratio:

$$TV_{min} = \text{(input speed)}/\text{(maximum output speed)} = 1\ 800/32.5 = 55.38$$
$$TV_{max} = \text{(input speed)}/\text{(minimum output speed)} = 1\ 800/31.5 = 57.14$$

Possible Ratio for Single Pair

The maximum ratio that any one pair of gears can produce occurs when the gear has 150 teeth and the pinion has 17 teeth (see Table 8-6). Then

$$VR_{max} = N_G/N_P = 150/17 = 8.82$$

This is too low.

Possible Train Value for Double-Reduction Train

If a double-reduction train is proposed, the train value will be

$$TV = (VR_1)(VR_2)$$

FIGURE 8–30
General layout of
proposed gear train

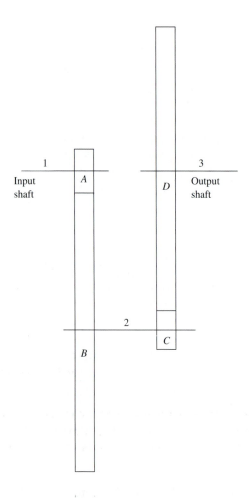

But the maximum value for either *VR* is 8.82. Then the maximum train value for the double-reduction train is

$$TV_{\max} = (8.82)(8.82) = (8.82)^2 = 77.9$$

It can be concluded that a double-reduction train should be practical.

Optional Designs

The general layout of the proposed train is shown in Figure 8–30. Its train value is

$$TV = (VR_1)(VR_2) = (N_B/N_A)(N_D/N_C)$$

We need to specify the number of teeth in each of the four gears to achieve a train value within the range just computed. Our approach is to specify two ratios, VR_1 and VR_2, such that their product is within the desired range. One possibility is to let the two ratios be equal. To produce the middle train value, each velocity ratio must be the square root of the target ratio, 56.25. That is,

$$VR_1 = VR_2 = \sqrt{56.25} = 7.50$$

TABLE 8–10

N_P	Computed $N_G = (7.5)(N_P)$	Nearest integer N_G	Actual VR: $VR = N_G/N_P$	Actual output speed (rpm): $n_G = n_P(VR)^2$
17	127.5	128	$128/17 = 7.529$	31.75
17	127.5	127	$127/17 = 7.470$	32.25
18	135	135	$135/18 = 7.500$	32.00
19	142.5	143	$143/19 = 7.526$	31.78
19	142.5	142	$142/19 = 7.474$	32.23
20	150	150	$150/20 = 7.500$	32.00

As shown in Table 8–10, we will use a process similar to that used in the previous example problem to select possible numbers of teeth.

Any of the possible designs shown in Table 8–10 would produce acceptable results. For example, we could specify

$$N_A = 18 \qquad N_B = 135 \qquad N_C = 18 \qquad N_D = 135$$

This combination would produce an output speed of exactly 32.0 rpm when the input speed was exactly 1 800 rpm.

Factoring Approach for Compound Gear Trains

Example Problem 8–12 Devise a gear train for a recorder for a precision measuring instrument. The input is a shaft that rotates at exactly 3 600 rpm. The output speed must be within the range of 11.0 and 11.5 rpm. Use 20°, full-depth, involute teeth; no fewer than 18 teeth; and no more than 150 teeth in any gear.

Solution **Nominal Target TV**

$$TV_{nom} = 3\ 600/11.25 = 320$$

Maximum TV

$$TV_{max} = 3\ 600/11.0 = 327.3$$

Minimum TV

$$TV_{min} = 3\ 600/11.5 = 313.0$$

Maximum Single VR

$$VR_{max} = 150/18 = 8.333$$

Maximum TV for Double Reduction

$$TV_{max} = (8.333)^2 = 69.4 \quad \text{(too low)}$$

Maximum TV for Triple Reduction

$$TV_{max} = (8.333)^3 = 578 \quad \text{(okay)}$$

FIGURE 8–31
Triple-reduction gear
train

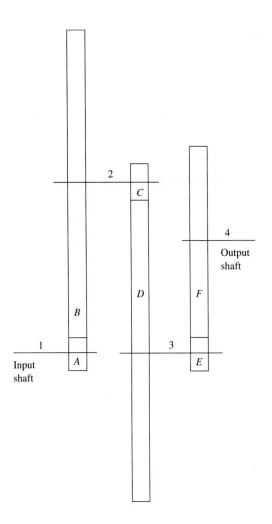

Design a triple-reduction gear train such as that shown in Figure 8–31. The train value is the product of the three individual velocity ratios:

$$TV = (VR_1)(VR_2)(VR_3)$$

If we can find three factors of 320 that are within the range of the possible ratio for a single pair of gears, they can be specified for each velocity ratio.

Factors of 320

One method is to divide by the smallest prime numbers that will divide evenly into given number, typically 2, 3, 5, or 7. For example,

$$320/2 = 160$$
$$160/2 = 80$$
$$80/2 = 40$$
$$40/2 = 20$$
$$20/2 = 10$$
$$10/2 = 5$$

Then the prime factors of 320 are 2, 2, 2, 2, 2, 2, and 5. We desire a set of three factors, which we can find by combining each set of three "2" factors into their product. That is,

$$(2)(2)(2) = 8$$

Then the three factors of 320 are

$$(8)(8)(5) = 320$$

Now let the number of teeth in the pinion of each pair be 18. The number of teeth in the gears will then be $(8)(18) = 144$ or $(5)(18) = 90$. Finally, we can specify

$$N_A = 18 \quad N_B = 144 \quad N_C = 18$$
$$N_D = 144 \quad N_E = 18 \quad N_F = 90$$

Residual Ratio

Example Problem 8–13   Devise a gear train for a conveyor drive. The drive motor rotates at 1 150 rpm, and it is desired that the output speed for the shaft that drives the conveyor be in the range of 24 to 28 rpm. Use a double-reduction gear train. Power transmission analysis indicates that it would be desirable for the reduction ratio for the first pair of gears to be somewhat greater than that for the second pair.

Solution   The start of this problem is similar to that for Example Problem 8–12.

Permissible Train Values

First let's compute the nominal train value that will produce an output speed of 26.0 rpm at the middle of the allowable range:

$$TV_{\text{nom}} = \text{(input speed)/(nominal output speed)} = 1\ 150/26 = 44.23$$

Now we can compute the minimum and maximum allowable speed ratio:

$$TV_{\text{min}} = \text{(input speed)/(maximum output speed)} = 1\ 150/24 = 47.92$$
$$TV_{\text{max}} = \text{(input speed)/(minimum output speed)} = 1\ 150/28 = 41.07$$

Possible Ratio for Single Pair

The maximum ratio that any one pair of gears can produce occurs when the gear has 150 teeth and the pinion has 17 teeth (see Table 8–6). Then

$$VR_{\text{max}} = N_G/N_P = 150/17 = 8.82 \quad \text{(too low)}$$

Possible Train Value for Double-Reduction Train

$$TV = (VR_1)(VR_2)$$

But the maximum value for either VR is 8.82. Then the maximum train value is

$$TV_{\text{max}} = (8.82)(8.82) = (8.82)^2 = 77.9$$

A double-reduction train is practical.

Optional Designs

The general layout of the proposed train is shown in Figure 8–30. Its train value is

$$TV = (VR_1)(VR_2) = (N_B/N_A)(N_D/N_C)$$

We need to specify the number of teeth in each of the four gears to achieve a train value within the range just computed. Our approach is to specify two ratios, VR_1 and VR_2, such that their product is within the desired range. If the two ratios were equal as before, each would be the square root of the target ratio, 44.23. That is,

$$VR_1 = VR_2 = \sqrt{44.23} = 6.65$$

But we want the first ratio to be somewhat larger than the second. Let's specify

$$VR_1 = 8.0 = (N_B/N_A)$$

If we let pinion A have 17 teeth, the number of teeth in gear B must be

$$N_B = (N_A)(8) = (17)(8) = 136$$

Then the second ratio should be approximately

$$(VR_2) = TV/(VR_1) = 44.23/8.0 = 5.53$$

This is the *residual ratio* left after the first ratio has been specified. Now if we specify 17 teeth for pinion C, gear D must be

$$(VR_2) = 5.53 = (N_D/N_C) = N_D/17$$
$$N_D = (5.53)(17) = 94.01$$

Rounding this off to 94 is likely to produce an acceptable result. Finally,

$$N_A = 17 \qquad N_B = 136 \qquad N_C = 17 \qquad N_D = 94$$

We should check the final design:

$$TV = (136/17)(94/17) = 44.235 = n_A/n_D$$

The actual output speed is

$$n_D = n_A/TV = (1\ 150\ \text{rpm})/44.235 = 26.0\ \text{rpm}$$

This is right in the middle of the desired range.

REFERENCES

1. American Gear Manufacturers Association. Standard 1012-F90. *Gear Nomenclature, Definitions of Terms with Symbols*. Alexandria, VA: American Gear Manufacturers Association, 1990.

2. American Gear Manufacturers Association. Standard 2002-B88. *Tooth Thickness Specification and Measurement*. Alexandria, VA: American Gear Manufacturers Association, 1988.

3. American Gear Manufacturers Association. Standard 2008-B90. *Standard for Assembling Bevel Gears*. Alexandria, VA: American Gear Manufacturers Association, 1990.

4. American Gear Manufacturers Association. Standard 370.01-1973 (R1978). *Design Manual for Fine-Pitch Gearing*. Alexandria, VA: American Gear Manufacturers Association, 1978.

5. American Gear Manufacturers Association. Standard 2005-C96. *Design Manual for Bevel Gearing*. Alexandria, VA: American Gear Manufacturers Association, 1996.

6. American Gear Manufacturers Association. Standard 6022-C93. *Design Manual for Cylindrical Wormgearing*. Alexandria, VA: American Gear Manufacturers Association, 1993.

7. American Gear Manufacturers Association. Standard 6001 D97. *Design and Selection of Components for Enclosed Gear Drives*. Alexandria, VA: American Gear Manufacturers Association, 1997.

8. American Gear Manufacturers Association. Standard 2000-A88. *Gear Classification and Inspection Handbook—Tolerances and Measuring Methods for Unassembled Spur and Helical Gears (Including Metric Equivalents).* Alexandria, VA: American Gear Manufacturers Association, 1988.

9. Drago, Raymond J. *Fundamentals of Gear Design.* Boston: Butterworths, 1988.

10. Dudley, Darle W. Handbook of Practical Gear Design. New York: McGraw-Hill, 1984.

11. Lipp, Robert. "Avoiding Tooth Interference in Gears." *Machine Design* 54, no. 1 (January 7, 1982).

12. Oberg, Erik, et al. *Machinery's Handbook.* 25th ed. New York: Industrial Press, 1996.

13. Shtipelman, Boris. *Design and Manufacture of Hypoid Gears.* John Wiley & Sons, 1978.

14. Wildhaber, Ernst. *Basic Relationship of Hypoid Gears.* American Machinist, 1946.

PROBLEMS

Gear Geometry

1. A gear has 44 teeth of the 20°, full-depth, involute form and a diametral pitch of 12. Compute the following:

   (a) Pitch diameter

   (b) Circular pitch

   (c) Equivalent module

   (d) Nearest standard module

   (e) Addendum

   (f) Dedendum

   (g) Clearance

   (h) Whole depth

   (i) Working depth

   (j) Tooth thickness

   (k) Outside diameter

   Repeat Problem 1 for the following gears:

2. $N = 34; P_d = 24$

3. $N = 45; P_d = 2$

4. $N = 18; P_d = 8$

5. $N = 22; P_d = 1.75$

6. $N = 20; P_d = 64$

7. $N = 180; P_d = 80$

8. $N = 28; P_d = 18$

9. $N = 28; P_d = 20$

   For Problems 10–17, repeat Problem 1 for the following gears in the metric module system. Replace Part (c) with equivalent P_d and Part (d) with nearest standard P_d.

10. $N = 34; m = 3$

11. $N = 45; m = 1.25$

12. $N = 18; m = 12$

13. $N = 22; m = 20$

14. $N = 20; m = 1$

15. $N = 180; m = 0.4$

16. $N = 28; m = 1.5$

17. $N = 28; m = 0.8$

18. Define *backlash,* and discuss the methods used to produce it.

19. For the gears of Problems 1 and 12, recommend the amount of backlash.

Velocity Ratio

20. An 8-pitch pinion with 18 teeth mates with a gear having 64 teeth. The pinion rotates at 2 450 rpm. Compute the following:

    (a) Center distance

    (b) Velocity ratio

    (c) Speed of gear

    (d) Pitch line speed

    Repeat Problem 20 for the following data:

21. $P_d = 4; \quad N_P = 20; N_G = 92; \quad n_P = 225$ rpm

22. $P_d = 20; \; N_P = 30; N_G = 68; \quad n_P = 850$ rpm

23. $P_d = 64; \; N_P = 40; N_G = 250; n_P = 3\,450$ rpm

24. $P_d = 12; \; N_P = 24; N_G = 88; \quad n_P = 1\,750$ rpm

25. $m = 2; \quad N_P = 22; N_G = 68; \quad n_P = 1\,750$ rpm

26. $m = 0.8; N_P = 18; N_G = 48; \quad n_P = 1\,150$ rpm

27. $m = 4; \quad N_P = 36; N_G = 45; \quad n_P = 150$ rpm

28. $m = 12; \; N_P = 15; N_G = 36; \quad n_P = 480$ rpm

    For Problems 29–32, all gears are made in standard 20°, full-depth, involute form. Tell what is wrong with the following statements:

29. An 8-pitch pinion having 24 teeth mates with a 10-pitch gear having 88 teeth. The pinion rotates at 1 750 rpm, and

the gear at approximately 477 rpm. The center distance is 5.900 in.

30. A 6-pitch pinion having 18 teeth mates with a 6-pitch gear having 82 teeth. The pinion rotates at 1 750 rpm, and the gear at approximately 384 rpm. The center distance is 8.3 in.

31. A 20-pitch pinion having 12 teeth mates with a 20-pitch gear having 62 teeth. The pinion rotates at 825 rpm, and the gear at approximately 160 rpm. The center distance is 1.850 in.

32. A 16-pitch pinion having 24 teeth mates with a 16-pitch gear having 45 teeth. The outside diameter of the pinion is 1.625 in. The outside diameter of the gear is 2.938 in. The center distance is 2.281 in.

Housing Dimensions

33. The gear pair described in Problem 20 is to be installed in a rectangular housing. Specify the dimensions X and Y as sketched in Figure 8–32 that would provide a minimum clearance of 0.10 in.

34. Repeat Problem 33 for the data of Problem 23.

35. Repeat Problem 33 for the data of Problem 26, but make the clearance 2.0 mm.

36. Repeat Problem 33 for the data of Problem 27, but make the clearance 2.0 mm.

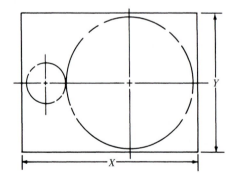

FIGURE 8–32   (Problems 33, 34, 35, and 36)

Analysis of Simple Gear Trains

For the gear trains sketched in the given figures, compute the output speed and the direction of rotation of the output shaft if the input shaft rotates at 1 750 rpm clockwise.

37. Use Figure 8–33.

38. Use Figure 8–34.

39. Use Figure 8–35.

40. Use Figure 8–36.

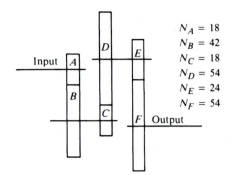

$N_A = 18$
$N_B = 42$
$N_C = 18$
$N_D = 54$
$N_E = 24$
$N_F = 54$

FIGURE 8–33   (Problem 37)

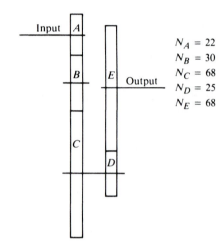

$N_A = 22$
$N_B = 30$
$N_C = 68$
$N_D = 25$
$N_E = 68$

FIGURE 8–34   (Problem 38)

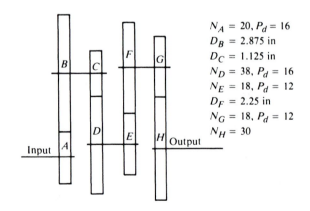

$N_A = 20, P_d = 16$
$D_B = 2.875$ in
$D_C = 1.125$ in
$N_D = 38, P_d = 16$
$N_E = 18, P_d = 12$
$D_F = 2.25$ in
$N_G = 18, P_d = 12$
$N_H = 30$

FIGURE 8–35   (Problem 39)

Helical Gearing

41. A helical gear has a transverse diametral pitch of 8, a transverse pressure angle of $14\frac{1}{2}°$, 45 teeth, a face width of 2.00 in, and a helix angle of 30°. Compute the circular

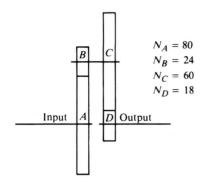

$N_A = 80$
$N_B = 24$
$N_C = 60$
$N_D = 18$

Input $\boxed{A}$ $\boxed{D}$ Output

FIGURE 8–36 (Problem 40)

pitch, normal circular pitch, normal diametral pitch, axial pitch, pitch diameter, and normal pressure angle. Then compute the number of axial pitches in the face width.

42. A helical gear has a normal diametral pitch of 12, a normal pressure angle of 20°, 48 teeth, a face width of 1.50 in, and a helix angle of 45°. Compute the circular pitch, normal circular pitch, transverse diametral pitch, axial pitch, pitch diameter, and transverse pressure angle. Then compute the number of axial pitches in the face width.

43. A helical gear has a transverse diametral pitch of 6, a transverse pressure angle of $14\frac{1}{2}°$, 36 teeth, a face width of 1.00 in, and a helix angle of 45°. Compute the circular pitch, normal circular pitch, normal diametral pitch, axial pitch, pitch, pitch diameter, and normal pressure angle. Then compute the number of axial pitches in the face width.

44. A helical gear has a normal diametral pitch of 24, a normal pressure angle of $14\frac{1}{2}°$, 72 teeth, a face width of 0.25 in, and a helix angle of 45°. Compute the circular pitch, normal circular pitch, transverse diametral pitch, axial pitch, pitch diameter, and transverse pressure angle. Then compute the number of axial pitches in the face width.

Bevel Gears

45. A straight bevel gear pair has the following data: $N_P = 15$; $N_G = 45$; $P_d = 6$; 20° pressure angle. Compute all of the geometric features from Table 8–7.

46. Draw the gear pair of Problem 45 to scale. The following additional dimensions are given (refer to Figure 8–22). Mounting distance (M_{dP}) for the pinion = 5.250 in; M_{dG} for the gear = 3.000 in; face width = 1.250 in. Supply any other needed dimensions.

47. A straight bevel gear pair has the following data: $N_P = 25$; $N_G = 50$; $P_d = 10$; 20° pressure angle. Compute all of the geometric features from Table 8–7.

48. Draw the gear pair of Problem 47 to scale. The following additional dimensions are given (refer to Figure 8–22). Mounting distance (M_{dP}) for the pinion = 3.375 in; M_{dG} for the gear = 2.625 in; face width = 0.700 in. Supply any other needed dimensions.

49. A straight bevel gear pair has the following data: $N_P = 18$; $N_G = 72$; $P_d = 12$; 20° pressure angle. Compute all of the geometric features from Table 8–7.

50. A straight bevel gear pair has the following data: $N_P = 16$; $N_G = 64$; $P_d = 32$; 20° pressure angle. Compute all of the geometric features from Table 8–7.

51. A straight bevel gear pair has the following data: $N_P = 12$; $N_G = 36$; $P_d = 48$; 20° pressure angle. Compute all of the geometric features from Table 8–7.

Wormgearing

52. A wormgear set has a single-thread worm with a pitch diameter of 1.250 in, a diametral pitch of 10, and a normal pressure angle of 14.5°. If the worm meshes with a wormgear having 40 teeth and a face width of 0.625 in, compute the lead, axial pitch, circular pitch, lead angle, addendum, dedendum, worm outside diameter, worm root diameter, gear pitch diameter, center distance, and velocity ratio.

53. Three designs are being considered for a wormgear set to produce a velocity ratio of 20 when the wormgear rotates at 90 rpm. All three have a diametral pitch of 12, a worm pitch diameter of 1.000 in, a gear face width of 0.500 in, and a normal pressure angle of 14.5°. One has a single-thread worm and 20 wormgear teeth; the second has a double-thread worm and 40 wormgear teeth; the third has a four-thread worm and 80 wormgear teeth. For each design, compute the lead, axial pitch, circular pitch, lead angle, gear pitch diameter, and center distance.

54. A wormgear set has a double-threaded worm with a normal pressure angle of 20°, a pitch diameter of 0.625 in, and a diametral pitch of 16. Its mating wormgear has 100 teeth and a face width of 0.3125 in. Compute the lead, axial pitch, circular pitch, lead angle, addendum, dedendum, worm outside diameter, center distance, and velocity ratio.

55. A wormgear set has a four-threaded worm with a normal pressure angle of $14\frac{1}{2}°$, a pitch diameter of 2.000 in, and a diametral pitch of 6. Its mating wormgear has 72 teeth and a face width of 1.000 in. Compute the lead, axial pitch, circular pitch, lead angle, addendum, dedendum, worm outside diameter, center distance, and velocity ratio.

56. A wormgear set has a single-threaded worm with a normal pressure angle of $14\frac{1}{2}°$, a pitch diameter of 4.000 in, and a diametral pitch of 3. Its mating wormgear has

54 teeth and a face width of 2.000 in. Compute the lead, axial pitch, circular pitch, lead angle, addendum, dedendum, worm outside diameter, center distance, and velocity ratio.

57. A wormgear set has a four-threaded worm with a normal pressure angle of 25°, a pitch diameter of 0.333 in, and a diametral pitch of 48. Its mating wormgear has 80 teeth and a face width of 0.156 in. Compute the lead, axial pitch, circular pitch, lead angle, addendum, dedendum, worm outside diameter, center distance, and velocity ratio.

Analysis of Complex Gear Trains

58. The input shaft for the gear train shown in Figure 8–37 rotates at 3 450 rpm. Compute the rotational speed of the output shaft.

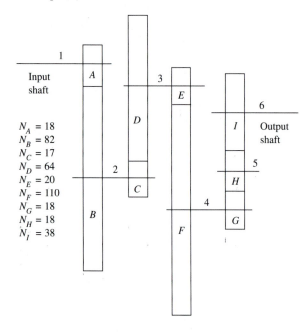

FIGURE 8–37 Gear train for Problem 58

59. The input shaft for the gear train shown in Figure 8–38 rotates at 12 200 rpm. Compute the rotational speed of the output shaft.

60. The input shaft for the gear train shown in Figure 8–39 rotates at 6 840 rpm. Compute the rotational speed of the output shaft.

61. The input shaft for the gear train shown in Figure 8–40 rotates at 2 875 rpm. Compute the rotational speed of the output shaft.

Kinematic Design of Gear Trains

62. Specify the numbers of teeth for the pinion and gear of a single gear pair to produce a velocity ratio of π as closely

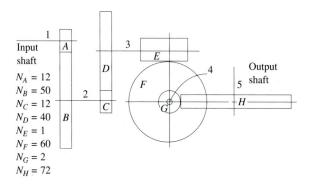

FIGURE 8–38 Gear train for Problem 59

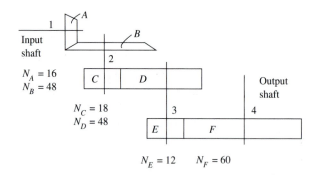

FIGURE 8–39 Gear train for Problem 60

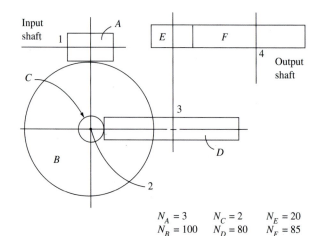

FIGURE 8–40 Gear train for Problem 61

as possible. Use no fewer than 16 teeth nor more than 24 teeth in the pinion.

63. Specify the numbers of teeth for the pinion and gear of a single gear pair to produce a velocity ratio of $\underline{e}$ 3 as closely as possible. Use no fewer than 16 teeth nor more than 24 teeth in the pinion.

64. Specify the numbers of teeth for the pinion and gear of a single gear pair to produce a velocity ratio of ℞ 38 as closely as possible. Use no fewer than 18 teeth nor more than 24 teeth in the pinion.

65. Specify the numbers of teeth for the pinion and gear of a single gear pair to produce a velocity ratio of 7.42 as closely as possible. Use no fewer than 18 teeth nor more than 24 teeth in the pinion.

66. Devise a gear train using all external spur gears. The input shaft rotates at exactly 1 800 rpm. The output shaft must rotate exactly 2.0 rpm. Use 20°, full-depth, involute teeth and no more than 150 teeth in any gear. Ensure that there is no interference.

67. Devise a gear train using all external spur gears. The input shaft rotates at exactly 1 800 rpm. The output shaft must rotate between 21.0 and 22.0 rpm. Use 25°, full-depth, involute teeth and no more than 150 teeth in any gear. Ensure that there is no interference.

68. Devise a gear train using all external spur gears. The input shaft rotates at exactly 3 360 rpm. The output shaft must rotate exactly 12.0 rpm. Use 20°, full-depth, involute teeth and no more than 150 teeth in any gear. Ensure that there is no interference.

69. Devise a gear train using all external spur gears. The input shaft rotates at exactly 4 200 rpm clockwise. The output shaft must rotate between 13.0 and 13.5 rpm clockwise. Use 20°, full-depth, involute teeth and no more than 150 teeth in any gear. Ensure that there is no interference.

70. Devise a gear train using any type of gears. The input shaft rotates at exactly 5 500 rpm. Two output shafts are to be provided. The first output shaft must rotate between 221 and 225 rpm. The second output shaft must rotate between 13 and 14 rpm. Use no more than 150 teeth in any gear.

9

Spur Gear Design

Spur Gear Design

Discussion Map

☐ A spur gear has involute teeth that are straight and parallel to the axis of the shaft that carries the gear.

Discover

Describe the action of the teeth of the driving gear on those of the driven gear. What kinds of stresses are produced?

How do the geometry of the gear teeth, the materials from which they are made, and the operating conditions affect the stresses and the life of the drive system?

This chapter will help you acquire the skills to perform the necessary analyses and to design safe spur gear drive systems that demonstrate long life.

A *spur gear* is one of the most fundamental types of gears. Its teeth are straight and parallel to the axis of the shaft that carries the gear. The teeth have the involute form that was described in Chapter 8. So, in general, the action of one tooth on a mating tooth is like that of two convex, curved members in contact: As the driving gear rotates, its teeth exert a force on the mating gear that is tangential to the pitch circles of the two gears. Because this force acts at a distance equal to the pitch radius of the gear, a torque is developed in the shaft that carries the gear. When the two gears rotate, they transmit power that is proportional to the torque. Indeed, that is the primary purpose of the spur gear drive system.

Consider the action described in the preceding paragraph:

- How does that action relate to the design of the gear teeth? Look back at Figure 8–1 in Chapter 8 as you consider this question and those that follow.
- As the force is exerted by the driving tooth on the driven tooth, what kinds of stresses are produced in the teeth? Consider both the point of contact of one tooth on the other and the whole tooth. Where are stresses a maximum?
- How could the teeth fail under the influence of these stresses?
- What material properties are critical to allow the gears to carry such loads safely and with a reasonable life span?
- What are the important geometric features that affect the level of stress produced in the teeth?
- How does the precision of the tooth geometry affect its operation?
- How does the nature of the application affect the gears? What if the machine that the gears drive is a rock crusher that takes large boulders and reduces them to gravel made up of small stones? How would that loading compare with that of a gear system that drives a fan providing ventilation air to a building?
- What is the influence of the driving machine? Would the design be different if an electric motor were the driver or if a gasoline engine were used?
- The gears are typically mounted on shafts that deliver power from the driver to the input gear of a gear train and that take power from the output gear and transmit it to the driven machine. Describe various ways that the gears can be attached to the shafts and located with respect to each other. How can the shafts be supported?

This chapter contains the kinds of information that you can use to answer such questions and to complete the analysis and design of spur gear power transmission systems. Later chapters cover similar topics for helical gears, bevel gears, and wormgearing, along with the design and specification of keys, couplings, seals, shafts, and bearings—all of which are needed to design a complete mechanical drive.

You Are the Designer

You have already made the design decision that a spur gear type of speed reducer is to be used for a particular application. How do you complete the design of the gears themselves?

This is a continuation of a design scenario that was started in Chapter 1 of this book when the original goals were stated and when an overview of the entire book was given. The introduction to Part II of the book continued this theme by indicating that the arrangement of the chapters is aligned with the design process that you could use to complete the design of the speed reducer.

Then in Chapter 8, you, as the designer, dealt with the kinematics of a gear reducer that would take power from the shaft of an electric motor rotating at 1 750 rpm and deliver it to a machine that was to operate at approximately 292 rpm. There you limited your interest to the decisions that affected motion and the basic geometry of the gears. It was decided that you would use a double-reduction gear train to reduce the speed of rotation of the drive system in two stages using two pairs of gears in series. You also learned how to specify the layout of the gear train, along with key design decisions such as the

numbers of teeth in all of the gears and the relationships among the diametral pitch, the number of teeth in the gears, the pitch diameters, and the distance between the centers of the shafts that carry those gears. For a chosen diametral pitch, you learned how to compute the dimensions of key features of the gear teeth such as the addendum, dedendum, and tooth width.

But the design is not complete until you have specified the material from which the gears are to be made and until you have verified that the gears will withstand the forces exerted on the gears as they transmit power and the corresponding torque. The teeth must not break, and they must have a sufficiently long life to meet the needs of the customer who uses the reducer.

To complete the design, you need more data: How much power is to be transmitted? To what kind of machine is the power from the output of the reducer being delivered? How does that affect the design of the gears? What is the anticipated duty cycle for the reducer in terms of the number of hours per day, days per week, and years of life expected? What options do you have for materials that are suitable for gears? Which material will you specify, and what will be its heat treatment?

You are the designer. The information in this chapter will help you complete the design.

9–1
OBJECTIVES OF THIS CHAPTER

After completing this chapter, you will be able to demonstrate the competencies listed below. They are presented in the order that they are covered in this chapter. The primary objectives are numbers 6, 7, and 8, which involve (a) the calculation of the bending strength and the ability of the gear teeth to resist pitting and (b) the design of gears to be safe with regard to both strength and pitting resistance. The competencies are as follows:

1. Compute the forces exerted on gear teeth as they rotate and transmit power.

2. Describe various methods for manufacturing gears and the levels of precision and quality to which they can be produced.

3. Specify a suitable level of quality for gears according to the use to which they are to be put.

4. Describe suitable metallic materials from which to make the gears, in order to provide adequate performance for both strength and pitting resistance.

5. Use the standards of the American Gear Manufacturers Association (AGMA) as the basis for completing the design of the gears.

6. Use appropriate stress analyses to determine the relationships among the applied forces, the geometry of the gear teeth, the precision of the gear teeth, and other factors specific to a given application, in order to make final decisions about those variables.

7. Perform the analysis of the tendency for the contact stresses exerted on the surfaces of the teeth to cause pitting of the teeth, in order to determine an adequate

hardness of the gear material that will provide an acceptable level of pitting resistance for the reducer.

8. Complete the design of the gears, taking into consideration both the stress analysis and the analysis of pitting resistance. The result will be a complete specification of the gear geometry, the material for the gear, and the heat treatment of the material.

9–2
CONCEPTS FROM PREVIOUS CHAPTERS

As you study this chapter, it is assumed that you are familiar with the geometry of gear features and the kinematics of one gear driving another as presented in Chapter 8 and illustrated in Figures 8–1, 8–8, 8–11, 8–12, 8–13, and 8–15. (See also References 3 and 19.) Key relationships that you should be able to use include the following:

$$\text{Pitch line speed} = v_t = R\omega = (D/2)\omega$$

where R = radius of the pitch circle
D = pitch diameter
ω = angular velocity of the gear

Because the pitch line speed is the same for both the pinion and the gear, values for $R, D,$ and ω can be for either. In the computation of stresses in gear teeth, it is usual to express the pitch line speed in the units of ft/min, while the size of the gear is given as its pitch diameter expressed in inches. Speed of rotation is typically given as n rpm, that is, n rev/min. Let's compute the unit-specific equation that gives pitch line speed in ft/min:

▷ **Pitch Line Speed**

$$v_t = (D/2)\omega = \frac{D \text{ in}}{2} \cdot \frac{n \text{ rev}}{\text{min}} \cdot \frac{2\pi \text{ rad}}{\text{rev}} \cdot \frac{1 \text{ ft}}{12 \text{ in}} = (\pi Dn/12) \text{ ft/min} \tag{9–1}$$

The velocity ratio can be expressed in many ways. For the particular case of a pinion driving a larger gear,

▷ **Velocity Ratio**

$$\text{Velocity ratio} = VR = \frac{\omega_P}{\omega_G} = \frac{n_P}{n_G} = \frac{R_G}{R_P} = \frac{D_G}{D_P} = \frac{N_G}{N_P} \tag{9–2}$$

A related ratio, m_G, called the *gear ratio*, is often used in analysis of the performance of gears. It is always defined as the ratio of the number of teeth in the larger gear to the number of teeth in the pinion, regardless of which is the driver. Thus, m_G is always greater than or equal to 1.0. When the pinion is the driver, as it is for a speed reducer, m_G is equal to VR. That is,

▷ **Gear Ratio**

$$\text{Gear ratio} = m_G = N_G/N_P \geq 1.0 \tag{9–3}$$

The diametral pitch, P_d, characterizes the physical size of the teeth of a gear. It is related to the pitch diameter and the number of teeth as follows:

▷ **Diametral Pitch**

$$P_d = N_G/D_G = N_P/D_P \tag{9–4}$$

The pressure angle, ϕ, is an important feature that characterizes the form of the involute curve that makes up the active face of the teeth of standard gears. See Figure 8–13. Also notice in Figure 8–12 that the angle between a normal to the involute curve and the tangent to the pitch circle of a gear is equal to the pressure angle.

9–3
FORCES ON
GEAR TEETH

To understand the method of computing stresses in gear teeth, consider the way power is transmitted by a gear system. For the simple single-reduction gear pair shown in Figure 9–1, power is received from the motor by the input shaft rotating at motor speed. Thus, there is a torque in the shaft that can be computed from the following equation:

▷ **Torque**
$$\text{Torque} = \text{power/rotational speed} = P/n \qquad (9\text{–}5)$$

The input shaft transmits the power from the coupling to the point where the pinion is mounted. The power is transmitted from the shaft to the pinion through the key. The teeth of the pinion drive the teeth of the gear and thus transmit the power to the gear. But again, power transmission actually involves the application of a torque during rotation at a given

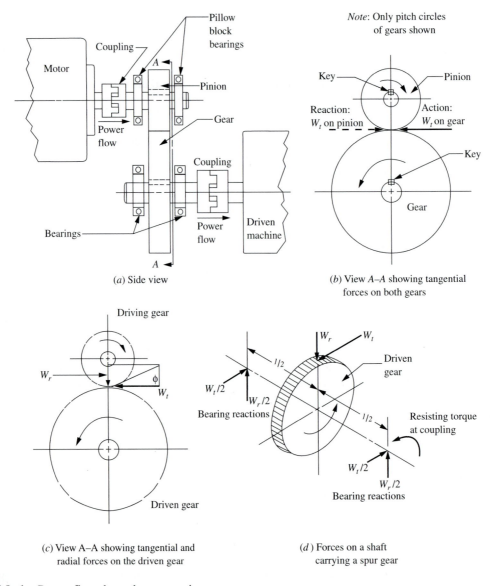

(a) Side view

(b) View A–A showing tangential forces on both gears

(c) View A–A showing tangential and radial forces on the driven gear

(d) Forces on a shaft carrying a spur gear

FIGURE 9–1   Power flow through a gear pair

speed. The torque is the product of the force acting tangent to the pitch circle of the pinion times the pitch radius of the pinion. We will use the symbol W_t to indicate the *tangential force.* As described, W_t is the force exerted *by the pinion teeth on the gear teeth.* But if the gears are rotating at a constant speed and are transmitting a uniform level of power, the system is in equilibrium. Therefore, there must be an equal and opposite tangential force exerted by the gear teeth back on the pinion teeth. This is an application of the principle of action and reaction.

To complete the description of the power flow, the tangential force on the gear teeth produces a torque on the gear equal to the product of W_t times the pitch radius of the gear. Because W_t is the same on the pinion and the gear, but the pitch radius of the gear is larger than that of the pinion, the torque on the gear (the output torque) is greater than the input torque. However, note that the power transmitted is the same or slightly less because of mechanical inefficiencies. The power then flows from the gear through the key to the output shaft and finally to the driven machine.

From this description of power flow, we can see that gears transmit power by exerting a force by the driving teeth on the driven teeth while the reaction force acts back on the teeth of the driving gear. Figure 9–2 shows a single gear tooth with the tangential force W_t acting on it. But this is not the total force on the tooth. Because of the involute form of the tooth, the total force transferred from one tooth to the mating tooth acts normal to the involute profile. This action is shown as W_n. The tangential force W_t is actually the horizontal component of the total force. To complete the picture, note that there is a vertical component of the total force acting radially on the gear tooth, indicated by W_r.

We will start the computation of forces with the transmitted force, W_t, because its value is based on the given data for power and speed. It is convenient to develop unit-specific equations for W_t because standard practice typically calls for the following units for key quantities pertinent to the analysis of gear sets:

Forces in pounds (lb)

Power in horsepower (hp) (Note that 1.0 hp = 550 lb·ft/s.)

FIGURE 9–2 Forces on gear teeth

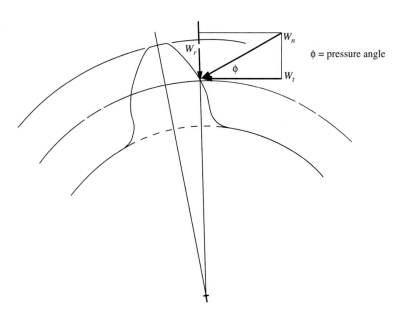

ϕ = pressure angle

Rotational speed in rpm, that is, rev/min

Pitch line speed in ft/min

Torque in lb·in

The torque exerted on a gear is the product of the transmitted load, W_t, and the pitch radius of the gear. The torque is also equal to the power transmitted divided by the rotational speed. Then

$$T = W_t(R) = W_t(D/2) = P/n$$

Then we can solve for the force, and the units can be adjusted as follows:

Tangential Force

$$W_t = \frac{2P}{Dn} = \frac{2P \text{ (hp)}}{D(\text{in}) \cdot n \text{ (rev/min)}} \cdot \frac{550 \text{ lb·ft/s}}{\text{hp}} \cdot \frac{1.0 \text{ rev}}{2\pi \text{ rad}} \cdot \frac{60 \text{ s/min}}{} \cdot \frac{12 \text{ in}}{\text{ft}}$$

$$W_t = (126\,000)(P)/(n\,D) \text{ lb} \tag{9–6}$$

Data for either the pinion or the gear can be used in this equation. Other relationships are now developed because they are needed in other parts of the process of analyzing the gears or the shafts that carry them.

Power is also the product of transmitted force, W_t, and the pitch line velocity:

$$P = W_t \cdot v_t$$

Then, solving for the force and adjusting units, we have

Tangential Force

$$W_t = \frac{P}{v_t} = \frac{P \text{ (hp)}}{v_t \text{ (ft/min)}} \cdot \frac{550 \text{ lb·ft/s}}{1.0 \text{ hp}} \cdot \frac{60 \text{ s/min}}{} = 33\,000 \, (P)/(v_t) \text{ lb} \tag{9–7}$$

We may also need to compute torque in lb·in:

$$T = \frac{P}{\omega} = \frac{P \text{ (hp)}}{n \text{ (rev/min)}} \cdot \frac{550 \text{ lb·ft/s}}{1.0 \text{ hp}} \cdot \frac{1.0 \text{ rev}}{2\pi \text{ rad}} \cdot \frac{60 \text{ s/min}}{} \cdot \frac{12 \text{ in}}{\text{ft}}$$

Torque

$$T = 63\,000 \, (P)/n \text{ lb·in} \tag{9–8}$$

These values can be computed for either the pinion or the gear by appropriate substitutions. Remember that the pitch line speed is the same for the pinion and the gear and that the transmitted loads on the pinion and the gear are the same, except that they act in opposite directions.

The normal force, W_n, and the radial force, W_r, can be computed from the known W_t by using the right triangle relations evident in Figure 9–2:

Radial Force

Normal Force

$$W_r = W_t \tan \phi \tag{9–9}$$

$$W_n = W_t/\cos \phi \tag{9–10}$$

where ϕ = pressure angle of the tooth form

In addition to causing the stresses in the gear teeth, these forces act on the shaft. In order to maintain equilibrium, the bearings that support the shaft must provide the reactions. Figure 9–1(d) shows the free-body diagram of the output shaft of the reducer.

9–4
GEAR
MANUFACTURE

The discussion of gear manufacture will begin with the method of producing the gear blank. Small gears are frequently made from wrought plate or bar, with the hub, web, spokes, and rim machined to final or near-final dimensions before the gear teeth are produced. The face width and the outside diameter of the gear teeth are also produced at this stage. Other gear blanks may be forged, sand cast, or die cast to achieve the basic form prior to machining. A few gears in which only moderate precision is required may be die cast with the teeth in virtually final form.

Large gears are frequently fabricated from components. The rim and the portion into which the teeth are machined may be rolled into a ring shape from a flat bar and then welded. The web or spokes and the hub are then welded inside the ring. Very large gears may be made in segments with the final assembly of the segments by welding or by mechanical fasteners.

The popular methods of machining the gear teeth are form milling, shaping, and hobbing. (See References 19 and 21.)

In *form milling* (Figure 9–3), a milling cutter that has the shape of the tooth space is used, and each space is cut completely before the gear blank is turned to the position of the next adjacent space. This method is used mostly for large gears, and great care is required to achieve accurate results.

Shaping (Figure 9–4) is a process in which the cutter reciprocates, usually on a vertical spindle. The shaping cutter rotates as it reciprocates and is fed into the gear blank. Thus, the involute-tooth form is generated gradually. This process is frequently used for internal gears.

Hobbing (Figures 9–5 and 9–6) is a process similar to milling except that the workpiece (the gear blank) and the cutter (the hob) rotate in a coordinated fashion. Here also, the tooth form is generated gradually as the hob is fed into the blank.

The gear teeth are finished to greater precision after form milling, shaping, or hobbing by the processes of grinding, shaving, and honing. Being products of secondary processes, they are expensive and should be used only where the operation requires high accuracy in the tooth form and spacing.

FIGURE 9–3 Form milling of gears (Illinois Tool Works, Chicago, IL)

FIGURE 9–4 Gear shaping [(a) and (b), Barber-Colman Company, Loves Park, IL; (c) and (d), Illinois Tool Works, Chicago, IL]

(a) Barber-Colman gear shaper

(b) Close-up of gear shaping

(c) Herringbone gear shaper cutters

(d) Spur gear shaper cutter

FIGURE 9–5 Gear hobbing (Barber-Colman Company, Loves Park, IL)

(a) Gear hobbing machine

(b) Closeup of gear hobbing process

FIGURE 9–6 Gear hobs [(a), Illinois Tool Works, Chicago, IL; (b), Barber-Colman Company, Loves Park, IL]

(*a*) Hob for small pitch gears having large teeth

(*b*) Hob for high pitch gears having small teeth

9–5 GEAR QUALITY

Quality in gearing is the precision of the individual gear teeth and the precision with which two gears rotate in relation to one another. The factors normally measured to determine quality are the following:

Runout: A measure of eccentricity and out-of-roundness

Tooth-to-tooth spacing: The difference in spacing between corresponding points on adjacent teeth

Profile: The variation of the actual tooth profile from the theoretically precise profile

Composite error is measured on a special device, shown in Figure 9–7(a), that places the test gear in tight mesh with a master gear of known precision. The two gears are rotated while in tight mesh, and the center of one gear is free to move. The variation in center distance for a complete revolution of the test gear is recorded. Figure 9–7(b) shows a typical recording with both *tooth-to-tooth composite error* and *total composite error* indicated. Thus, composite error is a measure of the combined effects of several types of gear-tooth errors.

The allowable amount of variation of the actual tooth form from the theoretical form, or the composite error, is specified by defining an AGMA quality number. Detailed charts giving the tolerances for many features are included in AGMA Standard 2000-A88. The quality numbers range from 5 to 15 with increasing precision. The actual tolerances are a function of the diametral pitch of the gear teeth and of the pitch diameter of the gear. Table 9–1 shows representative data for the total composite tolerance for several quality numbers.

Recommended Quality Numbers

These data should impress you with the precision that is normally exercised in the manufacture and installation of gears. The design of the entire gear system, including the shafts, bearings, and housing, must be consistent with this precision. Of course, the system should not be made more precise than necessary because of cost. For this reason, manufacturers have recommended quality numbers that will give satisfactory performance at a reasonable cost for a variety of applications. Table 9–2 lists several of these recommendations.

Also shown in Table 9–2 are recommendations for quality numbers for machine tool drives. Because this is such a wide range of specific applications, the recommended quality numbers are related to the *pitch line speed,* defined as the linear velocity of a point on

FIGURE 9–7

Recording of errors in gear geometry (Extracted from AGMA Standard 2000-A88, *Gear Classification and Inspection Handbook, Tolerances and Measuring Methods for Unassembled Spur and Helical Gears (Including Metric Equivalents),* with the permission of the publisher, American Gear Manufacturers Association, 1500 King Street, Suite 201, Alexandria, VA 22314)

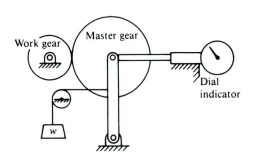

(*a*) Schematic diagram of a typical gear rolling fixture

(*b*) Chart of gear-tooth errors of a typical gear when run with a specified gear in a rolling fixture

the pitch circle of the gear. See Figure 8–15. Convenient relations for pitch line speed are adapted from Equation (9–1) and given here:

⇨ **Pitch Line Speed**
$$v_t = \pi D_G n_G / 12 = \pi D_P n_P / 12 \qquad (9\text{–}11)$$

where v_t = pitch line speed in feet per minute (fpm)
n_G = rotational speed of the gear (rpm)
n_P = rotational speed of the pinion (rpm)
D_G = pitch diameter of the gear (in)
D_P = pitch diameter of the pinion (in)

**9–6
ALLOWABLE
STRESS
NUMBERS**

Later in this chapter, design procedures are presented in which two forms of gear-tooth failure are considered.

A gear tooth acts like a cantilever beam in resisting the force exerted on it by the mating tooth. The point of highest tensile bending stress is at the root of the tooth where the involute curve blends with the fillet. The AGMA has developed a set of *allowable bending stress numbers,* called s_{at}, which are compared to computed bending stress levels in the tooth to rate the acceptability of a design.

TABLE 9–1 Selected values for total composite tolerance

AGMA quality number	Diametral pitch, P_d	Number of gear teeth				
		20	40	60	100	200
Q5	2	0.026 0	0.029 0	0.032 0	0.035 0	0.041 0
	8	0.012 0	0.013 0	0.014 0	0.015 0	0.017 0
	20	0.007 4	0.008 0	0.008 5	0.009 2	0.010 0
	32	0.006 0	0.006 4	0.006 8	0.007 3	0.008 0
Q8	2	0.009 4	0.011 0	0.012 0	0.013 0	0.015 0
	8	0.004 3	0.004 7	0.005 0	0.005 5	0.006 2
	20	0.002 7	0.002 9	0.003 1	0.003 4	0.003 7
	32	0.002 2	0.002 3	0.002 5	0.002 7	0.002 9
Q10	2	0.004 8	0.005 4	0.005 9	0.006 6	0.007 6
	8	0.002 2	0.002 4	0.002 6	0.002 8	0.003 2
	20	0.001 4	0.001 5	0.001 6	0.001 7	0.001 9
	32	0.001 1	0.001 2	0.001 3	0.001 4	0.001 5
Q12	2	0.002 5	0.002 8	0.003 0	0.003 4	0.003 9
	8	0.001 1	0.001 2	0.001 3	0.001 4	0.001 6
	20	0.000 71	0.000 77	0.000 81	0.000 87	0.000 97
	32	0.000 57	0.000 60	0.000 64	0.000 69	0.000 76
Q14	2	0.001 3	0.001 4	0.001 5	0.001 7	0.002 0
	8	0.000 57	0.000 62	0.000 67	0.000 73	0.000 82
	20	0.000 36	0.000 39	0.000 41	0.000 45	0.000 50
	32	0.000 29	0.000 31	0.000 33	0.000 35	0.000 39

SOURCE: Extracted from AGMA Standard 2000-A88, *Gear Classification and Inspection Handbook, Tolerances and Measuring Methods for Unassembled Spur and Helical Gears (Including Metric Equivalents)*, with the permission of the publisher, American Gear Manufacturers Association, 1500 King Street, Suite 201, Alexandria, VA 22314.

TABLE 9–2 Recommended AGMA quality numbers

Application	Quality number	Application	Quality number
Cement mixer drum drive	3–5	Small power drill	7–9
Cement kiln	5–6	Clothes washing machine	8–10
Steel mill drives	5–6	Printing press	9–11
Grain harvester	5–7	Computing mechanism	10–11
Cranes	5–7	Automotive transmission	10–11
Punch press	5–7	Radar antenna drive	10–12
Mining conveyor	5–7	Marine propulsion drive	10–12
Paper-box-making machine	6–8	Aircraft engine drive	10–13
Gas meter mechanism	7–9	Gyroscope	12–14

Machine tool drives and drives for other high-quality mechanical systems

Pitch line speed (fpm)	Quality number	Pitch line speed (m/s)
0–800	6–8	0–4
800–2 000	8–10	4–11
2 000–4 000	10–12	11–22
Over 4 000	12–14	Over 22

A second, independent form of failure is the pitting of the surface of the teeth, usually near the pitch line, where very high contact stresses occur. The transfer of force from the driving to the driven tooth theoretically occurs across a line contact because of the action of two convex curves on each other. Repeated application of these high contact stresses can cause a type of fatigue failure of the surface, resulting in local fractures and an actual loss of material. This is called *pitting.* The AGMA has developed a set of *allowable contact stress numbers,* called s_{ac}, which are compared to computed contact stress levels in the tooth to rate the acceptability of a design. (See References 8–10.)

Representative data for s_{at} and s_{ac} are given in the following section for general information and use in problems in this book. More extensive data are given in the AGMA standards listed at the end of the chapter. (See References 4–7.)

Many of the data given in this book for the design of spur and helical gears are taken from the AGMA Standard 2001-C95, *Fundamental Rating Factors and Calculation Methods for Involute Spur and Helical Gear Teeth,* with permission of the publisher, American Gear Manufacturers Association, 1500 King Street, Suite 201, Alexandria, VA 22314. That document should be consulted for details beyond the discussion in this book. Data in the U.S. Customary System only are given here. A separate document, AGMA Standard 2101-C95, has been published as a Metric Edition of AGMA 2001-C95. A brief summary of the differences between the terminology in these two standards is given later in this chapter.

9–7 GEAR MATERIALS

Gears can be made from a wide variety of materials to achieve properties appropriate to the application. From a mechanical design standpoint, strength and pitting resistance are the most important properties. But, in general, the designer should consider the producibility of the gear, taking into account all of the manufacturing processes involved, from the preparation of the gear blank, through the forming of the gear teeth, to the final assembly of the gear into a machine. Other considerations are weight, appearance, corrosion resistance, noise, and, of course, cost.

Steel Gear Materials

Through-hardened Steels. Gears for machine tool drives and many kinds of medium- to heavy-duty speed reducers and transmissions are typically made from medium-carbon steels. Among a wide range of carbon and alloy steels used are

AISI 1020	AISI 1040	AISI 1050
AISI 3140	AISI 4140	AISI 4150
AISI 4340	AISI 6150	AISI 8650

(See Reference 17.) AGMA Standard 2001-C95 gives data for the allowable bending stress number, s_{at}, and the allowable contact stress number, s_{ac}, for steels in the through-hardened condition. Figures 9–8 and 9–9 are graphs relating the stress numbers to the Brinell hardness number for the teeth. Notice that only knowledge of the hardness is required because of the direct relationship between hardness and the tensile strength of steels. See Appendix 19 for data that correlate the Brinell hardness number, HB, with the tensile strength of steel in Ksi. The range of hardnesses covered by the AGMA data is from 180 to 400 HB, corresponding to a tensile strength of approximately 87 to 200 Ksi. It is not recommended to use through-hardening above 400 HB because of inconsistent performance of the gears in service. Typically, case hardening is used when there is a desire to achieve a surface hardness above 400 HB. This is described later in this section.

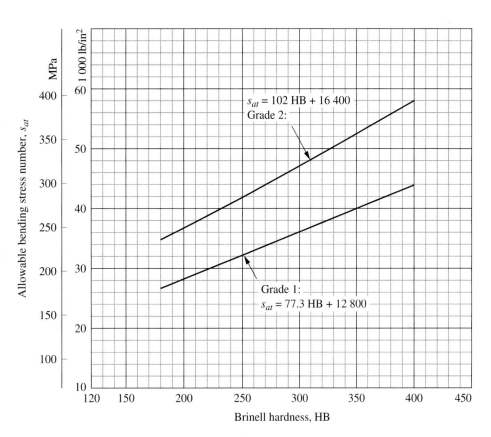

FIGURE 9–8
Allowable bending stress number for through-hardened steel gears, s_{at} (Extracted from AGMA 2001-C95 Standard, *Fundamental Rating Factors and Calculation Methods for Involute Spur and Helical Gear Teeth,* with permission of the publisher, American Gear Manufacturers Association, 1500 King Street, Suite 201, Alexandria, VA 22314)

$s_{at} = 102$ HB $+ 16\ 400$
Grade 2:

Grade 1:
$s_{at} = 77.3$ HB $+ 12\ 800$

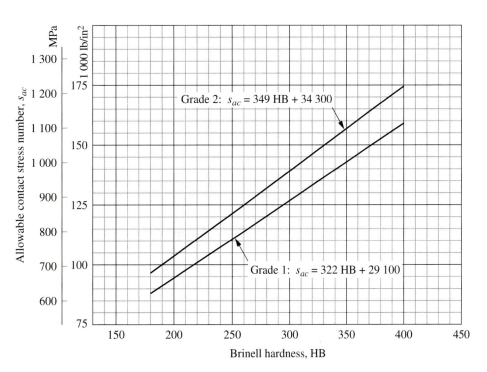

FIGURE 9–9
Allowable contact stress number for through-hardened steel gears, s_{ac} (Extracted from AGMA 2001-C95 Standard, *Fundamental Rating Factors and Calculation Methods for Involute Spur and Helical Gear Teeth,* with permission of the publisher, American Gear Manufacturers Association, 1500 King Street, Suite 201, Alexandria, VA 22314)

Grade 2: $s_{ac} = 349$ HB $+ 34\ 300$

Grade 1: $s_{ac} = 322$ HB $+ 29\ 100$

FIGURE 9–13
Allowable bending
stress numbers for
nitriding steel gears, s_{at}
(Extracted from
AGMA 2001-C95
Standard, *Fundamental
Rating Factors and
Calculation Methods
for Involute Spur and
Helical Gear Teeth,*
with permission of the
publisher, American
Gear Manufacturers
Association, 1500 King
Street, Suite 201,
Alexandria, VA 22314)

TABLE 9–4   Allowable stress numbers for iron and bronze gears

Material designation	Minimum hardness at surface (HB)	Allowable bending stress number		Allowable contact stress number	
		(Ksi)	(MPa)	(Ksi)	(MPa)
Gray cast iron, ASTM A48, as cast					
Class 20		5	35	50	345
Class 30	174	8.5	59	65	448
Class 40	201	13	90	75	517
Ductile (nodular) iron, ASTM A536					
60-40-18 annealed	140	22	152	77	530
80-55-06 quenched and tempered	179	22	152	77	530
100-70-03 quenched and tempered	229	27	186	92	634
120-90-02 quenched and tempered	269	31	214	103	710
Bronze, sand-cast, $s_{u\ min}$ = 40 Ksi (275 MPa)		5.7	39	30	207
Bronze, heat-treated, $s_{u\ min}$ = 90 Ksi (620 MPa)		23.6	163	65	448

SOURCE: Extracted from AGMA Standard 2001-C95 *Fundamental Rating Factors and Calculation Methods for Involute Spur and Helical Gear Teeth,* with the permission of the publisher, American Gear Manufacturers Association, 1500 King Street, Suite 201, Alexandria, VA 22314.

Plastic Gear Materials

Plastics perform well in applications where low weight, quiet operation, low friction, good corrosion resistance, and good wear properties are desired. Because their strengths are significantly lower than those of most metallic gear materials, plastics are used in relatively lightly loaded devices. Often plastic gears can be molded into final form without subse-

TABLE 9–5 Approximate allowable tooth bending stress in plastic gears

Material	Approximate allowable bending stress, Ksi (MPa)	
	Unfilled	Glass-filled
ABS	3 (21)	6 (41)
Acetal	5 (34)	7 (48)
Nylon	6 (41)	12 (83)
Polycarbonate	6 (41)	9 (62)
Polyester	3.5 (24)	8 (55)
Polyurethane	2.5 (17)	

Source: *Plastics Gearing.* Manchester, CN: ABA/PGT Publishing, 1976.

quent machining which gives production cost advantages. Some of the plastic materials used for gears are as follows:

Phenolic	Polyimide	Polyphenylene sulfide
Polycarbonate	Polyester	Polyurethane
Acetal	Nylon	Polyester elastomer
Styrene-acrylonitrile (SAN)		
Acrylonitrile-butadiene-styrene (ABS)		

These and other plastics can be produced in many formulations and can be filled with a variety of fillers to improve strength, wear resistance, impact resistance, temperature limit, moldability, and other properties. Therefore, standard allowable strengths have not been published. Table 9–5 shows a few selected data. Plastic material suppliers and companies specializing in plastic gear manufacture should be consulted for some of the special considerations in applying plastic gears. (See References 1 and 13.)

**9–8
STRESSES IN
GEAR TEETH**

The stress analysis of gear teeth is facilitated by consideration of the orthogonal force components, W_t and W_r, as shown in Figure 9–2.

The tangential force, W_t, produces a bending moment on the gear tooth similar to that on a cantilever beam. The resulting bending stress is maximum at the base of the tooth in the fillet that joins the involute profile to the bottom of the tooth space. Taking the detailed geometry of the tooth into account, Wilfred Lewis developed the equation for the stress at the base of the involute profile, which is now called the *Lewis equation:*

▷ **Lewis Equation
for Bending Stress
In Gear Teeth**

$$\sigma_t = \frac{W_t P_d}{FY}$$

(9–12)

where W_t = tangential force
P_d = diametral pitch of the tooth
F = face width of the tooth
Y = *Lewis form factor,* which depends on the tooth form, the pressure angle, the diametral pitch, the number of teeth in the gear, and the place where W_t acts

While the theoretical basis for the stress analysis of gear teeth is presented, the Lewis equation must be modified for practical design and analysis. One important limitation is that it does not take into account the stress concentration that exists in the fillet of the tooth. Figure 9–14 is a photograph of a photoelastic stress analysis of a model of a gear tooth. It indicates a stress concentration in the fillet at the root of the tooth as well as high contact stresses at the mating surface (the contact stress is discussed in the following section). Comparing the actual stress at the root with that predicted by the Lewis equation enables us to determine the stress concentration factor, K_t, for the fillet area. Placing this into equation (9–12) gives

$$\sigma_t = \frac{W_t P_d K_t}{FY} \tag{9–13}$$

The value of the stress concentration factor is dependent on the form of the tooth, the shape and size of the fillet at the root of the tooth, and the point of application of the force on the tooth. Note that the value of the Lewis form factor, Y, also depends on the tooth geometry. Therefore, the two factors are combined into one term, the *geometry factor, J,* where $J = Y/K_t$. The value of J also, of course, varies with the location of the point of application of the force on the tooth because Y and K_t vary.

Figure 9–15 shows graphs giving the values for the geometry factor for 20° and 25°, full-depth, involute teeth. The safest value to use is the one for the load applied at the tip of the tooth. However, this value is overly conservative because there is some load sharing by another tooth at the time that the load is initially applied at the tip of a tooth. The critical load on a given tooth occurs when the load is at the highest point of single-tooth contact, when the tooth carries the entire load. The upper curves in Figure 9–15 give the values for J for this condition.

Using the geometry factor, J, in the stress equation gives

$$\sigma_t = \frac{W_t P_d}{FJ} \tag{9–14}$$

The graphs in Figure 9–15 are taken from the former AGMA Standard 218.01 which has been superseded by the two new standards: AGMA 2001-C95, *Fundamental Rating Factors and Calculation Methods for Involute Spur and Helical Gear Teeth,* 1994, and AGMA 908-B89, *Geometry Factors for Determining the Pitting Resistance and Bending Strength of Spur, Helical and Herringbone Gear Teeth,* 1989. Standard 908-B89 includes

FIGURE 9–14
Photoelastic study of gear teeth under load (Measurements Group, Inc., Raleigh, NC)

an analytical method for calculating the geometry factor, *J*. But the values for *J* are unchanged from those in the former standard. Rather than graphs, the new standard reports values for *J* for a variety of tooth forms in tables. The graphs from the former standard are shown in Figure 9–15 so that you can visualize the variation of *J* with the number of teeth in the pinion and the gear.

FIGURE 9–15

Geometry factor, *J* (Extracted from AGMA 218.01 Standard, *Rating the Pitting Resistance and Bending Strength of Spur and Helical Involute Gear Teeth,* with the permission of the publisher, American Gear Manufacturers Association, 1500 King Street, Suite 201, Alexandria, VA 22314)

(*a*) 20° spur gear: standard addendum

(*b*) 25° spur gear: standard addendum

We can first use the principles from Section 9–3 to compute the transmitted load on the gear teeth:

$$D_p = N_p/P_d = 20/8 = 2.500 \text{ in}$$

$$v_t = \pi D_p n_P/12 = \pi(2.5)(1\,750)/12 = 1\,145 \text{ ft/min}$$

$$W_t = 33\,000(P)/v_t = (33\,000)(25)/(1\,145) = 720 \text{ lb}$$

From Figure 9–15, we find that $J_P = 0.335$ and $J_G = 0.420$.

The overload factor is found from Table 9–6. For a smooth, uniform electric motor driving an industrial saw generating moderate shock, $K_o = 1.50$ is a reasonable value.

The size factor $K_s = 1.00$ because the gear teeth with $P_d = 8$ are relatively small. See Table 9–7.

The load distribution factor, K_m, can be found from Equation (9–16) for commercial enclosed gear drives. For this design, $F = 1.50$ in, and

$$F/D = 1.50/2.50 = 0.60$$

$$C_{pf} = 0.04 \quad \text{(Figure 9–16)}$$

$$C_{ma} = 0.15 \quad \text{(Figure 9–17)}$$

$$K_m = 1.0 + C_{pf} + C_{ma} = 1.0 + 0.04 + 0.15 = 1.19$$

The rim thickness factor, K_B, can be taken as 1.00 because the gears are to be made from solid blanks.

The dynamic factor can be read from Figure 9–19. For $v_t = 1\,145$ ft/min and $Q_v = 6$, $K_v = 1.45$.

The stress can now be computed from Equation (9–15). We will compute the stress in the pinion first:

$$s_{tP} = \frac{(720)(8)}{(1.50)(0.335)} (1.50)(1.0)(1.19)(1.0)(1.45) = 29\,700 \text{ psi}$$

Notice that all factors in the stress equation are the same for the gear except the value of the geometry factor, J. Then the stress in the gear can be computed from

$$s_{tG} = \sigma_{tP}(J_P/J_G) = (29\,700)(0.335/0.420) = 23\,700 \text{ psi}$$

The stress in the pinion teeth will always be higher than the stress in the gear teeth because the value of J increases as the number of teeth increases.

9–9
SELECTION OF
GEAR MATERIAL
BASED ON
BENDING
STRESS

For safe operation, it is the designer's responsibility to specify a material that has an allowable bending stress greater than the computed stress due to bending from Equation (9–15). Recall that in Section 9–6, allowable stress numbers, s_{at}, were given for a variety of commonly used gear materials. Then it is necessary that

$$s_t < s_{at}$$

These data are valid for the following conditions:

Temperature less than 250°F

10^7 cycles of tooth loading

Reliability of 99%: Less than one failure in 100

Safety factor of 1.00

In this book we will assume that the operating temperature of the gears is less than 250°F. For higher temperatures, testing is recommended to determine the degree of reduction in the strength of the gear material.

Adjusted Allowable Bending Stress Numbers, s'_{at}

Data have been generated for different levels of expected life and reliability as discussed below. Designers may also choose to apply a factor of safety to the allowable bending stress number to account for uncertainties in the design analysis, material characteristics, or manufacturing tolerances, or to provide an extra measure of safety in critical applications. These factors are applied to the value of s_{at} to produce an *adjusted allowable bending stress number* which we will refer to as s'_{at}:

$$s'_{at} = s_{at} Y_N (SF \cdot K_P) \tag{9–17}$$

Stress Cycle Factor, Y_N

Figure 9–20 allows the determination of the life adjustment factor, Y_N, if the teeth of the gear being analyzed are expected to experience a number of cycles of loading much different from 10^7. Note that the general type of material is a factor in this chart for the lower number of cycles. For the higher number of cycles, a range is indicated by a shaded area. General design practice would use the upper line of this range. Critical applications where greater control of properties is exercised may use the lower part of the range.

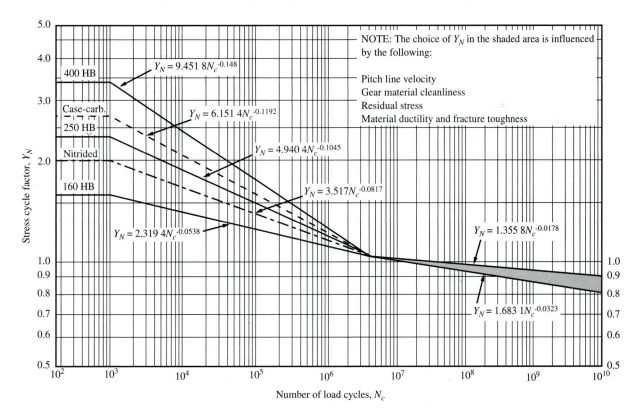

FIGURE 9–20   Bending strength stress cycle factor, Y_N (Extracted from AGMA 2001-C95 Standard, *Fundamental Rating Factors and Calculation Methods for Involute Spur and Helical Gear Teeth,* with permission of the publisher, American Gear Manufacturers Association, 1500 King Street, Suite 201, Alexandria, VA 22314)

Calculation of the expected number of cycles of loading can be done using

$$N_c = (60)(L)(n)(q) \qquad \textbf{(9–18)}$$

where N_c = expected number of cycles of loading

L = design life in hours

n = rotational speed of the gear in rpm

q = number of load applications per revolution

Design life is, indeed, a design decision based on the application. As a guideline, we will use a set of data created for use in bearing design and reported here as Table 9–8. Unless stated otherwise, we will use a design life of L = 20 000 h as listed for general industrial machines. The normal number of load applications per revolution for any given tooth is typically, of course, one. But consider the case of an idler gear that serves as both a driven and a driving gear in a gear train. It receives two cycles of load per revolution: one as it receives power from and one as it delivers power to its mating gears. Also, in certain types of gear trains, one gear may deliver power to two or more gears mating with it. Gears in a planetary gear train often have this characteristic.

As an example of the application of Equation (9–18), consider that the pinion in Example Problem 9–1 is designed to have a life of 20 000 h. Then

$$N_c = (60)(L)(n)(q) = (60)(20\ 000)(1\ 750)(1) = 2.1 \times 10^9 \text{ cycles}$$

Because this is higher than 10^7, an adjustment in the allowable bending stress number must be made.

Reliability Factor, K_R

Table 9–9 gives data that adjust for the design reliability desired. These data are based on statistical analyses of failure data.

Factor of Safety, SF

The factor of safety may be used to account for the following:

TABLE 9–8 Recommended design life

Application	Design life (h)
Domestic appliances	1 000–2 000
Aircraft engines	1 000–4 000
Automotive	1 500–5 000
Agricultural equipment	3 000–6 000
Elevators, industrial fans, multipurpose gearing	8 000–15 000
Electric motors, industrial blowers, general industrial machines	20 000–30 000
Pumps and compressors	40 000–60 000
Critical equipment in continuous 24-h operation	100 000–200 000

SOURCE: Eugene A. Avallone and Theodore Baumeister III, eds. *Marks' Standard Handbook for Mechanical Engineers.* 9th ed. New York: McGraw-Hill, 1986.

TABLE 9–9 Reliability factor, K_R

Reliability	K_R
0.90, one failure in 10	0.85
0.99, one failure in 100	1.00
0.999, one failure in 1 000	1.25
0.9999, one failure in 10 000	1.50

- Uncertainties in the design analysis
- Uncertainties in material characteristics
- Uncertainties in manufacturing tolerances

It may also be used to provide an extra measure of safety in critical applications.

No general guidelines are published, and designers must evaluate the conditions of each application. Note, however, that many of the factors often considered to be a part of a factor of safety in general design practice have already been included in the calculations for s_t and s'_{at}. Therefore, a modest value for factor of safety should suffice, say, between 1.00 and 1.50.

Procedure for Selecting Gear Materials for Bending Stress

The logic of the material selection process can be summarized as follows: The bending stress number from Equation (9–15) must be less than the adjusted allowable bending stress number from Equation (9–17). That is,

$$s_t < s'_{at}$$

Let's equate the expressions for these two values:

$$\frac{W_t P_d}{FJ} K_o K_s K_m K_B K_v = s_t < s_{at} \frac{Y_N}{SF \cdot K_R} \tag{9–19}$$

To use this relationship for material selection, it is convenient to solve for s_{at}:

$$\frac{K_R(SF)}{Y_N} s_t < s_{at} \tag{9–20}$$

We will use this equation for selecting gear materials based on bending stress. The list below summarizes the terms included in Equations (9–19) and (9–20) for your reference. You should review the more complete discussion of each before doing practice designs:

W_t = tangential force on gear teeth = $(63\,000)P/n$
P_d = diametral pitch of the gear
F = face width of the gear
J = geometry factor for bending stress   (see Figure 9–15)
K_o = overload factor   (see Table 9–6)
K_s = size factor   (see Table 9–7)
K_m = mesh alignment factor = $1.0 + C_{pf} + C_{ma}$ (see Figures 9–16 and 9–17)
K_B = rim thickness factor   (see Figure 9–18)
K_v = velocity factor   (see Figure 9–19)
K_R = reliability factor   (see Table 9–9)
SF = factor of safety   (design decision)
Y_N = bending strength stress cycle number   (see Figure 9–20)

Completing the calculation of the value of the left side of Equation (9–20) gives the required value for the allowable bending stress number, s_{at}. You should then go to the data in Section 9–7, "Gear Materials," to select a suitable material. Consider first whether the material should be steel, cast iron, bronze, or plastic. Then consult the related tables of data.

For steel materials, the following review should aid your selection:

1. Start by checking Figure 9–8 to see whether a through-hardened steel will give the needed s_{at}. If so, determine the required hardness. Then specify a steel material and its heat treatment by referring to Appendices 3 and 4.

2. If a higher s_{at} is needed, see Table 9–3 and Figures 9–12 and 9–13 for properties of case-hardened steels.

3. Appendix 5 will aid in the selection of carburized steels.

4. If flame or induction hardening is planned, specify a material with a good hardenability, such as AISI 4140 or 4340 or similar medium-carbon-alloy steels. See Appendix 4.

5. Refer to Figure 9–10 or 9–11 for recommended case depths for surface-hardened steels.

For cast iron or bronze, refer to Table 9–4. For plastics, refer to Table 9–5.

Example Problem 9–2

Specify suitable materials for the pinion and the gear from Example Problem 9–1. Design for a reliability of fewer than one failure in 10 000. The application is an industrial saw that will be fully utilized on a normal, one-shift, five-day-per-week operation.

Solution

The results of Example Problem 9–1 include the expected bending stress number for both the pinion and the gear as follows:

$$s_{tP} = 29\ 700 \text{ psi} \qquad s_{tG} = 23\ 700 \text{ psi}$$

We should consider the number of stress cycles, the reliability, and the safety factor to complete the calculation indicated in Equation (9–20).

Stress Cycle Factor, Y_N

From the problem statement, $n_P = 1\ 750$ rpm, $N_P = 20$ teeth, and $N_G = 70$ teeth. Let's use these data to determine the expected number of cycles of stress that the pinion and gear teeth will experience. The application conforms to common industry practice, calling for a design life of approximately 20 000 h as suggested in Table 9–8. The number of stress cycles for the pinion is

$$N_{cP} = (60)(L)(n_P)(q) = (60)(20\ 000)(1\ 750)(1) = 2.10 \times 10^9 \text{ cycles}$$

The gear rotates more slowly because of the speed reduction. Then

$$n_G = n_P\ (N_P/N_G) = (1\ 750 \text{ rpm})(20/70) = 500 \text{ rpm}$$

We can now compute the number of stress cycles for each gear tooth:

$$N_{cG} = (60)(L)(n_G)(q) = (60)(20\ 000)(500)(1) = 6.00 \times 10^8 \text{ cycles}$$

Because both values are above the nominal value of 10^7 cycles, a value of Y_N must be determined from Figure 9–20 for both the pinion and the gear:

$$Y_{NP} = 0.92 \qquad Y_{NG} = 0.96$$

Reliability Factor, K_R

For the design goal of fewer than one failure in 10 000, Table 9–9 recommends $K_R = 1.50$.

Factor of Safety

This is a design decision. Reviewing the discussion of factors throughout Example Problem 9–1 and this problem, we see that virtually all factors typically considered in adjusting stress on the teeth and strength of the material have been taken into account. Furthermore, when we select a material, it will likely have strength and hardness values somewhat above the minimum acceptable values. Therefore, as a design decision, let's use $SF = 1.00$.

Adjusted Value of s_{at}

We can now complete Equation (9–20) and use it for material selection. For the pinion,

$$\frac{K_R\,(SF)}{Y_{NP}}\,s_t = \frac{(1.50)(1.00)}{0.92}\,(29\,700\ \text{psi}) = 48\,450\ \text{psi} < s_{at}$$

For the gear,

$$\frac{K_R\,(SF)}{Y_{NG}}\,s_t = \frac{(1.50)(1.00)}{0.96}\,(23\,700\ \text{psi}) = 37\,050\ \text{psi} < s_{at}$$

Now, referring to Figure 9–8, and deciding to use Grade 1 steel, we find that the required allowable bending stress number for the pinion is greater than permitted for a through-hardened steel. But Table 9–3 indicates that a carburized, case-hardened steel with a case hardness of 55 to 64 HRC would be satisfactory, having a value of $s_{at} = 55$ Ksi = 55 000 psi. Referring to Appendix 5, we see that virtually any of the listed carburized materials could be used. Let's specify AISI 4320 SOQT 300, having a core tensile strength of 218 Ksi, 13% elongation, and a case hardness of 62 HRC.

For the gear, Figure 9–8 indicates that a through-hardened steel with a hardness of 320 HB would be satisfactory. From Appendix 3, let's specify AISI 4340 OQT 1000 having a hardness of 363 HB, a tensile strength of 171 Ksi, and 16% elongation.

Comments   These materials should provide satisfactory service, considering bending strength. The following section considers the other major failure mode: pitting resistance. It is possible, perhaps likely, that the requirements to meet that condition will be more severe than for bending.

**9–10
PITTING
RESISTANCE
OF GEAR
TEETH**

Gear teeth must be safe from failure of the teeth by breakage, and they must also be capable of running for the desired life without significant pitting of the tooth form. *Pitting* is the phenomenon in which small particles are removed from the surface of the tooth because of the high contact forces that are present between mating teeth. Pitting is actually the fatigue failure of the tooth surface. Prolonged operation after pitting begins causes the tooth form to change dramatically, causing vibration and noise. Thus, the design process should seek to prevent the start of pitting. (See Reference 9 for more information on pitting.)

The primary property of the gear tooth that provides resistance to pitting is the hardness of the tooth surface. The harder the surface, the higher is the compressive strength of the material. The force exerted by the driving tooth on the mating tooth (and the reaction back on the driving tooth) is theoretically applied to the line of contact between the tooth profiles. Actually, because of the elasticity of the material of the teeth, the force is spread over a small rectangular area. The resulting stress is highly concentrated around the area of application of the force and is called a *contact stress* or a *Hertz stress*.

The development of the equation for the contact stress on gear teeth is based on the analysis for two cylinders under a radial load. The radii of the cylinders are taken to be the radii of curvature of the involute-tooth forms of the mating teeth at the point of contact. The load on the teeth is the total normal load, found from

$$W_N = W_t/\cos\phi \qquad\qquad\textbf{(9–21)}$$

where W_N = load acting normal to the tooth surface
W_t = transmitted load acting tangential to the pitch line
ϕ = pressure angle

The resulting Hertz contact stress due to this load is

⇨ **Hertz Contact Stress on Gear Teeth**

$$\sigma_c = \sqrt{\frac{W_t}{F D_P} \frac{1}{\pi\left(\dfrac{1 - v_P^2}{E_P} + \dfrac{1 - v_G^2}{E_G}\right)} \frac{1}{\dfrac{\cos \phi \sin \phi}{2} \dfrac{m_G}{m_G + 1}}} \qquad (9\text{–}22)$$

where, in addition to the terms already defined,
D_P = pitch diameter of the pinion
v_P = Poisson's ratio for the pinion material
v_G = Poisson's ratio for the gear material
E_P = modulus of elasticity for the pinion material
E_G = modulus of elasticity for the gear material
m_G = gear ratio = N_G/N_P

(See Reference 2.)

The second term under the radical is dependent on the material properties and is given the name *elastic coefficient, C_p.* That is,

⇨ **Elastic Coefficient**

$$C_p = \sqrt{\frac{1}{\pi[(1 - v_P^2)/E_P + (1 - v_G^2)/E_G]}} \qquad (9\text{–}23)$$

For commonly used gear materials, the value of C_p can be found in Table 9–10.

The denominator of the last term under the radical in Equation (9–22) is called the *geometry factor, I,* and is dependent on the tooth geometry and on the gear ratio. Values can be found from Figure 9–21.

The graphs in Figure 9–21 are taken from the former AGMA Standard 218.01 which has been superseded by the two new standards: AGMA 2001-C95, *Fundamental Rating*

TABLE 9–10 Elastic coefficient, C_p

Pinion material	Modulus of elasticity, E_p, lb/in² (MPa)	Gear material and modulus of elasticity, E_G, lb/in² (MPa)					
		Steel 30 × 10⁶ (2 × 10⁵)	Malleable iron 25 × 10⁶ (1.7 × 10⁵)	Nodular iron 24 × 10⁶ (1.7 × 10⁵)	Cast iron 22 × 10⁶ (1.5 × 10⁵)	Aluminum bronze 17.5 × 10⁶ (1.2 × 10⁵)	Tin bronze 16 × 10⁶ (1.1 × 10⁵)
Steel	30 × 10⁶ (2 × 10⁵)	2 300 (191)	2 180 (181)	2 160 (179)	2 100 (174)	1 950 (162)	1 900 (158)
Mall. iron	25 × 10⁶ (1.7 × 10⁵)	2 180 (181)	2 090 (174)	2 070 (172)	2 020 (168)	1 900 (158)	1 850 (154)
Nod. iron	24 × 10⁶ (1.7 × 10⁵)	2 160 (179)	2 070 (172)	2 050 (170)	2 000 (166)	1 880 (156)	1 830 (152)
Cast iron	22 × 10⁶ (1.5 × 10⁵)	2 100 (174)	2 020 (168)	2 000 (166)	1 960 (163)	1 850 (154)	1 800 (149)
Al. bronze	17.5 × 10⁶ (1.2 × 10⁵)	1 950 (162)	1 900 (158)	1 880 (156)	1 850 (154)	1 750 (145)	1 700 (141)
Tin bronze	16 × 10⁶ (1.1 × 10⁵)	1 900 (158)	1 850 (154)	1 830 (152)	1 800 (149)	1 700 (141)	1 650 (137)

SOURCE: Extracted from AGMA Standard 2001-C95, *Fundamental Rating Factors and Calculation Methods for Involute Spur and Helical Gear Teeth,* with the permission of the publisher, American Gear Manufacturers Association, 1500 King Street, Suite 201, Alexandria, VA 22314.
NOTE: Poisson's ratio = 0.30; units for C_p are (lb/in²)^0.5 or (MPa)^0.5.

FIGURE 9–21

External spur pinion geometry factor, *I*, for standard center distances. All curves are for the lowest point of single-tooth contact on the pinion. (Extracted from AGMA Standard 218.01, *Rating the Pitting Resistance and Bending Strength of Spur and Helical Involute Gear Teeth,* with the permission of the publisher, American Gear Manufacturers Association, 1500 King Street, Suite 201, Alexandria, VA 22314)

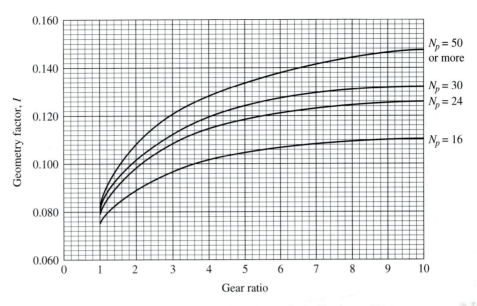

(*a*) 20° pressure angle, full-depth teeth (standard addendum = $1/P_d$)

(*b*) 25° pressure angle, full-depth teeth (standard addendum = $1/P_d$)

Factors and Calculation Methods for Involute Spur and Helical Gear Teeth, 1995, and AGMA 908-B89, *Geometry Factors for Determining the Pitting Resistance and Bending Strength of Spur, Helical and Herringbone Gear Teeth,* 1989. Standard 908-B89 includes an analytical method for calculating the geometry factor, *I*. But the values for *I* are unchanged from the former standard. Rather than using graphs, the new standard reports values for *I* for a variety of tooth forms in tables. The graphs from the former standard are shown in Figure 9–22 so that you can visualize the variation of *I* with the number of teeth in the pinion and the gear.

FIGURE 9–22   Pitting resistance stress cycle factor, Z_N (Extracted from AGMA Standard 2001-C95, *Fundamental Rating Factors and Calculation Methods for Involute Spur and Helical Gear Teeth,* with permission of the publisher, American Gear Manufacturers Association, 1500 King Street, Suite 201, Alexandria, VA 22314)

Note also that *I* factors for only two tooth forms are included in Figure 9–22 and that the values are valid only for those forms. Designers must ensure that *I* factors for the tooth form actually used are included in the contact stress analysis.

Substituting C_p and *I* into Equation (9–22) gives

$$\sigma_c = C_p \sqrt{\frac{W_t}{FD_pI}}$$

(9–24)

As with the equation for bending stress in gear teeth, several factors are added to the equation for contact stress as shown below. The resulting quantity is called the *contact stress number, s_c:*

▷ **Contact Stress Number**

$$s_c = C_p \sqrt{\frac{W_tK_oK_sK_mK_v}{FD_pI}}$$

(9–25)

This is the form of the contact stress equation that we will use in problem solutions.

The values for the overload factor, K_o; the size factor, K_s; the load-distribution factor, K_m; and the dynamic factor, K_v, can be taken to be the same as the corresponding values for the bending stress analysis in the preceding sections.

Example Problem 9–3 Compute the contact stress number for the gear pair described in Example Problem 9–1.

Solution Data from Example Problem 9–1 are summarized as follows:

$N_P = 20$ $N_G = 70,$ $F = 1.50$ in $W_t = 720$ lb $D_P = 2.500$ in

$K_o = 1.50$ $K_s = 1.00$ $K_m = 1.19$ $K_v = 1.45$

The gear teeth are 20°, full-depth, involute teeth. We also need the geometry factor for pitting resistance, I. From Figure 9–21(a), at a gear ratio of $m_G = N_G/N_P = 70/20 = 3.50$ and for $N_P = 20$, we read $I = 0.108$, approximately.

The design analysis for bending strength indicated that two steel gears should be used. Then, from Table 9–10, we find that $C_p = 2\,300$. Then the contact stress number is

$$s_c = C_p \sqrt{\frac{W_t K_o K_s K_m K_v}{F D_p I}} = 2\,300 \sqrt{\frac{(720)(1.50)(1.0)(1.19)(1.45)}{(1.50)(2.50)(0.108)}}$$

$$s_c = 156\,000 \text{ psi}$$

9–11 SELECTION OF GEAR MATERIAL BASED ON CONTACT STRESS

Because the pitting resulting from the contact stress is a different failure phenomenon from tooth failure due to bending, an independent specification for suitable materials for the pinion and the gear must now be made. In general, the designer must specify a material having an allowable contact stress number, s_{ac}, greater than the computed contact stress number, s_c. That is,

$$s_c < s_{ac}$$

In Section 9–6, values for s_{ac} were given for several materials that are valid for 10^7 cycles of loading at a reliability of 99% if the material temperature is under 250°F. For different life expectancy and reliability, other factors are added:

$$s_c > s_{ac} \frac{Z_N C_H}{(SF) K_R} \qquad (9\text{–}26)$$

Designs in this book are limited to applications where the operating temperature is less than 250°F and so no temperature factor is applied. Data for the reduction in hardness and strength as a function of temperature should be sought if higher temperatures are experienced.

The reliability factor, K_R, is the same as that for bending stress; it is given in Table 9–9. The other factors in Equation (9–26) are discussed below.

Pitting Resistance Stress Cycle Factor, Z_N

The term Z_N is the *pitting resistance stress cycle factor* and accounts for an expected number of contacts different from 10^7 as was assumed when the data were produced for the allowable contact stress number. Figure 9–22 shows values for Z_N where the solid-line curve is for most steels and the dashed-line curve is for nitrided steels. The number of cycles of contact is computed from Equation (9–18) and is the same as that used for bending. For the higher number of cycles, a range is indicated by the shaded area. General design practice would use the upper line of this range. Critical applications where greater control of properties is exercised may use the lower part of the range.

Factor of Safety, *SF*

The factor of safety is based on the same conditions as described for bending, and often the same value would be used for both bending and pitting resistance. Review that discussion in Section 9–9. However, if there are different levels of uncertainty, a different value should be chosen. No general guidelines are published. Because many factors are already considered in the pitting resistance calculations, a modest value for factor of safety should suffice, say, between 1.00 and 1.50.

Hardness Ratio Factor, C_H

Good gear design practice calls for making the pinion teeth harder than the gear teeth so that the gear teeth are smoothed and work-hardened during operation. This increases the gear capacity with regard to pitting resistance and is accounted for by the factor C_H. Figure 9–23 shows data for C_H for through-hardened gears that depend on the ratio of the hardness of the pinion and the hardness of the gear, expressed as the Brinell hardness number, and on the gear ratio where $m_G = N_G/N_P$. Use the given curves for hardness ratios between 1.2 and 1.7. For hardness ratios under 1.2, use $C_H = 1.00$. For hardness ratios over 1.7, use the value of C_H for 1.7, as no substantial additional improvement is gained.

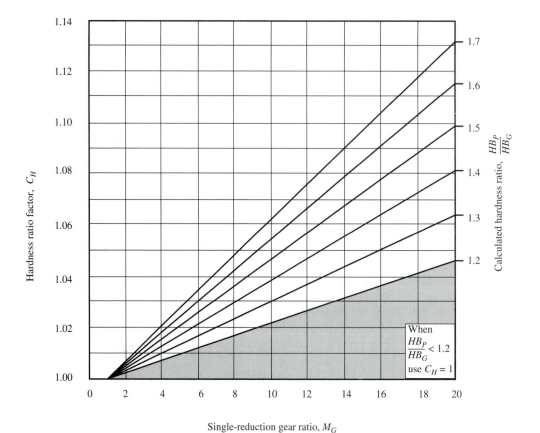

FIGURE 9–23 Hardness ratio factor, C_H (through-hardened) (Extracted from AGMA Standard 2001-C95, *Fundamental Rating Factors and Calculation Methods for Involute Spur and Helical Gear Teeth,* with permission of the publisher, American Gear Manufacturers Association, 1500 King Street, Suite 201, Alexandria, VA 22314)

Figure 9–24 shows data for C_H when pinions are surface-hardened to 48 HRC or higher and the gear is through-hardened up to 400 HB. The parameters are the Brinell hardness number for the gear and the surface finish of the pinion teeth, expressed as f_P and measured as the average roughness, R_a. Smoother teeth give a higher value of hardness factor and generally increase the pitting resistance of the gear teeth.

Note that C_H is applied for the calculations for only the gear, not the pinion.

Procedure for Selecting Gear Materials for Pitting Resistance

We can refer to the value on the right side of Equation (9–26) as the *modified allowable contact stress number*, as it accounts for nonstandard conditions under which the gears operate that are different from those assumed when the data for s_{ac} were determined as reported in Section 9–6.

Transposing the modifying factors from Equation (9–26) to the left side of the equation gives

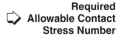

Required Allowable Contact Stress Number

$$\frac{K_R\,(SF)}{Z_N C_H}\, s_c < s_{ac} \qquad (9\text{--}27)$$

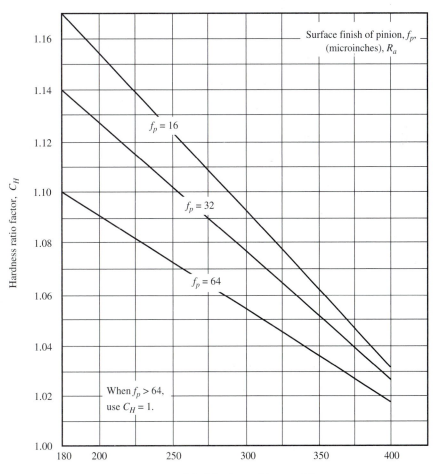

FIGURE 9–24

Hardness ratio factor, C_H (surface-hardened pinions) (Extracted from AGMA Standard 2001-C95, *Fundamental Rating Factors and Calculation Methods for Involute Spur and Helical Gear Teeth,* with permission of the publisher, American Gear Manufacturers Association, 1500 King Street, Suite 201, Alexandria, VA 22314)

This is the equation that we will use to determine the required properties of most metallic materials used for gears. We can summarize the procedure as follows.

Procedure for Determining the Required Properties of Most Metallic Materials

1. Solve for the contact stress number, s_c, from Equations (9–25) and (9–26) using the same factors as those used for bending stress number.

2. Use the value for K_R from the bending stress analysis, or evaluate it from Table 9–9.

3. Use Figure 9–22 to find Z_N.

4. Use Figure 9–23 to find that value for C_H for through-hardened gears, or use Figure 9–24 for a surface-hardened pinion and a through-hardened gear.

5. Specify a safety factor, typically between 1.00 and 1.50, considering the degree of uncertainty in material property data, the gear precision, the severity of the application, or the danger presented by the application.

6. Compute s_{ac} from Equation (9–27).

7. Refer to data in Section 9–7, "Gear Materials," to select a suitable material. You should consider first whether the material should be steel, cast iron, bronze, or plastic. Then consult the related tables of data.

The phenomenon of pitting does not apply directly to plastics. Refer to suppliers of plastic gears. For cast iron or bronze, refer to Table 9–4.

For steel materials, the following review should aid your selection:

1. Start by checking Figure 9–9 to see whether a through-hardened steel will give the needed s_{ac}. If so, determine the required hardness. Then specify a steel material and its heat treatment by referring to Appendices 3 and 4.

2. If a higher s_{ac} is needed, see Table 9–3 for properties of case-hardened steels.

3. Appendix 5 will aid in the selection of carburized steels.

4. If flame or induction hardening is planned, specify a material with a good hardenability, such as AISI 4140 or 4340 or similar medium-carbon-alloy steels. See Appendix 4.

5. Refer to Figure 9–10 or 9–11 for recommended case depths for surface-hardened steels.

Example Problem 9–4   Specify suitable materials for the pinion and the gear from Example Problem 9–3 based on contact stress. Application conditions are also described in Example Problems 9–1 and 9–2.

Solution   In Example Problem 9–3, we found that the expected contact stress number is $s_c = 156\,000$ psi. This should be modified as indicated in Equation (9–27).

In Example Problem 9–2, we determined that a carburized, case-hardened steel would be used for the pinion, and a through-hardened steel should be used for the gear. We must complete the material selection for pitting resistance independently from the bending stress analysis. However, we could not specify a material with lower properties than those specified in Example Problem 9–2 because then the bending strength would be inadequate.

In Example Problem 9–2, we used $K_R = 1.50$ for the desired reliability of fewer than one failure in 10 000. We decided to use $SF = 1.00$ because we anticipated no unusual factors in the application that have not already been taken into account by other factors.

We can find Z_N from Figure 9–22 using 2.10×10^9 cycles of loading on the pinion and 6.00×10^8 cycles for the gear as computed in Example Problem 9–2. We find, then, that

$$Z_{NP} = 0.88 \qquad Z_{NG} = 0.91$$

Let's complete the analysis for the pinion first. The hardness ratio factor does not apply to the pinion. Then Equation (9–27) gives

$$\frac{K_R(SF)}{Z_N} s_c = \frac{(1.50)(1.00)}{(0.88)} (156\ 000\ \text{psi}) = 265\ 900\ \text{psi} < s_{ac}$$

This value is quite high. Referring to Table 9–3, note that the only suitable listed material is a Grade 3 carburized and case-hardened steel having an allowable contact stress number of 275 Ksi. Let's complete the calculations for the gear materials and then discuss the results.

We can use Figure 9–24 to determine the hardness ratio factor. In Example Problem 9–2, the gear material was chosen to be a through-hardened steel with a hardness of 363 HB. With a very good surface finish on the pinion ($f_P = 16$), we read $C_H = 1.06$. Then Equation (9–27) gives

$$\frac{K_R(SF)}{Z_N C_H} s_c = \frac{(1.50)(1.00)}{(0.91)(1.06)} (156\ 000\ \text{psi}) = 242\ 600\ \text{psi} < s_{ac}$$

This value is also quite high, requiring the same Grade 3 carburized and case-hardened steel to provide adequate pitting resistance.

Comments and Design Decisions   Specifying the Grade 3 carburized and case-hardened steel would be expected to provide adequate strength and pitting resistance for this pair of gears. However, the design is very marginal, and it would be expensive because of the very special requirements of cleanliness for the material and other guarantees related to material composition and microstructure. Most designs are executed using Grade 1 steel. It is recommended that the gears be redesigned to produce a lower bending stress and contact stress. In general, that can be achieved by using larger teeth (a smaller value for diametral pitch, P_d), a larger diameter for each gear, and a larger face width. Greater precision in the manufacture of the gears, producing a higher quality number, Q_v, would lower the dynamic factor and therefore reduce the bending stress number and the contact stress number.

The next section outlines a design methodology for gears and provides guidelines for iterating on a design to produce alternatives from which we can select an optimum design for given conditions. In Section 9–14, a spreadsheet is developed that is useful for producing multiple iterations.

In Section 9–15, we use the same design requirements for the gear pair from Example Problems 9–1 through 9–4 as an example problem, making the necessary adjustments to our design decisions to ensure a satisfactory, economical gear design.

Then, in Chapter 15, this same design is used as the basis for a comprehensive discussion of the completion of the design of a power transmission. Considered there are the final selection of the gear design parameters, the design of the shafts for both the pinion and the gear (Chapter 12), the selection of two rolling contact bearings for each shaft (Chapter 14), and the enclosure of the transmission components in a suitable housing. Consideration is given for the inclusion of belt or chain drives for the input or output shafts of the gear reducer to provide more flexibility of its use. Chapter 15, then, is the culmination of virtually all of the design procedures presented in Part II of the book from Chapter 7 through Chapter 14.

9–12 DESIGN OF SPUR GEARS

In designs involving gear drives, normally the required speeds of rotation of the pinion and the gear and the amount of power that the drive must transmit are known. These factors are determined from the application. Also, the environment and operating conditions to which the drive will be subjected must be understood. It is particularly important to know the type of driving device and the driven machine, in order to judge the proper value for the application factor.

The designer must decide the type of gears to use; the arrangement of the gears on their shafts; the materials of the gears, including their heat treatment; and the geometry of the gears: numbers of teeth, diametral pitch, pitch diameters, tooth form, face width, and quality numbers.

This section presents a design procedure that accounts for the bending fatigue strength of the gear teeth and the pitting resistance, called *surface durability*. This procedure makes extensive use of the design equations presented in the preceding sections of the chapter and of the tables of material properties in Appendices 3 through 5, 8, and 12.

You should understand that there is no one best solution to a gear design problem; several good designs are possible. Your judgment and creativity and the specific requirements of the application will greatly affect the final design selected. The purpose here is to provide a means of approaching the problem to create a reasonable design.

Design Objectives

Some overall objectives of a design are listed below. The resulting drive should

> Be compact and small
>
> Operate smoothly and quietly
>
> Have long life
>
> Be low in cost
>
> Be easy to manufacture
>
> Be compatible with the other elements in the machine, such as bearings, shafts, the housing, the driver, and the driven machine

Design Procedure

The major objective of the design procedure is the determination of a safe and long-lasting gear drive. The procedure has these primary parts:

Procedure for Determining a Safe and Long-Lasting Gear Drive

1. A geometry that satisfies the required velocity ratio and application limitations, such as center distance and physical size, is proposed.

2. A tentative choice of the type of material to be used (steel, cast iron) is made.

3. A trial diametral pitch is chosen. Because of its strong effect on strength, pitting resistance, and geometry, the choice of diametral pitch is critical. The remaining parts of the procedure are directed at confirming that a reasonable design can be completed. If not, the results will help you choose the next trial pitch.

4. The loads, face width, and design factors are determined.

5. The bending stress on the pinion teeth is computed. If a reasonable value results, the procedure continues. Otherwise, a new pitch or revised geometry is selected.

6. The contact stress on the surface of the teeth is computed and the required material properties determined to ensure against pitting.

7. The final specifications of the materials for the pinion and the gear are made to satisfy the requirements of both strength and pitting resistance.

The detailed steps involved in the design of spur gears will be illustrated in Example Problem 9–5. It is not unusual to make several trials before settling on a design. If the process is set up in a computer program, on a spreadsheet, or on a programmable calculator, successive trials can be made very quickly (see Section 9–12).

Design Guidelines

Figure 9–25 shows a graph of power capacity of a pair of steel gears versus the speed of rotation of the pinion, with several values of diametral pitch shown. The hardnesses for the pinion and the gear used for these curves are in the middle range of possible values for steel (approximately HB 300). This should give you some feel for where to start. But note that the curves assume a uniform load ($K_o = 1.0$) and good alignment ($K_m = 1.0$). If your application has higher values for either factor, the entire set of curves is shifted downward. That is, you should choose a lower value for diametral pitch than indicated on the graph.

The face width, F, can be specified once the diametral pitch is chosen. Although a wide range of face widths is possible, the following limits are used for general machine drive gears:

Nominal Face Width

$$8/P_d < F < 16/P_d$$
$$\text{Nominal value of } F = 12/P_d \qquad \textbf{(9–28)}$$

Also, the face width normally is not greater than the pitch diameter of the pinion. An upper limit is placed on the face width to minimize problems with alignment. A very wide face width increases the chance for less than full face loading of the teeth. When the face width is less than the lower limit of Equation (9–28), it is probable that a more compact design can be achieved with a different pitch.

The following relationships should help you determine what changes in your design assumptions you should make after the first set of calculations to achieve a more optimum design:

- Decreasing the numerical value of the diametral pitch results in larger teeth and generally lower stresses. Also, the lower value of the pitch usually means a larger face width, which decreases stress and increases surface durability.

- Increasing the diameter of the pinion decreases the transmitted load, generally lowers the stresses, and improves the surface durability.

- Increasing the face width lowers the stress and improves the surface durability, but to a generally lesser extent than either the pitch or the pitch diameter changes discussed previously.

- Gears with more and smaller teeth tend to run more smoothly and quietly than gears with fewer and larger teeth.

- Standard values of diametral pitch should be used for ease of manufacture and lower cost (see Table 8–2).

- Using high-alloy steels with high surface hardness results in the most compact system, but the cost is higher.

FIGURE 9–25

Power vs. pinion speed

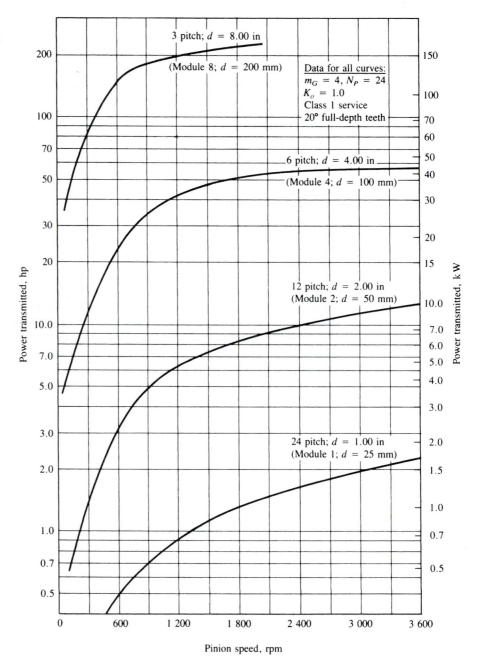

- Using very accurate gears (with ground or shaved teeth) results in lower dynamic loads and consequently lower stresses and improved surface durability, but the cost is higher.
- The number of teeth in the pinion should generally be as small as possible to make the system compact. But the possibility of interference is greater with fewer teeth. Check Table 8–6 to ensure that no interference will occur. (See also Reference 18.)

Example Problem 9–5 A pair of spur gears with 20°, full-depth teeth is to be designed as a part of the drive for a wood chipper to cut pulpwood for use in a paper mill. The power source is an electric motor that drives the pinion at 1 750 rpm. The gear must rotate between 460 and 465 rpm. The gears must transmit 3.0 hp. A compact design is desired, but in no case can the center distance exceed 5.00 in.

Complete the design that will be safe with a fatigue life of greater than 10 million cycles. Specify the diametral pitch, the number of teeth in the pinion and the gear, the pitch diameters of the pinion and the gear, the face width, and the material with its heat treatment.

Solution and General Design Procedure

Step 1. Compute the nominal velocity ratio.

At the midpoint of the given range of output speeds,

$$VR = n_p/n_G = (1\ 750\ \text{rpm})/(462.5\ \text{rpm}) = 3.78$$

Step 2. Specify the geometry for the first trial design.

Let's start by specifying the diametral pitch and the number of teeth in the pinion. Referring to Figure 9–25, you may think that a diametral pitch somewhat greater than 12 would work. But remember that those graphs are for uniform loads and good alignment. This design, for a wood chipper, has a high overload factor and will probably have fairly high misalignment. Therefore, let's choose $P_d = 12$. Because we want a compact design, let's choose the smallest number of teeth for the pinion that will ensure no interference. From Table 8–6, we can use $N_P = 18$.

Now, using the basic equations for gear pair geometry from Sections 8–4 and 8–6, we have

$$N_G = N_P(VR) = (18)(3.78) = 68.04$$

But because N_G must be an integer, use $N_G = 68$. The final velocity ratio is $VR = N_G/N_P = 68/18 = 3.778$. The actual output speed is $n_G = n_P/VR = (1\ 750\ \text{rpm})/(3.778) = 463\ \text{rpm}$. This is within the specified range.

Compute the pitch diameters:

$$D_P = N_P/P_d = 18/12 = 1.500\ \text{in}$$
$$D_G = N_G/P_d = 68/12 = 5.667\ \text{in}$$

Compute the center distance:

$$C = (N_P + N_G)/(2P_d) = (18 + 68)/(24) = 3.583\ \text{in}$$

The geometry is acceptable.

Step 3. Compute the pitch line speed:

$$v_t = \pi D_P n_P/12 = [\pi(1.500)(1\ 750)]/12 = 687\ \text{ft/min}$$

Step 4. Compute the transmitted load for 3.0 hp:

$$W_t = 33\ 000(P)/v_t = 33\ 000(3.0)/687 = 144\ \text{lb}$$

Step 5. Specify a quality number. Refer to Table 9–2 for guidelines. For the wood chipper, $Q_v = 6$ should be satisfactory.

Step 6. Determine the dynamic factor, K_v, from Figure 9–19. For $Q_v = 6$ and $v_t = 687\ \text{ft/min}$, we find that $K_v = 1.35$.

Step 7. Select the overload factor, K_o. From Table 9–6, with a uniform power source (an electric motor) and a heavy shock load (a wood chipper), use $K_o = 1.75$.

Step 8. Specify a size factor, K_s, from Table 9–7. For $P_d = 12$, $K_s = 1.00$.

Step 9. Specify the face width, F. Using the nominal value, we have

$$F = 12/P_d = 12/12 = 1.00\ \text{in}$$

Step 10. Specify a load distribution factor, K_m, from Equation (9–16).

For $F = 1.00$ in and $D_P = 1.50$ in, $F/D_P = 0.667$. The type of gear unit must be specified by the designer. In this case, let's use open gearing, supported on the frame of the wood chipper. Then $C_{pf} = 0.042$, found from Figure 9–16. From Figure 9–17, $C_{ma} = 0.264$. Then

$$K_m = 1.0 + C_{pf} + C_{ma} = 1.0 + 0.042 + 0.264 = 1.31$$

Step 11. Specify the rim thickness factor, K_B, using Figure 9–18.

Because the pinion is small, let's use a solid gear blank for which $K_B = 1.00$.

Step 12. Specify the tooth form to be used, and determine the geometry factors J_P and J_G from Figure 9–15. Normally, 20°, full-depth, involute teeth will be chosen. Let's choose this form for the wood cutter drive. Then $J_P = 0.325$ for $N_P = 18$ and $N_G = 68$. Also, $J_G = 0.410$.

Step 13. Compute the expected bending stress in the pinion teeth from Equation (9–15):

$$s_P = \frac{W_t P_d}{F J_P} K_o K_s K_m K_B K_v = \frac{(144)(12)}{(1.00)(0.325)} (1.75)(1.0)(1.31)(1.0)(1.35)$$

$$s_{tP} = 16\ 400 \text{ psi}$$

Step 14. Compute the expected bending stress in the gear from

$$s_{tG} = s_{tP}(J_P/J_G) = (16\ 400)(0.325/0.410) = 13\ 000 \text{ psi}$$

Step 15. Adjust the computed bending stress number if a reliability different from 99% or a life different from 10^7 cycles is desired. Refer to Equation (9–20) in Section 9–9. You may also apply a safety factor, *SF*. But no unusual conditions are expected beyond those already considered, so let's use $SF = 1.00$. Standard reliability and life are also satisfactory for this problem. Then no adjustments are necessary.

Step 16. Make a rough judgment about the suitability of the computed bending stresses. Can reasonable materials be specified to carry the stresses safely?

Check in Section 9–7:

Figure 9–8: Through-hardened steel

Table 9–3: Case-hardened steel

Table 9–4: Iron and bronze

Table 9–5: Plastics

If the stresses are reasonable, proceed to the contact stress analysis before finally specifying materials. Often the pitting resistance is the critical factor in material selection.

In this design problem, the computed bending stresses are safe, but low for steel materials. Some nodular iron materials could be considered. Let's proceed.

Step 17. Compute the expected contact stress on the surface of the teeth using Equation (9–25):

$$s_c = C_p \sqrt{\frac{W_t K_o K_s K_m K_v}{F D_P I}}$$

This requires that we decide on the general type of material to be used so that C_p can be determined from Table 9–10.

In this problem, let's tentatively decide to use steel for both the pinion and the gear. Then $C_p = 2\ 300$. We also need the geometry factor I from Figure 9–21. For $m_G = 3.778$ and $N_P = 18$, the value of I is approximately 0.104. From earlier steps we can use $K_o = 1.75$, $K_s = 1.00$, $K_m = 1.31$, $K_v = 1.35$, $F = 1.00$ in, $D_P = 1.50$ in, and $W_t = 144$ lb. Then

$$s_c = 2\,300 \sqrt{\frac{(144)(1.75)(1.0)(1.31)(1.35)}{(1.00)(1.50)(0.104)}} = 122\,900 \text{ psi}$$

Step 18. Apply reliability, safety, hardness ratio, and life factors to the computed contact stress if necessary. See Equation (9–27). In this problem we are using values of 1.00 for normal conditions.

Step 19. Determine the required hardness of the materials for the pinion and the gear to ensure safe operation *both* for bending stress and for contact stress.

In this problem, contact stress controls the design. From Figure 9–9, we find that a Grade 1 steel with a hardness of 300 HB will have an allowable contact stress s_{ac} of approximately 126 Ksi, slightly greater than the computed contact stress.

Step 20. Specify suitable materials and their treatments to meet the requirements identified in Steps 16 and 19.

For this problem, we could specify any steel that can be hardened to 300 HB. From Figure A4–4, we could specify the following:

For both the pinion and the gear: AISI 4140 OQT 1100, 300 HB minimum

Summary

Power transmitted = 3.0 hp from an electric motor

Pinion speed = 1 750 rpm; gear speed = 463 rpm

Velocity ratio = 3.778

Pinion: P_d = 12; N_P = 18; D_P = 1.500 in; face width = F = 1.00 in

Gear: P_d = 12; N_G = 68; D_G = 5.667 in; face width = F = 1.00 in

Pitch line speed = v_t = 687 ft/min

Transmitted load = W_t = 144 lb

Quality number = Q_v = 6; dynamic factor = K_v = 1.35

Overload factor = K_o = 1.75; size factor = K_s = 1.00

Load-distribution factor = K_m = 1.31

Rim thickness factor = K_B = 1.00

Life factor = K_L = 1.00 for 10^7 cycles

Reliability factor = K_R = 1.00 for 99% reliability

Bending stress number in pinion = s_{tP} = 16 600 psi

Bending stress number in gear = s_{tG} = 13 200 psi

For steel pinion and gear: Contact stress = σ_c = 122 900 psi

Required hardness in Grade 1 steel = 300 HB based on allowable contact stress

Material specification: AISI 4140 OQT 1100, 300 HB minimum

Other materials could be used.

9–13
GEAR DESIGN
FOR THE METRIC
MODULE SYSTEM

Section 8–4, "Gear Nomenclature and Gear-Tooth Features," described the metric module system of gearing and its relation to the diametral pitch system. As the design process was being developed in Sections 9–6 through 9–11, the data for stress analysis and surface durability analysis were taken from charts using U.S. Customary units (in, lb, hp, ft/min, and Ksi). Data for the metric module system were also available in the charts in units of millimeters (mm), newtons (N), kilowatts (kW), meters per second (m/s), and megapascals (MPa). But to use the SI data, we must modify some of the formulas.

The following example problem uses SI units. The procedure will be virtually the same as that used to design with U.S. Customary units. Those formulas that are converted to SI units are identified.

Example Problem 9–6   A gear pair is to be designed to transmit 15.0 kilowatts (kW) of power to a large meat grinder in a commercial meat processing plant. The pinion is attached to the shaft of an electric motor rotating at 575 rpm. The gear must operate at 270 to 280 rpm. The gear unit will be enclosed and of commercial quality. Commercially hobbed (quality number 5), 20°, full-depth, involute gears are to be used in the metric module system. The maximum center distance is to be 200 mm. Specify the design of the gears. Use $K_R = C_H = SF = Z_N = 1.00$.

Solution   The nominal velocity ratio is

$$VR = 575/275 = 2.09$$

From Figure 9–25, $m = 4$ is a reasonable trial module. Then

$$N_P = 18 \quad \text{(design decision)}$$
$$D_P = N_P m = (18)(4) = 72 \text{ mm}$$
$$N_G = N_P(VR) = (18)(2.09) = 37.6 \quad \text{(Use 38.)}$$
$$D_G = N_G m = (38)(4) = 152 \text{ mm}$$
$$\text{Final output speed} = n_G = n_P(N_P/N_G)$$
$$n_G = 575 \text{ rpm} \times (18/38) = 272 \text{ rpm} \quad \text{(okay)}$$
$$\text{Center distance} = C = (N_P + N_G)m/2 \quad \text{[Equation (8–18)]}$$
$$C = (18 + 38)(4)/2 = 112 \text{ mm} \quad \text{(okay)}$$

In SI units, the pitch line speed in meters per second (m/s) is

$$v_t = \pi D_P n_P/(60\,000)$$

where D_P is in mm and n_P is in revolutions per minute (rpm). Then

$$v_t = [(\pi)(72)(575)]/(60\,000) = 2.17 \text{ m/s}$$

In SI units, the transmitted load, W_t, is in newtons (N). If the power, P, is in kW, and v_t is in m/s,

$$W_t = 1\,000(P)/v_t = (1\,000)(15)/(2.17) = 6\,920 \text{ N}$$

In the U.S. Customary Unit System, it was recommended that the face width be approximately $F = 12/P_d$ in. The equivalent SI values are $F = 12(m)$ mm. For this problem, $F = 12(4) = 48$ mm. Let's use $F = 50$ mm.

Other factors are found as before.

$$K_s = K_B = 1.00$$
$$K_v = 1.34 \quad \text{(Figure 9–19)}$$
$$K_o = 1.50 \quad \text{(Table 9–6)}$$
$$K_m = 1.21 \quad \text{(Figures 9–16 and 9–17)} \quad (F/D_P = 50/72 = 0.69)$$
$$J_P = 0.315 \quad J_G = 0.380 \quad \text{[Figure 9–15(a)]}$$

Then the stress in the pinion is found from Equation (9–15), modified by letting $P_d = 1/m$:

$$s_{tP} = \frac{W_t K_o K_s K_B K_m K_v}{F m J_P} = \frac{(6\,920)(1.50)(1)(1)(1.21)(1.34)}{(50)(4)(0.315)} = 269 \text{ MPa}$$

This is a reasonable stress level. The required hardness of grade 1 material is HB 360, as found in Figure 9–8. Proceed with design for pitting resistance.

For two steel gears,

$$K_s = 1.0$$
$$C_p = 191 \quad \text{(Table 9–10)}$$
$$I = 0.092 \quad \text{(Figure 9–21)}$$
$$K_v = 1.34$$
$$K_o = 1.50$$
$$K_m = 1.21$$

The contact stress [Equation (9–25)] gives

$$s_c = C_P \sqrt{\frac{W_t K_o K_s K_m K_v}{F D_p I}} = 191 \sqrt{\frac{(6\,920)(1.50)(1.0)(1.21)(1.34)}{(50)(72)(0.092)}} = 1\,367 \text{ MPa}$$

Converting to Ksi gives

$$s_c = 1\,367 \text{ MPa} \times 1 \text{ Ksi}/6.895 \text{ MPa} = 198 \text{ Ksi}$$

From Table 9–3, the required surface hardness is HRC 58–64, case-carburized, Grade 2. Material selection from Appendix 5, for carburized steels, is as follows:

AISI 4320 SOQT 300; $s_u = 1\,500$ MPa; 13% elongation

Case-harden by carburizing to HRC 58 minimum

Case depth: 0.6 mm minimum (Figure 9–10)

**9–14
COMPUTER-
AIDED SPUR
GEAR DESIGN
AND ANALYSIS**

This section presents one approach to assisting the gear designer with the many calculations and judgments that must be made to produce an acceptable design. The spreadsheet shown in Figure 9–26 facilitates the completion of a prospective design for a pair of gears in a few minutes by an experienced designer. You must have studied all of the material in Chapters 8 and 9 in order to understand the data needed in the spreadsheet and to use it effectively.

The recommended use of the spreadsheet is to create a series of design iterations that allow you to progress toward an optimum design in a short amount of time. It follows the process outlined in Section 9–12 up to the point of computing the required allowable bending stress number and the allowable contact stress number for both the pinion and the gear. The designer must use those data to specify suitable materials for the gears and their heat treatments.

Given below is a discussion of the essential features of the spreadsheet. In general, it first calls for the input of basic performance data, allowing a proposed geometry to be specified. The final result is the completion of the stress analyses for bending and pitting resistance for both the pinion and the gear. Equations (9–19) and (9–20) are combined for the bending analysis. The analysis of pitting resistance uses Equations (9–25), (9–26), and (9–27). The designer must provide data for the several factors in those equations taken from appropriate figures and charts or based on design decisions. Virtually all computations are performed by the spreadsheet, allowing the designer to exercise judgment based on the intermediate results.

The format used for the spreadsheet helps the designer follow the process. After defining the problem at the top of the sheet, the first column at the left calls for several pieces of input data. Any value in italics within a gray-shaded area must be entered by the designer. White areas offer the results of calculations and provide guidance. The upper part of the second column also guides the designer in determining values for the several factors

DESIGN OF SPUR GEARS

Initial Input Data:

Input power:	$P=$	*3 hp*
Input speed:	$n_P=$	*1,750 rpm*
Diametral pitch:	$P_d=$	*12*
Number of pinion teeth:	$N_P=$	*18*
Desired output speed:	$n_G=$	*462.5 rpm*

Computed number of gear teeth: 68.1

Enter: Chosen no. of gear teeth: $N_G=$ *68*

Computed Data:

Actual output speed:	$n_G=$	463.2 rpm
Gear ratio:	$m_G=$	3.78
Pitch diameter, pinion:	$D_P=$	1.500 in
Pitch diameter, gear:	$D_G=$	5.667 in
Center distance:	$C=$	3.583 in
Pitch line speed:	$v_t=$	687 ft/min
Transmitted load:	$W_t=$	144 lb

Secondary Input Data:

	Min.	Nom.	Max.
Face width guidelines (in):	0.667	1.000	1.333

Enter: Face width: $F=$ *1.000 in*

Ratio: Face width/pinion diameter: $F/D_P=$ 0.67
Recommended range of ratio: $0.50 < F/D_P < 2.00$

Enter: Elastic coefficient: $C_p=$ *2,300*   Table 9–10

Enter: Quality number: $Q_v=$ Table 9–2

Enter: Bending geometry factors:
Pinion:	$J_P=$	*0.325*	Fig. 9–15
Gear:	$J_G=$	*0.410*	Fig. 9–15
Enter: Pitting geometry factor:	$I=$	*0.104*	Fig. 9–21

FIGURE 9–26   Spreadsheet solution for Example Problem 9–5

Application:	Wood chipper driven by an electric motor				
	Example Problem 9–5				

Factors in Design Analysis:					
Alignment factor, $K_m = 1.0 + C_{pf} + C_{ma}$	If $F < 1.0$	If $F > 1.0$			
Pinion proportion factor, $C_{pf} =$	0.042	0.042	[$0.50 < F/D_P < 2.00$]		
Enter: $C_{pf} =$	0.042		Fig. 9–16		
Type of gearing:	Open	Commer.	Precision	Ex. Prec.	
Mesh alignment factor, $C_{ma} =$	0.264	0.143	0.080	0.048	
Enter: $C_{ma} =$	0.264		Fig. 9–17		
Alignment factor: $K_m =$	1.31		[Computed]		
Overload factor: $K_o =$	1.75		Table 9–6		
Size factor: $K_s =$	1.00		Table 9–7: Use 1.00 if $P_d \geq 5$.		
Pinion rim thickness factor: $K_{BP} =$	1.00		Fig. 9–18: Use 1.00 if solid blank.		
Gear rim thickness factor: $K_{BG} =$	1.00		Fig. 9–18: Use 1.00 if solid blank.		
Dynamic factor: $K_v =$	1.35		[Computed: See Fig. 9–19.]		
Service factor: $SF =$	1.00		Use 1.00 if no unusual conditions.		
Hardness ratio factor: $C_H =$	1.00		Fig. 9–23 or 9–24; gear only		
Reliability factor: $K_R =$	1.00		Table 9–9: Use 1.00 for R = 0.99.		
Enter: Design life: =	3,000 hours		See Table 9–8.		

Pinion—Number of load cycles: $N_P = 3.2E+08$		Guidelines: Y_N, Z_N		
Gear—Number of load cycles: $N_G = 8.3E+07$		10^7 cycles	$>10^7$	$<10^7$
Bending stress cycle factor: $Y_{NP} =$	0.96	1.00	0.96	Fig. 9–20
Bending stress cycle factor: $Y_{NG} =$	0.98	1.00	0.98	Fig. 9–20
Pitting stress cycle factor: $Z_{NP} =$	0.92	1.00	0.92	Fig. 9–22
Pitting stress cycle factor: $Z_{NG} =$	0.95	1.00	0.95	Fig. 9–22

Stress Analysis: Bending

Pinion: Required $s_{at} =$	17,102 psi	See Fig. 9–8 or
Gear: Required $s_{at} =$	13,280 psi	Table 9–3, 9–4, or 9–5.

Stress Analysis: Pitting

Pinion: Required $s_{ac} =$	133,471 psi	See Fig. 9–9 or
Gear: Required $s_{ac} =$	129,256 psi	Table 9–3 or 9–4.

Specify materials, alloy and heat treatment, for most severe requirement.

needed to complete the stress analyses for bending and pitting resistance. The area at the lower right of the spreadsheet gives the primary output data on which the design decisions for materials and heat treatments are based.

The data in Figure 9–26 are taken from Example Problem 9–5 which was completed in the traditional manner in Section 9–12. Some enhancements are included with regard to the stress cycle factors, giving slightly different results.

Discussion of the Use of the Spur Gear Design Spreadsheet

1. ***Describing the application:*** In the heading of the sheet, the designer is asked to describe the application for identification purposes and to focus on the basic uses for the gears. Of particular interest is the nature of the prime mover and the driven machine.

2. ***Initial input data:*** It is assumed that designers begin with a knowledge of the power transmission requirement, the rotational speed of the pinion of the gear pair, and the desired output speed. Use Figure 9–25 to determine a trial value of the diametral pitch based on the power to be transmitted and the rotational speed of the pinion. The number of teeth in the pinion is a critical design decision because the size of the system depends on this value.

3. ***Number of gear teeth:*** The spreadsheet computes the approximate number of gear teeth to produce the desired output speed from $N_G = N_P(n_G/n_P)$. But, of course, the number of teeth in any gear must be an integer, and the actual value of N_G is entered by the designer.

4. ***Computed data:*** The seven values reported in the middle of the first column are all determined from the input data, and they allow the designer to evaluate the suitability of the geometry of the proposed design at this point. Changes to the input data can be made at this time if any value is out of the desired range in the judgment of the designer.

5. ***Secondary input data:*** When a suitable geometry for the gears is obtained, the designer enters the data called for at the lower part of the first column of the spreadsheet. The locations of data in pertinent tables and figures are listed.

6. ***Factors in design analysis:*** The stress analysis requires many factors to account for the unique situation of the design being pursued. Again, guidance is offered, but the designer must enter the values of the required factors. Many of the factors can have a value of 1.00 for normal conditions or to produce a conservative result.

7. ***Alignment factor:*** The alignment factor depends on two other factors: the pinion proportion factor and the mesh alignment factor as shown in Figures 9–16 and 9–17. The suggested values in the white areas are computed from the equations given in the figures. The designer must decide on the type of gearing to be used (open or closed) and the degree of precision to be designed into the system. The final result is computed from the input data.

8. ***Overload, size, and rim thickness factors:*** Consult Tables 9–6 and 9–7 along with Figure 9–18. Note that the rim thickness factor can be different for the pinion and the gear. Sometimes the smaller pinion is made from a solid blank while the larger gear can use a rim-and-spoke design.

9. ***Dynamic factor:*** The spreadsheet uses the equations included in Figure 9–19 to compute the dynamic factor using the quality number and pitch line speed found from data in the first column.

10. *Service factor:* This is a design decision as discussed in Section 9–12. Often a value of 1.00 is used if no unusual conditions are expected that are not already accounted for in other factors. Larger service factors allow for a higher degree of safety or to account for uncertainties.

11. *Hardness ratio factor:* This value depends on the ratio of the hardness of the pinion to the hardness of the gear teeth as shown in Figures 9–23 and 9–24. At the start of the design, these data are not known, and it is suggested that a value of $C_H = 1.00$ be used initially. Then, after one or more iterations are completed with tentative specifications for the pinion and gear materials, the value can be adjusted to refine the design. The factor C_H is applied only in the pitting resistance stress analysis for the gear, and it may allow the use of a less expensive or more ductile material with a lower hardness.

12. *Reliability factor:* The designer must select a value from Table 9–9 according to the desired level of reliability.

13. *Stress cycle factors:* Here the designer must specify the design life in hours of operation for the gear pair being designed. Table 9–8 provides suggestions according to the use of the system. The number of cycles of stress is then computed for both the pinion and the gear, assuming the normal case of one cycle of one-direction stress per revolution. If the gears operate in a reversing mode, as idlers, or in planetary gear trains, this calculation must be adjusted to account for the multiple cycles of stress experienced in each revolution. Guidelines recommend factors of 1.00 for 10^7 cycles for which the allowable stress numbers are computed. For a larger number of cycles, equations given in Figures 9–20 and 9–22 are used to compute the recommended factors. Because a variety of data are given for the case of fewer than 10^7 cycles, the designer is referred to the figures to determine the factors. In any case, the user of the spreadsheet must enter the selected values.

14. *Stress analyses for bending and pitting resistance:* Finally, the required allowable bending stress number and the required allowable contact stress number are computed using Equations (9–20) and (9–27), adjusted for the special values of factors for the pinion and the gear.

15. *Specification of the materials and their heat treatment:* The final step is left to the designer to use the computed values from the stress analyses and to specify materials that will provide an adequate strength and surface hardness of the gear teeth. Pertinent data are listed in Figures 9–8 and 9–9 and Tables 9–3, 9–4, and 9–5. The Appendices tables for material properties may also be consulted once the required hardnesses of the materials are determined.

9–15 USE OF THE SPUR GEAR DESIGN SPREADSHEET

The spreadsheet developed in Section 9–14 is a useful tool that aids the designer in the process of completing a design for a pair of gears to be safe with regard to bending stresses in the teeth of the gears and for pitting resistance. The use of the spreadsheet was demonstrated for the data in Example Problem 9–5 as shown in Figure 9–26.

A more important use for the spreadsheet is to propose and analyze several design alternatives and to work toward a goal of optimizing the design with regard to size, cost, or other parameters important to a particular design objective.

In this section, we use the data for Example Problems 9–1 through 9–4 to work toward an improved solution. The basic requirements were to produce a satisfactory design to transmit 25 hp with a pinion speed of 1 750 rpm and a gear speed of 500 rpm. The initial trial started in Example Problem 9–1 and carried through the other problems called for gears with a diametral pitch of 8 and 20 teeth in the pinion and 70 teeth in the gear. A quality number of 6 was chosen. Although those sound like reasonable choices, it was shown that the resulting stresses were significantly higher than desirable, particularly the required allowable contact stress number. It was shown in Example Problem 9–4 that a Grade 3 steel, case-hardened by carburizing, was required. This is a very expensive design because of the extreme controls on the material composition and cleanliness and because of the time-consuming heat treatment process.

Designers for typical machine and vehicle drives would plan to use Grade 1 steels and standard quenching and tempering heat treatments. Where small size is critical or where cost is not a major concern, case hardening by carburizing, induction or flame hardening, or nitriding can be used. Therefore, it is usually desirable to produce several design alternatives that can be analyzed for cost and manufacturability. Then the final selection can be made with assurance that a reasonably optimum design has been identified.

The spreadsheet in Figure 9–27 shows the combined results of Example Problems 9–1 through 9–4. The resulting stresses are slightly different because of small differences in the factors computed by the spreadsheet formulas and those read from charts. This summary can lead you to alternatives that meet the design goal of using only Grade 1 steels and achieving a cost-effective design.

Note that the most significant effects on stresses are produced by the following variables:

- **Diametral pitch, P_d.** A smaller value for diametral pitch represents a larger tooth size and correspondingly lower bending stress. Also, for a given number of teeth, the lower value for P_d results in a larger pinion diameter which lowers both the bending stress and the contact stress as discussed next.

- **Pitch diameter of the pinion, D_P.** For a given transmitted power and speed of rotation, a larger pinion results in a lower value of the transmitted load, W_t. Also, because this load is an important factor in both the bending stress and the contact stress equations, the stresses are significantly lowered. A larger pinion diameter results from either increasing the number of teeth for a given diametral pitch or using a smaller value for diametral pitch for a given number of teeth.

- **Face width, F:** A larger face width causes the transmitted load to be spread over more material, resulting in lower bending and contact stresses. This effect is moderate because practical designs would have a face width within a fairly small range as suggested in Equation (9–28).

- **Quality number, Q_v.** The quality number is a measure of the precision with which the teeth are machined and finished. A higher quality number provides for a smoother engagement and less dynamic impact as each tooth assumes the load from the mating tooth. This results in a lower dynamic factor as shown in Figure 9–19. A lower dynamic factor reduces both the bending and the contact stresses while producing a smoother, quieter drive. The effect on stresses is moderate, and the cost of producing the more accurate teeth should be considered.

Figures 9–28 , 9–29, and 9–30 show three additional trial designs for the system described in Example Problem 9–1. Each of these designs reaches the goal of producing a practical design that allows the use of Grade 1 steels. You should study these designs to ascertain the differences among the design decisions and the resulting required allowable stress numbers for both bending and pitting resistance. Figure 9–31 summarizes the major results from all of the trials.

Note that in all of the trials, the contact stress is critical. That is, the required hardness or case-hardening treatment for the gear teeth is most severe with regard to achieving adequate pitting resistance. The bending stress numbers are quite moderate. This situation is typical for steel gears, and designers often work first toward a satisfactory contact stress and then check to ensure that the bending stress is acceptable.

The trials shown in Figure 9–31 are arranged in order of increasing pinion diameter and increasing center distance, with the consequent decreasing of the bending and contact stresses. These are the most critical factors. Diametral pitch, face width, and quality numbers are also varied, but their effects are secondary.

The designs represented by Figures 9–28, 9–29, and 9–30 could all use the same material, provided that it could be hardened to the required levels. For example, specifying AISI 4340 Grade 1 steel or some similar medium-carbon-alloy steel would ensure that it could be flame or induction-hardened to a minimum of 50 HRC or through-hardened by quenching and tempering to the required 350 to 400 HB hardness on the Brinell scale.

9–16 POWER-TRANSMITTING CAPACITY

It is sometimes desirable to compute the amount of power that a gear pair can safely transmit after it has been completely defined. The *power-transmitting capacity* is the capacity when the tangential load causes the expected stress to equal the allowable stress number with all of the modifying factors considered. The capacity should be computed for both bending and pitting resistance and for both the pinion and the gear.

When similar materials are used for both the pinion and the gear, it is likely that the pinion will be critical for bending stress. But the most critical condition is usually pitting resistance. The following relationships can be used to compute the power-transmitting capacity. In this analysis, it is assumed that the operating temperature of the gears and their lubricants is 250°F and that gears are produced with the appropriate surface finish.

Bending

We start with Equation (9–19) in which the computed bending stress number is compared with the modified allowable bending stress number for the gear:

$$\frac{W_t P_d}{FJ} K_o K_s K_m K_B K_v = s_t < s_{at} \frac{Y_N}{(SF)K_R} \tag{9–19}$$

But solving for W_t gives

$$W_t = \frac{s_{at} Y_N FJ}{(SF)K_R K_o K_s K_m K_B K_v P_d} \tag{9–29}$$

It was shown in Equation (9–6) that

$$W_t = (126\,000)(P)/(n_P D_P) \tag{9–6}$$

Then substituting into Equation (9–29) gives

$$\frac{(126\,000)(P)}{n_P D_P} = \frac{s_{at} Y_N FJ}{(SF)K_R K_o K_s K_m K_B K_v P_d}$$

DESIGN OF SPUR GEARS

Initial Input Data:

Input power:	$P=$	25 hp
Input speed:	$n_P=$	1,750 rpm
Diametral pitch:	$P_d=$	8
Number of pinion teeth:	$N_P=$	20
Desired output speed:	$n_G=$	500 rpm

Computed number of gear teeth:	70.0

Enter: Chosen no. of gear teeth:	$N_G=$	70

Computed Data:

Actual output speed:	$n_G=$ 500.0 rpm
Gear ratio:	$m_G=$ 3.50
Pitch diameter, pinion:	$D_P=$ 2.500 in
Pitch diameter, gear:	$D_G=$ 8.750 in
Center distance:	$C=$ 5.625 in
Pitch line speed:	$v_t=$ 1,145 ft/min
Transmitted load:	$W_t=$ 720 lb

Secondary Input Data:

	Min.	Nom.	Max.
Face width guidelines (in):	1.000	1.500	2.000

Enter: Face width:	$F=$	1.500 in

Ratio: Face width/pinion diameter:	$F/D_P=$ 0.60
Recommended range of ratio:	$0.50 < F/D_P < 2.00$

Enter: Elastic coefficient:	$C_p=$ 2,300		Table 9–10

Enter: Quality number:	$Q_v=$	6	Table 9–2

Enter: Bending geometry factors:

Pinion:	$J_P=$ 0.335		Fig. 9–15
Gear:	$J_G=$ 0.420		Fig. 9–15
Enter: Pitting geometry factor:	$I=$ 0.108		Fig. 9–21

FIGURE 9–27 Spreadsheet solution for Example Problems 9–1 through 9–4

Application:	Industrial saw driven by an electric motor Example Problems 9–1 through 9–4			

Factors in Design Analysis:				
Alignment factor, $K_m = 1.0 + C_{pf} + C_{ma}$	If $F < 1.0$	If $F > 1.0$		
Pinion proportion factor, $C_{pf} =$	0.035	0.041	[$0.50 < F/D_P < 2.00$]	
Enter: $C_{pf} =$	*0.041*	Fig. 9–16		
Type of gearing:	Open	Commer.	Precision	Ex. Prec.
Mesh alignment factor, $C_{ma} =$	0.272	0.150	0.086	0.053
Enter: $C_{ma} =$	*0.15*	Fig. 9–17		
Alignment factor: $K_m =$	1.19	[Computed]		
Overload factor: $K_o =$	*1.50*	Table 9–6		
Size factor: $K_s =$	*1.00*	Table 9–7: Use 1.00 if $P_d \geq 5$.		
Pinion rim thickness factor: $K_{BP} =$	*1.00*	Fig. 9–18: Use 1.00 if solid blank.		
Gear rim thickness factor: $K_{BG} =$	*1.00*	Fig. 9–18: Use 1.00 if solid blank.		
Dynamic factor: $K_v =$	1.45	[Computed: See Fig. 9–19.]		
Service factor: $SF =$	*1.00*	Use 1.00 if no unusual conditions.		
Hardness ratio factor: $C_H =$	*1.06*	Fig. 9–23 or 9–24; gear only		
Reliability factor: $K_R =$	*1.50*	Table 9–9: Use 1.00 for R= 0.99.		
Enter: Design life:	*20,000 hours*	See Table 9–8.		

Pinion—Number of load cycles: $N_P =$ 2.1E+09		Guidelines: Y_N, Z_N		
Gear—Number of load cycles: $N_G =$ 6.0E+08		10^7 cycles	$> 10^7$	$< 10^7$
Bending stress cycle factor: $Y_{NP} =$	*0.93*	1.00	0.93	Fig. 9–20
Bending stress cycle factor: $Y_{NG} =$	*0.95*	1.00	0.95	Fig. 9–20
Pitting stress cycle factor: $Z_{NP} =$	*0.88*	1.00	0.88	Fig. 9–22
Pitting stress cycle factor: $Z_{NG} =$	*0.91*	1.00	0.91	Fig. 9–22

Stress Analysis: Bending

Pinion: Required $s_{at} =$	47,871 psi	See Fig. 9–8 or
Gear: Required $s_{at} =$	37,379 psi	Table 9–3, 9–4, or 9–5.

Stress Analysis: Pitting

Pinion: Required $s_{ac} =$	265,989 psi	See Fig. 9–9 or
Gear: Required $s_{ac} =$	242,660 psi	Table 9–3 or 9–4.

Specify materials, alloy and heat treatment, for most severe requirement.

DESIGN OF SPUR GEARS

Initial Input Data:

Input power:	$P=$	25 hp
Input speed:	$n_P=$	1,750 rpm
Diametral pitch:	$P_d=$	6
Number of pinion teeth:	$N_P=$	18
Desired output speed:	$n_G=$	500 rpm

Computed number of gear teeth:	63.0

Enter: Chosen no. of gear teeth:	$N_G=$	63

Computed Data:

Actual output speed:	$n_G=$	500.0 rpm
Gear ratio:	$m_G=$	3.50
Pitch diameter, pinion:	$D_P=$	3.000 in
Pitch diameter, gear:	$D_G=$	10.500 in
Center distance:	$C=$	6.750 in
Pitch line speed:	$v_t=$	1,374 ft/min
Transmitted load:	$W_t=$	600 lb

Secondary Input Data:

	Min.	Nom.	Max.
Face width guidelines (in):	1.333	2.000	2.667

Enter: Face width:	$F=$	2.750 in

Ratio: Face width/pinion diameter: $F/D_P=$	0.92
Recommended range of ratio: $0.50 < F/D_P < 2.00$	

Enter: Elastic coefficient:	$C_p=$	2,300	Table 9–10

Enter: Quality number:	$Q_v=$	8	Table 9–2

Enter: Bending geometry factors:			
Pinion:	$J_P=$	0.325	Fig. 9–15
Gear:	$J_G=$	0.410	Fig. 9–15
Enter: Pitting geometry factor:	$I=$	0.105	Fig. 9–21

FIGURE 9–28 Redesign for data of Example Problem 9–1, Trial 1

Application:	Industrial saw driven by an electric motor
	Redesign for data of Example Problem 9–1, Trial 1

Factors in Design Analysis:

Alignment factor, $K_m = 1.0 + C_{pf} + C_{ma}$		If $F < 1.0$	If $F > 1.0$	
Pinion proportion factor, $C_{pf}=$		0.067	0.089	$[0.50 < F/D_P < 2.00]$

Enter: $C_{pf}=$	0.089	Fig. 9–16	

Type of gearing:	Open	Commer.	Precision	Ex. Prec.
Mesh alignment factor, $C_{ma}=$	0.292	0.170 0.102	0.065	

Enter: $C_{ma}=$	0.17	Fig. 9–17

Alignment factor: $K_m=$	1.26	[Computed]

Overload factor: $K_o=$	1.50	Table 9–6
Size factor: $K_s=$	1.00	Table 9–7: Use 1.00 if $P_d \geq 5$.
Pinion rim thickness factor: $K_{BP}=$	1.00	Fig. 9–18: Use 1.00 if solid blank.
Gear rim thickness factor: $K_{BG}=$	1.00	Fig. 9–18: Use 1.00 if solid blank.

Dynamic factor: $K_v=$	1.30	[Computed: See Fig. 9–19.]

Service factor: $SF=$	1.00	Use 1.00 if no unusual conditions.
Hardness ratio factor: $C_H=$	1.00	Fig. 9–23 or 9–24; gear only
Reliability factor: $K_R=$	1.50	Table 9–9: Use 1.00 for R = 0.99.

Enter: Design life:	20,000 hours	See Table 9–8.

Pinion—Number of load cycles: $N_P = 2.1E+09$		Guidelines: Y_N, Z_N	
Gear—Number of load cycles: $N_G = 6.0E+08$		10^7 cycles $\quad >10^7$	$<10^7$
Bending stress cycle factor: $Y_{NP}=$	0.93	1.00 $\quad$ 0.93	Fig. 9–20
Bending stress cycle factor: $Y_{NG}=$	0.95	1.00 $\quad$ 0.95	Fig. 9–20
Pitting stress cycle factor: $Z_{NP}=$	0.88	1.00 $\quad$ 0.88	Fig. 9–22
Pitting stress cycle factor: $Z_{NG}=$	0.91	1.00 $\quad$ 0.91	Fig. 9–22

Stress Analysis: Bending

Pinion: Required $s_{at}=$	16,009 psi	See Fig. 9–8 or
Gear: Required $s_{at}=$	12,423 psi	Table 9–3, 9–4, or 9–5.

Stress Analysis: Pitting

Pinion: Required $s_{ac}=$	161,968 psi	See Fig. 9–9 or
Gear: Required $s_{ac}=$	156,629 psi	Table 9–3 or 9–4.

Specify materials, alloy and heat treatment, for most severe requirement.

DESIGN OF SPUR GEARS

Initial Input Data:

Input power:	$P=$	25 hp
Input speed:	$n_P=$	1,750 rpm
Diametral pitch:	$P_d=$	8
Number of pinion teeth:	$N_P=$	28
Desired output speed:	$n_G=$	500 rpm

Computed number of gear teeth:	98.0

Enter: Chosen no. of gear teeth:	$N_G=$	98

Computed Data:

Actual output speed:	$n_G=$	500.0 rpm
Gear ratio:	$m_G=$	3.50
Pitch diameter, pinion:	$D_P=$	3.500 in
Pitch diameter, gear:	$D_G=$	12.250 in
Center distance:	$C=$	7.875 in
Pitch line speed:	$v_t=$	1,604 ft/min
Transmitted load:	$W_t=$	514 lb

Secondary Input Data:

	Min.	Nom.	Max.
Face width guidelines (in):	1.000	1.500	2.000

Enter: Face width:	$F=$	2.000 in

Ratio: Face width/pinion diameter: $F/D_P=$	0.57
Recommended range of ratio: $0.50 < F/D_P < 2.00$	

Enter: Elastic coefficient:	$C_p=$	2,300	Table 9–10

Enter: Quality number:	$Q_v=$	8	Table 9–2

Enter: Bending geometry factors:

Pinion:	$J_P=$	0.380	Fig. 9–15
Gear:	$J_G=$	0.440	Fig. 9–15
Enter: Pitting geometry factor:	$I=$	0.115	Fig. 9–21

FIGURE 9–29 Redesign for data of Example Problem 9–1, Trial 2

Application:	Industrial saw driven by an electric motor
	Redesign for data of Example Problem 9–1, Trial 2

Factors in Design Analysis:

Alignment factor, $K_m = 1.0 + C_{pf} + C_{ma}$	If $F < 1.0$	If $F > 1.0$	
Pinion proportion factor, $C_{pf} =$	0.032	0.045	$[0.50 < F/D_P < 2.00]$
Enter: $C_{pf} =$	0.045	Fig. 9–16	

	Open	Commer.	Precision	Ex. Prec.
Type of gearing:				
Mesh alignment factor, $C_{ma} =$	0.280	0.158	0.093	0.058

Enter: $C_{ma} =$	0.158	Fig. 9–17
Alignment factor: $K_m =$	1.20	[Computed]

Overload factor: $K_o =$	1.50	Table 9–6
Size factor: $K_s =$	1.00	Table 9–7: Use 1.00 if $P_d \geq 5$.
Pinion rim thickness factor: $K_{BP} =$	1.00	Fig. 9–18: Use 1.00 if solid blank.
Gear rim thickness factor: $K_{BG} =$	1.00	Fig. 9–18: Use 1.00 if solid blank.
Dynamic factor: $K_v =$	1.33	[Computed: See Fig. 9–19.]

Service factor: $SF =$	1.00	Use 1.00 if no unusual conditions.
Hardness ratio factor: $C_H =$	1.00	Fig. 9–23 or 9–24; gear only
Reliability factor: $K_R =$	1.50	Table 9–9: Use 1.00 for $R = 0.99$.

Enter: Design life:	20,000 hours	See Table 9–8.

Pinion—Number of load cycles: $N_P = 2.1E+09$		Guidelines: Y_N, Z_N		
Gear—Number of load cycles: $N_G = 6.0E+08$		10^7 cycles	$> 10^7$	$< 10^7$
Bending stress cycle factor: $Y_{NP} =$	0.93	1.00	0.93	Fig. 9–20
Bending stress cycle factor: $Y_{NG} =$	0.95	1.00	0.95	Fig. 9–20
Pitting stress cycle factor: $Z_{NP} =$	0.88	1.00	0.88	Fig. 9–22
Pitting stress cycle factor: $Z_{NG} =$	0.91	1.00	0.91	Fig. 9–22

Stress Analysis: Bending

Pinion: Required $s_{at} =$	20,915 psi	See Fig. 9–8 or
Gear: Required $s_{at} =$	17,682 psi	Table 9–3, 9–4, or 9–5.

Stress Analysis: Pitting

Pinion: Required $s_{ac} =$	153,363 psi	See Fig. 9–9 or
Gear: Required $s_{ac} =$	148,307 psi	Table 9–3 or 9–4.

Specify materials, alloy and heat treatment, for most severe requirement.

DESIGN OF SPUR GEARS

Initial Input Data:

Input power:	$P=$	25 hp
Input speed:	$n_P=$	1,750 rpm
Diametral pitch:	$P_d=$	6
Number of pinion teeth:	$N_P=$	24
Desired output speed:	$n_G=$	500 rpm

Computed number of gear teeth: 84.0

Enter: Chosen no. of gear teeth: $N_G=$ 84

Computed Data:

Actual output speed:	$n_G=$	500.0 rpm
Gear ratio:	$m_G=$	3.50
Pitch diameter, pinion:	$D_P=$	4.000 in
Pitch diameter, gear:	$D_G=$	14.000 in
Center distance:	$C=$	9.000 in
Pitch line speed:	$v_t=$	1,833 ft/min
Transmitted load:	$W_t=$	450 lb

Secondary Input Data:

	Min.	Nom.	Max.
Face width guidelines (in):	1.333	2.000	2.667

Enter: Face width: $F=$ 2.000 in

Ratio: Face width/pinion diameter: $F/D_P=$ 0.50
Recommended range of ratio: $0.50 < F/D_P < 2.00$

Enter: Elastic coefficient: $C_p=$ 2,300 Table 9–10

Enter: Quality number: $Q_v=$ 6 Table 9–2

Enter: Bending geometry factors:
Pinion:	$J_P=$	0.360	Fig. 9–15
Gear:	$J_G=$	0.430	Fig. 9–15
Enter: Pitting geometry factor:	$I=$	0.112	Fig. 9–21

FIGURE 9–30 Redesign for data of Example Problem 9–1, Trial 3

Application:	Industrial saw driven by an electric motor Redesign for data of Example Problem 9–1, Trial 3			

Factors in Design Analysis:

	If $F < 1.0$	If $F > 1.0$		
Alignment factor, $K_m = 1.0 + C_{pf} + C_{ma}$				
Pinion proportion factor, $C_{pf} =$	0.025	0.038	[$0.50 < F/D_P < 2.00$]	

Enter: $C_{pf} =$	0.038	Fig. 9–16		

	Open	Commer.	Precision	Ex. Prec.
Type of gearing:				
Mesh alignment factor, $C_{ma} =$	0.280	0.158	0.093	0.058

Enter: $C_{ma} =$	0.158	Fig. 9–17

Alignment factor: $K_m =$	1.20	[Computed]

Overload factor: $K_o =$	1.50	Table 9–6
Size factor: $K_s =$	1.00	Table 9–7: Use 1.00 if $P_d \geq 5$.
Pinion rim thickness factor: $K_{BP} =$	1.00	Fig. 9–18: Use 1.00 if solid blank.
Gear rim thickness factor: $K_{BG} =$	1.00	Fig. 9–18: Use 1.00 if solid blank.

Dynamic factor: $K_v =$	1.56	[Computed: See Fig. 9–19.]

Service factor: $SF =$	1.00	Use 1.00 if no unusual conditions.
Hardness ratio factor: $C_H =$	1.00	Fig. 9–23 or 9–24; gear only
Reliability factor: $K_R =$	1.50	Table 9–9: Use 1.00 for $R = 0.99$.
Enter: Design life:	20,000 hours	See Table 9–8.

Pinion—Number of load cycles: $N_P = 2.1E+09$		Guidelines: Y_N, Z_N		
Gear—Number of load cycles: $N_G = 6.0E+08$		10^7 cycles	$>10^7$	$<10^7$
Bending stress cycle factor: $Y_{NP} =$	0.93	1.00	0.93	Fig. 9–20
Bending stress cycle factor: $Y_{NG} =$	0.95	1.00	0.95	Fig. 9–20
Pitting stress cycle factor: $Z_{NP} =$	0.88	1.00	0.88	Fig. 9–22
Pitting stress cycle factor: $Z_{NG} =$	0.91	1.00	0.91	Fig. 9–22

Stress Analysis: Bending

Pinion: Required $s_{at} =$	16,961 psi	See Fig. 9–8 or
Gear: Required $s_{at} =$	13,901 psi	Table 9–3, 9–4, or 9–5.

Stress Analysis: Pitting

Pinion: Required $s_{ac} =$	147,128 psi	See Fig. 9–9 or
Gear: Required $s_{ac} =$	142,277 psi	Table 9–3 or 9–4.

Specify materials, alloy and heat treatment, for most severe requirement.

COMPARISON OF DESIGN ALTERNATIVES FOR DATA OF EXAMPLE PROBLEMS 9–1 THROUGH 9–4

P= 25 hp   Pinion speed= 1 750 rpm   Gear speed= 500 rpm

	Original Solution and Alternative Trials				Comments:
	Problems 9–1 through 9–4	Alternative Trial 1	Alternative Trial 2	Alternative Trial 3	
Geometry, Quality, and Transmitted Load:					
N_P	20	18	28	24	All designs produce same gear ratio.
N_G	70	63	98	84	
P_d	8	6	8	6	Note change in diametral pitch.
D_P (in)	2.500	3.000	3.500	4.000	Increasing pinion diameter.
D_G (in)	8.750	10.500	12.250	14.000	Increasing gear diameter.
C (in)	5.625	6.750	7.875	9.000	Increasing center distance.
W_t (lb)	720	600	514	450	Decreasing transmitted load.
F (in)	1.50	2.75	2.00	2.00	Face width varies.
Q_v	6	8	8	6	Note change in quality number.
K_v	1.45	1.30	1.33	1.56	Note change in dynamic factor.
Stresses:					
s_{atP} (psi)	47,900	16,000	20,900	17,000	Moderate bending stress.
s_{atG} (psi)	37,400	12,450	17,700	13,900	Moderate bending stress.
s_{acP} (psi)	266,000	162,000	153,400	147,150	Contact stress critical.
s_{acG} (psi)	242,700	156,700	148,300	142,300	Contact stress critical.
Materials:					
Pinion	Grade 3 steel, carburized, case-hardened	Grade 1 steel, induction-hardened, 50 HRC	Grade 1 steel, through-hardened, 400 HB	Grade 1 steel, through-hardened, 400 HB	Design goal is to use only Grade 1 steels.
Gear	Grade 3 steel, carburized, case-hardened	Grade 1 steel, through-hardened, 400 HB	Grade 1 steel, through-hardened, 370 HB	Grade 1 steel, through-hardened, 350 HB	

FIGURE 9–31 Comparison of design alternative trials for data of Example Problem 9–1

Solving for P, we have

$$P = \frac{s_{at}\, Y_N\, FJ\, n_P D_P}{(126\,000)(P_d)\,(SF)\, K_R K_o K_s K_m K_B K_v} \qquad (9\text{–}30)$$

This equation should be solved for both the pinion and the gear. Most variables will be the same except for s_{at}, Y_N, J, and possibly K_B.

Pitting Resistance

Here we start with Equations (9–25), (9–26), and (9–27) in which the computed contact stress number is compared with the modified allowable contact stress number for the gear. Equation (9–26) can be expressed in the form

$$s_c = C_p \sqrt{\frac{W_t K_o K_s K_m K_v}{D_P F I}} = \frac{s_{ac} Z_N C_H}{(SF) K_R} \tag{9–19}$$

Squaring both sides of this equation and solving for W_t gives

$$\frac{W_t K_o K_s K_m K_v}{D_P F I} = \left[\frac{s_{ac} Z_N C_H}{(SF) K_R C_p} \right]^2$$

$$W_t = \frac{D_P F I}{K_o K_s K_m K_v} \left[\frac{s_{ac} Z_N C_H}{(SF) K_R C_p} \right]^2$$

Now substituting this into Equation (9–6) and solving for the power P gives

$$P = \frac{W_t D_P n_P}{126\,000} = \frac{D_P n_P D_P F I}{126\,000\, K_o K_s K_m K_v} \left[\frac{s_{ac} Z_N C_H}{(SF)\, K_R C_p} \right]^2$$

$$P = \frac{n_P F I}{126\,000\, K_o K_s K_m K_v} \left[\frac{s_{ac} D_P Z_N C_H}{(SF) K_R C_p} \right]^2 \tag{9–31}$$

Equations (9–30) and (9–31) should be used to compute the power-transmitting capacity for a pair of gears of known design with particular materials.

**9–17
PRACTICAL
CONSIDERATIONS
FOR GEARS AND
INTERFACES
WITH OTHER
ELEMENTS**

It is important to consider the design of the entire gear system when designing the gears because they must work in harmony with the other elements in the system. This section will briefly discuss some of these practical considerations and will show commercially available speed reducers.

Our discussion so far has been concerned primarily with the gear teeth, including the tooth form, pitch, face width, material selection, and heat treatment. Also to be considered is the type of gear blank. Figures 8–2 and 8–4 show several styles of blanks. Smaller gears and lightly loaded gears are typically made in the plain style. Gears with pitch diameters of approximately 5.0 in through 8.0 in are frequently made with thinned webs between the rim and the hub for lightening, with some having holes bored in the webs for additional lightening. Larger gears, typically with pitch diameters greater than 8.0 in, are made from cast blanks with spokes between the rim and the hub.

In many precision special machines and gear systems produced in large quantities, the gears are machined integral with the shaft carrying the gears. This, of course, eliminates some of the problems associated with mounting and location of the gears, but it may complicate the machining operations.

In general machine design, gears are usually mounted on separate shafts, with the torque transmitted from the shaft to the gear through a key. This setup provides a positive means of transmitting the torque while permitting easy assembly and disassembly. The axial location of the gear must be provided by another means, such as a shoulder on the shaft, a retaining ring, or a spacer (see Chapters 11 and 12).

Other considerations include the forces exerted on the shaft and the bearings that are due to the action of the gears. These subjects are discussed in Section 9–3. The housing design must provide adequate support for the bearings and protection of the interior components. Normally, it must also provide a means of lubricating the gears.

See References 10–15 for additional practical considerations.

Lubrication

The action of spur gear teeth is a combination of rolling and sliding. Because of the relative motion, and because of the high local forces exerted at the gear faces, adequate lubrication is critical to smoothness of operation and gear life. A continuous supply of oil at the pitch line is desirable for most gears unless they are lightly loaded or operate only intermittently.

In splash-type lubrication, one of the gears in a pair dips into an oil supply sump and carries the oil to the pitch line. At higher speeds, the oil may be thrown onto the inside surfaces of the case; then it flows down, in a controlled fashion, onto the pitch line. Simultaneously, the oil can be directed to the bearings that support the shafts. One difficulty with the splash type of lubrication is that the oil is churned; at high gear speeds, excessive heat can be generated, and foaming can occur.

A positive oil circulation system is used for high-speed and high-capacity systems. A separate pump draws the oil from the sump and delivers it at a controlled rate to the meshing teeth.

The primary functions of gear lubricants are to reduce friction at the mesh and to keep operating temperatures at acceptable levels. It is essential that a continuous film of lubricant be maintained between the mating tooth surfaces of highly loaded gears and that there be a sufficient flow rate and total quantity of oil to maintain cool temperatures. Heat is generated by the meshing gear teeth, by the bearings, and by the churning of the oil. This heat must be dissipated from the oil to the case or to some other external heat-exchange device in order to keep the oil itself below 160°F (approximately 70°C). Above this temperature, the lubricating ability of the oil, as indicated by its viscosity, is severely decreased. Also, chemical changes can be produced in the oil, decreasing its lubricity. Because of the wide variety of lubricants available and the many different conditions under which they must operate, it is recommended that suppliers of lubricants be consulted for proper selection. (See also Reference 8.)

Commercially Available Gear-Type Speed Reducers

By studying the design of commercially available gear-type speed reducers, you should get a better feel for these design details and the relationships among the component parts: the

FIGURE 9–32   Double-reduction spur gear reducer (Bison Gear & Engineering Corp., Downers Grove, IL)

FIGURE 9–33   Double-reduction gear reducer. First stage, helical gears; second stage, spur gears. (Bison Gear & Engineering Corp., Downers Grove, IL)

gears, the shafts, the bearings, the housing, the means of providing lubrication, and the coupling to the driving and driven machines.

Figure 9–32 shows a double-reduction spur gear speed reducer with an electric motor rigidly attached. Such a unit is often called a *gear motor*. Figure 9–33 is similar, except that one of the stages of reduction uses helical gears (discussed in the next chapter). The cross-sectional drawing shown with Figure 9–34 gives a clear picture of the several components of a reducer.

The planetary reducer in Figure 9–35 has quite a different design to accommodate the placement of the sun, planet, and ring gears. Figure 9–36 shows the eight-speed transmission from a large farm tractor and illustrates the high degree of complexity that may be involved in the design of transmissions.

(*a*) Cutaway of a concentric helical gear reducer

(*b*) Complete reducer

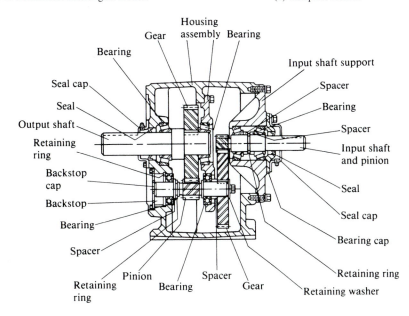

(c) Parts index

FIGURE 9–34   Concentric helical gear reducer (Winsmith Division, UMC Industries, Springville, NY)

(a)

(b) Schematic arrangement of planetary gearing

FIGURE 9–35   Planetary gear reducer (Rexnord, Milwaukee, WI)

FIGURE 9–36
Eight-speed tractor
transmission (Ford
Tractor Operations,
Troy, MI)

REFERENCES

1. Adams, Clifford E. *Plastics Gearing, Selection and Application.* New York: Marcel Dekker, 1986.

2. American Gear Manufacturers Association. Standard 908-B89. *Geometry Factors for Determining the Pitting Resistance and Bending Strength of Spur, Helical, and Herringbone Gear Teeth.* Alexandria, VA.: American Gear Manufacturers Association, 1989.

3. American Gear Manufacturers Association. Standard 1012-F90. *Gear Nomenclature, Definitions of Terms with Symbols.* Alexandria, VA: American Gear Manufacturers Association, 1990.

4. American Gear Manufacturers Association. Standard 2001-C95. *Fundamental Rating Factors and Calculation Methods for Involute Spur and Helical Gear Teeth.* Alexandria, VA: American Gear Manufacturers Association, 1995.

5. American Gear Manufacturers Association. Standard 2002-B88. *Tooth Thickness Specification and Measurement.* Alexandria, VA: American Gear Manufacturers Association, 1988.

6. American Gear Manufacturers Association. Standard 2004-B89. *Gear Materials and Heat Treatment Manual.* Alexandria, VA: American Gear Manufacturers Association, 1989.

7. American Gear Manufacturers Association. Standard 6010-E88. *Standard for Spur, Helical, Herringbone, and Bevel Enclosed Drives.* Alexandria, VA: American Gear Manufacturers Association, 1989.

8. American Gear Manufacturers Association. Standard 9005-D94. *Industrial Gear Lubrication.* Alexandria, VA: American Gear Manufacturers Association, 1994.

9. American Gear Manufacturers Association. Standard 1010-E95. *Nomenclature of Gear Tooth Failure Modes.* Alexandria, VA: American Gear Manufacturers Association, 1995.

10. American Gear Manufacturers Association. AGMA 427.01. *Information Sheet—Systems Considerations for Critical Service Gear Drives.* Alexandria, VA: American Gear Manufacturers Association, 1994.

11. American Society for Metals. *Source Book on Gear Design, Technology and Performance.* Metals Park, OH: American Society for Metals, 1980.

12. Drago, Raymond J. *Fundamentals of Gear Design.* Boston: Butterworths, 1988.

13. Dudley, Darle W. *Handbook of Practical Gear Design.* New York: McGraw-Hill, 1984.

14. Hosel, Theodor. *Comparison of Load Capacity Ratings for Involute Gears Due to ANSI/AGMA, ISO, DIN and Comecon Standards.* AGMA Technical Paper 89FTM4. Alexandria, VA: American Gear Manufacturers Association, 1989.

15. Lynwander, Peter. *Gear Drive Systems, Design and Application.* New York: Marcel Dekker, 1983.

16. Kern, Roy F. *Achievable Carburizing Specifications.* AGMA Technical Paper 88FTM1. Alexandria, VA: American Gear Manufacturers Association, 1988.

17. Kern, Roy F., and M. E. Suess. *Steel Selection.* New York: John Wiley & Sons, 1979.

18. Lipp, Robert. "Avoiding Tooth Interference in Gears." *Machine Design* 54, no. 1 (January 7, 1982).

19. Oberg, Erik, et al. *Machinery's Handbook.* 25th ed. New York: Industrial Press, 1996.

20. Shigley, Joseph E., and C. R. Mischke. *Mechanical Engineering Design.* 5th ed. New York: McGraw-Hill, 1989.

21. Society of Automotive Engineers. *Gear Design, Manufacturing and Inspection Manual.* Warrendale, PA: Society of Automotive Engineers, 1990.

PROBLEMS

Forces on Spur Gear Teeth

1. A pair of spur gears with 20°, full-depth, involute teeth transmits 7.5 hp. The pinion is mounted on the shaft of an electric motor operating at 1 750 rpm. The pinion has 20 teeth and a diametral pitch of 12. The gear has 72 teeth. Compute the following:
   a. The rotational speed of the gear
   b. The velocity ratio and the gear ratio for the gear pair
   c. The pitch diameter of the pinion and the gear
   d. The center distance between the shafts carrying the pinion and the gear
   e. The pitch line speed for both the pinion and the gear
   f. The torque on the pinion shaft and on the gear shaft
   g. The tangential force acting on the teeth of each gear
   h. The radial force acting on the teeth of each gear
   i. The normal force acting on the teeth of each gear

2. A pair of spur gears with 20°, full-depth, involute teeth transmits 50 hp. The pinion is mounted on the shaft of an electric motor operating at 1 150 rpm. The pinion has 18 teeth and a diametral pitch of 5. The gear has 68 teeth. Compute the following:
   a. The rotational speed of the gear
   b. The velocity ratio and the gear ratio for the gear pair
   c. The pitch diameter of the pinion and the gear
   d. The center distance between the shafts carrying the pinion and the gear
   e. The pitch line speed for both the pinion and the gear
   f. The torque on the pinion shaft and on the gear shaft
   g. The tangential force acting on the teeth of each gear

   h. The radial force acting on the teeth of each gear
   i. The normal force acting on the teeth of each gear

3. A pair of spur gears with 20°, full-depth, involute teeth transmits 0.75 hp. The pinion is mounted on the shaft of an electric motor operating at 3 450 rpm. The pinion has 24 teeth and diametral pitch of 24. The gear has 110 teeth. Compute the following:
   a. The rotational speed of the gear
   b. The velocity ratio and the gear ratio for the gear pair
   c. The pitch diameter of the pinion and the gear
   d. The center distance between the shafts carrying the pinion and the gear
   e. The pitch line speed for both the pinion and the gear
   f. The torque on the pinion shaft and on the gear shaft
   g. The tangential force acting on the teeth of each gear
   h. The radial force acting on the teeth of each gear
   i. The normal force acting on the teeth of each gear

4. For the data of Problem 1, repeat Parts (g), (h), and (i) if the teeth have 25° full depth instead of 20°.

5. For the data of Problem 2, repeat Parts (g), (h), and (i) if the teeth have 25° full depth instead of 20°.

6. For the data of Problem 3, repeat Parts (g), (h), and (i) if the teeth have 25° full depth instead of 20°.

Gear Manufacture and Quality

7. List three methods for producing gear teeth, and describe each method. Include a description of the cutter for each method along with its motion relative to the gear blank.

8. Specify a suitable quality number for the gears in the drive for a grain harvester. List the total composite tolerance for a pinion in the drive having a diametral pitch of 8 and 40 teeth and for its mating gear having 100 teeth.

9. Specify a suitable quality number for the gears in the drive for a high-speed printing press. List the total composite tolerance for a pinion in the drive having a diametral pitch of 20 and 40 teeth and for its mating gear having 100 teeth.

10. Specify a suitable quality number for the gears in the drive for an automotive transmission. List the total composite tolerance for a pinion in the drive having a diametral pitch of 8 and 40 teeth and its mating gear having 100 teeth.

11. Specify a suitable quality number for the gears in the drive for a gyroscope used in the guidance system for a spacecraft. List the total composite tolerance for a pinion in the drive having a diametral pitch of 32 and 40 teeth and its mating gear having 100 teeth.

12. Compare the values of total composite tolerance for the gears of Problems 8 and 10.

13. Compare the values of total composite tolerance for the gears of Problems 8, 9, and 11.

14. Specify a suitable quality number for the gears of Problem 1 if the drive is part of a precision machine tool.

15. Specify a suitable quality number for the gears of Problem 2 if the drive is part of a precision machine tool.

16. Specify a suitable quality number for the gears of Problem 3 if the drive is part of a precision machine tool.

Gear Materials

17. Identify the two major types of stresses that are created in gear teeth as they transmit power. Describe how the stresses are produced and where the maximum values of such stresses are expected to occur.

18. Describe the nature of the data contained in AGMA standards that relate to the ability of a given gear tooth to withstand the major types of stresses that it sees in operation.

19. Describe the general nature of steels that are typically used for gears, and list at least five examples of suitable alloys.

20. Describe the range of hardness that can typically be produced by through-hardening techniques and used successfully in steel gears.

21. Describe the general nature of the differences among steels produced as Grade 1, Grade 2, and Grade 3.

22. Suggest at least three applications in which Grade 2 or Grade 3 steel might be appropriate.

23. Describe three methods of producing gear teeth with strengths greater than can be achieved with through-hardening.

24. What AGMA standard should be consulted for data on the allowable stresses for steels used for gears?

25. In the AGMA standard identified in Problem 24, for what other materials besides steels are strength data provided?

26. Determine the allowable bending stress number and the allowable contact stress number for the following materials:
    a. Through-hardened, Grade 1 steel with a hardness of 200 HB
    b. Through-hardened, Grade 1 steel with a hardness of 300 HB
    c. Through-hardened, Grade 1 steel with a hardness of 400 HB
    d. Through-hardened, Grade 1 steel with a hardness of 450 HB
    e. Through-hardened, Grade 2 steel with a hardness of 200 HB
    f. Through-hardened, Grade 2 steel with a hardness of 300 HB
    g. Through-hardened, Grade 2 steel with a hardness of 400 HB

27. If the design of a steel gear indicates that an allowable bending stress number of 36 000 psi is needed, specify a suitable hardness level for Grade 1 steel. What hardness level would be required for Grade 2 steel?

28. What level of hardness can be expected for gear teeth that are case-hardened by carburizing?

29. Name three typical steels that are used in carburizing.

30. What is the level of hardness that can be expected for gear teeth that are case-hardened by flame or induction hardening?

31. Name three typical steels that are used for flame or induction hardening. What is an important property of such steels?

32. What level of hardness can be expected for gear teeth that are nitrided?

33. Determine the allowable bending stress number and the allowable contact stress number for the following materials:
    a. Flame-hardened AISI 4140 steel, Grade 1, with a surface hardness of 50 HRC
    b. Flame-hardened AISI 4140 steel, Grade 1, with a surface hardness of 54 HRC
    c. Carburized and case-hardened AISI 4620 Grade 1 steel, DOQT 300
    d. Carburized and case-hardened AISI 4620 Grade 2 steel, DOQT 300
    e. Carburized and case-hardened AISI 1118 Grade 1 steel, SWQT 350
    f. Nitrided, through-hardened, Grade 1 steel with a surface hardness of 84.5 HRN and a core hardness of 325 HB
    g. Nitrided, through-hardened, Grade 2 steel with a surface hardness of 84.5 HRN and a core hardness of 325 HB
    h. Nitrided, 2.5% chrome, Grade 3 steel with a surface hardness of 90.0 HRN and a core hardness of 325

i. Gray cast iron, class 20

j. Gray cast iron, class 40

k. Ductile iron, 100-70-03

l. Sand-cast bronze with a minimum tensile strength of 40 Ksi (275 MPa)

m. Heat-treated bronze with a minimum tensile strength of 90 Ksi (620 MPa)

n. Glass-filled nylon

o. Glass-filled polycarbonate

34. What depth should be specified for the case for a carburized gear tooth having a diametral pitch of 6?

35. What depth should be specified for the case for a carburized gear tooth having a metric module of 6?

Bending Stresses in Gear Teeth

For Problems 36–41, compute the bending stress number, s_t, using Equation (9–15). Assume that the gear blank is solid unless otherwise stated. (*Note that the data in these problems are used in later problems through Problem 59. You are advised to keep solutions to earlier problems accessible so that you can use data and results in later problems. The four problems that are keyed to the same set of data require the analysis of bending stress and contact stress and the corresponding specification of suitable materials based on those stresses. Later design problems, 60–70, use the complete analysis within each problem.*)

36. A pair of gears with 20°, full-depth, involute teeth transmits 10.0 hp while the pinion rotates at 1 750 rpm. The diametral pitch is 12, and the quality number is 6. The pinion has 18 teeth, and the gear has 85 teeth. The face width is 1.25 in. The input power is from an electric motor, and the drive is for an industrial conveyor. The drive is a commercial enclosed gear unit.

37. A pair of gears with 20°, full-depth, involute teeth transmits 40 hp while the pinion rotates at 1 150 rpm. The diametral pitch is 6, and the quality number is 5. The pinion has 20 teeth, and the gear has 48 teeth. The face width is 2.25 in. The input power is from an electric motor, and the drive is for a cement kiln. The drive is a commercial enclosed gear unit.

38. A pair of gears with 20°, full-depth, involute teeth transmits 0.50 hp while the pinion rotates at 3 450 rpm. The diametral pitch is 32, and the quality number is 10. The pinion has 24 teeth, and the gear has 120 teeth. The face width is 0.50 in. The input power is from an electric motor, and the drive is for a small machine tool. The drive is a precision enclosed gear unit.

39. A pair of gears with 25°, full-depth, involute teeth transmits 15.0 hp while the pinion rotates at 6 500 rpm. The diametral pitch is 10, and the quality number is 12. The pinion has 30 teeth, and the gear has 88 teeth. The face width is 1.50 in. The input power is from a universal electric motor, and the drive is for an actuator on an aircraft. The drive is an extra-precision, enclosed gear unit.

40. A pair of gears with 25°, full-depth, involute teeth transmits 125 hp while the pinion rotates at 2 500 rpm. The diametral pitch is 4, and the quality number is 8. The pinion has 32 teeth, and the gear has 76 teeth. The face width is 1.50 in. The input power is from a gasoline engine, and the drive is for a portable industrial water pump. The drive is a commercial enclosed gear unit.

41. A pair of gears with 25°, full-depth, involute teeth transmits 2.50 hp while the pinion rotates at 680 rpm. The diametral pitch is 20, and the quality number is 6. The pinion has 24 teeth, and the gear has 62 teeth. The face width is 0.25 in. The input power is from a vane-type fluid motor, and the drive is for a small lawn and garden tractor. The drive is a commercial enclosed gear unit.

Required Allowable Bending Stress Number

For Problems 42–47, compute the required allowable bending stress number, s_{at}, using Equation (9–20). Assume that no unusual conditions exist unless stated otherwise. That is, use a service factor, *SF*, of 1.00. Then specify a suitable steel and its heat treatment for both the pinion and the gear based on bending stress.

42. Use the data and results from Problem 36. Design for a reliability of 0.999 and a design life of 20 000 h.

43. Use the data and results from Problem 37. Design for a reliability of 0.99 and a design life of 8 000 h.

44. Use the data and results from Problem 38. Design for a reliability of 0.9999 and a design life of 12 000 h. Consider that the machine tool is a critical part of a production system calling for a service factor of 1.50 to avoid unexpected down time.

45. Use the data and results from Problem 39. Design for a reliability of 0.9999 and a design life of 4 000 h.

46. Use the data and results from Problem 40. Design for a reliability of 0.99 and a design life of 6 000 h.

47. Use the data and results from Problem 41. Design for a reliability of 0.90 and a design life of 2 000 h. The uncertainty of the actual use of the tractor calls for a service factor of 1.25. Consider using cast iron or bronze if the conditions permit.

Pitting Resistance

For Problems 48–53, compute the expected contact stress number, s_c, using Equation (9–25). Assume that both gears are to be steel unless stated otherwise.

48. Use the data and results from Problem 36.

49. Use the data and results from Problem 37.

50. Use the data and results from Problem 38.

51. Use the data and results from Problem 39.

52. Use the data and results from Problem 40.

53. Use the data and results from Problem 41. Use the materials selected in Problem 47.

Required Allowable Contact Stress Number

For Problems 54–59, compute the required allowable contact stress number, s_{ac}, using Equation (9–27). Use a service factor, *SF*, of 1.00 unless stated otherwise. Then specify suitable material for the pinion and the gear based on pitting resistance. Use steel unless an earlier decision has been made to use another material. Then evaluate whether the earlier decision is still valid. If not, specify a different material according to the most severe requirement. If no suitable material can be found, consider redesigning the original gears to enable reasonable materials to be used.

54. Use the data and results from Problems 36, 42, and 48.

55. Use the data and results from Problems 37, 43, and 49.

56. Use the data and results from Problems 38, 44, and 50.

57. Use the data and results from Problems 39, 45, and 51.

58. Use the data and results from Problems 40, 46, and 52.

59. Use the data and results from Problems 41, 47, and 53.

Design Problems

Problems 60–70, describe design situations. For each, design a pair of spur gears, specifying (at least) the diametral pitch, the number of teeth in each gear, the pitch diameters of each gear, the center distance, the face width, and the material from which the gears are to be made. Design for indefinite life with regard to both strength and pitting resistance. Work toward designs that are compact. Use standard values of diametral pitch, and avoid designs for which interference could occur. See Example Problem 9–5. Assume that the input to the gear pair is from an electric motor unless otherwise stated.

If the data are given in SI units, complete the design in the metric module system with dimensions in millimeters, forces in newtons, and stresses in megapascals. See Example Problem 9–6.

60. A pair of spur gears is to be designed to transmit 5.0 hp while the pinion rotates at 1 200 rpm. The gear must rotate between 385 and 390 rpm. The gear drives a reciprocating compressor.

61. A gear pair is to be a part of the drive for a milling machine requiring 20.0 hp with the pinion speed at 550 rpm and the gear speed to be between 180 and 190 rpm.

62. A drive for a punch press requires 50.0 hp with the pinion speed of 900 rpm and the gear speed of 225 to 230 rpm.

63. A single-cylinder gasoline engine has the pinion of a gear pair on its output shaft. The gear is attached to the shaft of a small cement mixer. The mixer requires 2.5 hp while rotating at approximately 75 rpm. The engine is governed to run at approximately 900 rpm.

64. A four-cylinder industrial engine runs at 2 200 rpm and delivers 75 hp to the input gear of a drive for a large wood chipper used to prepare pulpwood chips for paper making. The output gear must run between 4 500 and 4 600 rpm.

65. A small commercial tractor is being designed for chores such as lawn mowing and snow removal. The wheel drive system is to be through a gear pair in which the pinion runs at 600 rpm while the gear, mounted on the hub of the wheel, runs at 170 to 180 rpm. The wheel is 300 mm in diameter. The gasoline engine delivers 3.0 kW of power to the gear pair.

66. A water turbine transmits 75 kW of power to a pair of gears at 4 500 rpm. The output of the gear pair must drive an electric power generator at 3 600 rpm. The center distance for the gear pair must not exceed 150 mm.

67. A drive system for a large commercial band saw is to be designed to transmit 12.0 hp. The saw will be used to cut steel tubing for automotive exhaust pipes. The pinion rotates at 3 450 rpm, while the gear must rotate between 725 and 735 rpm. It has been specified that the gears are to be made from AISI 4340 steel, oil quenched and tempered. Case hardening is *not* to be used.

68. Repeat Problem 67, but consider a case-hardened carburized steel from Appendix 5. Try to achieve the smallest practical design. Compare the result with the design from Problem 67.

69. A gear drive for a special-purpose, dedicated machine tool is being designed to mill a surface on a rough steel casting. The drive must transmit 20 hp with a pinion speed of 650 rpm and an output speed between 110 and 115 rpm. The mill is to be used continuously, two shifts per day, six days per week, for at least five years. Design the drive to be as small as practical to permit its being mounted close to the milling head.

70. A cable drum for a crane is to rotate between 160 and 166 rpm. Design a gear drive for 25 hp in which the input pinion rotates at 925 rpm and the output rotates with the drum. The crane is expected to operate with a 50% duty cycle for 120 hours per week for at least 10 years. The pinion and the gear of the drive must fit within the 24-in inside diameter of the drum, with the gear mounted on the drum shaft.

Power-transmitting Capacity

71. Determine the power-transmitting capacity for a pair of spur gears having 20°, full-depth teeth, a diametral pitch of 10, a face width of 1.25 in, 25 teeth in the pinion, 60 teeth in the gear, and an AGMA quality class of 8. The pinion is made from AISI 4140 OQT 1000, and the gear is made from AISI 4140 OQT 1100. The pinion will rotate at 1 725 rpm on the shaft of an electric motor. The gear will drive a centrifugal pump.

72. Determine the power-transmitting capacity for a pair of spur gears having 20°, full-depth teeth, a diametral pitch of 6, 35 teeth in the pinion, 100 teeth in the gear, a face width of 2.00 in, and an AGMA quality class of 6. A gasoline engine drives the pinion at 1 500 rpm. The gear drives a conveyor for crushed rock in a quarry. The pinion is made from AISI 1040 WQT800. The gear is made from gray cast iron, ASTM A48-83, class 30.

73. It was found that the gear pair described in Problem 72 wore out when driven by a 25-hp engine. Propose a redesign that would be expected to give indefinite life under the conditions described.

Design of Double-Reduction Drives

74. Design a double-reduction gear train that will transmit 10.0 hp from an electric motor running at 1 750 rpm to an assembly conveyor whose drive shaft must rotate between 146 and 150 rpm. Note that this will require the design of two pairs of gears. Sketch the arrangement of the train, and compute the actual output speed.

75. A commercial food waste grinder in which the final shaft rotates at between 40 and 44 rpm is to be designed. The input is from an electric motor running at 850 rpm and delivering 0.50 hp. Design a double-reduction spur gear train for the grinder.

76. A small, powered hand drill is driven by an electric motor running at 3 000 rpm. The drill speed is to be approximately 550 rpm. Design the speed reduction for the drill. The power transmitted is 0.25 hp.

77. The output from the drill described in Problem 76 provides the drive for a small bench-scale band saw similar to the one in Figure 1–1. The saw blade is to move with a linear velocity of 375 ft/min. The saw blade rides on 9.0-in-diameter wheels. Design a spur gear reduction to drive the band saw. Consider using plastic gears.

78. Design a rack-and-pinion drive to lift a heavy access panel on a furnace. A fluid power motor rotating 1 500 rpm will provide 5.0 hp at the input to the drive. The linear speed of the rack is to be at least 2.0 ft/s. More than one stage of reduction may be used, but attempt to design with the fewest number of gears. The drive is expected to operate at least six times per hour for three shifts per day, seven days per week, for at least 15 years.

79. Design the gear drive for the wheels of an industrial lift truck. Its top speed is to be 20 mph. It has been decided that the wheels will have a diameter of 12.0 in. A DC motor supplies 20 hp at a speed of 3 000 rpm. The design life is 16 hours per day, six days per week, for 20 years.

10

Helical Gears, Bevel Gears, and Wormgearing

Helical Gears, Bevel Gears, and Wormgearing

Discussion Map

☐ The geometry of helical gears, bevel gears, and wormgearing was described in Chapter 8.

☐ The principles of stress analysis of gears were discussed in Chapter 9 for spur gears. Much of that information is applicable to the types of gears discussed in this chapter.

Discover

Review Chapters 8 and 9 now.

Recall some of the discussion at the beginning of Chapter 8 about uses for gears that you see in your world. Review that information now, and focus your discussion on helical gears, bevel gears, and wormgearing.

In this chapter, you acquire the skills to perform the necessary analyses to design safe gear drives that use helical gears, bevel gears, and wormgearing and that demonstrate long life.

Much was said in Chapters 8 and 9 about the kinematics of gears and about the stress analysis and design of spur gears. That information is important to the objectives of this chapter, in which we extend the application of those concepts to the analysis and design of helical gears, bevel gears, and wormgearing.

The basic geometry of helical gears was described in Section 8–7. The force system on helical gear teeth was also described, and that system is important to your understanding of the stresses and modes of potential failure for helical gears that we discuss in this chapter.

In Chapter 9, you learned how to analyze spur gears for bending strength and resistance to pitting of the surface of the teeth. This chapter modifies that same approach for application to the special geometry of helical gears. In fact, AGMA Standard 2001-C95 which we often referred to in Chapter 9 for spur gears is the same reference that we use for helical gears. So you will need to refer to Chapter 9 from time to time.

Similarly, the geometry of bevel gears was described in Section 8–8, and wormgearing was described in Sections 8–9 and 8–10. This chapter includes information about stresses in bevel gears and wormgearing.

Figure 10–1 shows an example of a large, commercially available, double-reduction, parallel-shaft reducer employing helical gears. Notice that the shafts are supported in tapered roller bearings that have the ability to withstand the thrust loads created by the helical gears. Chapter 14 elaborates on the selection of such bearings.

Figure 10–2 shows another form of helical gear reducer in which the drive motor is mounted above the reducer, and the shaft of the driven machine is inserted directly through the hollow output shaft of the reducer. This allows the reducer to be supported by the frame of the driven machine.

Refer to Figure 8–22 for drawings and photographs of bevel gear systems. The three-dimensional force system that acts between the teeth of bevel gears requires great care in the installation and alignment of the gears and requires the use of bearings that can withstand forces in all directions. The stress analysis is adapted from the approach described in Chapter 9, modified for the geometry of the bevel gear teeth.

Refer to Figure 8–24 for a photograph of a commercially available wormgear reducer. Notice the use of tapered roller bearings here also to withstand the thrust forces created by wormgearing. The lubrication of the gears is highlighted in the photograph. Lubrication is important because sliding action between the worm threads and the wormgear teeth

FIGURE 10–1
Parallel shaft reducer
(Emerson Power
Transmission
Corporation, Drive and
Component Division,
Ithaca, NY)

Splash lubrication system

Carburized and ground gearing

Timken ® tapered roller
bearings

Helical gearing

Welded steel housing

Magnetic drain plug

Output shaft

Input shaft

FIGURE 10–2
Helical shaft mount
reducer (Emerson
Power Transmission
Corporation, Drive and
Component Division,
Ithaca, NY)

Output gear

Tapered
roller
bearing

Hollow
output shaft

Shaft
seals

Ball
bearing

Pinions integral
with shafts

Cast housing

Reducer with motor
mounted to side of
a machine

inherently occurs, causing frictional heat to be generated. Consistent lubrication of the gear mesh is critical to the performance, efficiency, and life of the system.

This chapter presents more information about these factors in wormgearing system design. The design for strength and pitting resistance is presented for each type of gear, with an analysis of the geometry and the forces exerted on the gears.

You Are the Designer

The gear drives that were designed in Chapter 9 all assumed that spur gears would be used to accomplish the speed reduction or speed increase between the input and the output of the drive. But many other types of gears could have been used. Assume that you are the designer of the drive for the wood chipper described in Example Problem 9–5. How would the design be different if helical gears were used instead of spur gears? What forces

would be created and transferred to the shafts carrying the gears and to the bearings carrying the shafts? Would you be able to use smaller gears? How is the geometry of helical gears different from that of spur gears?

Rather than having the input and output shafts parallel as they were in designs up to this time, how can we design drives that deliver power to an output shaft at right angles to the input shaft? What special analysis techniques are applied to bevel gears and wormgearing?

The information in this chapter will help you answer these and other questions.

**10–1
OBJECTIVES OF
THIS CHAPTER**

After completing this chapter, you will be able to:

1. Describe the geometry of helical gears and compute the dimensions of key features.

2. Compute the forces exerted by one helical gear on its mating gear.

3. Compute the stress due to bending in helical gear teeth and specify suitable materials to withstand such stresses.

4. Design helical gears for surface durability.

5. Describe the geometry of bevel gears and compute the dimensions of key features.

6. Analyze the forces exerted by one bevel gear on another and show how those forces are transferred to the shafts carrying the gears.

7. Design and analyze bevel gear teeth for strength and surface durability.

8. Describe the geometry of worms and wormgears.

9. Compute the forces created by a wormgear drive system and analyze their effect on the shafts carrying the worm and the wormgear.

10. Compute the efficiency of wormgear drives.

11. Design and analyze wormgear drives to be safe for bending strength and wear. References 3, 4, 7, 8, and 17–22 are recommended for general guidelines for design and application of helical gears, bevel gears, and wormgearing.

**10–2
FORCES ON
HELICAL GEAR
TEETH**

Figure 10–3 shows a photograph of two helical gears in mesh and designed to be mounted on parallel shafts. This is the basic configuration that we analyze in this chapter. Refer to Figure 10–4 for a representation of the force system that acts between the teeth of two helical gears in mesh. In Chapter 8, using the same figure, we defined the following forces:

■ W_N is the *true normal force* that acts perpendicular to the face of the tooth in the plane normal to the surface of the tooth. The normal plane is shown in Part (d) of Figure 10–4. We seldom need to use the value of W_N because its three orthogonal components, defined next, are used in the analyses performed for helical gears. The

FIGURE 10–3
Helical gears. These gears have a 45° helix angle. (Emerson Power Transmission Corporation, Drive and Component Division, Ithaca, NY)

Left hand helix

Right hand helix

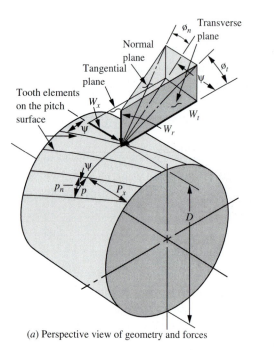

(a) Perspective view of geometry and forces

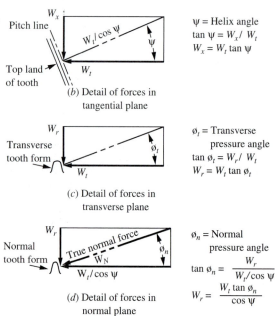

ψ = Helix angle
$\tan \psi = W_x / W_t$
$W_x = W_t \tan \psi$

(b) Detail of forces in tangential plane

ϕ_t = Transverse pressure angle
$\tan \phi_t = W_r / W_t$
$W_r = W_t \tan \phi_t$

(c) Detail of forces in transverse plane

ϕ_n = Normal pressure angle

$\tan \phi_n = \dfrac{W_r}{W_t / \cos \psi}$

$W_r = \dfrac{W_t \tan \phi_n}{\cos \psi}$

(d) Detail of forces in normal plane

FIGURE 10–4   Helical gear geometry and forces

values for the orthogonal components depend on the following three angles that help define the geometry of the helical gear teeth:

Normal pressure angle: ϕ_n

Transverse pressure angle: ϕ_t

Helix angle: ψ

For helical gears, the helix angle and one of the other two are specified. The third angle can be computed from

$$\tan \phi_n = \tan \phi_t \cos \psi \qquad \textbf{(10–1)}$$

- W_t is the *tangential force* that acts in the transverse plane and tangent to the pitch circle of the helical gear and that causes the torque to be transmitted from the driver to the driven gear. Therefore, this force is often called the *transmitted force*. It is functionally similar to W_t used in the analysis of spur gears in Chapter 8. We can compute its value from the same equations, as follows:

If the torque being transmitted (T) and the size of the gear (D) are known,

$$W_t = T/(D/2) \qquad \textbf{(10–2)}$$

If the power being transmitted (P) and the rotational speed (n) are known,

$$T = (P/n) \qquad \textbf{(10–3)}$$

For the unit-specific situation where power is expressed in horsepower and the rotational speed is in rpm, the torque in lb·in is

$$T = 63\,000(P)/n \qquad \textbf{(10–4)}$$

Then the tangential force can also be expressed as

$$W_t = 63\,000\,(P)/[(n)\,(D/2)] = 126\,000(P)/[(n)\,(D)] \qquad \textbf{(10–5)}$$

If the pitch line speed, v_t (ft/min) of the gear is known, along with the power being transmitted (hp), the tangential load is

$$W_t = 33\,000(P)/v_t \qquad \textbf{(10–6)}$$

The value of the tangential load is the most fundamental of the three orthogonal components of the true normal force. The calculation of the bending stress number and the pitting resistance of the gear teeth depends on W_t.

- W_r is the *radial force* that acts toward the center of the gear perpendicular to the pitch circle and to the tangential force. It tends to push the two gears apart. As can be seen in Figure 10–4(c),

$$W_r = W_t \tan \phi_t \qquad \textbf{(10–7)}$$

where ϕ_t = transverse pressure angle for the helical teeth

- W_x is the *axial force* that acts parallel to the axis of the gear and causes a thrust load that must be resisted by the bearings carrying the shaft. With the tangential force known, the axial force is computed from

$$W_x = W_t \tan \psi \qquad \textbf{(10–8)}$$

Example Problem 10–1   A helical gear has a normal diametral pitch of 8, a normal pressure angle of 20°, 32 teeth, a face width of 3.00 in, and a helix angle of 15°. Compute the diametral pitch, the transverse pressure angle, and the pitch diameter. If the gear is rotating at 650 rpm while transmitting 7.50 hp, compute the pitch line speed, the tangential force, the axial force, and the radial force.

Solution   **Diametral Pitch [Equation (8–28)]**

$$P_d = P_{nd} \cos \psi = 8 \cos(15) = 7.727$$

Transverse Pressure Angle [Equation (10–1)]

$$\phi_t = \tan^{-1}(\tan \phi_n / \cos \psi)$$
$$\phi_t = \tan^{-1}[\tan(20)/\cos(15)] = 20.65°$$

Pitch Diameter [Equation (8–27)]

$$D = N/P_d = 32/7.727 = 4.141 \text{ in}$$

Pitch Line Speed, v_t [Equation (9–1)]

$$v_t = \pi D n/12 = \pi(4.141)(650)/12 = 704.7 \text{ ft/min}$$

Tangential Force, W_t [Equation (10–6)]

$$W_t = 33\,000(P)/v_t = 33\,000(7.5)/704.7 = 351 \text{ lb}$$

Axial Force, W_x [Equation (10–8)]

$$W_x = W_t \tan \psi = 351 \tan(15) = 94 \text{ lb}$$

Radial Force, W_r [Equation (10–7)]

$$W_r = W_t \tan \phi_t = 351 \tan(20.65) = 132 \text{ lb}$$

10–3 STRESSES IN HELICAL GEAR TEETH

We will use the same basic equation for computing the bending stress number for helical gear teeth as we did for spur gear teeth in Chapter 9, given in Equation (9–15) and repeated here:

$$s_t = \frac{W_t P_d}{FJ} K_o K_s K_m K_B K_v \tag{9–15}$$

Figures 10–5, 10–6, and 10–7 show the values for the geometry factor, J, for helical gear teeth with 15°, 20°, and 22° normal pressure angles, respectively.[1] The K factors are the same as those used for spur gears. See References 9 and 18 and the following locations for values:

[1]Figures 10–5, 10–6, and 10–7:
Graphs for the geometry factor, J, for helical gears have been taken from AGMA Standard 218.01-1982, *Standard for Rating the Pitting Resistance and Bending Strength of Spur and Helical Involute Gear Teeth,* with the permission of the publisher, American Gear Manufacturers Association, 1500 King Street, Suite 201, Alexandria, VA 22314. This standard has been superseded by two standards: (1) Standard 908-B89, *Geometry Factors for Determining the Pitting Resistance and Bending Strength of Spur, Helical and Herringbone Gear Teeth,* 1989. (2) Standard 2001-C95, *Fundamental Rating Factors and Calculation Methods for Involute Spur and Helical Gear Teeth,* 1995. The method of calculating the value for J has not been changed. However, the new standards do not contain the graphs. Users are cautioned to ensure that geometry factors for a given design conform to the specific cutter geometry used to manufacture the gears. Standards 908-B89 and 2001-C95 should be consulted for the details of computing the values for J and for rating the performance of the gear teeth.

FIGURE 10–5

Geometry factor (J) for 15° normal pressure angle

(a) Geometry factor (J) for 15° normal pressure angle and indicated addendum

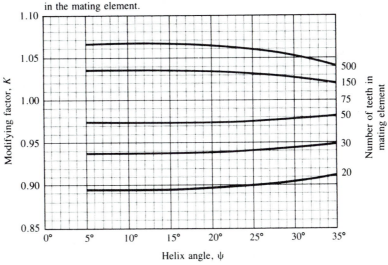

(b) J factor multipliers

K_o = overload factor   (Table 9–6)

K_s = size factor   (Table 9–7)

K_m = load-distribution factor   [Figures 9–16 and 9–17 and Equation (9–16)]

K_B = rim thickness factor   (Figure 9–18)

K_v = dynamic factor   (Figure 9–19)

FIGURE 10–6

Geometry factor (J) for 20° normal pressure angle

(*a*) Geometry factor (J) for 20° normal pressure angle, standard addendum, and finishing hob

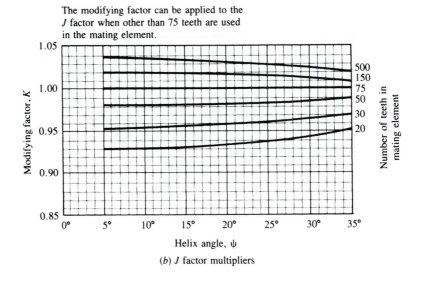

(*b*) J factor multipliers

For design, a material must be specified that has an allowable bending stress number, s_{at}, greater than the computed bending stress number, s_t. Design values of s_{at} can be found:

   Figure 9–8: Steel, through-hardened, Grades 1 and 2

   Table 9–3: Case-hardened steels

   Figures 9–12 and 9–13: Nitrided gears

FIGURE 10–7

Geometry factor (*J*) for
22° normal pressure
angle

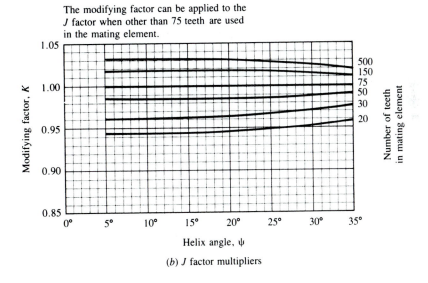

(*a*) Geometry factor (*J*), for 22° normal pressure angle, standard addendum, and pre-shave hob

(*b*) *J* factor multipliers

Table 9–4: Cast iron and bronze

Table 9–5: Plastics

(See also References 11, 16, and 21.) The data for steel, iron, and bronze apply to a design
life of 10^7 cycles at a reliability of 99% (fewer than one failure in 100). If other values for
design life or reliability are desired, the allowable stress can be modified using the procedure described in Section 9–9.

TABLE 10–1   Geometry factors for pitting resistance, *I*, for helical gears with 20° normal pressure angle and standard addendum

A. Helix angle $\psi = 15.0°$

Gear teeth	Pinion teeth				
	17	21	26	35	55
17	0.124				
21	0.139	0.128			
26	0.154	0.143	0.132		
35	0.175	0.165	0.154	0.137	
55	0.204	0.196	0.187	0.171	0.143
135	0.244	0.241	0.237	0.229	0.209

B. Helix angle $\psi = 25.0°$

Gear teeth	Pinion teeth					
	14	17	21	26	35	55
14	0.123					
17	0.137	0.126				
21	0.152	0.142	0.130			
26	0.167	0.157	0.146	0.134		
35	0.187	0.178	0.168	0.156	0.138	
55	0.213	0.207	0.199	0.189	0.173	0.144
135	0.248	0.247	0.244	0.239	0.230	0.210

SOURCE: Extracted from AGMA Standard 908-B89, *Geometry Factors for Determining the Pitting Resistance and Bending Strength of Spur, Helical and Herringbone Gear Teeth,* with the permission of the publisher, American Gear Manufacturers Association, 1500 King Street, Suite 201, Alexandria, VA 22314.

**10–4
PITTING
RESISTANCE
FOR HELICAL
GEAR TEETH**

Pitting resistance for helical gear teeth is evaluated using the same procedure as that discussed in Chapter 9 for spur gears. Equation (9–25) is repeated here:

$$s_c = C_p \sqrt{\frac{W_t K_o K_s K_m K_v}{F D_p I}}$$

$$(9–25)$$

All of the factors are the same for helical gears except the geometry factor for pitting resistance, *I*. The values for C_p are found in Table 9–10. Note that the other *K* factors have the same values as the *K* factors discussed and identified in Section 10–3.

Because of the larger variety of geometric features needed to define the form of helical gears, it is not reasonable to reproduce all of the necessary tables of values or the complete formulas for computing *I*. Values change with the gear ratio, the number of teeth in the pinion, the tooth form, the helix angle, and the specific values for addendum, whole depth, and fillet radius. See References 6 and 13 for extensive discussions about the procedures. To facilitate problem solving in this book, Tables 10–1 and 10–2 give a few values for *I*.

For design, when the computed contact stress number is known, a material must be specified that has an allowable contact stress number, s_{ac}, greater than s_c. Design values for s_{ac} can be found from the following:

Figure 9–9: Steel, through-hardened, Grades 1 and 2

Table 9–3: Steel, case-hardened, Grades 1, 2, and 3; flame- or induction-hardened, carburized, or nitrided

Table 9–4: Cast iron and bronze

TABLE 10–2 Geometry factors for pitting resistance, *I*, for helical gears with 25° normal pressure angle and standard addendum

A. Helix angle $\psi = 15.0°$

Gear teeth	Pinion teeth					
	14	17	21	26	35	55
14	0.130					
17	0.144	0.133				
21	0.160	0.149	0.137			
26	0.175	0.165	0.153	0.140		
35	0.195	0.186	0.175	0.163	0.143	
55	0.222	0.215	0.206	0.195	0.178	0.148
135	0.257	0.255	0.251	0.246	0.236	0.214

B. Helix angle $\psi = 25.0°$

Gear teeth	Pinion teeth						
	12	14	17	21	26	35	55
12	0.129						
14	0.141	0.132					
17	0.155	0.146	0.135				
21	0.170	0.162	0.151	0.138			
26	0.185	0.177	0.166	0.154	0.141		
35	0.203	0.197	0.188	0.176	0.163	0.144	
55	0.227	0.223	0.216	0.207	0.196	0.178	0.148
135	0.259	0.258	0.255	0.251	0.246	0.235	0.213

SOURCE: Extracted from AGMA Standard 908-B89, *Geometry Factors for Determining the Pitting Resistance and Bending Strength of Spur, Helical and Herringbone Gear Teeth,* with the permission of the publisher, American Gear Manufacturers Association, 1500 King Street, Suite 201, Alexandria, VA 22314.

The data from these sources apply to a design life of 10^7 cycles at a reliability of 99% (fewer than one failure in 100). If other values for design life or reliability are desired, or if a service factor is to be applied, the allowable contact stress number can be modified using the procedure described in Section 9–11.

10–5
DESIGN OF
HELICAL GEARS

The example problem that follows illustrates the procedure to design helical gears.

Example Problem 10–2

A pair of helical gears for a milling machine drive is to transmit 65 hp with a pinion speed of 3 450 rpm and a gear speed of 1 100 rpm. The power is from an electric motor. Design the gears.

Solution

Of course, there are several possible solutions. Here is one. Let's try a normal diametral pitch of 12, 24 teeth in the pinion, a helix angle of 15°, a normal pressure angle of 20°, and a quality number of 8.

Now compute the transverse diametral pitch, the axial pitch, the transverse pressure angle, and the pitch diameter. Then we will choose a face width that will give at least two axial pitches to ensure true helical action.

$$P_d = P_{dn} \cos \psi = 12 \cos(15°) = 11.59$$

$$P_x = \frac{\pi}{P_d \tan \psi} = \frac{\pi}{11.59 \tan(15°)} = 1.012 \text{ in}$$

$$\phi_t = \tan^{-1}(\tan \phi_n / \cos \psi) = \tan^{-1}[\tan(20°)/\cos(15°)] = 20.65°$$

$$d = N_P / P_d = 24/11.59 = 2.071 \text{ in}$$

$$F = 2P_x = 2(1.012) = 2.024 \text{ in} \quad \text{(nominal face width)}$$

Let's use 2.25 in, a more convenient value. The pitch line speed and the transmitted load are

$$v_t = \pi d n / 12 = \pi(2.071)(3\,450)/12 = 1\,871 \text{ ft/min}$$

$$W_t = 33\,000 \text{ (hp)}/v_t = 33\,000(65)/1\,871 = 1\,146 \text{ lb}$$

Now we can calculate the number of teeth in the gear:

$$VR = N_G / N_P = n_P / n_G = 3\,450/1\,100 = 3.14$$

$$N_G = N_P (VR) = 24(3.14) = 75 \text{ teeth} \quad \text{(integer value)}$$

The values for the factors in Equation (9–15) must now be determined to enable the calculation of the bending stress. The geometry factor for the pinion is found in Figure 10–6 for 24 teeth in the pinion and 75 teeth in the gear: $J_P = 0.48$. The value of J_G will be greater than the value of J_P, resulting in a lower stress in the gear.

The K factors are

K_o = overload factor = 1.5   (moderate shock)

K_s = size factor = 1.0

K_m = load-distribution factor = 1.26 for F/D = 1.09 and commercial-quality, enclosed gearing

K_B = rim thickness factor = 1.0   (solid gears)

K_v = dynamic factor = 1.35   for Q_v = 8 and v_t = 1 871 ft/min

The bending stress in the pinion can now be computed:

$$s_{tP} = \frac{W_t P_d}{F J_P} K_o K_s K_m K_B K_v \tag{9–15}$$

$$s_{tP} = \frac{(1\,146)(11.59)}{(2.25)(0.48)} (1.50)(1.0)(1.26)(1.0)(1.35) = 31\,400 \text{ psi}$$

From Figure 9–8, a Grade 1 steel with a hardness of approximately 250 HB would be required. Let's proceed to the design for pitting resistance.

Use Equation (9–25):

$$s_c = C_p \sqrt{\frac{W_t K_o K_s K_m K_v}{F D_p I}} \tag{9–25}$$

For two steel gears, $C_p = 2\,300$. Rough interpolation from the data in Table 10–1 for $N_P = 24$ and $N_G = 75$ gives $I = 0.202$. It is recommended that the computational procedure described in the AGMA standards be used to compute a more precise value for critical work. The contact stress is then

$$s_c = 2\,300 \sqrt{\frac{(1\,146)(1.50)(1.0)(1.26)(1.35)}{(2.25)(2.071)(0.202)}} = 128\,200 \text{ psi}$$

Figure 9–9 indicates that a Grade 1 steel with a hardness of 310 HB would be recommended. Assuming that standard life and reliability factors are acceptable, we could specify AISI 5150 OQT 1000, which has a hardness of 321 HB as listed in Appendix 3.

Comment   It is obvious that the contact stress governs this design. Let's adjust the solution for a higher reliability and to account for the expected number of cycles of operation. Certain design decisions must be made. For example, consider the following:

Design for a reliability of 0.999 (less than one failure in 1 000): $K_R = 1.25$ (Table 9–9).

Design life: Let's design for 10 000 h of life as suggested in Table 9–8 for multipurpose gearing. Then, using Equation (9–18), we can compute the number of cycles of loading. For the pinion rotating at 3 450 rpm with one cycle of loading per revolution,

$$N_c = (60)(L)(n)(q) = (60)(10\ 000)(3\ 450)(1.0) = 2.1 \times 10^9 \text{ cycles}$$

From Figure 9–22, we find that $Z_N = 0.89$.

No unusual conditions seem to exist in this application beyond those already considered in the various K factors. Therefore, we use a service factor, SF, of 1.00.

We can use Equation (9–27) to apply these factors. For the pinion, we use $C_H = 1.00$:

$$\frac{K_R(SF)}{Z_N C_H} s_c = s_{ac} = \frac{(1.25)(1.00)}{(0.89)(1.00)}(128\ 200 \text{ psi}) = 180\ 000 \text{ psi}$$

Table 9–3 indicates that Grade 1 steel, case-hardened by carburizing, would be suitable. From Appendix 5, let's specify AISI 4320 SOQT 450, having a case hardness of HRC 59 and a core hardness of 415 HB. This should be satisfactory for both bending and pitting resistance. Both the pinion and the gear should be of this material. There is a modest difference in the Z_N factor, but it should not lower the required allowable contact stress number below that requiring case hardening. Also, when both the pinion and the gear are case-hardened, the hardness ratio factor, C_H, is 1.00.

10–6
FORCES
ON STRAIGHT
BEVEL GEARS

Review Section 8–8 and Figure 8–22 for the geometry of bevel gears. Also see References 1, 5, and 12.

    Because of the conical shape of bevel gears and because of the involute-tooth form, a three-component set of forces acts on bevel gear teeth. Using notation similar to that for helical gears, we will compute the tangential force, W_t; the radial force, W_r; and the axial force, W_x. It is assumed that the three forces act concurrently at the midface of the teeth and on the pitch cone (see Figure 10–8). Although the actual point of application of the resultant force is a little displaced from the middle, no serious error results.

    The tangential force acts tangential to the pitch cone and is the force that produces the torque on the pinion and the gear. The torque can be computed from the known power transmitted and the rotational speed:

$$T = 63\ 000\ P/n$$

Then, using the pinion, for example, the transmitted load is

$$W_{tP} = T/r_m \tag{10–9}$$

where r_m = mean radius of the pinion

FIGURE 10–8

Forces on bevel gears

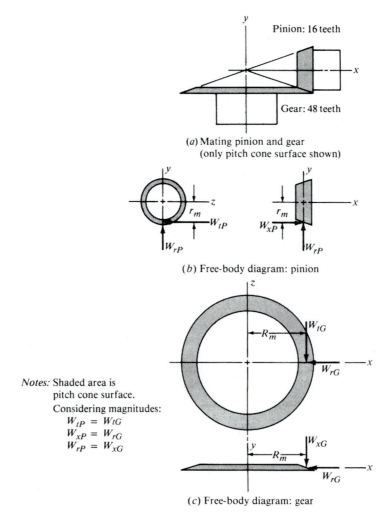

(a) Mating pinion and gear
(only pitch cone surface shown)

(b) Free-body diagram: pinion

Notes: Shaded area is pitch cone surface.
Considering magnitudes:
$W_{tP} = W_{tG}$
$W_{xP} = W_{rG}$
$W_{rP} = W_{xG}$

(c) Free-body diagram: gear

The value of r_m can be computed from

$$r_m = d/2 - (F/2)\sin \gamma \qquad \qquad \textbf{(10–10)}$$

Remember that the pitch diameter, d, is measured to the pitch line of the tooth at its large end.

The radial load acts toward the center of the pinion, perpendicular to its axis, causing bending of the pinion shaft. Thus,

$$W_{rP} = W_t \tan \phi \cos \gamma \qquad \qquad \textbf{(10–11)}$$

The axial load acts parallel to the axis of the pinion, tending to push it away from the mating gear. It causes a thrust load on the shaft bearings. It also produces a bending moment on the shaft because it acts at the distance from the axis equal to the mean radius of the gear. Thus,

$$W_{xP} = W_t \tan \phi \sin \gamma \qquad \qquad \textbf{(10–12)}$$

The values for the forces on the gear can be calculated by the same equations shown here for the pinion, if the geometry for the gear is substituted for that of the pinion. Refer to Figure 10–8 for the relationships between the forces on the pinion and the gear in both magnitude and direction.

Example Problem 10–3　　For the gear pair described in Example Problem 8–6, calculate the forces on the pinion and the gear if they are transmitting 2.50 hp with a pinion speed of 600 rpm. The geometry factors computed in Example Problem 8–6 apply. The data are summarized below.

Summary of Pertinent Results from Example Problem 8–6 and Given Data

Number of teeth in the pinion: $N_P = 16$

Number of teeth in the gear: $N_G = 48$

Diametral pitch: $P_d = 8$

Pitch diameter of pinion: $d = 2.000$ in

Pinion pitch cone angle: $\gamma = 18.43°$

Gear pitch cone angle: $\Gamma = 71.57°$

Face width: $F = 1.00$ in

Rotational speed of pinion: $n_P = 600$ rpm

Power transmitted: $P = 2.50$ hp

Solution　　Forces on the pinion are described by the following equations:

$$W_t = T/r_m$$

But

$$T_p = 63\ 000(P)/n_p = [63\ 000(2.50)]/600 = 263 \text{ lb·in}$$
$$r_m = d/2 - (F/2)\sin \gamma$$
$$r_m = (2.000/2) - (1.00/2)\sin(18.43°) = 0.84 \text{ in}$$

Then

$$W_t = T_p/r_m = 263 \text{ lb·in}/0.84 \text{ in} = 313 \text{ lb}$$
$$W_r = W_t \tan \phi \cos \gamma = 313 \text{ lb} \tan(20°)\cos(18.43°) = 108 \text{ lb}$$
$$W_x = W_t \tan \phi \sin \gamma = 313 \text{ lb} \tan(20°)\sin(18.43°) = 36 \text{ lb}$$

To determine the forces on the gear, first let's calculate the rotational speed of the gear:

$$n_G = n_P(N_P/N_G) = 600 \text{ rpm}(16/48) = 200 \text{ rpm}$$

Then

$$T_G = 63\ 000(2.50)/200 = 788 \text{ lb·in}$$
$$R_m = D/2 - (F/2)\sin \Gamma$$
$$R_m = 6.000/2 - (1.00/2)\sin(71.57°) = 2.53 \text{ in}$$
$$W_t = T_G/R_m = (788 \text{ lb·in})/(2.53 \text{ in}) = 313 \text{ lb}$$
$$W_r = W_t \tan \phi \cos \Gamma = 313 \text{ lb} \tan(20°)\cos(71.57°) = 36 \text{ lb}$$
$$W_x = W_t \tan \phi \sin \Gamma = 313 \text{ lb} \tan(20°)\sin(71.57°) = 108 \text{ lb}$$

Note from Figure 10–8 that the forces on the pinion and the gear form an *action-reaction pair.* That is, the forces on the gear are equal to those on the pinion, but they act in the opposite direction. Also, because of the 90° orientation of the shafts, the radial force on the pinion becomes the axial thrust load on the gear, and the axial thrust load on the pinion becomes the radial load on the gear.

**10–7
BEARING
FORCES ON
SHAFTS
CARRYING
BEVEL GEARS**

Because of the three-dimensional force system that acts on bevel gears, the calculation of the forces on shaft bearings can be cumbersome. An example is worked out here to show the procedure. In order to obtain numerical data, the arrangement shown in Figure 10–9 is proposed for the bevel gear pair that was the subject of Example Problems 8–6 and 10–3. The locations for the bearings are given with respect to the apex of the two pitch cones where the shaft axes intersect.

Note that both the pinion and the gear are *straddle-mounted;* that is, each gear is positioned between the supporting bearings. This is the most preferred arrangement because it usually provides the greatest rigidity and maintains the alignment of the teeth during power transmission. Care should be exercised to provide rigid mountings and stiff shafts when using bevel gears.

The arrangement of Figure 10–9 is designed so that the bearing on the right resists the axial thrust load on the pinion, and the lower bearing resists the axial thrust load on the gear.

Example Problem 10–4   Compute the reaction forces on the bearings that support the shafts carrying the bevel gear pair shown in Figure 10–9. The values of Example Problems 8–6 and 10–3 apply.

Solution   Referring to the results of Example Problem 10–3 and Figure 10–8, we have listed the forces acting on the gears:

Force	Pinion	Gear
Tangential	$W_{tP} = 313$ lb	$W_{tG} = 313$ lb
Radial	$W_{rP} = 108$ lb	$W_{rG} = 36$ lb
Axial	$W_{xP} = 36$ lb	$W_{xG} = 108$ lb

FIGURE 10–9
Layout of bevel gear
pair for Example
Problem 10–4

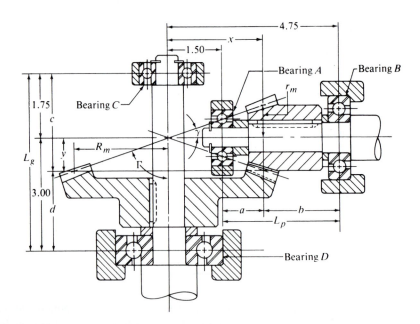

It is critical to be able to visualize the directions in which these forces are acting because of the three-dimensional force system. Notice in Figure 10–8 that a rectangular coordinate system has been set up. Figure 10–10 is an isometric sketch of the free-body diagrams of the pinion and the gear, simplified to represent the concurrent forces acting at the pinion/gear interface and at the bearing locations. Although the two free-body diagrams are separated for clarity, notice that you can bring them together by moving the point called *vertex* on each sketch together. This is the point in the actual gear system where the vertices of the two pitch cones lie at the same point. The two pitch points also coincide.

For setting up the equations of static equilibrium needed to solve for the bearing reactions, the distances $a, b, c, d, L_P,$ and L_G are needed, as shown in Figure 10–9. These require the two dimensions labeled x and y. Note from Example Problem 10–3 that

$$x = R_m = 2.53 \text{ in}$$
$$y = r_m = 0.84 \text{ in}$$

Then

$$a = x - 1.50 = 2.53 - 1.50 = 1.03 \text{ in}$$
$$b = 4.75 - x = 4.75 - 2.53 = 2.22 \text{ in}$$
$$c = 1.75 + y = 1.75 + 0.84 = 2.59 \text{ in}$$
$$d = 3.00 - y = 3.00 - 0.84 = 2.16 \text{ in}$$
$$L_P = 4.75 - 1.50 = 3.25 \text{ in}$$
$$L_G = 1.75 + 3.00 = 4.75 \text{ in}$$

FIGURE 10–10
Free-body diagrams for pinion and gear shafts

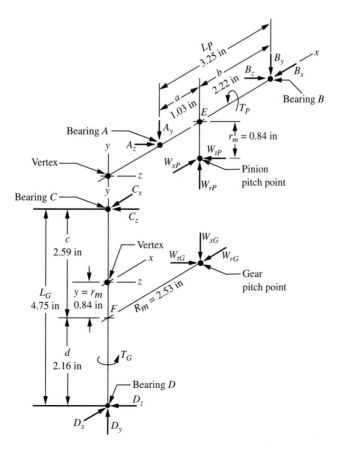

These values are shown on Figure 10–10.

To solve for the reactions, we need to consider the horizontal (x-z) and the vertical (x-y) planes separately. It may help you to look also at Figure 10–11, which breaks out the forces on the pinion shaft in these two planes. Then we can analyze each plane using the fundamental equations of equilibrium.

Bearing Reactions, Pinion Shaft: Bearings A and B

Step 1. To find B_z and A_z: In the x-z plane, only W_{tP} acts. Summing moments about A yields

$$0 = W_{tP}(a) - B_z(L_P) = 313(1.03) - B_z(3.25)$$
$$B_z = 99.2 \text{ lb}$$

Summing moments about B yields

$$0 = W_{tP}(b) - A_z(L_P) = 313(2.22) - A_z(3.25)$$
$$A_z = 214 \text{ lb}$$

Step 2. To find B_y and A_y: In the x-y plane, both W_{rP} and W_{xP} act. Summing moments about A yields

$$0 = W_{rP}(a) + W_{xP}(r_m) - B_y(L_P)$$
$$0 = 108(1.03) + 36(0.84) - B_y(3.25)$$
$$B_y = 43.5 \text{ lb}$$

Summing moments about B yields

$$0 = W_{rP}(b) - W_{xP}(r_m) - A_y(L_P)$$
$$0 = 108(2.22) - 36(0.84) - A_y(3.25)$$
$$A_y = 64.5 \text{ lb}$$

Step 3. To find B_x: Summing forces in the x-direction yields

$$B_x = W_{xP} = 36 \text{ lb}$$

FIGURE 10–11
Pinion shaft bending
moments

(a) Horizontal plane ($x-z$)   (b) Vertical plane ($x-y$)

This is the thrust force on bearing B.

Step 4. To find the total radial force on each bearing: Compute the resultant of the y- and z-components.

$$A = \sqrt{A_y^2 + A_z^2} = \sqrt{64.5^2 + 214^2} = 224 \text{ lb}$$

$$B = \sqrt{B_y^2 + B_z^2} = \sqrt{43.5^2 + 99.2^2} = 108 \text{ lb}$$

Bearing Reactions, Gear Shaft: Bearings C and D

Using similar methods, we can find the forces in Figure 10–12.

$$\left. \begin{array}{l} C_z = 142 \text{ lb} \\ C_x = 41.1 \text{ lb} \end{array} \right\} \quad C = 148 \text{ lb} \quad \text{(radial force on } C\text{)}$$

$$\left. \begin{array}{l} D_z = 171 \text{ lb} \\ D_x = 77.1 \text{ lb} \end{array} \right\} \quad D = 188 \text{ lb} \quad \text{(radial force on } D\text{)}$$

$$D_y = W_{xG} = 108 \text{ lb} \quad \text{(thrust force on } D\text{)}$$

FIGURE 10–12
Gear shaft bending moments

(a) x-y plane

(b) y-z plane

Summary   In selection of the bearings for these shafts, the following capacities are required:

> Bearing *A:* 224-lb radial
> Bearing *B:* 108-lb radial; 36-lb thrust
> Bearing *C:* 148-lb radial
> Bearing *D:* 188-lb radial; 108-lb thrust

**10–8
BENDING
MOMENTS ON
SHAFTS
CARRYING
BEVEL GEARS**

Because there are forces acting in two planes on bevel gears, as discussed in the preceding section, there is also bending in two planes. The analysis of the shearing force and bending moment diagrams for the shafts must take this into account.

Figures 10–11 and 10–12 show the resulting diagrams for the pinion and the gear shafts, respectively, for the gear pair used for Example Problems 8–6, 10–3, and 10–4. Notice that the axial thrust load on each shaft provides a concentrated moment to the shaft equal to the axial load times the distance that it is offset from the axis of the shaft. Also notice that the maximum bending moment for each shaft is the resultant of the moments in the two planes. On the pinion shaft, the maximum moment is 240 lb·in at *E,* where the lines of action for the radial and tangential forces intersect the shaft. Similarly, on the gear shaft, the maximum moment is 404 lb·in at *F.* These data are used in the shaft design (as discussed in Chapter 12).

**10–9
STRESSES IN
STRAIGHT BEVEL
GEAR TEETH**

The stress analysis for bevel gear teeth is similar to that already presented for spur and helical gear teeth. The maximum bending stress occurs at the root of the tooth in the fillet. This stress can be computed from

⇨ **Bending Stress
Number**

$$s_t = \frac{W_t P_d}{FJ} \frac{K_o K_s K_m}{K_v} \qquad (10\text{–}13)$$

The terms have all been used before. But there are minor differences in the manner of evaluating the factors, so they will be reviewed here.

Tangential Force, W_t

Contrary to the way W_t was computed in the preceding section, we compute it here by using the diameter of the gear at its large end, rather than the diameter at the middle of the tooth. This is more convenient, and the adjustment for the actual force distribution on the teeth is made in the value of the geometry factor, *J.* Then

⇨ **Tangential Force
on Bevel Gears**

$$W_t = \frac{T}{r} = \frac{63\,000(P)}{n_P} \frac{1}{d/2} \qquad (10\text{–}14)$$

where T = torque transmitted (lb·in)
  r = pitch radius of the pinion (in)
  P = power transmitted (hp)
  n_P = rotational speed of the pinion (rpm)
  d = pitch diameter of the pinion at its large end (in)

Dynamic Factor, K_v

The values for the dynamic factor for bevel gears are different from those for spur or helical gears. Factors affecting the dynamic factor include the accuracy of manufacture of the gear teeth (quality number Q); the pitch line speed, v_t; the tooth load; and the stiffness of the teeth. AGMA Standard 2003-A86 recommends the following procedure for computing K_v for the bending strength calculation and C_v for pitting resistance:

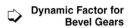

Dynamic Factor for
Bevel Gears

$$C_v = K_v = \left[\frac{K_z}{K_z + \sqrt{v_t}} \right]^u$$

(10–15)

where $u = \dfrac{8}{(2)^{0.5Q}} - s_{at} \left[\dfrac{125}{E_P + E_G} \right]$

$K_z = 85 - 10(u)$

If the equation for u results in a negative value, use $u = 0$. As a check on the specification of an appropriate value for the quality number, a minimum value for C_v should be computed from

$$C_{v\,min} = \frac{2}{\pi} \tan^{-1}(v_t/333)$$

The value of the result of the inverse tangent calculation must be in radians. If the actual value of C_v is less than $C_{v\,min}$, a higher quality number should be specified.

Size Factor, K_s

Use the values from Table 9–7.

Load-Distribution Factor, K_m

Values are highly dependent on the manner of mounting both the pinion and the gear. The preferred mounting is called *straddle mounting,* in which the gear is between its supporting bearings. Figure 10–9 shows straddle mounting of both the pinion and the gear. In addition, stiff, short shafts are recommended to minimize shaft deflections that cause misalignment of the gear teeth.

Refer to AGMA Standard 2003-A86 (see Reference 10) for general methods of evaluating K_m. For enclosed gearing with special care in mounting the gears and in controlling the form of the gear teeth, AGMA Standard 6010-E88 (Reference 13) recommends the values in Table 10–3. The gears should be tested under load to ensure an optimum tooth contact pattern. We will use these values in problem solutions.

Geometry Factor, J

Use Figure 10–13 if the pressure angle is 20° and the shaft angle is 90°.

The computed value for stress from Equation (10–13) can be compared with the allowable bending stress number from Tables 9–3 and 9–4 and Figure 9–8. A life factor, Y_N, or a reliability factor, K_R, can be applied as discussed in Chapter 9 if the design life is different from 10^7 cycles or if the desired reliability is different from 0.99.

TABLE 10–3 Load-distribution factors for bevel gears, K_m

Type of gearing	Both gears straddle-mounted	One gear straddle-mounted	Neither gear straddle-mounted
General commercial-quality	1.44	1.58	1.80
High-quality, commercial gearing	1.20	1.32	1.50

SOURCE: Extracted from AGMA 6010-E88, *Standard for Spur, Helical, Herringbone and Bevel Enclosed Drives,* with the permission of the publisher, American Gear Manufacturers Association, 1500 King Street, Suite 201, Alexandria, VA 22314.

FIGURE 10–13 Geometry factor, *J,* for straight bevel gears with 20° pressure angle and $0.120/P_d$ tool edge radius (Extracted from AGMA 6010-E88, *Standard for Spur, Helical, Herringbone and Bevel Enclosed Drives,* with permission of the publisher, American Gear Manufacturers Association, 1500 King Street, Suite 201, Alexandria, VA 22314)

Example Problem 10–5   Compute the bending stress in the teeth of the bevel pinion shown in Figure 10–9. The data from Example Problem 10–3 apply: $N_P = 16$; $N_G = 48$; $n_P = 600$ rpm; $P = 2.50$ hp; $P_d = 8$; $d = 2.000$ in; $F = 1.00$ in. Assume that the pinion is driven by an electric motor and that the load provides moderate shock. The quality number, Q_v, is to be 6.

Solution

$$W_t = \frac{T}{r} = \frac{63\,000(P)}{n_p}\frac{1}{d/2} = \frac{63\,000(2.50)}{600}\frac{1}{2.000/2} = 263 \text{ lb}$$

$$v_t = \pi d n_P/12 = \pi(2.000)(600)/12 = 314 \text{ ft/min}$$

$$K_o = 1.50 \quad \text{(from Table 9–6)}$$

$$K_s = 1.00$$

$$K_m = 1.44 \quad \text{(both gears straddle-mounted, general commercial-quality)}$$

$$J = 0.230 \quad \text{(from Figure 10–13)}$$

The value for the dynamic factor, K_v, must be computed from Equation (10–15), for $Q = 6$ and $v_t = 314$ ft/min. As a design decision, let's use two Grade 1 steel gears that are through-hardened at 300 HB with $s_{at} = 36\,000$ psi (Figure 9–8). The modulus of elasticity for both gears is 30×10^6 psi. Then

$$u = \frac{8}{(2)^{0.5(6)}} - (36\,000)\left[\frac{125}{60 \times 10^6}\right] = 0.85$$

$$K_z = 85 - 10(u) = 85 - 10(0.85) = 76.5$$

$$C_v = K_v = \left[\frac{K_z}{K_z + \sqrt{v_t}}\right]^u = \left[\frac{76.5}{76.5 + \sqrt{314}}\right]^{0.85} = 0.83$$

Checking $C_{v\,\min} = (2/\pi)\tan^{-1}(314/333) = 0.481$. The value of C_v is acceptable.
   Then, from Equation (12–22),

$$s_t = \frac{W_t P_d}{FJ}\frac{K_o K_s K_m}{K_v} = \frac{(263)(8)}{(1.00)(0.230)}\frac{(1.50)(1.00)(1.44)}{(0.83)} = 23\,800 \text{ psi}$$

Referring to Figure 9–8, note that this is a very modest stress level for steel gears, requiring only about HB 180 for the hardness. If stress were the only consideration, a redesign might be attempted to achieve a more compact system. However, normally the pitting resistance or surface durability of the teeth requires a harder material. The next section discusses pitting.

10–10
DESIGN OF BEVEL GEARS FOR PITTING RESISTANCE

The approach to the design of bevel gears for pitting resistance is similar to that for spur gears. The failure mode is fatigue of the surface of the teeth under the influence of the contact stress between the mating gears.
   The contact stress, called the *Hertz stress, s_c,* can be computed from

⇨ **Contact Stress Number**

$$s_c = C_p C_b \sqrt{\frac{W_t}{FdI}\frac{C_o C_m}{C_v}}$$

(10–16)

The factors C_o, C_v, and C_m are the same as K_o, K_v, and K_m, used for computing the bending stress in the preceding section. The terms W_t, F, and d have the same meanings, also. The

FIGURE 10–14
Geometry factors for straight and ZEROL® bevel gears (Extracted from AGMA 2003-A86, *Rating the Pitting Resistance and Bending Strength of Generated Straight Bevel, ZEROL® Bevel, and Spiral Bevel Gear Teeth,* with the permission of the publisher, American Gear Manufacturers Association, 1500 King Street, Suite 201, Alexandria, VA 22314)

factor C_p is the elastic coefficient and is the same as shown in Equation (9–23) and Table 9–10. For steel or cast iron gears,

$$C_p = 2\,300 \text{ for two steel gears} \quad (E = 30 \times 10^6 \text{ psi})$$
$$C_p = 1\,960 \text{ for two cast iron gears} \quad (E = 19 \times 10^6 \text{ psi})$$
$$C_p = 2\,100 \text{ for a steel pinion and a cast iron gear}$$

Using $C_b = 0.634$ allows the use of the same allowable contact stress as for spur and helical gears.

The factor I is the geometry factor for surface durability and can be found from Figure 10–14.

The Hertz contact stress, computed from Equation (10–16) should be compared with the allowable contact stress number, s_{ac}, from Figure 9–9 or Table 9–3 if the material is steel. For cast iron, use the values in Table 9–4.

Example Problem 10–6

Compute the Hertz stress for the gear pair in Figure 10–9 for the conditions used in Example Problem 10–5: $N_P = 16$; $N_G = 48$; $n_P = 600$ rpm; $P_d = 8$; $F = 1.00$ in; $d = 2.000$ in. Both gears are to be steel. Specify a suitable steel for the gears, along with its heat treatment.

Solution

From Example Problem 10–5: $W_t = 263$ lb, $C_o = 1.50$, $C_v = 0.83$, and $C_m = 1.44$. For two steel gears, $C_p = 2\,300$. From Figure 10–14, $I = 0.077$. Then

$$s_c = C_p C_b \sqrt{\frac{W_t}{FdI} \frac{C_o C_m}{C_v}} = (2\,300)(0.634) \sqrt{\frac{263}{(1.00)(2.000)(0.077)} \frac{(1.50)(1.44)}{0.83}}$$

$$s_c = 97\,200 \text{ psi}$$

Spiral bevel
gearing, first stage

Input shaft

Helical gearing
second and third stages

Fabricated steel housing

Fabricated steel housing

FIGURE 10–15   Spiral bevel, right-angle reducer (Sumitomo Machinery Corporation of America, Teterboro, NJ)

FIGURE 10–16
Final drive for a tractor
(Ford Tractor
Operations, Troy, MI)

Comparing this value with the allowable contact stress number from Figure 9–9 shows that a through-hardened, Grade 1 steel with a hardness of HB 220 is capable of withstanding this level of stress. Because this value is higher than that required for bending strength, it controls the design.

Practical Considerations for Bevel Gearing

Factors similar to those discussed for spur and helical gears should be considered in the design of systems using bevel gears. The accuracy of alignment and the accommodation of thrust loads discussed in the example problems are critical. Figures 10–15 and 10–16 show commercial applications.

**10–11
FORCES,
FRICTION, AND
EFFICIENCY IN
WORMGEAR
SETS**

See Chapter 8 for the geometry of wormgear sets. Also see References 2, 14, 15, and 17.

The force system acting on the worm/wormgear set is usually considered to be made of three perpendicular components as was done for helical and bevel gears. There are a tangential force, a radial force, and an axial force acting on the worm and the wormgear. We will use the same notation here as in the bevel gear system.

Figure 10–17 shows two orthogonal views (front and side) of a worm/wormgear pair, showing only the pitch diameters of the gears. The figure shows the separate worm and

FIGURE 10–17

Forces on a worm and a wormgear

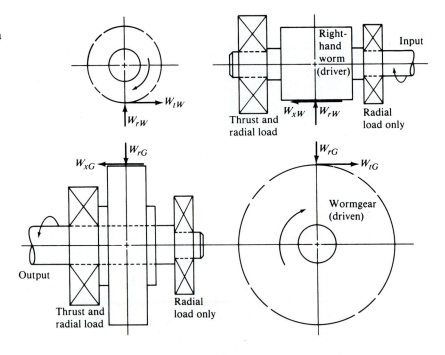

wormgear with the forces acting on each. Note that because of the 90° orientation of the two shafts,

> **Forces on Worms and Wormgears**

$$
\left.
\begin{array}{l}
W_{tG} = W_{xW} \\
W_{xG} = W_{tW} \\
W_{rG} = W_{rW}
\end{array}
\right\}
\tag{10–17}
$$

Of course, the directions of the paired forces are opposite because of the action/reaction principle.

The tangential force on the wormgear is computed first and is based on the required operating conditions of torque, power, and speed at the output shaft.

Coefficient of Friction, μ

Friction plays a major part in the operation of a wormgear set because there is inherently sliding contact between the worm threads and the wormgear teeth. The coefficient of friction is dependent on the materials used, the lubricant, and the sliding velocity. Based on the pitch line speed of the gear, the sliding velocity is

> **Sliding Velocity for Gear**

$$
v_s = v_{tG}/\sin \lambda
\tag{10–18}
$$

Based on the pitch line speed of the worm,

> **Sliding Velocity for Worm**

$$
v_s = v_{tW}/\cos \lambda
\tag{10–19}
$$

The AGMA (see Reference 15) recommends the following formulas to estimate the coefficient of friction for a hardened steel worm (58 HRC minimum), smoothly ground, or polished, or rolled, or with an equivalent finish, operating on a bronze wormgear. The

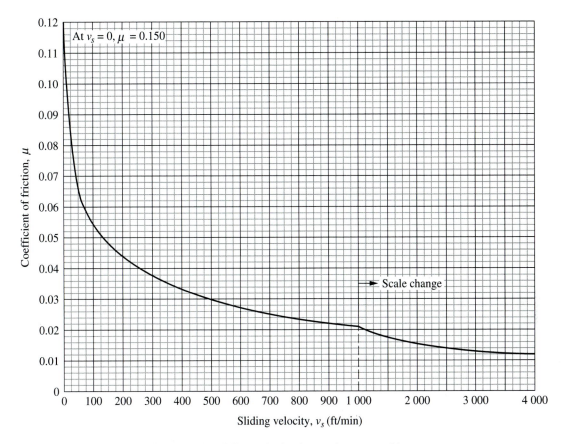

FIGURE 10–18 Coefficient of friction vs. sliding velocity for steel worm and bronze wormgear

choice of formula depends on the sliding velocity. *Note:* v_s must be in ft/min in the formulas; 1.0 ft/min = 0.0051 m/s.

Static Condition: $v_s = 0$

$$\mu = 0.150$$

Low Speed: $v_s < 10$ ft/min (0.051 m/s)

$$\mu = 0.124e^{(-0.074v_s^{0.645})} \qquad \textbf{(10–20)}$$

Higher Speed: $v_s > 10$ ft/min

$$\mu = 0.103e^{(-0.110v_s^{0.450})} + 0.012 \qquad \textbf{(10–21)}$$

Figure 10–18 is a plot of the coefficient μ versus the sliding velocity v_s.

Output Torque from Wormgear Drive, T_o

In most design problems for wormgear drives, the output torque and the rotating speed of the output shaft will be known from the requirements of the driven machine. Torque and speed are related to the output power by

**Output Torque
from Wormgear**

$$T_o = \frac{63\,000(P_o)}{n_G}$$

By referring to the end view of the wormgear in Figure 10–17, you can see that the output torque is

$$T_o = W_{tG} \cdot r_G = W_{tG}(D_G/2)$$

Then the following procedure can be used to compute the forces acting in a worm/wormgear drive system.

Procedure for Calculating the Forces on a Worm/Wormgear Set

Given:

Output torque, T_o, in lb·in

Output speed, n_G, in rpm

Pitch diameter of the wormgear, D_G, in inches

Lead angle, λ

Normal pressure angle, ϕ_n

Compute:

$$W_{tG} = 2T_o/D_G$$

$$W_{xG} = W_{tG} \frac{\cos \phi_n \sin \lambda + \mu \cos \lambda}{[\cos \phi_n \cos \lambda - \mu \sin \lambda]} \tag{10–22}$$

$$W_{rG} = \frac{W_{tG} \sin \phi_n}{\cos \phi_n \cos \lambda - \mu \sin \lambda} \tag{10–23}$$

The forces on the worm can be obtained by observation, using Equation (10–17). Equations (10–22) and (10–23) were derived using the components of both the tangential driving force on the wormgear and the friction force at the location of the meshing worm threads and wormgear teeth. The complete development of the equations is shown in Reference 20.

Friction Force, W_f

The friction force, W_f, acts parallel to the face of the worm threads and the gear teeth and depends on the tangential force on the gear, the coefficient of friction, and the geometry of the teeth:

$$W_f = \frac{\mu W_{tG}}{(\cos \lambda)(\cos \phi_n)} \tag{10–24}$$

Power Loss Due to Friction, P_L

Power loss is the product of the friction force and the sliding velocity at the mesh. That is,

$$P_L = \frac{v_s W_f}{33\,000}$$

In this equation, the power loss is in hp, v_s is in ft/min, and W_f is in in·lb.

Input Power, P_i

The input power is the sum of the output power and the power loss due to friction:

$$P_i = P_o + P_L$$

Efficiency, η

Efficiency is defined as the ratio of the output power to the input power:

$$\eta = P_o/P_i$$

Factors Affecting Efficiency

As can be seen in Equation (10–24), the lead angle, the normal pressure angle, and the coefficient of friction all affect the efficiency. The one that has the largest effect, and the one over which the designer has the most control, is the lead angle, λ. The larger the lead angle, the higher the efficiency, up to approximately $\lambda = 45°$. Now, looking back to the definition of the lead angle, note that the number of threads in the worm has a major effect on the lead angle. Therefore, to obtain a high efficiency, use multiple-threaded worms. But there is a disadvantage to this conclusion. More worm threads require more gear teeth to achieve the same ratio, resulting in a larger system overall. The designer is often forced to compromise.

Example Problem: Forces and Efficiency in Wormgearing

Review now the results of Example Problem 8–7, in which the geometry factors for a particular worm and wormgear set were computed. The following example problem extends the analysis to include the forces acting on the system for a given output torque.

Example Problem 10–7   The wormgear drive described in Example Problem 8–7 is transmitting an output torque of 4 168 lb·in. The transverse pressure angle is 20°. The worm is made from hardened and ground steel, and the wormgear is bronze. Compute the forces on the worm and the wormgear, the output power, the input power, and the efficiency.

Solution   Recall from Example Problem 8–7 that

$$\lambda = 14.04° \qquad D_G = 8.667 \text{ in} \qquad n_G = 101 \text{ rpm}$$
$$n_W = 1\,750 \text{ rpm} \qquad D_W = 2.000 \text{ in}$$

The normal pressure angle is required. From Equation (8–41),

$$\phi_n = \tan^{-1}(\tan \phi_t \cos \lambda) = \tan^{-1}(\tan 20° \cos 14.04°) = 19.45°$$

Because they recur in several formulas, let's compute the following:

$$\sin \phi_n = \sin 19.45° = 0.333$$
$$\cos \phi_n = \cos 19.45° = 0.943$$
$$\cos \lambda = \cos 14.04° = 0.970$$
$$\sin \lambda = \sin 14.04° = 0.243$$
$$\tan \lambda = \tan 14.04° = 0.250$$

We can now compute the tangential force on the wormgear:

$$W_{tG} = \frac{2T_o}{D_G} = \frac{(2)(4\ 168\ \text{lb·in})}{8.667\ \text{in}} = 962\ \text{lb}$$

The calculations of the axial and radial forces require a value for the coefficient of friction that, in turn, depends on the pitch line speed and the sliding velocity.

Pitch Line Speed of the Gear

$$v_{tG} = \pi D_G n_G/12 = \pi(8.667)(101)/12 = 229\ \text{ft/min}$$

Sliding Velocity

$$v_s = v_{tG}/\sin \lambda = 229/\sin 14.04° = 944\ \text{ft/min}$$

Coefficient of Friction

From Figure 10–18, at a sliding velocity of 944 ft/min, we can read $\mu = 0.022$.

Now the axial and radial forces on the wormgear can be computed from Equations (10–22) and (10–23).

Axial Force on the Wormgear

$$W_{xG} = 962\ \text{lb} \left[\frac{(0.943)(0.243) + (0.022)(0.970)}{(0.943)(0.970) - (0.022)(0.243)} \right] = 265\ \text{lb}$$

Radial Force on the Wormgear

$$W_{rG} = \left[\frac{(962)(0.333)}{(0.943)(0.970) - (0.022)(0.243)} \right] = 352\ \text{lb}$$

Now the output power, input power, and efficiency can be computed.

Output Power

$$P_o = \frac{T_o n_G}{63\ 000} = \frac{(4\ 168\ \text{lb·in})(101\ \text{rpm})}{63\ 000} = 6.68\ \text{hp}$$

The input power depends on the friction force and the consequent power loss due to friction.

Friction Force

$$W_f = \frac{\mu W_{tG}}{(\cos \lambda)(\cos \phi_n)} = \frac{(0.022)(962\ \text{lb})}{(0.970)(0.943)} = 23.1\ \text{lb}$$

Power Loss Due to Friction

$$P_L = \frac{v_s W_f}{33\ 000} = \frac{(944\ \text{ft/min})(23.1\ \text{lb})}{33\ 000} = 0.662\ \text{hp}$$

Input Power

$$P_i = P_o + P_L = 6.68 + 0.66 = 7.34\ \text{hp}$$

Efficiency

$$\eta = \frac{P_o}{P_i}(100\%) = \frac{6.68\ \text{hp}}{7.34\ \text{hp}}(100\%) = 91\%$$

Self-locking Wormgear Sets

Self-locking is the condition in which the worm drives the wormgear, but, if torque is applied to the gear shaft, the worm does not turn. It is locked! The locking action is produced by the friction force between the worm threads and the wormgear teeth, and this is highly dependent on the lead angle. It is recommended that a lead angle no higher than about 5.0° be used in order to ensure that self-locking will occur. This lead angle usually requires the use of a single-threaded worm; the low lead angle results in a low efficiency, possibly as low as 60% or 70%.

**10–12
STRESS IN
WORMGEAR
TEETH**

We present here an approximate method of computing the bending stress in the teeth of the wormgear. Because the geometry of the teeth is not uniform across the face width, it is not possible to generate an exact solution. However, the method given here should predict the bending stress with sufficient accuracy to check a design because most worm/wormgear systems are limited by pitting, wear, or thermal considerations rather than strength.

The AGMA, in its Standard 6034-A87, does not include a method of analyzing wormgears for strength. The method shown here was adapted from Reference 20. Only the wormgear teeth are analyzed because the worm threads are inherently stronger and are typically made from a stronger material.

The stress in the gear teeth can be computed from

$$\sigma = \frac{W_d}{yFp_n}$$

(10–25)

where W_d = dynamic load on the gear teeth
y = Lewis form factor   (see Table 10–4)
F = face width of the gear
p_n = normal circular pitch = $p \cos \lambda = \pi \cos \lambda / P_d$

(10–26)

The dynamic load can be estimated from

$$W_d = W_{tG}/K_v$$

(10–27)

and

$$K_v = 1\,200/(1\,200 + v_{tG})$$

(10–28)

$$v_{tG} = \pi D_G n_G/12 = \text{pitch line speed of the gear}$$

(10–29)

Only one value is given for the Lewis form factor for a given pressure angle because the actual value is very difficult to calculate precisely and does not vary much with the

TABLE 10–4   Approximate Lewis form factor for wormgear teeth

ϕ_n	y
$14\frac{1}{2}°$	0.100
20°	0.125
25°	0.150
30°	0.175

number of teeth. The actual face width should be used, up to the limit of two-thirds of the pitch diameter of the worm.

The computed value of tooth bending stress from Equation (10–25) can be compared with the fatigue strength of the material of the gear. For manganese gear bronze, use a fatigue strength of 17 000 psi; for phosphor gear bronze, use 24 000 psi. For cast iron, use approximately 0.35 times the ultimate strength, unless specific data are available for fatigue strength.

10–13
SURFACE
DURABILITY OF
WORMGEAR
DRIVES

AGMA Standard 6034-A87 (see Reference 15) gives a method for rating the surface durability of hardened steel worms operating with bronze gears. The ratings are based on the ability of the gears to operate without significant damage from pitting or wear.

The procedure calls for the calculation of a *rated tangential load, W_{tR},* from

▷ **Rated Tangential Load on Wormgears**

$$W_{tR} = C_s D_G^{0.8} F_e C_m C_v \qquad (10\text{–}30)$$

where C_s = materials factor   (from Figure 10–19)
         D_G = pitch diameter of the wormgear, in inches
         F_e = effective face width, in inches. Use the actual face width of the wormgear
            up to a maximum of 0.67 D_W.
         C_m = ratio correction factor   (from Figure 10–20)
         C_v = velocity factor   (from Figure 10–21)

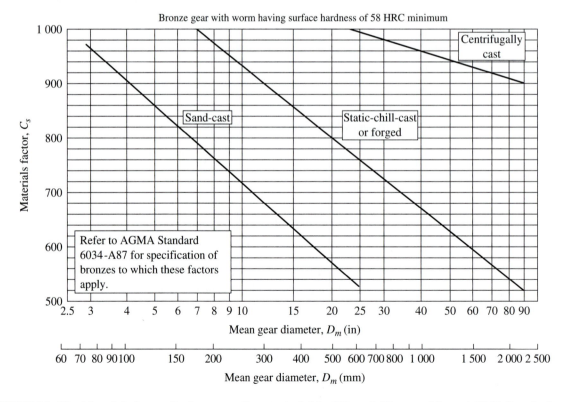

FIGURE 10–19   Materials factor, C_s, for center distance > 3.0 in (76 mm) (Extracted from AGMA Standard 6034-A87, *Practice for Enclosed Cylindrical Wormgear Speed Reducers and Gearmotors,* with the permission of publisher, American Gear Manufacturers Association, 1500 King Street, Suite 201, Alexandria, VA 22314)

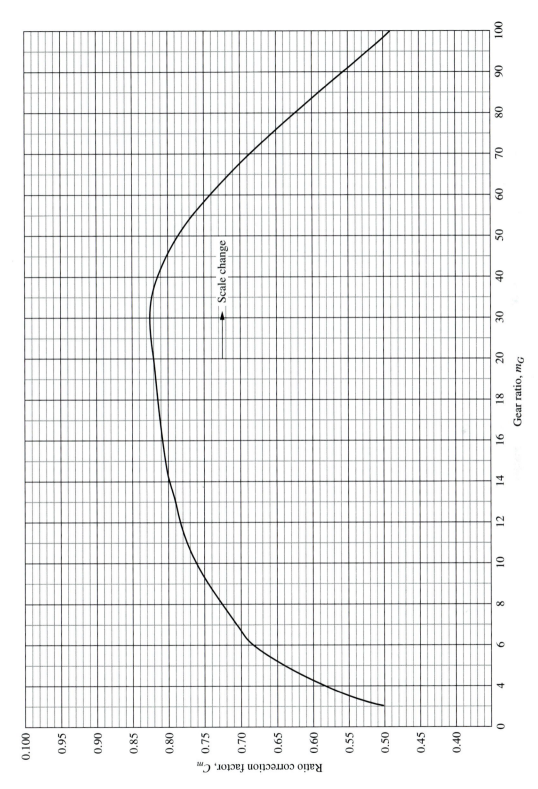

FIGURE 10–20 Ratio correction factor, C_m, vs. gear ratio, m_G

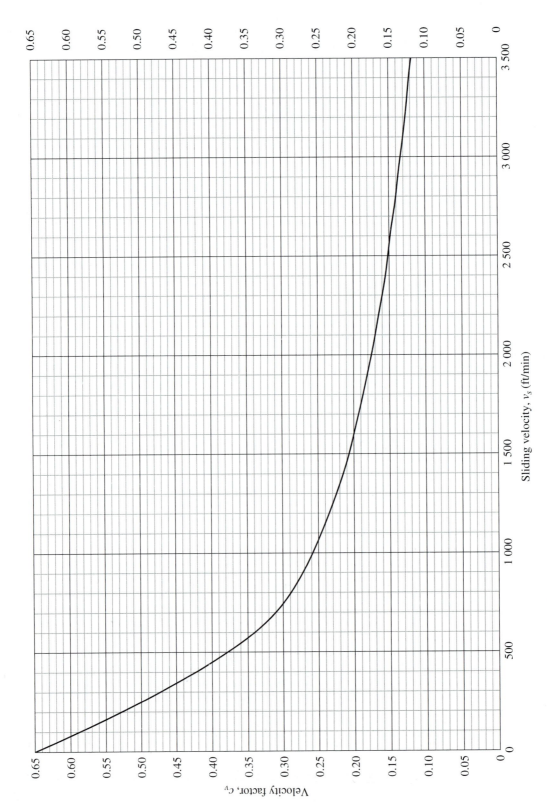

FIGURE 10–21 Velocity factor, C_v, vs. sliding velocity

Conditions on the Use of Equation (10–30)

1. The analysis is valid only for a hardened steel worm (58 HRC minimum) operating with gear bronzes specified in AGMA Standard 6034-A87. The classes of bronzes typically used are tin bronze, phosphor bronze, manganese bronze, and aluminum bronze. The materials factor, C_s, is dependent on the method of casting the bronze, as indicated in Figure 10–19. The values for C_s can be computed from the following formulas.

Sand-cast Bronzes:

For $D_G > 2.5$ in,

$$C_s = 1\,189.636 - 476.545 \log_{10}(D_G) \qquad \text{(10–31)}$$

For $D_G < 2.5$ in,

$$C_s = 1\,000$$

Static-Chill-cast or Forged Bronzes:

For $D_G > 8.0$ in,

$$C_s = 1\,411.651 - 455.825 \log_{10}(D_G) \qquad \text{(10–32)}$$

For $D_G < 8.0$ in,

$$C_s = 1\,000$$

Centrifugally Cast Bronzes:

For $D_G > 25$ in,

$$C_s = 1\,251.291 - 179.750 \log_{10}(D_G) \qquad \text{(10–33)}$$

For $D_G < 25$ in,

$$C_s = 1\,000$$

2. The wormgear diameter is the second factor in determining C_s. The *mean diameter* at the midpoint of the working depth of the gear teeth should be used. If standard addendum gears are used, the mean diameter is equal to the pitch diameter.

3. Use the actual face width, F, of the wormgear as F_e if $F < 0.667(D_W)$. For larger face widths, use $F_e = 0.67\,(D_W)$, because the excess width is not effective.

4. The ratio correction factor, C_m, can be computed from the following formulas.

For Gear Ratios, m_G, from 6 to 20
$$C_m = 0.020\,0(-m_G^2 + 40m_G - 76)^{0.5} + 0.46 \qquad \text{(10–34)}$$

For Gear Ratios, m_G, from 20 to 76
$$C_m = 0.010\,7(-m_G^2 + 56m_G + 5\,145)^{0.5} \qquad \text{(10–35)}$$

For $m_G > 76$

$$C_m = 1.148\,3 - 0.006\,58m_G \qquad (10\text{–}36)$$

5. The velocity factor depends on the sliding velocity, v_s, computed from Equation (10–18) or (10–19). Values for C_v can be computed from the following formulas.

For v_s from 0 to 700 ft/min

$$C_v = 0.659e^{(-0.001\,1v_s)} \qquad (10\text{–}37)$$

For v_s from 700 to 3 000 ft/min

$$C_v = 13.31v_s^{(-0.571)} \qquad (10\text{–}38)$$

For $v_s > 3\,000$ ft/min

$$C_v = 65.52v_s^{(-0.774)} \qquad (10\text{–}39)$$

6. The proportions of the worm and the wormgear must conform to the following limits defining the maximum and minimum pitch diameters of the worm in relation to the center distance, C, for the gear set. All dimensions are in inches.

$$\text{Maximum } D_W = C^{0.875}/1.6 \qquad (10\text{–}40)$$
$$\text{Minimum } D_W = C^{0.875}/3.0 \qquad (10\text{–}41)$$

7. The shaft carrying the worm must be sufficiently rigid to limit the deflection of the worm at the pitch point to the maximum value of $0.005\sqrt{P_x}$, where P_x is the axial pitch of the worm, which is numerically equal to the circular pitch, p, of the wormgear.

8. When you are analyzing a given wormgear set, the value of the rated tangential load, W_{tR}, must be greater than the actual tangential load, W_t, for satisfactory life.

9. Ratings given in this section are valid only for smooth systems such as fans or centrifugal pumps driven by an electric or hydraulic motor operating under 10 hours per day. More severe conditions, such as shock loading, internal combustion engine drives, or longer hours of operation, require the application of a service factor. Reference 15 lists several such factors based on field experience with specific types of equipment. For problems in this book, factors from Table 9–6 can be used.

Example Problem 10–8   Is the wormgear set described in Example Problem 8–7 satisfactory with regard to strength and wear when operating under the conditions of Example Problem 10–7? The wormgear has a face width of 1.25 in.

Solution   From previous problems and solutions,

$$W_{tG} = 962 \text{ lb} \qquad\qquad VR = m_G = 17.33$$
$$v_{tG} = 229 \text{ ft/min} \qquad\qquad v_s = 944 \text{ ft/min}$$
$$D_G = 8.667 \text{ in} \qquad\qquad D_W = 2.000 \text{ in}$$

Assume 58 HRC minimum for the steel worm. Assume that the bronze gear is sand-cast.

Stress

$$K_v = 1\,200/(1\,200 + v_{tG}) = 1\,200/(1\,200 + 229) = 0.84$$
$$W_d = W_{tG}/K_v = 962/0.84 = 1\,145 \text{ lb}$$
$$F = 1.25 \text{ in}$$
$$y = 0.125 \quad \text{(from Table 10–4)}$$
$$p_n = p \cos \lambda = (0.523\,6)\cos 14.04° = 0.508 \text{ in}$$

Then

$$\sigma = \frac{W_d}{yFp_n} = \frac{1\,145}{(0.125)(1.25)(0.508)} = 14\,430 \text{ psi}$$

The guidelines in Section 10–12 indicate that this stress level would be adequate for either manganese or phosphor gear bronze.

Surface Durability

Use Equation (10–30):

$$W_{tR} = C_s D_G^{0.8} F_e C_m C_v \tag{10–30}$$

C **Factors**

The values for the *C* factors can be found from Figures 10–19, 10–20, and 10–21. We find

$$C_s = 740 \quad \text{for sand-cast bronze} \quad \text{and} \quad D_G = 8.667 \text{ in}$$
$$C_m = 0.814 \quad \text{for } m_G = 17.33$$
$$C_v = 0.265 \quad \text{for } v_s = 944 \text{ ft/min}$$

We can use $F_e = F = 1.25$ in if this value is not greater than 0.67 times the worm diameter. For $D_W = 2.000$ in,

$$0.67D_W = (0.67)(2.00 \text{ in}) = 1.333 \text{ in}$$

Therefore, use $F_e = 1.25$ in. Then the rated tangential load is

$$W_{tR} = (740)(8.667)^{0.8}(1.25)(0.814)(0.265) = 1\,123 \text{ lb}$$

Because this value is greater than the actual tangential load of 962 lb, the design should be satisfactory, provided that the conditions defined for the application of Equation (10–30) are met.

REFERENCES

1. American Gear Manufacturers Association. Standard 2008-B90. *Standard for Assembling Bevel Gears*. Alexandria, VA: American Gear Manufacturers Association, 1990.

2. American Gear Manufacturers Association. Standard 6022-C93. *Design Manual for Cylindrical Wormgearing*. Alexandria, VA: American Gear Manufacturers Association, 1993.

3. American Gear Manufacturers Association. Standard AGMA 370.01-1973 (R1978). *Design Manual for Fine-Pitch Gearing*. Alexandria, VA: American Gear Manufacturers Association, 1978.

4. American Gear Manufacturers Association. Standard 6001-D97. *Design and Selection of Components for Enclosed Gear Drives*. Alexandria, VA: American Gear Manufacturers Association, 1997.

5. American Gear Manufacturers Association. Standard AGMA 390.03a (R1980). *Gear Handbook: Gear Classification, Materials and Measuring Methods for Bevel, Hypoid, Fine Pitch Wormgearing and Racks Only as Unassembled Gears*. Alexandria, VA: American Gear Manufacturers Association, 1988. Those portions of 390.03 not included in AGMA 2000-A88, June 1988.

6. American Gear Manufacturers Association. Standard AGMA 908-B89. *Geometry Factors for Determining the Pitting Resistance and Bending Strength of Spur, Helical, and Herringbone Gear Teeth.* Alexandria, VA: American Gear Manufacturers Association, 1989.

7. American Gear Manufacturers Association. Standard AGMA 1012-F90. *Gear Nomenclature, Definitions of Terms with Symbols.* Alexandria, VA: American Gear Manufacturers Association, 1990.

8. American Gear Manufacturers Association. Standard AGMA 2000-A88. *Gear Classification and Inspection Handbook—Tolerances and Measuring Methods for Unassembled Spur and Helical Gears (Including Metric Equivalents).* Alexandria, VA: American Gear Manufacturers Association, 1988. Partial replacement of AGMA 390.03.

9. American Gear Manufacturers Association. Standard AGMA 2001-C95. *Fundamental Rating Factors and Calculation Methods for Involute Spur and Helical Gear Teeth.* Alexandria, VA: American Gear Manufacturers Association, 1994.

10. American Gear Manufacturers Association. Standard AGMA 2003-A86. *Rating the Pitting Resistance and Bending Strength of Generated Straight Bevel, ZEROL® Bevel, and Spiral Bevel Gear Teeth.* Alexandria, VA: American Gear Manufacturers Association, 1988.

11. American Gear Manufacturers Association. Standard AGMA 2004-B89. *Gear Materials and Heat Treatment Manual.* Alexandria, VA: American Gear Manufacturers Association, 1989.

12. American Gear Manufacturers Association. Standard AGMA 2005-C96. *Design Manual for Bevel Gears.* Alexandria, VA: American Gear Manufacturers Association, 1996.

13. American Gear Manufacturers Association. Standard AGMA 6010-E88. *Standard for Spur, Helical, Herringbone, and Bevel Enclosed Drives.* Alexandria, VA: American Gear Manufacturers Association, 1988.

14. American Gear Manufacturers Association. Standard AGMA 6030-C87. *Design of Industrial Double-Enveloping Wormgears.* Alexandria, VA: American Gear Manufacturers Association, 1988.

15. American Gear Manufacturers Association. Standard AGMA 6034-B92. *Practice for Enclosed Cylindrical Wormgear Speed Reducers and Gearmotors.* Alexandria, VA: American Gear Manufacturers Association, 1992.

16. American Society for Metals. *Source Book on Gear Design, Technology and Performance.* Metals Park, OH: American Society for Metals, 1980.

17. Drago, Raymond J. *Fundamentals of Gear Design.* Boston: Butterworths, 1988.

18. Dudley, Darle W. *Handbook of Practical Gear Design.* New York: McGraw-Hill, 1984.

19. Lynwander, Peter. *Gear Drive Systems: Design and Applications.* New York: Marcel Dekker, 1983.

20. Shigley, Joseph E., and Charles R. Mischke. *Mechanical Engineering Design.* 5th ed. New York: McGraw-Hill, 1988.

21. Shigley, Joseph E., and Charles R. Mischke. *Gearing: A Mechanical Designers' Workbook.* New York: McGraw-Hill, 1990.

22. Society of Automotive Engineers. *Gear Design, Manufacturing and Inspection Manual AE-15.* Warrendale, PA: Society of Automotive Engineers, 1990.

PROBLEMS

Helical Gearing

1. A helical gear has a transverse diametral pitch of 8, a transverse pressure angle of $14\frac{1}{2}°$, 45 teeth, a face width of 2.00 in, and a helix angle of 30°.
   a. If the gear transmits 5.0 hp at a speed of 1 250 rpm, compute the tangential force, the axial force, and the radial force.
   b. If the gear operates with a pinion having 15 teeth, compute the bending stress in the pinion teeth. The power comes from an electric motor, and the drive is to a reciprocating pump. Specify a quality number for the teeth.
   c. Specify a suitable material for the pinion and the gear considering both strength and pitting resistance.

2. A helical gear has a normal diametral pitch of 12, a normal pressure angle of 20°, 48 teeth, a face width of 1.50 in, and a helix angle of 45°.
   a. If the gear transmits 2.50 hp at a speed of 1 750 rpm, compute the tangential force, the axial force, and the radial force.
   b. If the gear operates with a pinion having 16 teeth, compute the bending stress in the pinion teeth. The power comes from an electric motor, and the drive is to a centrifugal blower. Specify a quality number for the teeth.
   c. Specify a suitable material for the pinion and the gear considering both strength and pitting resistance.

3. A helical gear has a transverse diametral pitch of 6, a transverse pressure angle of $14\frac{1}{2}°$, 36 teeth, a face width of 1.00 in, and a helix angle of 45°.

a. If the gear transmits 15.0 hp at a speed of 2 200 rpm, compute the tangential force, the axial force, and the radial force.

b. If the gear operates with a pinion having 12 teeth, compute the bending stress in the pinion teeth. The power comes from a six-cylinder gasoline engine, and the drive is to a concrete mixer. Specify a quality number for the teeth.

c. Specify a suitable material for the pinion and the gear considering both strength and pitting resistance.

4. A helical gear has a normal diametral pitch of 24, a normal pressure angle of $14\frac{1}{2}°$, 72 teeth, a face width of 0.25 in, and a helix angle of 45°.

a. If the gear transmits 0.50 hp at a speed of 3 450 rpm, compute the tangential force, the axial force, and the radial force.

b. If the gear operates with a pinion having 16 teeth, compute the bending stress in the pinion teeth. The power comes from an electric motor, and the drive is to a winch that will experience moderate shock. Specify a quality number for the teeth.

c. Specify a suitable material for the pinion and the gear considering both strength and pitting resistance.

For Problems 5–13, complete the design of a pair of helical gears to operate under the stated conditions. Specify the geometry of the gears and the material and its heat treatment. Assume that the drive is from an electric motor unless otherwise specified. Consider both strength and pitting resistance.

5. A pair of helical gears is to be designed to transmit 5.0 hp while the pinion rotates at 1 200 rpm. The gear drives a reciprocating compressor and must rotate between 385 and 390 rpm.

6. A helical gear pair is to be a part of the drive for a milling machine requiring 20.0 hp with the pinion speed at 550 rpm and the gear speed to be between 180 and 190 rpm.

7. A helical gear drive for a punch press requires 50.0 hp with the pinion rotating at 900 rpm and the gear speed at 225 to 230 rpm.

8. A single-cylinder gasoline engine has the pinion of a helical gear pair on its output shaft. The gear is attached to the shaft of a small cement mixer. The mixer requires 2.5 hp while rotating at approximately 75 rpm. The engine is governed to run at approximately 900 rpm.

9. A four-cylinder industrial engine runs at 2 200 rpm and delivers 75 hp to the input gear of a helical gear drive for a large wood chipper used to prepare pulpwood chips for paper making. The output gear must run between 4 500 and 4 600 rpm.

10. A small commercial tractor is being designed for chores such as lawn mowing and snow removal. The wheel drive system is to be through a helical gear pair in which the pinion runs at 450 rpm while the gear, mounted on the hub of the wheel, runs at 75 to 80 rpm. The wheel has an 18-in diameter. The two-cylinder gasoline engine delivers 20.0 hp to the wheels.

11. A water turbine transmits 15.0 hp to a pair of helical gears at 4 500 rpm. The output of the gear pair must drive an electric power generator at 3 600 rpm. The center distance for the gear pair must not exceed 4.00 in.

12. Determine the power-transmitting capacity of a pair of helical gears having a normal pressure angle of 20°, a helix angle of 15°, a normal diametral pitch of 10, 20 teeth in the pinion, 75 teeth in the gear, and a face width of 2.50 in, if they are made from AISI 4140 OQT 1000 steel. They are of typical commercial quality. The pinion will rotate at 1 725 rpm on the shaft of an electric motor. The gear will drive a centrifugal pump.

13. Repeat Problem 12 with the gears made from AISI 4620 DOQT 300 carburized, case-hardened steel. Then compute the axial and radial forces on the gears.

Bevel Gears

14. A straight bevel gear pair has the following data: $N_P = 15$; $N_G = 45$; $P_d = 6$; 20° pressure angle. If the gear pair is transmitting 3.0 hp, compute the forces on both the pinion and the gear. The pinion speed is 300 rpm. The face width is 1.25 in. Compute the bending stress and the contact stress for the teeth, and specify a suitable material and heat treatment. The gears are driven by a gasoline engine, and the load is a concrete mixer providing moderate shock. Assume that neither gear is straddle-mounted.

15. A straight bevel gear pair has the following data: $N_P = 25$; $N_G = 50$; $P_d = 10$; 20° pressure angle. If the gear pair is transmitting 3.5 hp, compute the forces on both the pinion and the gear. The pinion speed is 1 250 rpm. The face width is 0.70 in. Compute the bending stress and the contact stress for the teeth, and specify a suitable material and heat treatment. The gears are driven by a gasoline engine, and the load is a conveyor providing moderate shock. Assume that neither gear is straddle-mounted.

16. Design a pair of straight bevel gears to transmit 5.0 hp at a pinion speed of 850 rpm. The gear speed should be approximately 300 rpm. Consider both strength and pitting resistance. The driver is a gasoline engine, and the driven machine is a heavy-duty conveyor.

17. Design a pair of straight bevel gears to transmit 0.75 hp at a pinion speed of 1 800 rpm. The gear speed should be approximately 475 rpm. Consider both strength and pitting resistance. The driver is an electric motor, and the driven machine is a reciprocating saw.

Wormgearing

18. A wormgear set has a single-thread worm with a pitch diameter of 1.250 in, a diametral pitch of 10, and a normal pressure angle of 14.5°. If the worm meshes with a wormgear having 40 teeth and a face width of 0.625 in, compute the gear pitch diameter, the center distance, and the velocity ratio. If the wormgear set is transmitting 924 lb·in of torque at its output shaft, which is rotating at 30 rpm, compute forces on the gear, efficiency, input speed, input power, and stress on the gear teeth. If the worm is hardened steel and the gear is chilled bronze, evaluate the rated load, and determine whether the design is satisfactory for pitting resistance.

19. Three designs are being considered for a wormgear set to produce a velocity ratio of 20 when the worm gear rotates at 90 rpm. All three have a diametral pitch of 12, a worm pitch diameter of 1.000 in, a gear face width of 0.500 in, and a normal pressure angle of 14.5°. One has a single-thread worm and 20 wormgear teeth; the second has a double-thread worm and 40 wormgear teeth; the third has a four-thread worm and 80 wormgear teeth. For each design, compute the rated output torque, considering both strength and pitting resistance. The worms are hardened steel, and the wormgears are chilled bronze.

20. For each of the three designs proposed in Problem 19, compute the efficiency.

The data for Problems 21, 22, and 23 are given in Table 10–5. Design a wormgear set to produce the desired velocity ratio when transmitting the given torque at the output shaft for the given output rotational speed.

TABLE 10–5 Data for Problems 21–23

Problem	VR	Torque (lb·in)	Output speed (rpm)
21.	7.5	984	80
22.	3	52.5	600
23.	40	4 200	45

24. Compare the two designs described in Table 10–6 when each is transmitting 1 200 lb·in of torque at its output shaft, which rotates at 20 rpm. Compute the forces on the worm and the wormgear, the efficiency, and the input power required.

TABLE 10–6 Designs for Problem 24

Design	P_d	N_t	N_G	D_w	F_G	Pressure angle
A	6	1	30	2.000	1.000	14.5°
B	10	2	60	1.250	0.625	14.5°

11

Keys, Couplings, and Seals

Keys, Couplings, and Seals

Discussion Map

- Keys and couplings connect functional parts of mechanisms and machines, allowing moving parts to transmit power or to locate parts relative to each other.

- Retaining rings hold assemblies together or hold parts on shafts, such as keeping a sprocket in position or holding a wheel on an axle.

- Seals protect critical components by excluding contaminants or by retaining fluids inside the housing of a machine.

Discover
Look around you and identify several examples of the use of keys, couplings, retaining rings, and seals in automobiles, trucks, home appliances, shop tools, gardening equipment, or bicycles.

This chapter will help you understand the functions and design requirements of such devices. In addition, you will learn to recognize commercially available designs and apply them properly.

Think of how two or more parts of a machine can be connected for the purpose of locating one part with respect to another. Now think about how that connection must be designed if the parts are moving and if power must be transmitted between them.

This chapter presents information on commercially available products that accomplish these functions. The generic categories of *keys, couplings,* and *seals* actually encompass numerous different designs.

A *key* is used to connect a drive member such as a belt pulley, chain sprocket, or gear to the shaft that carries it. Torque and power are transmitted across the key to or from the shaft. But how does the power get into or out of the shaft? One way might be that the output from the shaft of a motor or engine is connected to the input shaft of a transmission through a *flexible coupling* that reliably transmits power but allows for some misalignment between the shafts during operation because of the flexing of frame members or through progressive misalignment due to wear.

Seals may be difficult to see because they are typically encased in a housing or are covered in some way. Their function is to protect critical elements of a machine from contamination due to dust, dirt, or water or other fluids while allowing rotating or translating machine elements to move to accomplish their desired functions. Seals exclude undesirable materials from the inside of a mechanism, or they hold critical lubrication or cooling fluids inside the housing of the mechanism.

Look at machines that you interact with each day, and identify parts that fit the descriptions given here. Look in the engine compartment of a car or a truck. How are drive pulleys, linkages, latches, the hinges for the hood, the fan, the windshield wipers, and any other moving part connected to something else—the frame of the car, a rotating shaft, or some other moving part? If you are familiar with the inner workings of the engine and transmission, describe how those parts are connected. Look at the steering and suspension systems, the water pump, the fuel pump, the brake fluid reservoir, and the suspension struts or shock absorbers. Try to see where seals are used. Can you see the universal joints, sometimes called the *constant-velocity* (CV) joints, in the drive train? They should be connecting the output shaft from the transmission to the final parts of the drive train as the power is delivered to the wheels.

Find a small lawn or garden tractor at home or at a local home store. Typically their mechanisms are accessible, although protected from casual contact for safety reasons. Trace how power is transmitted from the engine, through a transmission, through a drive chain or belt, and all the way to the wheels or to the blade of a mower. How are functional parts connected together?

Look at home appliances, power tools in a home shop, and gardening equipment. Can you see parts that are held in place with *retaining rings?* These are typically thin, flat rings that are pressed onto shafts or inserted into grooves to hold a wheel on a shaft, to hold a gear or a pulley in position along the length of a shaft, or to simply hold some part of the device in place.

How are the *keys, couplings, seals, retaining rings,* or other connecting devices made? What materials are used? How are they installed? Can they be removed? What kinds of forces must they resist? How does their special geometry accomplish the desired function? How could they fail?

This chapter will help you become familiar with such mechanical components, with some of the manufacturers that offer commercial versions, and with correct application methods.

 You Are the Designer

In the first part of Chapter 8, you were the designer of a gear-type speed reducer whose conceptual design is shown in Figure 8–3. It has four gears, three shafts, and six bearings, all contained within a housing. How are the gears attached to the shafts? One way is to use *keys* at the interface between the hub of the gears and the shaft. You should be able to design the keys. How is the input shaft connected to the motor or engine that delivers the power? How is the output shaft connected to the driven machine? One way is to use *flexible couplings.* You should be able to specify commercially available couplings and apply them properly, considering the amount of torque they must transmit and how much misalignment they should permit.

How are the gears located axially along the shafts? Part of this function may be provided by shoulders machined on the shaft. But that works only on one side. On the other side, one type of locating means is a retaining ring that is installed into a groove in the shaft after the gear is in place. Rings may be used at the left of gears 1 and 2 and at the right of gears 3 and 4. Notice that the input and output shafts extend outside the housing. How can you keep contaminants from the outside from getting inside? How can you keep lubricating oil inside? Shaft seals can accomplish that function. Seals may also be provided on the bearings to retain lubricant inside and in full contact with the rotating balls or rollers of the bearing.

You should be familiar with the kinds of materials used for seals and with their special geometries. These concepts are discussed in this chapter.

11–1
OBJECTIVES OF THIS CHAPTER

After completing this chapter, you will be able to:

1. Describe several kinds of keys.

2. Specify a suitable size key for a given size shaft.

3. Specify suitable materials for keys.

4. Complete the design of keys and the corresponding keyways and keyseats, giving the complete geometries.

5. Describe *splines* and determine their torque capacity.

6. Describe several alternate methods of fastening machine elements to shafts.

7. Describe *rigid couplings* and *flexible couplings.*

8. Describe several types of flexible couplings.

9. Describe *universal joints.*

10. Describe *retaining rings* and other means of locating elements on shafts.

11. Specify suitable seals for shafts and other types of machine elements.

11–2 KEYS

A *key* is a machinery component placed at the interface between a shaft and the hub of a power-transmitting element for the purpose of transmitting torque [see Figure 11–1(a)]. The key is demountable to facilitate assembly and disassembly of the shaft system. It is installed in an axial groove machined into the shaft, called a *keyseat*. A similar groove in the hub of the power-transmitting element is usually called a *keyway,* but it is more properly also a keyseat. The key is typically installed into the shaft keyseat first; then the hub keyseat is aligned with the key, and the hub is slid into position.

Square and Rectangular Parallel Keys

The most common type of key for shafts up to $6\frac{1}{2}$ inches in diameter is the square key, as illustrated in Figure 11–1(b). The rectangular key, Figure 11–1(c), is recommended for larger shafts and is used for smaller shafts where the shorter height can be tolerated. Both the square and the rectangular keys are referred to as *parallel keys* because the top and bottom and the sides of the key are parallel.

Table 11–1 gives the preferred dimensions for parallel keys as a function of shaft diameter, as specified in ANSI Standard B17.1-1967. The width is nominally one-quarter of the diameter of the shaft. (See Reference 6.)

The keyseats in the shaft and the hub are designed so that exactly one-half of the height of the key is bearing on the side of the shaft keyseat and the other half on the side of the hub keyseat. Figure 11–2 shows the resulting geometry. The distance Y is the radial distance from the theoretical top of the shaft, before the keyseat is machined, to the top edge of the finished keyseat to produce a keyseat depth of exactly $H/2$. To assist in machining and inspecting the shaft or the hub, the dimensions S and T can be computed and shown on the part drawings. The equations are given in Figure 11–2. Tabulated values of Y, S, and T are available in References 6 and 9.

As discussed later in Chapter 12, keyseats in shafts are usually machined with either an end mill or a circular milling cutter, producing the profile or sled runner keyseat, respectively (refer to Figure 12–6). In general practice, the keyseats and keys are left with es-

FIGURE 11–1
Parallel keys

(*a*) Key and keyseat applied to a gear and shaft

(*b*) Square key

(*c*) Rectangular key

sentially square corners. But radiused keyseats and chamfered keys can be used to reduce the stress concentrations. Table 11–2 shows suggested values from ANSI Standard B17.1. When using chamfers on a key, be sure to take that factor into account when computing the bearing stress on the side of the key [see Equation (11–3)].

As alternates to the use of parallel keys, taper keys, gib head keys, pin keys, and Woodruff keys can be used to provide special features of installation or operation. Figure 11–3 shows the general geometry of these types of keys.

Taper Keys and Gib Head Keys

Taper keys are designed to be inserted from the end of the shaft after the hub is in position rather than installing the key first and then sliding the hub over the key as with parallel keys. The taper extends over at least the length of the hub, and the height, *H*, measured at the end of the hub, is the same as for the parallel key. The taper is typically 1/8 in per foot. Note that this design gives a smaller bearing area on the sides of the key, and the bearing stress must be checked.

The *gib head key* [Figure 11–3(c)] has a tapered geometry inside the hub that is the same as that of the plain taper key. But the extended head provides the means of extracting the key from the same end at which it was installed. This is very desirable if the opposite end is not accessible to drive the key out.

TABLE 11–1   Key size vs. shaft diameter

| Nominal shaft diameter | | Nominal key size | | |
| | | | Height, *H* | |
Over	To (incl.)	Width, *W*	Square	Rectangular
5/16	7/16	3/32	3/32	
7/16	9/16	1/8	1/8	3/32
9/16	7/8	3/16	3/16	1/8
7/8	$1\frac{1}{4}$	1/4	1/4	3/16
$1\frac{1}{4}$	$1\frac{3}{8}$	5/16	5/16	1/4
$1\frac{3}{8}$	$1\frac{3}{4}$	3/8	3/8	1/4
$1\frac{3}{4}$	$2\frac{1}{4}$	1/2	1/2	3/8
$2\frac{1}{4}$	$2\frac{3}{4}$	5/8	5/8	7/16
$2\frac{3}{4}$	$3\frac{1}{4}$	3/4	3/4	1/2
$3\frac{1}{4}$	$3\frac{3}{4}$	7/8	7/8	5/8
$3\frac{3}{4}$	$4\frac{1}{2}$	1	1	3/4
$4\frac{1}{2}$	$5\frac{1}{2}$	$1\frac{1}{4}$	$1\frac{1}{4}$	7/8
$5\frac{1}{2}$	$6\frac{1}{2}$	$1\frac{1}{2}$	$1\frac{1}{2}$	1
$6\frac{1}{2}$	$7\frac{1}{2}$	$1\frac{3}{4}$	$1\frac{3}{4}$	$1\frac{1}{2}$
$7\frac{1}{2}$	9	2	2	$1\frac{1}{2}$
9	11	$2\frac{1}{2}$	$2\frac{1}{2}$	$1\frac{3}{4}$
11	13	3	3	2
13	15	$3\frac{1}{2}$	$3\frac{1}{2}$	$2\frac{1}{2}$
15	18	4		3
18	22	5		$3\frac{1}{2}$
22	26	6		4
26	30	7		5

SOURCE: ANSI Standard B17.1-1967, *Keys and Keyseats* (American Society of Mechanical Engineers, New York).
NOTE: Values in nonshaded areas are preferred. Dimensions are in inches.

FIGURE 11–2

Dimensions for parallel keyseats

$$Y = \frac{D - \sqrt{D^2 - W^2}}{2}$$

(*a*) Chordal height

$$S = D - Y - \frac{H}{2} = \frac{D - H + \sqrt{D^2 - W^2}}{2}$$

(*b*) Depth of shaft keyseat

$$T = D - Y + \frac{H}{2} + C = \frac{D + H + \sqrt{D^2 - W^2}}{2} + C$$

Symbols

C = Allowance
  + 0.005-in clearance for parallel keys
  − 0.020-in interference for taper keys
D = Nominal shaft or bore dia., in
H = Nominal key height, in
W = Nominal key width, in
Y = Chordal height, in

(*c*) Depth of hub keyseat

TABLE 11–2   Suggested fillet radii and key chamfers

$H/2$, keyseat depth		Fillet radius	45° chamfer
Over	To (incl.)		
1/8	1/4	1/32	3/64
1/4	1/2	1/16	5/64
1/2	7/8	1/8	5/32
7/8	$1\frac{1}{4}$	3/16	7/32
$1\frac{1}{4}$	$1\frac{3}{4}$	1/4	9/32
$1\frac{3}{4}$	$2\frac{1}{2}$	3/8	13/32

SOURCE: ANSI Standard B17.1-1967, *Keys and Keyseats* (American Society of Mechanical Engineers, New York).
NOTE: All dimensions are given in inches.

Pin Keys

The *pin key,* shown in Figure 11–3(d), is a cylindrical pin placed in a cylindrical groove in the shaft and hub. Lower stress concentration factors result from this design as compared with parallel or taper keys. A close fit between the pin and the groove is required to ensure that the pin does not move and that the bearing is uniform along the length of the pin.

Woodruff Keys

Where light loading and relatively easy assembly and disassembly are desired, the *Woodruff key* should be considered. Figure 11–3(e) shows the standard configuration. The circular

FIGURE 11–3 Key types

(a) Plain taper key

(b) Alternate plain taper key

(c) Gib head taper key

(d) Pin key

(e) Woodruff keys

Note: Plain and gib head taper keys have a 1/8″ taper in 12″.

groove in the shaft holds the key in position while the mating part is slid over the key. The stress analysis for this type of key proceeds in the manner discussed below for the parallel key, taking into consideration the particular geometry of the Woodruff key. ANSI Standard B17.2-1967 lists the dimensions for a large number of standard Woodruff keys and their mating keyseats. (See Reference 7.) Table 11–3 provides a sampling. Note that the *key number* indicates the nominal key dimensions. The last two digits give the nominal diameter, B, in eighths of an inch, and the digits preceding the last two give the nominal width, W, in thirty-seconds of an inch. For example, key number 1210 has a diameter of 10/8 in ($1\frac{1}{4}$ in), and a width of 12/32 in (3/8 in). The actual size of the key is slightly smaller than half of the full circle, as shown in dimensions C and F in Table 11–3.

Selection and Installation of Keys and Keyseats

The key and the keyseat for a particular application are usually designed after the shaft diameter is specified by the methods of Chapter 12. Then, with the shaft diameter as a guide, the size of the key is selected from Table 11–1. The only remaining variables are the length of the key and its material. One of these can be specified, and the requirements for the other can then be computed.

TABLE 11–3   Woodruff key dimensions

Key number	Nominal key size, $W \times B$	Actual length, F	Height of key, C	Shaft keyseat depth	Hub keyseat depth
202	1/16 × 1/4	0.248	0.104	0.072 8	0.037 2
204	1/16 × 1/2	0.491	0.200	0.166 8	0.037 2
406	1/8 × 3/4	0.740	0.310	0.245 5	0.068 5
608	3/16 × 1	0.992	0.435	0.339 3	0.099 7
810	1/4 × $1\frac{1}{4}$	1.240	0.544	0.417 0	0.131 0
1210	3/8 × $1\frac{1}{4}$	1.240	0.544	0.354 5	0.193 5
1628	1/2 × $3\frac{1}{2}$	2.880	0.935	0.683 0	0.256 0
2428	3/4 × $3\frac{1}{2}$	2.880	0.935	0.558 0	0.381 0

SOURCE: ANSI Standard B17.2-1967, *Woodruff Keys and Keyseats* (American Society of Mechanical Engineers, New York).

NOTE: All dimensions are given in inches.

Typically the length of a key is specified to be a substantial portion of the hub length of the element in which it is installed to provide for good alignment and stable operation. But if the keyseat in the shaft is to be in the vicinity of other geometric changes, such as shoulder fillets and ring grooves, it is important to provide some axial clearance between them so that the effects of the stress concentrations are not compounded.

The key can be cut off square at its ends or provided with a radius at each end when installed in a profile keyseat to improve location. Square cut keys are usually used with the sled-runner-type keyseat.

The key is sometimes held in position with a set screw in the hub over the key. However, the reliability of this approach is questionable because of the possibility of the set screw's backing out with vibration of the assembly. Axial location of the assembly should be provided by more positive means, such as shoulders, retaining rings, or spacers.

**11–3
MATERIALS
FOR KEYS**

Keys are most often made from low-carbon, cold-drawn steel. For example, AISI 1020 CD is listed in Appendix 3 as having an ultimate tensile strength of 61 Ksi (420 MPa), a yield strength of 51 Ksi (352 MPa), and a 15% elongation. This is adequate strength and ductility for most applications. Standard key stock conforming to the dimensions of ANSI Standard B17.1 (Table 11–1) is available from industrial supply centers in similar materials. You should check the actual material and guaranteed strength of key stock used in critical applications.

If the low-carbon steel is not sufficiently strong, higher-carbon steel, such as AISI 1040 or 1045, could be used, also in the cold-drawn condition. Heat-treated steels could be used to gain even higher strength. However, the material should retain good ductility as indicated by a percent elongation value greater than about 10%, particularly when shock or impact loads could be encountered.

**11–4
STRESS
ANALYSIS
TO DETERMINE
KEY LENGTH**

There are two basic modes of potential failure for keys transmitting power: (1) shear across the shaft/hub interface and (2) compression failure due to the bearing action between the sides of the key and the shaft or hub material. The analysis for either failure mode requires an understanding of the forces that act on the key. Figure 11–4 shows the idealized case in which the torque on the shaft creates a force on the left side of the key. The key in turn exerts a force on the right side of the hub keyseat. The reaction force of the hub back on the key then produces a set of opposing forces that place the key in direct shear over its cross section, $W \times L$. The magnitude of the shearing force can be found from

FIGURE 11–4

Forces on a key

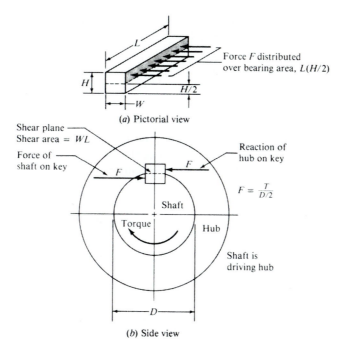

(a) Pictorial view

(b) Side view

$$F = T/(D/2)$$

The shearing stress is then

$$\tau = \frac{F}{A_s} = \frac{T}{(D/2)(WL)} = \frac{2T}{DWL} \tag{11–1}$$

In design, we can set the shearing stress equal to a design stress in shear for the maximum shear stress theory of failure:

$$\tau_d = 0.5s_y/N$$

Then the required length of the key is

$$L = \frac{2T}{\tau_d DW} \tag{11–2}$$

The failure in bearing is related to the compressive stress on the side of the key, the side of the shaft keyseat, or the side of the hub keyseat. The area in compression is the same for either of these zones, $L \times (H/2)$. Thus, the failure occurs on the surface with the lowest compressive yield strength. Let's define a *design stress for compression* as

$$\sigma_d = s_y/N$$

Then the compressive stress is

$$\sigma = \frac{F}{A_c} = \frac{T}{(D/2)(L)(H/2)} = \frac{4T}{DLH} \tag{11–3}$$

Letting this stress equal the design compressive stress allows the computation of the required length of the key for this mode of failure:

$$L = \frac{4T}{\sigma_d DH} \tag{11–4}$$

In typical industrial applications, $N = 3$ is adequate.

For the design of a square key in which the strength of the key material is lower than that of the shaft or the hub, Equations (11–2) and (11–4) produce the same result. Substituting the design stress into either equation would give

$$L = \frac{4TN}{DWs_y} \tag{11–5}$$

But be sure to evaluate the length from Equation (11–4) if either the shaft or the hub has a lower yield strength than the key.

Design Procedure for Parallel Keys

1. Complete the design of the shaft into which the key will be installed, and specify the actual diameter at the location of the keyseat.

2. Select the size of the key from Table 11–1. Use a square key with $W = H$ if the shaft size is 6.50 in or smaller. Use a rectangular key if the diameter is greater than 6.50 in. Then the key width W will be greater than the height H, requiring that both Equation (11–2) and (11–4) be used to determine the length.

3. Specify the material for the key, usually AISI 1020 CD steel. A higher-strength material can be used.

4. Determine the yield strength of the materials for the key, the shaft, and the hub.

5. If a square key is used and the key material has the lowest strength, use Equation (11–5) to compute the minimum required length of the key. This length will be satisfactory for both shear and bearing stress.

6. If a rectangular key is used, or if either the shaft or the hub has a lower strength than the key, use Equation (11–4) to compute the minimum required length of the key based on bearing stress. Also, use Equation (11–2) or (11–5) to compute the minimum required length based on shear of the key. The larger of the two computed lengths governs the design. Check to be sure that the computed length is shorter than the hub length. If not, a higher-strength material must be selected and the design process repeated. Alternatively, two keys or a spline can be used instead of a single key.

7. Specify the actual length of the key to be equal to or longer than the computed minimum length. A convenient standard size should be specified using the recommended basic sizes shown in Tables 13–1, 13–2, or 13–3. The key should extend over all or a substantial part of the length of the hub. But the keyseat should not run into other stress raisers such as shoulders or grooves.

8. Complete the design of the keyseat in the shaft and the keyway in the hub using the equations in Figure 11–2. ANSI Standard B17.1 should be consulted for standard tolerances on dimensions for the key and the keyseats.

Example Problem 11–1 A portion of a shaft where a gear is to be mounted is to be 2.00 in. The gear transmits 2 965 lb·in of torque. The shaft is to be made of AISI 1040 cold-drawn steel. The gear is made from AISI 8650 OQT 1000 steel. The width of the hub of the gear mounted at this location is 1.75 in. Design the key.

Solution From Table 11–1, the standard key dimension for a 2.00-in-diameter shaft would be 1/2 in square. See Figure 11–5 for the proposed design.

Material selection is a design decision. Let's choose AISI 1020 CD steel with $s_y = 51\,000$ psi.

A check of the yield strengths of the three materials in the key, the shaft, and the hub indicates that the key is the weakest material. Then Equation (11–5) can be used to compute the minimum required length of the key:

$$L = \frac{4TN}{DWs_y} = \frac{4(2\,965)(3)}{(2.00)(0.50)(51\,000)} = 0.698 \text{ in}$$

This length is well below the width of the hub of the gear. Notice that the design of the shaft includes retaining rings on both sides of the gear. It is desirable to keep the keyseat well clear of the ring grooves. Therefore, let's specify the length of the key to be 1.50 in.

Summary In summary, the key has the following characteristics:

Material: AISI 1020 CD steel

Width: 0.500 in

Height: 0.500 in

Length: 1.50 in.

Figure 11–5 shows some details of the completed design.

Shear and Bearing Areas for Woodruff Keys

The geometry of Woodruff keys makes it more difficult to determine the shear area and the bearing area for use in stress analyses. Figure 11–3(e) shows that the bearing area on the side of the key in the keyseat is a segment of a circle. The shear area is the product of the chord of that segment times the thickness of the key. The following equations describe the geometry:

Given

B = nominal diameter of the cylinder of which the key is a part

W = width (thickness) of the key

C = full height of the key

d_s = depth of the keyseat in the shaft

Results

$$\text{Shear area} = A_s = 2W\sqrt{d_s(B - d_s)} \tag{11–6}$$

To define the equations for the bearing areas on the side of the key in the shaft and in the hub, we first define three geometric variables, G, L, and J, as follows:

$$G = (\pi/180)B \cos^{-1}\{2[(B/2) - d_s]/B\}$$
$$L = 2\sqrt{d_s(B - d_s)}$$
$$J = (\pi/180)B \cos^{-1}\{2[(B/2) - C]/B\}$$

Shaft section at keyseat

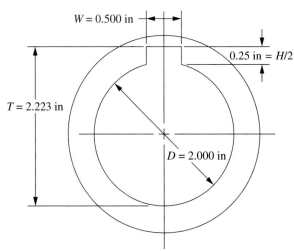

Hub section at keyseat

FIGURE 11–5   Details for proposed design of key and keyseats

Bearing Area in Shaft

$$A_{c\,shaft} = 0.5\{G(B/2) - L[(B/2) - d_s]\} \tag{11–7}$$

$$A_{c\,hub} = 0.5\{J(B/2) - F[(B/2) - C]\} - A_{c\,shaft} \tag{11–8}$$

**11–5
SPLINES**

A *spline* can be described as a series of axial keys machined into a shaft, with corresponding grooves machined into the bore of the mating part (gear, sheave, sprocket, and so on; see Figure 11–6). The splines perform the same function as a key in transmitting torque from the shaft to the mating element. The advantages of splines over keys are many. Because usually four or more splines are used, as compared with one or two keys, a more uniform transfer of the torque and a lower loading on a given part of the shaft/hub interface result. The splines are integral with the shaft, so no relative motion can occur as between a key and the shaft. Splines are accurately machined to provide a controlled fit between the mating internal and external splines. The surface of the spline is often hardened to resist wear and to facilitate its use in applications in which axial motion of the mating element is desired. Sliding motion between a standard parallel key and the mating element should not be permitted. Because of the multiple splines on the shaft, the mating element can be indexed to various positions.

Splines can be either straight-sided or involute. The involute form is preferred because it provides for self-centering of the mating element and because it can be machined with standard hobs used to cut gear teeth.

Straight-sided Splines

Straight splines are made according to the specifications of the Society of Automotive Engineers (SAE) and usually contain 4, 6, 10, or 16 splines. Figure 11–6 shows the six-spline version, from which you can see the basic design parameters of D (major diameter), d (mi-

FIGURE 11–6
Straight-sided spline

(*a*) General form of spline connection

(*b*) Internal spline

nor diameter), W (spline width), and h (spline depth). The dimensions for d, W, and h are related to the nominal major diameter D by the formulas given in Table 11–4. Note that the values of h and d differ according to the use of the spline. The permanent fit, A, is used when the mating part is not to be moved after installation. The B fit is used if the mating part will be moved along the shaft without a torque load. When the mating part must be moved under load, the C fit is used.

The torque capacity for SAE splines is based on the limit of 1 000-psi bearing stress on the sides of the splines, from which the following formula is derived:

$$T = 1\ 000NRh \qquad\qquad\qquad \textbf{(11–9)}$$

where N = number of splines

R = mean radius of the splines

h = depth of the splines (from Table 11–4)

The torque capacity is per inch of length of the spline. But note that

$$R = \frac{1}{2}\left[\frac{D}{2} + \frac{d}{2}\right] = \frac{D + d}{4}$$

$$h = \frac{1}{2}(D - d)$$

Then

$$T = 1\ 000N\,\frac{(D + d)}{4}\,\frac{(D - d)}{2} = 1\ 000N\,\frac{(D^2 - d^2)}{8} \qquad\qquad \textbf{(11–10)}$$

This equation can be further refined for each of the types of splines in Table 11–4 by substitution of the appropriate relationships for N and d. For example, for the six-spline version and the B fit, $N = 6$, $d = 0.850D$, and $d^2 = 0.722\ 5D^2$. Then

$$T = 1\ 000(6)\,\frac{[D^2 - 0.722\ 5D^2]}{8} = 208D^2$$

Thus, the required diameter to transmit a given torque would be

$$D = \sqrt{T/208}$$

TABLE 11–4   Formulas for SAE straight splines

No. of splines	W, for all fits	A: Permanent fit		B: To slide without load		C: To slide under load	
		h	d	h	d	h	d
Four	0.241D	0.075D	0.850D	0.125D	0.750D		
Six	0.250D	0.050D	0.900D	0.075D	0.850D	0.100D	0.800D
Ten	0.156D	0.045D	0.910D	0.070D	0.860D	0.095D	0.810D
Sixteen	0.098D	0.045D	0.910D	0.070D	0.860D	0.095D	0.810D

NOTE: These formulas give the maximum dimensions for W, h, and d.

In these formulas, dimensions are in inches and the torque is in pound-inches. We use this same approach to find the torque capacities and required diameters for the other versions of straight splines (Table 11–5).

The graphs in Figure 11–7 enable you to choose an acceptable diameter for a spline to carry a given torque, depending on the desired fit, *A, B,* or *C.* The data were taken from Table 11–5.

Involute Splines

Involute splines are typically made with pressure angles of 30°, 37.5°, or 45°. The 30° form is illustrated in Figure 11–8, showing the two types of fit that can be specified. The *major diameter fit* produces accurate concentricity between the shaft and the mating element. In the *side fit,* contact occurs only on the sides of the teeth, but the involute form tends to center the shaft in the mating splined hub.

TABLE 11–5 Torque capacity for straight splines per inch of spline length

Number of splines	Fit	Torque capacity	Required diameter
4	A	$139D^2$	$\sqrt{T/139}$
4	B	$219D^2$	$\sqrt{T/219}$
6	A	$143D^2$	$\sqrt{T/143}$
6	B	$208D^2$	$\sqrt{T/208}$
6	C	$270D^2$	$\sqrt{T/270}$
10	A	$215D^2$	$\sqrt{T/215}$
10	B	$326D^2$	$\sqrt{T/326}$
10	C	$430D^2$	$\sqrt{T/430}$
16	A	$344D^2$	$\sqrt{T/344}$
16	B	$521D^2$	$\sqrt{T/521}$
16	C	$688D^2$	$\sqrt{T/688}$

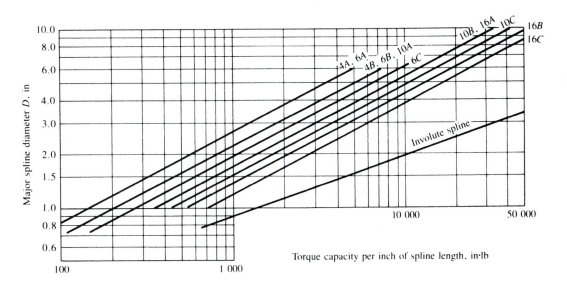

FIGURE 11–7 Torque capacity per inch of spline length, lb·in

FIGURE 11–8
30° involute spline

(*a*) Side-fit spline

Note chamfer on tips of
external spline teeth

(*b*) Major dia. fit spline

N = Number of spline teeth
P = Diametral pitch
$D = N/P$ = Pitch dia.
$p = \pi/P$ = Circular pitch

Minor dia.:
Internal: $\dfrac{N-1}{P}$
External: $\dfrac{N-1.35}{P}$

Major dia.:
Internal: $\dfrac{N+1.35}{P}$ side fit

           $\dfrac{N+1}{P}$ major dia. fit

External: $\dfrac{N+1}{P}$

Figure 11–8 also gives some of the basic formulas for key features of involute splines in the U.S. Customary Unit System with dimensions in inches. (See Reference 4.) The terms are similar to those for involute spur gears which are discussed more completely in Chapter 8. The basic spline size is governed by its *diametral pitch, P:*

$$P = N/D \qquad\qquad (11\text{–}11)$$

where N = number of spline teeth
       D = pitch diameter

The diametral pitch, then, is *the number of teeth per inch of pitch diameter.* Only even numbers of teeth from 6 to 60 are typically used. Up to 100 teeth are used on some 45° splines. Note that the pitch diameter lies *within* the tooth and is related to the major and minor diameters by the relationships shown in Figure 11–8.

The *circular pitch, p,* is the distance from one point on a tooth to the corresponding point on the next adjacent tooth, measured along the pitch circle. To find the nominal value of p, divide the circumference of the pitch circle by the number of spline teeth. That is,

$$p = \pi D/N \qquad\qquad (11\text{–}12)$$

But because $P = N/D$, we can also say

$$p = \pi/P \qquad\qquad (11\text{–}13)$$

The *tooth thickness, t,* is the thickness of the tooth measured along the pitch circle. Then the theoretical value is

$$t = p/2 = \pi/2P$$

The nominal value of the width of the tooth space is equal to t.

Standard Diametral Pitches. The following are the 17 standard diametral pitches in common use:

2.5	10	40
3	12	48
4	16	64
5	20	80
6	24	128
8	32	

The common designation for an involute spline is given as a fraction, P/P_s, where P_s is called the *stub pitch* and is always equal to $2P$. Thus, if a spline had a diametral pitch of 4, it would be called a 4/8 pitch spline. For convenience, we will use only the diametral pitch.

Length of Splines. Common designs use spline lengths from $0.75D$ to $1.25D$, where D is the pitch diameter of the spline. If these standards are used, the shear strength of the splines will exceed that of the shaft on which they are machined.

Metric Module Splines. The dimensions of splines made to metric standards are related to the *module, m,* where

$$m = D/N \qquad (11-14)$$

and both D and m are in millimeters. (See Reference 5.) Note that the symbol Z is used in place of N for the number of teeth in standards describing metric splines. Other features of metric splines can be found from the following formulas:

$$\text{Pitch diameter} = D = mN \qquad (11-15)$$
$$\text{Circular pitch} = p = \pi m \qquad (11-16)$$
$$\text{Basic tooth thickness} = t = \pi m/2 \qquad (11-17)$$

Standard Modules. There are 15 standard modules:

0.25	1.5	4
0.5	1.75	5
0.75	2	6
1	2.5	8
1.25	3	10

11–6 OTHER METHODS OF FASTENING ELEMENTS TO SHAFTS

The following discussion will acquaint you with some of the ways in which power-transmitting elements can be attached to shafts without keys or splines. (See Reference 10.) In most cases the designs have not been standardized, and analysis of individual cases, considering the forces exerted on the elements and the manner of loading of the fastening means, is necessary. In several of the designs, the analysis of shear and bearing will follow a procedure similar to that shown for keys. If a satisfactory analysis is not possible, testing of the assembly is recommended.

Pinning

With the element in position on the shaft, a hole can be drilled through both the hub and the shaft, and a pin can be inserted in the hole. Figure 11–9 shows three examples of this approach. The straight, solid, cylindrical pin is subjected to shear over two cross sections. If there is a force, F, on each end of the pin at the shaft/hub interface, and if the shaft diameter is D, then

$$T = 2F(D/2) = FD$$

or $F = T/D$. With the symbol d representing the pin diameter, the shear stress in the pin is

$$\tau = \frac{F}{A_s} = \frac{T}{D(\pi d^2/4)} = \frac{4T}{D(\pi d^2)} \qquad \textbf{(11–18)}$$

Letting the shear stress equal the design stress in shear as before, solving for d gives the required pin diameter:

$$d = \sqrt{\frac{4T}{D(\pi)(\tau_d)}} \qquad \textbf{(11–19)}$$

Sometimes the diameter of the pin is purposely made small to ensure that the pin will break if a moderate overload is encountered, in order to protect critical parts of a mechanism. Such a pin is called a *shear pin*.

One problem with a cylindrical pin is that fitting it adequately to provide precise location of the hub and to prevent the pin from falling out is difficult. The taper pin overcomes some of these problems, as does the split spring pin shown in Figure 11–9(c). For the split spring pin, the hole is made slightly smaller than the pin diameter so that a light force is required to assemble the pin in the hole. The spring force retains the pin in the hole and holds the assembly in position. But, of course, the presence of any of the pin-type connections produces stress concentrations in the shaft.

Clamping

A hub on a gear or other element can be slotted axially, and a clamp can be placed over the hub. When the clamp is drawn down tight, it forces the split hub into contact with the shaft. The pressure of the hub on the surface of the shaft permits transmission of torque. This type

FIGURE 11–9
Pinning

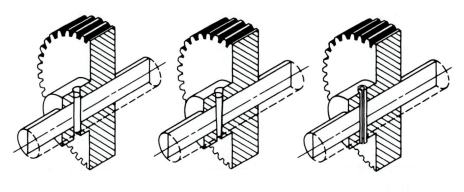

(*a*) Cylindrical pin   (*b*) Taper pin   (*c*) Spring pin of connection is difficult to control, and predicting the torque-transmitting capacity accurately is difficult.

Split Taper Bushing

The difficulties of simple clamping can be overcome by using a *split taper bushing,* available commercially on several belt sheaves and chain sprockets (see Figure 11–10). Whereas simple clamping relies on friction to transmit torque produced by pressure between the hub and the shaft, the split taper bushing uses a key that is a positive means of torque transmission; that is, no slipping can occur between the hub and the shaft. The split bushing has a small taper on the outer surface. When the bushing is pulled into a mating hub with a set of capscrews, the bushing is brought into tight contact with the shaft to hold the assembly in the proper axial position. The small taper locks the assembly together. Removal of the bushing is accomplished by removing the capscrews and using them in push-off holes to force the hub off the taper. The assembly can then be easily disassembled.

Set Screws

A *set screw* is a threaded fastener driven radially through a hub to bear on the outer surface of a shaft (see Figure 11–11). The point of the set screw is flat, oval, cone-shaped, cupped, or any of several proprietary forms. The point bears on the shaft or digs slightly into its surface. Thus, the set screw transmits torque by the friction between the point and the shaft or by the resistance of the material in shear. The capacity for torque transmission is somewhat variable, depending on the hardness of the shaft material and the clamping force created

Bushing

Bushings applied to belt sheaves

FIGURE 11–10   Split taper bushing (Emerson Power Transmission Corporation, Drive and Component Division, Ithaca, NY.)

FIGURE 11–11
Set screws

Set screws

FIGURE 11–12
Tapered shaft for
fastening machine
elements to shafts

(*a*) Taper and screw

(*b*) Taper and nut

when the screw is installed. Furthermore, the screw may loosen during operation because of vibration. For these reasons, set screws should be used with care. Some manufacturers provide set screws with plastic inserts in the side among the threads. When the set screw is screwed into a tapped hole, the plastic is deformed by the threads and holds the screw securely, resisting vibration. Using a liquid adhesive also helps resist loosening.

Another problem with using set screws is that the shaft surface is damaged by the point; this damage may make disassembly difficult. Machining a flat on the surface of the shaft may help reduce the problem and also produce a more consistent assembly.

When set screws are properly assembled on typical industrial shafting, their force capability is approximately as follows (see Reference 9):[1]

Screw Diameter	Holding Force
1/4 in	100 lb
3/8 in	250 lb
1/2 in	500 lb
3/4 in	1 300 lb
1 in	2 500 lb

Taper and Screw

The power-transmitting element (gear, sheave, sprocket, or other) that is to be mounted at the end of a shaft can be secured with a screw and a washer in the manner shown in Figure 11–12(a). The taper provides good concentricity and moderate torque transmission capac-

[1]Erik Oberg et al. *Machinery's Handbook.* 25th ed. New York: Industrial Press, 1997.

FIGURE 11–13
Methods of
mechanically locking
gears to shafts

Stake 3 or
more places

(c) Riveting

(b) Pinning

(d) Spinning

(a) Staking

ity. Because of the machining required, the connection is fairly costly. A modified form uses the tapered shaft with a threaded end for the application of a nut, as shown in Figure 11–12(b). The inclusion of a key lying in a keyseat machined parallel to the taper increases the torque-transmitting capacity greatly and ensures positive alignment.

Press Fit

Making the diameter of the shaft greater than the bore diameter of the mating element results in an interference fit. The resulting pressure between the shaft and the hub permits the transmission of torque at fairly high levels, depending on the degree of interference. This is discussed more fully in Chapter 13. Sometimes the press fit is combined with a key, with the key providing the positive drive and the press fit ensuring concentricity.

Molding

Plastic and die cast gears can be molded directly to their shafts. Often the gear is applied to a location that is knurled to improve the ability to transmit torque. A modification of this procedure is to take a separate gear blank with a prepared hub, locate it over the proper position on a shaft, and then cast zinc into the space between the shaft and the hub to lock them together.

Mechanical Locking

Some of the variations of methods of locking gears to shafts by mechanical means are illustrated in Figure 11–13. (See Reference 10.) Part (a), *staking,* is produced by deforming part of the hub into the gear. Part (b), *pegging* or *pinning,* is similar to the pin key shown in Figure 11–3. In Part (c), the gear is *riveted* to a flange on the shaft. Here other types of fasteners, such as standard nuts and bolts, can be used to permit removal of the gear. *Spinning* is used in Part (d) to deform part of the hub radially outward into intimate contact with the gear.

**11–7
COUPLINGS**

The term *coupling* refers to a device used to connect two shafts together at their ends for the purpose of transmitting power. There are two general types of couplings: rigid and flexible.

Rigid couplings are designed to draw two shafts together tightly so that no relative motion can occur between them. This design is desirable for certain kinds of equipment in which precise alignment of two shafts is required and can be provided. In such cases, the coupling must be designed to be capable of transmitting the torque in the shafts.

A typical rigid coupling is shown in Figure 11–14, in which flanges are mounted on the ends of each shaft and are drawn together by a series of bolts. The load path is then from the driving shaft to its flange, through the bolts, into the mating flange, and out to the driven shaft. The torque places the bolts in shear. The total shear force on the bolts depends on the radius of the bolt circle, $D_{bc}/2$, and the torque, T. That is,

$$F = T/(D_{bc}/2) = 2T/D_{bc}$$

Letting N be the number of bolts, the shear stress in each bolt is

$$\tau = \frac{F}{A_s} = \frac{F}{N(\pi d^2/4)} = \frac{2T}{D_{bc}N(\pi d^2/4)} \quad \textbf{(11–20)}$$

Letting the stress equal the design stress in shear and solving for the bolt diameter, we have

$$d = \sqrt{\frac{8T}{D_{bc}N\pi\tau_d}} \quad \textbf{(11–21)}$$

Notice that this analysis is similar to that for pinned connections in Section 11–6. The analysis assumes that the bolts are the weakest part of the coupling.

Rigid couplings should be used only when the alignment of the two shafts can be maintained very accurately, not only at the time of installation but also during operation of the machines. If significant angular, radial, or axial misalignment occurs, stresses that are difficult to predict and that may lead to early failure due to fatigue will be induced in the shafts. These difficulties can be overcome by the use of flexible couplings.

Flexible couplings are designed to transmit torque smoothly while permitting some axial, radial, and angular misalignment. The flexibility is such that when misalignment does occur, parts of the coupling move with little or no resistance. Thus, no significant axial or bending stresses are developed in the shaft.

FIGURE 11–14
Rigid coupling
(Rockwell
Automation/Reliance
Electric, Rockwell
International
Corporation, Costa
Mesa, CA.)

d = Bolt diameter

D_{bc}

Cutaway view

Many types of flexible couplings are available commercially, as discussed later in this section. Each is designed to transmit a given limiting torque. The manufacturer's catalog lists the design data from which you can choose a suitable coupling. Remember that torque equals power divided by rotational speed. So for a given size of coupling, as the speed of rotation increases, the amount of power that the coupling can transmit also increases, although not always in direct proportion. Of course, centrifugal effects determine the upper limit of speed.

The degree of misalignment that can be accommodated by a given coupling should be obtained from the manufacturer's catalog data, with values varying with the size and design of the coupling. Small couplings may be limited to parallel misalignment of 0.005 in, although larger couplings may allow 0.030 in or more. Typical allowable angular misalignment is ±3°. Axial movement allowed, sometimes called *end float,* is up to 0.030 in for many types of couplings.

Figures 11–15 through 11–23 show several commercially available flexible couplings. (Also see References 1–3.)

11–8 UNIVERSAL JOINTS

As noted in the preceding section, flexible couplings can accommodate up to approximately 3° of angular misalignment and approximately 0.030 in of parallel misalignment, depending on their design. Sometimes much more misalignment must be used; in these applications, the *universal joint* provides an effective coupling. Universal joints operate at angles of up to 45° at very slow speeds. At higher speeds, above approximately 10 rpm, the maximum recommended angle is 30°. Above 600 rpm, a 20° maximum angle is typical. Manufacturer's data should be consulted. Actual service conditions affect the permissible angle.

Figure 11–24 shows one of the smaller, simpler types of universal joints, consisting of two yokes, a center bearing block, and two pins that pass through the block at right angles to each other. Figure 11–25 shows a larger type, having two yokes and a center cross member, that is sometimes called a *Cardan joint.* The caps on the cross are equipped with needle bearings to permit low-friction relative motion between the members. Proper lubrication of the moving members of the joint is essential. Figure 11–26 shows another type of universal joint, which is used in heavy industrial applications.

When a single universal joint is driven by a shaft rotating at a uniform angular velocity, the output shaft has a nonuniform angular velocity. The varying angular velocity produces variable angular accelerations, which in turn cause vibrations and higher

FIGURE 11–15 Chain coupling. Torque is transmitted through a double roller chain. Clearances between the chain and the sprocket teeth on the two coupling halves accommodate misalignment. (Emerson Power Transmission Corporation, Drive and Component Division, Ithaca, NY.)

Elastomeric member bonded
to steel flanges and hubs

FIGURE 11–16 Ever-Flex coupling. The features of this coupling are that it
(1) generally minimizes torsional vibration; (2) cushions shock loads; (3) compensates
for parallel misalignment up to 1/32 in; (4) accommodates angular misalignment of ±3°;
and (5) provides adequate end float, ±1/32 in. (Emerson Power Transmission Corporation,
Drive and Component Division, Ithaca, NY.)

FIGURE 11–17 Grid-Flex coupling. Torque is transmitted through a flexible spring
steel grid. Flexing of the grid permits misalignment and makes it torsionally resilient to
resist shock loads. (Emerson Power Transmission Corporation, Drive and Component Division,
Ithaca, NY.)

stresses. Thus, the use of a single universal joint should be limited to slow-speed, low-
power applications.

Using two universal joints connected by an intermediate shaft eliminates the varying
angular velocity caused by one joint. With proper alignment, the angular velocity variation
of each joint is canceled by the other. Figure 11–27 shows a properly aligned double universal joint connecting a pair of parallel shafts having a large offset. If the shafts must be
at a specified angle, the two joints are installed so that the angle of each joint is one-half of
the total angle between the input and output shafts.

The intermediate shaft for the double universal joint is frequently made in two parts
with a spline connection between them. This arrangement allows for a variation in the ax-

FIGURE 11–18   Gear coupling. Torque is transmitted between crown-hobbed teeth from the coupling half to the sleeve. The crown shape on the gear teeth permits misalignment. (Emerson Power Transmission Corporation, Drive and Component Division, Ithaca, NY.)

FIGURE 11–19
Bellows coupling
(Stock Drive Products,
New Hyde Park, NY)

FIGURE 11–20
PARA-FLEX®
coupling. Using an
elastomeric element to
permit misalignment
and to cushion shocks
(Rockwell
Automation/Reliance
Electric, Rockwell
International
Corporation, Costa
Mesa, CA.)

ial location of the parts of the drive, even while the system is in operation. This feature is particularly desirable in vehicular applications because of the movement of the chassis components. In industrial machinery, such a design would permit easier, less critical installation of drive components.

Several proprietary designs for universal joints that have a uniform angular velocity relationship in a single joint are available. Examples are the Rzeppa and the Bendix-Weiss joints.

FIGURE 11–21
Dynaflex® coupling. Torque is transmitted through elastomeric material that flexes to permit misalignment and to attenuate shock loads. (Lord Corporation, Erie, PA)

FIGURE 11–22 Jaw-Type coupling (Emerson Power Transmission Corporation, Drive and Component Division, Ithaca, NY.)

Inserts

(a) Assembled coupling

| Neoprene (normal duty applications) | Bronze, oil impregnated (low-speed, high-torque applications) | Polyurethane (extra capacity at medium to high speed) |

(b) Types of inserts

FIGURE 11–23
FORM-FLEX® coupling. Torque is transmitted from hubs through laminated flexible elements to the spacer. (Formsprag Division, Dana Corp., Mt. Pleasant, MI)

FIGURE 11–24
Universal joint components (Curtis Universal Joint Co., Inc., Springfield, MA)

FIGURE 11–25
Automotive universal joints (Pontiac Motor Division, General Motors Corp., Pontiac, MI)

(*a*) Automotive universal joint components

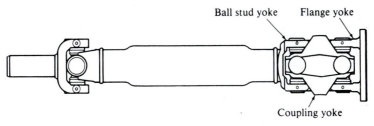

(*b*) Automotive propeller shaft with double Cardan universal joint

FIGURE 11–26
Industrial universal joint (Spicer Universal Joint Division, Dana Corp., Toledo, OH)

11–9 RETAINING RINGS AND OTHER MEANS OF AXIAL LOCATION

The preceding sections of this chapter have focused on means of connecting machine elements to shafts for the purpose of transmitting power. Therefore, they emphasized the ability of the elements to withstand a given torque at a given speed of rotation. It must be recognized that the axial location of the machine elements must also be ensured by the designer.

The choice of the means for axial location depends heavily on whether or not there is axial thrust transmitted by the element. Spur gears, V-belt sheaves, and chain sprockets

FIGURE 11–27
Double universal joint
(Curtis Universal Joints
Co., Inc., Springfield,
MA)

produce no significant thrust loads. Therefore, the need for axial location affects only inci-
dental forces due to vibration, handling, and shipping. Although not severe, these forces
should not be taken lightly. Movement of an element in relation to its mating element in an
axial direction can cause noise, excessive wear, vibration, or complete disconnection of the
drive. Any bicycle rider who has experienced the loss of a chain can appreciate the consequences of misalignment. Recall that for spur gears, the strength of the gear teeth and the
wear resistance are both directly proportional to the face width of the gear. Axial misalignment decreases the effective face width.

Some of the methods discussed in Section 11–6 for fastening elements to shafts for
the purpose of transmitting power also provide some degree of axial location. Refer to the
discussions of pinning, clamping, split taper bushings, set screws, the taper and screw,
the taper and nut, the press fit, molding, and the several methods of mechanically locking
the elements to the shaft.

Among the wide variety of other means available for axial location, we will discuss
the following:

- Retaining rings
- Collars
- Shoulders
- Spacers
- Locknuts

Retaining Rings

Retaining rings are placed on a shaft, in grooves in the shaft, or in a housing to prevent the
axial movement of a machine element. Figure 11–28 shows the wide variety of styles of
rings available from one manufacturer. The different designs allow for either internal or external mounting of the ring. The amount of axial thrust load capacity and the shoulder
height provided by the different ring styles also vary. The manufacturer's catalog provides

FIGURE 11–28
Standard Truarc® retaining ring series (Courtesy Waldes Truarc, Inc., Long Island City, NY)

INTERNAL	**BASIC** N5000 For housings and bores — Size Range 250—10.0 in. / 6.4—254.0 mm.	EXTERNAL	**BOWED** 5101* For shafts and pins — Size Range .188—1.750 in. / 4.8—44.4 mm.	EXTERNAL	**REINFORCED** 5115 For shafts and pins — Size Range 094—1.0 in. / •
INTERNAL	**BOWED** N5001* For housings and bores — Size Range .250—1.750 in. / 6.4—44.4 mm.	EXTERNAL	**BEVELED** 5102 For shafts and pins — Size Range 1.0—10.0 in. / 25.4—254.0 mm.	EXTERNAL	**BOWED E-RING** 5131 For shafts and pins — Size Range 110—1.375 in. / 2.8—34.9 mm.
INTERNAL	**BEVELED** *N5002/*N5003 For housings and bores — Size Range 1.0—10.0 in. / 25.4—254.0 mm. ° 1.56—2.81 in. / 39.7—71.4 mm.	EXTERNAL	**CRESCENT®** 5103 For shafts and pins — Size Range 125—2.0 in. / 3.2—50.8 mm.	EXTERNAL	**E-RING** 5133 For shafts and pins — Size Range 040—1.375 in. / 1.0—34.9 mm.
INTERNAL	**CIRCULAR** 5005 For housings and bores — Size Range 312—2.0 in. / •	EXTERNAL	**CIRCULAR** 5105 For shafts and pins — Size Range .094—1.0 in. / •	EXTERNAL	**RADIAL GRIPRING®** 5135 for shafts and pins — Size Range .094—.375 in. / 2.4—9.5 mm.
INTERNAL	**INVERTED** 5008 For housings and bores — Size Range .750—4.0 in. / 19.0—101.6 mm.	EXTERNAL	**INTERLOCKING** 5107* For shafts and pins — Size Range .469—3.375 in. / 11.9—85.7 mm.	EXTERNAL	**PRONG-LOCK®** 5139* For shafts and pins — Size Range .092—438 in. / •
EXTERNAL	**BASIC** 5100 For shafts and pins — Size Range .125—10.0 in. / 3.2—254.0 mm.	EXTERNAL	**INVERTED** 5108 For shafts and pins — Size Range 500—4.0 in. / 12.7—101.6 mm.	EXTERNAL	**REINFORCED E-RING** 5144 For shafts and pins — Size Range .094—.562 in. / 2.4—14.3 mm.

EXTERNAL	**HEAVY-DUTY** 5160 For shafts and pins — Size Range 394—2.0 in. / 10.0—50.8 mm.
EXTERNAL	**KLIPRING®** 5304 T-5304 For shafts and pins — Size Range .156—1.000 in. / 4.0—25.4 mm.
EXTERNAL	**GRIPRING®** 5555 For shafts and pins — Size Range .079—.750 in. / 2.0—19.0 mm.
EXTERNAL	**HIGH-STRENGTH** 5560* For shafts and pins — Size Range .101—328 in. / •
EXTERNAL	**PERMANENT SHOULDER** 5590* For shafts and pins — Size Range 250 .750 / 6.4 19.0 mm.
* Non-Stocking Ring Type: Available on special order only

FIGURE 11–29
Retaining rings applied to a water conditioner mechanism (Courtesy Waldes Truarc, Inc., Long Island City, NY)

data on the capacity and installation data for the rings. Figure 11–29 shows a few examples of retaining rings in use.

Collars

A *collar* is a ring slid over the shaft and positioned adjacent to a machine element for the purpose of axial location. It is held in position, typically, by set screws. Its advantage is that axial location can be set virtually anywhere along the shaft to allow adjustment of the position at the time of assembly. The disadvantages are chiefly those related to the use of set screws themselves (Section 11–6).

FIGURE 11–30
Application of spacers

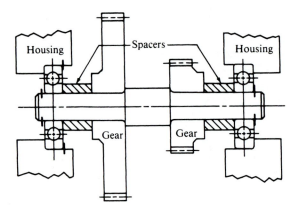

Shoulders

A *shoulder* is the vertical surface produced when a diameter change occurs on a shaft. Such a design is an excellent method for providing for the axial location of a machine element, at least on one side. Several of the shafts illustrated in Chapter 12 incorporate shoulders. The main design considerations are providing (1) a sufficiently large shoulder to locate the element effectively and (2) a fillet at the base of the shoulder that produces an acceptable stress concentration factor and that is compatible with the geometry of the bore of the mating element (see Chapter 12, Figures 12–2 and 12–7).

Spacers

A *spacer* is similar to a collar in that it is slid over the shaft against the machine element that is to be located. The primary difference is that set screws and the like are not necessary because the spacer is positioned *between* two elements and thus controls only the relative position between them. Typically, one of the elements is positively located by some other means, such as a shoulder or a retaining ring. Consider the shaft in Figure 11–30, on which two spacers locate each gear with respect to its adjacent bearing. The bearings are in turn located on one side by the housing supporting their outer races. The gears also seat against shoulders on one side.

Locknuts

When an element is located at the end of a shaft, a *locknut* can be used to retain it on one side. Figure 11–31 shows a bearing retainer type of locknut. These are available as stock items from bearing suppliers.

One practical consideration in the matter of axial location of machine elements is exercising care that elements are not *overrestrained.* Under certain conditions of differential thermal expansion or with an unfavorable tolerance stack, the elements may be forced together so tightly as to cause dangerous axial stresses. At times it may be desirable to locate only one bearing positively on a shaft and to permit the other to float slightly in the axial direction. The floating element may be held lightly with an axial spring force accommodating the thermal expansion without creating dangerous forces.

**11–10
TYPES OF SEALS**

Seals are an important part of machine design in situations where the following conditions apply:

FIGURE 11–31
Locknut and lockwasher for retaining bearings. Dimensions are given in manufacturer's catalog; part sizes are compatible with standard bearing sizes. (SKF Industries, Inc., King of Prussia, PA)

(a)

(b)

Locknut   Lockwasher

(c)

1. Contaminants must be excluded from critical areas of a machine.
2. Lubricants must be contained within a space.
3. Pressurized fluids must be contained within a component such as a valve or a hydraulic cylinder.

Some of the parameters affecting the choice of the type of sealing system, the materials used, and the details of its design are as follows:

1. The nature of the fluids to be contained or excluded.
2. Pressures on both sides of the seal.
3. The nature of any relative motion between the seal and the mating components.
4. Temperatures on all parts of the sealing system.
5. The degree of sealing required: Is some small amount of leakage permissible?
6. The life expectancy of the system.
7. The nature of the solid materials against which the seal must act: corrosion potential, smoothness, hardness, wear resistance.
8. Ease of service for replacement of worn sealing elements.

FIGURE 11–32
O-rings and T-rings
used as static seals

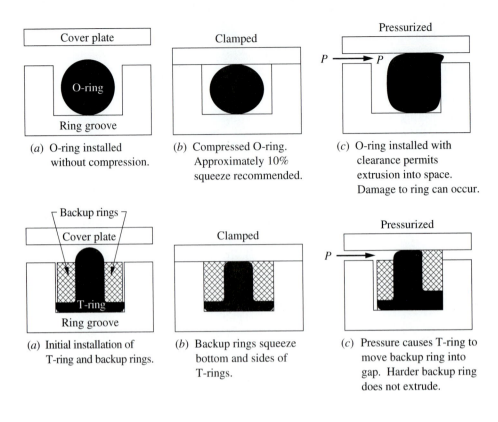

(*a*) O-ring installed
without compression.

(*b*) Compressed O-ring.
Approximately 10%
squeeze recommended.

(*c*) O-ring installed with
clearance permits
extrusion into space.
Damage to ring can occur.

(*a*) Initial installation of
T-ring and backup rings.

(*b*) Backup rings squeeze
bottom and sides of
T-rings.

(*c*) Pressure causes T-ring to
move backup ring into
gap. Harder backup ring
does not extrude.

The number of designs for sealing systems is virtually limitless, and only a brief overview will be presented here. More comprehensive coverage can be found in Reference 8. Often, designers rely on technical information provided by manufacturers of complete sealing systems or specific sealing elements. Also, in critical or unusual situations, testing of a proposed design is advised.

The choice of a type of sealing system depends on the duty that it must perform. Common conditions in which seals must operate are listed below, along with some of the types of seals used.

1. Static conditions such as sealing a closure on a pressurized container: elastomeric O-rings; T-rings; hollow metal O-rings; and sealants such as epoxies, silicones, and butyl caulking (Figure 11–32).

2. Sealing a closed container while allowing relative movement of some part, such as diaphragms, bellows, and boots (Figure 11–33).

3. Sealing around a continuously reciprocating rod or piston, such as in a hydraulic cylinder or a spool valve in a hydraulic system: lip seal; U-cup seal; V-packing; and split ring seals, sometimes called *piston rings* (Figure 11–34).

4. Sealing around a rotating shaft such as the input or output shafts of a speed reducer, transmission, or engine: lip seal; wipers and scrapers; and face seals (Figure 11–35).

5. Protection of rolling-element bearings supporting shafts to keep contaminants from the balls and rollers (Figure 11–36).

6. Sealing the active elements of a pump to retain the pumped fluid: face seals and V-packing.

FIGURE 11–33
Application of a diaphragm seal

Axial movement

Vent

Deflected diaphragm

Initial position of diaphragm

Pressure

7. Sealing infrequently moved elements such as a valve stem of a fluid-flow control valve: compression packings and V-packings.

8. Sealing between hard, rigid surfaces such as between a cylinder head and the block of an engine: resilient gaskets.

9. Circumferential seals, such as at the tips of turbine blades, and on large, high-speed rotating elements: labyrinth seals; abradable seals; and hydrostatic seals.

11–11
SEAL MATERIALS

Most seal materials are resilient to permit the sealing points to follow minor variations in the geometry of mating surfaces. Flexing of parts of the seal cross section also occurs in some designs, calling for resiliency in the materials. Alternatively, as in the case of hollow metal O-rings, the shape of the seal allows the flexing of hard materials to occur. Face seals require rigid, hard materials that can withstand constant sliding motion and that can be produced with fine accuracy, flatness, and smoothness.

Elastomers

Resilient seals such as O-rings, T-rings, and lip seals are often made from synthetic elastomers such as the following:

Neoprene	Butyl	Nitrile (Buna-*N*)
Fluorocarbon	Silicone	Fluorosilicone
Butadiene	Polyester	Ethylene propylene
Polysulfide	Polyurethane	Epichlorohydrin
Polyacrylate	PNF	
	(Phosphonitrilic fluoroelastomer)	

FIGURE 11–34 Lip seals, U-cup seal, wiper, and O-rings applied to a hydraulic actuator

Many proprietary formulations within these general classifications are available under trade names from seal producers and plastics manufacturers.

Properties needed in a given installation will limit the selection of candidate materials. The following list gives some of the more prevalent requirements for seals and some of the materials that meet those requirements:

Weather resistance: Silicone, fluorosilicone, fluorocarbon, ethylene propylene, polyurethane, polysulfide, polyester, neoprene, epichlorohydrin, and PNF.

Petroleum fluid resistance: Polyacrylate, polyester, PNF, nitrile, polysulfide, polyurethane, fluorocarbon, and epichlorohydrin.

Acid resistance: Fluorocarbon.

High-temperature operation: Ethylene propylene, fluorocarbon, polyacrylate, silicone, and PNF.

Cold-temperature operation: Silicone, fluorosilicone, ethylene propylene, and PNF.

Tensile strength: Butadiene, polyester, and polyurethane.

Abrasion resistance: Butadiene, polyester, and polyurethane.

Impermeability: Butyl, polyacrylate, polysulfide, and polyurethane.

FIGURE 11–35 Face seals

(*a*) Lip-type face seal

(*b*) Mechanical face seal

Rigid Materials

Face seals, and the parts of other types of sealing systems against which elastomers seal, require rigid materials that can withstand the sliding action and that are compatible with the environment around the seal. Some typical rigid materials used in sealing systems are described in the following list:

> *Metals:* Carbon steel, stainless steel, cast iron, nickel alloys, bronze, and tool steels.
> *Plastics:* Nylon, filled polytetrafluoroethylene (PTFE), and polyimide.
> *Carbon, ceramics, tungsten-carbide.*
> *Plating:* Chromium, cadmium, tin, nickel, and silver.
> *Flame-sprayed compounds.*

Packings

Packings for sealing shafts, rods, valve stems, and similar applications are made from a variety of materials, including leather, cotton, flax, several types of plastics, braided or twisted wire made from copper or aluminum, laminated cloth and elastomeric materials, and flexible graphite.

Gaskets

Common gasket materials are cork, cork and rubber compounds, filled rubber, paper, resilient plastics, and foams.

FIGURE 11–36 Seal for ball bearing

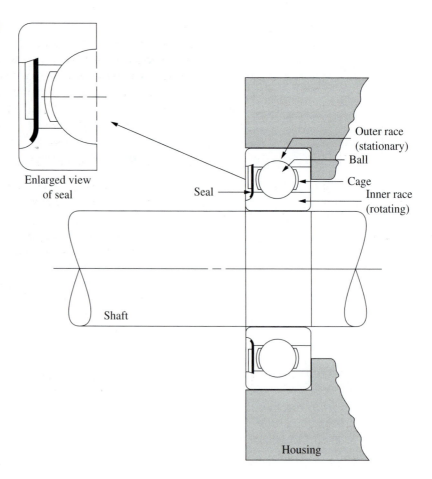

Enlarged view of seal

Outer race (stationary)
Ball
Cage
Inner race (rotating)
Seal
Shaft
Housing

Shafts

When radial lip seals are required around shafts, the shafts are typically steel. They should be hardened to HRC 30 to resist scoring of the surface. Tolerance on the diameter of the shaft on which the seal bears should conform to the following recommendations to ensure that the seal lip can follow the variations:

Shaft Diameter (in)	Tolerance (in)
$D \leq 4.000$	±0.003
$4.000 < D \leq 6.000$	±0.004
$D > 6.000$	±0.005

The surface of the shaft and any areas over which the seal must pass during installation should be free of burrs to protect against tearing the seal. A surface finish of 10 to 20 μin is recommended, with adequate lubrication to ensure full contact and to reduce friction between the seal and the shaft surface.

REFERENCES

1. American Gear Manufacturers Association. AGMA 510.03-1984. *Nomenclature for Flexible Couplings.* Alexandria, VA: American Gear Manufacturers Association, 1984.

2. American Gear Manufacturers Association. ANSI/AGMA 9002-A86. *Bores and Keyways for Flexible Couplings.* Alexandria, VA: American Gear Manufacturers Association, 1986.

3. American Gear Manufacturers Association. ANSI/AGMA 9001-A86. *Lubrication of Flexible Couplings.* Alexandria, VA: American Gear Manufacturers Association, 1986.

4. American National Standards Institute. ANSI B92.1-1970. *Involute Splines.* New York: American National Standards Institute, 1970.

5. American National Standards Institute. ANSI B92.2-1980. *Metric Module Involute Splines.* New York: American National Standards Institute, 1980.

6. American Society of Mechanical Engineers. ANSI B17.1-67. R73/89. *Keys and Keyseats.* New York: American Society of Mechanical Engineers, 1989.

7. American Society of Mechanical Engineers. ANSI B17.2-67. R78/90. *Woodruff Keys and Keyseats.* New York: American Society of Mechanical Engineers, 1990.

8. Lebeck, Alan O. *Principles and Design of Mechanical Face Seals.* New York: John Wiley & Sons, 1991.

9. Oberg, Erik, et al. *Machinery's Handbook.* 25th ed. New York: Industrial Press, 1996.

10. Penton Publishing Co. *Machine Design Magazine Power and Motion Control Volume.* Vol. 61, no. 12. Cleveland, OH: Penton Publishing, June 1989.

PROBLEMS

For Problems 1–4 and 7, determine the required key geometry: length, width, and height. Use AISI 1020 cold-drawn steel for the keys if a satisfactory design can be achieved. If not, use a higher-strength material. Unless otherwise stated, assume that the key material is weakest when compared with the shaft material or the mating elements.

1. Specify a key for a gear to be mounted on a shaft with a 2.00-in diameter. The gear transmits 21 000 lb·in of torque and has a hub length of 4.00 in.

2. Specify a key for a gear carrying 21 000 lb·in of torque if it is mounted on a 3.60-in-diameter shaft. The hub length of the gear is 4.00 in.

3. A V-belt sheave transmits 1 112 lb·in of torque to a 1.75-in-diameter shaft. The sheave is made from ASTM class 20 cast iron and has a hub length of 1.75 in.

4. A chain sprocket delivers 110 hp to a shaft at a rotational speed of 1 700 rpm. The bore of the sprocket is 2.50 inches in diameter. The hub length is 3.25 in.

5. Specify a suitable spline having a *B* fit for each of the applications in Problems 1–4.

6. Design a cylindrical pin to transmit the power, as in Problem 4. But design it so that it will fail in shear if the power exceeds 220 hp.

7. Specify a key for both the sprocket and the wormgear from Example Problem 12–4. Note the specifications for the final shaft diameters at the end of the problem.

8. Describe a Woodruff key no. 204.

9. Describe a Woodruff key no. 1628.

10. Make a detailed drawing of a Woodruff key connection between a shaft and the hub of a gear. The shaft has a diameter of 1.500 in. Use a no. 1210 Woodruff key. Dimension the keyseat in the shaft and the hub.

11. Repeat Problem 10, using a no. 406 Woodruff key in a shaft having a 0.500-in diameter.

12. Repeat Problem 10, using a no. 2428 Woodruff key in a shaft having a 3.250-in diameter.

13. Compute the torque that could be transmitted by the key of Problem 10 on the basis of shear and bearing if the key is made from AISI 1020 cold-drawn steel with a design factor of $N = 3$.

14. Repeat Problem 13 for the key of Problem 11.

15. Repeat Problem 13 for the key of Problem 12.

16. Make a drawing of a four-spline connection having a major diameter of 1.500 in and an *A* fit. Show critical dimensions.

17. Make a drawing of a 10-spline connection having a major diameter of 3.500 in and a *B* fit. Show critical dimensions.

18. Make a drawing of a 16-spline connection having a major diameter of 2.500 in and a *C* fit. Show critical dimensions.

19. Determine the torque capacity of the splines in Problems 16–18:

20. Describe the manner in which a set screw transmits torque if it is used in place of a key. Discuss the disadvantages of such an arrangement.

21. Describe a press fit as it would be used to secure a power transmission element to a shaft.

22. Describe the main differences between rigid and flexible couplings as they affect the stresses in the shafts that they connect.

23. Discuss a major disadvantage of using a single universal joint to connect two shafts with angular misalignment.

24. Describe five ways to locate power transmission elements positively on a shaft axially.

25. Describe three situations in which seals are applied in machine design.

26. List eight parameters that should be considered when selecting a type of seal and specifying a particular design.

27. Name three means of sealing a pressurized container under static conditions.

28. Name three methods of sealing a closed container while allowing relative movement of some part.

29. Name four types of seals used on reciprocating rods or pistons.

30. Name three types of seals applied to rotating shafts.

31. Describe the method of sealing a ball bearing from contaminants.

32. Describe an O-ring seal, and sketch its installation.

33. Describe a T-ring seal, and sketch its installation.

34. Describe some advantages of T-ring seals over O-rings.

35. Describe a diaphragm seal and the type of situation in which it is used.

36. Describe suitable methods of sealing the sides of a piston against the inner walls of the cylinder of a hydraulic actuator.

37. Describe the function of a scraper or wiper on a cylinder rod.

38. Describe the essential elements of a mechanical face seal.

39. Name at least six kinds of elastomers commonly used for seals.

40. Name at least three kinds of elastomers that are recommended for use when exposed to weather.

41. Name at least three kinds of elastomers that are recommended for use when exposed to petroleum-based fluids.

42. Name at least three kinds of elastomers that are recommended for use when exposed to cold-temperature operation.

43. Name at least three kinds of elastomers that are recommended for use when exposed to high-temperature operation.

44. A sealing application involves the following conditions: exposure to high-temperature petroleum fluids; impermeability. Specify a suitable elastomer for the seal.

45. A sealing application involves the following conditions: exposure to high temperature and weather; impermeability; and high strength and abrasion resistance. Specify a suitable elastomer for the seal.

46. Describe suitable shaft design details where elastomeric seals contact the shaft.

12

Shaft Design

The Big Picture

You Are the Designer

Shaft Design

Discussion Map

☐ A shaft is a rotating machine component that transmits power.

Discover

Identify examples of mechanical systems that incorporate power-transmitting shafts. Describe their geometry and the forces and torques that are exerted on them.

What kinds of stresses are produced in the shaft?

How are other elements mounted on the shaft? How does the shaft geometry accommodate them? How is the shaft supported? What kinds of bearings are used?

This chapter provides approaches that you can use to design shafts that are safe for their intended use. But you have the final responsibility for the design.

A *shaft* is the component of a mechanical device that transmits rotational motion and power. It is integral to any mechanical system in which power is transmitted from a prime mover, such as an electric motor or an engine, to other rotating parts of the system. Can you identify some kinds of mechanical systems incorporating rotating elements that transmit power?

Here are some examples: gear-type speed reducers, belt or chain drives, conveyors, pumps, fans, agitators, and many types of automation equipment. What others can you think of? Consider household appliances, lawn maintenance equipment, parts of a car, power tools, and machines around an office or in your workplace. Describe them and discuss how shafts are used. From what source is power delivered into the shaft? What kind of power-transmitting element, if any, is on the shaft itself? Or is the shaft simply transmitting the rotational motion and torque to some other element? If so, how is the shaft connected to that element?

Visualize the forces, torques, and bending moments that are created in the shaft during operation. In the process of transmitting power at a given rotational speed, the shaft is inherently subjected to a torsional moment, or *torque*. Thus, torsional shear stress is developed in the shaft. Also, a shaft usually carries power-transmitting components, such as gears, belt sheaves, or chain sprockets, which exert forces on the shaft in the transverse direction (perpendicular to its axis). These transverse forces cause bending moments to be developed in the shaft, requiring analysis of the stress due to bending. In fact, most shafts must be analyzed for combined stress.

Describe the specific geometry of shafts from some types of equipment that you can examine. Make a sketch of any variations in geometry that may occur, such as changes in diameter, to produce shoulders, grooves, keyseats, or holes. How are any power-transmitting elements held in position along the length of the shaft? How are the shafts supported? Typically, bearings are used to support the shaft while permitting rotation relative to the housing of the machine. What kinds of bearings are used? Do they have rolling elements such as ball bearings? Or are they smooth-surfaced bearings? What materials are used?

It is likely that you will find much diversity in the design of the shafts in different kinds of equipment. You should see that the functions of a shaft have a large influence on its design. Shaft geometry is greatly affected by the mating elements such as bearings, couplings, gears, chain sprockets, or other kinds of power-transmitting elements.

This chapter provides approaches that you can use to design shafts that are safe for their intended use. But the final responsibility for the design is yours because it is impractical to predict all conditions to which a given shaft will be subjected.

You Are the Designer

Consider the gear-type speed reducer shown in Figures 1–11 and 8–3. As you continue the design, three shafts must be designed. The input shaft carries the first gear in the gear train and rotates at the speed of the prime mover, typically an electric motor or an engine. The middle shaft carries two gears and rotates more slowly than the input shaft because of the first stage of speed reduction. The final gear in the train is carried by the third shaft, which also transmits the power to the driven machine. From what material should each shaft be made? What

torque is being transmitted by each shaft, and over what part of the shaft is it acting? How are the gears to be located on the shafts? How is the power to be transmitted from the gears to the shafts, or vice versa? What forces are exerted on the shaft by the gears, and what bending moments result? What forces must the bearings that support each shaft resist? What are the minimum acceptable diameters for the shafts at all sections to ensure safe operation? What should be the final dimensional specifications for the many features of the shafts, and what should be the tolerances on those dimensions? The material in this chapter will help you make these and other shaft design decisions.

**12–1
OBJECTIVES OF
THIS CHAPTER**

After completing this chapter, you will be able to:

1. Propose reasonable geometries for shafts to carry a variety of types of power-transmitting elements, providing for the secure location of each element and the reliable transmission of power.

2. Compute the forces exerted on shafts by gears, belt sheaves, and chain sprockets.

3. Determine the torque distribution on shafts.

4. Prepare shearing force and bending moment diagrams for shafts in two planes.

5. Account for stress concentration factors commonly encountered in shaft design.

6. Specify appropriate design stresses for shafts.

7. Apply the shaft design procedure recommended by the standard, ANSI B106.1M-1985, *Design of Transmission Shafting,* to determine the required diameter of shafts at any section to resist the combination of torsional shear stress and bending stress.

8. Specify reasonable final dimensions for shafts that satisfy strength requirements and installation considerations and that are compatible with the elements mounted on the shafts.

**12–2
SHAFT DESIGN
PROCEDURE**

Because of the simultaneous occurrence of torsional shear stresses and normal stresses due to bending, the stress analysis of a shaft virtually always involves the use of a combined stress approach. The recommended approach for shaft design and analysis is the *distortion energy theory of failure*. This theory was introduced in Chapter 5 and will be discussed more fully in Section 12–5. Vertical shear stresses and direct normal stresses due to axial loads also occur at times, but they typically have such a small effect that they can be neglected. On very short shafts or on portions of shafts where no bending or torsion occurs, such stresses may be dominant. For a shaft that is not rotating, the discussions in Chapters 3, 4, and 5 explain the appropriate analysis.

The specific tasks to be performed in the design and analysis of a shaft depend on the shaft's proposed design in addition to the manner of loading and support. With this in mind, the following is a recommended procedure for the design of a shaft.

Procedure for Design of a Shaft

1. Determine the rotational speed of the shaft.

FIGURE 12–1

Intermediate shaft for a
double-reduction, spur
gear–type speed
reducer

Spur gear: 20 in
20° F D teeth

Spur gear: 10 in
20° F D teeth

Shaft to be
designed,
mating gears
not shown in
this view

← 10 in → ← 15 in → ← 10 in →

Pitch circle

Outside
diameter
of gear

Rotation

Gear pairs as viewed
from right end of shaft

Input: Gear *P* drives gear *A*   *Output:* Gear *C* drives gear *Q*

2. Determine the power or the torque to be transmitted by the shaft.

3. Determine the design of the power-transmitting components or other devices that will be mounted on the shaft, and specify the required location of each device.

4. Specify the location of bearings to support the shaft. The reactions on bearings supporting radial loads are assumed to act at the midpoint of the bearings. For example, if a single-row ball bearing is used, the load path is assumed to pass directly through the balls. If thrust (axial) loads exist in the shaft, you must specify which bearing is to be designed to react against the thrust load. Then the bearing that does not resist the thrust should be permitted to move slightly in the axial direction to ensure that no unexpected and undesirable thrust load is exerted on that bearing.

   Another important concept is that normally two and only two bearings are used to support a shaft. They should be placed on either side of the power-transmitting elements if possible to provide stable support for the shaft and to produce reasonably well-balanced loading of the bearings. The bearings should be placed close to the power-transmitting elements to minimize bending moments. Also, the overall length of the shaft should be kept small to keep deflections at reasonable levels.

5. Propose the general form of the geometry for the shaft, considering how each element on the shaft will be held in position axially and how power transmission from each element to the shaft is to take place. For example, consider the shaft in Figure 12–1, which is to carry two gears as the intermediate shaft in a double-reduction, spur gear–type speed reducer. Gear *A* accepts power from gear *P* by way of the input shaft. The power is transmitted from gear *A* to the shaft through the key at the interface between the gear hub and the shaft. The power is then transmitted down the shaft to point *C,* where it passes through another key into gear *C.* Gear *C* then transmits the power to gear *Q* and thus to the output shaft.

FIGURE 12–2

Proposed geometry for the shaft in Figure 12–1. Sharp fillets at r_3, r_5; well-rounded fillets at r_1, r_2, r_4; profile keyseats at A, C.

Sharp fillets at r_3, r_5
Well-rounded fillets at r_1, r_2, r_4
Profile keyseats at A, C

It is now decided that the bearings will be placed at points B and D on the shaft to be designed. But how will the bearings and the gears be located so as to ensure that they stay in position during operation, handling, shipping, and so forth? Of course, there are many ways to do this. One way is proposed in Figure 12–2. Shoulders are to be machined in the shaft to provide surfaces against which to seat the bearings and the gears on one side of each element. The gears are restrained on the other side by retaining rings snapped into grooves in the shaft. The bearings will be held in position by the housing acting on the outer races of the bearings. Keyseats will be machined in the shaft at the location of each gear. This proposed geometry provides for positive location of each element.

6. Determine the magnitude of torque that the shaft sees at all points. It is recommended that a torque diagram be prepared, as will be shown later.

7. Determine the forces that are exerted on the shaft, both radially and axially.

8. Resolve the radial forces into components in perpendicular directions, usually vertically and horizontally.

9. Solve for the reactions on all support bearings in each plane.

10. Produce the complete shearing force and bending moment diagrams to determine the distribution of bending moments in the shaft.

11. Select the material from which the shaft will be made, and specify its condition: cold-drawn, heat-treated, and so on.

12. Determine an appropriate design stress, considering the manner of loading (smooth, shock, repeated and reversed, or other).

13. Analyze each critical point of the shaft to determine the minimum acceptable diameter of the shaft at that point in order to ensure safety under the loading at that point. In general, the critical points are several and include those where a change of diameter takes place, where the higher values of torque and bending moment occur, and where stress concentrations occur.

14. Specify the final dimensions for each point on the shaft. Normally, the results from Step 13 are used as a guide, and convenient values are then chosen. Design details such as tolerances, fillet radii, shoulder heights, and keyseat dimensions

must also be specified. Sometimes the size and the tolerance for a shaft diameter are dictated by the element to be mounted there. For example, ball bearing manufacturers' catalogs give recommended limits for bearing seat diameters on shafts.

This process will be demonstrated after the concepts of force and stress analysis are presented.

12–3 FORCES EXERTED ON SHAFTS BY MACHINE ELEMENTS

Gears, belt sheaves, chain sprockets, and other elements typically carried by shafts exert forces on the shaft that cause bending moments. The following is a discussion of the methods for computing these forces for some cases. In general, you will have to use the principles of statics and dynamics to determine the forces for any particular element.

Spur Gears

The force exerted on a gear tooth during power transmission acts normal (perpendicular) to the involute-tooth profile, as discussed in Chapter 9 and shown in Figure 12–3. It is convenient for the analysis of shafts to consider the rectangular components of this force acting in the radial and tangential directions. It is most convenient to compute the tangential force, W_t, directly from the known torque being transmitted by the gear. For U.S. Customary units,

 Torque

$$T = 63\ 000(P)/n \qquad \text{(12–1)}$$

Tangential Force

$$W_t = T/(D/2) \qquad \text{(12–2)}$$

where P = power being transmitted in hp
n = rotational speed in rpm
T = torque on the gear in lb·in
D = pitch diameter of the gear in inches

FIGURE 12–3
Forces on gear teeth

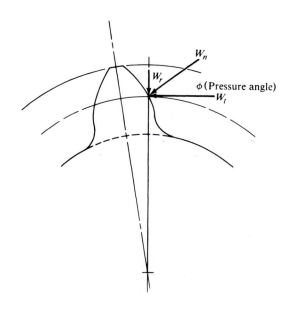

The angle between the total force and the tangential component is equal to the pressure angle, ϕ, of the tooth form. Thus, if the tangential force is known, the radial force can be computed directly from

▷ **Radial Forces**

$$W_r = W_t \tan \phi \qquad (12\text{–}3)$$

and there is no need to compute the total force at all. For gears, the pressure angle is typically $14\frac{1}{2}°$, $20°$, or $25°$.

Helical Gears

In addition to the tangential and radial forces encountered with spur gears, helical gears produce an axial force (as discussed in Chapter 10). First compute the tangential force from Equations (12–1) and (12–2). Then, if the helix angle of the gear is ψ, and if the normal pressure angle is ϕ_n, the radial load can be computed from

▷ **Radial Force**

$$W_r = W_t \tan \phi_n / \cos \psi \qquad (12\text{–}4)$$

The axial load is

▷ **Axial Force**

$$W_a = W_t \tan \psi \qquad (12\text{–}5)$$

Bevel Gears

Refer to Chapter 10 to review the formulas for the three components of the total force on bevel gear teeth in the tangential, radial, and axial directions. Example Problem 10–4 gives a comprehensive analysis of the forces, torques, and bending moments on shafts carrying bevel gears.

Worms and Wormgears

Chapter 10 also gives the formulas for computing the forces on worms and wormgears in the tangential, radial, and axial directions. See Example Problem 10–10.

Chain Sprockets

Figure 12–4 shows a pair of chain sprockets transmitting power. The upper part of the chain is in tension and produces the torque on either sprocket. The lower part of the chain, referred to as the *slack side,* exerts no force on either sprocket. Therefore, the total bending force on the shaft carrying the sprocket is equal to the tension in the tight side of the chain. If the torque on a certain sprocket is known,

▷ **Force in Chain**

$$F_c = T/(D/2) \qquad (12\text{–}6)$$

where D = pitch diameter of that sprocket

Notice that the force, F_c, acts along the direction of the tight side of the belt. Because of the size difference between the two sprockets, that direction is at some angle from the

FIGURE 12–4

Forces on chain sprockets

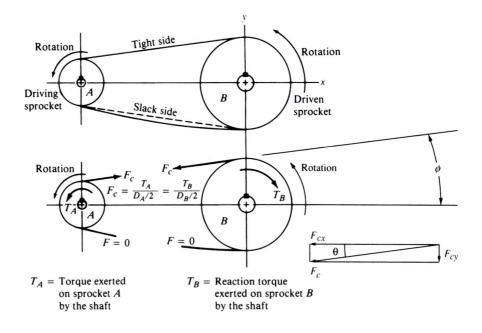

T_A = Torque exerted on sprocket A by the shaft

T_B = Reaction torque exerted on sprocket B by the shaft

centerline between the shaft centers. A precise analysis would call for the force, F_c, to be resolved into components parallel to the centerline and perpendicular to it. That is,

$$F_{cx} = F_c \cos \theta \quad \text{and} \quad F_{cy} = F_c \sin \theta$$

where the x-direction is parallel to the centerline

the y-direction is perpendicular to it

the angle θ is the angle of inclination of the tight side of the chain with respect to the x-direction

These two components of the force would cause bending in both the x-direction and the y-direction. Alternatively, the analysis could be carried out in the direction of the force, F_c, in which single plane bending occurs.

If the angle θ is small, little error will result from the assumption that the entire force, F_c, acts along the x-direction. Unless stated otherwise, this book will use this assumption.

V-Belt Sheaves

The general appearance of the V-belt drive system looks similar to the chain drive system. But there is one important difference: Both sides of the V-belt are in tension, as indicated in Figure 12–5. The tight side tension, F_1, is greater than the "slack side" tension, F_2, and thus there is a net driving force on the sheaves equal to

▷ **Net Driving Force**

$$F_N = F_1 - F_2 \tag{12–7}$$

The magnitude of the net driving force can be computed from the torque transmitted:

▷ **Net Driving Force**

$$F_N = T/(D/2) \tag{12–8}$$

FIGURE 12–5

Forces on belt sheaves or pulleys

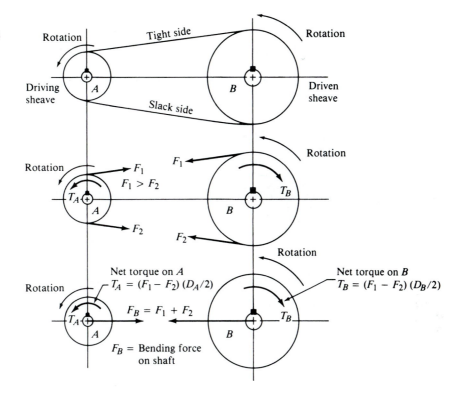

But notice that the bending force on the shaft carrying the sheave is dependent on the *sum*, $F_1 + F_2 = F_B$. To be more precise, the components of F_1 and F_2 parallel to the line of centers of the two sprockets should be used. But unless the two sprockets are radically different in diameter, little error will result from $F_B = F_1 + F_2$.

To determine the bending force, F_B, a second equation involving the two forces F_1 and F_2 is needed. This is provided by assuming a ratio of the tight side tension to the slack side tension. For V-belt drives, the ratio is normally taken to be

$$F_1/F_2 = 5 \qquad\qquad (12\text{–}9)$$

It is convenient to derive a relationship between F_N and F_B of the form

$$F_B = CF_N \qquad\qquad (12\text{–}10)$$

where C = constant to be determined

$$C = \frac{F_B}{F_N} = \frac{F_1 + F_2}{F_1 - F_2} \qquad\qquad (12\text{–}11)$$

But from Equation (12–9), $F_1 = 5F_2$. Then

$$C = \frac{F_1 + F_2}{F_1 - F_2} = \frac{5F_2 + F_2}{5F_2 - F_2} = \frac{6F_2}{4F_2} = 1.5$$

Equation (12–10) then becomes, for V-belt drives,

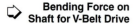

⇨ **Bending Force on Shaft for V-Belt Drive**

$$F_B = 1.5F_N = 1.5T/(D/2) \qquad \textbf{(12–12)}$$

It is customary to consider the bending force, F_B, to act as a single force in the direction along the line of centers of the two sheaves as shown in Figure 12–5.

Flat-Belt Pulleys

The analysis of the bending force exerted on shafts by flat-belt pulleys is identical to that for V-belt sheaves except that the ratio of the tight side to the slack side tension is typically taken to be 3 instead of 5. Using the same logic as with V-belt sheaves, we can compute the constant C to be 2.0. Then, for flat-belt drives,

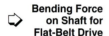

⇨ **Bending Force on Shaft for Flat-Belt Drive**

$$F_B = 2.0F_N = 2.0T/(D/2) \qquad \textbf{(12–13)}$$

Flexible Couplings

More detailed discussion of flexible couplings was presented in Chapter 11, but it is important to observe here how the use of a flexible coupling affects the design of a shaft.

A flexible coupling is used to transmit power between shafts while accommodating minor misalignments in the radial, angular, or axial directions. Thus, the shafts adjacent to the couplings are subjected to torsion, but the misalignments cause no axial or bending loads.

12–4 STRESS CONCENTRATIONS IN SHAFTS

In order to mount and locate the several types of machine elements on shafts properly, a final design typically contains several diameters, keyseats, ring grooves, and other geometric discontinuities that create stress concentrations. The shaft design proposed in Figure 12–2 is an example of this observation.

These stress concentrations must be taken into account during the design analysis. But a problem exists because the true design values of the stress concentration factors, K_t, are unknown at the start of the design process. Most of the values are dependent on the diameters of the shaft and on the fillet and groove geometries, and these are the objectives of the design.

You can overcome this dilemma by establishing a set of preliminary design values for commonly encountered stress concentration factors, which can be used to produce initial estimates for the minimum acceptable shaft diameters. Then, after the refined dimensions are selected, you can analyze the final geometry to determine the real values for stress concentration factors. Comparing the final values with the preliminary values will enable you to judge the acceptability of the design. The charts from which the final values of K_t can be determined are found in Figures A15–1 and A15–4.

Preliminary Design Values for K_t

Considered here are the types of geometric discontinuities most often found in power-transmitting shafts: keyseats, shoulder fillets, and retaining ring grooves. In each case, a suggested design value is relatively high in order to produce a conservative result for the first approximation to the design. Again it is emphasized that the final design should be checked for safety. That is, if the final value is less than the original design value, the design is still safe. Conversely, if the final value is higher, the stress analysis for the design must be rechecked.

Keyseats. A *keyseat* is a longitudinal groove cut into a shaft for the mounting of a key, permitting the transfer of torque from the shaft to a power-transmitting element, or vice versa. The detail design of keys was covered in Chapter 11.

Two types of keyseats are most frequently used: profile and sled runner (see Figure 12–6). The profile keyseat is milled into the shaft, using an end mill having a diameter equal to the width of the key. The resulting groove is flat-bottomed and has a sharp, square corner at its end. The sled runner keyseat is produced by a circular milling cutter having a width equal to the width of the key. As the cutter begins or ends the keyseat, it produces a smooth radius. For this reason, the stress concentration factor for the sled runner keyseat is lower than that for the profile keyseat. Normally used design values are

$$K_t = 2.0 \quad \text{(profile)}$$
$$K_t = 1.6 \quad \text{(sled runner)}$$

Each of these is to be applied to the bending stress calculation for the shaft, using the full diameter of the shaft. The factors take into account both the reduction in cross section and the effect of the discontinuity. Consult the references listed for more detail about stress concentration factors for keyseats. (See Reference 5.) If the torsional shear stress is fluctuating rather than steady, the stress concentration factor is also applied to that.

Shoulder Fillets. When a change in diameter occurs in a shaft to create a shoulder against which to locate a machine element, a stress concentration dependent on the ratio of the two diameters and on the radius in the fillet is produced (see Figure 12–7). It is recommended that the fillet radius be as large as possible to minimize the stress concentration, but at times the design of the gear, bearing, or other element affects the radius that can be

FIGURE 12–6
Keyseats

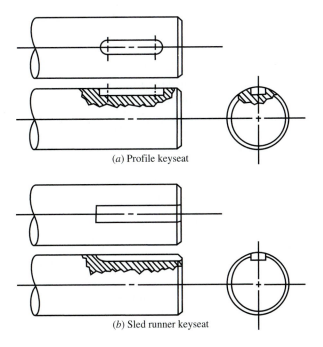

(*a*) Profile keyseat

(*b*) Sled runner keyseat

FIGURE 12–7

Fillets on shafts

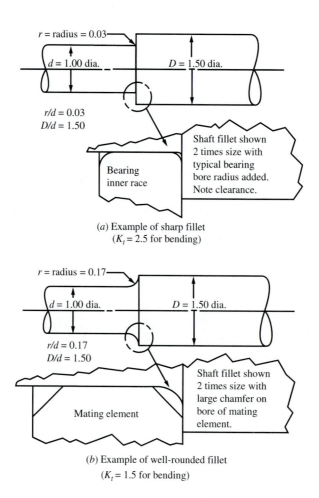

(*a*) Example of sharp fillet
($K_t = 2.5$ for bending)

(*b*) Example of well-rounded fillet
($K_t = 1.5$ for bending)

used. For the purpose of design, we will classify fillets into two categories: sharp and well-rounded.

The term *sharp* here does not mean truly sharp, without any fillet radius at all. Such a shoulder configuration would have a very high stress concentration factor and should be avoided. Instead, *sharp* describes a shoulder with a relatively small fillet radius. One situation in which this is likely to occur is where a ball or roller bearing is to be located. The inner race of the bearing has a factory-produced radius, but it is small. The fillet radius on the shaft must be smaller yet in order for the bearing to be seated properly against the shoulder. When an element with a large chamfer on its bore is located against the shoulder, or when nothing at all seats against the shoulder, the fillet radius can be much larger (*well-rounded*), and the corresponding stress concentration factor is smaller. We will use the following values for design for bending:

$$K_t = 2.5 \quad \text{(sharp fillet)}$$
$$K_t = 1.5 \quad \text{(well-rounded fillet)}$$

Referring to the chart for stress concentration factors in Figure A15–1, you can see that these values correspond to ratios of *r/d* of approximately 0.03 for the sharp fillet case and 0.17 for the well-rounded fillet for a *D/d* ratio of 1.50.

Retaining Ring Grooves.   Retaining rings are used for many types of locating tasks in shaft applications. The rings are installed in grooves in the shaft after the element to be retained is in place. The geometry of the groove is dictated by the ring manufacturer. Its usual configuration is a shallow groove with straight side walls and bottom and a small fillet at the base of the groove. The behavior of the shaft in the vicinity of the groove can be approximated by considering two sharp-filleted shoulders positioned close together. Thus, the stress concentration factor for a groove is fairly high.

When bending exists, we will use $K_t = 3.0$ for preliminary design as an estimate to account for the fillets and the reduction in diameter at the groove to determine the nominal shaft diameter before the groove is cut. When torsion exists along with bending, or when only torsion exists at a section of interest, the stress concentration factor is not applied to the torsional shear stress component because it is steady. To account for the decrease in diameter at the groove, however, increase the resulting computed diameter by approximately 6%, a typical value for commercial retaining ring grooves. But after the final shaft diameter and groove geometry are specified, the stress in the groove should be computed with the appropriate stress concentration factor for the groove geometry.

12–5
DESIGN
STRESSES
FOR SHAFTS

In a given shaft, several different stress conditions can exist at the same time. For any part of the shaft that transmits power, there will be a torsional shear stress, while bending stress is usually present on the same parts. There may be other parts where only bending stresses occur. Some points may not be subjected to either bending or torsion but will experience vertical shearing stress. Axial tensile or compressive stresses may be superimposed on the other stresses. Then there may be some points where no significant stresses at all are created.

Thus, the decision of what design stress to use depends on the particular situation at the point of interest. In many shaft design and analysis projects, computations must be done at several points to account completely for the variety of loading and geometry conditions that exist.

Several cases discussed in Chapter 5 for computing design factors, N, are useful for determining design stresses for shaft design. The bending stresses will be assumed to be completely reversed and repeated because of the rotation of the shaft. Because ductile materials perform better under such loads, it will be assumed that the material for the shaft is ductile. It will also be assumed that the torsional loading is relatively constant and acting in one direction. If other situations exist, consult the appropriate case from Chapter 5.

The symbol τ_d will be used for the design stress when a shear stress is the basis for the design. The symbol σ_d will be used when a normal stress is the basis.

Design Shear Stress—Steady Torque

It was stated in Chapter 5 that the best predictor of failure in ductile materials due to a steady shear stress was the distortion energy theory in which the design shear stress is computed from

$$\tau_d = s_y/(N\sqrt{3}) = (0.577s_y)/N \qquad \text{(12–14)}$$

We will use this value for steady torsional shear stress, vertical shear stress, or direct shear stress in a shaft.

Design Shear Stress—Reversed Vertical Shear

Points on a shaft where no torque is applied and where the bending moments are zero or very low are often subjected to significant vertical shearing forces which then govern the design analysis. This typically occurs where a bearing supports an end of a shaft and where no torque is transmitted in that part of the shaft.

Figure 12–8(a) shows the distribution of vertical shearing stresses on such a circular cross section. Note that the maximum shearing stress is at the neutral axis of the shaft, that is, at the diameter. The stress decreases in a roughly parabolic manner to zero at the outer surface of the shaft. Recall from strength of materials that the maximum vertical shearing stress for the special case of a solid circular cross section can be computed from

$$\tau_{max} = 4V/3A$$

where V = vertical shearing force
A = area of the cross section

Where stress concentration factors are to be considered,

$$\tau_{max} = K_t(4V/3A)$$

Also note, as shown in Figure 12–8, that the rotation of the shaft causes any point at the outer part of the cross section to experience a reversing shearing stress that varies from $+\tau_{max}$ to zero to $-\tau_{max}$ to zero in each revolution. Then the stress analysis should be completed using Equation (5–15) in Case E in Section 5–9 of Chapter 5 dealing with design stresses:

$$N = s'_{sn}/\tau_{max} \qquad\qquad (5\text{–}15)$$

We recommend using the distortion energy theory. Then the endurance strength in shear is

$$s'_{sn} = 0.577s_n$$

Thus, Equation (5–15) can be written in the form

$$N = 0.577s'_n/\tau_{max}$$

Expressed as a design stress, this is

$$\tau_d = 0.577s'_n/N$$

Now letting $\tau_{max} = \tau_d = K_t(4V)/3A$ gives

FIGURE 12–8
Shearing stress in a
rotating shaft due to
vertical shearing
force, V

(a) Distribution of shearing
stress on a circular
cross section of a shaft

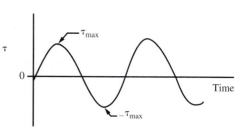

(b) Variation of shearing stress
on a given element at the
surface of a circular rotating shaft

$$\frac{K_t(4V)}{3A} = \frac{0.577s'_n}{N}$$

Solving for N gives

$$N = \frac{0.577s'_n(3A)}{K_t(4V)} = \frac{0.433s'_n(A)}{K_t(V)} \qquad (12\text{--}15)$$

Equation (12–15) is useful if the goal is to evaluate the design factor for a given magnitude of loading, a given geometry for the shaft, and given material properties.

Now, solving for the required area gives

$$A = \frac{K_t(V)N}{0.433s'_n} = \frac{2.31K_t(V)N}{s'_n}$$

But our usual objective is the design of the shaft to determine the required diameter. By substituting

$$A = \pi D^2/4$$

we can solve for D:

**Required Shaft
Diameter**

$$D = \sqrt{2.94K_t(V)N/s'_n} \qquad (12\text{--}16)$$

This equation should be used to compute the required diameter for a shaft where a vertical shearing force V is the only significant loading present. In most shafts, the resulting diameter will be much smaller than that required at other parts of the shaft where significant values of torque and bending moment occur. Also, practical considerations may require that the shaft be somewhat larger than the computed minimum to accommodate a reasonable bearing at the place where the shearing force is equal to the radial load on the bearing.

Implementation of Equations (12–15) and (12–16) has the complication that values for the stress concentration factor under conditions of vertical shearing stress are not well known. Published data, such as that in the Appendices of this book and in Reference 5, report values for stress concentration factors for axial normal stress, bending normal stress, and torsional shear stress. But values for vertical shearing stress are rarely reported. As an approximation, we will use the values for K_t for torsional shear stress when using these equations.

Design Normal Stress—Fatigue Loading

For the repeated, reversed bending in a shaft caused by transverse loads applied to the rotating shaft, the design stress is related to the endurance strength of the shaft material. The actual conditions under which the shaft is *manufactured* and *operated* should be considered when specifying the design stress. The following procedure will be used in this book.

Procedure for Computing Estimated
Endurance Strength and Design Stress

1. Determine the ultimate tensile strength of the material, s_u, from test results, supplier specifications, or published data. The most accurate and reliable data available should be used. When there is doubt about the accuracy of the data, larger-than-average design factors should be used.

FIGURE 12–9 Size factor for shaft design

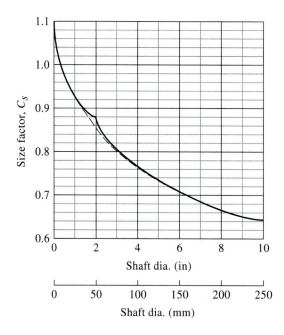

2. Determine the estimated endurance strength, s_n, of the material from Figure 5–9. Recall that the data in this figure consider the manner in which the shaft is produced in addition to the relationship between the basic endurance strength and the ultimate strength. If the ultimate strength is higher than the limit of Figure 5–9 (220 Ksi or 1 500 MPa), use the values for $s_u = 220$ Ksi.

3. Apply a size factor C_s to account for the stress gradient within the material and for the probability of a given section having a damaging occlusion that can serve as a place for initiation of a fatigue crack. Reference 1 recommends the following:

For Diameters Less than 2.0 in (D in Inches)
$$C_s = (D/0.3)^{-0.068}$$

For Diameters Less than 50 mm (D in mm)
$$C_s = (D/7.6)^{-0.068}$$

For Diameters over 2.0 in to 10.0 in (D in Inches)
$$C_s = D^{-0.19}$$

For Diameters over 50 mm to 250 mm (D in mm)
$$C_s = 1.85D^{-0.19}$$

Figure 12–9 shows a plot of these formulas. The two sets of formulas are taken from different sources, and there is some small discontinuity of slope near $D = 2.0$ in (50 mm). The short, dashed line blends the curves and gives slightly conservative values.

TABLE 12–1 Reliability factors, C_R

Desired reliability	Reliability factor, C_R
0.50	1.00
0.90	0.90
0.99	0.81
0.999	0.75

4. Apply a reliability factor, C_R. Endurance strength data typically reported are average values over many tests, thus implying a reliability of 0.50 (50%). Assuming that the actual failure data follow a normal distribution, the factors from Table 12–1 can be used to adjust for higher levels of reliability.

   Note that any stress concentration factor will be accounted for in the design equation developed later. Other factors, not considered here, that could have an adverse effect on the endurance strength of the shaft material, and therefore on the design stress, are temperatures above approximately 400°F (200°C); variation in peak stress levels above the nominal endurance strength for some periods of time; vibration; residual stresses; case hardening; interference fits; corrosion; thermal cycling; plating or surface coatings; and stresses not accounted for in the basic stress analysis. Testing on actual components is recommended for such conditions.

5. Compute $s'_n = s_n C_s C_R$.

6. For parts of the shaft subjected to only reversed bending, let the design stress be

$$\sigma_d = s'_n/N \tag{12–17}$$

Design Factor, N

Under typical industrial conditions, the design factor of $N = 3$ is recommended. If the application is very smooth, a value as low as $N = 2$ may be justified. Under conditions of shock or impact, $N = 4$ or higher should be used, and careful testing is advised.

**12–6
SHAFTS IN
BENDING AND
TORSION ONLY**

Examples of shafts subjected to bending and torsion only are those carrying spur gears, V-belt sheaves, or chain sprockets. The power being transmitted causes the torsion, and the transverse forces on the elements cause bending. In the general case, the transverse forces do not all act in the same plane. In such cases, the bending moment diagrams for two perpendicular planes are prepared first. Then the resultant bending moment at each point of interest is determined. The process will be illustrated in Example Problem 12–1.

A design equation is now developed based on the assumption that the bending stress in the shaft is repeated and reversed as the shaft rotates, but that the torsional shear stress is nearly uniform. The design equation is based on the principle shown graphically in Figure 12–10 in which the vertical axis is the ratio of the reversed bending stress to the endurance strength of the material. (See Reference 7.) The horizontal axis is the ratio of the torsional shear stress to the yield strength of the material in shear. The points having the value of 1.0 on these axes indicate impending failure in pure bending or pure torsion, respectively. Experimental data show that failure under combinations of bending and torsion roughly follows the curve connecting these two points, which obeys the following equation:

$$(\sigma/s'_n)^2 + (\tau/s_{ys})^2 = 1 \tag{12–18}$$

FIGURE 12–10
Basis for shaft design
equation for repeated
reversed bending stress
and steady torsional
shear stress

We will use $s_{ys} = s_y/\sqrt{3}$ for the distortion energy theory. Also, a design factor can be introduced to each term on the left side of the equation to yield an expression based on *design stresses:*

$$(N\sigma/s'_n)^2 + (N\tau\sqrt{3}/s_y)^2 = 1$$

Now we can introduce a stress concentration factor for bending in the first term only, because this stress is repeated. No factor is needed for the torsional shear stress term because it is assumed to be steady, and stress concentrations have little or no effect on the failure potential. Then

$$(K_tN\sigma/s'_n)^2 + (N\tau\sqrt{3}/s_y)^2 = 1 \tag{12–19}$$

For rotating, solid, circular shafts, the bending stress due to a bending moment, *M,* is

$$\sigma = M/S \tag{12–20}$$

where $S = \pi D^3/32$ is the rectangular section modulus. The torsional shear stress is

$$\tau = T/Z_p \tag{12–21}$$

where $Z_p = \pi D^3/16$ is the polar section modulus

Note that $Z_p = 2S$ and that, therefore,

$$\tau = T/(2S)$$

Substituting these relationships into Equation (12–19) gives

$$\left[\frac{K_tNM}{Ss'_n}\right]^2 + \left[\frac{NT\sqrt{3}}{2Ss_y}\right]^2 = 1 \tag{12–22}$$

Now the terms *N* and *S* can be factored out, and the terms $\sqrt{3}$ and 2 can be brought outside the bracket in the torsion term:

$$\left[\frac{N}{S}\right]^2 \left[\left[\frac{K_t M}{s'_n}\right]^2 + \frac{3}{4}\left[\frac{T}{s_y}\right]^2\right] = 1$$

We now take the square root of the entire equation:

$$\frac{N}{S}\sqrt{\left[\frac{K_t M}{s'_n}\right]^2 + \frac{3}{4}\left[\frac{T}{s_y}\right]^2} = 1$$

Let $S = \pi D^3/32$.

$$\frac{32N}{\pi D^3}\sqrt{\left[\frac{K_t M}{s'_n}\right]^2 + \frac{3}{4}\left[\frac{T}{s_y}\right]^2} = 1 \tag{12–23}$$

Now we can solve for the diameter D:

➪ **Design Equation for Shaft Design**

$$D = \left[\frac{32N}{\pi}\sqrt{\left[\frac{K_t M}{s'_n}\right]^2 + \frac{3}{4}\left[\frac{T}{s_y}\right]^2}\right]^{1/3} \tag{12–24}$$

Equation (12–24) is used for shaft design in this book. It is compatible with the standard ANSI B106.1M-1985. (See Reference 1.)

12–7
SHAFT DESIGN
EXAMPLE

Design Example 12–1

Design the shaft shown in Figures 12–1 and 12–2. It is to be machined from AISI 1144 OQT 1000 steel. The shaft is part of the drive for a large blower system supplying air to a furnace. Gear A receives 200 hp from gear P. Gear C delivers the power to gear Q. The shaft rotates at 600 rpm.

Solution

First determine the properties of the steel for the shaft. From Figure A4–2, $s_y = 83\,000$ psi, $s_u = 118\,000$ psi, and the percent elongation is 19%. Thus, the material has good ductility. Using Figure 5–9, we can estimate $s_n = 42\,000$ psi.

A size factor should be applied to the endurance strength because the shaft will be quite large to be able to carry 200 hp. Although we do not know the actual size at this time, we might select $C_s = 0.80$ from Figure 12–9 as an estimate.

A reliability factor should also be specified. This is a design decision. For this problem, let's design for a reliability of 0.99 and use $C_R = 0.81$. Now we can compute the modified endurance strength:

$$s'_n = s_n C_s C_R = (42\,000)(0.80)(0.81) = 27\,200 \text{ psi}$$

The design factor is taken to be $N = 3$. The blower is not expected to present any unusual shock or impact.

Now we can compute the torque in the shaft from Equation (12–1):

$$T = 63\,000(P)/n = 63\,000(200)/600 = 21\,000 \text{ lb·in}$$

Note that only that part of the shaft from A to C is subjected to this torque. There is zero torque from the right of gear C over to bearing D.

Forces on the Gears

Figure 12–11 shows the two pairs of gears with the forces acting *on gears* A *and* C *shown.* Observe that gear A is driven by gear P, and gear C drives gear Q. It is very important for the directions of these forces to be correct. The values of the forces are found from Equations (12–2) and (12–3).

$$W_{tA} = T_A/(D_A/2) = 21\ 000/(20/2) = 2\ 100\ \text{lb}$$
$$W_{rA} = W_{tA} \tan(\phi) = 2\ 100 \tan(20°) = 764\ \text{lb}$$
$$W_{tC} = T_C/(D_C/2) = 21\ 000/(10/2) = 4\ 200\ \text{lb}$$
$$W_{rC} = W_{tC} \tan(\phi) = 4\ 200 \tan(20°) = 1\ 529\ \text{lb}$$

Forces on the Shaft

The next step is to show these forces on the shaft in their proper planes of action and in the proper direction. The reactions at the bearings are computed, and the shearing force and bending moment diagrams are prepared. The results are shown in Figure 12–12.

    We continue the design by computing the minimum acceptable diameter of the shaft at several points along the shaft. At each point, we will observe the magnitude of torque and the bending moment that exist at the point, and we will estimate the value of any stress concentration factors. If more than one stress concentration exist in the vicinity of the point of interest, the larger value is used for design. This assumes that the geometric discontinuities themselves do not interact, which is good practice. For example, at point A, the keyseat should end well before the shoulder fillet begins.

1. *Point A:* Because gear A is located at point A, the shaft from A and to the right is subjected to torque. To the left of A, where there is a retaining ring, there are no forces, moments, or torques. Because it is the free end of the shaft, there is no

FIGURE 12–11

Forces on gears A and C

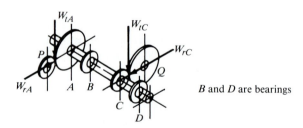

(*a*) Pictorial view of forces on gears A and C

B and D are bearings

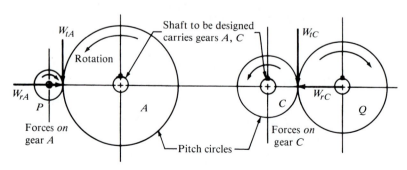

Input: Gear P drives gear A   *Output*: Gear C drives gear Q

(*b*) Forces on gear A   (*c*) Forces on gear C bending moment at A. Stress concentrations can be neglected because the torque is steady. Then, from Equations (12–14) and (12–21),

$$\tau_d = 0.577s_y/N = 0.577(83\ 000)/3 = 15\ 960 \text{ psi}$$

Letting $\tau = \tau_d$ in Equation (12–21) and solving for Z_p gives

$$Z_p = T/\tau_d = 21\ 000/15\ 960 = 1.315 \text{ in}^3$$

But because $Z_p = \pi D_1^3/16$,

$$D_1 = \sqrt[3]{16(Z_p)/\pi} = \sqrt[3]{16(1.315)/\pi} = 1.89 \text{ in}$$

2. **Point B:** Point B is the location of a bearing with a sharp fillet to the right of B and a well-rounded fillet to the left. It is desirable to make D_2 at least slightly smaller than D_3 at the bearing seat to permit the bearing to be slid easily onto the shaft up to the place where it is pressed to its final position. There is usually a light press fit between the bearing bore and the shaft seat.

To the left of B (diameter D_2),

$$T = 21\ 000 \text{ lb·in}$$

The bending moment at B is the resultant of the moment in the x- and y-planes from Figure 12–12:

$$M_B = \sqrt{M_{Bx}^2 + M_{By}^2} = \sqrt{(7\ 640)^2 + (21\ 000)^2} = 22\ 350 \text{ lb·in}$$

$$K_t = 1.5 \quad \text{(well-rounded fillet)}$$

Using Equation (12–24) because of the combined stress condition,

$$D_2 = \left[\left(\frac{32N}{\pi}\right)\sqrt{\left(\frac{K_tM}{s_n'}\right)^2 + \frac{3}{4}\left(\frac{T}{s_y}\right)^2}\right]^{1/3} \tag{12–24}$$

$$D_2 = \left[\frac{32(3)}{\pi}\sqrt{\left[\frac{1.5(22\ 350)}{27\ 200}\right]^2 + \frac{3}{4}\left[\frac{21\ 000}{83\ 000}\right]^2}\right]^{1/3} = 3.37 \text{ in}$$

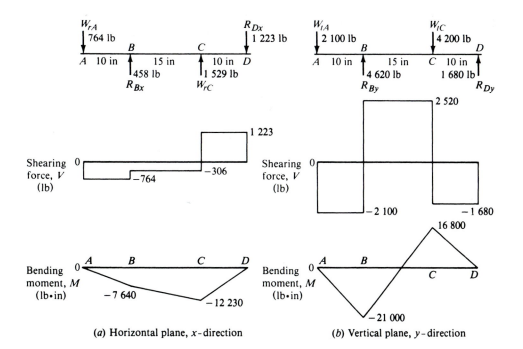

(a) Horizontal plane, x-direction   (b) Vertical plane, y-direction

FIGURE 12–12 Load, shear, and moment diagrams for the shaft in Figure 12–11

At B and to the right of B (diameter D_3), everything is the same, except the value of $K_t = 2.5$ for the sharp fillet. Then

$$D_3 = \left[\frac{32(3)}{\pi} \sqrt{ \left[\frac{2.5(22\ 350)}{27\ 200} \right]^2 + \frac{3}{4} \left[\frac{21\ 000}{83\ 000} \right]^2 } \right]^{1/3} = 3.98 \text{ in}$$

Notice that D_4 will be larger than D_3 in order to provide a shoulder for the bearing. Therefore, it will be safe. Its actual diameter will be specified after we have completed the stress analysis and selected the bearing at B. The bearing manufacturer's catalog will specify the minimum acceptable diameter to the right of the bearing to provide a suitable shoulder against which to seat the bearing.

3. **Point C:** Point C is the location of gear C with a well-rounded fillet to the left, a profile keyseat at the gear, and a retaining ring groove to the right. The use of a well-rounded fillet at this point is actually a design decision that requires that the design of the bore of the gear accommodate a large fillet. Usually this means that a chamfer is produced at the ends of the bore. The bending moment at C is

$$M_c = \sqrt{M_{Cx}^2 + M_{Cy}^2} = \sqrt{(12\ 230)^2 + (16\ 800)^2} = 20\ 780 \text{ lb·in}$$

To the left of C the torque of 21 000 lb·in exists with the profile keyseat giving $K_t = 2.0$. Then

$$D_5 = \left[\frac{32(3)}{\pi} \sqrt{ \left[\frac{2.0(20\ 780)}{27\ 200} \right]^2 + \frac{3}{4} \left[\frac{21\ 000}{83\ 000} \right]^2 } \right]^{1/3} = 3.61 \text{ in}$$

To the right of C there is no torque, but the ring groove suggests $K_t = 3.0$ for design, and there is reversed bending. Using Equations (12–17) and (12–20), we have

$$\sigma_d = s_n'/N = 27\ 200/3 = 9\ 067 \text{ psi}$$

Then let $\sigma = \sigma_d$ in Equation (12–20), and compute the required S:

$$S = K_t M_C/\sigma_d = 3.0(20\ 780)/9\ 067 = 6.88 \text{ in}^3$$

But $S = \pi D_5^3/32$. Then

$$D_5 = \sqrt[3]{32S/\pi} = \sqrt[3]{32(6.88)/\pi} = 4.12 \text{ in}$$

This value is higher than that computed for the left of C, so it governs the design at point C.

4. **Point D:** Point D is the seat for bearing D, and there is no torque or bending moment here. However, there is a vertical shearing force equal to the reaction at the bearing. Using the resultant of the x- and y-plane reactions, the shearing force is

$$V_D = \sqrt{(1\ 223)^2 + (1\ 680)^2} = 2\ 078 \text{ lb}$$

We can use Equation (12–16) to compute the required diameter for the shaft at this point:

$$D = \sqrt{2.94 K_t(V)N/s_n'} \qquad\qquad \textbf{(12–16)}$$

Referring to Figure 12–2, we see a sharp fillet near this point on the shaft. Then a stress concentration factor of 2.5 should be used:

$$D_6 = \sqrt{\frac{2.94(2.5)(2\ 078)(3)}{27\ 200}} = 1.30 \text{ in}$$

This is very small compared to the other computed diameters, and it will usually be so. In reality, the diameter at D will probably be made much larger than this computed value because of the size of a reasonable bearing to carry the radial load of 2 078 lb.

Summary
The computed minimum required diameters for the various parts of the shaft in Figure 12–2 are as follows:

$$D_1 = 1.89 \text{ in}$$
$$D_2 = 3.37 \text{ in}$$
$$D_3 = 3.98 \text{ in}$$
$$D_5 = 4.12 \text{ in}$$
$$D_6 = 1.3 \text{ in}$$

Also, D_4 must be somewhat greater than 4.12 in in order to provide adequate shoulders for gear C and bearing B.

12–8 RECOMMENDED BASIC SIZES FOR SHAFTS

When mounting a commercially available element, of course, follow the manufacturer's recommendation for the basic size of the shaft and the tolerance.

In the U.S. Customary Unit System, diameters are usually specified to be common fractions or their decimal equivalents. Appendix 2 lists the preferred basic sizes that you can use for dimensions over which you have control in decimal-inch, fractional-inch, and metric units. (See Reference 2.)

When commercially available, unmounted bearings are to be used on a shaft, it is likely that their bores will be in metric dimensions. Typical sizes available and their decimal equivalents are listed in Table 14–3.

Design Example 12–2
Specify convenient decimal-inch dimensions for the six diameters from Design Example 12–1. Choose the bearing seat dimensions from Table 14–3. Choose all other dimensions from Appendix 2.

Solution
Table 12–2 shows one possible set of recommended diameters.

Diameters D_3 and D_6 are the decimal equivalents of the metric diameters of the inner races of bearings from Table 14–3. The procedures in Chapter 14 would have to be used to determine whether bearings having those diameters are suitable to carry the given radial loads. Also, D_4 would have to be checked to see whether it provides a sufficiently high shoulder against which to seat the bearing mounted at point B on the shaft. Then detailed specifications for fillet radii, lengths, keyseats, and retaining ring grooves would have to be defined. The actual values for stress concentration factors and the size factor should then

TABLE 12–2 Recommended diameters

Mating part	Diameter number (from Design Example 12–1)	Minimum diameter	Specified diameter (basic size)
Gear	D_1	1.89 in	2.000 in
Nothing	D_2	3.37 in	3.400 in
Bearing	D_3	3.98 in	4.133 9 in (105 mm)
Nothing	D_4	$>D_3$ or D_5	4.600 in
Gear	D_5	4.12 in	4.200 in
Bearing	D_6	1.30 in	3.149 6 in (80 mm)

be determined. Finally, the stress analysis should be repeated to ensure that the resulting design factor is acceptable. Equation (12–23) can be solved for N and evaluated for actual conditions.

**12–9
ADDITIONAL
DESIGN
EXAMPLES**

Two additional design examples are given in this section. The first is for a shaft that contains three different types of power transmission devices: a V-belt sheave, a chain sprocket, and a spur gear. Some of the forces are acting at angles other than vertical and horizontal, requiring the resolution of the bending forces on the shaft into components before the shearing force and bending moment diagrams are created. Design Example 12–4 involves a shaft carrying a wormgear and a chain sprocket. The axial force on the wormgear presents a slight modification of the design procedure. Except for these differences, the design procedure is the same as that for Design Example 12–1. Thus, much of the detailed manipulation of formulas is omitted.

Design Example 12–3

The shaft shown in Figure 12–13 receives 110 hp from a water turbine through a chain sprocket at point C. The gear pair at E delivers 80 hp to an electrical generator. The V-belt sheave at A delivers 30 hp to a bucket elevator that carries grain to an elevated hopper. The shaft rotates at 1 700 rpm. The sprocket, sheave, and gear are located axially by retaining rings. The sheave and gear are keyed with sled runner keyseats, and there is a profile key-

(a) Side view of shaft

(b) Orientation of elements A, C, and E as viewed from right end of shaft

FIGURE 12–13   Shaft design

seat at the sprocket. Use AISI 1040 cold-drawn steel for the shaft. Compute the minimum acceptable diameters D_1 through D_7 as defined in Figure 12–13.

Solution First, the material properties for the AISI 1040 cold-drawn steel are found from Appendix 3:

$$s_y = 71\,000 \text{ psi} \qquad s_u = 80\,000 \text{ psi}$$

Then from Figure 5–9, $s_n = 30\,000$ psi. Let's design for a reliability of 0.99 and use $C_R = 0.81$. The shaft size should be moderately large, so we can assume $C_s = 0.85$ as a reasonable estimate. Then the modified endurance strength is

$$s'_n = s_n C_s C_R = (30\,000)(0.85)(0.81) = 20\,650 \text{ psi}$$

This application is fairly smooth: a turbine drive and a generator and a conveyor at the output points. A design factor of $N = 3$ should be satisfactory.

Torque Distribution in the Shaft

Recalling that all of the power comes into the shaft at C, we can then observe that 30 hp is delivered down the shaft from C to the sheave at A. Also, 80 hp is delivered down the shaft from C to the gear at E. From these observations, the torque in the shaft can be computed:

$$T = 63\,000(30)/1\,700 = 1\,112 \text{ lb·in} \quad \text{from } A \text{ to } C$$

$$T = 63\,000(80)/1\,700 = 2\,965 \text{ lb·in} \quad \text{from } C \text{ to } E$$

Figure 12–14 shows a plot of the torque distribution *in the shaft* superimposed on the sketch of the shaft. When designing the shaft at C, we will use 2 965 lb·in *at C and to the right,*

FIGURE 12–14   Torque distribution in the shaft

but we can use 1 112 lb·in *to the left of C.* Notice that no part of the shaft is subjected to the full 110 hp that comes into the sprocket at *C.* The power splits into two parts as it enters the shaft. When analyzing the sprocket itself, we must use the full 110 hp and the corresponding torque:

$$T = 63\,000(110)/1\,700 = 4\,076\ \text{lb·in}\quad\text{(torque in the sprocket)}$$

Forces

We will compute the forces at each element separately and show the component forces that act in the vertical and horizontal planes, as in Design Example 12–1. Figure 12–15 shows the directions of the applied forces and their components for each element.

1. **Forces on sheave A:** Use Equations (12–7), (12–8), and (12–12):

$$F_N = F_1 - F_2 = T_A/(D_A/2) = 1\,112/3.0 = 371\ \text{lb}\quad\text{(net driving force)}$$

$$F_A = 1.5F_N = 1.5(371) = 556\ \text{lb}\quad\text{(bending force)}$$

The bending force acts upward and to the left at an angle of 60° from the horizontal. As shown in Figure 12–15, the components of the bending force are

$$F_{Ax} = F_A \cos(60°) = (556)\cos(60°) = 278\ \text{lb}\quad\text{(toward the left)}$$

$$F_{Ay} = F_A \sin(60°) = (556)\sin(60°) = 482\ \text{lb}\quad\text{(upward)}$$

2. **Forces on sprocket C:** Use Equation (12–6):

$$F_C = T_C/(D_C/2) = 4\,076/5.0 = 815\ \text{lb}$$

This is the bending load on the shaft. The components are

$$F_{Cx} = F_C \sin(40°) = (815)\sin(40°) = 524\ \text{lb}\quad\text{(to the left)}$$

$$F_{Cy} = F_C \cos(40°) = (815)\cos(40°) = 624\ \text{lb}\quad\text{(downward)}$$

(a) Pictorial view showing forces

(b) Forces on shaft at sheave A

(c) Forces on shaft at sprocket C

(d) Forces on gear E

FIGURE 12–15   Forces resolved into *x*- and *y*-components

3. *Forces on gear E:* The transmitted load is found from Equation (12–2), and the radial load from Equation (12–3). The directions are shown in Figure 12–15.

$$F_{Ey} = W_{tE} = T_E/(D_E/2) = 2\,965/6.0 = 494 \text{ lb} \quad \text{(upward)}$$

$$F_{Ex} = W_{rE} = W_{tE} \tan(\phi) = (494)\tan(20°) = 180 \text{ lb} \quad \text{(to the left)}$$

Load, Shear, and Moment Diagrams

Figure 12–16 shows the forces acting on the shaft at each element, the reactions at the bearings, and the shearing force and bending moment diagrams for both the horizontal (*x*-) and vertical (*y*-) planes. In the figure, the computations of the resultant bending moment at points *B, C,* and *D* are also shown.

Design of the Shaft

We will use Equation (12–24) to determine the minimum acceptable diameter of the shaft at each point of interest. Because the equation requires a fairly large number of individual operations, and because we will be using it at least seven times, it may be desirable to write a computer program just for that operation. Or the use of a spreadsheet would be nearly ideal. Note that we can use Equation (12–24) even though there is only torsion or only bending by entering zero for the missing value.

Equation (12–24) is repeated here for reference. In the solution below, the data used for each design point are listed. You may want to verify the calculations for the required minimum diameters. The design factor of *N* = 3 has been used.

$$D = \left[\frac{32N}{\pi} \sqrt{\left(\frac{K_t M}{s'_n}\right)^2 + \frac{3}{4}\left(\frac{T}{s_y}\right)^2} \right]^{1/3}$$

(12–24)

1. *Point A:* Torque = 1 112 lb·in; moment = 0. The sheave is located with retaining rings. Because the torque is steady, we will not use a stress concentration factor in this calculation, as discussed in Section 12–4. But then we will find the nominal

FIGURE 12–16
Load, shear, and moment diagrams

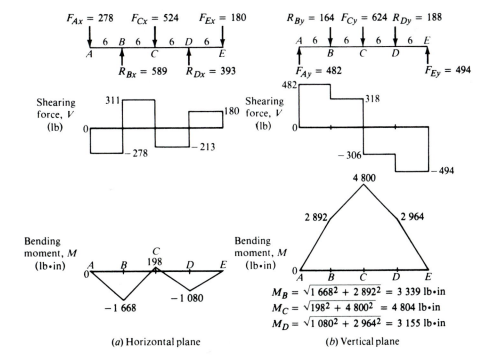

$$M_B = \sqrt{1\,668^2 + 2\,892^2} = 3\,339 \text{ lb·in}$$
$$M_C = \sqrt{198^2 + 4\,800^2} = 4\,804 \text{ lb·in}$$
$$M_D = \sqrt{1\,080^2 + 2\,964^2} = 3\,155 \text{ lb·in}$$

(*a*) Horizontal plane   (*b*) Vertical plane

diameter at the groove by increasing the computed result by about 6%. The result should be conservative for typical groove geometries.

Using Equation (12–24), $D_1 = 0.75$ in. Increasing this by 6% gives $D_1 = 0.79$ in.

2. **To the left of point B:** This is the relief diameter leading up to the bearing seat. A well-rounded fillet radius will be specified for the place where D_2 joins D_3. Thus,

$$\text{Torque} = 1\ 112\ \text{lb·in} \qquad \text{Moment} = 3\ 339\ \text{lb·in} \qquad K_t = 1.5$$

Then $D_2 = 1.95$ in.

3. **At point B and to the right:** This is the bearing seat with a shoulder fillet at the right, requiring a fairly sharp fillet:

$$\text{Torque} = 1\ 112\ \text{lb·in} \qquad \text{Moment} = 3\ 339\ \text{lb·in} \qquad K_t = 2.5$$

Then $D_3 = 2.31$ in.

4. **At point C:** It is planned that the diameter be the same all the way from the right of bearing B to the left of bearing D. The worst condition is at the right of C, where there is a ring groove and the larger torque value is

$$\text{Torque} = 2\ 965\ \text{lb·in} \qquad \text{Moment} = 4\ 804\ \text{lb·in} \qquad K_t = 3.0$$

Then $D_4 = 2.77$ in.

5. **At point D and to the left:** This is a bearing seat similar to that at B:

$$\text{Torque} = 2\ 965\ \text{lb·in} \qquad \text{Moment} = 3\ 155\ \text{lb·in} \qquad K_t = 2.5$$

Then $D_5 = 2.27$ in.

6. **To the right of point D:** This is a relief diameter similar to D_2:

$$\text{Torque} = 2\ 965\ \text{lb·in} \qquad \text{Moment} = 3\ 155\ \text{lb·in} \qquad K_t = 1.5$$

Then $D_6 = 1.92$ in.

7. **At point E:** The gear is mounted with retaining rings on each side:

$$\text{Torque} = 2\ 965\ \text{lb·in} \qquad \text{Moment} = 0 \qquad D_7 = 1.03$$

Summary with Convenient Values Specified

Using Appendix 2, we specify convenient fractions at all places, including bearing seats (see Table 12–3). It is assumed that inch bearings of the pillow block type will be used.

We decided to make the diameters D_1, D_2, D_6, and D_7 the same to minimize machining and to provide a little extra safety factor at the ring grooves. Again the bearing bore sizes would have to be checked against the load rating of the bearings. The size of D_4 would have to be checked to see that it provides a sufficient shoulder for the bearings at B and D.

TABLE 12–3 Specification of values

Mating part	Diameter number	Minimum diameter	Specified diameter	
			Fraction	Decimal
Sheave	D_1	0.79	2	2.000
Nothing	D_2	1.95	2	2.000
Bearing	D_3	2.31	$2\frac{1}{2}$	2.500
Sprocket	D_4	2.77	3	3.000
Bearing	D_5	2.27	$2\frac{1}{2}$	2.500
Nothing	D_6	1.92	2	2.000
Gear	D_7	1.03	2	2.000

Design Example 12–4   A wormgear is mounted at the end of the shaft as shown in Figure 12–17. The gear has the same design as that discussed in Chapter 10 and delivers 6.68 hp to the shaft at a speed of 101 rpm. The magnitudes and directions of the forces on the gear are given in the figure. Notice that there is a system of three orthogonal forces acting on the gear. The power is transmitted by a chain sprocket at B to drive a conveyor removing cast iron chips from a machining system. Design the shaft.

Solution   The torque on the shaft from the wormgear at point D to the chain sprocket at B is

$$T = W_{tG}(D_G/2) = 962(4.333) = 4\ 168\ \text{lb·in}$$

The force on the chain sprocket is

$$F_c = T/(D_s/2) = 4\ 168/(6.71/2) = 1\ 242\ \text{lb}$$

This force acts horizontally toward the right as viewed from the end of the shaft.

Bending Moment Diagrams

Figure 12–18 shows the forces acting on the shaft in both the vertical and the horizontal planes and the corresponding shearing force and bending moment diagrams. You should review these diagrams, especially that for the vertical plane, to grasp the effect of the axial force of 270 lb. Notice that because it acts above the shaft, it creates a bending moment at the end of the shaft of 1 165 lb·in. It also affects the reactions at the bearings. The resultant bending moments at B, C, and D are also shown in the figure.

In the design of the entire system, we must decide which bearing will resist the axial force. For this problem, let's specify that the bearing at C will transfer the axial thrust force to the housing. This decision places a compressive stress in the shaft from C to D and requires that means be provided to transmit the axial force from the wormgear to the bearing. The geometry proposed in Figure 12–17 accomplishes this, and it will be adopted for the following stress analysis. The procedures are the same as those used in Design Examples 12–1 through 12–3, and only summary results will be shown. The consideration of the axial compressive stress is discussed, along with the computations at point C on the shaft.

FIGURE 12–17
Shaft design

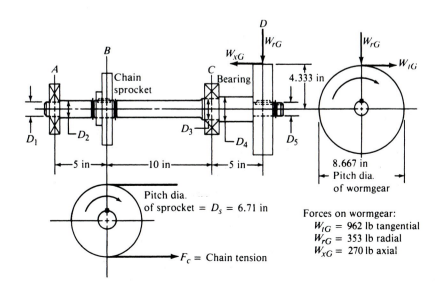

Forces on wormgear:
$W_{tG} = 962$ lb tangential
$W_{rG} = 353$ lb radial
$W_{xG} = 270$ lb axial

12–10 SPREADSHEET AID FOR SHAFT DESIGN

A spreadsheet is useful to organize the data required to compute the minimum required shaft diameter at various points along a shaft and to complete the calculation using Equations (12–16) and (12–24). Note that Equation (12–24) can be used for bending only, torsion only, or the combination of bending and torsion.

Figure 12–19 shows a typical example, using U.S. Customary units, for data from Design Example 12–1. Describe the application in the upper panel for future reference. Then complete the following steps:

- Enter the shaft material specification along with its ultimate and yield strength properties found from tables in the Appendices.

- Find the basic endurance strength from Figure 5–9 considering the ultimate tensile strength and the manner of production (ground, machined, etc.).

- Enter values for the size factor and the reliability factor. The spreadsheet then computes the modified endurance strength, s'_n.

- Enter the stress concentration factor for the point of interest.

- Enter the design factor.

- After analysis such as that shown in Design Example 12–1, enter the torque and the components of the bending moment in the x- and y-planes that exist at the point of interest along the shaft. The spreadsheet computes the combined bending moment.

- Enter the components of the vertical shearing force in the x- and y-planes. The spreadsheet computes the combined shearing force.

The minimum acceptable shaft diameters from both Equation (12–16) (vertical shear only) and Equation (12–24) (torsion and/or bending) are computed. You must observe which required diameter is larger.

REFERENCES

1. American Society of Mechanical Engineers. ANSI Standard B106.1M-1985. *Design of Transmission Shafting.* New York: American Society of Mechanical Engineers, 1985.

2. American Society of Mechanical Engineers. ANSI Standard B4.1-67. *Preferred Limits and Fits for Cylindrical Parts.* (R87/94). New York: American Society of Mechanical Engineers, 1994.

3. Avallone, E., and T. Baumeister, et al. *Marks' Standard Handbook for Mechanical Engineers.* 10th ed. New York: McGraw-Hill, 1996.

4. Oberg, Erik, et al. *Machinery's Handbook.* 25th ed. New York: Industrial Press, 1996.

5. Pilkey, Walter D. *Peterson's Stress Concentration Factors.* 2d ed. New York: John Wiley & Sons, 1997.

6. Shigley, J. E., and C. R. Mischke. *Mechanical Engineering Design.* 5th ed. New York: McGraw-Hill, 1989.

7. Soderberg, C. R. "Working Stresses." *Journal of Applied Mechanics* 57 (1935): A–106.

DESIGN OF SHAFTS

Application: Design Example 12–1, Drive for a Blower System
Diameter D_3—To right of point B—Bending and torsion

This design aid computes the mininum acceptable diameter for shafts using Equation (12–24) for shafts subjected to steady torsion and/or rotating bending.

Equation (12–16) is used when only vertical shear stress is present.

Input Data:		(Insert values in italics.)	

Shaft material specification:		AISI 1144 OQT 1000 Steel	
Tensile strength:	$s_u=$	118,000 psi	
Yield strength:	$s_y=$	83,000 psi	
Basic endurance strength:	$s_n=$	42,000 psi	From Figure 5–9
Size factor:	$C_s=$	0.8	From Figure 12–9
Reliability factor:	$C_R=$	0.81	From Table 12–1
Modified endurance strength:	$S'_n=$	27,216 psi	**Computed**
Stress concentration factor:	$K_t=$	2.5	Sharp fillet
Design factor:	$N=$	3	Nominal $N=3$

Shaft Loading Data: Bending and Torsion			

Bending moment components:	$M_x=$	21,000 lb·in	$M_y=7{,}640$ lb·in
Combined bending moment:	$M=$	22,347 lb·in	**Computed**
Torque:	$T=$	21,000 lb·in	

Minimum shaft diameter:	**D=**	**3.98** in	**Computed from Eq. (12–24)**

Shaft Loading Data: Vertical Shearing Force Only			

Shearing force components:	$V_x=$	764 lb	$V_y=2{,}520$ lb
Combined shearing force:	$V=$	2,633 lb	**Computed**

Minimum shaft diameter:	**D=**	**1.461** in	**Computed from Eq. (12–16)**

FIGURE 12–19 Spreadsheet aid for shaft design

PROBLEMS

Forces and Torque on Shafts—Spur Gears

1. See Figure 12–20. The shaft rotating at 550 rpm carries a spur gear *B* having 96 teeth with a diametral pitch of 6. The teeth are of the 20°, full-depth, involute form. The gear receives 30 hp from a pinion directly above it. Compute the torque delivered to the shaft and the tangential and radial forces exerted on the shaft by the gear.

2. See Figure 12–21. The shaft rotating at 200 rpm carries a spur gear *C* having 80 teeth with a diametral pitch of 8. The teeth are of the 20°, full-depth, involute form. The gear delivers 6 hp to a pinion directly below it. Compute the torque delivered by the shaft to gear *C* and the tangential and radial forces exerted on the shaft by the gear.

3. See Figure 12–22. The shaft rotating at 480 rpm carries a spur pinion *B* having 24 teeth with a diametral pitch of 8. The teeth are of the 20°, full-depth, involute form. The pinion delivers 5 hp to a gear directly below it. Compute the torque delivered by the shaft to pinion *B* and the tangential and radial forces exerted on the shaft by the pinion.

4. See Figure 12–23. The shaft rotating at 120 rpm carries a spur gear *A* having 80 teeth with a diametral pitch of 5. The teeth are of the 20°, full-depth, involute form. The gear receives 40 hp from a pinion to the right as shown. Compute the torque delivered to the shaft and the tangential and radial forces exerted on the shaft by the gear.

FIGURE 12–20   (Problems 1, 14, and 24)

FIGURE 12–21   (Problems 2, 12, 13, and 25)

5. See Figure 12–24. The shaft rotating at 240 rpm carries a spur gear *D* having 48 teeth with a diametral pitch of 6. The teeth are of the 20°, full-depth, involute form. The gear receives 15 hp from pinion *Q* located as shown. Compute the torque delivered to the shaft and the tangential and radial forces exerted on the shaft by the gear. Resolve the tangential and radial forces into their horizontal and vertical components, and determine the net forces acting on the shaft at *D* in the horizontal and vertical directions.

6. See Figure 12–25. The shaft rotating at 310 rpm carries a spur pinion *E* having 36 teeth with a diametral pitch of 6. The teeth are of the 20°, full-depth, involute form. The pinion delivers 20 hp to a gear to the left as shown. Com-

pute the torque delivered by the shaft to pinion *E* and the tangential and radial forces exerted on the shaft by the pinion. Include the weight of the pinion.

7. See Figure 12–26. The shaft rotating at 480 rpm carries a spur gear *C* having 50 teeth with a diametral pitch of 5. The teeth are of the 20°, full-depth, involute form. The gear receives 50 hp from a pinion directly below it. Compute the torque delivered to the shaft and the tangential and radial forces exerted on the shaft by the gear.

8. See Figure 12–26. The shaft rotating at 480 rpm carries a spur pinion *A* having 30 teeth with a diametral pitch of 6. The teeth are of the 20°, full-depth, involute form. The

FIGURE 12–22   (Problems 3, 15, 16, and 26)

FIGURE 12–23   (Problems 4, 19, and 27)

FIGURE 12–24 (Problems 5, 20, 21, and 28)

FIGURE 12–25 (Problems 6 and 29)

pinion delivers 30 hp to a gear to the left as shown. Compute the torque delivered by the shaft to pinion A and the tangential and radial forces exerted on the shaft by the pinion.

9. See Figure 12–27. The shaft rotating at 220 rpm carries a spur pinion C having 60 teeth with a diametral pitch of 10. The teeth are of the 20°, full-depth, involute form. The pin-

ion delivers 5 hp to a gear directly above it. Compute the torque delivered by the shaft to pinion C and the tangential and radial forces exerted on the shaft by the pinion.

10. See Figure 12–27. The shaft rotating at 220 rpm carries a spur gear D having 96 teeth with a diametral pitch of 8. The teeth are of the 20°, full-depth, involute form. The

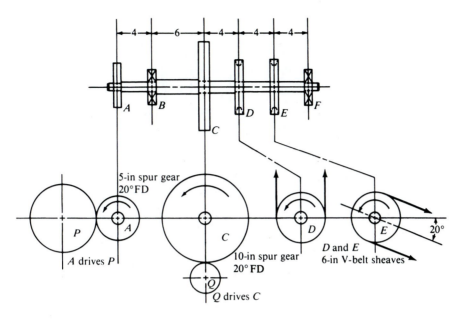

FIGURE 12–26 (Problems 7, 8, and 30)

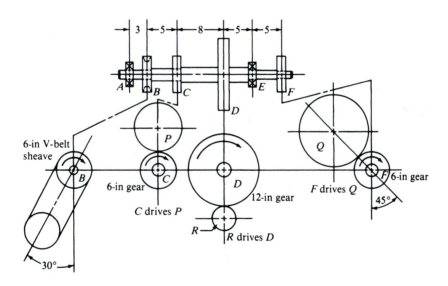

FIGURE 12–27 (Problems 9, 10, 11, and 31)

gear receives 12.5 hp from a pinion directly below it. Compute the torque delivered to the shaft and the tangential and radial forces exerted on the shaft by the gear.

11. See Figure 12–27. The shaft rotating at 220 rpm carries a spur pinion *F* having 60 teeth with a diametral pitch of 10. The teeth are of the 20°, full-depth, involute form. The pinion delivers 5 hp to a gear located as shown. Compute the torque delivered by the shaft to pinion *F* and the tangential and radial forces exerted on the shaft by the pinion. Resolve the forces into their horizontal and vertical

components, and determine the net forces acting on the shaft at *F* in the horizontal and vertical directions.

Forces and Torque on Shafts—Sprockets and Belt Sheaves

12. See Figure 12–21. The shaft rotating at 200 rpm carries a 6-in-diameter chain sprocket *D* that delivers 4 hp to a mating sprocket above. Compute the torque delivered by the shaft to sprocket *D* and the force exerted on the shaft by the sprocket.

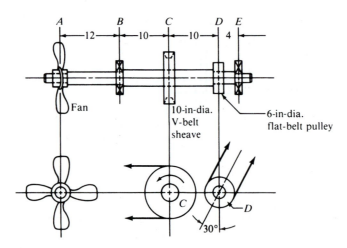

FIGURE 12–28   (Problems 17, 18, and 32)

13. See Figure 12–21. The shaft rotating at 200 rpm carries a 20-in-diameter flat-belt pulley at *A* that receives 10 hp from below. Compute the torque delivered by the pulley to the shaft and the force exerted on the shaft by the pulley.

14. See Figure 12–20. The shaft rotating at 550 rpm carries a 10-in-diameter V-belt sheave at *D* that delivers 30 hp to a mating sheave as shown. Compute the torque delivered by the shaft to the sheave and the total force exerted on the shaft by the sheave. Resolve the force into its horizontal and vertical components, and show the net forces acting on the shaft at *D* in the horizontal and vertical directions.

15. See Figure 12–22. The shaft rotating at 480 rpm carries a 10-in-diameter chain sprocket at *C* that receives 11 hp from a mating sprocket below and to the left as shown. Compute the torque delivered to the shaft by the sprocket and the total force exerted on the shaft by the sprocket. Resolve the force into its horizontal and vertical components, and show the net forces acting on the shaft at *C* in the horizontal and vertical directions.

16. See Figure 12–22. The shaft rotating at 480 rpm carries two 4-in-diameter sheaves at *D* and *E* that each deliver 3 hp to mating sheaves as shown. Compute the torque delivered by the shaft to each sheave and the total force exerted on the shaft by each sheave. Resolve the force at *E* into its horizontal and vertical components, and show the net forces acting on the shaft at *E* in the horizontal and vertical directions.

17. See Figure 12–28. The shaft rotating at 475 rpm carries a 10-in V-belt sheave at *C* that receives 15.5 hp from a mating sheave to the left as shown. Compute the torque delivered to the shaft at *C* by the sheave and the total force exerted on the shaft at *C* by the sheave.

18. See Figure 12–28. The shaft rotating at 475 rpm carries a 6-in-diameter flat-belt pulley at *D* that delivers 3.5 hp to a mating pulley above and to the right as shown. Compute the torque delivered to the shaft by the pulley and the total force exerted on the shaft by the pulley. Resolve the force into its horizontal and vertical components, and show the net forces acting on the shaft at *D* in the horizontal and vertical directions.

19. See Figure 12–23. The shaft rotating at 120 rpm carries two identical 14-in-diameter chain sprockets at *C* and *D*. Each sprocket delivers 20 hp to mating sprockets to the left as shown. Compute the torque delivered by the shaft to each sprocket and the total force exerted on the shaft by each sprocket.

20. See Figure 12–24. The shaft rotating at 240 rpm carries a 12-in-diameter V-belt sheave at *A* that delivers 10 hp to a mating sheave directly below. Compute the torque delivered by the shaft to the sheave and the total force exerted on the shaft at *A* by the sheave.

21. See Figure 12–24. The shaft rotating at 240 rpm carries a 6-in-diameter chain sprocket at *E* that delivers 5.0 hp to a mating sprocket to the right and above as shown. Compute the torque delivered by the shaft to the sprocket and the total force exerted on the shaft by the sprocket. Resolve the force into its horizontal and vertical components, and show the net forces acting on the shaft at *E* in the horizontal and vertical directions.

Forces and Torque on Shafts—Helical Gears

22. See Figure 12–29. The shaft is rotating at 650 rpm, and it receives 7.5 hp through a flexible coupling. The power is delivered to an adjacent shaft through a single helical gear *B* having a normal pressure angle of 20° and a helix angle of 15°. The pitch diameter for the gear is 4.141 in. Review the discussion of forces on helical gears in Chapter 10, and use Equations (12–1), (12–2), (12–4), and (12–5) in this chapter to verify the values for the forces given in the figure. Draw the complete free-body diagrams for the shaft in both the vertical plane and the horizontal plane. Then draw the complete shearing force and bending moment diagrams for the shaft in both planes.

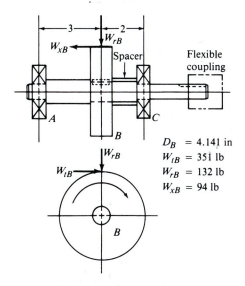

D_B = 4.141 in
W_{tB} = 351 lb
W_{rB} = 132 lb
W_{xB} = 94 lb

FIGURE 12–29 (Problems 22 and 33)

Forces and Torque on Shafts—Wormgears

23. See Figure 12–30. The shaft is rotating at 1 750 rpm, and it receives 7.5 hp through a 5.00-in V-belt drive from a mating belt sheave directly below. The power is delivered by a worm with a pitch diameter of 2.00 in. The forces on the worm were computed in Design Example 12–4 and are shown in the figure. Review the description of these forces and the method of computing them. Draw the complete free-body diagrams for the shaft in both the vertical plane and the horizontal plane. Then draw the complete shearing force and bending moment diagrams for the shaft in both planes.

Comprehensive Shaft Design Problems

For each of the following problems, it will be necessary to do the following:

(a) Determine the magnitude of the torque in the shaft at all points

(b) Compute the forces acting on the shaft at all power-transmitting elements

(c) Compute the reactions at the bearings

(d) Draw the complete load, shear, and bending moment diagrams

Neglect the weight of the elements on the shafts, unless otherwise noted.

The objective of any problem, at the discretion of the instructor, could be any of the following:

■ Design the complete shaft, including the specification of the overall geometry and the consideration of stress concentration factors. The analysis would show the minimum acceptable diameter at each point on the shaft to be safe from the standpoint of strength.

■ Given a suggested geometry of one part of the shaft, specify the minimum acceptable diameter for the shaft at that point.

■ Specify the required geometry at any selected element on the shaft: a gear, sheave, bearing, or other.

■ Make a working drawing of the design for the shaft, following the appropriate stress analysis, and specify the final dimensions.

■ Suggest how the given shaft can be redesigned by moving or reorienting the elements on the shaft to improve the design to produce lower stresses, smaller shaft size, more convenient assembly, and so on.

■ Incorporate the given shaft into a more comprehensive machine, and complete the design of the entire

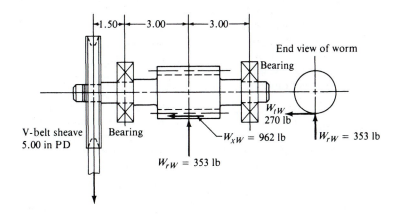

FIGURE 12–30 (Problems 23 and 34)

machine. In most problems, the type of machine for which the shaft is being designed is suggested.

24. The shaft in Figure 12–20 is part of a drive for an automated transfer system in a metal stamping plant. Gear Q delivers 30 hp to gear B. Sheave D delivers the power to its mating sheave as shown. The shaft carrying B and D rotates at 550 rpm. Use AISI 1040 cold-drawn steel.

25. The shaft in Figure 12–21 rotates at 200 rpm. Pulley A receives 10 hp from below. Gear C delivers 6 hp to the mating gear below it. Chain sprocket D delivers 4 hp to a shaft above. Use AISI 1117 cold-drawn steel.

26. The shaft in Figure 12–22 is part of a special machine designed to reclaim scrap aluminum cans. The gear at B delivers 5 hp to a chopper that cuts the cans into small pieces. The V-belt sheave at D delivers 3 hp to a blower that draws air through the chopper. V-belt sheave E delivers 3 hp to a conveyor that raises the shredded aluminum to an elevated hopper. The shaft rotates at 480 rpm. All power comes into the shaft through the chain sprocket at C. Use AISI 1137 OQT 1300 steel for the shaft. The elements at B, C, D, and E are held in position with retaining rings and keys in profile keyseats. The shaft is to be of uniform diameter, except at its ends, where the bearings are to be mounted. Determine the required diameter.

27. The shaft in Figure 12–23 is the drive shaft for a large, bulk material conveyor. The gear receives 40 hp and rotates at 120 rpm. Each chain sprocket delivers 20 hp to one side of the conveyor. Use AISI 1020 cold-drawn steel.

28. The shaft in Figure 12–24 is part of a conveyor drive system that feeds crushed rock into a railroad car. The shaft rotates at 240 rpm and is subjected to moderate shock during operation. All power is input to the gear at D. The V-belt sheave at A delivers 10.0 hp vertically downward. Chain sprocket E delivers 5.0 hp. Note the position of the gear Q, which drives gear D.

29. Figure 12–25 illustrates an intermediate shaft of a punch press that rotates at 310 rpm while transmitting 20 hp from the V-belt sheave to the gear. The flywheel is not absorbing or giving up any energy at this time. Consider the weight of all elements in the analysis.

30. The shaft in Figure 12–26 is part of a material handling system aboard a ship. All power comes into the shaft through gear C, which rotates at 480 rpm. Gear A delivers 30 hp to a hoist. V-belt sheaves D and E each deliver 10 hp to hydraulic pumps. Use AISI 3140 OQT 1000 steel.

31. The shaft in Figure 12–27 is part of an automatic machining system. All power is input through gear D. Gears C and F drive two separate tool feed devices requiring 5.0 hp each. The V-belt sheave at B requires 2.5 hp to drive a coolant pump. The shaft rotates at 220 rpm. All gears are

spur gears with 20°, full-depth teeth. Use AISI 1020 cold-drawn steel.

32. The shaft in Figure 12–28 is part of a grain-drying system. At A is a propeller-type fan that requires 12 hp when rotating at 475 rpm. The fan weighs 34 lb, and its weight should be included in the analysis. The flat-belt pulley at D delivers 3.5 hp to a screw conveyor handling the grain. All power comes into the shaft through the V-belt sheave at C. Use AISI 1144 cold-drawn steel.

33. Figure 12–29 shows a helical gear mounted on a shaft that rotates at 650 rpm while transmitting 7.5 hp. The gear is also analyzed in Example Problem 10–1, and the tangential, radial, and axial forces on it are shown in the figure. The pitch diameter of the gear is 4.141 in. The power is delivered from the shaft through a flexible coupling at its right end. A spacer is used to position the gear relative to bearing C. The thrust load is taken at bearing A.

34. The shaft shown in Figure 12–30 is the input shaft of a wormgear drive. The V-belt sheave receives 7.5 hp from directly downward. The worm rotates at 1 750 rpm and has a pitch diameter of 2.000 in. This is the driving worm for the wormgear described in Design Example 12–4. The tangential, radial, and axial forces are shown in the figure. The worm is to be machined integrally with the shaft, and it has a root diameter of 1.614 in. Assume that the geometry of the root area presents a stress concentration factor of 1.5 for bending. Analyze the stress in the area of the worm thread root, and specify a suitable material for the shaft.

35. The double-reduction, helical gear reducer shown in Figure 12–31 transmits 5.0 hp. Shaft 1 is the input, rotating at 1 800 rpm and receiving power directly from an electric motor through a flexible coupling. Shaft 2 rotates at 900 rpm. Shaft 3 is the output, rotating at 300 rpm. A chain sprocket is mounted on the output shaft as shown and delivers the power upward. The data for the gears are given in Table 12–5.

Each gear has a $14\frac{1}{2}°$ normal pressure angle and a 45° helix angle. The combinations of left- and right-hand helixes are arranged so that the axial forces oppose each other on shaft 2 as shown. Use AISI 4140 OQT 1200 for the shafts.

36. Complete the design of the shafts carrying the bevel gear and pinion shown in Figure 10–8. The forces on the gears, the bearing reactions, and the bending moment diagrams are developed in Example Problems 10–3 and 10–4 and are shown in Figures 10–8 through 10–12. Assume that the 2.50 hp enters the pinion shaft from the right through a flexible coupling. The power is delivered by way of the lower extension of the gear shaft through another flexible coupling. Use AISI 1040 OQT 1200 for the shafts.

37. The vertical shaft shown in Figure 12–32 is driven at a speed of 600 rpm with 4.0 hp entering through the bevel

FIGURE 12–31 (Problem 35)

TABLE 12–5

Gear	Diametral pitch	Pitch diameter	Number of teeth	Face width
P	8	1.500 in	12	0.75 in
B	8	3.000 in	24	0.75 in
C	6	2.000 in	12	1.00 in
Q	6	6.000 in	36	1.00 in

Gear A	Gear B
18 teeth	54 teeth
1.80-in dia.	5.40-in dia.

Gear C	Gear D
24 teeth	48 teeth
4.00-in dia.	8.00-in dia.

gear. Each of the two chain sprockets delivers 2.0 hp to the side to drive mixer blades in a chemical reactor vessel. The bevel gear has a diametral pitch of 5, a pitch diameter of 9.000 in, a face width of 1.31 in, and a pressure angle of 20°. Use AISI 4140 OQT 1000 steel for the shaft. See Chapter 10 for the methods for computing the forces on the bevel gear.

38. Figure 12–33 shows a sketch for a double-reduction spur gear train. Shaft 1 rotates at 1 725 rpm, driven directly by an electric motor that transmits 15.0 hp to the reducer. The gears in the train all have 20°, full-depth teeth and the following numbers of teeth and pitch diameters:

Note that the speed reduction for each gear pair is proportional to the ratio of the number of teeth. Therefore, shaft 2 rotates at 575 rpm, and shaft 3 rotates at 287.5 rpm. Assume that all shafts carry 15.0 hp. The distance from the middle of each bearing to the middle of the face of the nearest gear is 3.00 in. Shaft 2 has a length of 5.00 in between the two gears, making the total distance between the middle of the two bearings equal to 11.00 in. The extensions of the input and the output shafts carry torque, but no bending loads are exerted on them. Complete the design of all three shafts. Provide for profile keyseats for each gear and sled-runner keyseats at the outboard ends of

FIGURE 12–32   Bevel gears shown in section for Problem 37. See Chapter 10.

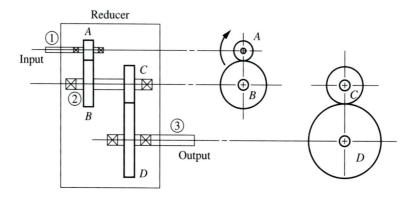

FIGURE 12–33   Gear reducer for Problems 38 and 39

the input and output shafts. Provide for the location of each gear and bearing on the shaft.

39. Figure 12–34 shows a speed reducer with a V-belt drive delivering power to the input shaft and a chain drive taking power from the output shaft and delivering it to a conveyor. The drive motor supplies 12.0 hp and rotates 1 150 rpm. The speed reductions of the V-belt drive and the chain drive are proportional to the ratio of the diameter of the driving and driven sheaves or sprockets. The arrangement of the gears in the reducer is the same as that described in Figure 12–33 and Problem 38. Determine the

forces applied to the motor shaft, to each of the three shafts of the reducer, and to the conveyor drive shaft. Then complete the design of the three shafts of the reducer, assuming that all shafts transmit 12.0 hp.

40. Figure 12–35 shows a drive for a system to crush coal and deliver it by conveyor to a railroad car. Gear A delivers 15 kW to the crusher, and gear E delivers 7.5 kW to the conveyor. All power enters the shaft through gear C. The shaft carrying gears A, C, and E rotates at 480 rpm. Design that shaft. The distance from the middle of each bearing to the middle of the face of the nearest gear is 100 mm.

FIGURE 12–34 Drive system for Problem 39

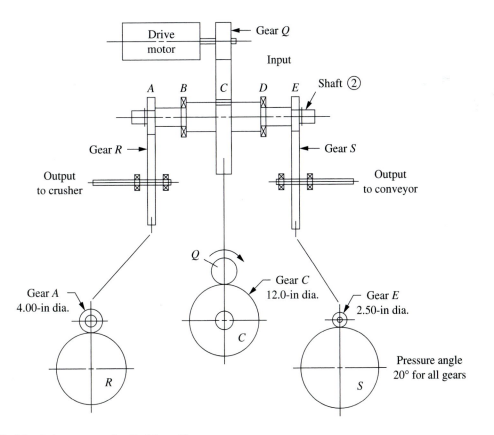

FIGURE 12–35 Drive system for Problem 40

FIGURE 12–36 Shaft and levers for windshield wiper system for Problem 41

41. A shaft is to be designed for a linkage of a windshield wiper mechanism for a truck (see Figure 12–36). A 20-N force is applied to lever 1 by an adjacent link. The reaction force, F_2, on lever 2 is transmitted to another link. The distance, d, between elements is 20.0 mm. Bearings A and C are straight, cylindrical, bronze bushings, 10.0 mm long. Design the shaft and levers 1 and 2.

13

Tolerances and Fits

The Big Picture

You Are the Designer

Tolerances and Fits

Discussion Map

☐ A *tolerance* is the permissible variation on key dimensions of mechanical parts. You, as a designer, must specify the tolerance for each dimension, considering how it will perform and how it will be manufactured.

☐ Two or more parts may be assembled together either with a *clearance fit,* allowing free relative motion, or with an *interference fit,* where the two parts must be pressed together and therefore will not move during operation of the device.

Discover

Look for examples of mechanical parts that have clearance fits and some that have interference fits.

Describe the reasons why the designer may have made the design in this way.

This chapter will help you acquire the competency to specify suitable fits for mating parts and the dimensional tolerances that will produce the desired fit.

The objective of most of the analysis procedures discussed in this book is to determine the minimum acceptable geometric size at which a component is safe and performs properly under specified conditions. As a designer, you must also then specify the final dimensions for the components, including the *tolerances* on those dimensions.

The term *tolerance* refers to the permissible deviation of a dimension from the specified basic size. The proper performance of a machine can depend on the tolerances specified for its parts, particularly those that must fit together for location or for suitable relative motion.

The term *fit* usually refers to the *clearances* that are permissible between mating parts in a mechanical device that must assemble easily and that must often move relative to each other during normal operation of the device. Such fits are usually called *running* or *sliding fits. Fit* can also refer to the amount of *interference* that exists when the inside part should be larger than the outside part. Interference fits are used to ensure that mating parts do not move relative to each other.

Look for examples of parts for mechanical devices that have clearance fits. Any machine having shafts that rotate in plane surface bearings should have such parts. This type of bearing is called a *journal bearing,* and there must be a small but reliable clearance between the shaft and the bearing to permit smooth rotation of the shaft. But the clearance cannot be too large, or the operation of the machine will appear to be too crude and rough.

Consider also any assembly having hinges that allow one part to rotate relative to the other, such as an access door or panel or the cover of a container. The hinge parts will have a clearance fit. Many types of measuring equipment, such as calipers, dial indicators, and electronic probes, that have moving parts must be designed carefully to maintain the expected precision of the measurements while allowing reliable motion. On the other end of the spectrum is the rather loose fit that is typical of toys and some other types of recreation equipment. The fit of the wheel of a child's wagon onto its axle is typically quite large to allow free rotation and to permit easy assembly. We will show in this chapter that the wide range of fits that are encountered in practice are designed by specifying a *class of fit* and then determining the allowable range of key dimensions of mating parts.

Look also for examples where two parts are assembled so that they cannot move relative to one another. They are held tightly together because the inside part is larger than the outside part. Perhaps some parts of your car are like that.

This chapter will help you design parts that must have clearance or interference fits. When an interference fit occurs, it is often desirable to predict the level of stress that the mating parts are subjected to. That topic will also be covered in this chapter.

You Are the Designer

Assume that you are responsible for the design of the gear-type speed reducer discussed in Chapter 1 and sketched in Figure 1-11. The sketch is repeated here as Figure 13-1. The input and output shafts each carry one gear while being supported on two bearings mounted in the housing. The gears are keyed to the shafts to permit the transmission of torque from the shaft to the gear, or vice versa. The material in Chapter 12 will help you to design the shafts themselves. Chapter 14 discusses bearing selection and application. The results of these design decisions include the calculation of the minimum acceptable diameter for the shaft at any section and the specification of suitable bearings to carry the loads applied with a reasonable life. The bearing manufacturer will specify the tolerances on the various dimensions of the bearing. Your job is to specify the final dimensions for the shaft at all points, including the tolerances on those dimensions.

Consider the part of the input shaft where the first gear in the train is mounted. What is a convenient nominal size to specify for the shaft diameter at the location of the gear? Do you want to be able to slide the gear easily over the left end of the shaft and up against the shoulder that locates it? If so, how much clearance should be allowed between the shaft and the bore of the gear to ensure ease of assembly while still providing accurate location of the gear and smooth operation? When the shaft is machined, what range of dimensions will you allow the manufacturing staff to produce? What surface finish should be specified for the shaft, and what manufactur-

FIGURE 13-1 Conceptual design for a speed reducer

ing process is required to produce that finish? What is the relative cost of the manufacturing operation? Similar questions must be answered for the bore of the gear.

Rolling contact bearings, such as the familiar ball bearing, are designed to be installed on the shaft with an *interference fit*. That is, the inside diameter of the bearing is smaller than the outside diameter of the shaft where the bearing is to be seated. A significant force is required to press the bearing onto the shaft. What dimensions do you specify for the shaft at the bearing seat? How much interference should be specified? How much stress is produced in the shaft because of the interference fit? You as the designer must answer these questions.

13-1 OBJECTIVES OF THIS CHAPTER

After completing this chapter, you will be able to:

1. Define the terms *tolerance, allowance, unilateral tolerance,* and *bilateral tolerance.*
2. Describe the relationships among tolerances, production processes, and cost.
3. Specify basic sizes for dimensions according to a set of preferred sizes.
4. Use the ANSI Standard B4.1, *Preferred Limits and Fits for Cylindrical Parts,* to specify tolerances, fits, and clearances.
5. Specify transitional, interference, and force fits.
6. Compute the pressure created between parts subjected to interference fits and the resulting stresses in the mating members.
7. Use a spreadsheet to assist in the calculation of stresses for interference fits.

**13–2
FACTORS
AFFECTING
TOLERANCES
AND FITS**

Consider the plain surface bearings designed in Chapter 16. A critical part of the design is the specification of the diametral clearance between the journal and the bearing. The typical value is just a few thousandths of an inch. But some variation must be allowed on both the journal outside diameter and the bearing inside diameter for reasons of economy of manufacture. Thus, there will be a variation of the actual clearance in production devices, depending on where the individual mating components fall within their own tolerance bands. Such variations must be accounted for in the analysis of the bearing performance. Too small a clearance could cause seizing. Conversely, too large a clearance would reduce the precision of the machine and adversely affect the lubrication.

The mounting of power transmission elements onto shafts is another situation in which tolerances and fits must be considered. A chain sprocket for general-purpose mechanical drives is usually produced with a bore that will easily slide into position on the shaft during assembly. But once in place, it will transmit power smoothly and quietly only if it is not excessively loose. A high-speed turbine rotor must be installed on its shaft with an interference fit, eliminating any looseness that would cause vibrations at the high rotational speeds.

When relative motion between two parts is required, a clearance fit is necessary. But here, too, there are differences. Some measuring instruments have parts that must move with no perceptible looseness (sometimes called *play*) between mating parts which would adversely affect the accuracy of measurement. An idler sheave in a belt drive system must rotate on its shaft reliably, without the tendency to seize, but with only a small amount of play. The requirement for the mounting of a wheel of a child's wagon on its axle is much different. A loose clearance fit is satisfactory for the expected use of the wheel, and it permits wide tolerances on the wheel bore and on the axle diameter for economy.

**13–3
TOLERANCES,
PRODUCTION
PROCESSES,
AND COST**

A *unilateral tolerance* deviates in only one direction from the basic size. A *bilateral tolerance* deviates both above and below the basic size. The *total tolerance* is the difference between the maximum and minimum permissible dimensions.

Geometric tolerancing should be used to control the size, form, location, and finish of manufactured parts. It is especially useful in quantity production of interchangeable parts. The complete definition of the standards for geometric dimensioning and tolerancing is found in the Standard ANSI Y14.5M-1995. (See Reference 3; Reference 6 shows some applications.)

The term *allowance* refers to an intentional difference between the maximum material limits of mating parts. For example, a positive allowance for a hole/shaft pair would define the minimum *clearance* between the mating parts from the largest shaft mating with the smallest hole. A negative allowance would result in the shaft being larger than the hole (*interference*).

The term *fit* refers to the relative looseness (clearance fit) or tightness (interference fit) of mating parts, especially as it affects the motion of the parts or the force between them after assembly. Specifying the degree of clearance or interference is one of the tasks of the designer.

It is costly to produce components with very small tolerances on dimensions. It is the designer's responsibility to set the tolerances at the highest possible level that results in satisfactory operation of the machine. Certainly judgment and experience must be exercised in such a process. In quantity production situations, it may be cost-effective to test prototypes with a range of tolerances to observe the limits of acceptable performance.

In general, the production of parts with small tolerances on their dimensions requires multiple processing steps. A shaft may first have to be turned on a lathe and then ground to produce the final dimensions and surface finish. In extreme cases, lapping may be required.

Each subsequent step in the manufacture adds cost. Even if different operations are not required, the holding of small tolerances on a single machine, such as a lathe, may require several passes, ending with a finish cut. Cutting tool changes must be more frequent, also, as wear of the tool takes the part out of tolerance.

The production of part features with small tolerances usually involves finer surface finishes, as well. Figure 13–2 shows the general relationship between the surface finish and the cost of producing a part. The typical total tolerance produced by the processes described is included in the figure. The increase in cost is dramatic for the small tolerances and fine finishes.

Figure 13–3 presents the relationship between the surface finish and the machining operations available to produce it.

The basic reference for tolerances and fits in the United States is the ANSI Standard B4.1-1967, *Preferred Limits and Fits for Cylindrical Parts.* Metric dimensions should conform to ANSI B4.2-1978, *Preferred Metric Limits and Fits.* The most recent version should be consulted. (See References 1 and 2.)

An international body, the International Organization for Standardization (ISO), gives metric data for limits and fits in Recommendation 286 (ISO R286), used in Europe and many other countries.

The term *tolerance grade* refers to a set of tolerances that can be produced with an approximately equal production capability. The actual total tolerance allowed within each grade depends on the nominal size of the dimension. Smaller tolerances can be achieved for smaller dimensions, and vice versa. Standards ISO R286 and ANSI B4.1

FIGURE 13–2

Machining costs versus surface finish specified [Reprinted with permission from the Association for Integrated Manufacturing Technology (formerly the Numerical Control Society)]

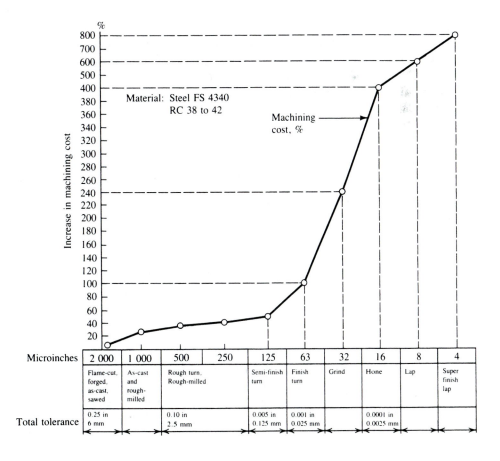

FIGURE 13–3
Finishes produced by
various techniques
(roughness
average, R_a)

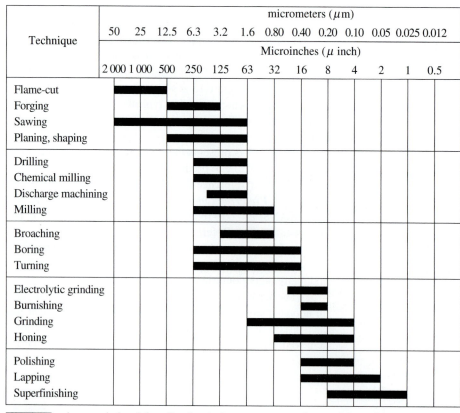

= Average industrial application (values may vary under special conditions)

TABLE 13–1   Tolerance grades

Application	Tolerance grades								
Measuring tools	01	0	1	2	3	4	5	6	7
Fits of machined parts	4	5	6	7	8	9	10	11	
Material, as supplied	8	9	10	11	12	13	14		
Rough forms (casting, sawing, forging, etc.)	12	13	14	15	16				

include complete data for tolerance grades from 01 through 16 as shown in Table 13–1. Tolerances are smaller for the smaller grade number.

A sampling of tolerance data for a few grades and ranges of size is given in Table 13–2. Figure 13–2 shows the capability of selected manufacturing processes to produce work within the given tolerance grades.

**13–4
PREFERRED
BASIC SIZES**

The first step in specifying a dimension for a part is to decide on the basic size, that dimension to which the tolerances are applied. The analysis for strength, deflection, or performance of the part determines the nominal or minimum size required. Unless special conditions exist, the basic size should be chosen from the following lists of preferred basic sizes: Table 13–3 for fractional-inch sizes, Table 13–4 for decimal-inch sizes, and Table 13–5 for metric sizes from the SI. If possible, select from the first-choice column. If

TABLE 13–2 Tolerances for some tolerance grades

				Tolerance grade				
	4	5	6	7	8	9	10	11
Nominal size (in)				Tolerances in thousandths of an inch				
0.24–0.40	0.15	0.25	0.4	0.6	0.9	1.4	2.2	3.5
0.40–0.71	0.20	0.3	0.4	0.7	1.0	1.6	2.8	4.0
0.71–1.19	0.25	0.4	0.5	0.8	1.2	2.0	3.5	5.0
1.19–1.97	0.3	0.4	0.6	1.0	1.6	2.5	4.0	6.0
1.97–3.15	0.3	0.5	0.7	1.2	1.8	3.0	4.5	7.0
3.15–4.73	0.4	0.6	0.9	1.4	2.2	3.5	5.0	9.0
4.73–7.09	0.5	0.7	1.0	1.6	2.5	4.0	6.0	10.0

TABLE 13–3 Preferred basic sizes

Fractional (in)			
1/64	0.015 625	5	5.000 0
1/32	0.031 25	$5\frac{1}{4}$	5.250 0
1/16	0.062 5	$5\frac{1}{2}$	5.500 0
3/32	0.093 75	$5\frac{3}{4}$	5.750 0
1/8	0.125 0	6	6.000 0
5/32	0.156 25	$6\frac{1}{2}$	6.500 0
3/16	0.187 5	7	7.000 0
1/4	0.250 0	$7\frac{1}{2}$	7.500 0
5/16	0.312 5	8	8.000 0
3/8	0.375 0	$8\frac{1}{2}$	8.500 0
7/16	0.437 5	9	9.000 0
1/2	0.500 0	$9\frac{1}{2}$	9.500 0
9/16	0.562 5	10	10.000 0
5/8	0.625 0	$10\frac{1}{2}$	10.500 0
11/16	0.687 5	11	11.000 0
3/4	0.750 0	$11\frac{1}{2}$	11.500 0
7/8	0.875 0	12	12.000 0
1	1.000 0	$12\frac{1}{2}$	12.500 0
$1\frac{1}{4}$	1.250 0	13	13.000 0
$1\frac{1}{2}$	1.500 0	$13\frac{1}{2}$	13.500 0
$1\frac{3}{4}$	1.750 0	14	14.000 0
2	2.000 0	$14\frac{1}{2}$	14.500 0
$2\frac{1}{4}$	2.250 0	15	15.000 0
$2\frac{1}{2}$	2.500 0	$15\frac{1}{2}$	15.500 0
$2\frac{3}{4}$	2.750 0	16	16.000 0
3	3.000 0	$16\frac{1}{2}$	16.500 0
$3\frac{1}{4}$	3.250 0	17	17.000 0
$3\frac{1}{2}$	3.500 0	$17\frac{1}{2}$	17.500 0
$3\frac{3}{4}$	3.750 0	18	18.000 0
4	4.000 0	$18\frac{1}{2}$	18.500 0
$4\frac{1}{4}$	4.250 0	19	19.000 0
$4\frac{1}{2}$	4.500 0	$19\frac{1}{2}$	19.500 0
$4\frac{3}{4}$	4.750 0	20	20.000 0

SOURCE: ANSI Standard B4.1-1967 (Revised 1979), *Preferred Limits and Fits for Cylindrical Parts* (American Society of Mechanical Engineers, New York, NY).

TABLE 13–4 Preferred basic sizes

Decimal (in)		
0.010	2.00	8.50
0.012	2.20	9.00
0.016	2.40	9.50
0.020	2.60	10.00
0.025	2.80	10.50
0.032	3.00	11.00
0.040	3.20	11.50
0.05	3.40	12.00
0.06	3.60	12.50
0.08	3.80	13.00
0.10	4.00	13.50
0.12	4.20	14.00
0.16	4.40	14.50
0.20	4.60	15.00
0.24	4.80	15.50
0.30	5.00	16.00
0.40	5.20	16.50
0.50	5.40	17.00
0.60	5.60	17.50
0.80	5.80	18.00
1.00	6.00	18.50
1.20	6.50	19.00
1.40	7.00	19.50
1.60	7.50	20.00
1.80	8.00	

SOURCE: ANSI Standard B4.1-1967 (Revised 1979), *Preferred Limits and Fits for Cylindrical Parts* (American Society of Mechanical Engineers, New York, NY).

TABLE 13–5   Preferred metric sizes

First	Second	First	Second	First	Second
	Nominal size (mm)
1			13		55
	1.1	14		60	
1.2			15		65
	1.4	16			70
1.6			17		75
	1.8	18		80	
2			19		90
	2.2	20		100	
2.5			21		110
	2.8	22		120	
3			23		130
	3.5		24	140	
4		25			150
	4.5		26	160	
5			28		170
	5.5	30		180	
6			32		190
	6.5	35		200	
	7		38		220
8		40		250	
	9		42		280
10		45		300	
	11		48		320
12		50		350	

SOURCE: ANSI Standard B4.1-1967, *Preferred Metric Limits and Fits* (American Society of Mechanical Engineers, New York, NY).

a size between two first-choice sizes is required, the second-choice column should be used. This will limit the number of sizes typically encountered in the manufacture of products and will lead to cost-effective standardization. The choice of system depends on the policies of the company and the market for the product. Appendix 2 also includes the tables of preferred basic sizes for convenience.

**13–5
CLEARANCE FITS**

When there must always be a clearance between mating parts, a clearance fit is specified. The designation for standard clearance fits from ANSI Standard B4.1 for members that must move together is the *running* or *sliding clearance fit (RC)*. Within this standard, there are nine classes, RC1 through RC9, with RC1 providing the smallest clearance, and RC9 the largest. The following descriptions for the individual members of this class should help you to decide which is most appropriate for a given application.

RC1 (close sliding fit): Accurate location of parts that must assemble without perceptible play.

RC2 (sliding fit): Parts that will move and turn easily but are not intended to run freely. Parts may seize with small temperature changes, especially in the larger sizes.

RC3 (precision running fit): Precision parts operating at slow speeds and light loads that must run freely. Changes in temperature may cause difficulties.

RC4 (close running fit): Accurate location with minimum play for use under moderate loads and speeds. A good choice for accurate machinery.

RC5 (medium running fit): Accurate machine parts for higher speeds and/or loads than for RC4.

RC6 (medium running fit): Similar to RC5 for applications in which larger clearance is desired.

RC7 (free running fit): Reliable relative motion under wide temperature variations in applications where accuracy is not critical.

RC8 (loose running fit): Permits large clearances, allowing the use of parts with commercial, "as-received" tolerances.

RC9 (loose running fit): Similar to RC8, with approximately 50% larger clearances.

The complete standard ANSI B4.1 lists the tolerances on the mating parts and the resulting limits of clearances for all nine classes and for sizes from zero to 200 in. Table 13–6 is abstracted from the standard. Let RC2 represent the precision fits (RC1, RC2, RC3); let RC5 represent the accurate, reliable running fits (RC4 to RC7); and let RC8 represent the loose fits (RC8, RC9).

The numbers in Table 13–6 are in thousandths of an inch. Thus, a clearance of 2.8 from the table means a difference in size between the inside and outside parts of 0.002 8 in. The tolerances on the hole and the shaft are to be applied to the basic size to determine the limits of size for that dimension.

In applying tolerances and the allowance to the dimensions of mating parts, designers use the basic hole system and the basic shaft system. In the *basic hole system,* the design size of the hole is the basic size, and the allowance is applied to the shaft; the basic size is the minimum size of the hole. In the *basic shaft system,* the design size of the shaft is the basic size, and the allowance is applied to the hole; the basic size is the maximum size of the shaft. The basic hole system is preferred.

Figure 13–4 shows a graphical display of the tolerances and fits for all nine RC classes when applied to a shaft/hole combination in which the basic size is 2.000 in and the basic hole system is used. Note that such a diagram shows the total tolerance on both the shaft and the hole. The tolerance for the hole always starts at the basic size, while the shaft tolerance is offset below the basic size to provide for the minimum clearance (smallest hole combined with the largest shaft). The maximum clearance combines the largest hole with the smallest shaft. This figure also shows the dramatic range of clearances provided by the nine classes within the RC system.

The codes within the tolerance bars in Figure 13–4 refer to the tolerance grades mentioned earlier. The capital H combined with a tolerance grade number is used for the hole in the basic hole system for which there is no fundamental deviation from the basic size. The lowercase letters in the shaft tolerance bars indicate some fundamental deviation of the shaft size from the basic size. Then the tolerance is added to the fundamental deviation. The size of the tolerance is indicated by the number.

ISO standard R286 for limits and fits also uses the letter/number codes. For example, a specification for a shaft/hole combination with a basic size of 50 mm that is to provide a free running fit (similar to RC7) could be defined on a drawing as

Hole: ∅50 H9   Shaft: ∅50 d8

No tolerance values need to be given on the drawing.

TABLE 13–6   Clearance fits (RC)

| Nominal size range (in) | Class RC2 | | | Class RC5 | | | Class RC8 | | | Nominal size range (in) |
| | Limits of clearance | Standard limits | | Limits of clearance | Standard limits | | Limits of clearance | Standard limits | | |
Over To		Hole	Shaft		Hole	Shaft		Hole	Shaft	Over To
0–0.12	0.1 0.55	+0.25 0	−0.1 −0.3	0.6 1.6	+0.6 −0	−0.6 −1.0	2.5 5.1	+1.6 0	−2.5 −3.5	0–0.12
0.12–0.24	0.15 0.65	+0.3 0	−0.15 −0.35	0.8 2.0	+0.7 −0	−0.8 −1.3	2.8 5.8	+1.8 0	−2.8 −4.0	0.12–0.24
0.24–0.40	0.2 0.85	+0.4 0	−0.2 −0.45	1.0 2.5	+0.9 −0	−1.0 −1.6	3.0 6.6	+2.2 0	−3.0 −4.4	0.24–0.40
0.40–0.71	0.25 0.95	+0.4 0	−0.25 −0.55	1.2 2.9	+1.0 −0	−1.2 −1.9	3.5 7.9	+2.8 0	−3.5 −5.1	0.40–0.71
0.71–1.19	0.3 1.2	+0.5 0	−0.3 −0.7	1.6 3.6	+1.2 −0	−1.6 −2.4	4.5 10.0	+3.5 0	−4.5 −6.5	0.71–1.19
1.19–1.97	0.4 1.4	+0.6 0	−0.4 −0.8	2.0 4.6	+1.6 −0	−2.0 −3.0	5.0 11.5	+4.0 0	−5.0 −7.5	1.19–1.97
1.97–3.15	0.4 1.6	+0.7 0	−0.4 −0.9	2.5 5.5	+1.8 −0	−2.5 −3.7	6.0 13.5	+4.5 0	−6.0 −9.0	1.97–3.15
3.15–4.73	0.5 2.0	+0.9 0	−0.5 −1.1	3.0 6.6	+2.2 −0	−3.0 −4.4	7.0 15.5	+5.0 0	−7.0 −10.5	3.15–4.73
4.73–7.09	0.6 2.3	+1.0 0	−0.6 −1.3	3.5 7.6	+2.5 −0	−3.5 −5.1	8.0 18.0	+6.0 0	−8.0 −12.0	4.73–7.09
7.09–9.85	0.6 2.6	+1.2 0	−0.6 −1.4	4.0 8.6	+2.8 −0	−4.0 −5.8	10.0 21.5	+7.0 0	−10.0 −14.5	7.09–9.85
9.85–12.41	0.7 2.8	+1.2 0	−0.7 −1.6	5.0 10.0	+3.0 −0	−5.0 −7.0	12.0 25.0	+8.0 0	−12.0 −17.0	9.85–12.41

Source: ANSI Standard B4.1-1967 (Revised 1979), *Preferred Limits and Fits for Cylindrical Parts* (American Society of Mechanical Engineers, New York, NY).

Note: Limits are in thousandths of an inch.

Example Problem 13–1   A shaft carrying an idler sheave for a belt drive system is to have a nominal size of 2.00 in. The sheave must rotate reliably on the shaft, but with the smoothness characteristic of accurate machinery. Specify the limits of size for the shaft and the sheave bore, and list the limits of clearance that will result. Use the basic hole system.

Solution   An RC5 fit should be satisfactory in this application. From Table 13–6, the hole tolerance limits are +1.8 and −0. The sheave hole then should be within the following limits:

Sheave Hole

$$2.000\ 0 + 0.001\ 8 = 2.001\ 8 \text{ in}\quad \text{(largest)}$$
$$2.000\ 0 - 0.000\ 0 = 2.000\ 0 \text{ in}\quad \text{(smallest)}$$

FIGURE 13–4 RC fits for basic hole size of 2.00 in showing tolerances for hole and shaft, minimum clearance and maximum clearance

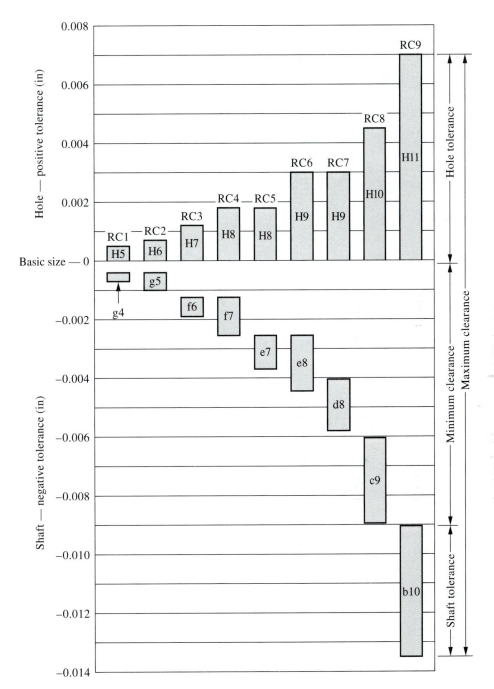

Notice that the smallest hole is the basic size.

The shaft tolerance limits are −2.5 and −3.7. The resulting size limits are as follows:

Shaft Diameter

$$2.000\ 0 - 0.002\ 5 = 1.997\ 5 \text{ in} \quad \text{(largest)}$$
$$2.000\ 0 - 0.003\ 7 = 1.996\ 3 \text{ in} \quad \text{(smallest)}$$

Figure 13–5 illustrates these results.

FIGURE 13–5 An RC5 fit using the basic hole system

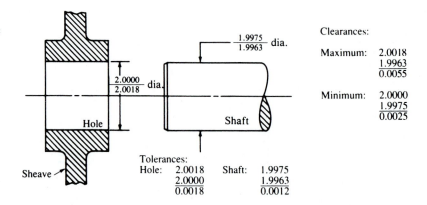

Combining the smallest shaft with the largest hole gives the largest clearance. Conversely, combining the largest shaft with the smallest hole gives the smallest clearance. Therefore, the limits of clearance are

$$2.001\ 8 - 1.996\ 3 = 0.005\ 5 \text{ in} \quad \text{(largest)}$$
$$2.000\ 0 - 1.997\ 5 = 0.002\ 5 \text{ in} \quad \text{(smallest)}$$

These values check with the limits of clearance in Table 13–6. Notice that the total tolerance for the shaft is 0.001 2 in, and for the hole 0.001 8 in, both relatively small values.

Locational Clearance Fits

Another clearance fit system is available for parts for which control of the location is desired although the parts will normally not move in relation to each other in operation. Called *locational clearance (LC)* fits, they include eleven classes. The first four, LC1 to LC4, have a zero clearance (size to size) as the lower limit of the fit, regardless of size or class. The upper limit of fit increases with both the size of the parts and the class number. Classes LC5 through LC11 provide some positive clearance for all sizes, increasing with the size of the parts and with the class. Numerical values for the tolerances and fits for these classes have been published. (See Reference 1.)

**13–6
INTERFERENCE
FITS**

Interference fits are those in which the inside member is larger than the outside member, requiring the application of force during assembly. There is some deformation of the parts after assembly, and a pressure exists at the mating surfaces.

Force fits are designed to provide a controlled pressure between mating parts throughout the range of sizes for a given class. They are used where forces or torques are transmitted through the joint. Instead of being assembled through the application of force, similar fits are obtained by *shrink fitting,* in which one member is heated to expand it while the other remains cool. The parts are then assembled with little or no force. After cooling, the same dimensional interference exists as for the force fit. *Locational interference fits* are used for location only. There is no movement between parts after assembly, but there is no special requirement for the resulting pressure between mating parts.

Force Fits (FN)

As shown in Table 13–7, five classes of force fits are defined in ANSI Standard B4.1. (See Reference 1.)

TABLE 13–7 Force and shrink fits (FN)

Nominal size range (in) Over To	Class FN1 Limits of interference	Standard limits Hole	Shaft	Class FN2 Limits of interference	Standard limits Hole	Shaft	Class FN3 Limits of interference	Standard limits Hole	Shaft	Class FN4 Limits of interference	Standard limits Hole	Shaft	Class FN5 Limits of interference	Standard limits Hole	Shaft
0–0.12	0.05 0.5	+0.25 −0	+0.5 +0.3	0.2 0.85	+0.4 −0	+0.85 +0.6				0.3 0.95	+0.4 −0	+0.95 +0.7	0.3 1.3	+0.6 −0	+1.3 +0.9
0.12–0.24	0.1 0.6	+0.3 −0	+0.6 +0.4	0.2 1.0	+0.5 −0	+1.0 +0.7				0.4 1.2	+0.5 −0	+1.2 +0.9	0.5 1.7	+0.7 −0	+1.7 +1.2
0.24–0.40	0.1 0.75	+0.4 −0	+0.75 +0.5	0.4 1.4	+0.6 −0	+1.4 +1.0				0.6 1.6	+0.6 −0	+1.6 +1.2	0.5 2.0	+0.9 −0	+2.0 +1.4
0.40–0.56	0.1 0.8	+0.4 −0	+0.8 +0.5	0.5 1.6	+0.7 −0	+1.6 +1.2				0.7 1.8	+0.7 −0	+1.8 +1.4	0.6 2.3	+1.0 −0	+2.3 +1.6
0.56–0.71	0.2 0.9	+0.4 −0	+0.9 +0.6	0.5 1.6	+0.7 −0	+1.6 +1.2				0.7 1.8	+0.7 −0	+1.8 +1.4	0.8 2.5	+1.0 −0	+2.5 +1.8
0.71–0.95	0.2 1.1	+0.5 −0	+1.1 +0.7	0.6 1.9	+0.8 −0	+1.9 +1.4				0.8 2.1	+0.8 −0	+2.1 +1.6	1.0 3.0	+1.2 −0	+3.0 +2.2
0.95–1.19	0.3 1.2	+0.5 −0	+1.2 +0.8	0.6 1.9	+0.8 −0	+1.9 +1.4	0.8 2.1	+0.8 −0	+2.1 +1.6	1.0 2.3	+0.8 −0	+2.3 +1.8	1.3 3.3	+1.2 −0	+3.3 +2.5
1.19–1.58	0.3 1.3	+0.6 −0	+1.3 +0.9	0.8 2.4	+1.0 −0	+2.4 +1.8	1.0 2.6	+1.0 −0	+2.6 +2.0	1.5 3.1	+1.0 −0	+3.1 +2.5	1.4 4.0	+1.6 −0	+4.0 +3.0
1.58–1.97	0.4 1.4	+0.6 −0	+1.4 +1.0	0.8 2.4	+1.0 −0	+2.4 +1.8	1.2 2.8	+1.0 −0	+2.8 +2.2	1.8 3.4	+1.0 −0	+3.4 +2.8	2.4 5.0	+1.6 −0	+5.0 +4.0
1.97–2.56	0.6 1.8	+0.7 −0	+1.8 +1.3	0.8 2.7	+1.2 −0	+2.7 +2.0	1.3 3.2	+1.2 −0	+3.2 +2.5	2.3 4.2	+1.2 −0	+4.2 +3.5	3.2 6.2	+1.8 −0	+6.2 +5.0
2.56–3.15	0.7 1.9	+0.7 −0	+1.9 +1.4	1.0 2.9	+1.2 −0	+2.9 +2.2	1.8 3.7	+1.2 −0	+3.7 +3.0	2.8 4.7	+1.2 −0	+4.7 +4.0	4.2 7.2	+1.8 −0	+7.2 +6.0
3.15–3.94	0.9 2.4	+0.9 −0	+2.4 +1.8	1.4 3.7	+1.4 −0	+3.7 +2.8	2.1 4.4	+1.4 −0	+4.4 +3.5	3.6 5.9	+1.4 −0	+5.9 +5.0	4.8 8.4	+2.2 −0	+8.4 +7.0
3.94–4.73	1.1 2.6	+0.9 −0	+2.6 +2.0	1.6 3.9	+1.4 −0	+3.9 +3.0	2.6 4.9	+1.4 −0	+4.9 +4.0	4.6 6.9	+1.4 −0	+6.9 +6.0	5.8 9.4	+2.2 −0	+9.4 +8.0
4.73–5.52	1.2 2.9	+1.0 −0	+2.9 +2.2	1.9 4.5	+1.6 −0	+4.5 +3.5	3.4 6.0	+1.6 −0	+6.0 +5.0	5.4 8.0	+1.6 −0	+8.0 +7.0	7.5 11.6	+2.5 −0	+11.6 +10.0
5.52–6.30	1.5 3.2	+1.0 −0	+3.2 +2.5	2.4 5.0	+1.6 −0	+5.0 +4.0	3.4 6.0	+1.6 −0	+6.0 +5.0	5.4 8.0	+1.6 −0	+8.0 +7.0	9.5 13.6	+2.5 −0	+13.6 +12.0

SOURCE: ANSI Standard B4.1-1967 (Revised 1979), *Preferred Limits and Fits for Cylindrical Parts* (American Society of Mechanical Engineers, New York, NY).

NOTE: Limits are in thousandths of an inch.

FN1 (light drive fit): Only light pressure required to assemble mating parts. Used for fragile parts and where no large forces must be transmitted across the joint.

FN2 (medium drive fit): General-purpose class used frequently for steel parts of moderate cross section.

TABLE 13–8 Coefficient of thermal expansion

Material	Coefficient of thermal expansion, α	
	in/in·°F	mm/mm·°C
Steel:		
AISI 1020	6.5×10^{-6}	11.7×10^{-6}
AISI 1050	6.1×10^{-6}	11.0×10^{-6}
AISI 4140	6.2×10^{-6}	11.2×10^{-6}
Stainless steel:		
AISI 301	9.4×10^{-6}	16.9×10^{-6}
AISI 430	5.8×10^{-6}	10.4×10^{-6}
Aluminum:		
2014	12.8×10^{-6}	23.0×10^{-6}
6061	13.0×10^{-6}	23.4×10^{-6}
Bronze:	10.0×10^{-6}	18.0×10^{-6}

FN3 (heavy drive fit): Used for heavy steel parts.

FN4 (force fit): Used for high-strength assemblies where high resulting pressures are required.

FN5 (force fit): Similar to FN4 for higher pressures.

The use of shrink fit methods is desirable in most cases of interference fits and is virtually required in the heavier classes and larger-size parts. The temperature increase required to produce a given expansion for assembly can be computed from the basic definition of the coefficient of thermal expansion:

$$\delta = \alpha L (\Delta t) \qquad (13\text{–}1)$$

where δ = total deformation desired (in or mm)

α = coefficient of thermal expansion (in/in·°F or mm/mm·°C)

L = nominal length of member being heated (in or mm)

Δt = temperature difference (°F or °C)

For cylindrical parts, L is the diameter and δ is the change in diameter required. Table 13–8 gives the values for α for several materials. (See Reference 7 for more data.)

**13–7
TRANSITION FITS**

The *location transition (LT)* fit is used where accuracy of location is important, but where a small amount of clearance or a small amount of interference is acceptable. There are six classes, LT1 to LT6. In any class there is an overlap in the tolerance limits for both the hole and the shaft so that possible combinations produce a small clearance, a small interference, or even a size-to-size fit. Complete tables of data for these fits are published in Reference 1.

**13–8
ROBUST
PRODUCT
DESIGN**

Careful control of the tolerances on the dimensions of machine parts is the responsibility of both the product designer and the manufacturing staff. The goal is to ensure the functionality of the product while permitting economical manufacture. Often these goals seem incompatible.

Robust product design is a tool that can help. (See Reference 5.) It is a technique in which a sequence of experiments is undertaken to determine the variables in a product's design that most affect its performance. Then the optimum limits on those variables are defined. The concepts can be used for dimensions, material properties, control settings, and many other factors.

The design of the experiments used to implement a robust product is critical to the success of the design. An initial screening experiment is used to select the most significant variables. Then additional experiments are run to determine the response of the system to combinations of these variables. Computer-generated plots of results establish limits for the significant variables.

Robust product design is based on the *Taguchi method,* an important element in the process of improving manufacturing quality developed by Dr. Genichi Taguchi. He showed how design of experiments can be used to produce products of consistently high quality that are insensitive to the operating environment.

Another tool that can assist in the allocation of tolerances is called *variation simulation analysis,* or *VSA,* developed by the Applied Computer Solutions Company. (See Reference 4.) VSA allows the designer to model the statistical variation expected for several key variables in a product made from several parts. It uses statistical techniques to predict the resulting assembly configurations considering the "tolerance stack" of the individual components. The computer-assisted simulation allows the designer to tailor tolerance specifications to achieve an economical yet reliable product.

13–9 STRESSES FOR FORCE FITS

When force fits are used to secure mechanical parts, the interference creates a pressure acting at the mating surfaces. The pressure causes stresses in each part. Under heavy force fits or even with lighter fits in fragile parts, the stresses developed can be great enough to yield ductile materials. A permanent set results, which normally destroys the usefulness of the assembly. With brittle materials such as cast iron, actual fracture may result.

The stress analysis applicable to force fits is related to the analysis of thick-walled cylinders. The outer member expands under the influence of the pressure at the mating surface, with the tangential tensile stress developed being a maximum at the mating surface. There is a radial stress equal to the pressure itself. Also, the inner member contracts because of the pressure and is subjected to a tangential compressive stress along with the radial compressive stress equal to the pressure. (See Reference 8.)

The usual objective of the analysis is to determine the magnitude of the pressure due to a given interference fit that would be developed at the mating surfaces. Then the stresses due to this pressure in the mating members are computed. The following procedure can be used:

Procedure for Computing Stresses for Force Fits

1. Determine the amount of interference from the design of the parts. For standard force fits, Table 13–7 can be used. The maximum limit of interference would, of course, give the maximum stresses for the parts. Note that the interference values are based on the total interference on the diameter, which is the sum of the expansion of the outer ring plus the contraction of the inner member (see Figure 13–6).

2. Compute the pressure at the mating surface from Equation (13–2) if both members are of the same material:

Pressure Created by Force Fit

$$p = \frac{E\delta}{2b}\left[\frac{(c^2 - b^2)(b^2 - a^2)}{2b^2(c^2 - a^2)}\right]$$

(13–2)

The tensile stress in the steel sleeve is

$$\sigma_o = p\left(\frac{c^2 + b^2}{c^2 - b^2}\right) = 3\,517\left(\frac{1.75^2 + 1.25^2}{1.75^2 - 1.25^2}\right) = 10\,846 \text{ psi}$$

The compressive stress in the bronze bushing is

$$\sigma_i = -p\left(\frac{b^2 + a^2}{b^2 - a^2}\right) = -3\,517\left(\frac{1.25^2 + 1.00^2}{1.25^2 - 1.00^2}\right) = -16\,025 \text{ psi}$$

The increase in the diameter of the sleeve is

$$\delta_o = \frac{2bp}{E_o}\left[\frac{c^2 + b^2}{c^2 - b^2} + v_o\right]$$

$$\delta_o = \frac{2(1.25)(3\,517)}{30 \times 10^6}\left[\frac{1.75^2 + 1.25^2}{1.75^2 - 1.25^2} + 0.27\right] = 0.000\,98 \text{ in}$$

The decrease in the diameter of the bushing is

$$\delta_i = -\frac{2bp}{E_i}\left[\frac{b^2 + a^2}{b^2 - a^2} - v_i\right]$$

$$\delta_i = \frac{2(1.25)(3\,517)}{17 \times 10^6}\left[\frac{1.25^2 + 1.00^2}{1.25^2 - 1.00^2} + 0.27\right] = 0.002\,22 \text{ in}$$

Note that the sum of δ_o and δ_i equals 0.003 2, the total interference, δ.

The spreadsheet to analyze force fits is shown in Figure 13–7. The data shown are from Example Problem 13–2.

STRESSES FOR FORCE FITS Refer to Figure 13–6 for geometry.		*Data from:* Example Problem 13–2	
Input Data:		**Numerical values in *italics* must be inserted for each problem.**	
Inside radius of inner member:	*a=*	*1.0000 in*	
Outside radius of inner member:	*b=*	*1.2500 in*	
Outside radius of outer member:	*c=*	*1.7500 in*	
Total interference:	*d=*	*0.0032 in*	
Modulus of outer member:	*E_o=*	*3.00E+07 psi*	
Modulus of inner member:	*E_i=*	*1.70E+07 psi*	
Poisson's ratio for outer member:	*v_o=*	*0.27*	
Poisson's ratio for inner member:	*v_i=*	*0.27*	
Computed Results			
Pressure at mating surface: ***p=***		**3,518 psi**	Using Equation (13–3)
Tensile stress in outer member: $\sigma_o=$		**10,846 psi**	Using Equation (13–4)
Compressive stress in inner member: $\sigma_i=$		**−16,025 psi**	Using Equation (13–5)
Increase in diameter of outer member: $\delta_o=$		**0.00098 in**	Using Equation (13–6)
Decrease in diameter of inner member: $\delta_i=$		**0.00222 in**	Using Equation (13–7)

FIGURE 13–7 Spreadsheet solution for pressure, stresses, and deformations in cylindrical members mating with a force fit

REFERENCES

1. The American Society of Mechanical Engineers. ANSI Standard B4.1-1967 (R87/94). *Preferred Limits and Fits for Cylindrical Parts.* New York: American Society of Mechanical Engineers.

2. The American Society of Mechanical Engineers. ANSI Standard B4.2-1978 (R84). *Preferred Metric Limits and Fits.* New York: American Society of Mechanical Engineers.

3. The American Society of Mechanical Engineers. ANSI Standard Y14.5M. *Dimensioning and Tolerancing.* New York: American Society of Mechanical Engineers, 1995.

4. Applied Computer Solutions. *Variation Simulation Analysis, 3-D VSA* (Computer Program). St. Clair Shores, MI: Applied Computer Solutions Company, 1989.

5. Donnelly, T. A. "Robust Product Design." *Machine Design Magazine* (October 8, 1987).

6. Jensen, C., and R. Hines. *Interpreting Engineering Drawings.* 4th ed. Albany, NY: Delmar Publishers, 1989.

7. Oberg, E., et al. *Machinery's Handbook.* 25th ed. New York: Industrial Press, 1996.

8. Shigley, J. E., and C. R. Mischke. *Mechanical Engineering Design.* 5th ed. New York: McGraw-Hill, 1989.

PROBLEMS

1. Specify the class of fit, the limits of size, and the limits of clearance for the bore of a slow-moving but heavily loaded conveyor roller that must rotate freely on a stationary shaft under heavy load. The nominal diameter of the shaft is 3.500 in. Use the basic hole system.

2. A sine plate is a measuring device that pivots at its base, allowing the plate to assume different angles that are set with gage blocks for precision. The pin in the pivot has a nominal diameter of 0.500 0 in. Specify the class of fit, the limits of size, and the limits of clearance for the pivot. Use the basic hole system.

3. A child's toy wagon has an axle with a nominal diameter of 5/8 in. For the wheel bore and axle, specify the class of fit, the limits of size, and the limits of clearance. Use the basic shaft system.

4. The planet gear of an epicyclic gear train must rotate reliably on its shaft while maintaining accurate position with respect to the mating gears. For the planet gear bore and its shaft, specify the class of fit, the limits of size, and the limits of clearance. Use the basic hole system. The nominal shaft diameter is 0.800 in.

5. The base of a hydraulic cylinder is mounted to the frame of a machine by means of a clevis joint, which allows the cylinder to oscillate during operation. The clevis must provide reliable motion, but some play is acceptable. For the clevis holes and the pin, which have a nominal diameter of 1.25 in, specify the class of fit, the limits of size, and the limits of clearance. Use the basic hole system.

6. A heavy door on a furnace swings upward to permit access to the interior of the furnace. During various modes of operation, the door and its hinge assembly see temperatures of 50°F to 500°F. The nominal diameter of each hinge pin is 4.00 in. For the hinge and its pin, specify the class of fit, the limits of size, and the limits of clearance. Use the basic shaft system.

7. The stage of an industrial microscope pivots to permit the mounting of a variety of shapes. The stage must move with precision and reliability under widely varying temperatures. For the pin mount for the stage, having a nominal diameter of 0.750 in, specify the class of fit, the limits of size, and the limits of clearance. Use the basic hole system.

8. An advertising sign is suspended from a horizontal rod and is allowed to swing under wind forces. The rod is to be commercial bar stock, nominally 1.50 inches in diameter. For the mating hinges on the sign, specify the class of fit, the limits of size, and the limits of clearance. Use the basic shaft system.

9. A spacer made of AISI 1020 hot-rolled steel is in the form of a hollow cylinder with a nominal inside diameter of 3.25 in and an outside diameter of 4.000 in. It is to be mounted on a solid steel shaft with a heavy force fit. Specify the dimensions for the shaft and sleeve, and compute the stress in the sleeve after installation. Use the basic hole system.

10. A bronze bushing, with an inside diameter of 3.50 in and a nominal outside diameter of 4.00 in, is pressed into a steel sleeve having an outside diameter of 4.50 in. For an FN3 class of fit, specify the dimensions of the parts, the limits of interference, and the stresses created in the bushing and the sleeve. Use the basic hole system.

11. It is proposed to install a steel rod with a nominal diameter of 3.00 inches in a hole of an aluminum cylinder with an outside diameter of 5.00 in, with an FN5 force fit. Would this be satisfactory?

12. The allowable compressive stress in the wall of an aluminum tube is 8 500 psi. Its outside diameter is 2.000 in, and the wall thickness is 0.065 in. What is the maximum amount of interference between this tube and a steel sleeve that can be tolerated? The sleeve has an outside diameter of 3.00 in.

13. To what temperature would the spacer of Problem 9 have to be heated so that it would slide over the shaft with a clearance of 0.002 in? The ambient temperature is 75°F.

14. For the bronze bushing and the steel sleeve of Problem 10, the ambient temperature is 75°F. How much would the bronze bushing shrink if placed in a freezer at −20°F? Then to what temperature would the steel sleeve have to be heated to provide a clearance of 0.004 in for assembly on the cold bushing?

15. For the bronze bushing of Problem 10, what would be the final inside diameter after assembly into the sleeve if it started with an inside diameter of 3.500 0 in?

16. For any of Problems 1–8, make a diagram of the tolerances, fits, and clearances for your specifications, using a method similar to that in Figure 13–4.

14

Rolling Contact Bearings

Rolling Contact Bearings

Discussion Map

☐ *Bearings* are used to support a load while permitting relative motion between two elements of a machine.

☐ Some bearings use rolling elements, such as spherical balls or cylindrical or tapered rollers. This results in a very low coefficient of friction.

Discover

Look for examples of bearings on machines, cars, trucks, bicycles, and consumer products.

Describe the bearings, including how they are installed and what kinds of forces are exerted on them.

This chapter presents information about such bearings and gives methods for analyzing bearings and for selecting commercially available bearings.

The purpose of a bearing is to support a load while permitting relative motion between two elements of a machine. The term *rolling contact bearings* refers to the wide variety of bearings that use spherical balls or some other type of roller between the stationary and the moving elements. The most common type of bearing supports a rotating shaft, resisting purely radial loads or a combination of radial and axial (thrust) loads. Some bearings are designed to carry only thrust loads. Most bearings are used in applications involving rotation, but some are used in linear motion applications.

The components of a typical rolling contact bearing are the inner race, the outer race, and the rolling elements. Figure 14–1 shows the common single-row, deep-groove ball bearing. Usually the outer race is stationary and is held by the housing of the machine. The inner race is pressed onto the rotating shaft and thus rotates with it. Then the balls roll between the outer and inner races. The load path is from the shaft, to the inner race, to the balls, to the outer race, and finally to the housing. The presence of the balls allows a very smooth, low-friction rotation of the shaft. The typical coefficient of friction for a rolling contact bearing is approximately 0.001 to 0.005. These values reflect only the rolling elements themselves and the means of retaining them in the bearing. The presence of seals, excessive lubricant, or unusual loading increases these values.

Look for consumer products, industrial machinery, or transportation equipment (cars, trucks, bicycles, and so on), and identify any uses of rolling contact bearings. Be sure that the machine is turned off, and then try to gain access to the mechanical drive shafts used to transmit power from a motor or an engine to some moving parts of the machine. Are these shafts supported in ball or roller bearings? Or are they supported in plain surface bearings in which the shaft passes through cylindrical members called *bushings* or *bearings,* typically with lubricants present between the rotating shaft and the stationary bearing? Plain surface bearings are discussed in Chapter 16.

For the shafts that are supported in ball or roller bearings, describe the bearing. How is it mounted on the shaft? How is it mounted in the housing of the machine? Can you identify what kinds of forces are acting on the bearing and in what directions they act? Are the forces directed radially toward the centerline of the shaft? Is there any force that acts parallel to the axis of the shaft? Compare the bearings that you find with the photographs that are in this chapter. Which varieties of bearings have you found? Measure or estimate the physical size of the bearings, particularly the diameter of the bore that is in contact with the shaft, the outside diameter, and the width. Can you see the rolling elements—balls or rollers? If so, sketch them and estimate their diameters and/or lengths. Are any of the rolling elements tapered rollers like those shown later in this chapter in Figure 14–7?

FIGURE 14–1   Single-row, deep-groove ball bearing (Hoover-NSK Bearing Company, Ann Arbor, MI)

When you complete this chapter, you will be able to identify several kinds of rolling contact bearings and specify suitable bearings to carry specified loads. You will also be able to apply such bearings properly, planning for their installation on shafts and in housings.

 You Are the Designer

In Chapter 12, you were the designer of a shaft that was rotating at 600 rpm carrying two gears as a part of a power transmission system. Figures 12–1 and 12–2 showed the basic layout that was proposed. The shaft was designed to be supported on two bearings at points B and D. Then, in Example Problem 12–1, we completed the force analysis by computing the forces applied *to* the shaft by the gears and then computing the *reactions at the bearings.* Figure 12–12 showed the results, summarized here:

$$R_{Bx} = 458 \text{ lb} \qquad R_{By} = 4\ 620 \text{ lb}$$
$$R_{Dx} = 1\ 223 \text{ lb} \qquad R_{Dy} = 1\ 680 \text{ lb}$$

where x refers to the horizontal direction and y refers to the vertical direction. All forces on the bearings are in the radial direction. Your task now is to specify suitable rolling contact bearings for the shaft to withstand those forces and transfer them from the shaft to the housing of the speed reducer.

What kind of bearing should be selected? How do the forces just identified affect the choice? What life expectancy is reasonable for the bearings, and how does that affect the selection of the bearings? What size should be specified? How are the bearings to be installed on the shaft, and how does that affect the detailed design of the shaft? What limit dimensions should be defined for the bearing seats on the shaft? How is the bearing to be located axially on the shaft? How is it to be installed in the housing and located there? How is lubrication to be provided for the bearings? Is there a need for shields and seals to keep contaminants out of the bearings? The information in this chapter will help you make these and other design decisions.

TABLE 14–3 Bearing selection data for single-row, deep-groove, Conrad-type ball bearings

A. Series 6200

| Bearing number | Nominal bearing dimensions | | | | | | r^* | Preferred shoulder diameter | | Bearing weight | Basic static load rating, C_o | Basic dynamic load rating, C |
| | d | | D | | B | | | Shaft | Housing | | | |
	mm	in	mm	in	mm	in	in	in	in	lb	lb	lb
6200	10	0.3937	30	1.1811	9	0.3543	0.024	0.500	0.984	0.07	520	885
6201	12	0.4724	32	1.2598	10	0.3937	0.024	0.578	1.063	0.08	675	1 180
6202	15	0.5906	35	1.3780	11	0.4331	0.024	0.703	1.181	0.10	790	1 320
6203	17	0.6693	40	1.5748	12	0.4724	0.024	0.787	1.380	0.14	1 010	1 660
6204	20	0.7874	47	1.8504	14	0.5512	0.039	0.969	1.614	0.23	1 400	2 210
6205	25	0.9843	52	2.0472	15	0.5906	0.039	1.172	1.811	0.29	1 610	2 430
6206	30	1.1811	62	2.4409	16	0.6299	0.039	1.406	2.205	0.44	2 320	3 350
6207	35	1.3780	72	2.8346	17	0.6693	0.039	1.614	2.559	0.64	3 150	4 450
6208	40	1.5748	80	3.1496	18	0.7087	0.039	1.811	2.874	0.82	3 650	5 050
6209	45	1.7717	85	3.3465	19	0.7480	0.039	2.008	3.071	0.89	4 150	5 650
6210	50	1.9685	90	3.5433	20	0.7874	0.039	2.205	3.268	1.02	4 650	6 050
6211	55	2.1654	100	3.9370	21	0.8268	0.059	2.441	3.602	1.36	5 850	7 500
6212	60	2.3622	110	4.3307	22	0.8661	0.059	2.717	3.996	1.73	7 250	9 050
6213	65	2.5591	120	4.7244	23	0.9055	0.059	2.913	4.390	2.18	8 000	9 900
6214	70	2.7559	125	4.9213	24	0.9449	0.059	3.110	4.587	2.31	8 800	10 800
6215	75	2.9528	130	5.1181	25	0.9843	0.059	3.307	4.783	2.64	9 700	11 400
6216	80	3.1496	140	5.5118	26	1.0236	0.079	3.504	5.118	3.09	10 500	12 600
6217	85	3.3465	150	5.9055	28	1.1024	0.079	3.740	5.512	3.97	12 300	14 600
6218	90	3.5433	160	6.2992	30	1.1811	0.079	3.937	5.906	4.74	14 200	16 600
6219	95	3.7402	170	6.6929	32	1.2598	0.079	4.213	6.220	5.73	16 300	18 800
6220	100	3.9370	180	7.0866	34	1.3386	0.079	4.409	6.614	6.94	18 600	21 100
6221	105	4.1339	190	7.4803	36	1.4173	0.079	4.606	7.008	8.15	20 900	23 000
6222	110	4.3307	200	7.8740	38	1.4961	0.079	4.803	7.402	9.59	23 400	24 900
6224	120	4.7244	215	8.4646	40	1.5748	0.079	5.197	7.992	11.4	26 200	26 900

TABLE 14–3 *(continued)*

A. *Series 6200, continued*

Bearing number	d (mm)	d (in)	D (mm)	D (in)	B (mm)	B (in)	r* (in)	Shaft (in)	Housing (in)	Bearing weight (lb)	Basic static load rating, C_o (lb)	Basic dynamic load rating, C (lb)
6226	130	5.118 1	230	9.055 1	40	1.574 8	0.098	5.669	8.504	12.7	29 100	28 700
6228	140	5.511 8	250	9.842 5	42	1.653 5	0.098	6.063	9.291	19.6	29 300	28 700
6230	150	5.905 5	270	10.629 9	45	1.771 7	0.098	6.457	10.079	25.3	32 500	30 000
6232	160	6.299 2	290	11.417 3	48	1.889 8	0.098	6.850	10.886	32.0	35 500	32 000
6234	170	6.692 9	310	12.204 7	52	2.047 2	0.118	7.362	11.535	38.5	43 000	36 500
6236	180	7.086 6	320	12.598 4	52	2.047 2	0.118	7.758	11.929	41.0	46 500	39 000
6238	190	7.480 3	340	13.385 8	55	2.165 4	0.118	8.150	12.717	50.5	54 500	44 000
6240	200	7.874 0	360	14.173 2	58	2.283 5	0.118	8.543	13.504	61.5	60 000	46 500

B. *Series 6300*

Bearing number	d (mm)	d (in)	D (mm)	D (in)	B (mm)	B (in)	r* (in)	Shaft (in)	Housing (in)	Bearing weight (lb)	Basic static load rating, C_o (lb)	Basic dynamic load rating, C (lb)
6300	10	0.393 7	35	1.378 0	11	0.433 1	0.024	0.563	1.181	0.12	805	1 400
6301	12	0.472 4	37	1.456 7	12	0.472 4	0.039	0.656	1.220	0.13	990	1 680
6302	15	0.590 6	42	1.653 5	13	0.511 8	0.039	0.781	1.417	0.18	1 200	1 980
6303	17	0.669 3	47	1.850 4	14	0.551 2	0.039	0.875	1.614	0.25	1 460	2 360
6304	20	0.787 4	52	2.047 2	15	0.590 6	0.039	1.016	1.772	0.32	1 730	2 760
6305	25	0.984 3	62	2.440 9	17	0.669 3	0.039	1.220	2.165	0.52	2 370	3 550
6306	30	1.181 1	72	2.834 6	19	0.748 0	0.039	1.469	2.559	0.76	3 150	4 600
6307	35	1.378 0	80	3.149 6	21	0.826 8	0.059	1.688	2.795	1.01	4 050	5 800
6308	40	1.574 8	90	3.543 3	23	0.905 5	0.059	1.929	3.189	1.40	5 050	7 050
6309	45	1.771 7	100	3.937 0	25	0.984 3	0.059	2.126	3.583	1.84	6 800	9 150
6310	50	1.968 5	110	4.330 7	27	1.063 0	0.079	2.362	3.937	2.42	8 100	10 700
6311	55	2.165 4	120	4.724 4	29	1.141 7	0.079	2.559	4.331	2.98	9 450	12 300
6312	60	2.362 2	130	5.118 1	31	1.220 5	0.079	2.835	4.646	3.75	11 000	14 100
6313	65	2.559 1	140	5.511 8	33	1.299 2	0.079	3.031	5.039	4.63	12 600	16 000
6314	70	2.755 9	150	5.905 5	35	1.378 0	0.079	3.228	5.433	5.51	14 400	18 000
6315	75	2.952 8	160	6.299 2	37	1.456 7	0.079	3.425	5.827	6.61	16 300	19 600

TABLE 14–3 (*continued*)

B. *Series 6300, continued*

Bearing number	Nominal bearing dimensions							Preferred shoulder diameter		Bearing weight	Basic static load rating, C_o	Basic dynamic load rating, C
	d		D		B		r^*	Shaft	Housing			
	mm	in	mm	in	mm	in	in	in	in	lb	lb	lb
6316	80	3.149 6	170	6.692 9	39	1.535 4	0.079	3.622	6.220	7.93	18 300	21 300
6317	85	3.346 5	180	7.086 6	41	1.614 2	0.098	3.898	6.535	9.37	20 400	22 900
6318	90	3.543 3	190	7.480 3	43	1.692 9	0.098	4.094	6.929	10.8	22 500	24 700
6319	95	3.740 2	200	7.874 0	45	1.771 7	0.098	4.291	7.323	12.5	24 900	26 400
6320	100	3.937 0	215	8.464 6	47	1.850 4	0.098	4.488	7.913	15.3	29 800	30 000
6321	105	4.133 9	225	8.858 3	49	1.929 1	0.098	4.685	8.307	17.9	32 500	31 700
6322	110	4.330 7	240	9.448 8	50	1.968 5	0.098	4.882	8.898	21.0	38 000	35 500
6324	120	4.724 4	260	10.236 2	55	2.165 4	0.098	5.276	9.685	27.6	38 500	36 000
6326	130	5.118 1	280	11.023 6	58	2.283 5	0.118	5.827	10.315	40.8	44 500	39 500
6328	140	5.511 8	300	11.811 0	62	2.440 9	0.118	6.220	11.102	48.5	51 000	43 500
6330	150	5.905 5	320	12.598 4	65	2.559 1	0.118	6.614	11.890	57.3	58 000	47 500
6332	160	6.299 2	340	13.385 8	68	2.677 2	0.118	7.008	12.677	58	58 500	48 000
6334	170	6.692 9	360	14.173 2	72	2.834 6	0.118	7.402	13.465	84	73 500	56 500
6336	180	7.086 6	380	14.960 6	75	2.952 8	0.118	7.795	14.252	98	84 000	61 500
6338	190	7.480 3	400	15.748 0	78	3.070 9	0.157	8.346	14.882	112	84 000	61 500
6340	200	7.874 0	420	16.535 4	80	3.149 6	0.157	8.740	15.669	127	91 500	65 500

SOURCE: Hoover-NSK Bearing Company, Ann Arbor, MI.

*Maximum fillet that corner radius will clear.

You should be aware that different manufacturers use other bases for the rated life. For example, some use 90 million cycles as the rated life and determine the rated load for that life. Also, some will report *average life,* for which 50% of the bearings will not survive. Thus, average life can be called the L_{50} life, not the L_{10}. Note that the average life is approximately five times as long as the L_{10} life. (See Reference 13.) Be sure that you understand the basis for the ratings in any given catalog. (See References 1, 3, 7, 9, 12, and 15 for additional analysis of roller bearing performance.)

Example Problem 14–1

A catalog lists the basic dynamic load rating for a ball bearing to be 7 050 lb for a rated life of 1 million rev. What would be the expected L_{10} life of the bearing if it were subjected to a load of 3 500 lb?

Solution

In Equation (14–1),

$$P_1 = C = 7\ 050\ \text{lb} \quad \text{(basic dynamic load rating)}$$
$$P_2 = P_d = 3\ 500\ \text{lb} \quad \text{(design load)}$$
$$L_1 = 10^6\ \text{rev} \quad (L_{10}\ \text{life at load } C)$$
$$k = 3 \quad \text{(ball bearing)}$$

Then letting the life, L_2, be called the *design life, L_d,* at the design load,

$$L_2 = L_d = L_1\left(\frac{P_1}{P_2}\right)^k = 10^6\left(\frac{7\ 050}{3\ 500}\right)^{3.00} = 8.17 \times 10^6\ \text{rev}$$

This must be interpreted as the L_{10} life at a load of 3 500 lb.

**14–8
DESIGN LIFE**

We will use the method developed in Example Problem 14–1 to refine the procedure for computing the required basic dynamic load rating C for a given design load P_d and a given design life L_d. If the reported load data in the manufacturer's literature is for 10^6 revolutions, Equation (14–1) can be written as

⇨ **Design Life**

$$L_d = (C/P_d)^k(10^6) \tag{14–2}$$

The required C for a given design load and life would be

⇨ **Basic Dynamic Load Rating**

$$C = P_d(L_d/10^6)^{1/k} \tag{14–3}$$

Most people do not think in terms of the number of revolutions that a shaft makes. Rather, they consider the speed of rotation of the shaft, usually in rpm, and the design life of the machine, usually in hours of operation. The design life is specified by the designer, considering the application. As a guide, Table 14–4 can be used. Now, for a specified design life in hours, and a known speed of rotation in rpm, the number of design revolutions for the bearing would be

$$L_d = (h)(rpm)(60\ min/h)$$

Example Problem 14–2

Compute the required basic dynamic load rating, C, for a ball bearing to carry a radial load of 650 lb from a shaft rotating at 600 rpm that is part of an assembly conveyor in a manufacturing plant.

TABLE 14–4   Recommended design life for bearings

Application	Design life L_{10}, h
Domestic appliances	1 000–2 000
Aircraft engines	1 000–4 000
Automotive	1 500–5 000
Agricultural equipment	3 000–6 000
Elevators, industrial fans, multipurpose gearing	8 000–15 000
Electric motors, industrial blowers, general industrial machines	20 000–30 000
Pumps and compressors	40 000–60 000
Critical equipment in continuous, 24-h operation	100 000–200 000

SOURCE: Eugene A. Avallone and Theodore Baumeister III, eds., *Marks' Standard Handbook for Mechanical Engineers,* 9th ed. New York: McGraw-Hill, 1986.

Solution   From Table 14–4, let's select a design life of 30 000 h. Then L_d is

$$L_d = (30\ 000\ \text{h})(600\ \text{rpm})(60\ \text{min/h}) = 1.08 \times 10^9\ \text{rev}$$

From Equation (14–3),

$$C = 650(1.08 \times 10^9/10^6)^{1/3} = 6\ 670\ \text{lb}$$

To facilitate calculations, some manufacturers provide charts or tables of life factors and speed factors that make computing the number of revolutions unnecessary. Note that the rated life of 1 million rev would be achieved by a shaft rotating at $33\frac{1}{3}$ rpm for 500 h. If the actual speed or desired life is different from these two values, a speed factor, f_N, and a life factor, f_L, can be determined from charts such as those shown in Figure 14–12. The factors account for the load/life relationship of Equation (14–1). The required basic dynamic load rating, C, for a bearing to carry a design load, P_d, would then be

⇨ **Required Basic Dynamic Load Rating**

$$C = P_d f_L / f_N \tag{14–4}$$

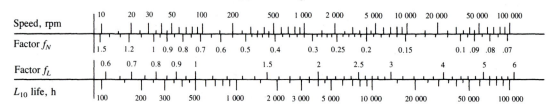

FIGURE 14–12   Life and speed factors for ball and roller bearings

Other catalogs use different approaches, but they are all based on the load/life relationship of Equation (14–1).

If we solved Example Problem 14–2 by using the charts of Figure 14–12, the following would result:

$$f_N = 0.381 \quad \text{(for 600 rpm)}$$
$$f_L = 3.90 \quad \text{(for 30 000-h life)}$$
$$C = 650(3.90)/(0.381) = 6\ 654 \text{ lb}$$

This compares closely with the value of 6 670 lb found previously.

14–9 BEARING SELECTION: RADIAL LOADS ONLY

The selection of a bearing takes into consideration the load capacity, as discussed, and the geometry of the bearing that will ensure that it can be installed conveniently in the machine. We will first consider unmounted bearings carrying radial loads only. Then we will consider unmounted bearings carrying a combination of radial and thrust loads. The term *unmounted* refers to the case in which the designer must provide for the proper application of the bearing onto the shaft and into a housing.

The bearing is normally selected after the shaft design has progressed to the point where the required minimum diameter of the shaft has been determined, using the techniques presented in Chapter 12. The radial loads are also known, along with the orientation of the bearings with respect to other elements in the machine.

Procedure for Selecting a Bearing

1. Specify the design load on the bearing, usually called *equivalent load*. The method of determining the equivalent load when only a radial load, *R,* is applied takes into account whether the inner or the outer race rotates.

Equivalent Load, Radial Load Only

$$P = VR \qquad \text{(14–5)}$$

The factor *V* is called a *rotation factor* and takes the value of 1.0 if the inner race of the bearing rotates, the usual case. Use $V = 1.2$ if the outer race rotates.

2. Determine the minimum acceptable diameter of the shaft that will limit the bore size of the bearing.

3. Select the type of bearing, using Table 14–1 as a guide.

4. Specify the design life of the bearing, using Table 14–4.

5. Determine the speed factor and the life factor if such tables are available for the selected type of bearing. We will use Figure 14–12.

6. Compute the required basic dynamic load rating, *C,* from Equation (14–1), (14–3), or (14–4).

7. Identify a set of candidate bearings that have the required basic dynamic load rating.

8. Select the bearing having the most convenient geometry, also considering its cost and availability.

9. Determine mounting conditions, such as shaft seat diameter and tolerance, housing bore diameter and tolerance, means of locating the bearing axially, and special needs such as seals or shields.

Example Problem 14–3   Select a single-row, deep-groove ball bearing to carry 650 lb of pure radial load from a shaft that rotates at 600 rpm. The design life is to be 30 000 h. The bearing is to be mounted on a shaft with a minimum acceptable diameter of 1.48 in.

Solution   Note that this is a pure radial load and the inner race is to be pressed onto the shaft and rotate with it. Therefore, the factor $V = 1.0$ in Equation (14–5), and the design load is equal to the radial load. These are the same data used in Example Problem 14–2, where we found the required basic dynamic load rating, C, to be 6 670 lb. From Table 14–3, giving design data for two classes of bearings, we find that we could use a bearing 6211 or a bearing 6308. Either has a rated C of just over 6 670 lb. But note that the 6211 has a bore of 55 mm (2.165 4 in), and the 6308 has a bore of 40 mm (1.574 8 in). The 6308 is more nearly in line with the desired shaft size.

Summary of Data for the Selected Bearing

   Bearing number: 6308, single-row, deep-groove ball bearing

   Bore: $d = 40$ mm (1.574 8 in)

   Outside diameter: $D = 90$ mm (3.543 3 in)

   Width: $B = 23$ mm (0.905 5 in)

   Maximum fillet radius: $r = 0.059$ in

   Basic dynamic load rating: $C = 7\ 050$ lb

14–10
BEARING SELECTION: RADIAL AND THRUST LOADS COMBINED

⇨ **Equivalent Load with Radial and Thrust Loads**

When both radial and thrust loads are exerted on a bearing, the equivalent load is the constant radial load that would produce the same rated life for the bearing as the combined loading. The method of computing the equivalent load, P, for such cases is presented in the manufacturer's catalog and takes the form

$$P = VXR + YT \tag{14–6}$$

where $P =$ equivalent load
$V =$ rotation factor (as defined)
$R =$ applied radial load
$T =$ applied thrust load
$X =$ radial factor
$Y =$ thrust factor

The values of X and Y vary with the specific design of the bearing and with the magnitude of the thrust load relative to the radial load. For relatively small thrust loads, $X = 1$ and $Y = 0$, so the equivalent load equation reverts to the form of Equation (14–5) for pure radial loads. To indicate the limiting thrust load for which this is the case, manufacturers list a factor called e. If the ratio $T/R > e$, Equation (14–6) must be used to compute P. If $T/R < e$, Equation (14–5) is used. Table 14–5 shows one set of data for a single-row, deep-groove ball bearing. Note that both e and Y depend on the ratio T/C_o, where C_o is the static load rating of a particular bearing. This presents a difficulty in bearing selection because the value of C_o is not known until the bearing has been selected. Therefore, a simple

TABLE 14–5   Radial and thrust factors for single-row, deep-groove ball bearings

e	T/C_o	Y	e	T/C_o	Y
0.19	0.014	2.30	0.34	0.170	1.31
0.22	0.028	1.99	0.38	0.280	1.15
0.26	0.056	1.71	0.42	0.420	1.04
0.28	0.084	1.55	0.44	0.560	1.00
0.30	0.110	1.45			

NOTE: $X = 0.56$ for all values of Y.

trial-and-error method is applied. If a significant thrust load is applied to a bearing along with a radial load, perform the following steps:

1. Assume a value of Y from Table 14–5. The value $Y = 1.50$ is reasonable, being at about the middle of the range of possible values.
2. Compute $P = VXR + YT$.
3. Compute the required basic dynamic load rating C from Equation (14–1), (14–3), or (14–4).
4. Select a candidate bearing having a value of C at least equal to the required value.
5. For the selected bearing, determine C_o.
6. Compute T/C_o.
7. From Table 14–5, determine e.
8. If $T/R > e$, then determine Y from Table 14–5.
9. If the new value of Y is different from that assumed in Step 1, repeat the process.
10. If $T/R < e$, use Equation (14–5) to compute P, and proceed as for a pure radial load.

Example Problem 14–4   Select a single-row, deep-groove ball bearing from Table 14–3 to carry a radial load of 1 850 lb and a thrust load of 675 lb. The shaft is to rotate at 1 150 rpm, and a design life of 20 000 h is desired. The minimum acceptable diameter for the shaft is 3.10 in.

Solution   Use the procedure outlined above.

Step 1. Assume $Y = 1.50$.

Step 2. $P = VXR + YT = (1.0)(0.56)(1\ 850) + (1.50)(675) = 2\ 049$ lb.

Step 3. From Figure 14–12, the speed factor $f_N = 0.30$, and the life factor $f_L = 3.41$. Then the required basic dynamic load rating C is

$$C = Pf_L/f_N = 2\ 049(3.41)/(0.30) = 23\ 300 \text{ lb}$$

Step 4. From Table 14–3, we could use either bearing number 6222 or 6318. The 6318 has a bore of 3.543 3 in and is well suited to this application.

Step 5. For bearing number 6318, $C_o = 22\ 500$ lb.

Step 6. $T/C_o = 675/22\ 500 = 0.03$.

Step 7. From Table 14–5, $e = 0.22$ (approximately).

Step 8. $T/R = 675/1\ 850 = 0.36$. Because $T/R > e$, we can find $Y = 1.97$ from Table 14–5 by interpolation based on $T/C_o = 0.03$.

Step 9. Recompute $P = (1.0)(0.56)(1\ 850) + (1.97)(675) = 2\ 366$ lb:

$$C = 2\ 366(3.41)/(0.30) = 26\ 900\ \text{lb}$$

The bearing number 6318 is not satisfactory at this load. Let's choose bearing number 6320 and repeat the process from Step 5.

Step 5. $C_o = 29\ 800$ lb.

Step 6. $T/C_o = 675/29\ 800 = 0.023$.

Step 7. $e = 0.20$.

Step 8. $T/R > e$. Then $Y = 2.10$.

Step 9. $P = (1.0)(0.56)(1\ 850) + (2.10)(675) = 2\ 454$ lb. Thus,

$$C = 2\ 454(3.41)/(0.30) = 27\ 900\ \text{lb}$$

Because bearing number 6320 has a value of $C = 30\ 000$ lb, it is satisfactory.

14–11
MOUNTING OF
BEARINGS

Up to this point, we have considered the load-carrying capacity of the bearings and the bore size in selecting a bearing for a given application. Although these are the most critical parameters, the successful application of a bearing must consider its proper mounting. Bearings are precision machine elements. Great care must be exercised in their handling, mounting, installation, and lubrication.

The primary considerations in the mounting of a bearing are as follows:

- The shaft seat diameter and tolerances
- The housing internal bore and tolerances
- The shaft shoulder diameter against which the inner race of the bearing will be located
- The housing shoulder diameter provided for locating the outer race
- The radius of the fillets at the base of the shaft and housing shoulders
- The means of retaining the bearing in position

In a typical installation, the bore of the bearing makes a light interference fit on the shaft, and the outside diameter of the outer race makes a close clearance fit in the housing bore. To ensure proper operation and life, the mounting dimensions must be controlled to a total tolerance of only *a few ten-thousandths of an inch*. Most catalogs specify the limit dimensions for both the shaft seat diameter and the housing bore diameter.

Likewise, the catalog will specify the desirable shoulder diameters for the shaft and the housing that will provide a secure surface against which to locate the bearing while ensuring that the shaft shoulder contacts only the inner race and the housing shoulder contacts only the outer race. Table 14–3 includes these values.

The fillet radius specified in the catalog (see *r* in Table 14–3) is the maximum permissible radius *on the shaft and in the housing* that will clear the external radius on the bearing races. Using too large a radius would not permit the bearing to seat tightly against the shoulder. Of course, the actual fillet radius should be made as large as possible up to the maximum to minimize the stress concentration at the shoulder.

Bearings can be retained in the axial direction by many of the means described in Chapter 11. Three popular methods are retaining rings, end caps, and locknuts. Figure 14–13 shows one possible arrangement. Note that for the left bearing, the shaft diameter is slightly smaller to the left of the bearing seat. This allows the bearing to be slid easily over the shaft up to the place where it must be pressed on. The retaining ring for the outer race could be supplied as a part of the outer race instead of as a separate piece.

The right bearing is held on the shaft with a locknut threaded on the end of the shaft. See Figure 14–14 for the design of standard locknuts. The internal tab on the lockwasher engages a groove in the shaft, and one of the external tabs is bent into a groove on the nut

FIGURE 14–13
Bearing mounting
illustration

FIGURE 14–14
Locknut and
lockwasher for
retaining bearings
(SKF Industries, Inc.,
King of Prussia, PA)

after it is seated to keep the nut from backing off. The external cap not only protects the bearing but also retains the outer race in position.

Care must be exercised to ensure that the bearings are not overly constrained. If both bearings are held tightly, any changes in dimensions due to thermal expansion or unfavorable tolerance stackup would cause binding and could lead to dangerous unexpected loads on the bearings. It is desirable to give one bearing complete location while allowing the other bearing to float axially.

14–12
TAPERED
ROLLER
BEARINGS

The taper on the rollers of tapered roller bearings, evident in Figure 14–7, results in a load path different from that for the bearings discussed thus far. Figure 14–15 shows two tapered roller bearings supporting a shaft with a combination of a radial load and a thrust load. The design of the shaft is such that the thrust load is resisted by the left bearing. But a peculiar feature of this type of bearing is that a radial load on one of the bearings creates a thrust on the opposing bearing, also; this feature must be considered in analysis of the bearing.

The location of the radial reaction must also be determined with care. Part (b) of Figure 14–15 shows a dimension a that is found by the intersection of a line perpendicular to the axis of the roller and the centerline of the shaft. The radial reaction at the bearing acts through this point. The distance a is reported in the tables of data for the bearings.

The Anti-Friction Bearings Manufacturers' Association (AFBMA) recommends the following approach in computing the equivalent loads on a tapered roller bearing:

Equivalent Load
for Tapered
Roller Bearing

$$P_A = 0.4F_{rA} + 0.5\frac{Y_A}{Y_B}F_{rB} + Y_AT_A \qquad (14\text{–}7)$$

$$P_B = F_{rB} \qquad (14\text{–}8)$$

where P_A = equivalent radial load on bearing A
    P_B = equivalent radial load on bearing B
    F_{rA} = applied radial load on bearing A
    F_{rB} = applied radial load on bearing B
    T_A = thrust load on bearing A
    Y_A = thrust factor for bearing A from tables
    Y_B = thrust factor for bearing B from tables

FIGURE 14–15
Example of tapered
roller bearing
installation

(*a*) Shaft loading and support   (*b*) Bearing details

Table 14–6 shows an abbreviated set of data from a catalog to illustrate the method of computing equivalent loads.

For the several hundred designs of standard tapered roller bearings available commercially, the value of the thrust factor varies from as small as 1.07 to as high as 2.26. In design problems, a trial-and-error procedure is usually necessary. Example Problem 14–5 illustrates one approach.

Example Problem 14–5   The shaft shown in Figure 14–15 carries a transverse load of 6 800 lb and a thrust load of 2 500 lb. The thrust is resisted by bearing *A*. The shaft rotates at 350 rpm and is to be used in a piece of agricultural equipment. Specify suitable tapered roller bearings for the shaft.

Solution   The radial loads on the bearings are

$$F_{rA} = 6\ 800(4\ \text{in}/10\ \text{in}) = 2\ 720\ \text{lb}$$

$$F_{rB} = 6\ 800(6\ \text{in}/10\ \text{in}) = 4\ 080\ \text{lb}$$

$$T_A = 2\ 500\ \text{lb}$$

To use Equation (14–6), we must assume values of Y_A and Y_B. Let's use $Y_A = Y_B = 1.75$. Then

$$P_A = 0.40(2\ 720) + 0.5\ \frac{1.75}{1.75}\ 4\ 080 + 1.75(2\ 500) = 7\ 503\ \text{lb}$$

$$P_B = F_{rB} = 4\ 080\ \text{lb}$$

Using Table 14–4 as a guide, let's select 4 000 h as a design life. Then the number of revolutions would be

$$L_d = (4\ 000\ \text{h})(350\ \text{rpm})(60\ \text{min/h}) = 8.4 \times 10^7\ \text{rev}$$

The required basic dynamic load rating can now be calculated from Equation (14–3), using $k = 3.33$:

$$C_A = P_A(L_d/10^6)^{1/k}$$

$$C_A = 7\ 503(8.4 \times 10^7/10^6)^{0.30} = 28\ 400\ \text{lb}$$

Similarly,

$$C_B = 4\ 080(8.4 \times 10^7/10^6)^{0.30} = 15\ 400\ \text{lb}$$

From Table 14–6, we can choose the following bearings.

TABLE 14–6   Tapered roller bearing data

Bore	Outside diameter	Width	*a*	Thrust factor, *Y*	Basic dynamic load rating, *C*
1.000 0	2.500 0	0.812 5	0.583	1.71	8 370
1.500 0	3.000 0	0.937 5	0.690	1.98	12 800
1.750 0	4.000 0	1.250 0	0.970	1.50	21 400
2.000 0	4.375 0	1.500 0	0.975	2.02	26 200
2.500 0	5.000 0	1.437 5	1.100	1.65	29 300
3.000 0	6.000 0	1.625 0	1.320	1.47	39 700
3.500 0	6.375 0	1.875 0	1.430	1.76	47 700

NOTE: Dimensions are in inches. Load *C* is in pounds for an L_{10} life of 1 million rev.

Bearing A

$$d = 2.500\ 0 \text{ in} \qquad D = 5.000\ 0 \text{ in}$$
$$C = 29\ 300 \text{ lb} \qquad Y_A = 1.65$$

Bearing B

$$d = 1.750\ 0 \text{ in} \qquad D = 4.000\ 0 \text{ in}$$
$$C = 21\ 400 \text{ lb} \qquad Y_B = 1.50$$

We can now recompute the equivalent loads:

$$P_A = 0.40(2\ 720) + 0.5\ \frac{1.65}{1.50}\ 4\ 080 + 1.65(2\ 500) = 7\ 457 \text{ lb}$$

$$P_B = F_{rB} = 4\ 080 \text{ lb}$$

From these, the new values of $C_A = 28\ 200$ lb and $C_B = 15\ 400$ lb are still satisfactory for the selected bearings.

One caution must be observed in using the equations for equivalent loads for tapered roller bearings. If, from Equation (14–7), the equivalent load on bearing A is less than the applied radial load, the following equation should be used.

If $P_A < F_{rA}$, then let $P_A = F_{rA}$ and compute P_B.

$$P_B = 0.4F_{rB} + 0.5\frac{Y_B}{Y_A}F_{rA} - Y_B T_A \qquad \textbf{(14–9)}$$

A similar analysis is used for angular contact ball bearings in which the design of the races results in a load path similar to that for tapered roller bearings. Figure 14–3 shows an angular contact bearing and the angle through the pressure center. This is equivalent to the line perpendicular to the axis of the tapered roller bearing. The radial reaction on the bearing acts through the intersection of this line and the axis of the shaft. Also, a radial load on one bearing induces a thrust load on the opposing bearing, requiring the application of the equivalent load formulas of the type used in Equations (14–7) and (14–9). The angle of the load line in commercially available angular contact bearings ranges from 15° to 40°.

14–13
PRACTICAL
CONSIDERATIONS
IN THE
APPLICATION
OF BEARINGS

This section will discuss lubrication of bearings, installation, preloading, stiffness, operation under varying loads, sealing, limiting speeds, and standards and bearing tolerance classes related to the manufacture and application of bearings.

Lubrication

The functions of lubrication in a bearing unit are as follows:

1. To provide a low-friction film between the rolling elements and the races of the bearing and at points of contact with cages, guiding surfaces, retainers, and so on.

2. To protect the bearing components from corrosion.

3. To help dissipate heat from the bearing unit.

4. To carry heat away from the bearing unit.

5. To help dispel contaminants and moisture from the bearing.

Rolling contact bearings are usually lubricated with either grease or oil. Under normal ambient temperatures (approximately 70°F) and relatively slow speeds (under 500 rpm), grease is satisfactory. At higher speeds or higher ambient temperatures, oil lubrication applied in a continuous flow is required, possibly with external cooling of the oil.

Oils used in bearing lubrication are usually clean, stable mineral oils. Under lighter loads and lower speeds, light oil is used. Heavier loads and/or higher speeds require heavier oils up to SAE 30. A recommended upper limit for lubricant temperature is 160°F. The choice of the correct oil or grease depends on many factors, so each application should be discussed with the bearing manufacturer. In general, a viscosity of 70 to 100 Saybolt Universal Seconds (SUS) should be maintained at the operating temperature of the lubricant in the bearing.

In some critical applications such as bearings in jet engines and very-high-speed devices, lubricating oil is pumped under pressure to an enclosed housing for the bearing where the oil is directed at the rolling elements themselves. A controlled return path is also provided. The temperature of the oil in the sump is monitored and controlled with heat exchangers or refrigeration to maintain oil viscosity within acceptable limits. Such systems provide reliable lubrication and ensure the removal of heat from the bearing.

See Section 14–14 for additional discussion about the importance of oil film thickness in bearings. Reference 7 contains an extensive amount of information on this subject.

Greases used in bearings are mixtures of lubricating oils and thickening agents, usually soaps such as lithium or barium. The soaps act as carriers for the oil which is drawn out at the point of need within the bearing. Additives to resist corrosion or oxidation of the oil itself are sometimes added. Classifications of greases specify the operating temperatures to which the greases will be exposed, as defined by the Anti-Friction Bearing Manufacturers' Association (AFBMA), and outlined in Table 14–7.

Installation

It has already been stated that most bearings should be installed with a light interference fit between the bore of the bearing and the shaft to preclude the possibility of rotation of the inner race of the bearing with respect to the shaft. Such a condition would result in uneven wear of the bearing elements and early failure. To install the bearing then requires rather heavy forces applied axially. Care must be exercised so that the bearing is not damaged during installation. The installation force should be applied directly to the inner race of the bearing.

If the force were applied through the outer race, the load would be transferred through the rolling elements to the inner race. Because of the small contact area, it is likely that such transfer of forces would overstress some element, exceeding the static load capacity. Brinelling would result, along with the noise and rapid wear that accompany this condition. For large bearings it may be necessary to heat the bearing to expand its diameter in order

TABLE 14–7   Types of greases used for bearing lubrication

Group	Type of grease	Operation temperature range (°F)
I	General-purpose	−40–250
II	High-temperature	0–300
III	Medium-temperature	32–200
IV	Low-temperature	−67–225
V	Extremely high-temperature	Up to 450

to keep the installation forces within reason. Removal of bearings intended for reuse requires similar precautions. Bearing pullers are available to facilitate this task.

Preloading

Some bearings are made with internal clearances that must be taken up in a particular direction to ensure satisfactory operation. In such cases, preloading must be provided, usually in the axial direction. On horizontal shafts, springs are typically used, with axial adjustment of the spring deflection sometimes provided to adjust the amount of preload. When space is limited, the use of Belleville washers is desirable because they provide high forces with small deflections. Shims can be used to adjust the actual deflection and preload obtained (see Chapter 19). On vertical shafts the weight of the shaft assembly itself may be sufficient to provide the required preload.

Bearing Stiffness

Stiffness is the deflection that a given bearing undergoes when carrying a given load. Usually the radial stiffness is most important because the dynamic behavior of the rotating shaft system is affected. The critical speed and the mode of vibration are both functions of the bearing stiffness. Generally speaking, the softer the bearing (low stiffness), the lower the critical speed of the shaft assembly. Stiffness is measured in the units used for springs, such as pounds per inch or newtons per millimeter. Of course, the stiffness values are quite high, with values of 500 000 to 1 000 000 lb/in reasonable. The manufacturer should be consulted when such information is needed, because it is rarely included in standard catalogs.

Operation under Varying Loads

The load/life relationships used thus far assume that the load is reasonably constant in magnitude and direction. If the load varies considerably, an effective mean load must be used for determining the expected life of the bearing. (See References 5 and 13.) Oscillating loads also require special analysis because only a few of the rolling elements share the load. See Section 14–15 for additional information about life prediction under varying loads.

Sealing

When the bearing is to operate in dirty or moist environments, special shields and seals are usually specified. They can be provided on either or both sides of the rolling elements. Shields are typically metal and are fixed to the stationary race, but they remain clear of the rotating race. Seals are made of elastomeric materials and do contact the rotating race. Bearings fitted with both seals and shields and precharged at the factory with grease are sometimes called *permanently lubricated*. Although such bearings are likely to give many years of satisfactory service, extreme conditions can produce a degradation of the lubricating properties of the grease. The presence of seals also increases the friction in a bearing. Sealing can be provided outside the bearing in the housing or at the shaft/housing interface. On high-speed shafts, a *labyrinth seal,* consisting of a noncontacting ring around the shaft with a few thousandths of an inch radial clearance, is frequently used. Grooves, sometimes in the form of a thread, are machined in the ring; the relative motion of the shaft with respect to the ring creates the sealing action.

Limiting Speeds

Most catalogs list limiting speeds for each bearing. Exceeding these limits may result in excessively high operating temperatures due to friction between the cages supporting the rolling elements. Generally, the limiting speed is lower for larger bearings than for smaller bearings. Also, a given bearing will have a lower limiting speed as loads increase. With special care, either in the fabrication of the bearing cage or in the lubrication of the bearing, bearings can be operated at higher speeds than those listed in the catalog. The manufacturer should be consulted in such applications. The use of ceramic rolling elements with their lower mass can result in higher limiting speeds.

Standards

Several groups are involved in standard setting for the bearing industry. A partial list is given here:

Anti-Friction Bearing Manufacturers' Association	(AFBMA)
Annular Bearing Engineers Committee	(ABEC)
Roller Bearing Engineers Committee	(RBEC)
Ball Manufacturers Engineers Committee	(BMEC)
American National Standards Institute	(ANSI)
International Standards Organization	(ISO)

Many standards are co-listed by the ANSI and AFBMA organizations. The catalog of ANSI lists 22 different standards related to bearings. Three of them are listed here:

Terminology and Definitions for Ball and Roller Bearings and Parts, ANSI B3.7

Load Ratings and Fatigue Life for Ball Bearings, ANSI/AFBMA 9

Load Ratings and Fatigue Life for Roller Bearings, ANSI/AFBMA 11

Tolerances

Several different tolerance classes are recognized in the bearing industry to accommodate the needs of the wide variety of equipment using rolling contact bearings. In general, of course, all bearings are precision machine elements and should be treated as such. As noted before, the general range of tolerances is of the order of a few ten-thousandths of an inch. The standard tolerance classes are defined by ABEC, as identified below.

ABEC 1:	Standard radial ball and roller bearings
ABEC 3:	Semiprecision instrument ball bearings
ABEC 5:	Precision radial ball and roller bearings
ABEC 5P:	Precision instrument ball bearings
ABEC 7:	High-precision radial ball bearings
ABEC 7P:	High-precision instrument ball bearings

Most machine applications would use ABEC 1 tolerances, the data for which are usually provided in the catalogs. Machine tool spindles requiring extra smooth and accurate running would use ABEC 5 or ABEC 7 classes.

**14–14
IMPORTANCE
OF OIL FILM
THICKNESS IN
BEARINGS**

In bearings operated at heavy loads or high speeds, it is critical to maintain a film of lubricating oil at the surface of the rolling elements. A steady supply of clean lubricant with an adequate viscosity is required. With careful analysis of the geometry of the bearing, the speed of rotation, and the properties of the lubricant, it is possible to estimate the thickness of the oil film between the rolling elements and the races. Although the film thickness may be only a few microinches, it has been shown that oil starvation of the contact area is a chief cause of premature failure of rolling contact bearings. Conversely, if a film thickness significantly greater than the height of surface roughness features can be maintained, the expected life will be several times greater than published data from catalogs. (See References 7 and 15.)

The nature of the lubrication at the interface between rolling elements and the races is called *elastohydrodynamic lubrication* because it depends on the specific elastic deformation of the mating surfaces under the influence of high contact stresses and the creation of a pressurized film of lubricant by the dynamic action of the rolling elements.

The data needed to evaluate film thickness for ball bearings are given in the following lists:

Bearing Geometry Factors

Ball diameter

Number of balls

Radius of curvature of the inner race groove in both the circumferential and the axial directions

Pitch diameter of the bearing; the average of the bore diameter and the outside diameter

Surface roughness of the balls and the races

Contact angle for angular contact bearings

Bearing Material Factors

Modulus of elasticity of balls and races

Poisson's ratio

Lubricant Factors

Dynamic viscosity at the operating temperature *within* the bearing

Pressure coefficient of viscosity; the change of viscosity with pressure

Operational Factors

Rotational speed of both the inner and the outer races

Radial load

Thrust load

The details of the analysis are given in Reference 7. A simplified approach with helpful charts to ease the determination of critical parameters is given in Reference 15.

The results of the analysis include the following:

The minimum film thickness of the lubricant, h_o

The composite roughness of the balls and the races, S

The ratio $\Lambda = h_o/S$

The service life of the bearing is dependent on the value of Λ:

- If $\Lambda < 0.90$, a life less than the manufacturer's rated life is to be expected because of surface damage due to inadequate lubricant film.
- If Λ is in the range of 0.90 to 1.50, rated service life can be expected.
- If Λ is in the range of 1.50 to 3.0, an increase of life of up to three times the rated life is possible.
- If Λ is greater than 3.0, a life of up to six times the rated life is possible.

General Recommendations for Achieving Long Life for Bearings

1. Choose a bearing with an adequate rated life using the procedures outlined in this chapter.
2. Ensure that the bearing has a fine surface finish and is not damaged by rough handling, poor installation practices, corrosion, vibration, or exposure to electrical current flow.
3. Ensure that operating loads are within the design values.
4. Supply the bearing with a copious flow of clean lubricant at an adequate viscosity at the operating temperature within the bearing according to the bearing manufacturer's recommendations. Provide external cooling for the lubricant if necessary. For an existing system, this is the factor over which you have the most control without a significant redesign of the system itself.
5. If redesign is possible, design the system to operate at as low a speed as possible.

14–15
LIFE PREDICTION
UNDER VARYING
LOADS

The design and analysis procedures used so far in this book assumed that the bearing would operate at a single design load throughout its life. It is possible to predict the life of the bearing under such conditions fairly accurately using manufacturers' data published in catalogs. If loads vary with time, a modified procedure is required.

One procedure that is recommended by bearing manufacturers is called the *Palmgren-Miner rule*, or sometimes simply *Miner's rule*. References 11 and 12 describe their work, and Reference 10 discusses a modified approach more tailored to bearings.

The basis of Miner's rule is that if a given bearing is subjected to a series of loads of different magnitudes for known lengths of time, each load contributes to the eventual failure of the bearing in proportion to the ratio of the load to the expected life that the bearing would have under that load. Then the cumulative effect of the series of loads must account for all such contributions to failure.

A similar approach, described in Reference 7, introduces the concept of *mean effective load, F_m*:

Mean Effective
Load under
Varying Loads

$$F_m = \left(\frac{\sum_i (F_i)^p N_i}{N} \right)^{1/p} \qquad \textbf{(14–10)}$$

where F_i = individual load among a series of i loads

$\quad N_i$ = number of revolutions at which F_i operates

$\quad N$ = total number of revolutions in a complete cycle

$\quad p$ = exponent on the load/life relationship; $p = 3$ for ball bearings, and $p = 10/3$ for rollers

Alternatively, if the bearing is rotating at a constant speed, and because the number of revolutions is proportional to the time of operation, N_i can be the number of minutes of operation at F_i, and N is the sum of the number of minutes in the total cycle. That is,

$$N = N_1 + N_2 + \cdots + N_i$$

Then the total expected life, in millions of revolutions of the bearing, would be

$$L = \left(\frac{C}{F_m}\right)^P \qquad \textbf{(14–11)}$$

Example Problem 14–6   A single-row, deep-groove ball bearing number 6308 is subjected to the following set of loads for the given times:

Condition	F_i	Time
1	650 lb	30 min
2	750 lb	10 min
3	250 lb	20 min

This cycle of 60 min is repeated continuously throughout the life of the bearing. The shaft carried by the bearing rotates at 600 rpm. Estimate the total life of the bearing.

Solution   Using Equation (14–10), we have

$$F_m = \left(\frac{\Sigma_i (F_i)^P N_i}{N}\right)^{1/p} \qquad \textbf{(14–10)}$$

$$F_m = \left(\frac{30(650)^3 + 10(750)^3 + 20(250)^3}{30 + 10 + 20}\right)^{1/3} = 597 \text{ lb}$$

Now use Equation (14–11):

$$L = \left(\frac{C}{F_m}\right)^P \qquad \textbf{(14–11)}$$

From Table 14–3, for the 6308 bearing, we find that $C = 7\,050$ lb. Then

$$L = \left(\frac{7\,050}{597}\right)^3 = 1\,647 \text{ million rev}$$

At a rotational speed of 600 rpm, the number of hours of life would be

$$L = \frac{1\,647 \times 10^6 \text{ rev}}{1} \cdot \frac{\text{min}}{600 \text{ rev}} \cdot \frac{\text{h}}{60 \text{ min}} = 45\,745 \text{ h}$$

Note that this is the same bearing used in Example Problem 14–3 which was selected to operate at least 30 000 h when carrying a steady load of 650 lb.

REFERENCES

1. Avallone, Eugene A., and Theodore Baumeister III, eds. *Marks' Standard Handbook for Mechanical Engineers.* 10th ed. New York: McGraw-Hill, 1996.

2. Dodge/Rockwell Automation. *Dodge Mounted Bearings Bulletin.* Greenville, SC: Rockwell International Corp.

3. Eschmann, Paul, Ludwig Hasbargen, Karl Weisgand, and Johannes Brandlein. *Ball and Roller Bearings: Theory, Design, and Application.* 2d ed. New York: John Wiley & Sons, 1985.

4. FAG Bearings Corporation. *Precision Bearings Catalog.* Stamford, CT: FAG Bearings Company.

5. FMC Corporation. *Bearing Technical Journal.* Indianapolis, IN: FMC Bearing Division, FMC Corporation.

6. Hannoosh, J. G. "Ceramic Bearings Enter the Mainstream." *Design News* (November 23, 1988).

7. Harris, Tedric A. *Rolling Bearing Analysis.* 3d ed. New York: John Wiley & Sons, 1991.

8. Hoover-NSK Bearing Company. *Catalog No. AM-1, Ball, Tapered Roller, Cylindrical Roller & Spherical Roller Bearings.* Ann Arbor, MI: Hoover-NSK Bearing Company.

9. Juvinall, Robert C., and Kurt M. Marshek. *Fundamentals of Machine Component Design.* 2d ed. New York: John Wiley & Sons, 1991.

10. Kauzlarich, James J. "The Palmgren-Miner Rule Derived." *Proceedings of the 15th Leeds/Lyon Symposium of Tribology.* (Leeds, United Kingdom, September 6–9, 1988).

11. Miner, M. A. "Cumulative Damage in Fatigue." *Journal of Applied Mechanics* 67 (1945): A159–A164.

12. Palmgren, A. *Ball and Roller Bearing Engineering.* 3d ed. Philadelphia, PA: Burbank, 1959.

13. SKF Industries, Inc. *SKF Engineering Data.* Philadelphia, PA: SKF Industries, Inc.

14. The Timken Company. *Tapered Roller Bearing Engineering Journal.* Canton, OH: The Timken Company.

15. Werner, J. M. "Predicting Service Life for Ball Bearings." *Machine Design Magazine* (March 7, 1974):102–107.

PROBLEMS

1. A radial ball bearing has a basic dynamic load rating of 2 350 lb for a rated (L_{10}) life of 1 million rev. What would its L_{10} life be when operating at a load of 1 675 lb?

2. Determine the required basic dynamic load rating for a bearing to carry 1 250 lb from a shaft rotating at 880 rpm if the design life is to be 20 000 h.

3. A catalog lists the basic dynamic load rating for a ball bearing to be 3 150 lb for a rated life of 1 million rev. What would be the L_{10} life of the bearing if it were subjected to a load of (a) 2 200 lb and (b) 4 500 lb?

4. Compute the required basic dynamic load rating, *C,* for a ball bearing to carry a radial load of 1 450 lb at a shaft speed of 1 150 rpm for an industrial fan.

5. Specify suitable bearings for the shaft of Example Problem 12–1. Note the data contained in Figures 12–1, 12–2, 12–11, and 12–12.

6. Specify suitable bearings for the shaft of Example Problem 12–3. Note the data contained in Figures 12–13, 12–15, and 12–16.

7. Specify suitable bearings for the shaft of Example Problem 12–4. Note the data contained in Figures 12–17 and 12–18.

8. For any of the bearings specified in Problems 2–7, make a scale drawing of the shaft, the bearings, and the part of the housing to support the outer races of the bearings. Be sure to consider fillet radii and axial location of the bearings.

9. A bearing is to carry a radial load of 455 lb and no thrust load. Specify a suitable bearing from Table 14–3 if the shaft rotates at 1 150 rpm and the design life is 20 000 h.

For each of the problems in Table 14–8, repeat Problem 9 with the new data.

TABLE 14–8

Problem number	Radial load	Thrust load	rpm	Design life, h
10.	857 lb	0	450	30 000
11.	1 265 lb	645 lb	210	5 000
12.	235 lb	88 lb	1 750	20 000
13.	2 875 lb	1 350 lb	600	15 000
14.	3.8 kN	0	3 450	15 000
15.	5.6 kN	2.8 kN	450	2 000
16.	10.5 kN	0	1 150	20 000
17.	1.2 kN	0.85 kN	860	20 000

18. In Chapter 12, Figures 12–20 through 12–35 showed shaft design exercises related to the problems at the end of the chapter. For each bearing of each shaft, specify a suitable bearing from Table 14–3. If the shaft design has already been completed to the point that the minimum acceptable diameter of the shaft at the bearing seat is known, consider that diameter when specifying the bearing. Note the Chapter 12 problem statements for the speed of the shaft and the loading data.

19. Bearing number 6332 from Table 14–3 is carrying the set of loads in Table 14–9 while rotating at 600 rpm. Compute the expected L_{10} life for the bearing under these conditions if the cycle repeats continuously.

TABLE 14–9

Condition	Load, F_i	Time, N_i
1	4 500 lb	25 min
2	2 500 lb	15 min

20. Bearing number 6318 from Table 14–3 is carrying the set of loads in Table 14–10 while rotating at 600 rpm. Compute the expected L_{10} life for the bearing under these conditions if the cycle repeats continuously.

TABLE 14–10

Condition	Load, F_i	Time, N_i
1	2 500 lb	25 min
2	1 500 lb	15 min

21. Bearing number 6211 from Table 14–3 is carrying the set of loads in Table 14–11 while rotating at 1 700 rpm. Compute the expected L_{10} life for the bearing under these conditions if the cycle repeats continuously.

TABLE 14–11

Condition	Load, F_i	Time, N_i
1	600 lb	480 min
2	200 lb	115 min
3	100 lb	45 min

22. Bearing number 6211 from Table 14–3 is carrying the set of loads in Table 14–12 while rotating at 1 700 rpm. Compute the expected L_{10} life for the bearing under these conditions if the cycle repeats continuously.

TABLE 14–12

Condition	Load, F_i	Time, N_i
1	450 lb	480 min
2	180 lb	115 min
3	50 lb	45 min

23. Bearing number 6206 from Table 14–3 is carrying the set of loads in Table 14–13 while rotating at 101 rpm for one 8-h shift. Compute the expected L_{10} life for the bearing under these conditions if the cycle repeats continuously. If the machine operates two shifts per day, six days per week, in how many weeks would you expect to have to replace the bearing?

TABLE 14–13

Condition	Load, F_i	Time, N_i
1	500 lb	6.75 h
2	800 lb	0.40 h
3	100 lb	0.85 h

24. Bearing number 6212 from Table 14–3 is carrying the set of loads in Table 14–14 while rotating at 101 rpm for one 8-h shift. Compute the expected L_{10} life for the bearing under these conditions if the cycle repeats continuously. If the machine operates two shifts per day, six days per week, in how many weeks would you expect to have to replace the bearing?

TABLE 14–14

Condition	Radial load	Thrust load	Time, N_i
1	1 750 lb	350 lb	6.75 h
2	600 lb	250 lb	0.40 h
3	280 lb	110 lb	0.85 h

15

Completion of the Design of a Power Transmission

Completion of the Design of a Power Transmission

Discussion Map

☐ Now we bring together the concepts and design procedures from the last eight chapters to complete the design of the power transmission.

Discover

Consider how all of the machine elements that you have studied in Chapters 7–14 fit together.

Think also about the entire life cycle of the transmission, from its design to its disposal.

This chapter presents a summary of the steps that you should take to complete the design. Some procedures are quite detailed. You should be able to apply this experience to any design you are responsible for in the future.

This is where all of the work of Part II of this book comes together. In Chapters 7–14, you learned important concepts and design procedures for many kinds of machine elements that could all be part of a given power transmission. In each case, we mentioned how the elements need to work together. Now we show the approach to completing the design of a power transmission which demonstrates an integrated approach and illustrates the title of this book, *Machine Elements in Mechanical Design.* The emphasis is on the whole design.

The lesson of this chapter is that you as a designer must constantly keep in mind how the part you are currently working on fits with other parts and how its design can affect the design of other parts. You must also consider how the part is to be manufactured, how it may be serviced and repaired as necessary, and how it will eventually be taken out of service. What will happen to the materials in the product when they have served their useful life as part of your current project?

Although we are using a power transmission in this example, the skills and insights that you will acquire should be transferable to the design of almost any other mechanical device or system.

**15–1
OBJECTIVES OF
THIS CHAPTER**

After completing this chapter, you will be able to:

1. Bring together the individual components of a mechanical, gear-type power transmission into a unified, complete system.

2. Resolve the interface questions where two components fit together.

3. Establish reasonable tolerances and limit dimensions on key dimensions of components, especially where assembly and operation of the components are critical.

4. Verify that the final design is safe and suitable for its intended purpose.

5. Add details to some of the components that were not considered in earlier analyses.

**15–2
DESCRIPTION OF
THE POWER
TRANSMISSION
TO BE DESIGNED**

The project to be completed in this chapter is the design of a single-reduction speed reducer that uses spur gears. We will use the data from Example Problem 9–1 in which the gears for the drive for an industrial saw were designed. You should review that problem now and note that it was carried forward through Example Problems 9–1 through 9–4. It was considered again in Section 9–15 where we refined the design using the spreadsheet developed in Section 9–14.

We will also use elements of the mechanical design process that were first outlined in Chapter 1 in Sections 1–4 through 1–6. We will state the functions and design requirements for the power transmission, establish a set of criteria for evaluating design decisions, and implement the design tasks that were outlined in Section 1–6. References 4, 7, and 8 provide additional approaches that you may find useful for other design projects.

Basic Statement of the Problem

We will design a power transmission for an industrial saw that will be used to cut tubing for vehicle exhaust pipes to length prior to the forming processes. The saw will receive 25 hp from the shaft of an electric motor rotating at 1 750 rpm. The drive shaft for the saw should rotate at approximately 500 rpm.

Functions, Design Requirements, and Selection Criteria for the Power Transmission

Functions.   The functions of the power transmission are as follows:

1. To receive power from an electric motor through a rotating shaft.
2. To transmit the power through machine elements that reduce the rotational speed to a desired value.
3. To deliver the power at the lower speed to an output shaft which ultimately drives the saw.

Design Requirements.   Additional information is presented here for the specific case of the industrial saw. You would normally be responsible for acquiring the necessary information and for making design decisions at this point in the design process. You would be involved in the design of the saw and would be able to discuss its desirable features with colleagues in marketing, sales, manufacturing planning, and production management and field service, and, perhaps, with customers. The kinds of information that you should seek are illustrated in the following list:

1. The reducer must transmit 25 hp.
2. The input is from an electric motor whose shaft rotates at a full-load speed of 1 750 rpm. It has been proposed to use a NEMA frame 284T motor having a shaft diameter of 1.875 in and a keyway to accommodate a 1/2 × 1/2 in key. See Chapter 21, Figure 21–18 and Table 21–3, for more data on the dimensions of the motor.
3. The output of the reducer delivers power to the saw through a shaft that rotates between 495 and 505 rpm. The speed reduction ratio should then be in the range of 3.46 to 3.53.
4. A mechanical efficiency of greater than 95% is desirable.
5. The minimum torque delivered to the saw should be 2 950 lb·in.
6. The saw is a band saw. The cutting operation is generally smooth, but moderate shock may be encountered as the saw blade engages the tubes and if there is any binding of the blade in the cut.
7. The speed reducer will be mounted on a rigid plate that is part of the base of the saw. The means of mounting the reducer should be specified.
8. It has been decided that flexible couplings may be used to connect the motor shaft to the input shaft of the reducer and to connect the output shaft directly to

the shaft of the main drive wheel for the band saw. The design for the shaft for the band-saw drive has not yet been completed. It is likely that its diameter will be the same as that of the output shaft of the reducer.

9. Whereas a small, compact size for the reducer is desirable, space in the machine base should be able to accommodate most reasonable designs.

10. The saw is expected to operate 16 hours per day, 5 days per week, with a design life of 5 years. This is approximately 20 000 hours of operation.

11. The machine base will be enclosed and will prevent any casual contact with the reducer. However, the functional components of the reducer should be enclosed in their own rigid housing to protect them from contaminants and to provide for the safety of those who work with the equipment.

12. The saw will operate in a factory environment and should be capable of operation in the temperature range of 50°F to 100°F.

13. The saw is expected to be produced in quantities of 5 000 units per year.

14. A moderate cost is critical to the marketing success of the saw.

Selection Criteria.   The list of criteria should be produced by an interdisciplinary team composed of people having broad experience with the market for and use of such equipment. The details will vary according to the specific design. As an illustration of the process, the following criteria are suggested for the present design:

1. *Safety:* The speed reducer should operate safely and provide a safe environment for people near the machine.

2. *Cost:* Low cost is desirable so that the saw appeals to a large set of customers.

3. *Small size.*

4. *High reliability.*

5. *Low maintenance.*

6. *Smooth operation; low noise; low vibration.*

**15–3
DESIGN
ALTERNATIVES
AND SELECTION
OF THE DESIGN
APPROACH**

There are many ways that the speed reduction for the saw can be accomplished. Figure 15–1 shows four possibilities: (a) belt drive, (b) chain drive, (c) gear-type drive connected through flexible couplings, and (d) gear-type drive with a belt drive on the input side and connected to the saw with a flexible coupling.

Reference 6 includes a more extensive set of alternatives and a more detailed analysis.

Selection of the Basic Design Approach

Table 15–1 shows an example of the rating that could be done to select the type of design to be produced for the speed reducer for the saw. A 10-point scale is used, with 10 being the highest rating. Of course, with more information about the actual application, a different design approach could be selected. Also, it may be desirable to proceed with more than one design to determine more of the details, thus allowing a more rational decision. A modification of the design decision matrix calls for weighting factors to be assigned to each criterion to reflect its relative importance. See References 3 and 7 for extensive discussions of rational decision analysis techniques.

On the basis of this decision analysis, let's proceed with (c), the design of a gear-type speed reducer using flexible couplings to connect with the drive motor and the driven shaft of the saw. It is considered to have a higher level of safety for operators and maintenance peo-

(a) Belt drive option

(b) Chain drive option

(c) Single-stage gear-reducer option

(d) Belt drive with single-stage gear-reducer option

FIGURE 15–1   Options for speed reduction for saw drive design project

TABLE 15–1   Decision analysis chart

			Alternatives	
			(c)	(d)
			Gear with	Gear
	(a)	(b)	flexible	with input
Criteria	Belt	Chain	couplings	belt drive
1. Safety	6	6	9	7
2. Cost	9	8	7	6
3. Size	5	6	9	6
4. Reliability	7	6	10	7
5. Maintenance	6	5	9	6
6. Smoothness	8	6	9	8
Totals:	41	31	53	40

ple because its rotating components are enclosed. The input and output shafts and the couplings can be covered at the time of installation. Reliability is expected to be higher because precision metallic parts are used and the drive is enclosed in a sealed housing. The flexing of belts and the significant number of moving parts in a chain drive are judged to provide lower reliability. Initial cost may be higher than for belt or chain drives. However, it is expected that maintenance will be somewhat less, leading to lower cost overall. The space taken by the design should be small, simplifying the design of other parts of the saw. Design alternative (d) is attractive if there is some interest in providing a variable speed operation in the future. By using different belt drive ratios, we can achieve different cutting speeds for the saw. A further alternative would be to consider a variable-speed electric drive motor, either to replace the need for a reducer at all or to be used in conjunction with the gear-type reducer.

**15–4
DESIGN
ALTERNATIVES
FOR THE GEAR-
TYPE REDUCER**

Now that we have selected the gear-type reducer, we need to decide which type to use. Here are some alternatives:

1. *Single-reduction spur gears:* The nominal ratio of 3.50 : 1 is reasonable for a single pair of gears. Spur gears produce only radial loads which simplify selection of the bearings that support the shafts. Efficiency should be greater than 95% with reasonable precision of the gears, bearings, and seals. Spur gears are relatively inexpensive to produce. Shafts would be parallel and should be fairly easy to align with the motor and the drive shaft for the saw.

2. *Single-reduction helical gears:* These gears are equally practical as spur gears. Shaft alignment is similar. A smaller size should be possible because of the greater capacity of helical gears. However, axial thrust loads would be created which must be accommodated by the bearings and the housing. Cost is likely to be somewhat higher.

3. *Bevel gears:* These gears produce a right-angle drive which may be desirable, but not necessary in the present design. They are also somewhat more difficult to design and assemble to achieve adequate precision.

4. *Worm and wormgear drive:* This drive also produces a right-angle drive. It is typically used to achieve a higher reduction ratio than 3.50 : 1. Efficiency is usually much lower than the 95% called for in the design requirements. Heat generation could be a problem with 25 hp and the lower efficiency. A larger motor could be required to overcome the loss of power and still provide the required torque at the output shaft.

Design Decision for the Gear Type

For the present design, we choose the single-reduction spur gear reducer. Its simplicity is desirable, and the final cost is likely to be lower than that of the other proposed designs. The smaller size of the helical reducer is not considered to be of high priority.

**15–5
GENERAL
LAYOUT AND
DESIGN DETAILS
OF THE
REDUCER**

Figure 15–2 shows the proposed arrangement of the components for the single-reduction spur gear–type speed reducer. Note that the illustration in Part (b) is the top view. The design involves the following tasks:

1. Design a pinion and gear to transmit 25 hp with a pinion speed of 1 750 rpm and a gear speed in the range from 495 to 505 rpm. The nominal ratio is 3.50 : 1. Design for both strength and pitting resistance to achieve approximately 20 000 hours of life and a reliability of at least 0.999.

(*a*) Pictorial view

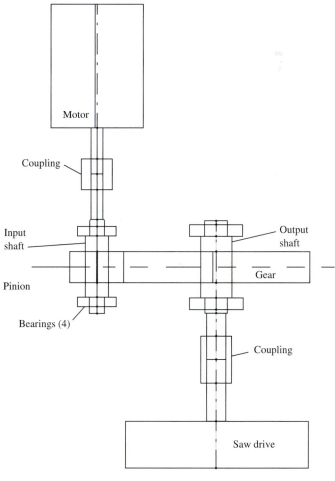

Motor

Coupling

Input
shaft

Pinion

Bearings (4)

Output
shaft

Gear

Coupling

Saw drive

(*b*) Top view

FIGURE 15–2   General layout of saw drive with single-stage gear reducer

2. Design two shafts, one for the pinion and one for the gear. Provide positive axial location for the gears on the shaft. The input shaft must be designed to extend beyond the housing to enable the motor shaft to be coupled to it. The output shaft must accommodate a coupling that mates with the drive shaft of the saw. Use a design reliability of 0.999.

3. Design six keys: one for each gear; one for the motor; one for the input shaft at the coupling; one for the output shaft at the coupling; and one for the drive shaft for the saw.

4. Specify two flexible couplings: one for the input shaft and one for the output shaft.

5. Specify four commercially available rolling contact bearings, two for each shaft. The L_{10} design life should be 20 000 hours.

6. Design a housing to enclose the gears and the bearings and to support them rigidly.

7. Provide a means of lubricating the gears within the housing.

8. Provide seals for the input and output shafts at the place where they pass through the housing. We will not specify the particular seals because of lack of data in this book. However, refer to Chapter 11 for suggestions for the types of seals that may be suitable.

Gear Design

The conditions for this design are the same as those considered in Example Problems 9–1 through 9–4. Several design iterations were produced in Section 9–12 using the aid of the gear design spreadsheet. The design alternatives all used a reliability factor of 1.50 to gain an expected reliability of 0.9999, fewer than one failure in 10 000. This is more conservative than the suggested reliability of 0.999. An overload factor of 1.50 was used to account for the moderate shock expected from the operation of the saw. We will use the design documented in Figure 9–29 having the following major features:

- Diametral pitch: $P_d = 8$; 20°, full-depth, involute teeth
- Number of teeth in the pinion: $N_P = 28$
- Number of teeth in the gear: $N_G = 98$
- Diameter of the pinion: $D_P = 3.500$ in
- Diameter of the gear: $D_G = 12.250$ in
- Center distance: $C = 7.875$ in
- Face width: $F = 2.00$ in
- Quality number: $Q_v = 8$
- Tangential force: $W_t = 514$ lb
- Required bending stress number for pinion: $s_{at} = 20\ 900$ psi
- Required contact stress number for pinion: $s_{ac} = 153\ 000$ psi; requires 370 HB steel
- Material specified: AISI 4340 OQT 900; 388 HB; $s_u = 190$ Ksi; 15% elongation

Shaft Design

1. *Forces:* Figure 15–3(a) shows the proposed configuration for the input shaft which carries the pinion and which connects to the motor shaft through a flexible coupling. Figure 15–3(b) shows the gear, similarly configured. The only active forces on the shafts are the tangential force and the radial force from the gear teeth.

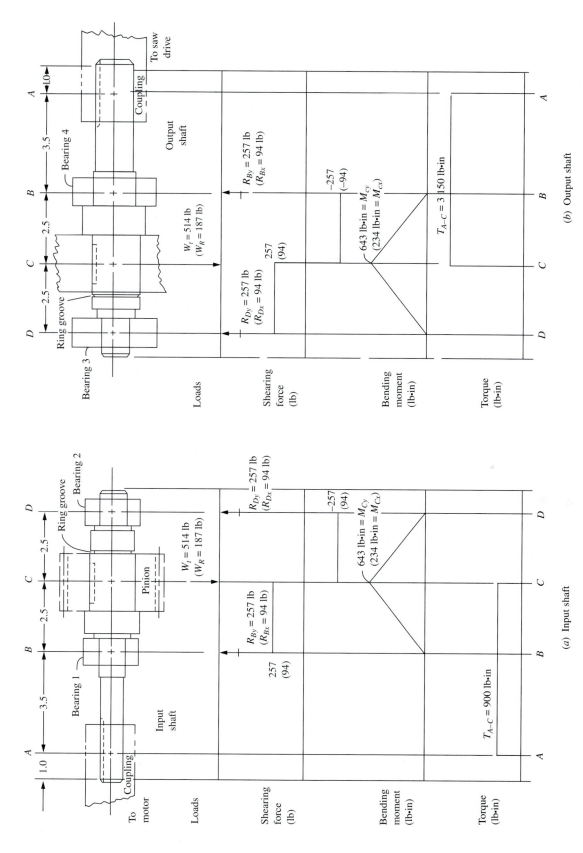

FIGURE 15–3 Diagrams of shaft configurations, shearing force, bending moment, and torque.

The flexible couplings at the ends of the shafts allow torque transmission, but no radial or axial forces are transmitted when the alignment of the shafts is within the recommended limits for the coupling. See Chapter 11. Without the flexible couplings, it is very likely that significant radial loads would be produced, requiring somewhat larger diameters for the shaft and larger bearings. See Chapter 12.

The spreadsheet analysis for the gears gives the tangential force as $W_t = 514$ lb. It acts downward in the vertical plane on the pinion and upward on the gear. The radial force is

$$W_r = W_t \tan \phi = (514 \text{ lb})\tan(20°) = 187 \text{ lb}$$

The radial force acts horizontally toward the left on the pinion, tending to separate the pinion from the gear. It acts toward the right on the gear.

2. ***Torque values:*** The torque on the input shaft is

$$T_1 = (63\ 000)(P)/n_P = (63\ 000)(25)/1\ 750 = 900 \text{ lb·in}$$

This value acts from the coupling at the left end of the shaft to the pinion where the power is delivered through the key to the pinion and thus to the mating gear.

The torque on the output shaft is computed next, assuming that no power is lost. The resulting value of torque is conservative for use in the shaft design:

$$T_2 = (63\ 000)(P)/n_G = (63\ 000)(25)/500 = 3\ 150 \text{ lb·in}$$

The torque acts in the output shaft from the gear to the coupling at the right end of the shaft. Assuming that the system is 95% efficient, the actual output torque is approximately

$$T_o = T_2 (0.95) = 2\ 992 \text{ lb·in}$$

This value is within the required range, as indicated in the design requirements, item 5.

3. ***Shearing force and bending moment diagrams:*** Figure 15–3 also shows the shearing force and bending moment diagrams for the two shafts. Because the active loading occurs only at the gears, the form of each diagram is the same in the vertical and horizontal directions. The first number given is the value for load, the shearing force, or the bending moment in the vertical plane. The second number in parentheses is the value in the horizontal plane. The maximum bending moment in each shaft occurs where the gears are mounted. The values are

$$M_y = 643 \text{ lb·in} \qquad M_x = 234 \text{ lb·in}$$

The resultant moment is $M_{max} = 684$ lb·in.

The bending moment is zero at the bearings and in the extensions for the input and output shafts.

4. ***Support reactions—bearing forces:*** The reactions at all bearings are the same for this example because of the simplicity of the loading pattern and the symmetry of the design. The horizontal and vertical components are

$$F_y = 257 \text{ lb} \qquad F_x = 93.5 \text{ lb}$$

The resultant force is the radial force that must be carried by the bearings: $F_r = 274$ lb. This value also produces vertical shearing stress in the shaft at the bearings.

5. **Material selection for shafts:** Each shaft will have a series of different diameters, shoulders with fillets, keyseats, and a ring groove as shown in Figure 15–3. Thus, much machining will be required. The shafts will be subjected to a combination of steady torque and reversed, repeated bending during normal use as the saw cuts steel tubing for vehicular exhaust systems. Occasional moderate shock loading is expected as the saw engages the tubing and when binding occurs in the cut due to dullness of the blade or unusually hard steel in the tubing.

   These conditions call for a steel that has moderately high strength, good fatigue resistance, good ductility, and good machinability. Such shafts are typically made from a medium-carbon-alloy steel (0.30% to 0.60% carbon) in either the cold-drawn or the oil-quenched and tempered condition. Good machinability is obtained from a steel having moderately high sulfur content, a characteristic of the 1100 series. Where good hardenability is also desired, higher manganese content is used.

   An example of such an alloy is AISI 1144 which has 0.40% to 0.48% carbon, 1.35% to 1.65% manganese, and 0.24% to 0.33% sulfur. It is called a *resulfurized, free-machining* grade of steel. Figure A4–2 shows the range of properties available for this material when it is oil-quenched and tempered. We select a tempering temperature of 1 000°F which produces good balance between strength and ductility.

   In summary, the material specified is

   AISI 1144 OQT 1000 steel: $s_u = 118\ 000$ psi; $s_y = 83\ 000$ psi; 20% elongation

   The endurance strength of the material can be estimated using the method outlined in Chapters 5 and 12:

   Basic endurance strength: $s_n = 43\ 000$ psi (from Figure 5–9 for a machined surface)

   Size factor: $C_s = 0.86$ (from Figure 12–9 with an estimate of a 2.0-in diameter)

   Reliability factor: $C_R = 0.75$ (desired reliability of 0.999)

   Modified endurance strength: $s_n' = s_n(C_s)C_R = (43\ 000\ \text{psi})(0.86)(0.75) = 27\ 700$ psi

6. **Design factor N:** The choice of a design factor N should consider the many factors discussed in Chapter 5 where a nominal value of $N = 3$ was suggested for general machine design. With the expectation of moderate shock and impact loading, let's specify $N = 4$ for extra safety.

7. **Minimum allowable shaft diameters:** The minimum allowable shaft diameters are now computed at several sections along the shaft using Equation (12–24) if there is any combination of torsion or bending loads at the section of interest. For those sections subjected only to vertical shearing loads, such as at the bearings labeled D in Figure 15–3, Equation (12–16) is used. Table 15–2 summarizes the data used in these equations for each section and reports the computed minimum diameter. The spreadsheet from Section 12–10 was used to complete the analysis.

TABLE 15–2   Summary of the results of shaft diameter calculations for the preliminary design sizing for the shaft

A. Input shaft

Section	Diameter (and related component)	Torque (lb·in)	M_x (lb·in)	M_y (lb·in)	V_x (lb)	V_y (lb)	K_t	Feature	Minimum	Design
			Bending moments		Shearing forces				Diameter (in)	
A	D_1 (*coupling*)	900	0	0	0	0	1.60	S.R. keyseat	0.73	1.000
B (to right)	D_2 (*bearing*)	900	0	0	0	0	2.50	Sharp fillet	0.73	*
Note: D_3 must be greater than D_2 or D_4 to provide shoulders for bearing and gear.										2.000
C	D_4 (*gear*)	900	234	643	94	257	2.00	Prof. keyseat	1.35	1.750
C (to right)	D_4 (*gear*)	0	234	643	94	257	3.00	Ring groove	1.53	1.750
D	D_5 (*bearing*)	0	0	0	94	257	2.50	Sharp fillet	0.54	*

B. Output shaft

Section	Diameter (and related component)	Torque (lb·in)	M_x (lb·in)	M_y (lb·in)	V_x (lb)	V_y (lb)	K_t	Feature	Minimum	Design
			Bending moments		Shearing forces				Diameter (in)	
A	D_1 (*coupling*)	3 150	0	0	0	0	1.60	S.R. keyseat	1.10	1.250
B (to left)	D_2 (*bearing*)	3 150	0	0	0	0	2.50	Sharp fillet	1.10	*
Note: D_3 must be greater than D_2 or D_4 to provide shoulders for bearing and gear.										2.000
C	D_4 (*gear*)	3 150	234	643	94	257	2.00	Prof. keyseat	1.34	1.750
C (to left)	D_4 (*gear*)	0	234	643	94	257	3.00	Ring groove	1.53	1.750
D	D_5 (*bearing*)	0	0	0	94	257	2.50	Sharp fillet	0.54	*

NOTE: Bearing seat diameters denoted by * are to be specified. S.R. = sled runner; Prof. = profile.

The last column in Table 15–2 also lists some *preliminary* design decisions for convenient diameters at the given locations. These will be re-evaluated and refined as the design is completed. The diameters suggested for the shaft extensions at A on both the input and the output shafts have been set to standard values available for the bores of flexible couplings. The actual flexible couplings are discussed later.

Note that the diameters for the bearing seats at sections B and D have not been given. The reason is that the next task in the design project is to specify commercially available rolling contact bearings to carry the radial loads with suitable life. The diameters of the shafts must be specified according to the limit dimensions recommended by the bearing manufacturer. Therefore, we will leave Table 15–2 as it is for now and revisit it after completing the bearing selection process.

Bearing Selection

We will use the method outlined in Section 14–9 to select commercially available, single-row, deep-groove ball bearings from the data given in Table 14–3. The design load is equal to the radial load, and the value can be found from the shaft analysis shown in Figure 15–3.

The reactions at the supports for each shaft are, in fact, the radial loads to which the bearings are subjected.

Because of the symmetry of the design of this system, and because there are no radial loads on the shaft except those produced by the action of the gear teeth, the radial loads on each of the four bearings in this design are the same. Earlier in this chapter, in the shaft design section, we determined that the bearing load is 274 lb.

Recall that design life for the bearings, L_d, is the total number of revolutions expected in service. Therefore, it is dependent on both the rotational speed of the shaft and the design life in hours. We are using a design life of 20 000 hours for all bearings. Shaft 1, the input shaft, rotates at 1 750 rpm, resulting in a total number of revolutions of

$$L_d = (20\ 000\ \text{h})(1\ 750\ \text{rev/min})(60\ \text{min/h}) = 2.10 \times 10^9\ \text{rev}$$

Shaft 2, the output shaft, rotates at 500 rpm. Then its design life is

$$L_d = (20\ 000\ \text{h})(500\ \text{rev/min})(60\ \text{min/h}) = 6.0 \times 10^8\ \text{rev}$$

Data for the bearings in Table 14–3 are for a life of 1.0 million rev (10^6 rev).

Now we will use Equation (14–3) with $k = 3$ to compute the required basic dynamic load rating C for each ball bearing. For the bearings on shaft 1,

$$C = P_d(L_d/10^6)^{1/k} = (274\ \text{lb})(2.10 \times 10^9/10^6)^{1/3} = 3\ 510\ \text{lb}$$

Similarly, for the bearings on shaft 2,

$$C = P_d(L_d/10^6)^{1/k} = (274\ \text{lb})(6.0 \times 10^8/10^6)^{1/3} = 2\ 310\ \text{lb}$$

Listed in Table 15–3 are candidate bearings for each shaft having basic dynamic load ratings at least as high as those just computed. We also need to refer to Table 15–2 to determine the minimum acceptable diameters for the shafts at each bearing seat to ensure that the inner race diameter for the bearing is compatible.

We have selected the smallest bearing for each location on shaft 1 that had a suitable value for the basic dynamic load rating. For shaft 2, we decided that the diameter of the shaft extension should be 1.25 in and that the bearing bore must be larger. Bearing 6207 provides a suitable bore and an additional safety factor on the load rating.

Note that the dimensions for the bearings are actually listed in mm by the manufacturer as indicated in Table 14–3. The decimal-inch equivalents are somewhat inconvenient, but they must be used. The dimensions in mm are listed below:

Inch Dimension	**mm**
0.551 2	14
0.590 6	15
0.669 3	17
0.984 3	25
1.378 0	35
1.850 4	47
2.047 2	52
2.440 9	62
2.834 6	72
0.039	1.00-mm fillet radius

TABLE 15–3   Candidate bearings for shafts 1 and 2

A. For shaft 1: Minimum bore = 0.60 in at section B; 0.54 in at section D

Bearing no.	C (lb)	d (in)	D (in)	B (in)	r_{max} (in)	Comment
6207	4 450	1.378 0	2.834 6	0.669 3	0.039	Large for both B and D
6305	3 550	0.984 3	2.440 9	0.669 3	0.039	Specify for both B and D

B. For shaft 2: Minimum bore = 1.10 in at section B; 0.54 in at section D

Bearing no.	C (lb)	d (in)	D (in)	B (in)	r_{max} (in)	Comment
6205	2 430	0.984 3	2.047 2	0.590 6	0.039	Small for B; specify for D
6207	4 450	1.378 0	2.834 6	0.669 3	0.039	Specify for B

Bearing Mounting on the Shafts and in the Housing

With the specifications for the bearings, we can finalize the basic dimensions for the shaft diameters. Table 15–4 is an update of the data in Table 15–2 with the bearing bore dimensions given. A few other changes are included as well, which is typical of the iterative nature of design. For example, the diameter of the input shaft at the coupling (section A) was made slightly smaller than the bearing seat diameter. This permits the bearing to be slid onto the shaft easily to the point where it is then pressed into position on its seat at section B and against the shoulder. Another check needs to be made at section D for both shafts where a 1.750-in diameter steps down to the bearing seat diameter of 0.984 in. There is the possibility that the step is too large and that it may interfere with the outer race of the bearing. That will be checked as we complete the details of mounting the bearings. If interference does occur, it should be a simple matter to provide another small step to make the bearing shoulder an acceptable height.

Mounting of ball and roller bearings on shafts and into housings requires very careful consideration of limit dimensions on all mating parts to ensure proper fits as defined by the bearing manufacturer. The total tolerances on shaft diameters are only a few ten-thousandths of an inch in sizes up to about 6.00 in. Total tolerances on housing bore diameters range from about 0.001 to 0.004 in for sizes from about 1.00 in to over 16.0 in. Violation of the recommended fits will likely cause unsatisfactory performance and possibly early failure of the bearing.

The bore of a bearing is typically pressed onto the shaft seat with a light interference fit to ensure that the inner race rotates with the shaft. The OD of the bearing is typically a close sliding fit in the housing, with the minimum clearance being zero. This facilitates installation and allows some slight movement of the bearing as thermal deformation occurs during operation. Tighter fits than those recommended by the manufacturer may cause the rolling elements to bind between the inner and outer races, resulting in higher loads and higher friction. Looser fits may permit the outer race to rotate relative to the housing, a very undesirable situation.

Only one of the two bearings on a shaft should be located and held fixed axially in the housing to provide proper alignment of functional components such as the gears in this design. The second bearing should be installed in a way that allows some small axial movement during operation. If the second bearing is held fixed also, it is likely that extra axial loads will be developed for which the bearing has not been designed.

We first discuss the specification of shaft limit dimensions at the bearing seats.

Bearing Seat Diameters.   Because most commercially available bearings are produced to metric dimensions, the fits are specified according to the tolerance system of the

TABLE 15–4 Summary of the results of the shaft diameter calculations for the revised design sizing for the shaft

A. Input shaft

Section	Diameter (and related component)	Torque (lb·in)	M_x (lb·in)	M_y (lb·in)	V_x (lb)	V_y (lb)	K_t	Feature	Minimum	Design
			Bending moments		Shearing forces				Diameter (in)	
A	D_1 (coupling)	900	0	0	0	0	1.60	S.R. keyseat	0.73	0.875
B (to right)	D_2 (bearing)	900	0	0	0	0	2.50	Sharp fillet	0.73	0.984
Note: D_3 must be greater than D_2 or D_4 to provide shoulders for bearing and gear.										2.000
C	D_4 (gear)	900	234	643	94	257	2.00	Prof. keyseat	1.35	1.750
C (to right)	D_4 (gear)	0	234	643	94	257	3.00	Ring groove	1.53	1.750
D	D_5 (bearing)	0	0	0	94	257	2.50	Sharp fillet	0.54	0.984

B. Output shaft

Section	Diameter (and related component)	Torque (lb·in)	M_x (lb·in)	M_y (lb·in)	V_x (lb)	V_y (lb)	K_t	Feature	Minimum	Design
			Bending moments		Shearing forces				Diameter (in)	
A	D_1 (coupling)	3 150	0	0	0	0	1.60	S.R. keyseat	1.10	1.250
B (to left)	D_2 (bearing)	3 150	0	0	0	0	2.50	Sharp fillet	1.10	1.378
Note: D_3 must be greater than D_2 or D_4 to provide shoulders for bearing and gear.										2.000
C	D_4 (gear)	3 150	234	643	94	257	2.00	Prof. keyseat	1.34	1.750
C (to left)	D_4 (gear)	0	234	643	94	257	3.00	Ring groove	1.53	1.750
D	D_5 (bearing)	0	0	0	94	257	2.50	Sharp fillet	0.54	0.984

NOTE: S.R. = sled runner; Prof. = profile.

International Standards Organization (ISO). Only a sampling of the data are listed here to illustrate the process of specifying limit dimensions for shafts and housings to accommodate bearings. Manufacturers' catalogs include much more extensive data.

For bearings carrying moderate to heavy loads such as those in this example design, the following tolerance grades are recommended for the bearing seats on shafts and housing bore fits with the outer race:

Bearing Bore Diameter Range	Tolerance Grade
10–18 mm	j5
20–100 mm	k5
105–140 mm	m5
150–200 mm	m6
Housing bore (any)	H8

Table 15–5 shows representative data for the actual limit dimensions for these grades over the size ranges included for the bearings listed in Table 14–3. Note that the bearing bore and the bearing *OD* dimensions are those expected from the bearing manufacturer. You must

TABLE 15–5 Shaft and housing fits for bearings

A. Shaft fits

Bearing bore			ISO tolerance grade	Shaft diameter		Limits of fit	
Nominal (mm)	Maximum (in)	Minimum (in)		Maximum (in)	Minimum (in)	Minimum (in)	Maximum (in)
10	0.3937	0.3934	j5	0.3939	0.3936	0.0001L	0.0005T
12	0.4724	0.4721	j5	0.4726	0.4723	0.0001L	0.0005T
15	0.5906	0.5903	j5	0.5908	0.5905	0.0001L	0.0005T
17	0.6693	0.6690	j5	0.6695	0.6692	0.0001L	0.0005T
20	0.7874	0.7870	k5	0.7878	0.7875	0.0001T	0.0008T
25	0.9843	0.9839	k5	0.9847	0.9844	0.0001T	0.0008T
30	1.1811	1.1807	k5	1.1815	1.1812	0.0001T	0.0008T
35	1.3780	1.3775	k5	1.3785	1.3781	0.0001T	0.0010T
40	1.5748	1.5743	k5	1.5753	1.5749	0.0001T	0.0010T
45	1.7717	1.7712	k5	1.7722	1.7718	0.0001T	0.0010T
50	1.9685	1.9680	k5	1.9690	1.9686	0.0001T	0.0010T
55	2.1654	2.1648	k5	2.1660	2.1655	0.0001T	0.0012T
60	2.3622	2.3616	k5	2.3628	2.3623	0.0001T	0.0012T
65	2.5591	2.5585	k5	2.5597	2.5592	0.0001T	0.0012T
70	2.7559	2.7553	k5	2.7565	2.7560	0.0001T	0.0012T
75	2.9528	2.9522	k5	2.9534	2.9529	0.0001T	0.0012T
80	3.1496	3.1490	k5	3.1502	3.1497	0.0001T	0.0012T
85	3.3465	3.3457	k5	3.3472	3.3466	0.0001T	0.0015T
90	3.5433	3.5425	k5	3.5440	3.5434	0.0001T	0.0015T
95	3.7402	3.7394	k5	3.7409	3.7403	0.0001T	0.0015T
100	3.9370	3.9362	k5	3.9377	3.9371	0.0001T	0.0015T
105	4.1339	4.1331	m5	4.1350	4.1344	0.0005T	0.0019T
110	4.3307	4.3299	m5	4.3318	4.3312	0.0005T	0.0019T
115	4.5276	4.5268	m5	4.5287	4.5281	0.0005T	0.0019T
120	4.7244	4.7236	m5	4.7255	4.7249	0.0005T	0.0019T
125	4.9213	4.9203	m5	4.9226	4.9219	0.0006T	0.0023T
130	5.1181	5.1171	m5	5.1194	5.1187	0.0006T	0.0023T
140	5.5118	5.5108	m5	5.5131	5.5124	0.0006T	0.0023T
150	5.9055	5.9045	m6	5.9071	5.9061	0.0006T	0.0026T
160	6.2992	6.2982	m6	6.3008	6.2998	0.0006T	0.0026T
170	6.6929	6.6919	m6	6.6945	6.6935	0.0006T	0.0026T
180	7.0866	7.0856	m6	7.0882	7.0872	0.0006T	0.0026T
190	7.4803	7.4791	m6	7.4821	7.4810	0.0007T	0.0030T
200	7.8740	7.8728	m6	7.8758	7.8747	0.0007T	0.0030T

control the shaft diameter and the housing bore to the specified minimum and maximum dimensions. The table also lists the minimum and maximum fits that result. The symbol L indicates that there is a net clearance (loose) fit; T indicates an interference (tight) fit. So bearings must be pressed onto the shaft seat. Sometimes heating of the bearing and cooling of the shaft are used to produce a clearance to facilitate assembly. When the parts return to normal temperatures, the final fit is produced.

We now show the determination of the limit dimensions for the shaft at each bearing seat.

TABLE 15–5 (*continued*)

B. Housing Fits

Bearing *OD*			ISO tolerance grade	Housing bore		Limits of fit	
Nominal (mm)	Maximum (in)	Minimum (in)		Maximum (in)	Minimum (in)	Minimum (in)	Maximum (in)
30	1.1811	1.1807	H8	1.1811	1.1824	0	0.0017L
32	1.2598	1.2594	H8	1.2598	1.2613	0	0.0019L
35	1.3780	1.3776	H8	1.3780	1.3795	0	0.0019L
37	1.4567	1.4563	H8	1.4567	1.4582	0	0.0019L
40	1.5748	1.5744	H8	1.5748	1.5763	0	0.0019L
42	1.6535	1.6531	H8	1.6535	1.6550	0	0.0019L
47	1.8504	1.8500	H8	1.8504	1.8519	0	0.0019L
52	2.0472	2.0467	H8	2.0472	2.0490	0	0.0023L
62	2.4409	2.4404	H8	2.4409	2.4427	0	0.0023L
72	2.8346	2.8341	H8	2.8346	2.8364	0	0.0023L
80	3.1496	3.1491	H8	3.1496	3.1514	0	0.0023L
85	3.3465	3.3459	H8	3.3465	3.3486	0	0.0027L
90	3.5433	3.5427	H8	3.5433	3.5454	0	0.0027L
100	3.9370	3.9364	H8	3.9370	3.9391	0	0.0027L
110	4.3307	4.3301	H8	4.3307	4.3328	0	0.0027L
120	4.7244	4.7238	H8	4.7244	4.7265	0	0.0027L
125	4.9213	4.9206	H8	4.9213	4.9238	0	0.0032L
130	5.1181	5.1174	H8	5.1181	5.1206	0	0.0032L
140	5.5118	5.5111	H8	5.5118	5.5143	0	0.0032L
150	5.9055	5.9048	H8	5.9055	5.9080	0	0.0032L
160	6.2992	6.2982	H8	6.2992	6.3017	0	0.0035L
170	6.6929	6.6919	H8	6.6929	6.6954	0	0.0035L
180	7.0866	7.0856	H8	7.0866	7.0891	0	0.0035L
190	7.4803	7.4791	H8	7.4803	7.4831	0	0.0040L
200	7.8740	7.8728	H8	7.8740	7.8768	0	0.0040L
215	8.4646	8.4634	H8	8.4646	8.4674	0	0.0040L
225	8.8583	8.8571	H8	8.8583	8.8611	0	0.0040L
230	9.0551	9.0539	H8	9.0551	9.0579	0	0.0040L
240	9.4488	9.4476	H8	9.4488	9.4516	0	0.0040L
250	9.8425	9.8413	H8	9.8425	9.8453	0	0.0040L
260	10.2362	10.2348	H8	10.2362	10.2394	0	0.0046L
270	10.6299	10.6285	H8	10.6299	10.6331	0	0.0046L
280	11.0236	11.0222	H8	11.0236	11.0268	0	0.0046L
290	11.4173	11.4159	H8	11.4173	11.4205	0	0.0046L
300	11.8110	11.8096	H8	11.8110	11.8142	0	0.0046L
310	12.2047	12.2033	H8	12.2047	12.2079	0	0.0046L
320	12.5984	12.5968	H8	12.5984	12.6019	0	0.0051L
340	13.3858	13.3842	H8	13.3858	13.3893	0	0.0051L
360	14.1732	14.1716	H8	14.1732	14.1767	0	0.0051L
380	14.9606	14.9590	H8	14.9606	14.9641	0	0.0051L
400	15.7480	15.7464	H8	15.7480	15.7515	0	0.0051L
420	16.5354	16.5336	H8	16.5354	16.5392	0	0.0056L

NOTE: L = loose; T = tight.

Shaft 1: Input Shaft.   Both bearings 1 and 2 are number 6305.

Nominal bore = 25 mm (0.984 3 in)

From Table 15–5: k5 ISO tolerance grade on the shaft seat; limits of 0.984 7–0.984 4 in

Resulting fit between bearing bore and shaft seat: 0.000 1 in tight to 0.000 8 in tight

OD of the outer race = 62 mm (2.440 9 in)

From Table 15–5: H8 ISO tolerance grade on the housing bore; limits of 2.440 9– 2.442 7 in

Resulting fit between outer race and housing bore: 0.0 to 0.002 3 in loose

Shaft 2: Output Shaft.   Bearing 3 at D is number 6205.

Nominal bore = 25 mm (0.984 3 in)

From Table 15–5: k5 ISO tolerance grade on the shaft seat; limits of 0.984 7–0.984 4 in

Resulting fit between bearing bore and shaft seat: 0.000 1 in tight to 0.000 8 in tight

OD of the outer race = 52 mm (2.047 2 in)

From Table 15–5: H8 ISO tolerance grade on the housing bore; limits of 2.047 2– 2.049 0 in

Resulting fit between outer race and housing bore: 0.0 to 0.002 3 in loose

Shaft 2: Output Shaft.   Bearing 4 at B is number 6207.

Nominal bore = 35 mm (1.378 0 in)

From Table 15–5: k5 ISO tolerance grade on the shaft seat; limits of 1.378 5–1.378 1 in

Resulting fit between bearing bore and shaft seat: 0.000 1 in tight to 0.001 0 in tight

OD of the outer race = 72 mm (2.834 6 in)

From Table 15–5: H8 ISO tolerance grade on the housing bore; limits of 2.834 6– 2.836 4 in

Resulting fit between outer race and housing bore: 0.0 to 0.002 3 in loose

Shaft and Housing Shoulder Diameters.   Each of the bearings in this design is to be seated against a shoulder on one side of the bearing. The shaft shoulder must be suffi-ciently large to provide a solid, flat surface against which to seat the side of the inner race. But the shoulder must not be so high that it contacts the outer race because the inner race rotates at shaft speed and the outer race is stationary.

Similarly, a shoulder in the housing must provide for the solid location of the outer race but not be such that it contacts the inner race.

Bearing manufacturers' catalogs provide data such as those shown in Table 15–6 to guide you in specifying suitable shoulder heights. The value of S is the minimum shaft shoulder diameter. The nominal maximum diameter is the mean diameter for the bearing at the middle of the balls. The value of H is the maximum housing shoulder diameter, with the nominal minimum diameter being the mean diameter for the bearing.

For example, in the present design, the minimum shoulder diameter at each bearing on shaft 1 should be 1.14 in as indicated for the bearing number 305 in Table 15–6. (Note that the bearing number 6305 specified for the shaft is of the same *series* as the number 305, indicating that it would have similar dimensions.) The maximum housing shoulder diameter for the number 305 bearing is 2.17 in, where the outer race is to seat against a shoulder.

TABLE 15–6   Shaft and housing shoulder diameters

200 series bearings						300 series bearings					
No.	S	H	No.	S	H	No.	S	H	No.	S	H
200	0.50	0.98	216	3.55	5.12	300	0.50	1.18	316	3.62	6.22
201	0.58	1.06	217	3.75	5.51	301	0.63	1.22	317	3.90	6.54
202	0.69	1.18	218	3.94	5.91	302	0.75	1.42	318	4.09	6.93
203	0.77	1.34	219	4.21	6.22	303	0.83	1.61	319	4.29	7.32
204	0.94	1.61	220	4.41	6.61	304	0.94	1.77	320	4.49	7.91
205	1.14	1.81	221	4.61	7.01	305	1.14	2.17	321	4.69	8.31
206	1.34	2.21	222	4.80	7.40	306	1.34	2.56	322	4.88	8.90
207	1.53	2.56	224	5.20	7.99	307	1.69	2.80	324	5.28	9.69
208	1.73	2.87	226	5.67	8.50	308	1.93	3.19	326	5.83	10.32
209	1.94	3.07	228	6.06	9.29	309	2.13	3.58	328	6.22	11.10
210	2.13	3.27	230	6.46	10.08	310	2.36	3.94	330	6.61	11.89
211	2.41	3.68	232	6.85	10.87	311	2.56	4.33	332	7.01	12.68
212	2.67	3.98	234	7.40	11.50	312	2.84	4.65	334	7.40	13.47
213	2.86	4.37	236	7.80	11.89	313	3.03	5.04	336	7.80	14.25
214	3.06	4.57	238	8.19	12.68	314	3.23	5.43	338	8.35	14.88
215	3.25	4.76	240	8.58	13.47	315	3.43	5.83	340	8.74	15.67

NOTES:

S = minimum shaft shoulder diameter
    Maximum diameter should not exceed the mean diameter of the bearing at the middle of the balls.

H = maximum housing shoulder diameter
    Minimum diameter should not be less than the mean diameter of the bearing at the middle of the balls.

On shaft 2, the shoulder for bearing 6205 should also be at least 1.14 in, and the shoulder for bearing 6207 should be 1.53 in minimum. The maximum housing shoulder diameter for the 6205 bearing on shaft 2 is 1.81. For the 6207 bearing, the maximum housing shoulder diameter is 2.56 in.

Table 15–7 shows the pertinent data used to decide on the values for the shoulder diameters and, in the last two columns, gives the specified values. Where the specified shoulder diameter is less than the preliminary value shown in Table 15–4, another step in the shaft will be used to provide the proper shoulder for the bearing and for the gear. This can be seen in the drawings of the shafts given at the end of this chapter.

The use of a 1.75-in diameter for the shoulder at D on shaft 1 was specified because that is the diameter of the shaft chosen earlier. It is a bit higher than the mean diameter of the bearing, but it should still be lower than that of the outer race. More complete data in a manufacturer's catalog indicates the diameter of the inner surface of the outer race to be 2.00 in, so the 1.75-in diameter is acceptable.

TABLE 15–7 Shaft and housing shoulder diameters for bearings in the present design

A. Shaft 1

Bearing no.	Bore	OD	Mean diameter	S Minimum	H Maximum	Specified shaft shoulder	Specified housing shoulder
6305 at B	0.9843	2.4409	1.713	1.14	2.17	1.50	2.00
6305 at D	0.9843	2.4409	1.713	1.14	2.17	1.75	2.00

B. Shaft 2

Bearing no.	Bore	OD	Mean diameter	S Minimum	H Maximum	Specified shaft shoulder	Specified housing shoulder
6207 at B	1.3780	2.8346	2.106	1.53	2.56	2.00	2.25
6205 at D	0.9843	2.0472	1.516	1.14	1.81	1.50	1.75

Fillet Radii. Each of the bearings specified for the reducer calls for the maximum fillet radius at the shoulder that locates the bearing to be 0.039 in. See Table 14–3. Let's specify the limits on the radius to be 0.039 to 0.035. Before committing to the design, we will check the stress concentration factor at each shoulder.

Flexible Couplings. The use of flexible couplings on both the input and the output shafts has been taken into account in the shaft design and analysis. They allow the transmission of torque between two shafts but do not exert significant radial or axial forces on the shaft. In the present design, the use of flexible couplings made the shaft design simpler and decreased the loads on bearings compared with having a device such as a belt sheave or a chain sprocket on the shaft.

Now we specify suitable couplings for the input and output shafts. Chapter 11 showed many examples for such couplings, and you should review them now. It is impractical to reproduce data for all couplings in this book. As you read this section, it would be good for you to seek a copy of the catalog of one of the manufacturers of couplings and study their recommended selection procedures.

We have selected couplings of the type pictured in Figure 11–16, called the *Browning Ever-flex Coupling* from Emerson Power Transmission, a division of the Emerson Electric Company. Rubber flex members are permanently bonded to steel hubs, and the flexing of the rubber accommodates parallel misalignment of the mating shafts up to 0.032 in, angular misalignment of ±3°, and axial end float of the shafts of up to ±0.032 in. It is important for you to design the drive system for the saw to provide this alignment of the input shaft to the drive motor and of the output shaft to the drive shaft of the saw.

The selection of a suitable coupling relies on the power transmission rating of the various sizes available. But the power rating must be correlated to the speed of rotation because the real variable is the torque to which the coupling is subjected. Both the input and the output couplings transmit nominally 25 hp in this design for the drive for the saw. But the input shaft rotates at 1 750 rpm, and the output shaft rotates at 500 rpm. Because torque is inversely proportional to speed of rotation, the torque experienced by the coupling on the output shaft is approximately 3.5 times higher than that on the input shaft. The coupling catalog data also call for the use of a *service factor* based on the kind of machine being driven, and some suggested values are included in the catalog. We judge that a service factor of 1.5 is suitable for the saw which will see mostly smooth power transmission with occasional moderate shock loading.

The service factor is applied to the nominal power being transmitted to compute a value for the *normal rating* for the couplings. Then

$$\text{Normal rating} = \text{power input} \times \text{service factor} = 25 \text{ hp}(1.5) = 37.5 \text{ hp}$$

The catalog tables list coupling number CFR6 with a suitable normal rating at 1 750 rpm for the input shaft and number CFR9 for the output shaft at 500 rpm.

We specify hubs for the couplings that have machined bores and keyways with a range of bores allowed. Each coupling half can have a different bore according to the shaft size on which it is to be mounted. For the input shaft, we have specified the diameter to be 0.875 in (7/8 in), and this will be the specification for the bore of that half for the CFR6 coupling. The keyway is 3/16 × 3/32 to accept a 3/16-in square key. The nominal maximum length of shaft inside each half of the coupling is 2.56 in.

The other half of the CFR6 coupling mounts on the motor shaft. Recall that the design requirements listed at the beginning of this design process specified a 25-hp motor with a NEMA Frame 284T. Table 21–3 gives the shaft diameter for this motor to be 1.875 in ($1\frac{7}{8}$ in) with a 1/2 × 1/4 in keyway to accept a 1/2-in square key. This will be the bore specified for the motor half of the CFR6 coupling. The reason for the large difference in the sizes of the shafts for the motor and for our reducer is that the general-purpose motor must be designed to carry a significant side load, and our shaft does not.

The output shaft of the reducer at the coupling has a diameter of 1.250 in, and the input shaft for the saw will have the same size. Therefore, both halves of the CFR9 coupling will have that bore with a 1/4 × 1/8 in keyway to accept a 1/4-in square key. The nominal maximum shaft length inside each half of the coupling is 3.125 in.

Keys and Keyseats.   A total of six keys need to be specified: two for each half of the flexible couplings on the input and output shafts, and one for each gear in the reducer. The methods of Chapter 11 are used to verify the suitability of the keys and to specify the required length using Equation (11–5). We will use standard key sizes made from AISI 1020 CD steel having a yield strength of 51 000 psi.

1. *Keys for CFR6 coupling on the input shaft:* First let's check the keys inside the couplings because their sizes have already been specified by the coupling manufacturer. The coupling half that mounts on the input shaft is critical because its bore diameter of 0.875 in is smaller, resulting in larger forces on the key when transmitting the torque of 900 lb·in which was computed earlier during the shaft design. The key is 3/16 in square (0.188 in). We use a design factor N of 4 as we did in the shaft design. Then, from Equation (11–5),

$$L = \frac{4TN}{DWs_y} = \frac{4(900 \text{ lb·in})(4)}{(0.875 \text{ in})(0.188 \text{ in})(51\ 000 \text{ psi})} = 1.72 \text{ in}$$

We can specify a key length of 2.50 in for extra safety and to match the length of the hub of the CFR6 coupling. The 1/2-in key for the motor shaft should be made to match the length of the coupling hub also, and it should be very safe because of the larger key size and the larger shaft size carrying the same torque.

2. *Keys for the CFR9 coupling on the output shaft:* For the output shaft and the drive shaft for the saw,

$$T = 3\ 150\ \text{lb·in}$$

$$D = 1.25\ \text{in}$$

$$W = 0.250\ \text{in}\quad\text{(key width)}$$

$$L = \frac{4TN}{DWs_y} = \frac{4(3\ 150\ \text{lb·in})(4)}{(1.25\ \text{in})(0.250\ \text{in})(51\ 000\ \text{psi})} = 3.16\ \text{in}$$

We will make the key length 3.125 in (3 1/8 in), the full length of the hub of the CFR9 coupling. The conservative design factor of 4 should make this length acceptable.

3. *Key for the pinion on shaft 1:* The bore of the pinion is to be nominally 1.75 in as determined in the shaft design and shown in Table 15–4. The key size for this diameter should be 3/8 in square according to Table 11–1. The torque being transmitted is 900 lb·in. Then Equation (11–5) gives

$$L = \frac{4TN}{DWs_y} = \frac{4(900\ \text{lb·in})(4)}{(1.75\ \text{in})(0.375\ \text{in})(51\ 000\ \text{psi})} = 0.430\ \text{in}$$

The face width of the gear is 2.00 in. Let's use a key length of 1.50 in and center the profile keyseat at section *C* on the shaft so that the keyseat does not significantly interact with the ring groove to the right or with the shoulder fillet to the left.

4. *Key for the gear on shaft 2:* The bore of the gear is to be nominally 1.75 in as determined in the shaft design and shown in Table 15–4. The key size for this diameter should be 3/8 in square according to Table 11–1. The torque being transmitted is 3 150 lb·in. Then Equation (11–5) gives

$$L = \frac{4TN}{DWs_y} = \frac{4(3\ 150\ \text{lb·in})(4)}{(1.75\ \text{in})(0.375\ \text{in})(51\ 000\ \text{psi})} = 1.50\ \text{in}$$

The face width of the gear is 2.00 in. Let's use a key length of 1.50 in for this key also.

The key designs are summarized in the following list:

Summary of Key Designs

Motor shaft: 1/2-in square key × 2.50 in long

Input shaft of reducer at coupling: 3/16-in square key × 2.50 in long; sled runner keyseat

Input shaft at pinion: 3/8-in square key × 1.50 in long; profile shaft keyseat

Output shaft at gear: 3/8-in square key × 1.50 in long; profile shaft keyseat

Output shaft at coupling: 1/4-in square key × 3.125 in long; sled runner keyseat

Drive shaft for saw at coupling: 1/4-in square key × 3.125 in long; sled runner keyseat

The tolerances for the keys and keyseats are summarized as follows: Standard square bar stock in AISI 1020 or 1030 steel is available to be used for keys. Typical tolerances are

TABLE 15–8 Tolerances and fits for keys and keyseats

Key size (width in inches)	Tolerance on key (all +0.000)	Tolerance on keyseat (all −0.000)	Fit range
Up to 1/2	−0.002	+0.002	0.000–0.004
Over 1/2 to 3/4	−0.002	+0.003	0.000–0.005
Over 3/4 to 1	−0.003	+0.003	0.000–0.006
Over 1 to $1\frac{1}{2}$	−0.003	+0.004	0.000–0.007
Over $1\frac{1}{2}$ to $2\frac{1}{2}$	−0.004	+0.004	0.000–0.008

given in Table 15–8. Also given are recommended tolerances on the keyseat width dimension and the resulting fit between the key and the keyseat. A small clearance fit is desirable to permit easy assembly while not allowing the key to rock noticeably when installed.

Tolerances on Other Shaft Dimensions.   It is the designer's responsibility to establish tolerances on each dimension of each component in a mechanical device. The tolerances must ensure that the component fulfills its function. But it should also be as large as practical to permit economical manufacture. This pair of conflicting principles must be dealt with. You should review the discussion in Chapter 13 on tolerances and fits. See also References 1, 2, and 5, along with other comprehensive texts on technical drawing and interpretation of engineering drawings.

Special attention should be paid to the features of a component which mate with other components and with which they must operate reliably or with which they must be accurately located. The fit of the inner races of the bearings on the shafts is an example of such features. Others in this reducer are the clearances between parts that must be assembled together easily but that must not have large relative motion during operation. The fit of the bore of the gears on the shafts or the fit of the *ID* of the couplings on the ends of the shafts are examples. Either a close sliding fit or a close locational fit is recommended for such components, as discussed in Chapter 13. Data are given in Table 13–6 for the RC2, RC5, and RC8 fits. More complete data are available in technical drawing textbooks and in References 1, 2, 3, and 7 in Chapter 13. We will apply the RC5 fit to accurate mating parts that must assemble easily but where little perceptible play between the parts is desired. The RC fits use the *basic hole system,* as illustrated in Chapter 13.

Where no other component mates with certain features of a given component, the tolerances should be as large as practical such that they could be produced with basic machining, molding, or casting processes without the need for subsequent finishing. It is often recommended that blanket tolerances be given for such dimensions and that the precision with which the basic size is stated on the drawing implies a certain tolerance. For decimal dimensions in U.S. Customary units, a note similar to the following is often given:

DIMENSIONS IN inches. TOLERANCES ARE AS
FOLLOWS UNLESS OTHERWISE STATED.

XX.X = ± 0.050
XX.XX = ±0.010
XX.XXX = ±0.005
XX.XXXX = ±0.0005
ANGLES: ±0.50°

where X represents a specified digit

For example, if a given dimension has a basic size of 2.5 inches, the dimension can be stated on the drawing in any of four ways with different interpretations:

2.5	means	2.5 ± 0.050 or limits of 2.550 to 2.450 in
2.50	means	2.50 ± 0.010 or limits of 2.510 to 2.490 in
2.500	means	2.500 ± 0.005 or limits of 2.505 to 2.495 in
2.5000	means	2.5000 ± 0.0005 or limits of 2.5005 to 2.4995 in

Any other desired tolerance must be specified on the dimension. Of course, you may select different standard tolerances according to the needs of the system being designed.

Similar data for metric drawings would appear as follows:

DIMENSIONS IN mm. TOLERANCES ARE AS
FOLLOWS UNLESS OTHERWISE STATED.

XX.X = ± 1.0
XX.XX = ± 0.25
XX.XXX = ± 0.15
XX.XXXX = ± 0.012
ANGLES: $\pm 0.50°$

Some tolerance notes also relate the degree of precision to the nominal size of the feature, with tighter tolerances on smaller dimensions and looser tolerances on larger dimensions. The international tolerance (IT) grades, discussed in Chapter 13, use this approach.

Geometric Tolerances.   Geometric tolerancing is used to control location, form, profile, orientation, and runout on a dimensioned feature. Its purpose is to ensure proper assembly and/or operation of parts. Figure 15–4 shows some of the geometric symbols used. Some of the more commonly used geometric tolerances are for straightness, flatness, cylindricity, concentricity, perpendicularity, parallelism, and position. Each of these feature control notes contains a tolerance and a datum or reference feature.

Surface Finish.   The designer should also control the surface finish of all features critical to the performance of the device being designed. This includes the mating surfaces which have already been discussed. But also any surface that experiences relatively high stresses, particularly reversed bending, should have a smooth surface.

See Figure 5–9 for a very rough indication of the effect of surface finishes between, for example, *machined* and *ground* surfaces on the basic endurance strength of steels. In general *ground* surfaces have an average roughness, R_a, of 16 μin (0.4 μm). Figure 13–3 shows the expected range of surface finish for many kinds of machining processes. Note that turning is reported to be capable of producing that level of surface finish, but it is at the limit of its capability and will likely require very fine finish cuts with sharp-edged tooling having a broad radius nose. The more nominal surface finish from turning, milling, broaching, and boring is 63 μin (1.6 μm). This would correspond to the *machined* category in Figure 5–9.

Bearing seats on shafts for accurate machinery are typically ground, particularly in the smaller sizes under 3.0 in (80 mm), with a maximum allowable average roughness of 16 μin (0.4 μm). Above that size and up to 20 in (500 mm), 32 μin (0.8 μm) is allowed. See manufacturers' catalogs. The shaft drawings (Figures 15–5 and 15–6) show surface finish specifications at the bearing seats and in the general notes.

Tolerance	Characteristic	Symbol
Form	Straightness	—
	Flatness	▱
	Circularity	○
	Cylindricity	⌭
Profile	Profile of a line	⌒
	Profile of a surface	⌓
Orientation	Angularity	∠
	Perpendicularity	⊥
	Parallelism	//
Location	Position/symmetry	⌖
	Concentricity	◎
Runout	Circular runout	↗
	Total runout	⌰

(a) Geometric symbols

Geometric characteristic ⌐   Dia. ⌐   Tolerance ⌐

◎ Ø 0.008 C

Datum ⌐

(b) Feature control frame

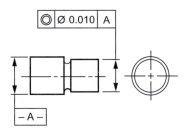

◎ Ø 0.010 A

— A —

(c) Concentricity

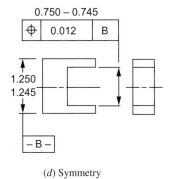

0.750 − 0.745

⌖ 0.012 B

1.250
1.245

— B —

(d) Symmetry

— 0.004

Ø 0.750
0.745

−0.004 tolerance

(e) Straightness

▱ 0.002

−0.002 tolerance

(f) Flatness

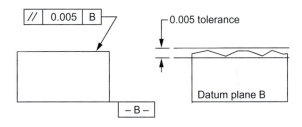

// 0.005 B

0.005 tolerance

Datum plane B

— B —

(g) Parallelism: planes

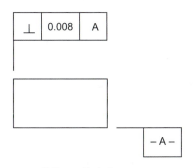

// Ø 0.005 A

— A —

(h) Parallelism: cylinders

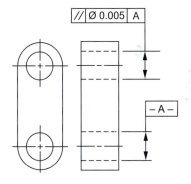

⊥ 0.008 A

— A —

(i) Perpendicularity

FIGURE 15–4   Examples of geometric tolerancing

**15–6
FINAL DESIGN
DETAILS FOR
THE SHAFTS**

Figures 15–5 and 15–6 show the final design for the input and output shafts. Data from throughout this chapter have been used to specify pertinent dimensions. Where fillets are specified, a final check on the stress condition has been made to ensure that the estimated stress concentration factors used in the earlier design analysis are satisfactory and that the final stress levels are safe.

Keyseat details have been shown in sections below the main dimensioned shaft drawings. See Chapter 11 for computing the vertical dimension from the bottom of the shaft to the bottom of the keyseat.

The retaining ring grooves are drawn to the dimensions specified for a basic external ring (Type 5100) for a 1.75-in-diameter shaft from the Waldes Truarc Company as pictured in Figure 11–28.

Four shaft diameters are sized to the RC5 fit in Figures 15–5 and 15–6. On shaft 1 they are at the extension where the coupling mounts and at the pinion. On shaft 2 they are at the gear and at the coupling location on the output extension. Table 15–9 summarizes the data for the fits. You should verify these using the procedure shown in Chapter 13. Note that the limit dimensions for both the shaft diameter and the bore of the mating element are given and that the total tolerance on each dimension is small, less than 0.002 in for any dimension. Note also the small variation of the clearance on mating parts as indicated in the last column called "Fit."

For the shafts in this project, we have specified geometric tolerances for concentricity of four critical diameters for each shaft. Figure 15–6(b) documents the approach for the output shaft. Because of the similarity of the two shafts, the nature of the callouts would be the same for the input shaft. The datum or reference diameter is specified at the gear. Then the diameters at the two bearing seats and at the end of the shaft where the coupling mounts are controlled with concentricity feature control blocks. Shoulders for locating bearings and gears are controlled for perpendicularity to the axis of the shaft as represented by the gear diameter. The keyseat is controlled for parallelism to the axis of the shaft.

**15–7
ASSEMBLY
DRAWING**

Figure 15–7 is an assembly drawing for the reducer with all features drawn to scale. The housing has been shown as a rectangular box shape for simplicity. Bearings are held in bearing retainers which are then fastened to the housing walls. Special attention to the alignment of the retainers is required and would be an important task for the detailing of the housing, not shown here.

The assembly of all components into the housing is facilitated by having the right side removable. Again, alignment of the cover piece with the main housing is critical.

Seals have been shown in the bearing retainers where the shafts penetrate the side walls of the housing. See Chapter 11 for more information on seals.

TABLE 15–9   Calculations and specifications for fits of elements on the shafts of the transmission

Location	Nominal diameter	Bore limit dimensions for external element	OD limit dimensions for shaft	Fit
Shaft 1:				
Input at coupling	0.875 0	0.876 2/0.875 0	0.873 4/0.872 6	+0.002 6/+0.003 6
Pinion	1.750 0	1.751 6/1.750 0	1.748 0/1.747 0	+0.002 0/+0.004 6
Shaft 2:				
Gear	1.750 0	1.751 6/1.750 0	1.748 0/1.747 0	+0.002 0/+0.004 6
Output at coupling	1.250 0	1.251 6/1.250 0	1.248 0/1.247 0	+0.002 0/+0.004 6

NOTE: RC5 fit is used for all locations; dimensions are in inches.

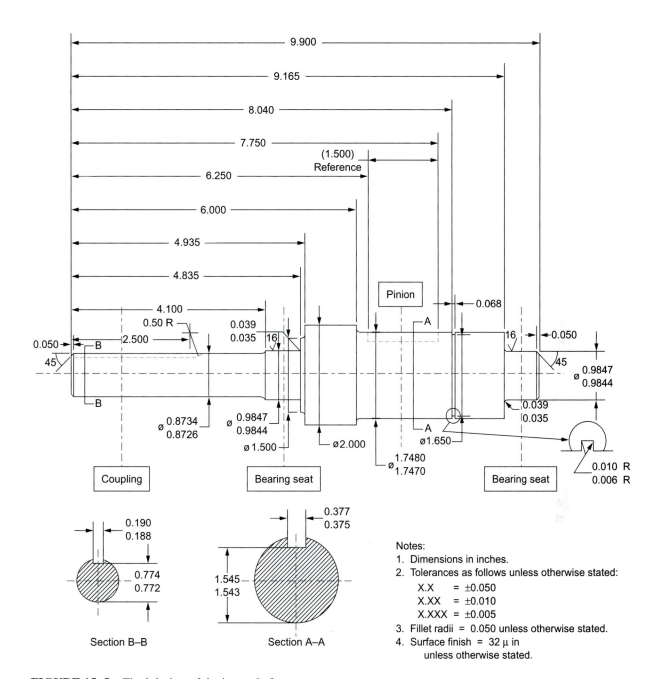

FIGURE 15–5 Final design of the input shaft

Notes:
1. Dimensions in inches.
2. Tolerances as follows unless otherwise stated:

    X.X   = ±0.050
    X.XX = ±0.010
    X.XXX = ±0.005

3. Fillet radii = 0.050 unless otherwise stated.
4. Surface finish = 32 μ in unless otherwise stated.

(a) Dimensioned drawing

FIGURE 15–6   Final design of the output shaft

594

Keyseat

(b) Geometric tolerances

FIGURE 15–6 (*continued*)

FIGURE 15–7 Assembly drawing for the reducer

Critique of the Design

The design shown in Figures 15–5, 15–6, and 15–7 present a design that meets the basic design requirements established at the beginning of this chapter. It is likely that refinements could be made if more detail were available about the saw for which the reducer is being designed.

It appears that the length of the shafts could be made somewhat shorter. The distances between the center of the gears and the bearings was arbitrarily set at 2.50 in at the start of the design process when dimensions for any components were unknown. Now that the nominal size of the gears, bearings, and couplings are known, further iterations on the design could result in a smaller package.

You should look at other commercially available gear-type speed reducers for other features that might be built into this design. Note particularly Figures 9–32, 9–33, and 9–34 in Chapter 9 and Figures 10–1, 10–2, and 10–15 in Chapter 10.

REFERENCES

1. American Society of Mechanical Engineers. ASME Standard Y14.5M. *Dimensioning and Tolerancing.* New York: American Society of Mechanical Engineers, 1995.

2. Earle, James H. *Graphics Technology.* Reading, MA: Addison-Wesley, 1995.

3. Kepner, Charles H., and Benjamin B. Tregoe. *The New Rational Manager.* Princeton, NJ: Princeton Research Press, 1981.

4. National Research Council. *Improving Engineering Design: Designing for Competitive Advantage.* Washington, DC: National Academy Press, 1991.

5. Oberg, Erik, et al. *Machinery's Handbook.* 25th ed. New York: Industrial Press, 1996.

6. Peerless-Winsmith, Inc. *The Speed Reducer Book: A Practical Guide to Enclosed Gear Drives.* Springville, NY: Peerless-Winsmith, 1980.

7. Pugh, Stuart. *Total Design: Integrated Methods for Successful Product Engineering.* Reading, MA: Addison-Wesley, 1991.

8. Ullman, David G. *The Mechanical Design Process.* 2nd ed. New York: McGraw-Hill, 1997.

16

Plain Surface Bearings

Plain Surface Bearings

Discussion Map

☐ The purpose of a bearing is to support a load while permitting relative motion between two elements of a machine. This chapter discusses *plain surface* bearings where the two parts that move relative to each other do not have rolling elements between them.

Discover

Look around your home and in your car for products that have plain surface bearings. Find some that experience rotational motion and some that have linear sliding contact. Consider simple items such as hinges, door locks, latches, or the wheels of a lawn mower. Most of these experience boundary lubrication.

Now see if you can find information about the crankshaft bearings in the engine of your car. Such bearings usually employ full-film hydrodynamic lubrication. *What can you discover about them?*

How do you think a large telescope or an astronomical radio antenna is supported to permit it to be moved easily and positioned precisely? One method is called hydrostatic bearings. *What can you discover about such bearings?*

This chapter will help you explore all of these kinds of bearings and complete the basic design analyses required to ensure satisfactory operation.

The purpose of a bearing is to support a load while permitting relative motion between two elements of a machine. This chapter discusses *plain surface* bearings where the two parts that move relative to each other do not have rolling elements between them. Note that *rolling contact bearings* were discussed in Chapter 14.

Plain surface bearings for rotating parts are, of course, cylindrical with a typical arrangement as shown in Figure 16–1. The inner member, called the *journal,* is usually that

FIGURE 16–1

Bearing geometry

Radial clearance

$$C_r = \frac{D_b - D_j}{2}$$

Bearing dia. D_b

Journal dia. D_j

Diametral clearance

$$C_d = D_b - D_j = 2C_r$$

Bearing

Clearance

Journal

Housing

part of a shaft where any radial reaction forces are transferred to the base of the machine. The stationary member that mates with the journal is called simply the *bearing*.

Where have you seen such plain surface bearings in operation? As you did in Chapter 14 on rolling contact bearings, look for consumer products, industrial machinery, or transportation equipment (cars, trucks, bicycles, and so on—any device where there are rotating shafts). If the bearings are not visible from the outside (and that is often the case), you may need to partially disassemble the product to see inside.

But there are some obvious examples that should be accessible to you. The wheels of a household lawn mower, wheelbarrow, or simple cart are typically mounted directly on axles to form the plain surface bearing. Hand tools such as grass and hedge clippers, pruning shears, pliers, and adjustable or ratchet wrenches use plain surface bearings at points where one part must rotate relative to another.

Look at almost any hinge for a door. It incorporates a solid pin riding inside a cylindrical outer member. That is an example of a plain surface bearing. Garage doors often have rollers that ride in channels, but the shafts that carry the rollers are operating in plain cylindrical bearings, often with very large clearances. The cable system that connects to the counterbalance springs runs over pulleys having plain surface bearings rotating on stationary shafts. If the door has an electric door opener, several of its components probably use plain surface bearings. Get a ladder and climb up to inspect them. *But be careful of the moving parts!*

Besides the rotational motion of the kinds of applications described above, some plain surface bearings have linear sliding motion. Look inside a printer for a computer of the ink-jet or dot-matrix-pin type. Identify the part that delivers the actual image to the paper as it traverses across the page. It typically slides on a polished rod with great precision. Many linkages contain parts that slide. Look at locking mechanisms, latches, staplers, switches, car seat adjusters, gear shifters for cars, parts of window-lift devices in cars, and coin-operated machines. Can you identify parts that have linear sliding motion?

If you visit a factory, you will likely see numerous examples of both rotational and linear sliding action. Transfer devices, reciprocating machinery, machine tool slides, and packaging equipment employ many plain surface bearings.

Most of the examples just mentioned have intermittent and relatively slow motion between the mating parts, whether the motion is rotational or linear. In such applications, there is often a lubricant between the moving parts. Or the materials are carefully chosen to produce low friction and reliable, smooth motion. This kind of bearing experiences *boundary lubrication* as discussed in this chapter.

You may be familiar with the bearings on the crankshaft of an internal combustion engine. They are often called *journal bearings* because they have the classic configuration shown in Figure 16–1. But consider their general operation. After the engine has been started, the crank rotates at several thousand revolutions per minute, typically from about 1 500 to 6 000 rpm. This is one of the most important applications of plain surface bearings using *full-film hydrodynamic lubrication*. In such bearings, there is a continuous film of lubricant, usually oil, that actually lifts the rotating shaft journal off the stationary bearing. Thus, there is no metal-to-metal contact between the members. Much of this chapter is devoted to this kind of bearing and the analysis of the conditions required to maintain the supporting oil film. Engine bearings are some of the most highly engineered mechanical devices.

Can you identify other examples of bearings that use *full-film hydrodynamic lubrication?*

Another example of a class of plain surface bearings is called *hydrostatic*. Pumps create high pressure in a fluid such as oil that is delivered by carefully shaped pads where the pressure lifts the part to be moved. Then it can be moved easily, either slowly or rapidly.

Think about a huge telescope or radio antenna that must be positioned accurately and moved with ease. Such applications are rare but very important, justifying the large engineering and technical effort of designing, installing, and maintaining them.

This chapter will help you understand boundary lubrication, full-film hydrodynamic lubrication, and hydrostatic lubrication.

You Are the Designer

Your company is designing a conveyor system to transport products in a shipping and receiving system of a major manufacturer. The design has progressed to the point where it has been decided to use a flexible belt conveyor driven by flat pulleys at either end. The shafts that carry the pulleys must be supported in bearings in the side frames of the conveyor. Your task is to design the bearings.

First you must decide what type of bearing to use: plain surface or journal bearings as discussed in this chapter, or rolling contact bearings which are discussed in Chapter 14. For the plain surface bearings, you must determine whether full-film hydrodynamic lubrication can be achieved with its advantages of low friction and long life. Or will the shaft operate in the bearing with boundary lubrication? What materials will the bearing and the journal be made from? What dimensions will be specified for all components? What lubricant should be used? These and other questions are discussed in this chapter.

16–1 OBJECTIVES OF THIS CHAPTER

After completing this chapter, you will be able to:

1. Describe the three modes of operation of a plain surface bearing (boundary, mixed-film, and full-film hydrodynamic lubrication), and discuss the conditions under which each will normally occur.

2. Discuss the significance of the bearing parameter, $\mu n/p$.

3. List the decisions that a bearing designer must make to completely define a plain surface bearing system.

4. List the materials often used for journals and bearings, and describe their important properties.

5. Define the *pV factor* and use it in the design of boundary-lubricated bearings.

6. Describe the operation of full-film hydrodynamically lubricated bearings.

7. Complete the design of full-film bearings, defining the size of the journal and bearing, the diametral clearance, the bearing length, the minimum film thickness, the surface finish, the lubricant, and the resulting frictional performance of the bearing system.

8. Describe a hydrostatic bearing system and complete the basic design of such bearings.

16–2 THE BEARING DESIGN TASK

The term *plain surface bearing* refers to the kind of bearing in which two surfaces move relative to each other without the benefit of rolling contact. Thus, there is sliding contact. The actual shape of the surfaces can be anything that permits the relative motion. The most common shapes are flat surfaces and concentric cylinders.

The name *journal bearing* is sometimes used for plain surface bearings. This term is derived from the terminology of the components of the complete bearing system. For the case of a bearing on a rotating shaft, the portion of the rotating shaft at the bearing is called the *journal.* The stationary part that supports the load is the *bearing.* Figure 16–1 shows the basic geometry of a journal bearing.

A given bearing system can operate with any of three types of lubrication:

Boundary lubrication: There is actual contact between the solid surfaces of the moving and the stationary parts of the bearing system, although a film of lubricant is present.

Mixed-film lubrication: There is a transition zone between boundary and full-film lubrication.

Full-film lubrication: The moving and stationary parts of the bearing system are separated by a complete film of lubricant that supports the load. The term *hydrodynamic lubrication* is often used to describe this type.

All of these types of lubrication can be encountered in a bearing without external pressurization of the bearing. If lubricant under pressure is supplied to the bearing, it is called a *hydrostatic bearing,* which is discussed separately. Running dry surfaces together is not recommended unless there is inherently good lubricity between the mating materials. Some plastics are used dry, as discussed in Sections 16–4 and 16–5.

Bearing design involves so many design decisions that it is not possible to develop a procedure that produces the single best design. Thus, several feasible designs could be proposed, and the designer must make judgments based on knowledge of the application and on the principles of bearing operation to define the final design. The following lists identify the information needed to design a bearing system and the types of design decisions that must be made. (See Reference 14.) In this discussion, it is assumed that the bearing will be cylindrical, like that used to support a rotating shaft. Modified lists could be made for linear sliding surfaces or some other geometry.

Bearing Requirements

Magnitude, direction, and degree of variation of the radial load

Magnitude and direction of the thrust load, if any

Rotational speed of the journal (shaft)

Frequency of starts and stops, and duration of idle periods

Magnitude of the load when the system is stopped and when it is started

Life expectancy of the bearing system

Environment in which the bearing will operate

Design Decisions

Materials for the journal and the bearing

Diameters, including tolerances, of the journal and the bearing

Nominal value and range of journal clearance in the bearing

Surface finish for the journal and the bearing

Length of the bearing

Method of manufacturing the bearing system

Type of lubricant to be used and the means of supplying it

Operating temperature of the bearing system and of the lubricant

Method of maintaining the lubricant cleanliness and temperature

Analyses Required

Type of lubrication: boundary, mixed-film, full-film

Coefficient of friction

Frictional power loss

Minimum film thickness

Thermal expansion

Heat dissipation required and the means of accomplishing it

Shaft stiffness and slope of the shaft in the bearing

16–3 BEARING PARAMETER, $\mu n/p$

The performance of a bearing differs radically, depending on which type of lubrication occurs. There is a marked decrease in the coefficient of friction when the operation changes from boundary to full-film lubrication. Wear also decreases with full-film lubrication. Thus, it is desirable to understand the conditions under which one or the other type of lubrication occurs.

The creation of full-film lubrication, the most desirable type, is encouraged by light loads, high relative speed between the moving and stationary parts, and presence of a high-viscosity lubricant in the bearing in copious supply. For a rotating journal bearing, the combined effect of these three factors, as it relates to the friction in the bearing, can be evaluated by computing the *bearing parameter*, $\mu n/p$. The viscosity of the lubricant is indicated by μ, the rotational speed by n, and the bearing load by the pressure, p. To compute the pressure, divide the applied radial load on the bearing by the *projected area* of the bearing, that is, the product of the length times the diameter.

Although we will not use it for design or analysis, it is instructional to note that the bearing parameter $\mu n/p$ is dimensionless when each term is expressed in consistent units. Some unit systems that can be used are listed here:

Unit System	Viscosity, μ	Rotational Speed, n	Pressure, p
SI metric	$N \cdot s/m^2$ or $Pa \cdot s$	rev/s	N/m^2 or Pa
English	$lb \cdot s/in^2$ or reyn	rev/s	lb/in^2
Old metric (obsolete)	$dyne \cdot s/cm^2$ or poise	rev/s	$dynes/cm^2$

The effect of the bearing parameter is shown in Figure 16–2, which is a plot of the coefficient of friction, f, for the bearing versus the value of $\mu n/p$. At low values of $\mu n/p$, boundary lubrication occurs, and the coefficient of friction is high. For example, with a steel shaft sliding slowly in a lubricated bronze bearing (boundary lubrication), the value of f would be approximately 0.08 to 0.14. At high values of $\mu n/p$, the full hydrodynamic film is created, and the value of f is normally in the range of 0.001 to 0.005. Note that this compares favorably with precision rolling contact bearings. Between boundary and full-film lubrication, the *mixed-film* type, which is some combination of the other two, occurs.

It is recommended that designers avoid the mixed-film zone because it is virtually impossible to predict the performance of the bearing system. Also notice that the curve is very steep in this zone. Thus, a small change in any of the three factors, μ, n, or p, produces a large change in f, resulting in erratic performance of the machine.

The value of $\mu n/p$ at which full-film lubrication is produced is very difficult to predict. Besides the individual factors of speed, pressure (load), and viscosity (a function of

FIGURE 16–2
Bearing performance and types of lubrication related to the bearing parameter, $\mu n/p$

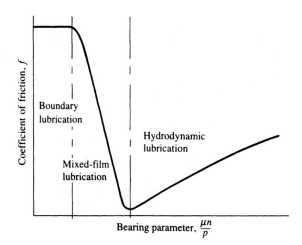

the type of lubricant and its temperature), the variables affecting the production of the film include the quantity of lubricant supplied, the adhesion of the lubricant to the surfaces, the materials of the journal and the bearing, the structural rigidity of the journal and the bearing, and the surface roughness of the journal and the bearing. After completing the design process presented later in this chapter, you are advised to test the design.

In general, boundary lubrication should be expected for slow-speed operation with a surface speed less than about 10 ft/min (0.05 m/s). Reciprocating or oscillating motion or a combination of a light lubricant and high pressure would also produce boundary lubrication.

The design of bearings to produce full-film lubrication is described in Section 16–6. In general, it requires a surface speed of greater than 25 ft/min (0.13 m/s) continuously in one direction with an adequate supply of oil at a proper viscosity.

16–4 BEARING MATERIALS

In rotational applications, the journal on the shaft is frequently steel. The stationary bearing may be made of any of a wide variety of materials, including the following:

> Bronze
>
> Babbitt
>
> Aluminum
>
> Porous metals
>
> Plastics (nylon, TFE, PTFE, phenolic, acetal,
>
>   polycarbonate, filled polyimide)

The properties desirable for materials used for plain bearings are unique, and compromises must frequently be made. The following list discusses these properties.

1. *Strength:* The function of the bearing is to carry the applied load and deliver it to the supporting structure. At times, the loads are varied, requiring fatigue strength as well as static strength.

2. *Embeddability:* The property of embeddability relates to the ability of the material to hold contaminants in the bearing without causing damage to the rotating journal. Thus, a relatively soft material is desirable.

3. *Corrosion resistance:* The total environment of the bearing must be considered, including journal material, lubricant, temperature, air-borne particulates, and corrosive gases or vapors.

4. *Cost:* Always an important factor, cost includes not only the material cost but also processing and installation costs.

A brief discussion of the performance of some of the bearing materials follows.

Cast Bronze

The name *bronze* refers to several alloys of copper with tin, lead, zinc, or aluminum, either singly or in combination. The leaded bronzes contain 25% to 35% lead, which gives them good embeddability and resistance to seizing under conditions of boundary lubrication. However, their strength is relatively low. The cast bearing bronze, SAE CA932, has 83% copper, 7% tin, 7% lead, and 3% zinc. It possesses a good combination of properties for such uses as pumps, machinery, and appliances. Tin and aluminum bronzes have higher strength and hardness and can carry greater loads, particularly in impact situations. But they rate lower on embeddability.

Babbitt

Babbitts may be lead-based or tin-based, having nominally 80% of the parent metal. Various alloy compositions of copper and antimony (as well as lead and tin) can tailor the properties to suit a particular application. Because of their softness, babbitts have outstanding embeddability and resistance to seizure, important properties in applications in which boundary lubrication occurs. They have rather low strength, however, and are often applied as liners in steel or cast iron housings.

Aluminum

With the highest strength of the commonly used bearing materials, aluminum is used in severe applications in engines, pumps, and aircraft. The high hardness of aluminum bearings results in poor embeddability, requiring clean lubricants.

Porous Metals

Products of the powder metal industry, porous metals are sintered from powders of bronze, iron, and aluminum; some are mixed with lead or copper. The sintering leaves a large number of voids in the bearing material into which lubricating oil is forced. Then, during operation, the oil migrates out of the pores, supplying the bearing. Such bearings are particularly good for slow-speed, reciprocating, or oscillating motions.

Plastics

Generally referred to as *self-lubricating materials,* plastics used in bearing applications have inherently low friction characteristics. They can be operated dry, but most improve in performance with a lubricant present. Embeddability is usually good, as is resistance to seizure. But many have low strength, limiting their load-carrying capacity. Backing with metal sleeves is frequently done to improve load-carrying capacity. Major advantages are corrosion resistance and, when operated dry, freedom from contamination. These properties are particularly important in the processing of foodstuffs and chemical products.

with D in mm and n in rpm. Then the units for pV are

$$pV = (\text{kPa})(\text{m/s})$$

A useful conversion to the U.S. Customary System is

$$1.0 \text{ psi-fpm} = 0.035 \text{ kPa·m/s}$$

Again, the units for pV can be reformatted to reflect the rate of energy transfer per unit area:

$$pV = \text{kPa·m/s} = \frac{\text{kN}}{\text{m}^2} \cdot \frac{\text{m}}{\text{s}} = \frac{\text{kW}}{\text{m}^2}$$

where $1 \text{ kW} = 1 \text{ kN·m/s}$

Operating Temperature

Most plastics are limited to approximately 200°F (93°C). However, PTFE can operate at 500°F (260°C). Babbitt is limited to 300°F (150°C), while tin-bronze and aluminum can operate at 500°F (260°C). A major advantage of carbon-graphite bearings is their ability to operate at up to 750°F (400°C).

Wear Factor

The wear factor, K, is measured under fixed conditions with the material loaded as a thrust washer. When equilibrium is reached, wear is measured as volumetric loss of material as a function of time. The load and the velocity also affect wear; thus, K is defined as

Wear Factor for Bearing Material

$$K = W/FVT \qquad\qquad (16\text{--}1)$$

where W = wear, volume of material lost (in^3)
  F = applied load (lb)
  V = linear velocity (ft/min)
  T = time (h)

Then the units for K are

$$K = \frac{\text{in}^3}{\text{lb}} \frac{1}{\text{ft/min}} \frac{1}{\text{h}} = \frac{\text{in}^3 \cdot \text{min}}{\text{lb·ft·h}}$$

But because wear is a very slow phenomenon, the values are very small. Note in Table 16–1 that the values are to be multiplied by 10^{-10}. Whereas the K factor and Equation (16–1) may not be used to predict actual wear in a given application, the K factor allows the comparison of relative wear among alternative materials.

Design Procedure

The following is one method of completing the preliminary design of boundary-lubricated plain surface bearings.

Given information: Radial load on the bearing, F (lb or N); speed of rotation, n (rpm); nominal minimum shaft diameter, D_{min} (in or mm) (based on stress or deflection analysis).

Objectives of the design process: To specify the nominal diameter and length of the bearing and a material that will have a safe value of pV.

Procedure for Designing Boundary-lubricated Plain Surface Bearings

1. Specify a trial diameter, D, for the journal and the bearing.
2. Specify a ratio of bearing length to diameter, L/D, typically in the range of 0.5 to 2.0. For nonlubricated (dry-rubbing) or oil-impregnated porous bearings, $L/D = 1$ is recommended. (See Reference 16.) For carbon-graphite bearings, $L/D = 1.5$ is recommended.
3. Compute $L = D(L/D) =$ nominal length of the bearing.
4. Specify a convenient value for L.
5. Compute the bearing pressure (lb/in^2 or Pa):

$$p = F/LD$$

6. Compute the linear speed of the journal surface:

U.S. Customary units:	$V = \pi Dn/12$ ft/min or fpm
SI metric units:	$V = \pi Dn/(60\ 000)$ m/s
Also note:	1.0 ft/min = 0.005 08 m/s
	1.0 m/s = 197 ft/min

7. Compute pV (psi-fpm or Pa·m/s or kW/m^2).
8. Multiply $2(pV)$ to obtain a design value for pV.
9. Specify a material from Table 16–1 with a rated value of pV equal to or greater than the design value.
10. Complete the design of the bearing system considering diametral clearance, lubricant selection, lubricant supply, surface finish specification, thermal control, and mounting considerations. Often the supplier of the bearing material provides recommendations for many of these design decisions.
11. Nominal diametral clearance: Many factors affect the final specification for clearance, such as need for precision, thermal expansion of all parts of the bearing system, load variations, shaft deflection expected, means of providing lubricant, and manufacturing capability. One long-used rule of thumb is to provide 0.001 in of clearance per inch of journal diameter. Figure 16–3 shows recommended minimum values for clearance based on journal diameter and rotational speed under steady loads. These values apply to the smallest clearance under any combination of tolerances on the dimensions of the bearing system to avoid problems of bearing heating and eventual seizure. The operating clearance will then be higher than these values because of manufacturing tolerances. Performance over the entire range of clearances should be evaluated, preferably by testing.

FIGURE 16–3

Minimum recommended diametral clearance for bearings considering journal diameter and rotational speed (Taken from R. J. Welsh. *Plain Bearing Design Handbook.* London: Butterworths, 1983)

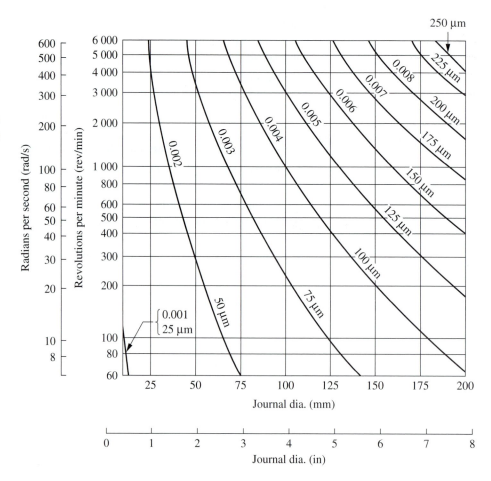

Example Problem 16–1   A bearing is to be designed to carry a radial load of 150 lb from a shaft having a minimum acceptable diameter of 1.50 in and rotating at 500 rpm. Design the bearing to operate under boundary-lubrication conditions.

Solution   We will use the design procedure previously outlined.

>   **Step 1.** Trial diameter: Try $D = D_{min} = 1.50$ in.
>
>   **Steps 2–4.** Try $L/D = 1.0$. Then $L = D = 1.50$ in.
>
>   **Step 5.** Bearing pressure:
>
> $$p = F/LD = (150 \text{ lb})/(1.50 \text{ in})(1.50 \text{ in}) = 66.7 \text{ psi}$$
>
>   **Step 6.** Journal speed:
>
> $$V = \pi Dn/12 = \pi(1.50)(500)/12 = 196 \text{ ft/min}$$
>
>   **Step 7.** pV factor:
>
> $$pV = (66.7 \text{ psi})(196 \text{ fpm}) = 13\ 100 \text{ psi-fpm}$$
>
>   **Step 8.** Design value of $pV = 2(13\ 100) = 26\ 200$ psi-fpm.
>
>   **Step 9.** From Table 16–1, we could use a bearing made from PPS filled with PTFE
and glass having a rated value of pV of 30 000 psi-fpm.

Steps 10–11. Nominal diametral clearance: From Figure 16–3, we can recommend a minimum $C_d = 0.002$ in based on $D = 1.50$ in and $n = 500$ rpm. Other design details are dependent on the details of the system into which the bearing will be placed.

Example Problem 16–2   Design a boundary-lubricated plain surface bearing to carry a radial load of 2.50 kN from a shaft rotating at 1 150 rpm. The nominal minimum diameter of the journal is 65 mm.

Solution   We will use the design procedure previously outlined.

> **Step 1.** Trial diameter: Try $D = 75$ mm.
>
> **Steps 2–4.** Try $L/D = 1.0$. Then $L = D = 75$ mm.
>
> **Step 5.** Bearing pressure:
> $$p = F/LD = (2\ 500\ \text{N})/(75\ \text{mm})(75\ \text{mm}) = 0.444\ \text{N/mm}^2$$
> Converting to kPa, we have
> $$p = 0.444\ \text{N/mm}^2\ (10^3\ \text{kPa})/(\text{N/mm}^2) = 444\ \text{kPa}$$
>
> **Step 6.** Journal speed:
> $$V = \pi Dn/(60\ 000) = \pi(75)(1\ 150)/(60\ 000) = 4.52\ \text{m/s}$$
>
> **Step 7.** pV factor:
> $$pV = (444\ \text{kPa})(4.52\ \text{m/s}) = 2\ 008\ \text{kPa·m/s}$$
>
> **Step 8.** Design value for $pV = 2(2\ 008) = 4\ 016$ kPa·m/s.

This value exceeds the capability of any of the materials listed in Table 16–1. Let's return to Step 1.

> **Step 1.** Try $D = 150$ mm.
>
> **Step 2.** Let $L/D = 1.25$.
>
> **Step 3.** Then
> $$L = D(L/D) = (150\ \text{mm})(1.25) = 187.5\ \text{mm}$$
>
> **Step 4.** Let's use the more convenient value of 175 mm for L.
>
> **Step 5.** Bearing pressure:
> $$p = F/LD = (2\ 500\ \text{N})/(175\ \text{mm})(150\ \text{mm}) = 0.095\ \text{N/mm}^2 = 95\ \text{kPa}$$
>
> **Step 6.** Journal speed:
> $$V = \pi Dn/(60\ 000) = \pi(150)(1\ 150)/(60\ 000) = 9.03\ \text{m/s}$$
>
> **Step 7.** pV factor:
> $$pV = (95\ \text{kPa})(9.03\ \text{m/s}) = 860\ \text{kPa·m/s} = 860\ \text{kW/m}^2$$
>
> **Step 8.** Design value of $pV = 2(860) = 1\ 720$ kW/m^2.
>
> **Step 9.** From Table 16–1, we can select the bronze bearing material having a limiting pV value of 1 750 kW/m^2.
>
> **Steps 10–11.** Nominal diametral clearance: From Figure 16–3, we can recommend a minimum $C_d = 150$ μm based on $D = 150$ mm and $n = 1\ 150$ rpm. Other design details are dependent on the details of the system into which the bearing will be placed.

FIGURE 16–4

Position of the journal relative to the bearing as a function of mode of operation

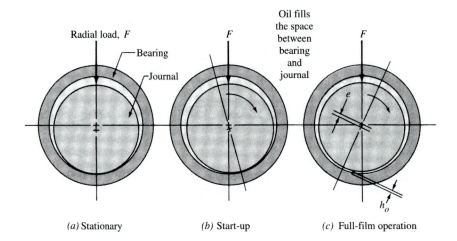

(a) Stationary   *(b)* Start-up   *(c)* Full-film operation

16–6
FULL-FILM
HYDRODYNAMIC
BEARINGS

In the *full-film hydrodynamic bearing,* the load on the bearing is supported on a continuous film of lubricant, usually oil, so that no contact between the bearing and the rotating journal occurs. A pressure must be developed in the oil in order to carry the load. With proper design, the motion of the journal inside the bearing creates the necessary pressure.

Figure 16–4 shows the progressive action in a plain surface bearing from start-up to the steady-state hydrodynamic operation. Observe that boundary lubrication and mixed-film lubrication precede the establishment of full-film hydrodynamic lubrication. At start-up, the radial load applied through the journal to the bearing forces the journal off center in the direction of the load, taking up all clearance [Figure 16–4(a)]. At the initial slow rotational speeds, the friction between the journal and the bearing causes the journal to climb up the bearing wall, somewhat as shown in Figure 16–4(b). Because of the viscous shear stresses developed in the oil, the moving journal draws oil into the converging, wedge-shaped area above the region of contact. The resulting pumping action produces a pressure in the oil film; when the pressure is sufficiently high, the journal is lifted from the bearing. The frictional forces are greatly reduced under this operating condition, and the journal moves eventually to the steady-state position shown in Figure 16–4(c). Note that the journal is offset from the direction of the load; that there is a certain eccentricity, *e,* between the geometric center of the bearing and the center of the journal; and that there is a point of minimum film thickness, h_o, at the nose of the wedge-shaped pressurized zone.

Figure 16–5 illustrates the general form of the pressure distribution within a full-film hydrodynamically lubricated bearing. The clearance between the bearing and the journal is highly exaggerated. Part (a) of the figure shows the rise of pressure as the rotating shaft draws oil into the converging wedge, approaching the point of minimum film thickness. The maximum pressure occurs there and then falls rapidly to nearly zero as the space between the journal and the bearing diverges again. The integrated effect of the pressure distribution is a force sufficient to support the shaft on a film of oil without metal-to-metal contact.

Figure 16–5(b) shows the pressure distribution axially along the shaft through the line of minimum film thickness or maximum pressure. The highest pressure value occurs in the middle of the bearing length, and it falls rapidly as the ends are approached because the pressure outside the bearing is ambient pressure, typically atmospheric pressure. There is continual leakage flow from both ends of the bearing. This illustrates the importance of providing a means of continually supplying oil to the bearing to maintain the full-film operation. Without a steady, adequate supply of oil, the system would not be able to create the

FIGURE 16–5
Pressure distribution in the oil film for hydrodynamic lubrication

(a) Section through middle of bearing

(b) Diametral section through line of minimum film thickness

pressurized film to carry the shaft, and boundary lubrication would result. The significantly higher frictional forces thus created would cause rapid heating of the bearing/journal interface, and seizure would likely occur very rapidly.

**16–7
DESIGN OF
FULL-FILM
HYDRO-
DYNAMICALLY
LUBRICATED
BEARINGS**

The following discussion presents several guidelines for bearing design for typical industrial applications.

Surface Roughness

A ground journal with a surface roughness of 16 to 32 microinches (μin) root mean square (RMS) is recommended for bearings of good quality. The bearing should be equally smooth or made from one of the softer materials so that "wearing in" can smooth the high spots, creating a good fit between the journal and the bearing. In high-precision equipment, polishing or lapping can be used to produce a surface finish of the order of 8 to 16 μin RMS.

Minimum Film Thickness

The limiting acceptable value of the minimum film thickness depends on the surface roughness of the journal and the bearing because the film must be thick enough to eliminate any solid contact during expected operating conditions. The suggested design value also depends on the size of the journal. For ground journals, the following relationship can be used to estimate the design value:

$$h_o = 0.000\,25D \qquad\qquad (16–2)$$

where D = diameter of the bearing

Diametral Clearance

The clearance between the journal and the bearing depends on the nominal diameter of the bearing, the precision of the machine for which the bearing is being designed, the speed of rotation, and the surface roughness of the journal. The coefficient of thermal expansion for both the journal and the bearing must also be considered to ensure a satisfactory clearance under all expected operating conditions. An overall guideline of making the clearance in the range of 0.001 to 0.002 times the bearing diameter can be used. Figure 16–3 shows a graph of the recommended minimum diametral clearance as a function of the bearing diameter and the rotational speed. Some variation above the values from the curve is permissible.

Ratio of Bearing Length to Diameter

Because the journal is a part of the shaft itself, its diameter is usually limited by stress and deflection considerations of the type discussed in Chapter 12. Then the length of the bearing is specified to provide for a suitable level of bearing pressure. General-purpose industrial machinery bearings typically operate at approximately 200 to 500 psi (1.4 to 3.4 MPa) bearing pressure, based on the projected area of the bearing [$p = $ load/(LD)]. The pressure may range from as low as 50 psi (0.34 MPa) for light-duty equipment to 2 000 psi (13.4 MPa) for heavy machinery under varying loads, such as in internal combustion engines. The leakage of oil from the bearing is also affected by the bearing length. The typical range of length to diameter ratio (L/D) for full-film hydrodynamic bearings is from 0.35 to 1.5. But many successful bearings operate outside this range. Reference 16 recommends $L/D = 0.60$ for most industrial applications.

Lubricant Temperature

The viscosity of the oil is a critical parameter in the performance of a bearing. Figure 16–6 shows the great variation of viscosity with temperature, which indicates that temperature control is advisable. Also, most petroleum lubricating oils should be limited to approximately 160°F (70°C) in order to retard oxidation. The temperature of interest is, of course, that inside the bearing. Frictional energy or thermal energy from the equipment itself can increase the oil temperature above that in the supply reservoir. In design examples we will select the lubricant that will ensure a satisfactory viscosity at 160°F, unless otherwise noted. If the actual operating temperature is lower, the resulting film thickness will be greater than the design value—a conservative result. It is incumbent on the designer to ensure that the limiting temperature is not exceeded, using forced cooling if necessary.

Lubricant Viscosity

The specification of the lubricant for the bearing is one of the final decisions to be made in the design procedure that follows. In the calculations, it is the dynamic viscosity, μ, which is used. In the U.S. Customary Unit (lb-in-s) System, the dynamic viscosity is expressed in lb·s/in^2, which is given the name *reyn* in honor of Osbourne Reynolds, who performed much significant work in fluid flow. In SI units, the standard unit is N·s/m^2 or Pa·s. Past work frequently used the now-obsolete unit of poise or centipoise, derived from the dyne-cm-s metric system. Some useful conversions are

$$1.0 \text{ reyn} = 6\,895 \text{ Pa·s}$$
$$1.0 \text{ Pa·s} = 1\,000 \text{ centipoise}$$

Other viscosity conversions may also be useful. (See Reference 11.)

Sommerfeld Number

The combined effect of many of the variables involved in the operation of a bearing under hydrodynamic lubrication can be characterized by the dimensionless number, *S,* called the

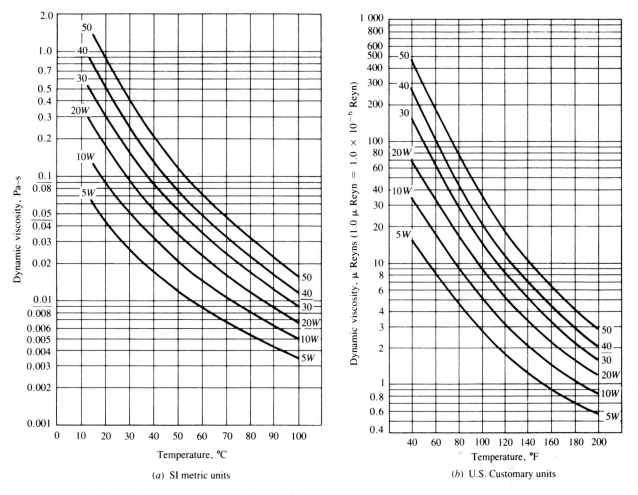

FIGURE 16–6 Viscosity vs. temperature for SAE oils

Sommerfeld number. In fact, some refer to this as the *bearing characteristic number,* defined as follows:

⇨ **Bearing Characteristic Number**

$$S = \frac{\mu n_s (R/C_r)^2}{p} \qquad (16\text{–}3)$$

Note that S is similar to the bearing parameter, $\mu n/p$, discussed in Section 16–3, in that it involves the combined effect of viscosity, rotational speed, and bearing pressure. In order for S to be dimensionless, the following units should be used for the factors:

	U.S. Customary Units	**SI Units**
μ	lb·s/in^2 (reyns)	Pa·s (N·s/m^2)
n_s	rev/s	rev/s
p	lb/in^2 (psi)	Pa (N/m^2)
R, C_r	in	m or mm

Any consistent units can be used. Figure 16–7, adapted from Reference 1, shows the relationship between the Sommerfeld number and the film thickness ratio, h_o/C_r. Figure 16–8

5. Cast Bronze Institute. *Cast Bronze Hydrostatic Bearing Design Manual.* Chicago: Cast Bronze Institute, 1975.

6. Clauss, F. J. *Solid Lubricants and Self-Lubricating Solids.* New York: Academic Press, 1972.

7. Faires, V. M. *Design of Machine Elements.* New York: Macmillan, 1965.

8. Fuller, D. D. *Theory and Practice of Lubrication for Engineers.* New York: John Wiley & Sons, 1956.

9. Juvinall, R. C., and Kurt M. Marshek. *Fundamentals of Machine Component Design.* 2d ed. New York: John Wiley & Sons, 1991.

10. *Machine Design Magazine Mechanical Drives Reference Issue.* Cleveland: Penton/IPC, 1987.

11. Mott, R. L. *Applied Fluid Mechanics.* 4th ed. Upper Saddle River, New Jersey: Prentice Hall, 1994.

12. Someya, Tsunen, ed. *Journal Bearing Handbook.* New York: Springer-Verlag, 1989.

13. Stolarski, T. A. *Tribology in Machine Design.* Oxford, England: Industrial Press, 1990.

14. Welsh, R. J. *Plain Bearing Design Handbook.* London: Butterworths, 1983.

15. Wills, J. George. *Lubrication Fundamentals.* New York: Marcel Dekker, 1980.

16. Wolverton, M. P., et al. "How Plastic Composites Wear at High Temperatures." *Machine Design Magazine* (February 10, 1983).

PROBLEMS

For Problems 1–8 and the data listed in Table 16–2, design a plain surface bearing, using the boundary-lubricated approach of Section 16–5. Use an L/D ratio for the bearing in the range of 0.50 to 1.50. Compute the pV factor, and specify a material from Table 16–1.

TABLE 16–2

Problem number	Radial load (lb)	Shaft diameter (in)	Shaft speed (rpm)
1.	225	3.00	1 750
2.	100	1.50	1 150
3.	200	1.25	850
4.	75	0.50	600
5.	850	4.50	625
6.	500	3.75	450
7.	800	3.00	350
8.	60	0.75	750

TABLE 16–3

Problem number	Radial load	Min. shaft diameter	Shaft speed (rpm)	Application
9.	1 250 lb	2.60 in	1 750	Electric motor
10.	2 250 lb	3.50 in	850	Conveyor
11.	1 875 lb	2.25 in	1 150	Air compressor
12.	1 250 lb	1.75 in	600	Precision spindle
13.	500 lb	1.15 in	2 500	Precision spindle
14.	850 lb	1.45 in	1 200	Idler sheave
15.	4 200 lb	4.30 in	450	Shaft for chain drive
16.	18.7 kN	100 mm	500	Conveyor
17.	2.25 kN	25 mm	2 200	Machine tool
18.	5.75 kN	65 mm	1 750	Printer

For Problems 9–18 and the data listed in Table 16–3, design a hydrodynamically lubricated bearing, using the method outlined in Section 16–7. Specify the journal nominal diameter, the bearing length, the diametral clearance, the minimum film thickness of the lubricant during operation, the surface finish of the journal and bearing, the lubricant, and its maximum operating temperature. For your design, compute the coefficient of friction, the friction torque, and the power dissipated as the result of friction.

For Problems 19–28 and the data listed in Table 16–4, design a hydrostatic bearing of circular shape. Specify the pad diameter, the recess diameter, the recess pressure, the film thickness, the lubricant and its temperature, the oil flow rate, and the pumping power. The load specified is for a single bearing. You may choose to use multiple bearings. (The supply pressure is the maximum available at the pump.)

TABLE 16–4

Problem number	Load	Supply pressure
19.	1 250 lb	300 psi
20.	5 000 lb	300 psi
21.	3 500 lb	500 psi
22.	750 lb	500 psi
23.	250 lb	150 psi
24.	500 lb	150 psi
25.	22.5 kN	2.0 MPa
26.	1.20 kN	750 kPa
27.	8.25 kN	1.5 MPa
28.	12.5 kN	1.5 MPa

17

Power Screws and Ball Screws

Power Screws and Ball Screws

Discussion Map

☐ *Power screws* and *ball screws* are designed to convert rotary motion to linear motion and to exert the necessary force to move a machine element along a desired path. They use the principle of a screw thread and its mating nut.

Discover

Visit a machine shop and see if you can identify power screws or ball screws. Look in particular at the lathes and the milling machines. Manual machines likely use power screws. Those using computer numerical control should have ball screws. Describe the shape of the threads. How are the screws powered? How are they fastened to other machine parts?

Can you find other equipment that uses power screws or ball screws? Look in laboratories where materials are tested or where large forces must be generated.

This chapter will help you learn to analyze power screws and ball screw drives and to specify suitable sizes for a given application.

Power screws and ball screws are designed to convert rotary motion to linear motion and to exert the necessary force to move a machine element along a desired path. Power screws operate on the classic principle of the screw thread and its mating nut. If the screw is supported in bearings and rotated while the nut is restrained from rotating, the nut will translate along the screw. If the nut is made an integral part of a machine, for example, the tool holder for a lathe, the screw will drive the tool holder along the bed of the machine to take a cut. Conversely, if the nut is supported while it is rotating, the screw can be made to translate. The screw jack uses this approach.

A ball screw is similar in function to a power screw, but the configuration is different. The nut contains many small, spherical balls that make rolling contact with the threads of the screw, giving very low friction and very high efficiencies when compared with power screws. Modern machine tools, automation equipment, vehicle steering systems, and actuators on aircraft use ball screws for high precision, fast response, and smooth operation.

Visit a machine shop where there are metal-cutting machine tools. Look for examples of power screws that convert rotary motion to linear motion. They are likely to be on manual lathes moving the tool holder. Or look at the table drive for a milling machine. Inspect the form of the threads of the power screw. Are they of a form similar to that of a screw thread with sloped sides? Or are the sides of the threads straight? Compare the screw threads with those shown in Figure 17–1 for square, Acme, and buttress forms.

While in the shop, do you see any type of material-testing equipment or a device called an *arbor press* that exerts large axial forces? Such machines often employ square-thread power screws to produce the axial force and motion from rotational input, through either a hand crank or an electric motor drive. If they are not in the machine shop, look for them in the metallurgy lab or another room where materials testing is done.

Now look further in the machine shop. Are there machines that use digital readouts to indicate position of the table or the tool? Are there computer numerical control machine tools? Any of these types of machines should have ball screws rather than the traditional power screws because ball screws require significantly less power and torque to drive them against a given load. They can also be moved faster and positioned more accurately than power screws. You may or may not be able to see the recirculating balls in the nut of the power screw, as illustrated in Figure 17–2. But you should be able to see the different-shaped threads looking like grooves with circular bottoms in which the spherical balls roll.

Have you seen such power screws or ball screws anywhere outside a machine shop? Some garage door openers employ a screw drive, but others use chain drives. Perhaps your

FIGURE 17–1 Forms of power screw threads [(b), ANSI Standard B1.5-1973; (c), ANSI Standard B1.9-1973]

Square   Acme

2ϕ
$29°$
ϕ
$p = $ pitch
$7°$
$0.5p$
$0.5p$
$0.163p$
$.5\,p$
$45°$
p
p
$0.663p$

(a) Square thread   (b) Acme thread
(Ref.: ANSI B1.5-1973)   (c) Buttress thread
(Ref.: ANSI B1.9-1973)

FIGURE 17–2 Ball bearing screw (Thomson Industries, Inc., Port Washington, NY)

Return tube
Ball screw
Bearing balls
Ball nut

home has a screw jack or a scissors jack for raising the car to change a tire. Both use power screws. Have you ever sat in a seat on an airplane where you can see the mechanisms that actuate the flaps on the rear edge of the wings? Try it sometime, and observe the actuators during takeoff or landing. It is likely that you will see a ball screw in action.

This chapter will help you learn the methods of analyzing the performance of power screws and ball screws and to specify the proper size for a given application.

You Are the Designer

You are a member of a plant engineering team for a large steel processing plant. One of the furnaces in the plant in which steel is heated prior to final heat treatment is installed beneath the floor, and the large ingots are lowered vertically into it. While the ingots are soaking in the furnace, a large, heavy hatch is placed over the opening to minimize the escape of heat and provide more uniform temperature. The hatch weighs 25 000 lb.

You are asked to design a system that will permit the hatch to be raised at least 15 in above the floor within 12.0 s and to lower it again within 12.0 s.

What design concept would you propose? Of course, there are many feasible concepts, but suppose that you proposed a system like that sketched in Figure 17–3. An overhead support structure is suggested on which a worm/wormgear drive system would be mounted. One shaft would be driven directly by the gear drive while a second shaft would be driven simultaneously by a chain drive. The shafts are power screws, supported in bearings

FIGURE 17–3 An Acme screw–driven system for raising a hatch

at the top and bottom. A yoke is connected to the hatch and is mounted on the screws with the nuts that mate with the screw integral with the yoke. Therefore, as the screw rotates, the nuts carry the yoke and the hatch vertically upward or downward.

As the designer of the hatch lift system, you must make several decisions. What size screw is required to ensure that it can safely raise the 25 000-lb hatch? Notice that the screws are placed in tension as they are supported on the collars on the upper support system. What diameter, thread type, and thread size should be used?

The sketch suggests the Acme thread form. What other styles are available? At what speed must the screws be rotated to raise the hatch in three minutes or less? How much power is required to drive the screws? What safety concerns exist while the system is handling this heavy load? What advantage would there be to using a ball screw rather than a power screw?

The material in this chapter will help you make these decisions, along with providing methods of computing stresses, torques, and efficiencies.

17–1
OBJECTIVES OF THIS CHAPTER

After completing this chapter, you will be able to:

1. Describe the operation of a power screw and the general form of *square threads, Acme threads,* and *buttress threads* as they are applied to power screws.

2. Compute the torque that must be applied to a power screw to raise or lower a load.

3. Compute the efficiency of power screws.

4. Compute the power required to drive a power screw.

5. Describe the design of a ball screw and its mating nut.

6. Specify suitable ball screws for a given set of requirements of load, speed, and life.

7. Compute the torque required to drive a ball screw, and compute its efficiency.

17–2
POWER SCREWS

Figure 17–1 shows three types of power screw threads: the square thread, the Acme thread, and the buttress thread. Of these, the square and buttress threads are the most efficient. That is, they require the least torque to move a given load along the screw. However, the Acme

TABLE 17–1 Preferred Acme screw threads

Nominal major diameter (in)	Threads per in, n	Pitch, $p = 1/n$ (in)	Minimum minor diameter (in)	Minimum pitch diameter (in)	Tensile stress area (in^2)	Shear stress area (in^2)[a]
1/4	16	0.062 5	0.161 8	0.204 3	0.026 32	0.335 5
5/16	14	0.071 4	0.214 0	0.261 4	0.044 38	0.434 4
3/8	12	0.083 3	0.263 2	0.316 1	0.065 89	0.527 6
7/16	12	0.083 3	0.325 3	0.378 3	0.097 20	0.639 6
1/2	10	0.100 0	0.359 4	0.430 6	0.122 5	0.727 8
5/8	8	0.125 0	0.457 0	0.540 8	0.195 5	0.918 0
3/4	6	0.166 7	0.537 1	0.642 4	0.273 2	1.084
7/8	6	0.166 7	0.661 5	0.766 3	0.400 3	1.313
1	5	0.200 0	0.750 9	0.872 6	0.517 5	1.493
$1\frac{1}{8}$	5	0.200 0	0.875 3	0.996 7	0.688 1	1.722
$1\frac{1}{4}$	5	0.200 0	0.999 8	1.121 0	0.883 1	1.952
$1\frac{3}{8}$	4	0.250 0	1.071 9	1.218 8	1.030	2.110
$1\frac{1}{2}$	4	0.250 0	1.196 5	1.342 9	1.266	2.341
$1\frac{3}{4}$	4	0.250 0	1.445 6	1.591 6	1.811	2.803
2	4	0.250 0	1.694 8	1.840 2	2.454	3.262
$2\frac{1}{4}$	3	0.333 3	1.857 2	2.045 0	2.982	3.610
$2\frac{1}{2}$	3	0.333 3	2.106 5	2.293 9	3.802	4.075
$2\frac{3}{4}$	3	0.333 3	2.355 8	2.542 7	4.711	4.538
3	2	0.500 0	2.432 6	2.704 4	5.181	4.757
$3\frac{1}{2}$	2	0.500 0	2.931 4	3.202 6	7.388	5.700
4	2	0.500 0	3.430 2	3.700 8	9.985	6.640
$4\frac{1}{2}$	2	0.500 0	3.929 1	4.199 1	12.972	7.577
5	2	0.500 0	4.428 1	4.697 3	16.351	8.511

[a]Per inch of length of engagement.

thread is not greatly less efficient, and it is easier to machine. The buttress thread is desirable when force is to be transmitted in only one direction.

Table 17–1 gives the preferred combinations of basic major diameter, D, and number of threads per inch, n, for Acme screw threads. The pitch, p, is the distance from a point on one thread to the corresponding point on the adjacent thread, and $p = 1/n$.

Other pertinent dimensions listed in Table 17–1 include the minimum minor diameter and the minimum pitch diameter of a screw with an external thread. When you are performing stress analyses on the screw, the safest approach is to compute the area corresponding to the minor diameter for tensile or compressive stresses. However, a more accurate stress computation results from using the *tensile stress area* (listed in Table 17–1), found from

> **Tensile Stress Area for Screw Threads**

$$A_t = \frac{\pi}{4}\left[\frac{D_r + D_p}{2}\right]^2$$

(17–1)

This is the area corresponding to the average of the minor (or root) diameter, D_r, and the pitch diameter, D_p. The data reflect the minimums for commercially available screws according to recommended tolerances.

The shear stress area, A_s, listed in Table 17–1 is also found in published data and represents the area in shear approximately at the pitch line of the threads for a 1.0-in length of engagement. Other lengths would require that the area be modified by the ratio of the actual length to 1.0 in.

Torque Required to Move a Load

When using a power screw to exert a force, as with a jack raising a load, you need to know how much torque must be applied to the nut of the screw to move the load. The parameters involved are the force to be moved, F; the size of the screw, as indicated by its pitch diameter, D_p; the lead of the screw, L; and the coefficient of friction, f. Note that the *lead* is defined as the axial distance that the screw would move in one complete revolution. For the usual case of a single-threaded screw, the lead is equal to the pitch and can be read from Table 17–1 or computed from $L = p = 1/n$.

In the development of Equation (17–2) for the torque required to turn the screw, Figure 17–4(a), which depicts a load being pushed up an inclined plane against a friction force, is used. This is a reasonable representation for a square thread if you think of the thread as being unwrapped from the screw and laid flat. The torque for an Acme thread is slightly different from this due to the thread angle. The revised equation for the Acme thread will be shown later.

The torque computed from Equation (17–2) is called T_u, implying that the force is applied to move a load up the plane, that is, to raise the load. This observation is completely appropriate if the load is raised vertically, as with a jack. If, however, the load is horizontal or at some angle, Equation (17–2) is still valid if the load is to be advanced along the screw "up the thread." Equation (17–4) shows the required torque, T_d, to lower a load or move a load "down the thread."

The torque to move a load up the thread is

Torque Required to Move a Load up a Power Screw with a Square Thread ⇨

$$T_u = \frac{FD_p}{2}\left[\frac{L + \pi f D_p}{\pi D_p - fL}\right] \qquad (17\text{–}2)$$

This equation accounts for the force required to overcome friction between the screw and the nut in addition to the force required just to move the load. If the screw or the nut bears against a stationary surface while rotating, there will be an additional friction torque developed at that surface. For this reason, many jacks and similar devices incorporate antifriction bearings at such points.

The coefficient of friction for use in Equation (17–2) depends on the materials used and the manner of lubricating the screw. For well-lubricated steel screws acting in steel nuts, $f = 0.15$ should be conservative.

An important factor in the analysis for torque is the angle of inclination of the plane. In a screw thread, the angle of inclination is referred to as the *lead angle,* λ. It is the angle

FIGURE 17–4

Screw thread force analysis

P = Force required to move the load
F_f = Friction force
N = Normal force
λ = Lead angle
D_p = Pitch diameter

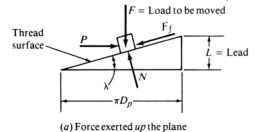

(a) Force exerted *up* the plane

(b) Force exerted *down* the plane

between the tangent to the helix of the thread and the plane transverse to the axis of the screw. It can be seen from Figure 17–4 that

$$\tan \lambda = L/(\pi D_p) \tag{17–3}$$

where πD_p = circumference of the pitch line of the screw

Then if the rotation of the screw tends to raise the load (move it up the incline), the friction force opposes the motion and acts down the plane.

Conversely, if the rotation of the screw tends to lower the load, the friction force will act up the plane, as shown in Figure 17–4(b). The torque analysis changes, producing Equation (17–4):

Torque Required to Lower a Load Down a Power Screw with a Square Thread

$$T_d = \frac{FD_p}{2} \left[\frac{\pi f D_p - L}{\pi D_p + fL} \right] \tag{17–4}$$

If the screw thread is very steep (that is, if it has a high lead angle), the friction force may not be able to overcome the tendency for the load to "slide" down the plane, and the load will fall due to gravity. In most cases for power screws with single threads, however, the lead angle is rather small, and the friction force is large enough to oppose the load and keep it from sliding down the plane. Such a screw is called *self-locking,* a desirable characteristic for jacks and similar devices. Quantitatively, the condition that must be met for self-locking is

$$f > \tan \lambda \tag{17–5}$$

The coefficient of friction must be greater than the tangent of the lead angle. For $f = 0.15$, the corresponding value of the lead angle is 8.5°. For $f = 0.1$, for very smooth, well-lubricated surfaces, the lead angle for self-locking is 5.7°. The lead angles for the screw designs listed in Table 17–1 range from 1.94° to 5.57°. Thus, it is expected that all would be self-locking. However, operation under vibration should be avoided, as this may still cause movement of the screw.

Efficiency of a Power Screw

Efficiency for the transmission of a force by a power screw can be expressed as the ratio of the torque required to move the load without friction to that with friction. Equation (17–2) gives the torque required with friction, T_u. Letting $f = 0$, the torque required without friction, T', is

$$T' = \frac{FD_p}{2} \frac{L}{\pi D_p} = \frac{FL}{2\pi} \tag{17–6}$$

Then the efficiency, *e,* is

Efficiency of a Power Screw

$$e = \frac{T'}{T_u} = \frac{FL}{2\pi T_u} \tag{17–7}$$

Alternate Forms of the Torque Equations

Equations (17–2) and (17–4) can be expressed in terms of the lead angle, rather than the lead and the pitch diameter, by noting the relationship in Equation (17–3). With this substitution, the torque required to move the load would be

⇨ **Torque to Raise a Load with a Square Thread**

$$T_u = \frac{FD_p}{2} \left[\frac{(\tan \lambda + f)}{(1 - f \tan \lambda)} \right] \qquad \text{(17–8)}$$

and the torque required to lower the load is

⇨ **Torque to Lower a Load with a Square Thread**

$$T_d = \frac{FD_p}{2} \left[\frac{(f - \tan \lambda)}{(1 + f \tan \lambda)} \right] \qquad \text{(17–9)}$$

Adjustment for Acme Threads

The difference between Acme threads and square threads is the presence of the thread angle, ϕ. Note from Figure 17–1 that $2\phi = 29°$, and therefore $\phi = 14.5°$. This changes the direction of action of the forces on the thread from that depicted in Figure 17–4. Figure 17–5 shows that F would have to be replaced by $F/\cos\phi$. Carrying this through, the analysis for torque would give the following modified forms of Equations (17–8) and (17–9). The torque to move the load up the thread is

⇨ **Torque to Raise a Load with an Acme Thread**

$$T_u = \frac{FD_p}{2} \left[\frac{(\cos \phi \tan \lambda + f)}{(\cos \phi - f \tan \lambda)} \right] \qquad \text{(17–10)}$$

and the torque to move the load down the thread is

⇨ **Torque to Lower a Load with an Acme Thread**

$$T_d = \frac{FD_p}{2} \left[\frac{(f - \cos \phi \tan \lambda)}{(\cos \phi + f \tan \lambda)} \right] \qquad \text{(17–11)}$$

Power Required to Drive a Power Screw

If the torque required to rotate the screw is applied at a constant rotational speed, n, then the power in horsepower to drive the screw is

$$P = \frac{Tn}{63\,000}$$

FIGURE 17–5   Force on an Acme thread

(a) Force normal to a square thread

(b) Force normal to an Acme thread

References 1 and 2 include more detail about the formulas that characterize the performance of power screws.

Example Problem 17–1   Two Acme-threaded power screws are to be used to raise a heavy access hatch, as sketched in Figure 17–3. The total weight of the hatch is 25 000 lb, divided equally between the two screws. Select a satisfactory screw from Table 17–1 on the basis of tensile strength, limiting the tensile stress to 10 000 psi. Then determine the required thickness of the yoke that acts as the nut on the screw to limit the shear stress in the threads to 5 000 psi. For the screw thus designed, compute the lead angle, the torque required to raise the load, the efficiency of the screw, and the torque required to lower the load. Use a coefficient of friction of 0.15.

Solution   The load to be lifted places each screw in direct tension. Therefore, the required tensile stress area is

$$A_t = \frac{F}{\sigma_a} = \frac{12\ 500\ \text{lb}}{10\ 000\ \text{lb/in}^2} = 1.25\ \text{in}^2$$

From Table 17–1, a $1\frac{1}{2}$-in-diameter Acme thread screw with four threads per inch would provide a tensile stress area of 1.266 in^2.

For this screw, each inch of length of a nut would provide 2.341 in^2 of shear stress area in the threads. The required shear area is then

$$A_s = \frac{F}{\tau_a} = \frac{12\ 500\ \text{lb}}{5\ 000\ \text{lb/in}^2} = 2.50\ \text{in}^2$$

Then the required length of the yoke would be

$$h = 2.5\ \text{in}^2 \left[\frac{1.0\ \text{in}}{2.341\ \text{in}^2} \right] = 1.07\ \text{in}$$

For convenience, let's specify $h = 1.25$ in.

The lead angle is (remember that $L = p = 1/n = 1/4 = 0.250$ in)

$$\lambda = \tan^{-1} \frac{L}{\pi D_p} = \tan^{-1} \frac{0.250}{\pi(1.342\ 9)} = 3.39°$$

The torque required to raise the load can be computed from Equation (17–10):

$$T_u = \frac{FD_p}{2} \left[\frac{(\cos \phi \tan \lambda + f)}{(\cos \phi - f \tan \lambda)} \right] \qquad \textbf{(17–10)}$$

Using $\cos \phi = \cos(14.5°) = 0.968$, and $\tan \lambda = \tan(3.39°) = 0.059\ 2$, we have

$$T_u = \frac{(12\ 500\ \text{lb})(1.342\ 9\ \text{in})}{2} \frac{[(0.968)(0.059\ 2) + 0.15]}{[0.968 - (0.15)(0.059\ 2)]} = 1\ 809\ \text{lb·in}$$

The efficiency can be computed from Equation (17–7):

$$e = \frac{FL}{2\pi T_u} = \frac{(12\ 500\ \text{lb})(0.250\ \text{in})}{2(\pi)(1\ 809\ \text{lb·in})} = 0.275 \quad \text{or} \quad 27.5\%$$

The torque required to lower the load can be computed from Equation (17–11):

$$T_d = \frac{FD_p}{2} \left[\frac{(f - \cos \phi \tan \lambda)}{(\cos \phi + f \tan \lambda)} \right] \qquad \textbf{(17–11)}$$

$$T_d = \frac{(12\ 500\ \text{lb})(1.342\ 9\ \text{in})}{2} \frac{[0.15 - (0.968)(0.059\ 2)]}{[0.968 + (0.15)(0.059\ 2)]} = 796\ \text{lb·in}$$

Example Problem 17–2   It is desired to raise the hatch in Figure 17–3 a total of 15.0 inches in no more than 12.0 s. Compute the required rotational speed for the screws and the power required.

Solution   The screw selected in the solution for Example Problem 17–1 was a $1\frac{1}{2}$-in Acme-threaded screw with four threads per inch. Thus, the load would be moved 1/4 in with each revolution. The linear speed required is

$$V = \frac{15.0 \text{ in}}{12.0 \text{ s}} = 1.25 \text{ in/s}$$

The required rotational speed would be

$$n = \frac{1.25 \text{ in}}{\text{s}} \frac{1 \text{ rev}}{0.25 \text{ in}} \frac{60 \text{ s}}{\text{min}} = 300 \text{ rpm}$$

Then the power required to drive each screw would be

$$P = \frac{Tn}{63\,000} = \frac{(1\,809 \text{ lb·in})(300 \text{ rpm})}{63\,000} = 8.61 \text{ hp}$$

17–3
BALL SCREWS

The basic action of using screws to produce linear motion from rotation was described in Section 17–2 on power screws. A special adaptation of this action that minimizes the friction between the screw threads and the mating nut is the *ball screw.*

Figure 17–2 shows a cutaway view of a commercially available ball screw. It replaces the sliding friction of the conventional power screw with the rolling friction of bearing balls. The bearing balls circulate in hardened steel races formed by concave helical grooves in the screw and the nut. All reactive loads between the screw and the nut are carried by the bearing balls which provide the only physical contact between these members. As the screw and the nut rotate relative to each other, the bearing balls are diverted from one end and are carried by the ball-guide return tubes to the opposite end of the ball nut. This recirculation permits unrestricted travel of the nut in relation to the screw. (See Reference 3.)

Applications of ball screws occur in automotive steering systems, machine tool tables and linear actuators, jacking and positioning mechanisms, aircraft controls such as flap actuating devices, packaging equipment, and instruments. Figure 17–6 shows a machine with a ball screw installed to move a component along the ways of the bed.

The application parameters to be considered in selecting a ball screw include the following:

The axial load to be exerted by the screw during rotation

The rotational speed of the screw

The maximum static load on the screw

The direction of the load

The manner of supporting the ends of the screw

The length of the screw

The expected life

The environmental conditions

When transmitting a load, a ball screw experiences stresses similar to those on a ball bearing, as discussed in Chapter 14. The load is transferred from the screw to the balls, from the balls to the nut, and from the nut to the driven device. The contact stress between the

FIGURE 17–6
Application of a ball screw (Thomson Industries, Inc., Port Washington, NY)

balls and the races in which they roll eventually causes fatigue failure, indicated by pitting of the balls or the races.

Thus, the rating of ball screws gives the load capacity of the screw for a given life which 90% of the screws of a given design will survive. This is similar to the L_{10} life of ball bearings. Because ball screws are typically used as linear actuators, the most pertinent life parameter is the distance traveled by the nut relative to the screw.

Manufacturers usually report the rated load that a given screw can exert for 1 million inches (25.4 km) of cumulative travel. The relationship between load, P, and life, L, is also similar to that for a ball bearing:

Relationship between Bearing Load and Life

$$\frac{L_2}{L_1} = \left(\frac{P_1}{P_2}\right)^3$$

(17–12)

Thus, if the load on a ball screw is doubled, the life is reduced to one-eighth of the original life. If the load is cut in half, the life is increased by eight times. Figure 17–7 shows the nominal performance of ball screws of a small variety of sizes. Many more sizes and styles are available.

In addition to the load/life considerations, the proper application of ball screws must take into account the vibration tendencies and the tendency for the screw to buckle. Because of the low friction inherent in a ball screw, many are applied at a high rotational speed. Care must be exercised that the operational speed is not near a critical speed of the screw. The critical speed is dependent on the length (actually the length squared!), the screw diameter, and the manner of supporting the ends. Thus, one of the worst cases is a long, small-diameter

FIGURE 17–7 Ball screw performance

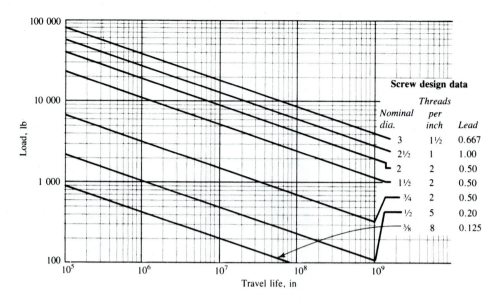

Nominal dia.	Threads per inch	Lead
3	1½	0.667
2½	1	1.00
2	2	0.50
1½	2	0.50
¾	2	0.50
½	5	0.20
⅜	8	0.125

screw, supported in one simple bearing at each end. Such a system would have a relatively low critical speed. One of the best designs, resulting in a high critical speed, is a short, large-diameter screw, mounted rigidly in a pair of widely spaced bearings on each end. Catalogs give specific design data for critical speeds. As long as the operating speed is below 0.80 times the critical speed, the operation should be satisfactory.

Column buckling should be checked whenever a ball screw is loaded in compression. The parameters are the same as those discussed in Chapter 6. The good and bad design approaches for column loading are similar to those described for critical speed. Of course, whenever possible, the load on the screw should be tensile. Then column buckling is not a factor.

Torque and Efficiency

The efficiency of a ball bearing screw is typically taken to be 90%. This far exceeds the efficiency for power screws without rolling contact that are typically in the range of 20% to 30%. Thus, far less torque is required to exert a given load with a given size of screw. Power is correspondingly reduced. The computation of the torque to turn is adapted from Equation (17–7), relating efficiency to torque:

⇨ **Efficiency of a Ball Screw**

$$e = \frac{FL}{2\pi T_u}$$

(17–7)

Then, using $e = 0.90$,

⇨ **Torque to Drive a Ball Screw**

$$T_u = \frac{FL}{2\pi e} = 0.177FL$$

(17–13)

Because of the low friction, ball screws are virtually never self-locking. In fact, designers also use this property to advantage by purposely using the applied load on the nut to rotate the screw. This is called *backdriving;* backdriving torque can be computed from

⇨ **Backdriving Torque for a Ball Screw**

$$T_b = \frac{FLe}{2\pi} = 0.143FL$$

(17–14)

Example Problem 17–3   Select suitable ball screws for the application described in Example Problem 17–1 and illustrated in Figure 17–3. The hatch must be lifted 15.0 in to open it eight times per day, and then it must be closed. The design life is 10 years. The lifting or lowering is to be completed in no more than 12.0 s.

For the screw selected, compute the torque to turn the screw, the power required, and the actual expected life.

Solution   The data required to select a screw from Figure 17–7 are the load and the travel of the nut on the screw over the desired life. The load is 12 500 lb on each screw:

$$\text{Travel} = \frac{15.0 \text{ in}}{\text{stroke}} \frac{2 \text{ strokes}}{\text{cycle}} \frac{8 \text{ cycles}}{\text{day}} \frac{365 \text{ days}}{\text{year}} \frac{10 \text{ years}}{} = 8.76 \times 10^5 \text{ in}$$

From Figure 17–7, the 2-in screw with two threads per inch and a lead of 0.50 in is satisfactory.

The torque required to turn the screw is

$$T_u = 0.177 FL = 0.177(12\,500)(0.50) = 1\,106 \text{ lb·in}$$

The rotational speed required is

$$n = \frac{1 \text{ rev}}{0.50 \text{ in}} \frac{15.0 \text{ in}}{12.0 \text{ s}} \frac{60 \text{ s}}{\text{min}} = 150 \text{ rpm}$$

The power required for each screw is

$$P = \frac{Tn}{63\,000} = \frac{(1\,106)(150)}{63\,000} = 2.63 \text{ hp}$$

Compare this with the 8.61 hp required for the Acme screw in Example Problem 17–1.

The actual travel life expected for this screw at a 12 500-lb load would be approximately 3.2×10^6 in, using Figure 17–7. This is 3.65 times longer than required.

REFERENCES

1. Faires, V. M. *Design of Machine Elements*. 5th ed. New York: Macmillan, 1965.

2. Shigley, J. E., and C. R. Mischke. *Mechanical Engineering Design*. 5th ed. New York: McGraw-Hill, 1988.

3. Thomson Industries, Inc. *Industrial Ball Screw Catalog*. Port Washington, NY: Thomson Industries, Inc.

4. Warner Electric Brake & Clutch Company. *Ball Bearing Screws Catalog*. Beloit, WI: Warner Electric Brake & Clutch Company.

PROBLEMS

1. Name three types of threads used for power screws.

2. Make a scale drawing of an Acme thread having a major diameter of $1\frac{1}{2}$ in and four threads per inch. Draw a section 2.0 in long.

3. Repeat Problem 2 for a buttress thread.

4. Repeat Problem 2 for a square thread.

5. If an Acme-thread power screw is loaded in tension with a force of 30 000 lb, what size screw from Table 17–1 should be used to maintain a tensile stress below 10 000 psi?

6. For the screw chosen in Problem 5, what would be the required axial length of the nut on the screw that transfers the load to the frame of the machine if the shear stress in the threads must be less than 6 000 psi?

7. Compute the torque required to raise the load of 30 000 lb with the Acme screw selected in Problem 5. Use a coefficient of friction of 0.15.

8. Compute the torque required to lower the load with the screw from Problem 5.

9. If a square-thread screw having a major diameter of 3/4 in and six threads per inch is used to lift a load of 4 000 lb, compute the torque required to rotate the screw. Use a coefficient of friction of 0.15.

10. For the screw of Problem 9, compute the torque required to rotate the screw when lowering the load.

11. Compute the lead angle for the screw of Problem 9. Is it self-locking?

12. Compute the efficiency of the screw of Problem 9.

13. If the load of 4 000 lb is lifted by the screw described in Problem 9 at the rate of 0.5 in/s, compute the required rotational speed of the screw and the power required to drive it.

14. A ball screw for a machine table drive is to be selected. The axial force to be transmitted by the screw is 600 lb. The table moves 24 in per cycle, and it is expected to cycle 10 times per hour for a design life of 10 years. Select an appropriate screw.

15. For the screw selected in Problem 14, compute the torque required to drive the screw.

16. For the screw selected in Problem 14, the normal travel speed of the table is 10.0 in/min. Compute the power required to drive the screw.

17. If the cycle time for the machine in Problem 14 were decreased to obtain 20 cycles/h instead of 10, what would be the expected life in years of the screw originally selected?

18

Fasteners

Fasteners

Discussion Map

☐ *Fasteners* connect or join two or more components. Common types are *bolts* and *screws* such as those illustrated in Figures 18–1 through 18–4.

Discover
Look for examples of bolts and screws. List how many types you have found. For what functions were they being used? What kinds of forces are the fasteners subjected to? What materials are used for the fasteners?

In this chapter you will learn to analyze the performance of fasteners and to select suitable types and sizes.

A *fastener* is any device used to connect or join two or more components. Literally hundreds of fastener types and variations are available. The most common are threaded fasteners referred to by many names, among them *bolts, screws, nuts, studs, lag screws,* and *set screws.*

A *bolt* is a threaded fastener designed to pass through holes in the mating members and to be secured by tightening a nut from the end opposite the head of the bolt. See Figure 18–1(a), called a *hex head bolt.* Several other types of bolts are shown in Figure 18–2.

A *screw* is a threaded fastener designed to be inserted through a hole in one member to be joined and into a threaded hole in the mating member. See Figure 18–1(b). The threaded hole may have been preformed, for example, by tapping, or it may be formed by the screw itself as it is forced into the material. *Machine screws,* also called *cap screws,* are precision fasteners with straight-threaded bodies that are turned into tapped holes (see Figure 18–3). *Sheet-metal screws, lag screws, self-tapping screws,* and *wood screws* usually form their own threads. Figure 18–4 shows a few styles.

Search for examples where the kinds of fasteners illustrated in Figures 18–1 through 18–4 are used. How many can you find? Make a list using the names for the fasteners in the

FIGURE 18–1
Comparison of a bolt with a screw (R. P. Hoelscher et al., *Graphics for Engineers,* New York: John Wiley & Sons, 1968)

(*a*) Hex bolt

(*b*) Hex head cap screw

FIGURE 18–2 Bolt styles. See also the hex bolt in Figure 18–1. (R. P. Hoelscher et al., *Graphics for Engineers,* New York: John Wiley & Sons, 1968)

(a) Carriage bolt   (b) Elevator bolt   (c) Countersunk bolt   (d) Plow bolt   (e) Track bolt   (f) Stud bolt   (g) Stove bolt   (h) Stove bolt

FIGURE 18–3 Cap screws or machine screws. See also the hex head cap screw in Figure 18–1. (R. P. Hoelscher et al., *Graphics for Engineers,* New York: John Wiley & Sons, 1968)

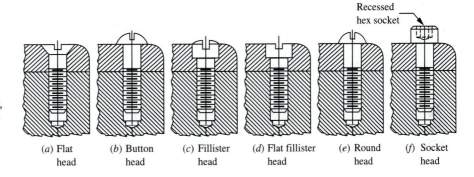

Recessed hex socket

(a) Flat head   (b) Button head   (c) Fillister head   (d) Flat fillister head   (e) Round head   (f) Socket head

FIGURE 18–4 Sheet-metal and lag screws (R. P. Hoelscher et al., *Graphics for Engineers,* New York: John Wiley & Sons, 1968)

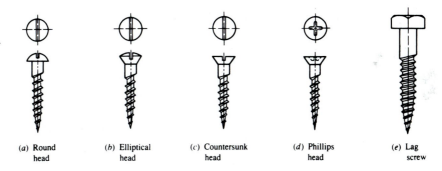

(a) Round head   (b) Elliptical head   (c) Countersunk head   (d) Phillips head   (e) Lag screw

figures. Describe the application. What function is the fastener performing? What kinds of forces are exerted on each fastener during service? How large is the fastener? Measure as many dimensions as you can. What material is each fastener made from?

Look in your car, particularly under the hood in the engine compartment. If you can, also look under the chassis to see where fasteners are used to hold different components onto the frame or some other structural member.

Look also at bicycles, lawn and garden equipment, grocery carts, display units in a store, hand tools, kitchen appliances, toys, exercise equipment, and furniture. If you have access to a factory, you should be able to identify hundreds or thousands of examples. Try to get some insight about where certain types of fasteners are used and for what purposes.

In this chapter, you will learn about many of the types of fasteners that you will encounter, including how to analyze their performance.

You Are the Designer

Review Figure 15–7 which shows the assembly of the gear-type power transmission that was designed in that chapter. Fasteners are called for in several places on the housing for the transmission, but they were not specified in that chapter. The four *bearing retainers* are to be fastened to the housing and the cover by threaded fasteners. The cover itself is to be attached to the housing by fasteners. Finally, the mounting base has provisions for using fasteners to hold the entire transmission to a support structure.

You are the designer. What kinds of fasteners would you consider for these applications? What material

should be used to make them? What strength should the material have? If threaded fasteners are used, what size should the threads be, and how long must they be? What head style would you specify? How much torque should be applied to the fastener to ensure that there is sufficient clamping force between the joined members? How does the design of the gasket between the cover and the housing affect the choice of the fasteners and the specification of the tightening torque for them? What alternatives are there to the use of threaded fasteners to hold the components together and to still allow disassembly?

This chapter presents information that you can use to make such design decisions. The references at the end of the chapter give other valuable sources of information from the large body of knowledge about fasteners.

18–1
OBJECTIVES OF
THIS CHAPTER

After completing this chapter, you will be able to:

1. Describe a bolt in comparison with a machine screw.

2. Name and describe nine styles of heads for bolts.

3. Name and describe six styles of heads for machine screws.

4. Describe sheet-metal screws and lag screws.

5. Describe six styles of set screws and their application.

6. Describe nine types of locking devices that restrain a nut from becoming loose on a bolt.

7. Use tables of data for various grades of steel materials used for bolts as published by the Society of Automotive Engineers (SAE) and the American Society for Testing and Materials (ASTM), and for standard metric grades.

8. List at least 10 materials other than steel that are used for fasteners.

9. Use tables of data for standard screw threads in the American Standard and metric systems for dimensions and stress analysis.

10. Define *proof load, clamping load,* and *tightening torque* as applied to bolts and screws, and compute design values.

11. Compute the effect of adding an externally applied force on a bolted joint, including the final force on the bolts and the clamped members.

12. List and describe 16 different coating and finishing techniques that are used for metal fasteners.

13. Describe rivets, quick-operating fasteners, welding, brazing, and adhesives, and contrast them with bolts and screws for fastening applications.

18–2
OTHER TYPES OF
FASTENERS AND
ACCESSORIES

Most bolts and screws have enlarged heads that bear down on the part to be clamped and thus exert the clamping force. *Set screws* are headless, are inserted into tapped holes, and are designed to bear directly on the mating part, locking it into place. Figure 18–5 shows several styles of points and drive means for set screws. Caution must be used with set screws, as with any threaded fastener, so that vibration does not loosen the screw.

FIGURE 18–5 Set screws with different head and point styles applied to hold a collar on a shaft (R. P. Hoelscher et al., *Graphics for Engineers,* New York: John Wiley & Sons, 1968)

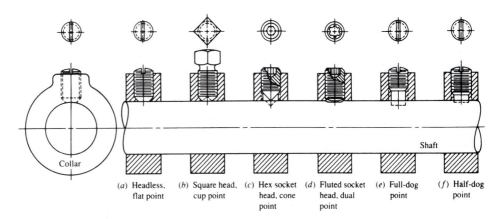

(a) Headless, flat point   (b) Square head, cup point   (c) Hex socket head, cone point   (d) Fluted socket head, dual point   (e) Full-dog point   (f) Half-dog point

A *washer* may be used under either or both the bolt head and the nut to distribute the clamping load over a wide area and to provide a bearing surface for the relative rotation of the nut. The basic type of washer is the plain flat washer, a flat disc with a hole in it through which the bolt or screw passes. Other styles, called *lockwashers,* have axial deformations or projections that produce axial forces on the fastener when compressed. These forces keep the threads of the mating parts in intimate contact and decrease the probability that the fastener will loosen in service.

Figure 18–6 shows several means of using washers and other types of locking devices. Part (a) is a jam nut tightened against the regular nut. Part (b) is the standard lockwasher. Part (c) is a locking tab that keeps the nut from turning. Part (d) is a cotter inserted through a hole drilled through the bolt. Part (e) uses a cotter, but it also passes through slots in the nut. Part (f) is one of several types of thread-deformation techniques used. Part (g), an *elastic stop nut,* uses a plastic insert to keep the threads of the nut in tight contact with the bolt. This may be used on machine screws as well. In part (h), the elastic stop nut is riveted to a thin plate, allowing a mating part to be bolted from the opposite side. The thin metal device in (i) bears against the top of the nut and grips the threads, preventing axial motion of the nut.

A *stud* is like a stationary bolt attached permanently to a part of one member to be joined. The mating member is then placed over the stud, and a nut is tightened to clamp the parts together.

Additional variations occur when these types of fasteners are combined with different head styles. Several of these are shown in the figures already discussed. Others are listed next:

Square	Hex	Heavy hex	Hex jam
Hex castle	Hex flat	Hex slotted	12-point
High crown	Low crown	Round	T-head
Pan	Truss	Hex washer	Flat countersunk
Plow	Cross recess	Fillister	Oval countersunk
Hex socket	Spline socket	Button	Binding

Additional combinations are created by consideration of the American National Standards or British Standard (metric); material grades; finishes; thread sizes; lengths; class (tolerance grade); manner of forming heads (machining, forging, cold heading); and the manner of forming threads (machining, die cutting, tapping, rolling, and plastic molding).

A high strength-to-weight ratio is the chief advantage of *titanium* alloys used for fasteners in aerospace applications. Appendix 11 gives a list of properties of several alloys.

Plastics are used widely because of their light weight, corrosion resistance, insulating ability, and ease of manufacture. Nylon 6/6 is the most frequently used material, but others include ABS, acetal, TFE fluorocarbons, polycarbonate, polyethylene, polypropylene, and polyvinylchloride. Appendix 13 lists several plastics and their properties. In addition to being used in screws and bolts, plastics are used extensively where the fastener is designed specially for the particular application.

Coatings and *finishes* are provided for metallic fasteners to improve appearance or corrosion resistance. Some also lower the coefficient of friction for more consistent results relating tightening torque to clamping force. Steel fasteners can be finished with black oxide, bluing, bright nickel, phosphate, and hot-dip zinc. Plating can be used to deposit cadmium, copper, chromium, nickel, silver, tin, and zinc. Various paints, lacquers, and chromate finishes are also used. Aluminum is usually anodized.

Socket Head Cap Screws

A very popular type of machine screw is the socket head cap screw. The usual configuration, shown in Figure 18–3(f), has a cylindrical head with a recessed hex socket. Also readily available are flat head styles for countersinking to produce a flush surface, button head styles for a low profile appearance, and shoulder screws providing a precision bearing surface for location or pivoting.

Socket head cap screws of the *1960 Series* are made from a heat-treated alloy steel having the following strengths:

Size Range	Tensile Strength (Ksi)	Yield Strength (Ksi)
0–5/8	190	170
3/4–3	180	155

Roughly equivalent performance is obtained from metric socket head cap screws made to the metric strength grade 12.9. The same geometry is available in corrosion-resistant stainless steel, typically 18-8, at somewhat lower strength levels. Consult the manufacturers.

18–4 THREAD DESIGNATIONS

Table 18–4 shows pertinent dimensions for threads in the American Standard styles, and Table 18–5 gives SI metric styles. For consideration of strength and size, the designer must know the basic major diameter, the pitch of the threads, and the area available to resist tensile loads. Note that the pitch is equal to $1/n$, where n is the number of threads per inch in the American Standard system. In the SI, the pitch in millimeters is designated directly. The tensile stress area listed in Tables 18–4 and 18–5 takes into account the actual area cut by a transverse plane. Because of the helical path of the thread on the screw, such a plane will cut near the root on one side of the screw but will cut near the major diameter on the other. The equation for the tensile stress area for American Standard threads is

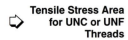
Tensile Stress Area for UNC or UNF Threads

$$A_t = (0.785\ 4)(D - 0.974\ 3p)^2 \qquad (18\text{–}1)$$

where D = major diameter
p = pitch of the thread

TABLE 18–4 American Standard thread dimensions

A. Numbered sizes

Size	Basic major diameter (in)	Coarse threads: UNC		Fine threads: UNF	
		Threads per in	Tensile stress area (in^2)	Threads per in	Tensile stress area (in^2)
0	0.060 0			80	0.001 80
1	0.073 0	64	0.002 63	72	0.002 78
2	0.086 0	56	0.003 70	64	0.003 94
3	0.099 0	48	0.004 87	56	0.005 23
4	0.112 0	40	0.006 04	48	0.006 61
5	0.125 0	40	0.007 96	44	0.008 30
6	0.138 0	32	0.009 09	40	0.010 15
8	0.164 0	32	0.014 0	36	0.014 74
10	0.190 0	24	0.017 5	32	0.020 0
12	0.216 0	24	0.024 2	28	0.025 8

B. Fractional sizes

Size	Basic major diameter (in)	Threads per in	Tensile stress area (in^2)	Threads per in	Tensile stress area (in^2)
1/4	0.250 0	20	0.031 8	28	0.036 4
5/16	0.312 5	18	0.052 4	24	0.058 0
3/8	0.375 0	16	0.077 5	24	0.087 8
7/16	0.437 5	14	0.106 3	20	0.118 7
1/2	0.500 0	13	0.141 9	20	0.159 9
9/16	0.562 5	12	0.182	18	0.203
5/8	0.625 0	11	0.226	18	0.256
3/4	0.750 0	10	0.334	16	0.373
7/8	0.875 0	9	0.462	14	0.509
1	1.000	8	0.606	12	0.663
$1\frac{1}{8}$	1.125	7	0.763	12	0.856
$1\frac{1}{4}$	1.250	7	0.969	12	1.073
$1\frac{3}{8}$	1.375	6	1.155	12	1.315
$1\frac{1}{2}$	1.500	6	1.405	12	1.581
$1\frac{3}{4}$	1.750	5	1.90		
2	2.000	$4\frac{1}{2}$	2.50		

For metric threads, the tensile stress area is

Tensile Stress Area for Metric Threads

$$A_t = (0.785\ 4)(D - 0.938\ 2p)^2 \tag{18–2}$$

For most standard screw thread sizes, at least two pitches are available: the *coarse* series and the *fine thread* series. Both are included in Tables 18–4 and 18–5.

The smaller American Standard threads use a number designation from 0 to 12. The corresponding major diameter is listed in Table 18–4(A). The larger sizes use fractional-inch designations. The decimal equivalent for the major diameter is shown in Table 18–4(B). Metric threads list the major diameter and the pitch in millimeters, as shown in Table 18–5. Samples of the standard designations for a thread are given next.

TABLE 18–5   Metric thread dimensions

Basic major diameter (mm)	Coarse threads		Fine threads	
	Pitch (mm)	Tensile stress area (mm^2)	Pitch (mm)	Tensile stress area (mm^2)
1	0.25	0.460		
1.6	0.35	1.27	0.20	1.57
2	0.4	2.07	0.25	2.45
2.5	0.45	3.39	0.35	3.70
3	0.5	5.03	0.35	5.61
4	0.7	8.78	0.5	9.79
5	0.8	14.2	0.5	16.1
6	1	20.1	0.75	22.0
8	1.25	36.6	1	39.2
10	1.5	58.0	1.25	61.2
12	1.75	84.3	1.25	92.1
16	2	157	1.5	167
20	2.5	245	1.5	272
24	3	353	2	384
30	3.5	561	2	621
36	4	817	3	865
42	4.5	1 121		
48	5	1 473		

American Standard: Basic size followed by number of threads per inch and the thread series designation.

10–24 UNC   10–32 UNF

1/2–13 UNC   1/2–20 UNF

$1\frac{1}{2}$–6 UNC $1\frac{1}{2}$–12 UNF

Metric: M (for "metric"), followed by the basic major diameter and then the pitch in millimeters.

M3 × 0.5   M3 × 0.35   M10 × 1.5

18–5
PERFORMANCE
OF BOLTED
JOINTS

Clamping Load

When a bolt or a screw is used to clamp two parts, the force exerted between the parts is the *clamping load*. The designer is responsible for specifying the clamping load and for ensuring that the fastener is capable of withstanding the load. The maximum clamping load is often taken to be 0.75 times the proof load, where the proof load is the product of the proof stress times the tensile stress area of the bolt or screw.

Tightening Torque

The clamping load is created in the bolt or the screw by exerting a tightening torque on the nut or on the head of the screw. An approximate relationship between the torque and the axial tensile force in the bolt or screw (the clamping force) is

⟲ **Tightening Torque** $$T = KDP \qquad (18\text{–}3)$$

where T = torque, lb·in
D = nominal outside diameter of threads, in
P = clamping load, lb
K = constant dependent on the lubrication present

For average commercial conditions, use $K = 0.15$ if any lubrication at all is present. Even cutting fluids or other residual deposits on the threads will produce conditions consistent with $K = 0.15$. If the threads are well cleaned and dried, $K = 0.20$ is better. Of course, these values are approximate, and variations among seemingly identical assemblies should be expected. Testing and statistical analysis of the results are recommended.

Example Problem 18–1   A set of three bolts is to be used to provide a clamping force of 12 000 lb between two components of a machine. The load is shared equally among the three bolts. Specify suitable bolts, including the grade of the material, if each is to be stressed to 75% of its proof strength. Then compute the required tightening torque.

Solution   The load on each screw is to be 4 000 lb. Let's specify a bolt made from SAE Grade 5 steel, having a proof strength of 85 000 psi. Then the allowable stress is

$$\sigma_a = 0.75(85\ 000\ \text{psi}) = 63\ 750\ \text{psi}$$

The required tensile stress area for the bolt is then

$$A_t = \frac{\text{load}}{\sigma_a} = \frac{4\ 000\ \text{lb}}{63\ 750\ \text{lb/in}^2} = 0.062\ 7\ \text{in}^2$$

From Table 18–4(B), we find that the 3/8–16 UNC thread has the required tensile stress area. The required tightening torque will be

$$T = KDP = 0.15(0.375\ \text{in})(4\ 000\ \text{lb}) = 225\ \text{lb·in}$$

Externally Applied Force on a Bolted Joint

The analysis shown in Example Problem 18–1 considers the stress in the bolt only under static conditions and only for the clamping load. It was recommended that the tension on the bolt be very high, approximately 75% of the proof load for the bolt. Such a load will use the available strength of the bolt efficiently and will prevent the separation of the connected members.

When a load is applied to a bolted joint over and above the clamping load, special consideration must be given to the behavior of the joint. Initially, the force on the bolt (in tension) is equal to the force on the clamped members (in compression). Then some of the additional load will act to stretch the bolt beyond its length assumed after the clamping load was applied. Another portion will result in a *decrease* in the compressive force in the clamped member. Thus, only part of the applied force is carried by the bolt. The amount is dependent on the relative stiffness of the bolt and the clamped members.

If a very stiff bolt is clamping a flexible member, such as a resilient gasket, most of the added force will be taken by the bolt because it takes little force to change the compression in the gasket. In this case, the bolt design must take into account not only the initial clamping force but also the added force.

Conversely, if the bolt is relatively flexible compared with the clamped members, virtually all of the externally applied load will initially go to decreasing the clamping force

FIGURE 19–6
Notation for lengths
and forces

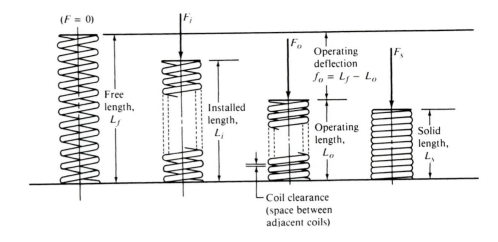

Forces

We will use the symbol F to indicate forces exerted by a spring, with various subscripts to specify which level of force is being considered. The subscripts correspond to those used for the lengths. Thus,

F_s = force at solid length, L_s; the maximum force that the spring ever sees.

F_o = force at operating length, L_o; the maximum force the spring sees in *normal operation.*

F_i = force at installed length, L_i; the force varies between F_o and F_i for a reciprocating spring.

F_f = force at free length, L_f; this force is zero.

Spring Rate

The relationship between the force exerted by a spring and its deflection is called its *spring rate, k.* Any change in force divided by the corresponding change in deflection can be used to compute the spring rate:

 Spring Rate

$$k = \Delta F / \Delta L \qquad (19\text{--}1)$$

For example,

$$k = \frac{F_o - F_i}{L_i - L_o} \qquad (19\text{--}1a)$$

or

$$k = \frac{F_o}{L_f - L_o} \qquad (19\text{--}1b)$$

or

$$k = \frac{F_i}{L_f - L_i} \qquad (19\text{--}1c)$$

FIGURE 19–7 Pitch angle

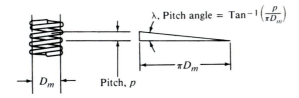

λ, Pitch angle $= \text{Tan}^{-1}\left(\frac{p}{\pi D_m}\right)$

πD_m

D_m Pitch, p

In addition, if the spring rate is known, the force at any deflection can be computed. For example, if a spring had a rate of 42.0 lb/in, the force exerted at a deflection from free length of 2.25 in would be

$$F = k(L_f - L) = (42.0 \text{ lb/in})(2.25 \text{ in}) = 94.5 \text{ lb}$$

Spring Index

The ratio of the mean diameter of the spring to the wire diameter is called the *spring index, C:*

☞ **Spring Index**

$$C = D_m/D_w$$

It is recommended that C be greater than 5.0, with typical machinery springs having C values ranging from 5 to 12. For C less than 5, the forming of the spring will be very difficult, and the severe deformation required may create cracks in the wire. The stresses and the deflections in springs are dependent on C, and a larger C will help to eliminate the tendency for a spring to buckle.

Number of Coils

The total number of coils in a spring will be called N. But in the calculation of stress and deflections for a spring, some of the coils are inactive and are neglected. For example, in a spring with squared and ground ends or simply squared ends, each end coil is inactive, and the number of *active coils, N_a,* is $N - 2$. For plain ends, all coils are active: $N_a = N$. For plain coils with ground ends, $N_a = N - 1$.

Pitch

Pitch, p, refers to the axial distance from a point on one coil to the corresponding point on the next adjacent coil. The relationships among the pitch, free length, wire diameter, and number of active coils are given next:

Squared and ground ends:	$L_f = pN_a + 2D_w$
Squared ends only:	$L_f = pN_a + 3D_w$
Plain and ground ends:	$L_f = p(N_a + 1)$
Plain ends:	$L_f = pN_a + D_w$

Pitch Angle

Figure 19–7 shows the pitch angle, λ; note that the larger the pitch angle, the steeper the coils appear to be. Most practical spring designs produce a pitch angle less than about 12°. If the angle is greater than 12°, undesirable compressive stresses develop in the wire, and the formulas presented later are inaccurate. The pitch angle can be computed by the formula

$$\lambda = \tan^{-1}\left[\frac{p}{\pi D_m}\right]$$

(19–2)

You can see the logic of this formula by taking one coil of a spring and unwrapping it onto a flat surface, as illustrated in Figure 19–7. The horizontal line is the mean circumference of the spring, and the vertical line is the pitch, p.

Installation Considerations

Frequently, a spring is installed in a cylindrical hole or around a rod. When it is, adequate clearances must be provided. When a compression spring is compressed, its diameter gets larger. Thus, the inside diameter of a hole enclosing the spring must be greater than the outside diameter of the spring to eliminate rubbing. An initial diametral clearance of one-tenth of the wire diameter is recommended for springs having a diameter of 0.50 in (12 mm) or greater. If a more precise estimate of the actual outside diameter of the spring is required, the following formula can be used for the OD at the solid length condition:

$$OD_s = \sqrt{D_m^2 + \frac{p^2 - D_w^2}{\pi^2}} + D_w$$

(19–3)

Even though the spring ID gets larger, it is also recommended that the clearance at the ID be approximately $0.1D_w$.

Springs with squared ends or squared and ground ends are frequently mounted on a button-type seat or in a socket with a depth equal to the height of just a few coils for the purpose of locating the spring.

Coil Clearance. The term *coil clearance* refers to the space between adjacent coils when the spring is compressed to its operating length, L_o. The actual coil clearance, *cc*, can be estimated from

 Coil Clearance

$$cc = (L_o - L_s)/N_a$$

One guideline is that the coil clearance should be greater than $D_w/10$, especially in springs loaded cyclically. Another recommendation relates to the overall deflection of the spring:

$$(L_o - L_s) > 0.15(L_f - L_s)$$

Materials Used for Springs

Virtually any elastic material can be used for a spring. However, most mechanical applications use metallic wire—high-carbon steel (most common), alloy steel, stainless steel, brass, bronze, beryllium copper, or nickel-base alloys. Most spring materials are made according to specifications of the ASTM. Table 19–3 lists some common types.

Types of Loading and Allowable Stresses

The allowable stress to be used for a spring depends on the type of loading, the material, and size of the wire. Loading is usually classified into three types:

TABLE 19–3 Spring materials

Material type	ASTM no.	Relative cost	Temperature limits, °F
A. *High-carbon steels*			
Hard-drawn	A227	1.0	0–250
General-purpose steel with 0.60%–0.70% carbon; low cost			
Music wire	A228	2.6	0–250
High-quality steel with 0.80%–0.95% carbon; very high strength; excellent surface finish; hard-drawn; good fatigue performance; used mostly in smaller sizes up to 0.125 in			
Oil-tempered	A229	1.3	0–350
General-purpose steel with 0.60%–0.70% carbon; used mostly in larger sizes above 0.125 in; not good for shock or impact			
B. *Alloy steels*			
Chromium-vanadium	A231	3.1	0–425
Good strength, fatigue resistance, impact strength, high-temperature performance; valve-spring quality			
Chromium-silicon	A401	4.0	0–475
Very high strength and good fatigue and shock resistance			
C. *Stainless steels*			
Type 302	A313(302)	7.6	<0–550
Very good corrosion resistance and high-temperature performance; nearly nonmagnetic; cold-drawn; types 304 and 316 also fall under this ASTM class and have improved workability but lower strength			
Type 17-7 PH	A313(631)	11.0	0–600
Good high-temperature performance			
D. *Copper alloys:* All have good corrosion resistance and electrical conductivity.			
Spring brass	B134	High	0–150
Phosphor bronze	B159	8.0	<0–212
Beryllium copper	B197	27.0	0–300
E. *Nickel-base alloys:* All are corrosion-resistant, have good high- and low-temperature properties, and are nonmagnetic or nearly nonmagnetic (trade names of the International Nickel Company).			
Monel™			−100–425
K-Monel™			−100–450
Inconel™			Up to 700
Inconel-X™		44.0	Up to 850

SOURCE: Associated Spring, Barnes Group, Inc. *Engineering Guide to Spring Design.* Bristol, CT, 1987. Carlson, Harold. *Spring Designer's Handbook.* New York: Marcel Dekker, 1978. Oberg, E., et al. *Machinery's Handbook.* 25th ed. New York: Industrial Press, 1996.

- ■ ***Light service:*** Static loads or up to 10 000 cycles of loading with a low rate of loading (nonimpact).
- ■ ***Average service:*** Typical machine design situations; moderate rate of loading and up to 1 million cycles.
- ■ ***Severe service:*** Rapid cycling for above 1 million cycles; possibility of shock or impact loading; engine valve springs are a good example.

The strength of a given material is greater for the smaller sizes. Figures 19–8 through 19–13 show the design stresses for six different materials. Note that some curves can be used for more than one material by the application of a factor. As a conservative approach to design, we will use the average service curve for most design examples, unless true high cycling conditions exist. We will use the light service curve as the upper limit on stress when the spring is compressed to its solid height. If the stress exceeds the light service value by a small amount, the spring will undergo permanent set because of yielding.

19–4
STRESSES AND
DEFLECTION
FOR HELICAL
COMPRESSION
SPRINGS

As a compression spring is compressed under an axial load, the wire is twisted. Therefore, the stress developed in the wire is *torsional shear stress,* and it can be derived from the classical equation $\tau = Tc/J$.

When the equation is applied specifically to a helical compression spring, some modifying factors are needed to account for the curvature of the spring wire and for the direct shear stress created as the coils resist the vertical load. Also, it is convenient to express the shear stress in terms of the design variables encountered in springs. The resulting equation for stress is attributed to Wahl. (See Reference 7.) The maximum shear stress, which occurs at the inner surface of the wire, is

⇨ **Shear Stress**
in a Spring

$$\tau = \frac{8KFD_m}{\pi D_w^3} = \frac{8KFC}{\pi D_w^2} \tag{19–4}$$

These are two forms of the same equation as the definition of $C = D_m/D_w$ demonstrates. The shear stress for any applied force, F, can be computed. Normally we will be concerned about the stress when the spring is compressed to solid length under the influence of F_s and when the spring is operating at its normal maximum load, F_o. Notice that the stress is inversely proportional to the *cube* of the wire diameter. This illustrates the great effect that variation in the wire size has on the performance of the spring.

The Wahl factor, K, in Equation (19–4) is the term that accounts for the curvature of the wire and the direct shear stress. Analytically, K is related to C:

⇨ **Wahl Factor**

$$K = \frac{4C - 1}{4C - 4} + \frac{0.615}{C} \tag{19–5}$$

Figure 19–14 shows a plot of K versus C for round wire. Recall that $C = 5$ is the recommended minimum value of C. The value of K rises rapidly for $C < 5$.

FIGURE 19–14
Wahl factor vs. spring index for round wire

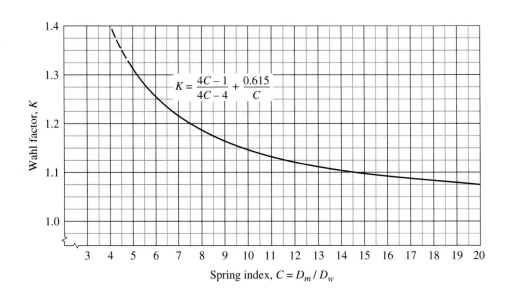

Deflection

Because the primary manner of loading on the wire of a helical compression spring is torsion, the deflection is computed from the angle of twist formula:

$$\theta = TL/GJ$$

where θ = angle of twist in radians
T = applied torque
L = length of the wire
G = modulus of elasticity of the material in shear
J = polar moment of inertia of the wire

Again, for convenience, we will use a different form of the equation in order to calculate the linear deflection, f, of the spring from the typical design variables of the spring. The resulting equation is

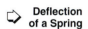
Deflection of a Spring

$$f = \frac{8FD_m^3 N_a}{GD_w^4} = \frac{8FC^3 N_a}{GD_w} \tag{19–6}$$

Recall that N_a is the number of *active* coils, as discussed in Section 19–3. Table 19–4 lists the values for G for typical spring materials. Note again, in Equation (19–6), that the wire diameter has a strong effect on the performance of the spring.

Buckling

The tendency for a spring to buckle increases as the spring becomes tall and slender, much as for a column. Figure 19–15 shows plots of the critical ratio of deflection to the free length versus the ratio of free length to the mean diameter for the spring. Three different end conditions are described in the figure. As an example of the use of this figure, consider a spring

TABLE 19–4 Spring wire modulus of elasticity in shear (G) and tension (E)

Material and ASTM no.	Shear modulus, G		Tension modulus, E	
	(psi)	(GPa)	(psi)	(GPa)
Hard-drawn steel: A227	11.5×10^6	79.3	28.6×10^6	197
Music wire: A228	11.85×10^6	81.7	29.0×10^6	200
Oil-tempered: A229	11.2×10^6	77.2	28.5×10^6	196
Chromium-vanadium: A231	11.2×10^6	77.2	28.5×10^6	196
Chromium-silicon: A401	11.2×10^6	77.2	29.5×10^6	203
Stainless steels: A313				
Types 302, 304, 316	10.0×10^6	69.0	28.0×10^6	193
Type 17-7 PH	10.5×10^6	72.4	29.5×10^6	203
Spring brass: B134	5.0×10^6	34.5	15.0×10^6	103
Phosphor bronze: B159	6.0×10^6	41.4	15.0×10^6	103
Beryllium copper: B197	7.0×10^6	48.3	17.0×10^6	117
Monel and K-Monel	9.5×10^6	65.5	26.0×10^6	179
Inconel and Inconel-X	10.5×10^6	72.4	31.0×10^6	214

NOTE: Data are average values. Slight variations with wire size and treatment may occur.

FIGURE 19–15
Spring buckling criteria. If the actual ratio of f_o/L_f is greater than the critical ratio, the spring will buckle at operating deflection.

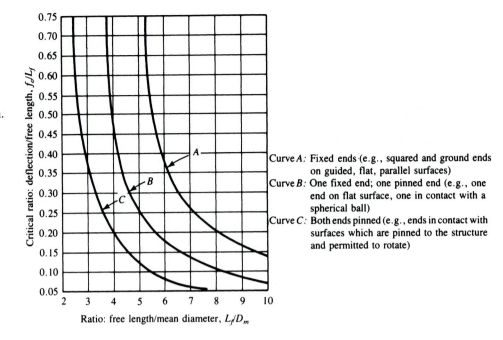

Curve A: Fixed ends (e.g., squared and ground ends on guided, flat, parallel surfaces)
Curve B: One fixed end; one pinned end (e.g., one end on flat surface, one in contact with a spherical ball)
Curve C: Both ends pinned (e.g., ends in contact with surfaces which are pinned to the structure and permitted to rotate)

having squared and ground ends, a free length of 6.0 in, and a mean diameter of 0.75 in. We want to know what deflection would cause the spring to buckle. First compute

$$\frac{L_f}{D_m} = \frac{6.0}{0.75} = 8.0$$

Then, from Figure 19–15, the critical deflection ratio is 0.20. From this we can compute the critical deflection:

$$\frac{f_o}{L_f} = 0.20 \quad \text{or} \quad f_o = 0.20(L_f) = 0.20(6.0 \text{ in}) = 1.20 \text{ in}$$

That is, if the spring is deflected more than 1.20 in, the spring should buckle.

Additional development and discussion of formulas for stresses and deflection of helical compression springs can be found in References 4 and 6. Reference 3 provides valuable information on the analysis of spring failures.

**19–5
ANALYSIS
OF SPRING
CHARACTERISTICS**

This section demonstrates the use of the concepts developed in previous sections to analyze the geometry and the performance characteristics of a given spring. Assume that you have found a spring but there are no performance data available for it. By making a few measurements and computations, you should be able to determine those characteristics. One piece of information you *would* have to know is the material from which the spring is made so that you could evaluate the acceptability of calculated stress levels.

The method of analysis is presented in Example Problem 19–1.

Example Problem 19–1   A spring is known to be made from music wire, ASTM A228 steel, but no other data are known. You are able to measure the following features using simple measurement tools:

Free length $= L_f = 1.75$ in

Outside diameter $= OD = 0.561$ in

Wire diameter $= D_w = 0.055$ in

The ends are squared and ground.

The *total* number of coils is 10.0.

This spring will be used in an application where the normal operating load is to be 14.0 lb. Approximately 300 000 cycles of loading are expected.

For this spring, compute and/or do the following:

1. The music wire gage number, mean diameter, inside diameter, spring index, and Wahl factor.

2. The expected stress at the operating load of 14.0 lb.

3. The deflection of the spring under the 14.0-lb load.

4. The operating length, solid length, and spring rate.

5. The force on the spring when it is at its solid length and the corresponding stress at solid length.

6. The design stress for the material; then compare it with the actual operating stress.

7. The maximum permissible stress; then compare it with the stress at solid length.

8. Check the spring for buckling and coil clearance.

9. Specify a suitable diameter for a hole in which to install the spring.

Solution   The solution is presented in the same order as the requested items just listed. The formulas used are found in the preceding sections of this chapter.

Step 1. The wire is 24-gage music wire (Table 19–2). Thus,

$$D_m = OD - D_w = 0.561 - 0.055 = 0.506 \text{ in}$$

$$ID = D_m - D_w = 0.506 - 0.055 = 0.451 \text{ in}$$

$$\text{Spring index} = C = D_m/D_w = 0.506/0.055 = 9.20$$

$$\text{Wahl factor} = K = (4C - 1)/(4C - 4) + 0.615/C$$

$$K = [4(9.20) - 1]/[4(9.20) - 4] + 0.615/9.20$$

$$K = 1.158$$

Step 2. Stress in spring at $F = F_o = 14.0$ lb [Equation (19–4)]:

$$\tau_o = \frac{8KF_oC}{\pi D_w^2} = \frac{8(1.158)(14.0)(9.20)}{\pi(0.055)^2} = 125\,560 \text{ psi}$$

Step 3. Deflection at operating force [Equation (19–6)]:

$$f_o = \frac{8F_oC^3N_a}{GD_w} = \frac{8(14.0)(9.20)^3(8.0)}{(11.85 \times 10^6)(0.055)} = 1.071 \text{ in}$$

Note that the number of active coils for a spring with squared and ground ends is $N_a = N - 2 = 10.0 - 2 = 8.0$. Also, the spring wire modulus, G, was found in Table 19–4. The value of f_o is the deflection *from free length* to the operating length.

Step 4. Operating length: We compute operating length as

$$L_o = L_f - f_o = 1.75 - 1.071 = 0.679 \text{ in}$$

$$\text{Solid length} = L_s = D_w(N) = 0.055(10.0) = 0.550 \text{ in}$$

Spring Index: We use Equation (19–1).

$$k = \frac{\Delta F}{\Delta L} = \frac{F_o}{L_f - L_o} = \frac{F_o}{f_o} = \frac{14.0 \text{ lb}}{1.071 \text{ in}} = 13.07 \text{ lb /in}$$

Step 5. We can find the force at solid length by multiplying the spring rate times the deflection from the free length to the solid length. Then

$$F_s = k(L_f - L_s) = (13.07 \text{ lb/in})(1.75 \text{ in} - 0.550 \text{ in}) = 15.69 \text{ lb}$$

The stress at solid length, τ_s, could be found from Equation (19–4), using $F = F_s$. However, an easier method is to recognize that the stress is directly proportional to the force on the spring and that all of the other data in the formula are the same as those used to compute the stress under the operating force, F_o. We can then use the simple proportion

$$\tau_s = \tau_o(F_s/F_o) = (125\ 560 \text{ psi})(15.69/14.0) = 140\ 700 \text{ psi}$$

Step 6. Design stress, τ_d: From Figure 19–9, in the graph of design stress versus spring wire diameter for ASTM A228 steel, we can use the *average service* curve based on the expected number of cycles of loading. We read $\tau_d = 135\ 000$ psi for the 0.055-in wire. Because the actual operating stress, τ_o, is less than this value, it is satisfactory.

Step 7. Maximum allowable stress, τ_{max}: It is recommended that the *light service* curve be used to determine this value. For $D_w = 0.055$, $\tau_{max} = 150\ 000$ psi. The actual expected maximum stress that occurs at solid length ($\tau_s = 140\ 700$ psi) is less than this value, and therefore the design is satisfactory with regard to stresses.

Step 8. Buckling: To evaluate buckling, we must compute

$$L_f/D_m = (1.75 \text{ in})/(0.506 \text{ in}) = 3.46$$

Referring to Figure 19–15 and using curve *A* for squared and ground ends, we see that the critical deflection ratio is very high and that buckling should not occur. In fact, for any value of $L_f/D_m < 5.2$, we can conclude that buckling will not occur.

Coil clearance, *cc:* We evaluate *cc* as follows:

$$cc = (L_o - L_s)/N_a = (0.679 - 0.550)/(8.0) = 0.016 \text{ in}$$

Comparing this to the recommended minimum clearance of

$$D_w/10 = (0.055 \text{ in})/10 = 0.005\ 5 \text{ in}$$

we can judge this clearance to be acceptable.

Step 9. Hole diameter: It is recommended that the hole into which the spring is to be installed should be greater in diameter than the *OD* of the spring by the amount of $D_w/10$. Then

$$D_{hole} > OD + D_w/10 = 0.561 \text{ in} + (0.055 \text{ in})/10 = 0.567 \text{ in}$$

A diameter of 5/8 in (0.625 in) would be a satisfactory standard size.

This completes the example problem.

19–6
DESIGN OF
HELICAL
COMPRESSION
SPRINGS

The objective of the design of helical compression springs is to specify the geometry for the spring to operate under specified limits of load and deflection, possibly with space limitations, also. We will specify the material and the type of service by considering the environment and the application.

A typical problem statement follows. Then two solution procedures are shown, and each is implemented with the aid of a spreadsheet.

Example Problem 19–2 A helical compression spring is to exert a force of 8.0 lb when compressed to a length of 1.75 in. At a length of 1.25 in, the force must be 12.0 lb. The spring will be installed in a machine that cycles slowly, and approximately 200 000 cycles total are expected. The temperature will not exceed 200°F. The spring will be installed in a hole having a diameter of 0.75 in.

For this application, specify a suitable material, wire diameter, mean diameter, *OD, ID,* free length, solid length, number of coils, and type of end condition. Check the stress at the maximum operating load and at the solid length condition.

The first of two solution procedures will be shown. The numbered steps can be used as a guide for future problems and as a kind of algorithm for the spreadsheet approach that follows the manual solution.

Solution Method 1 The procedure works directly toward the overall geometry of the spring by specifying the mean diameter to meet the space limitations. The process requires that the designer have tables of data available for wire diameters (such as Table 19–2) and graphs of design stresses for the material from which the spring will be made (such as Figures 19–8 through 19–13). We must make an initial estimate for the design stress for the material by consulting the charts of design stress versus wire diameter to make a reasonable choice. In general, more than one trial must be made, but the results of early trials will help you decide the values to use for later trials.

Step 1. Specify a material and its shear modulus of elasticity, *G.*

For this problem, several standard spring materials can be used. Let's select ASTM A231 chromium-vanadium steel wire, having a value of $G = 11\,200\,000$ psi (See Table 19–4.)

Step 2. From the problem statement, identify the operating force, F_o; the operating length at which that force must be exerted, L_o; the force at some other length, called the *installed force,* F_i; and the installed length, L_i.

Remember, F_o is the maximum force that the spring experiences under normal operating conditions. Many times, the second force level is not specified. In that case, let $F_i = 0$, and specify a design value for the free length, L_f, in place of L_i.

For this problem, $F_o = 12.0$ lb; $L_o = 1.25$ in; $F_i = 8.0$ lb; and $L_i = 1.75$ in.

Step 3. Compute the spring rate, k using Equation (19–1a):

$$k = \frac{F_o - F_i}{L_i - L_o} = \frac{12.0 - 8.0}{1.75 - 1.25} = 8.00 \text{ lb/in}$$

Step 4. Compute the free length, L_f:

$$L_f = L_i + F_i/k = 1.75 \text{ in} + [(8.00 \text{ lb})/(8.00 \text{ lb/in})] = 2.75 \text{ in}$$

The second term in the preceding equation is the amount of deflection from free length to the installed length in order to develop the installed force, F_i. Of course, this step becomes unnecessary if the free length is specified in the original data.

Step 5. Specify an initial estimate for the mean diameter, D_m.

Keep in mind that the mean diameter will be smaller than the *OD* and larger than the *ID.* Judgment is necessary to get started. For this problem, let's specify $D_m = 0.60$ in. This should permit the installation into the 0.75-in-diameter hole.

Step 6. Specify an initial design stress.

The charts for the design stresses for the selected material can be consulted, considering also the service. In this problem, we should use average service. Then for the ASTM A231 steel, as shown in Figure 19–11, a nominal design stress would be 130 000 psi. This is strictly an estimate based on the strength of the material. The process includes a check on stress later.

Step 7. Compute the trial wire diameter by solving Equation (19–4) for D_w. Notice that everything else in the equation is known except the Wahl factor, K, because it depends on the wire diameter itself. But K varies only little over the normal range of spring indexes, C. From Figure 19–14, note that $K = 1.2$ is a nominal value. This, too, will be checked later. With the assumed value of K, some simplification can be done:

$$D_w = \left[\frac{8KF_oD_m}{\pi\tau_d}\right]^{1/3} = \left[\frac{(8)(1.2)(F_o)(D_m)}{(\pi)(\tau_d)}\right]^{1/3}$$

Combining constants gives

▷ Trial Wire Diameter

$$D_w = \left[\frac{8KF_oD_m}{\pi\tau_d}\right]^{1/3} = \left[\frac{(3.06)(F_o)(D_m)}{(\tau_d)}\right]^{1/3} \tag{19–7}$$

For this problem,

$$D_w = \left[\frac{(3.06)(F_o)(D_m)}{\tau_d}\right]^{1/3} = \left[\frac{(3.06)(12)(0.6)}{130\ 000}\right]^{0.333}$$

$$D_w = 0.055\ 3\ \text{in}$$

Step 8. Select a standard wire diameter from the tables, and then determine the design stress and the maximum allowable stress for the material at that diameter. The design stress will normally be for average service, unless high cycling rates or shock indicate that severe service is warranted. The light service curve should be used with care because it is very near to the yield strength. In fact, we will use the light service curve as an estimate of the maximum allowable stress.

For this problem, the next larger standard wire size is 0.062 5 in, no. 16 on the U.S. Steel Wire Gage chart. For this size, the curves in Figure 19–11 for ASTM A231 steel wire show the design stress to be approximately 145 000 psi for average service, and the maximum allowable stress to be 170 000 psi from the light service curve.

Step 9. Compute the actual values of C and K, the spring index and the Wahl factor:

$$C = \frac{D_m}{D_w} = \frac{0.60}{0.062\ 5} = 9.60$$

$$K = \frac{4C-1}{4C-4} + \frac{0.615}{C} = \frac{4(9.60)-1}{4(9.60)-4} + \frac{0.615}{9.60} = 1.15$$

Step 10. Compute the actual expected stress due to the operating force, F_o, from Equation (19–4):

$$\tau_o = \frac{8KF_oD_m}{\pi D_w^3} = \frac{(8)(1.15)(12.0)(0.60)}{(\pi)(0.062\ 5)^3} = 86\ 450\ \text{psi}$$

Comparing this with the design stress of 145 000 psi, we see that it is safe.

Step 11. Compute the number of active coils required to give the proper deflection characteristics for the spring. Using Equation (19–6) and solving for N_a, we have

$$f = \frac{8FC^3N_a}{GD_w}$$

▷ Number of Active Coils

$$N_a = \frac{fGD_w}{8FC^3} = \frac{GD_w}{8kC^3} \quad (\textit{Note: F/f = k, the spring rate.}) \tag{19–8}$$

Then, for this problem,

$$N_a = \frac{GD_w}{8kC^3} = \frac{(11\ 200\ 000)(0.062\ 5)}{(8)(8.0)(9.60)^3} = 12.36\ \text{coils}$$

Notice that $k = 8.0$ lb/in is the spring rate. Do not confuse this with K, the Wahl factor.

Step 12. Compute the solid length, L_s; the force on the spring at solid length, F_s; and the stress in the spring at solid length, τ_s. This computation will give the maximum stress that the spring will receive.

The solid length occurs when all of the coils are touching, but recall that there are two inactive coils for springs with squared and ground ends. Thus,

$$L_s = D_w(N_a + 2) = 0.062\ 5(14.36) = 0.898 \text{ in}$$

The force at solid length is the product of the spring rate times the deflection to solid length ($L_f - L_s$):

$$F_s = k(L_f - L_s) = (8.0 \text{ lb/in})(2.75 - 0.898) \text{ in} = 14.8 \text{ lb}$$

Because the stress in the spring is directly proportional to the force, a simple method of computing the solid length stress is

$$\tau_s = (\tau_o)(F_s/F_o) = (86\ 450 \text{ psi})(14.8/12.0) = 106\ 750 \text{ psi}$$

When this value is compared with the maximum allowable stress of 170 000 psi, we see that it is safe, and the spring will not yield when compressed to solid length.

Step 13. Complete the computations of geometric features, and compare them with space and operational limitations:

$$OD = D_m + D_w = 0.60 + 0.062\ 5 = 0.663 \text{ in}$$
$$ID = D_m - D_w = 0.60 - 0.062\ 5 = 0.538 \text{ in}$$

These dimensions are satisfactory for installation in a hole having a diameter of 0.75 in.

Step 14. The tendency to buckle is checked, along with the coil clearance.

This procedure completes the design of one satisfactory spring for this application. It may be desirable to make other trials to attempt to find a more nearly optimum spring.

Spreadsheet for Spring Design Method 1

The 14 steps required to complete one trial design using Method 1 which was demonstrated in Example Problem 19–2 are fairly involved, tedious, and time-consuming. Furthermore, it is highly likely that several iterations are required to produce an optimum solution that meets application considerations of the physical size of the spring, acceptable levels of stress under all loads, cost, and other factors. You might want to investigate the use of different materials for the same basic design goals for the forces, lengths, and spring rate.

For these and many other reasons, it is recommended that computer-assisted design approaches be developed to perform most of the calculations and to guide you through the solution procedure. This could be done with a computer program, a spreadsheet, mathematical analysis software, or a programmable calculator. Once written, the program or spreadsheet can be used for any similar design problem in the future by yourself or others.

Figure 19–16 shows one approach using a spreadsheet with the data from Example Problem 19–2 used for illustration. This approach is attractive because the entire solution is presented on one page, and the user is guided through the solution procedure. Let's summarize the use of this spreadsheet. As you read this, you should compare the spreadsheet entries with the details of the solution of Example Problem 19–2. The various formulas used there are programmed into the appropriate cells of the spreadsheet.

1. The process begins, as with virtually any spring design procedure, with the specification of the relationship between force and length for two separate conditions,

DESIGN OF HELICAL COMPRESSION SPRING—METHOD 1
Specify mean diameter and design stresses. Compute wire diameter and number of coils.

1. Enter data for forces and lengths.
2. Specify material; shear modulus, G; and an estimate for design stress.
3. Enter trial mean diameter for spring, considering space available.
4. Check computed values for spring rate, free length, and new trial wire diameter.
5. Enter your choice for wire diameter of a standard size.
6. Enter design stress and maximum allowable stress from Figures 19–8 through 19–13 for new D_w.

Numerical values in italics in shaded areas must be inserted for each problem.	Problem ID:	Example Problem 19–2

Initial Input Data:			Comments
Maximum operating force =	F_o =	*12.0 lb*	
Operating length =	L_o =	*1.25 in*	
Installed force =	F_i =	*8.0 lb*	*Note:* F_i = 0 if:
Installed length =	L_i =	*1.75 in*	L_i = free length
Trial mean diameter =	D_m =	*0.60 in*	
Type of spring wire:	*ASTM A231 steel*		See Figures 19–8 through 19–13.
Shear modulus of elasticity of spring wire =	G =	*1.12E + 07 psi*	From Table 19–4
Initial estimate of design stress =	τ_{di} =	*130,000 psi*	From design stress graph

Computed Values:		
Computed spring rate =	k =	8.00 lb/in
Computed free length =	L_f =	2.75 in
Computed trial wire diameter =	D_{wt} =	0.055 in

Secondary Input Data:			
Standard wire diameter =	D_w =	*0.062 5 in*	From Table 19–2
Design stress =	τ_d =	*145,000 psi*	From design stress graph
Maximum allowable stress =	τ_{max} =	*170,000 psi*	Use light service curve.

Computed Values:			
Outside diameter =	D_o =	0.663 in	
Inside diameter =	D_i =	0.538 in	
Number of active coils =	N_a =	12.36	
Spring index =	C =	9.60	**Should not be <5.0**
Wahl factor =	K =	1.15	
Stress at operating force =	τ_o =	86,459 psi	**Cannot be >145,000**
Solid length =	L_s =	0.898 in	**Cannot be >1.25**
Force at solid length =	F_s =	14.82 lb	
Stress at solid length =	τ_s =	106,768 psi	**Cannot be >170,000**

Check Buckling, Coil Clearance, and Hole Size:			
Buckling: Ratio =	L_f/D_m =	4.58	Check Figure 19–15 if >5.2.
Coil clearance =	cc =	0.029 in	Should be >0.006 25
If installed in hole, minimum hole diameter =	D_{hole} >	0.669 in	For side clearance

FIGURE 19–16   Spreadsheet for Spring Design Method 1 of Example Problem 19–2

typically called the *operating force-length* and the *installed force-length*. From these data, the spring rate is effectively specified as well. Sometimes the free length is known, which then equals the installed length, and the installed force equals zero. Also, the designer would know roughly the space into which the spring is to be installed.

2. The heading for the spreadsheet gives a brief overview of the design approach built into Method 1. The designer specifies a target mean diameter for the spring to fit a particular application. The spring material is specified, and the graph of the strength of that material is used as a guide for estimating the design stress. This requires the user to make a rough estimate also of the wire diameter, but a specific size is not yet chosen. The service (light, average, or severe) must also be specified at this time.

3. The spreadsheet then computes the resulting spring rate, the free length, and a trial wire diameter to produce an acceptable stress. Equation (19–7) is used to compute wire size.

4. The designer then enters a standard wire size, typically greater than the computed value. Table 19–2 is a listing of standard wire sizes. At this time, also, the designer must look again at the design stress graph for the chosen material and must determine a revised value for the design stress corresponding with the new wire diameter. The maximum allowable stress is also entered, found from the light service curve for the material at the specified wire size.

5. With the data entered, the spreadsheet completes the entire set of remaining calculations. The spring must be manufactured with exactly the specified diameters and number of active coils. The end condition for the spring, from the possibilities shown in Figure 19–3, should also be specified.

6. The designer's task is to evaluate the suitability of the results for basic geometry, stresses, potential for buckling, coil clearance, and installation of the spring into a hole. The **Comments** section to the right includes several prompts. But this is the designer's responsibility: to make judgments and design decisions.

Notice the advantage of using a spreadsheet: The designer does the thinking, and the spreadsheet does the computing. For subsequent design iterations, only the values that change need to be entered. For example, if the designer wants to try a different wire diameter for the same spring material, only the three data values in the section called **Secondary Input Data** need to be changed, and a new result is produced immediately. Many design iterations can be completed in a short time with this approach.

You might see ways to enhance the utility of the spreadsheet, and you are encouraged to do that.

Spreadsheet for Spring Design Method 2

An alternative spring design method is shown in Figure 19–17. This method allows the designer more freedom in manipulating the parameters. The data used are basically the same as those of Example Problem 19–2 except that there is no requirement for installing the spring in a certain size hole. We refer to this modified problem statement as Example Problem 19–3.

DESIGN OF HELICAL COMPRESSION SPRING—METHOD 2
Specify wire diameter, design stresses, and number of coils. Compute mean diameter.

1. Enter data for forces and lengths.
2. Specify material; shear modulus, G; and an estimate for design stress.
3. Enter trial wire diameter.
4. Check computed values for spring rate, free length, and new trial wire diameter.
5. Enter your choice for wire diameter of a standard size.
6. Enter design stress and maximum allowable stress from Figures 19–8 through 19–13 for new D_w.
7. Check computed maximum number of coils. Enter selected actual number of coils.

Numerical values in italics in shaded areas must be inserted for each problem.	Problem ID:	Example Problem 19–3, Trial 1:

Initial Input Data:			Comments
Maximum operating force =	$F_o =$	12.0 lb	
Operating length =	$L_o =$	1.25 in	
Installed force =	$F_i =$	8.0 lb	*Note:* $F_i = 0$ if:
Installed length =	$L_i =$	1.75 in	$L_i =$ free length
Type of spring wire:	ASTM A231 steel		See Figures 19–8 through 19–13.
Shear modulus of elasticity of spring wire =	$G =$	1.12E + 07 psi	From Table 19–4
Initial estimate of design stress =	$\tau_{di} =$	144,000 psi	From design stress graph
Initial trial wire diameter =	$D_w =$	0.06 in	

Computed Values:		
Computed spring rate =	$k =$	8.00 lb/in
Computed free length =	$L_f =$	2.75 in
Computed trial wire diameter =	$D_{wt} =$	0.042 in

Secondary Input Data:			
Standard wire diameter =	$D_w =$	0.047 5 in	From Table 19–2
Design stress =	$\tau_d =$	149,000 psi	From design stress graph
Maximum allowable stress =	$\tau_{max} =$	174,000 psi	Use light service curve.

Computed: Maximum number of coils =	$N_{max} =$	24.32	

Input: Number of active coils =	$N_a =$	22	Suggest using an integer.

Computed Values:			
Spring index =	$C =$	7.23	**Should not be** <**5.0**
Wahl factor =	$K =$	1.21	
Mean diameter =	$D_m =$	0.343 in	
Outside diameter =	$D_o =$	0.391 in	
Inside diameter =	$D_i =$	0.296 in	
Solid length =	$L_s =$	1.140 in	**Cannot be** >**1.25**
Stress at operating force =	$\tau_o =$	118 030 psi	**Cannot be** >**149,000**
Force at solid length =	$F_s =$	12.88 lb	
Strength at solid length =	$\tau_s =$	126 685 psi	**Cannot be** >**174,000**

Check Buckling, Coil Clearance, and Hole Size:			
Buckling: Ratio =	$L_f/D_m =$	8.01	Check Figure 19–15 if >5.2.
Coil clearance =	$cc =$	0.005 0 in	Should be >0.004 75
If installed in hole, minimum hole diameter =	$D_{hole} >$	0.396 in	For side clearance

FIGURE 19–17   Spreadsheet for Spring Design Method 2 of Example Problem 19–3, Trial 1

Example Problem 19–3	The process is similar to that of Method 1, so only the major differences are discussed below. Follow along with the spreadsheet in Figure 19–17 as we describe Method 2, Trials 1 and 2.

Solution Method 2, Trial 1

1. The general procedure is outlined at the top of the spreadsheet. The designer selects a material, estimates the wire diameter as an initial trial, and inputs a corresponding estimate of the design stress.

2. The spreadsheet then computes a new trial wire diameter using a formula derived from the fundamental equation for shear stress in a helical compression spring, Equation (19–4). The development is described here.

$$\tau = \frac{8KFC}{\pi D_w^2} \tag{19–4}$$

Let $F = F_o$ and $\tau = \tau_d$ (the design stress). Solving for the wire diameter gives

$$D_w = \sqrt{\frac{8KF_oC}{\pi\tau_d}} \tag{19–9}$$

The values of K and C are not yet known, but a good estimate for the wire diameter can be computed if the spring index is assumed to be approximately 7.0, a reasonable value. The corresponding value for the Wahl factor is $K = 1.2$, from Equation (19–5). Combining these assumed values with the other constants in the preceding equation gives

$$D_w = \sqrt{21.4(F_o)/(\tau_d)} \tag{19–10}$$

This formula is programmed into the spreadsheet in the cell to the right of the one labeled D_{wt}, the computed trial wire diameter.

3. The designer then enters a standard wire size and determines revised values for the design stress and the maximum allowable stress from the graphs of material properties, Figures 19–8 through 19–13.

4. The spreadsheet then computes the maximum permissible number of active coils for the spring. The logic here is that *the solid length must be less than the operating length.* The solid length is the product of the wire diameter and the total number of coils. For squared and ground ends, this is

$$L_s = D_w(N_a + 2)$$

Note that different relationships are used for the total number of coils for springs with other end conditions. See the discussion "Number of Coils" in Section 19–3. Now letting $L_s = L_o$ as a limit and solving for the number of coils gives

$$(N_a)_{max} = (L_o - 2D_w)/D_w \tag{19–11}$$

This is the formula programmed into the spreadsheet cell to the right of the one labeled N_{max}.

5. The designer now has the freedom to choose any number of active coils less than the computed maximum value. Note the effects of that decision. Choosing a small number of coils will provide more clearance between adjacent coils and will use less wire per spring. However, the stresses produced for a given load will be higher, so there is a practical limit. One approach is to try progressively fewer coils until the stress approaches the design stress. Whereas any number of coils can be produced, even fractional numbers, we suggest trying integer values for the convenience of the manufacturer.

6. After the selected number of coils has been entered, the spreadsheet can complete the remaining calculations. One additional new formula is used in this spreadsheet

to compute the value of the spring index, C. It is developed from the second form of Equation (19–6) relating the deflection for the spring, f, to a corresponding applied force, F, the value of C, and other parameters that are already known. First we solve for C^3:

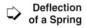

**Deflection
of a Spring**

$$f = \frac{8FD_m^3 N_a}{GD_w^4} = \frac{8FC^3 N_a}{GD_w} \qquad (19\text{–}6)$$

$$C^3 = \frac{fGD_w}{8FN_a}$$

Now notice that we have the force F in the denominator and the corresponding deflection f in the numerator. But the spring rate k is defined as the ratio of F/f. Then we can substitute k into the denominator and solve for C:

$$C = \left[\frac{GD_w}{8kN_a} \right]^{1/3} \qquad (19\text{–}12)$$

This formula is programmed into the spreadsheet cell to the right of the one labeled $C =$.

7. Recall that C is defined as the ratio, D_m/D_w. We can now solve for the mean diameter:

$$D_m = CD_w$$

This is used to compute the mean diameter in the cell to the right of $D_m =$.

8. The remaining calculations use equations already developed and used earlier. Again, the designer is responsible for evaluating the suitability of the results and for performing any additional iterations to search for an optimum result.

**Solution Method 1,
Trial 2**

Now notice that the solution obtained in Trial 1 and shown in Figure 19–17 is far from optimum. Its free length of 2.75 in is quite long compared with the mean diameter of 0.343 in. The buckling ratio of $L_f/D_m = 8.01$ indicates that the spring is long and slender. Checking Figure 19–15, we can see that buckling is predicted.

One way to work toward a more suitable geometry is to increase the wire diameter and reduce the number of coils. The net result will be a larger mean diameter, improving the buckling ratio.

Figure 19–18 shows the result of several iterations, finally using $D_w = 0.062\,5$ in (larger than the former value of 0.047 5 in) and 16 active coils (down from 22 in the first trial). The buckling ratio is down to 4.99, indicating that buckling is unlikely. The stress at operating force is comfortably lower than the design stress. The other geometrical features also appear to be satisfactory.

This example should demonstrate the value of using spreadsheets or other computer-based computational aids. You should also have a better feel for the kinds of design decisions that can work toward a more optimum design.

**19–7
EXTENSION
SPRINGS**

Extension springs are designed to exert a pulling force and to store energy. They are made from closely coiled helical coils similar in appearance to helical compression springs. Most extension springs are made with adjacent coils touching in such a manner that an initial force must be applied to separate the coils. Once the coils are separated, the force is linearly proportional to the deflection, as it is for helical compression springs. Figure 19–19 shows a typical extension spring, and Figure 19–20 shows the characteristic type of load-deflection curve. By convention, the initial force is found by projecting the straight line portion of the curve back to zero deflection.

DESIGN OF HELICAL COMPRESSION SPRING—METHOD 2
Specify wire diameter, design stresses, and number of coils. Compute mean diameter.

1. Enter data for forces and lengths.
2. Specify material; shear modulus, G; and an estimate for design stress.
3. Enter trial wire diameter.
4. Check computed values for spring rate, free length, and new trial wire diameter.
5. Enter your choice for wire diameter of a standard size.
6. Enter design stress and maximum allowable stress from Figures 19–8 through 19–13 for new D_w.
7. Check computed maximum number of coils. Enter selected actual number of coils.

Numerical values in italics in shaded areas must be inserted for each problem.	Problem ID:	*Example Problem 19–3, Trial 2: D_w = 0.062 5*

Initial Input Data:			Comments
Maximum operating force =	F_o =	*12.0 lb*	
Operating length =	L_o =	*1.25 in*	
Installed force =	F_i =	*8.0 lb*	*Note: F_i = 0 if:*
Installed length =	L_i =	*1.75 in*	L_i = free length
Type of spring wire:		*ASTM A231 steel*	See Figures 19–8 through 19–13.
Shear modulus of elasticity of spring wire =	G =	*1.12E+07 psi*	From Table 19–4
Initial estimate of design stress =	τ_{di} =	*144,000 psi*	From design stress graph
Initial trial wire diameter =	D_w =	*0.06 in*	

Computed Values:			
Computed spring rate =	k =	8.00 lb/in	
Computed free length =	L_f =	2.75 in	
Computed trial wire diameter =	D_{wt} =	0.042 in	

Secondary Input Data:			
Standard wire diameter =	D_w =	*0.062 5 in*	From Table 19–2
Design stress =	τ_d =	*142,600 psi*	From design stress graph
Maximum allowable stress =	τ_{max} =	*167,400 psi*	Use light service curve.

Computed: Maximum number of coils =	N_{max} =	18.00	

Input: Number of active coils =	N_a =	*16*	Suggest using an integer.

Computed Values:			
Spring index =	C =	8.81	**Should not be <5.0**
Wahl factor =	K =	1.17	
Mean diameter =	D_m =	0.551 in	
Outside diameter =	D_o =	0.613 in	
Inside diameter =	D_i =	0.488 in	
Solid length =	L_s =	1.125 in	**Cannot be >1.25**
Stress at operating force =	τ_o =	80,341 psi	**Cannot be >142,600**
Force at solid length =	F_s =	13.00 lb	
Stress at solid length =	τ_s =	87,036 psi	**Cannot be >167,400**

Check Buckling, Coil Clearance, and Hole Size:			
Buckling: Ratio =	L_f/D_m =	4.99	Check Figure 19–15 if >5.2.
Coil clearance =	cc =	0.007 8 in	Should be >0.006 25
If installed in hole, minimum hole diameter =	D_{hole} >	0.619 in	For side clearance

FIGURE 19–18 Spreadsheet for Spring Design Method 1 of Example Problem 19–3, Trial 2

FIGURE 19–19
Extension spring

FIGURE 19–20
Load-deflection curve
for an extension spring

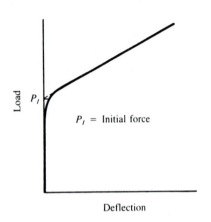

The stresses and deflections for an extension spring can be computed by using the formulas used for compression springs. Equation (19–4) is used for the torsional shear stress, Equation (19–5) for the Wahl factor to account for the curvature of the wire and the direct shear stress, and Equation (19–6) for the deflection characteristics. All coils in an extension spring are active. In addition, since the end loops or hooks deflect, their deflection may affect the actual spring rate.

The initial tension in an extension spring is typically 10% to 25% of the maximum design force. Figure 19–21 shows one manufacturer's recommendation of the preferred torsional stress due to initial tension as a function of the spring index.

End Configurations for Extension Springs

A wide variety of end configurations may be obtained for attaching the spring to mating machine elements, some of which are shown in Figure 19–4. The cost of the spring can be greatly affected by its end type, and it is recommended that the manufacturer be consulted before ends are specified.

Frequently, the weakest part of an extension spring is its end, especially in fatigue loading cases. The loop end shown in Figure 19–22, for example, has a high bending stress at point A and a torsional shear stress at point B. Approximations for the stresses at these points can be computed as follows:

Bending Stress at A

$$\sigma_A = \frac{16 D_m F_o K_1}{\pi D_w^3} + \frac{4 F_o}{\pi D_w^2} \qquad \text{(19–13)}$$

$$K_1 = \frac{4 C_1^2 - C_1 - 1}{4 C_1 (C_1 - 1)} \qquad \text{(19–14)}$$

$$C_1 = 2 R_1 / D_w$$

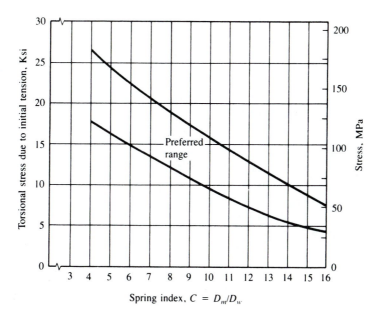

FIGURE 19–22
Stresses in ends of
extension spring

(a) Bending stress at A   (b) Torsion stress at B

Torsional Stress at B

$$\tau_B = \frac{8D_m F_o K_2}{\pi D_w^3} \qquad (19\text{–}15)$$

$$K_2 = \frac{4C_2 - 1}{4C_2 - 4} \qquad (19\text{–}16)$$

$$C_2 = 2R_2/D_w$$

The ratios C_1 and C_2 relate to the curvature of the wire and should be large, typically greater than 4, to avoid high stresses.

Allowable Stresses for Extension Springs

The torsional shear stress in the coils of the spring and in the end loops can be compared to the curves in Figures 19–8 through 19–13. Some designers reduce these allowable stresses by about 10%. The bending stress in the end loops, such as that from Equation (19–13), should be compared to the allowable bending stresses for torsion springs, as presented in the next section.

Example Problem 19–4 A helical extension spring is to be installed in a latch for a large, commercial laundry machine. When the latch is closed, the spring must exert a force of 16.25 lb at a length between attachment points of 3.50 in. As the latch is opened, the spring is pulled to a length of 4.25 in with a maximum force of 26.75 lb. An outside diameter of 5/8 in (0.625 in) is desired. The latch will be cycled only about 10 times per day, and thus the design stress will be based on average service. Use ASTM A227 steel wire. Design the spring.

Solution As before, the suggested design procedure will be given as numbered steps, followed by the calculations for this set of data.

Step 1. Assume a trial mean diameter and a trial design stress for the spring.

Let the mean diameter be 0.500 in. For ASTM A227 wire under average service, a design stress of 110 000 psi is reasonable (from Figure 19–8).

Step 2. Compute a trial wire diameter from Equation (19–4) for the maximum operating force, the assumed mean diameter and design stress, and an assumed value for K of about 1.20.

$$D_w = \left[\frac{8KF_oD_m}{\pi\tau_d}\right]^{1/3} = \left[\frac{(8)(1.20)(26.75)(0.50)}{(\pi)(110\,000)}\right]^{1/3} = 0.072 \text{ in}$$

A standard wire size in the U.S. Steel Wire Gage system is available in this size. Use 15-gage.

Step 3. Determine the actual design stress for the selected wire size.

From Figure 19–8, at a wire size of 0.072 in, the design stress is 120 000 psi.

Step 4. Compute the actual values for outside diameter, mean diameter, inside diameter, spring index, and the Wahl factor, K. These factors are the same as those defined for helical compression springs.

Let the outside diameter be as specified, 0.625 in. Then

$$D_m = OD - D_w = 0.625 - 0.072 = 0.553 \text{ in}$$

$$ID = OD - 2D_w = 0.625 - 2(0.072) = 0.481 \text{ in}$$

$$C = D_m/D_w = 0.553/0.072 = 7.68$$

$$K = \frac{4C - 1}{4C - 4} + \frac{0.615}{C} = \frac{4(7.68) - 1}{4(7.68) - 4} + \frac{0.615}{7.68} = 1.19$$

Step 5. Compute the actual expected stress in the spring wire under the operating load from Equation (19–4):

$$\tau_o = \frac{8KF_oD_m}{\pi D_w^3} = \frac{8(1.19)(26.75)(0.553)}{\pi(0.072)^3} = 120\,000 \text{ psi} \quad \text{(okay)}$$

Step 6. Compute the required number of coils to produce the desired deflection characteristics. Solve Equation (19–6) for the number of coils, and substitute $k = $ force/deflection $= F/f$:

$$k = \frac{26.75 - 16.25}{4.25 - 3.50} = 14.0 \text{ lb/in}$$

$$N_a = \frac{GD_w}{8C^3k} = \frac{(11.5 \times 10^6)(0.072)}{8(7.68)^3(14.0)} = 16.3 \text{ coils}$$

Step 7. Compute the body length for the spring, and propose a trial design for the ends.

$$\text{Body length} = D_w(N_a + 1) = (0.072)(16.3 + 1) = 1.25 \text{ in}$$

Let's propose to use a full loop at each end of the spring, adding a length equal to the *ID* of the spring at each end. Then the total free length is

$$L_f = \text{body length} + 2(ID) = 1.25 + 2(0.481) = 2.21 \text{ in}$$

Step 8. Compute the deflection from free length to operating length:

$$f_o = L_o - L_f = 4.25 - 2.21 = 2.04 \text{ in}$$

Step 9. Compute the initial force in the spring at which the coils just begin to separate. This is done by subtracting the amount of force due to the deflection, f_o:

$$F_I = F_o - kf_o = 26.75 - (14.0)(2.04) = -1.81 \text{ lb}$$

The negative force resulting from a free length that is too small for the specified conditions is clearly impossible.

Let's try $L_f = 2.50$ in, which will require a redesign of the end loops. Then

$$f_o = 4.25 - 2.50 = 1.75 \text{ in}$$

$$F_I = 26.75 - (14.0)(1.75) = 2.25 \text{ lb} \quad \text{(reasonable)}$$

Step 10. Compute the stress in the spring under the initial tension, and compare it with the recommended levels in Figure 19–21.

Because the stress is proportional to the load,

$$\tau_I = \tau_o(F_I/F_o) = (120\,000)(2.25/26.75) = 10\,100 \text{ psi}$$

For $C = 7.68$, this stress is in the preferred range from Figure 19–21.

At this point, the coiled portion of the spring is satisfactory. The final configuration of the end loops must be completed and analyzed for stress.

19–8 HELICAL TORSION SPRINGS

Many machine elements require a spring that exerts a rotational moment, or torque, instead of a push or a pull force. The helical torsion spring is designed to satisfy this requirement. The spring has the same general appearance as either the helical compression or the extension spring, with round wire wrapped into a cylindrical shape. Usually the coils are close together but with a small clearance, allowing no initial tension in the spring as there is for extension springs. Figure 19–23 shows a few examples of torsion springs with a variety of end treatments.

The familiar clip-type clothespin uses a torsion spring to provide the gripping force. Many cabinet doors are designed to close automatically under the influence of a torsion spring. Some timers and switches use torsion springs to actuate mechanisms or

FIGURE 19–23
Torsion springs showing a variety of end types

close contacts. Torsion springs often provide counterbalancing of machine elements that are mounted to a hinged plate.

The following are some of the special features and design guides for torsion springs:

1. The moment applied to a torsion spring should always act in a direction that causes the coils to wind more tightly, rather than to open the spring up. This takes advantage of favorable residual stresses in the wire after forming.

2. In the free (no-load) condition, the definition of mean diameter, outside diameter, inside diameter, wire diameter, and spring index are the same as those used for compression springs.

3. As the load on a torsion spring increases, its mean diameter, D_m, decreases, and its length, L, increases, according to the following relations:

$$D_m = D_{mI}N_a/(N_a + \theta) \qquad\qquad (19\text{–}17)$$

where D_{mI} = initial mean diameter at the free condition
N_a = number of active coils in the spring (to be defined later)
θ = angular deflection of the spring from the free condition, expressed in revolutions or a fraction of a revolution

$$L = D_w(N_a + 1 + \theta) \qquad\qquad (19\text{–}18)$$

This equation assumes that all coils are touching. If any clearance is provided, as is often desirable to reduce friction, a length of N_a times the clearance must be added.

4. Torsion springs must be supported at three or more points. They are usually installed around a rod to provide location and to transfer reaction forces to the structure. The rod diameter should be approximately 90% of the *ID* of the spring at the maximum load.

Stress Calculations

The stress in the coils of a helical torsion spring is *bending stress* created because the applied moment tends to bend each coil into a smaller diameter. Thus, the stress is computed from a form of the flexure formula, $\sigma = Mc/I$, modified to account for the curved wire. Also, because most torsion springs are made from round wire, the section modulus I/c is $S = \pi D_w^3/32$. Then

$$\sigma = \frac{McK_b}{I} = \frac{MK_b}{S} = \frac{MK_b}{\pi D_w^3/32} = \frac{32MK_b}{\pi D_w^3} \qquad\qquad (19\text{–}19)$$

K_b is the curvature correction factor and is reported by Wahl (see Reference 7) to be

$$K_b = \frac{4C^2 - C - 1}{4C(C - 1)} \qquad\qquad (19\text{–}20)$$

where C = spring index

Deflection, Spring Rate, and Number of Coils

The basic equation governing deflection is

$$\theta' = ML_w/EI$$

where θ' = angular deformation of the spring in radians (rad)
  M = applied moment, or torque
  L_w = length of wire in the spring
  E = tensile modulus of elasticity
  I = moment of inertia of the spring wire

We can substitute equations for L_w and I and convert θ' (radians) to θ (revolutions) to produce a more convenient form for application to torsion springs:

$$\theta = \frac{ML_w}{EI} = \frac{M(\pi D_m N_a)}{E(\pi D_w^4/64)} \frac{1 \text{ rev}}{2\pi \text{ rad}} = \frac{10.2 M D_m N_a}{E D_w^4} \qquad \textbf{(19–21)}$$

To compute the spring rate, k_θ (moment per revolution), solve for M/θ:

$$k_\theta = \frac{M}{\theta} = \frac{E D_w^4}{10.2 D_m N_a} \qquad \textbf{(19–22)}$$

Friction between coils and between the ID of the spring and the guide rod may decrease the rate slightly from this value.

The number of coils, N_a, is composed of a combination of the number of coils in the body of the spring, called N_b, and the contribution of the ends as they are subjected to bending, also. Calling N_e the contribution of the ends having lengths L_1 and L_2, we have

$$N_e = (L_1 + L_2)/(3\pi D_m) \qquad \textbf{(19–23)}$$

Then compute $N_a = N_b + N_e$.

Design Stresses

Because the stress in a torsion spring is bending and not torsional shear, the design stresses are different from those used for compression and extension springs. Figures 19–24 through 19–29 include six graphs of the design stress versus wire diameter for the same alloys used earlier for compression springs.

Design Procedure for Torsion Springs

A general procedure for designing torsion springs is presented and illustrated with Example Problem 19–5.

FIGURE 19–28

Design bending stresses for torsion springs ASTM A401 steel wire, chromium-silicon alloy, oil-tempered (Reprinted from Harold Carlson, *Spring Designer's Handbook,* p. 148, by courtesy of Marcel Dekker, Inc.)

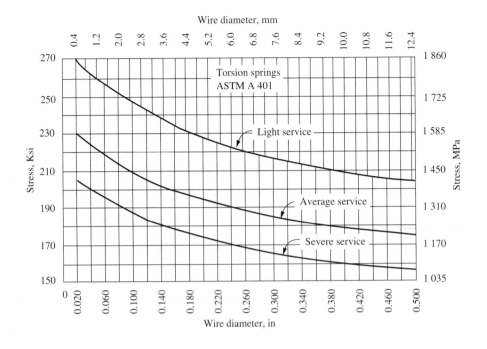

FIGURE 19–29

Design bending stresses for torsion springs ASTM A313 stainless steel wire, types 302 and 304, corrosion-resistant (Reprinted from Harold Carlson, *Spring Designer's Handbook,* p. 150, by courtesy of Marcel Dekker, Inc.)

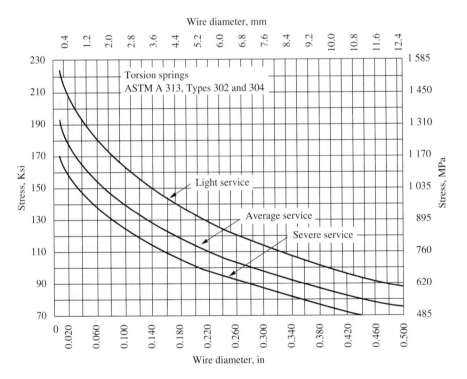

Example Problem 19–5   A timer incorporates a mechanism to close a switch after the timer rotates one complete revolution. The switch contacts are actuated by a torsion spring that is to be designed. A cam on the timer shaft slowly moves a lever attached to one end of the spring to a point where the maximum torque on the spring is 3.00 lb·in. At the end of the revolution, the cam permits the lever to rotate 60° suddenly with the motion produced by the energy stored in the spring. At this new position, the torque on the spring is 1.60 lb·in. Because of space limitations, the *OD* of the spring should not be greater than 0.50 in, and the length should not be greater than 0.75 in. Use music wire, ASTM A228 steel wire. The number of cycles of the spring will be moderate, so use design stresses for average service.

Solution   **Step 1.** Assume a trial value for the mean diameter and an estimate for the design stress.

Let's use a mean diameter of 0.400 in and estimate the design stress for A228 music wire, average service, to be 180 000 psi (Figure 19–25).

Step 2. Solve Equation (19–19) for the wire diameter, compute a trial size, and select a standard wire size. Let $K_b = 1.15$ as an estimate. Also use the largest applied torque:

$$D_w = \left[\frac{32MK_b}{\pi\sigma_d}\right]^{1/3} = \left[\frac{32(3.0)(1.15)}{\pi(180\ 000)}\right]^{1/3} = 0.058 \text{ in}$$

From Table 19–2, we can choose 25-gage music wire with a diameter of 0.059 in. For this size wire, the actual design stress for average service is 178 000 psi.

Step 3. Compute the *OD, ID,* spring index, and the new K_b:

$$OD = D_m + D_w = 0.400 + 0.059 = 0.459 \text{ in} \quad \text{(okay)}$$

$$ID = D_m - D_w = 0.400 - 0.059 = 0.341 \text{ in}$$

$$C = D_m/D_w = 0.400/0.059 = 6.78$$

$$K_b = \frac{4C^2 - C - 1}{4C(C - 1)} = \frac{4(6.78)^2 - 6.78 - 1}{4(6.78)(6.78 - 1)} = 1.123$$

Step 4. Compute the actual expected stress from Equation (19–19):

$$\sigma = \frac{32MK_b}{\pi D_w^3} = \frac{32(3.0)(1.123)}{(\pi)(0.059)^3} = 167\ 000 \text{ psi} \quad \text{(okay)}$$

Step 5. Compute the spring rate from the given data.

The torque exerted by the spring decreases from 3.00 to 1.60 lb·in as the spring rotates 60°. Convert 60° to a fraction of a revolution (rev):

$$\theta = \frac{60}{360} = 0.167 \text{ rev}$$

$$k_\theta = \frac{M}{\theta} = \frac{3.00 - 1.60}{0.167} = 8.40 \text{ lb·in/rev}$$

Step 6. Compute the required number of coils by solving for N_a from Equation (19–22):

$$N_a = \frac{ED_w^4}{10.2D_m k_\theta} = \frac{(29 \times 10^6)(0.059)^4}{(10.2)(0.400)(8.40)} = 10.3 \text{ coils}$$

Step 7. Compute the equivalent number of coils due to the ends of the spring from Equation (19–23).

This requires some design decisions. Let's use straight ends, 2.0 in long on one side and 1.0 in long on the other. These ends will be attached to the structure of the timer during operation. Then

$$N_e = (L_1 + L_2)/(3\pi D_m) = (2.0 + 1.0)/[3\pi(0.400)] = 0.80 \text{ coil}$$

Step 8. Compute the required number of coils in the body of the spring:

$$N_b = N_a - N_e = 10.3 - 0.8 = 9.5 \text{ coils}$$

Step 9. Complete the geometric design of the spring, including the size of the rod on which it will be mounted.

We first need the total angular deflection for the spring from the free condition to the maximum load. In this case, we know that the spring rotates 60° during operation. To this we must add the rotation from the free condition to the initial torque, 1.60 lb·in. Thus,

$$\theta_I = M_I/k_\theta = 1.60 \text{ lb·in}/(8.4 \text{ lb·in/rev}) = 0.19 \text{ rev}$$

Then the total rotation is

$$\theta_t = \theta_I + \theta_o = 0.19 + 0.167 = 0.357 \text{ rev}$$

From Equation (19–15), the mean diameter at the maximum operating torque is

$$D_m = D_{ml}N_a/(N_a + \theta_t) = [(0.400)(10.3)]/[(10.3 + 0.357)] = 0.387 \text{ in}$$

The minimum inside diameter is

$$ID_{\min} = 0.387 - D_w = 0.387 - 0.059 = 0.328 \text{ in}$$

The rod diameter on which the spring is mounted should be approximately 0.90 times this value. Then

$$D_r = 0.9(0.328) = 0.295 \text{ in} \quad (\text{say, } 0.30 \text{ in})$$

The length of the spring, assuming that all coils are originally touching, is computed from Equation (19–18):

$$L_{\max} = D_w(N_a + 1 + \theta_t) = (0.059)(10.3 + 1 + 0.356) = 0.688 \text{ in} \quad (\text{okay})$$

This value for length is the maximum space required in the direction along the axis of the coil when the spring is fully actuated. The specifications allow an axial length of 0.75 in, so this design is acceptable.

REFERENCES

1. Associated Spring, Barnes Group, Inc. *Engineering Guide to Spring Design.* Bristol, CT: Associated Spring, 1987.

2. Carlson, Harold. *Spring Designer's Handbook.* New York: Marcel Dekker, 1978.

3. Carlson, Harold. *Springs—Troubleshooting and Failure Analysis.* New York: Marcel Dekker, 1980.

4. Faires, V. M. *Design of Machine Elements.* 4th ed. New York: Macmillan, 1965.

5. Oberg, E., et al. *Machinery's Handbook.* 25th ed. New York: Industrial Press, 1996.

6. Shigley, J. E., and C. R. Mischke. *Mechanical Engineering Design.* 5th ed. New York: McGraw-Hill, 1989.

7. Wahl, A. M. *Mechanical Springs.* New York: McGraw-Hill, 1963.

PROBLEMS

Compression Springs

1. A spring has an overall length of 2.75 in when it is not loaded and a length of 1.85 in when carrying a load of 12.0 lb. Compute its spring rate.

2. A spring is loaded initially with a load of 4.65 lb and has a length of 1.25 in. The spring rate is given to be 18.8 lb/in. What is the free length of the spring?

3. A spring has a spring rate of 76.7 lb/in. At a load of 32.2 lb, it has a length of 0.830 in. Its solid length is 0.626 in. Compute the force required to compress the spring to solid height. Also compute the free length of the spring.

4. A spring has an overall length of 63.5 mm when it is not loaded and a length of 37.1 mm when carrying a load of 99.2 N. Compute its spring rate.

5. A spring is loaded initially with a load of 54.05 N, and it has a length of 39.47 mm. The spring rate is given to be 1.47 N/mm. What is the free length of the spring?

6. A spring has a spring rate of 8.95 N/mm. At a load of 134 N, it has a length of 29.4 mm. Its solid length is 21.4 mm. Compute the force required to compress the spring to solid height. Also compute the free length of the spring.

7. A helical compression spring with squared and ground ends has an outside diameter of 1.100 in, a wire diameter of 0.085 in, and a solid height of 0.563 in. Compute the *ID,* the mean diameter, the spring index, and the approximate number of coils.

8. The following data are known for a spring:

> Total number of coils = 19
>
> Squared and ground ends
>
> Outside diameter = 0.560 in
>
> Wire diameter = 0.059 in (25-gage music wire)
>
> Free length = 4.22 in

For this spring, compute the spring index, the pitch, the pitch angle, and the solid length.

9. For the spring of Problem 8, compute the force required to reduce its length to 3.00 in. At that force, compute the stress in the spring. Would that stress be satisfactory for average service?

10. Would the spring of Problems 8 and 9 tend to buckle when compressed to the length of 3.00 in?

11. For the spring of Problem 8, compute the estimate for the outside diameter when it is compressed to solid length.

12. The spring of Problem 8 must be compressed to solid length to install it. What force is required to do this? Compute the stress at solid height. Is that stress satisfactory?

13. A support bar for a machine component is suspended to soften applied loads. During operation, the load on each spring varies from 180 to 220 lb. The position of the rod is to move no more than 0.500 in as the load varies. Design a compression spring for this application. Several million cycles of load application are expected. Use ASTM A229 steel wire.

14. Design a helical compression spring to exert a force of 22.0 lb when compressed to a length of 1.75 in. When its length is 3.00 in, it must exert a force of 5.0 lb. The spring will be cycled rapidly, with severe service required. Use ASTM A401 steel wire.

15. Design a helical compression spring for a pressure relief valve. When the valve is closed, the spring length is 2.0 in, and the spring force is to be 1.50 lb. As the pressure on the valve increases, a force of 14.0 lb causes the valve to open and compress the spring to a length of 1.25 in. Use corrosion-resistant ASTM A313, type 302 steel wire, and design for average service.

16. Design a helical compression spring to be used to return a pneumatic cylinder to its original position after being actuated. At a length of 10.50 in, the spring must exert a force of 60 lb. At a length of 4.00 in, it must exert a force of 250 lb. Severe service is expected. Use ASTM A231 steel wire.

17. Design a helical compression spring using music wire that will exert a force of 14.0 lb when its length is 0.68 in. The free length is to be 1.75 in. Use average service.

18. Design a helical compression spring using stainless steel wire, ASTM A313, type 316, for average service, which will exert a force of 8.00 lb after deflecting 1.75 in from a free length of 2.75 in.

19. Repeat Problem 18 with the additional requirement that the spring must operate around a rod having a diameter of 0.625 in.

20. Repeat Problem 17 with the additional requirement that the spring is to be installed in a hole having a diameter of 0.750 in.

21. Design a helical compression spring using ASTM A231 steel wire for severe service. The spring will exert a force

of 45.0 lb at a length of 3.05 in and a force of 22.0 lb at a length of 3.50 in.

22. Design a helical compression spring using round steel wire, ASTM A227. The spring will actuate a clutch and must withstand several million cycles of operation. When the clutch discs are in contact, the spring will have a length of 2.50 in and must exert a force of 20 lb. When the clutch is disengaged, the spring will be 2.10 in long and must exert a force of 35 lb. The spring will be installed around a round shaft having a diameter of 1.50 in.

23. Design a helical compression spring using round steel wire, ASTM A227. The spring will actuate a clutch and must withstand several million cycles of operation. When the clutch discs are in contact, the spring will have a length of 60 mm and must exert a force of 90 N. When the clutch is disengaged, the spring will be 50 mm long and must exert a force of 155 N. The spring will be installed around a round shaft having a diameter of 38 mm.

24. Evaluate the performance of a helical compression spring made from 17-gage ASTM A229 steel wire that has an outside diameter of 0.531 in. It has a free length of 1.25 in, squared and ground ends, and a total of 7.0 coils. Compute the spring rate and the deflection and stress when carrying 10.0 lb. At this stress level, for what service (light, medium, or severe) would the spring be suitable?

Extension Springs

For Problems 25–31, be sure that the stress in the spring under initial tension is within the range suggested in Figure 19–21.

25. Design a helical extension spring using music wire to exert a force of 7.75 lb when the length between attachment points is 2.75 in, and a force of 5.25 lb at a length of 2.25 in. The outside diameter must be less than 0.300 in. Use severe service.

26. Design a helical extension spring for average service using music wire to exert a force of 15.0 lb when the length between attachment points is 5.00 in, and a force of 5.20 lb at a length of 3.75 in. The outside diameter must be less than 0.75 in.

27. Design a helical extension spring for severe service using music wire to exert a maximum force of 10.0 lb at a length of 3.00 in. The spring rate should be 6.80 lb/in. The outside diameter must be less than 0.75 in.

28. Design a helical extension spring for severe service using music wire to exert a maximum force of 10.0 lb at a length

of 6.00 in. The spring rate should be 2.60 lb/in. The outside diameter must be less than 0.75 in.

29. Design a helical extension spring for average service using music wire to exert a maximum force of 10.0 lb at a length of 9.61 in. The spring rate should be 1.50 lb/in. The outside diameter must be less than 0.75 in.

30. Design a helical extension spring for average service using stainless steel wire, ASTM A313, type 302, to exert a maximum force of 162 lb at a length of 10.80 in. The spring rate should be 38.0 lb/in. The outside diameter should be approximately 1.75 in.

31. An extension spring has an end similar to that shown in Figure 19–22. The pertinent data are as follows: U.S. Wire Gage no. 19; mean diameter = 0.28 in; $R_1 = 0.25$ in; $R_2 = 0.094$ in. Compute the expected stresses at points A and B in the figure for a force of 5.0 lb. Would those stresses be satisfactory for ASTM A227 steel wire for average service?

Torsion Springs

32. Design a helical torsion spring for average service using stainless steel wire, ASTM A313, type 302, to exert a torque of 1.00 lb·in after a deflection of 180° from the free condition. The outside diameter of the coil should be no more than 0.500 in. Specify the diameter of a rod on which to mount the spring.

33. Design a helical torsion spring for severe service using stainless steel wire, ASTM A313, type 302, to exert a torque of 12.0 lb·in after a deflection of 270° from the free condition. The outside diameter of the coil should be no more than 1.250 in. Specify the diameter of a rod on which to mount the spring.

34. Design a helical torsion spring for severe service using music wire to exert a maximum torque of 2.50 lb·in after a deflection of 360° from the free condition. The outside diameter of the coil should be no more than 0.750 in. Specify the diameter of a rod on which to mount the spring.

35. A helical torsion spring has a wire diameter of 0.038 in; an outside diameter of 0.368 in; 9.5 coils in the body; one end 0.50 in long; the other end 1.125 in long; and a material of ASTM A401 steel. What torque would cause the spring to rotate 180°? What would the stress be then? Would it be safe?

20

Machine Frames, Bolted Connections, and Welded Joints

Machine Frames, Bolted Connections, and Welded Joints

Discussion Map

☐ As you develop the design of machine elements (as you have throughout this book), you must also design the housing, frame, or structure that supports them and protects them from the elements.

Discover

Select a variety of products, machines, vehicles, and even toys. Observe how they are built. What is the basic shape of the structure that holds everything together? Why was that shape chosen by the designer? What functions are performed by the frame?

What kinds of forces, bending moments, and torsional moments (torques) are produced when the product is operating? How are they managed and controlled? What is the load path *that delivers them to the ultimate structure?*

This chapter will help you identify some efficient approaches to the design of structures and frames and to analyze the performance of fasteners and welded joints loaded in many different ways.

Throughout this book, you have studied individual machine elements while continuously considering how those elements must work together in a more comprehensive machine. As the design progresses, there comes a time when *you must put it all together.* But then you are faced with the questions, "What do I put it in? How do I hold all of the functional components safely, allowing assembly and service while providing a secure, rigid structure?"

It is impractical to generate a completely general approach to the design of a structure or a frame for a machine, a vehicle, a consumer product, or even a toy. Each is different with regard to its functions; the number, size, and type of components in the product; the intended use; and the expected demand for an aesthetic design. For example, toys often exhibit clever design approaches because the manufacturer wants to provide a safe, functional toy while minimizing the material used and the amount of personnel time required to produce the toy.

In this chapter you will explore some basic concepts for creating a satisfactory frame design, considering the shape of structural components, material properties, the use of fasteners such as bolts, and the fabrication of welded assemblies. You will learn some of the techniques for analyzing and designing bolted assemblies to consider loads on the bolts in several directions. The design of welded joints to be safe and rigid is discussed.

Chapter 18 covered part of the story, that dealing with bolts loaded in pure tension, as in a clamping function. This chapter extends that chapter to consider eccentrically loaded joints: those that must resist a combination of direct shear and a bending moment on a bolt pattern.

The ability of a welded joint to carry a variety of loads is discussed with the objective of designing the weld. Here both uniformly loaded and eccentrically loaded joints are treated.

To appreciate the value of such study, select a variety of products, machines, and vehicles, and observe how they are built. What is the basic shape of the structure that holds everything together? Where are forces, bending moments, and torsional moments (torques) generated? What kinds of stress do they create? Consider the *load path* by following a force from the place where it is generated through all of the means by which that force or its effects are passed to a series of members and to the point where it is supported by a basic frame of the machine or where it is delivered out of the system of interest. Considering the load paths for all forces that exist in a mechanical device should give you an understanding of the desirable characteristics of the structure and should help you know how to produce a design that optimizes the management of the forces exerted on it.

Take apart a variety of mechanical devices to observe their structure. How did shape contribute to the rigidity and safety of the device? Is it rather stiff? Or is it more flexible? Are there ribs or thickened sections to provide special strength or rigidity for certain parts?

This chapter will help you identify some efficient approaches to the design of structures and frames and to analyze the performance of fasteners and welded joints loaded in many ways.

The subject of machine frames and structures is quite complex. It is discussed from the standpoint of general principles and guidelines, rather than specific design techniques. Critical frames are typically designed with computerized finite-element analysis. Also, experimental stress analysis techniques are often used to verify designs.

 You Are the Designer

In Chapter 16, you were the designer of bearings for conveyor systems for a large product distribution center. Now, how do you design the frame and the structure of the conveyor system? What general form is desirable? What materials and shape should be used for the structural elements? Are the elements loaded in tension, compression, bending, shear, torsion, or some combination of these types of stress? How do the manner of loading and the nature of the stress in the structure affect these decisions? Should the frame be fabricated from standard structural shapes and bolted together? Or should it be made from steel plate and welded? How about using aluminum? Or should it be cast from cast iron or cast steel? Can it be molded from plastic? Can composites be used? How would the weight of the structure be affected? How much rigidity is desirable for this kind of structure? What shapes of load-carrying elements contribute to a stiff, rigid structure, as well as one that is safe in resisting applied stresses?

If the structure involves bolting or welding, how are the joints to be designed? What forces, both magnitude and direction, must be carried by the fasteners or the welds?

The material in this chapter will help you make some of these design decisions. Much of the information is general in nature rather than giving you specific design procedures. You must exercise judgment and creativity not only in the design of the conveyor frame but also in the analysis of its components. Because the frame design could evolve into a form too complex for analysis using traditional techniques of stress analysis, you may have to employ finite-element modeling to determine whether the design is adequate or, perhaps, overdesigned. Perhaps one or more prototypes should be built for testing.

**20–1
OBJECTIVES OF
THIS CHAPTER**

After completing this chapter, you will be able to:

1. Apply the principles of stress and deflection analysis to propose a reasonable and efficient shape for a structure or frame and for the components involved.

2. Specify materials that are well suited to the demands of a given design, given certain conditions of load, environment, fabrication requirements, safety, and aesthetics.

3. Analyze eccentrically loaded bolted joints.

4. Design welded joints to carry many types of loading patterns.

**20–2
MACHINE
FRAMES AND
STRUCTURES**

The design of machine frames and structures is largely art in that the components of the machine must be accommodated. The designer is often restricted in where supports can be placed in order not to interfere with the operation of the machine or in order to provide access for assembly or service.

Step 7. Determine the resultant of all forces acting on each bolt. A vector summation can be performed either analytically or graphically, or each force can be resolved into horizontal and vertical components. The components can be summed and then the resultant can be computed.

Let's use the latter approach for this problem. The shear force acts directly downward, in the y-direction. The x- and y-components of F_1 are

$$F_{1x} = F_1 \sin \theta = (4\ 200\ \text{lb})\sin(36.9°) = 2\ 520\ \text{lb}$$

$$F_{1y} = F_1 \cos \theta = (4\ 200)\cos(36.9°) = 3\ 360\ \text{lb}$$

The total force in the y-direction is then

$$F_{1y} + F_s = 3\ 360 + 875 = 4\ 235\ \text{lb}$$

Then the resultant force on bolt 1 is

$$R_1 = \sqrt{(2\ 520)^2 + (4\ 235)^2} = 4\ 928\ \text{lb}$$

Step 8. Specify the bolt material; compute the required area for the bolt; and select an appropriate size. For this problem, let's specify ASTM A325 steel for the bolts having an allowable shear stress of 17 500 psi from Table 20–1. Then the required area for the bolt is

$$A_s = \frac{R_1}{\tau_a} = \frac{4\ 928\ \text{lb}}{17\ 500\ \text{lb/in}^2} = 0.282\ \text{in}^2$$

FIGURE 20–5

Geometry of bolted joint and forces on bolt 1

(a) Proposed bolt pattern

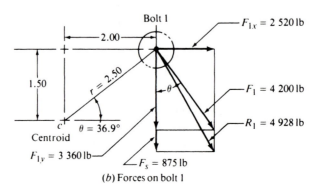

(b) Forces on bolt 1

The required diameter would be

$$D = \sqrt{\frac{4A_s}{\pi}} = \sqrt{\frac{4(0.282 \text{ in}^2)}{\pi}} = 0.599 \text{ in}$$

Let's specify a 5/8-in bolt having a diameter of 0.625 in.

20–4
WELDED JOINTS

The design of welded joints requires consideration of the manner of loading on the joint, the types of materials in the weld and in the members to be joined, and the geometry of the joint itself. The load may be either uniformly distributed over the weld such that all parts of the weld are stressed to the same level, or the load may be eccentrically applied. Both are discussed in this section.

The materials of the weld and the parent members determine the allowable stresses. Table 20–2 lists several examples for steel and aluminum. The allowables listed are for shear on fillet welds. For steel, welded by the electric arc method, the type of electrode is an indication of the tensile strength of the filler metal. For example, the E70 electrode has a minimum tensile strength of 70 Ksi (483 MPa). Additional data are available in publications of the American Welding Society (AWS), the American Institute for Steel Construction (AISC), and the Aluminum Association (AA).

Types of Joints

Joint type refers to the relationship between mating parts, as illustrated in Figure 20–6. The butt weld allows a joint to be the same nominal thickness as the mating parts and is usually loaded in tension. If the joint is properly made with the appropriate weld metal, the joint

TABLE 20–2 Allowable shear stresses on fillet welds

A. Steel

Electrode type	Typical metals joined (ASTM grade)	Allowable shear stress
E60	A36, A500	18 Ksi (124 MPa)
E70	A242, A441	21 Ksi (145 MPa)
E80	A572, Grade 65	24 Ksi (165 MPa)
E90		27 Ksi (186 MPa)
E100		30 Ksi (207 MPa)
E110		33 Ksi (228 MPa)

B. Aluminum

Metal joined	Filler Alloy							
	1100		4043		5356		5556	
	Allowable Shear Stress							
	Ksi	MPa	Ksi	MPa	Ksi	MPa	Ksi	MPa
1100	3.2	22	4.8	33				
3003	3.2	22	5.0	34				
6061			5.0	34	7.0	48	8.5	59
6063			5.0	34	6.5	45	6.5	45

FIGURE 20–6   Types of weld joints

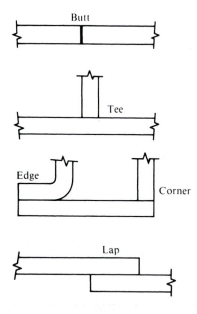

FIGURE 20–7   Some types of welds showing edge preparation

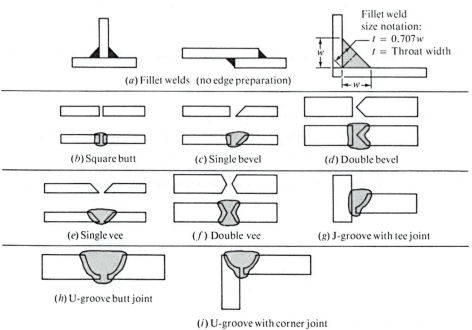

(a) Fillet welds (no edge preparation)

Fillet weld size notation:
$t = 0.707w$
t = Throat width

(b) Square butt

(c) Single bevel

(d) Double bevel

(e) Single vee

(f) Double vee

(g) J-groove with tee joint

(h) U-groove butt joint

(i) U-groove with corner joint

will be stronger than the parent metal. Thus, no special analysis of the joint is required if the joined members themselves are shown to be safe. Caution is advised, however, when the materials to be joined are adversely affected by the heat of the welding process. Heat-treated steels and many aluminum alloys are examples. The other types of joints in Figure 20–6 are assumed to place the weld in shear.

Types of Welds

Figure 20–7 shows several types of welds named for the geometry of the edges of the parts to be joined. Note the special edge preparation required, especially for thick plates, to permit the welding rod to enter the joint and build a continuous weld bead.

TABLE 20–3 Allowable shear stresses and forces on welds

Base metal ASTM grade	Electrode	Allowable shear stress	Allowable force per inch of leg
Building-type structures:			
A36, A441	E60	13 600 psi	9 600 lb/in
A36, A441	E70	15 800 psi	11 200 lb/in
Bridge-type structures:			
A36	E60	12 400 psi	8 800 lb/in
A441, A242	E70	14 700 psi	10 400 lb/in

Size of Weld

The five types of groove-type welds in Figure 20–7 are made as complete penetration welds. Then, as indicated before for butt welds, the weld is stronger than the parent metals, and no further analysis is required.

Fillet welds are typically made as equal-leg right triangles, with the size of the weld indicated by the length of the leg. A fillet weld loaded in shear would tend to fail along the shortest dimension of the weld that is the line from the root of the weld to the theoretical face of the weld and normal to the face. The length of this line is found from simple trigonometry to be $0.707w$, where w is the leg dimension.

The objectives of the design of a fillet welded joint are to specify the length of the legs of the fillet; the pattern of the weld; and the length of the weld. Presented here is the method that treats the weld as a line having no thickness. The method involves determining the *maximum force per inch* of weld leg length. Comparing the actual force with an allowable force allows the calculation of the required leg length.

Table 20–3 gives data for the allowable shear stress and the allowable force per inch for some combinations of base metal and welding electrode. In general, the allowables for building-type structures are for steady loads. The values for bridge-type loading accounts for the cyclic effects. For true fatigue-type repeated loading, refer to the literature. (See References 2 and 3.)

Method of Treating Weld as a Line

Four different types of loading are considered here: (1) direct tension or compression, (2) direct vertical shear, (3) bending, and (4) twisting. The method allows the designer to perform calculations in a manner very similar to that used to design the load-carrying members themselves. In general, the weld is analyzed separately for each type of loading to determine the force per inch of weld size due to each load. The loads are then combined vectorially to determine the maximum force. This maximum force is compared with the allowables from Table 20–3 to determine the size of the weld required.

The relationships used are summarized next:

Type of Loading	Formula (and Equation Number) for Force per Inch of Weld	
Direct tension or compression	$f = P/A_w$	(20–4)
Direct vertical shear	$f = V/A_w$	(20–5)
Bending	$f = M/S_w$	(20–6)
Twisting	$f = Tc/J_w$	(20–7)

FIGURE 20–8

Geometry factors for weld analysis

TABLE 20–4 Minimum weld sizes for thick plates

Plate thickness (in)	Minimum leg size for fillet weld (in)
≤1/2	3/16
>1/2–3/4	1/4
>3/4–$1\frac{1}{2}$	5/16
>$1\frac{1}{2}$–$2\frac{1}{4}$	3/8
>$2\frac{1}{4}$–6	1/2
>6	5/8

In these formulas, the geometry of the weld is used to evaluate the terms A_w, S_w, and J_w, using the relationships shown in Figure 20–8. Note the similarity between these formulas and those used to perform the stress analysis. Note, also, the similarity between the geometry factors for welds and the properties of areas used for the stress analysis. Because the weld is treated as a line having no thickness, the units for the geometry factors are different from those of the area properties, as indicated in Figure 20–8.

The use of this method of weld analysis will be demonstrated with example problems. In general, the method requires the following steps:

1. Propose the geometry of the joint and the design of the members to be joined.
2. Identify the types of stresses to which the joint is subjected (bending, twisting, vertical shear, direct tension, or compression).
3. Analyze the joint to determine the magnitude and the direction of the force on the weld due to each type of load.
4. Combine the forces vectorially at the point or points of the weld where the forces appear to be maximum.
5. Divide the maximum force on the weld by the allowable force from Table 20–3 to determine the required leg size for the weld. Note that when thick plates are welded, there are minimum acceptable sizes for the welds as listed in Table 20–4.

Example Problem 20–2 Design a bracket similar to that in Figure 20–4, but use welding to attach the bracket to the column. The bracket is 6.00 in high and is made from ASTM A36 steel having a thickness of 1/2 in. The column is also made from A36 steel and is 8.00 in wide.

Solution *Step 1.* The proposed geometry is a design decision and may have to be subjected to some iteration to achieve an optimum design. For a first trial, let's use the C-shaped weld pattern shown in Figure 20–9.

Step 2. The weld will be subjected to direct vertical shear and twisting caused by the 3 500-lb load on the bracket.

Step 3. To compute the forces on the weld, we must know the geometry factors A_w and J_w. Also, the location of the centroid of the weld pattern must be computed [see Figure 20–9(b)]. Use Case 5 in Figure 20–8.

$$A_w = 2b + d = 2(4) + 6 = 14 \text{ in}$$

$$J_w = \frac{(2b + d)^3}{12} - \frac{b^2(b + d)^2}{(2b + d)} = \frac{(14)^3}{12} - \frac{16(10)^2}{14} = 114.4 \text{ in}^3$$

$$\bar{x} = \frac{b^2}{2b + d} = \frac{16}{14} = 1.14 \text{ in}$$

Force Due to Vertical Shear

$$V = P = 3\,500 \text{ lb}$$

$$f_s = P/A_w = (3\,500 \text{ lb})/14 \text{ in} = 250 \text{ lb/in}$$

This force acts vertically downward on all parts of the weld.

Forces Due to the Twisting Moment

$$T = P[8.00 + (b - \bar{x})] = 3\,500[8.00 + (4.00 - 1.14)]$$

$$T = 3\,500(10.86) = 38\,010 \text{ lb·in}$$

The twisting moment causes a force to be exerted on the weld that is perpendicular to a radial line from the centroid of the weld pattern to the point of interest. In this case, the end of the weld to the upper right experiences the greatest force. It is most convenient to break the force down into horizontal and vertical components and then subsequently recombine all such components to compute the resultant force:

$$f_{th} = \frac{Tc_v}{J_w} = \frac{(38\,010)(3.00)}{114.4} = 997 \text{ lb/in}$$

$$f_{tv} = \frac{Tc_h}{J_w} = \frac{(38\,010)(2.86)}{114.4} = 950 \text{ lb/in}$$

Step 4. The vectorial combination of the forces on the weld is shown in Figure 20–9(c). Thus, the maximum force is 1 560 lb/in.

FIGURE 20–9
C-shaped weld bracket

(a) Basic design of bracket   (b) Dimensions of bracket

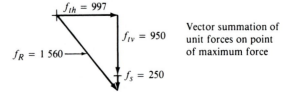

(c) Analysis of forces

Step 5. Selecting an E60 electrode for the welding, we find that the allowable force per inch of weld leg size is 9 600 lb/in (Table 20–3). Then the required weld leg size is

$$w = \frac{1\ 560\ \text{lb/in}}{9\ 600\ \text{lb/in per in of leg}} = 0.163\ \text{in}$$

Table 20–4 shows that the minimum size weld for a 1/2-in plate is 3/16 in (0.188 in). That size should be specified.

Example Problem 20–3   A steel strap, 1/4 in thick, is to be welded to a rigid frame to carry a dead load of 12 500 lb, as shown in Figure 20–10. Design the strap and its weld.

Solution   The basic objectives of the design are to specify a suitable material for the strap, the welding electrode, the size of the weld, and the dimensions W and h, as shown in Figure 20–10.

Let's specify that the strap is to be made from ASTM A441 structural steel and that it is to be welded with an E70 electrode, using the minimum size weld, 3/16 in. Appendix 7 gives the yield strength of the A441 steel as 42 000 psi. Using a design factor of 2, we can compute an allowable stress of

$$\sigma_a = 42\ 000/2 = 21\ 000\ \text{psi}$$

Then the required area of the strap is

$$A = \frac{P}{\sigma_a} = \frac{12\ 500\ \text{lb}}{21\ 000\ \text{lb/in}^2} = 0.595\ \text{in}^2$$

But the area is $W \times t$, where $t = 0.25$ in. Then the required width W is

$$W = A/t = 0.595/0.25 = 2.38\ \text{in}$$

Let's specify that $W = 2.50$ in.

To compute the required length of the weld h, we need the allowable force on the 3/16-in weld. Table 20–3 indicates the allowable force on the A441 steel welded with an E70 electrode to be 11 200 lb/in per in of leg size. Then

$$f_a = \frac{11\ 200\ \text{lb/in}}{1.0\ \text{in leg}} \times 0.188\ \text{in leg} = 2\ 100\ \text{lb/in}$$

The actual force on the weld is

$$f_a = P/A_w = P/2h$$

FIGURE 20–10
Steel strap

P = 12 500 lb

Then solving for h gives

$$h = \frac{P}{2(f_a)} = \frac{12\ 500\ \text{lb}}{2(2\ 100\ \text{lb/in})} = 2.98\ \text{in}$$

Let's specify $h = 3.00$ in.

Example Problem 20–4 Evaluate the design shown in Figure 20–11 with regard to stress in the welds. All parts of the assembly are made of ASTM A36 structural steel and are welded with an E60 electrode. The 2 500-lb load is a dead load.

Solution The critical point would be the weld at the top of the tube where it is joined to the vertical surface. At this point, there is a three-dimensional force system acting on the weld as shown in Figure 20–12. The offset location of the load causes a twisting on the weld that produces a force f_t on the weld toward the left in the y-direction. The bending produces a force f_b, acting outward along the x-axis. The vertical shear force f_s acts downward along the z-axis.

From statics, the resultant of the three force components would be

$$f_R = \sqrt{f_t^2 + f_b^2 + f_s^2}$$

Now each component force on the weld will be computed.

FIGURE 20–11
Bracket assembly

FIGURE 20–12
Force vectors

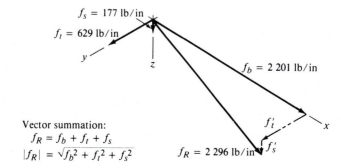

Vector summation:
$$f_R = f_b + f_t + f_s$$
$$|f_R| = \sqrt{f_b^2 + f_t^2 + f_s^2}$$

Twisting Force, f_t

$$f_t = \frac{Tc}{J_w}$$

$$T = (2\,500\text{ lb})(8.00\text{ in}) = 20\,000\text{ lb·in}$$
$$c = OD/2 = 4.500/2 = 2.25\text{ in}$$
$$J_w = (\pi)(OD)^3/4 = (\pi)(4.500)^3/4 = 71.57\text{ in}^3$$

Then

$$f_t = \frac{Tc}{J_w} = \frac{(20\,000)(2.25)}{71.57} = 629\text{ lb/in}$$

Bending Force, f_b

$$f_b = \frac{M}{S_w}$$

$$M = (2\,500\text{ lb})(14.00\text{ in}) = 35\,000\text{ lb·in}$$
$$S_w = (\pi)(OD)^2/4 = (\pi)(4.500)^2/4 = 15.90\text{ in}^2$$

Then

$$f_b = \frac{M}{S_w} = \frac{35\,000}{15.90} = 2\,201\text{ lb/in}$$

Vertical Shear Force, f_s

$$f_s = \frac{P}{A_w}$$

$$A_w = (\pi)(OD) = (\pi)(4.500\text{ in}) = 14.14\text{ in}$$
$$f_s = \frac{P}{A_w} = \frac{2\,500}{14.14} = 177\text{ lb/in}$$

Now the resultant can be computed:

$$f_R = \sqrt{f_t^2 + f_b^2 + f_s^2}$$
$$f_R = \sqrt{629^2 + 2\,201^2 + 177^2} = 2\,296\text{ lb/in}$$

Comparing this with the allowable force on a 1.0-in weld gives

$$w = \frac{2\,296\text{ lb/in}}{9\,600\text{ lb/in per in of leg size}} = 0.239\text{ in}$$

The 1/4-in fillet specified in Figure 20–11 is satisfactory.

REFERENCES

1. American Institute of Steel Construction. *Manual of Steel Construction.* 9th ed. New York: American Institute of Steel Construction, 1989.

2. Blodgett, Omer W. *Design of Welded Structures.* Cleveland, OH: James F. Lincoln Arc Welding Foundation, 1966.

3. Blodgett, Omer W. *Design of Weldments.* Cleveland, OH: James F. Lincoln Arc Welding Foundation, 1963.

4. Marshek, Kurt M. *Design of Machine and Structural Parts.* New York: John Wiley & Sons, 1987.

5. Mott, Robert L. *Applied Strength of Materials.* 3d ed. Upper Saddle River, NJ: Prentice-Hall, 1996.

6. Slocum, Alexander H. *Precision Machine Design.* Upper Saddle River, NJ: Prentice-Hall, 1992.

7. Weiser, Peter F., ed. *Steel Casting Handbook.* 5th ed. Rocky River, OH: Steel Founder's Society of America, 1980.

PROBLEMS

For Problems 1–6, design a bolted joint to join the two members shown in the appropriate figure. Specify the number of bolts, the pattern, the bolt grade, and the bolt size.

1. Figure 20–13

FIGURE 20–13   (Problems 1, 7, and 13)

2. Figure 20–14

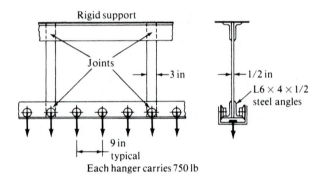

FIGURE 20–14   (Problem 2)

3. Figure 20–15

FIGURE 20–15   (Problem 3)

4. Figure 20–16

FIGURE 20–16   (Problems 4 and 8)

5. Figure 20–17

6. Figure 20–18

For Problems 7–12, design a welded joint to join the two members shown in the appropriate figure. Specify the weld pattern, the type of electrode to be used, and the size of weld. In Problems 7–9, the members are ASTM A36 steel. In Problems 10–12, the members are ASTM A441 steel. Use the method of treating the joint as a line, and use the allowable forces per inch of leg for building-type structures from Table 20–3.

7. Figure 20–13

8. Figure 20–16

9. Figure 20–17

10. Figure 20–18

11. Figure 20–19

12. Figure 20–19 (but $P_2 = 0$)

For Problems 13–16, design a welded joint to join the two aluminum members shown in the appropriate figure. Specify the weld pattern, the type of filler alloy, and the size of weld. The types of materials joined are listed in the problems.

13. Figure 20–13: 6061 alloy (but $P = 4\ 000$ lb)

14. Figure 20–20: 6061 alloy

15. Figure 20–21: 6063 alloy

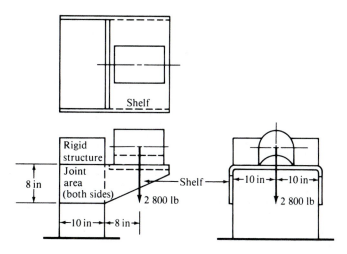

FIGURE 20–17 (Problems 5 and 9)

Load shared equally by *four* brackets (only two shown)

FIGURE 20–18 (Problems 6 and 10)

FIGURE 20–19 (Problems 11 and 12)

FIGURE 20–20 (Problem 14)

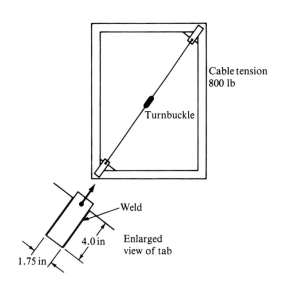

FIGURE 20–21 (Problem 15)

16. Figure 20–22: 3003 alloy

FIGURE 20–22   (Problem 16)

17. Compare the weight of a tensile rod carrying a dead load of 4 800 lb if it is made from (a) AISI 1020 HR steel; (b) AISI 5160 OQT 1300 steel; (c) aluminum 2014-T6; (d) aluminum 7075-T6; (e) titanium 6A1-4V, annealed; and (f) titanium 3A1-13V-11Cr, aged. Use $N = 2$ based on yield strength.

21

Electric Motors

Electric Motors

Discussion Map

☐ Electric motors provide the power for a huge array of products in homes, factories, schools, commercial facilities, transportation equipment, and many portable devices. The two major classifications of motors are *alternating current (AC)* and *direct current (DC)*. Some can operate on either type of power.

Discover

Some motors that will operate well in the United States will run differently or will not run at all in other countries. Why? What are the electric voltage and frequency standards in various parts of the world?

Look for a variety of machines and products that are powered by electric motors. Choose both large and small devices; some portable and some that plug into standard electrical outlets. Look around your home, at work, and in a factory.

Try to find the nameplate for each motor, and copy down as much information as you can. How are the nameplate data related to the performance characteristics of the motor? Is it an AC motor or a DC motor? How fast does it run? What is its electrical rating in terms of voltage and current?

This chapter will help you identify many kinds of motors, understand the general performance characteristics of each, and apply them properly.

The electric motor is very widely used for providing the prime power to industrial machinery, consumer products, and business equipment. This chapter will describe the various types of motors and will discuss their performance characteristics. The objective is to provide you with the background needed to specify motors and to communicate with vendors to acquire the proper motor for a given application.

Types of motors discussed in this chapter are direct current (DC), alternating current (AC, both single-phase and three-phase), universal motors, stepper motors, and variable-speed AC motors. Motor controls are also discussed.

Look for a variety of machines and products that are powered by electric motors. Choose both large and small devices; some portable and some that plug into standard electrical outlets. Look around your home, at work, and in a factory if possible.

Try to find the nameplate for each motor and copy down as much information as you can. What does each piece of information mean? Are some electrical wiring diagrams included? How are the nameplate data related to the performance characteristics of the motor? Is it an AC motor or a DC motor? How fast does it run? What is its electrical rating in terms of voltage and current? For the motors in portable devices, what kind of power supply is used? What kinds of batteries are required? How many? How are they connected? In series? In parallel? How do those factors relate to the voltage rating of the motor? What is the relationship between the kind of work done by the motor and the time between charging sessions?

Compare some of the larger motors with the photographs in Section 21–8 of this chapter. Can you identify the frame type? What company manufactured the motor?

See if you can find more information about the motors or their manufacturers from the Internet. Look up some of the company names in this chapter to learn more about the lines of motors that they offer. Do a broader search to find as many different types as you can.

What is the rated power for each motor? What units are used for the power rating? Convert all of the power ratings to watts. (See Appendix 18.) Then convert them all to horsepower to gain an appreciation for the physical size of various motors and to visualize the comparison between the SI unit of watts and the older U.S. Customary unit of horsepower.

Separate the motors you find into those that are AC and those that are DC. Do you see any significant differences between the two classes of motors? Within each class, what variations do you see? Do any of the nameplates describe the type of motor, such as *synchronous, universal, split-phase, NEMA design C,* or other designations?

What kinds of controls are connected to the motors? Switches? Starters? Speed controls? Protection devices?

This chapter will help you identify many kinds of motors, understand the general performance characteristics of each, and apply them properly.

 You Are the Designer

Consider a conveyor system that you are to design. One possibility for driving the conveyor system is to use an electric motor. What type should be used? At what speed will it operate? What type of electric power is available to supply the motor? What power is required? What type of housing and mounting style should be specified? What are the dimensions of the motor? How is the motor connected to the drive pulley of the conveyor system? The information in this chapter will help you to answer these and other questions.

21–1
OBJECTIVES OF
THIS CHAPTER

After completing this chapter, you will be able to:

1. Describe the factors that must be specified to select a suitable motor.
2. Describe the principles of operation of AC motors.
3. Identify the typical classifications of AC electric motors according to rated power.
4. Identify the common voltages and frequencies of AC power and the speed of operation of AC motors operating on those systems.
5. Describe single-phase and three-phase AC power.
6. Describe the typical frame designs, sizes, and enclosure styles of AC motors.
7. Describe the general form of a motor performance curve.
8. Describe the comparative performance of *shaded-pole, permanent-split capacitor, split-phase,* and *capacitor-start,* single-phase AC motors.
9. Describe three-phase, squirrel-cage AC motors.
10. Describe the comparative performance of *NEMA design B, NEMA design C, NEMA design D,* and *wound-rotor* three-phase AC motors.
11. Describe *synchronous motors.*
12. Describe *universal motors.*
13. Describe three means of producing DC power and the common voltages produced.
14. Describe the advantages and disadvantages of DC motors compared with AC motors.
15. Describe four basic designs of DC motors—*shunt-wound, series-wound, compound-wound,* and *permanent magnet*—and describe their performance curves.
16. Describe torque motors, servomotors, stepper motors, brushless DC motors, and printed circuit motors.

17. Describe motor control systems for system protection, speed control, starting, stopping, and overload protection.

18. Describe speed control of AC motors.

19. Describe the control of DC motors.

21–2
MOTOR
SELECTION
FACTORS

As a minimum, the following items must be specified for motors:

- Motor type: DC, AC, single-phase, three-phase, and so on
- Power rating and speed
- Operating voltage and frequency
- Type of enclosure
- Frame size
- Mounting details

In addition, there may be many special requirements that must be communicated to the vendor. The primary factors to be considered in selecting a motor include the following:

- Operating torque, operating speed, and power rating. Note that these are related by the equation

$$Power = torque \times speed$$

- Starting torque.
- Load variations expected and corresponding speed variations that can be tolerated.
- Current limitations during the running and starting phases of operation.
- Duty cycle: how frequently the motor is to be started and stopped.
- Environmental factors: temperature, presence of corrosive or explosive atmospheres, exposure to weather or to liquids, availability of cooling air, and so on.
- Voltage variations expected: Most motors will tolerate up to ± 10% variation from the rated voltage. Beyond this, special designs are required.
- Shaft loading, particularly side loads and thrust loads which can affect the life of shaft bearings.

Motor Size

A rough classification of motors by size is used to group motors of similar design. Horsepower (hp) is currently used most frequently, with the metric unit of watts or kilowatts also used at times. The conversion is

$$1.0 \, hp = 0.746 \, kW = 746 \, W$$

The classifications are as follows:

- *Subfractional horsepower:* 1 to 40 millihorsepower (mhp), where 1 mhp = 0.001 hp. Thus, this range includes 0.001 to 0.040 hp (0.75 to 30 W, approximately).
- *Fractional horsepower:* 1/20 to 1.0 hp (37 to 746 W, approximately).
- *Integral horsepower:* 1.0 hp (0.75 kW) and larger.

References 1–6 provide additional information about the selection and application of electric motors.

21–3 AC POWER AND GENERAL INFORMATION ABOUT AC MOTORS

Alternating current (AC) power is produced by the electric utility and delivered to the industrial, commercial, or residential consumer in a variety of forms. In the United States, AC power has a frequency of 60 hertz (Hz) or 60 cycles/s. In many other countries, 50 Hz is used. Some aircraft use 400-Hz power from an on-board generator.

AC power is also classified as single-phase or three-phase. Most residential units and light commercial installations have only single-phase power, carried by two conductors plus ground. The waveform of the power would appear like that in Figure 21–1, a single continuous sine wave at the system frequency whose amplitude is the rated voltage of the power. Three-phase power is carried on a three-wire system and is composed of three distinct waves of the same amplitude and frequency, with each phase offset from the next by 120°, as illustrated in Figure 21–2. Industrial and large commercial installations use three-phase power for the larger electrical loads because smaller motors are possible and there are economies of operation.

AC Voltages

Some of the more popular voltage ratings available in AC power are listed in Table 21–1. Given are the nominal system voltage and the typical motor voltage rating for that system in both single-phase and three-phase. In most cases, the highest voltage available should be

FIGURE 21–1
Single-phase AC power

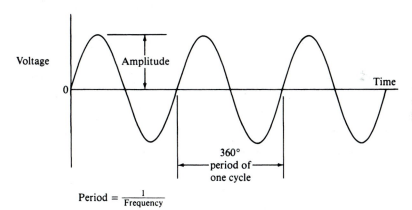

FIGURE 21–2
Three-phase AC power

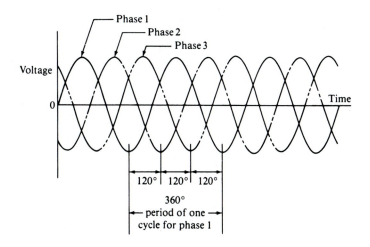

TABLE 21–1   AC motor voltages

| System voltage | Motor voltage ratings | |
	Single-phase	Three-phase
120	115	115
120/208	115	200
240	230	230
480		460
600		575

TABLE 21–2   AC motor speeds

Number of poles	Synchronous speed (rpm)	Full-load[a] speed (rpm)
2	3 600	3 450
4	1 800	1 725
6	1 200	1 140
8	900	850
10	720	690
12	600	575

[a]Approximately 95% of synchronous speed (normal slip).

used because the current flow for a given power is smaller. This allows smaller conductors to be used.

Speeds of AC Motors

An AC motor at zero load would tend to operate at or near its *synchronous speed, n_s,* which is related to the frequency, *f*, of the AC power and to the number of electrical poles, *p*, wound into the motor, according to the equation

⇨ **Synchronous Speed**

$$n_s = \frac{120f}{p} \text{ rev/min} \qquad\qquad (21\text{–}1)$$

Motors have an even number of poles, usually from 2 to 12, resulting in the synchronous speeds listed in Table 21–2 for 60-Hz power. But the induction motor, the most widely used type, operates at a speed progressively slower than its synchronous speed as the load (torque) demand increases. When the motor is delivering its rated torque, it will be operating near its rated or full-load speed, also listed in Table 21–2. Note that the full-load speed is not precise and that those listed are for motors with normal slip of approximately 5%. Some motors described later are "high-slip" motors having lower full-load speeds. Some 4-pole motors are rated at 1 750 rpm at full load, indicating only about 3% slip. *Synchronous motors* operate precisely at the synchronous speed with no slip.

**21–4
PRINCIPLES OF
OPERATION OF
AC INDUCTION
MOTORS**

Later in this chapter we will discuss the particular details of several different types of AC motors, but the most common of these is the *induction motor.* The two active parts of an induction motor are the *stator,* or stationary element, and the *rotor,* or rotating element. Figure 21–3 shows a longitudinal cross section of an induction motor showing the stator, in the form of a hollow cylinder, fixed in the housing. The rotor is positioned inside the stator and is carried on the shaft. The shaft, in turn, is supported by bearings in the housing.

The stator is constructed of many thin, flat discs of steel, called *laminations,* stacked together and insulated from one another. Figure 21–4 shows the shape of the laminations, including a series of slots around the inside. These slots are typically aligned as the stator laminations are stacked, thus forming channels along the length of the stator core. Several layers of copper wire are passed through the channels and are looped around to form a set of continuous coils, called *windings.* The pattern of the coils in the stator determines the number of poles for the motor, usually 2, 4, 6, 8, 10, or 12. Table 21–2 shows that the rotational speed of the motor depends on the number of poles.

FIGURE 21–3
Longitudinal section
through an induction
motor

FIGURE 21–4
Induction motor
laminations

The rotor also has a lamination stack with channels along the length. The channels are typically filled with solid bars made from a good electrical conductor such as copper or aluminum, with the ends of all bars connected to continuous rings at each end. In some smaller rotors, the complete set of bars and end rings are cast from aluminum as a unit. As shown in Figure 21–5, if this casting were viewed without the laminations, it would appear similar to a squirrel cage. Thus, induction motors are often called squirrel-cage motors. The combination of the squirrel cage and the laminations is fixed onto the motor shaft with good precision to assure concentric alignment with the stator and good dynamic balance while rotating. When the rotor is installed in the supporting bearings and is inserted inside the stator, there is a small gap of approximately 0.020 in (0.50 mm) between the outer surface of the rotor and the inner surface of the stator.

Three-Phase Motors

The principles of operation of AC motors is first discussed for three-phase induction motors. Single-phase motor designs are discussed later. The three-phase electrical power, shown schematically in Figure 21–2, is connected to the stator windings. As the current flows in the windings, electromagnetic fields are created that are exposed to the conductors

FIGURE 21–5
Squirrel cage

in the rotor. Because the three phases of power are displaced from each other in time, the effect created is a set of fields rotating around the stator. A conductor placed in a moving magnetic field has a current induced in it, and a force is exerted perpendicular to the conductor. The force acts near the periphery of the rotor, thus creating a torque to rotate the rotor.

It is the production of the induced current in the rotor that leads to calling such motors *induction motors.* Note also that there are no direct electrical connections to the rotor, thus greatly simplifying the design and construction of the motor and contributing to its high reliability.

21–5
AC MOTOR
PERFORMANCE

The performance of electric motors is usually displayed on a graph of speed versus torque, as shown in Figure 21–6. The *vertical axis* is the rotational speed of the motor as a percentage of synchronous speed. The *horizontal axis* is the torque developed by the motor as a percentage of the full-load or rated torque. When exerting its full-load torque, the motor operates at its full-load speed and delivers the rated power. See Table 21–2 for a listing of synchronous speeds and full-load speeds.

The torque at the bottom of the curve where the speed is zero is called the *starting torque* or *locked-rotor torque.* It is the torque available to initially get the load moving and begin its acceleration. This is one of the most important selection parameters for motors, as will be discussed in the descriptions of the individual types of motors.

The "knee" of the curve, called the *breakdown torque,* is the maximum torque developed by the motor during acceleration. The slope of the speed/torque curve in the vicinity of the full-load operating point is an indication of *speed regulation.* A flat curve (a low slope) indicates good speed regulation with little variation in speed as load varies. Conversely, a steep curve (a high slope) indicates poor speed regulation, and the motor will exhibit wide swings in speed as load varies. Such motors produce a "soft" acceleration of a load which may be an advantage in some applications. But where fairly constant speed is desired, a motor with good speed regulation should be selected.

FIGURE 21–6

General form of motor performance curve

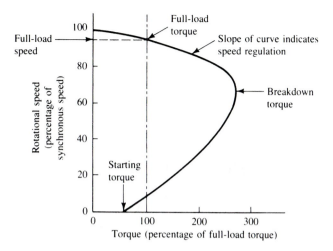

21–6
THREE-PHASE, SQUIRREL-CAGE INDUCTION MOTORS

Three of the most commonly used three-phase AC motors are simply designated as designs B, C, and D by the National Electrical Manufacturers Association (NEMA). They differ primarily in the value of starting torque and in the speed regulation near full load. Figure 21–7 shows the performance curves for these three designs for comparison. Each of these designs employs the solid squirrel-cage type of rotor, and thus there is no electrical connection to the rotor.

The 4-pole design with a synchronous speed of 1 800 rpm is the most common and is available in virtually all power ratings from 1/4 hp to 500 hp. Certain sizes are available in 2-pole (3 600 rpm), 6-pole (1 200 rpm), 8-pole (900 rpm), 10-pole (720 rpm), and 12-pole (600 rpm) designs.

NEMA Design B

The performance of the three-phase design B motor is similar to that of the single-phase split-phase motor described later. It has a moderate starting torque (about 150% of full-load torque) and good speed regulation. The breakdown torque is high, usually 200% of full-load torque or more. Starting current is fairly high, at approximately six times full-load current. The starting circuit must be selected to be able to handle this current for the short time required to bring the motor up to speed.

Typical uses for the design B motor are centrifugal pumps, fans, blowers, and machine tools such as grinders and lathes.

NEMA Design C

High starting torque is the main advantage of the design C motor. Loads requiring 200% to 300% of full-load torque to start can be driven. Starting current is typically lower than for the design B motor for the same starting torque. Speed regulation is good and is about the same as for the design B motor. Reciprocating compressors, refrigeration systems, heavily loaded conveyors, and ball-and-rod mills are typical uses.

NEMA Design D

The design D motor has a high starting torque, about 300% of full-load torque. But it also has poor speed regulation, which results in large speed changes with varying loads. Sometimes called a *high-slip motor,* it operates at 5% to 13% slip at full load, whereas

FIGURE 21–9

Performance curves for
a universal motor

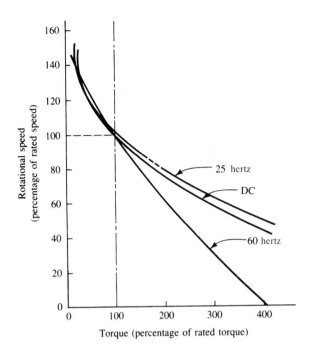

power and DC power. Note that performance near the rated load is similar, regardless of the nature of the input power. Note also that these motors have poor speed regulation; that is, the speed varies greatly with load.

21–7
SINGLE-PHASE
MOTORS

The four most common types of single-phase motors are the *split-phase, capacitor-start, permanent-split capacitor,* and *shaded-pole.* Each is unique in its physical construction and in the manner in which the electrical components are connected to provide for starting and running of the motor. The emphasis here is not on how to design the motors, but rather on the performance, so that a suitable motor can be selected.

Figure 21–10 shows the performance characteristics of these four types of motors so that they can be compared. The special features of the performance curves for the four types of motors are discussed in later sections.

In general, the construction of single-phase motors is similar to that for three-phase motors, consisting of a fixed stator, a solid rotor, and a shaft carried on bearings. The induction principle discussed earlier applies also to single-phase motors. Differences occur because single-phase power does not inherently rotate around the stator to create a moving field. Each type uses a different scheme for initially starting the motor. See Figure 21–11.

Single-phase motors are usually in the subfractional or fractional horsepower range from 1/50 hp (15 W) to 1.0 hp (750 W), although some are available up to 10 hp (7.5 kW).

Split-Phase Motors

The stator of the split-phase motor [Figure 21–11(b)] has two windings: the *main winding,* which is continuously connected to the power line, and the *starting winding,* which is connected only during the starting of the motor. The starting winding creates a slight phase shift that creates the initial torque to start and accelerate the rotor. After the rotor reaches approximately 75% of its synchronous speed, the starting winding is cut out by a centrifugal switch, and the rotor continues to run on the main winding.

FIGURE 21–10
Performance curves of four types of single-phase electric motors

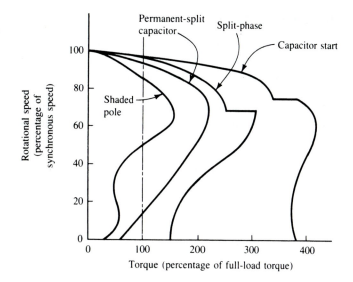

The performance curve for the split-phase motor is shown in Figure 21–10. It has moderate starting torque, approximately 150% of full-load torque. It has good efficiency and is designed for continuous operation. Speed regulation is good. One of the disadvantages is that it requires a centrifugal switch to cut out the starting winding. The step in the speed/torque curve indicates this cutout.

These characteristics make the split-phase motor one of the most popular types, used in business machines, machine tools, centrifugal pumps, electric lawn mowers, and similar applications.

Capacitor-Start Motors

Like the split-phase motor, the capacitor-start motor [Figure 21–11(c)] also has two windings: a *main* or *running winding* and a *starting winding.* But in it, a capacitor is connected in series with the starting winding, giving a much higher starting torque than with the split-phase motor. A starting torque of 250% of full load or higher is common. Again a centrifugal switch is used to cut out the starting winding and the capacitor. The running characteristics of the motor are then very similar to those of the split-phase motor: good speed regulation and good efficiency for continuous operation.

Disadvantages include the switch and the relatively bulky capacitor. Frequently the capacitor is mounted right on top of the motor and is very evident (see Figure 21–17). It may also be integrated into a package containing the starting switch, a relay, or other control elements.

Uses for the capacitor-start motor include the many types of machines that need the high starting torque. Examples include heavily loaded conveyors, refrigeration compressors, and pumps and agitators for heavy fluids.

Permanent-Split Capacitor Motors

A capacitor is connected in series with the starting winding at all times. The starting torque of the permanent-split capacitor motor [Figure 21–11(d)] is typically quite low, approximately 40% or less of full-load torque. Thus, only low-inertia loads such as fans and blowers are usually used. An advantage is that you can tailor the running performance and the speed regulation to match the load by selecting the appropriate capacitor value. Also, no centrifugal switch is required.

Stopping the Motor

Where no special conditions exist when the system is shut down, the motor can be permitted to coast to a stop after the power is interrupted. The time required to stop will depend on the inertia and the friction in the system. If controlled, rapid stopping is required, external brakes can be used. *Brake motors,* which have a brake integral with the motor, are available. Typically, the design is of the "fail-safe" nature, in which the brake is disengaged by an electromagnetic coil when the motor is energized. When the motor is de-energized, either on purpose or because of power failure, the brake is actuated by mechanical spring force.

On circuits with reversing starters, *plug stopping* can be used. When it is desired to stop the motor running in the forward direction, the control can be switched immediately to reverse. There would then be a decelerating torque applied to the rotor, stopping it quickly. Care must be exercised to cut out the reversing circuit when the motor is at rest to prevent it from continuing in the reverse direction.

Overload Protection

The chief cause of failure in electric motors is overheating of the wound coils due to excessive current. The current is dependent on the load on the motor. A short circuit, of course, would cause a virtually instantaneously high current of a damaging level.

The protection against a short circuit can be provided by *fuses,* but careful application of fuses to motors is essential. A fuse contains an element that literally melts when a particular level of current flows through it. Thus, the circuit is opened. Reactivating the circuit would require replacing the fuse. Time-delay fuses, or "slow-blowing" fuses, are needed for motor circuits to prevent the fuses from blowing when the motor starts, drawing the relatively high starting current that is normal and not damaging. After the motor starts, the fuse will blow at a set value of overcurrent.

Fuses are inadequate for larger or more critical motors because they provide protection at only one level of overcurrent. Each motor design has a characteristic *overheating curve,* as shown in Figure 21–25. This indicates that the motor could withstand different levels of overcurrent for different periods of time. For example, for the motor heating curve of Figure 21–25, a current twice as high as the full-load current (200%) could exist for up

FIGURE 21–25
Motor heating curve and response curve of a typical overload protector (Square D Company, Palatine, IL)

to 9 min before a damaging temperature was produced in the windings. But a 400% overload would cause damage in less than 2 min. An ideal overload protection device would parallel the overheating curve of the given motor, always cutting out the motor at a safe current level, as shown in Figure 21–25. Devices are available commercially to provide this protection. Some use special melting alloys, bimetallic strips similar to a thermostat, or magnetic coils that are sensitive to the current flowing in them. Most large motor starters include overload protection integral with the starter.

Another type of overload protection uses a temperature-sensitive device inserted in the windings of the motor when it is made. Then it opens the motor circuit when the windings reach a dangerous temperature, regardless of the reason.

Enclosures for Motor Controls

As stated before, one of the functions of a motor control system is to protect personnel from contact with dangerous parts of the electrical system. Also, protection of the system from the environment must be provided. These functions are accomplished by the enclosure.

NEMA has established standards for enclosures for the variety of environments encountered by motor controls. The most frequent types are described in Table 21–6.

AC Variable-Speed Drives

Standard AC motors operate at a fixed speed for a given load if powered by AC power at a fixed frequency, for example, 60 Hz. Variable-speed operation can be obtained with a control system that produces a variable frequency power. Two such control types are in common use: the *six-step method* and the *pulse-width modulation (PWM) method*. Either system takes the 60-Hz line voltage and first rectifies it to a DC voltage. The six-step method then uses an inverter to produce a series of square waves that provides a voltage to the motor winding which varies both voltage and frequency in six steps per cycle. In the PWM system, the DC voltage is input to an inverter that produces a series of pulses of variable width. The rate of polarity reversals determines the frequency applied to the motor. See Figures 21–26 and 21–27.

TABLE 21–6   Motor control enclosures

NEMA design number	Description
1	General-purpose: indoor use; not dusttight
3	Dusttight, raintight: outdoor-weather-resistant
3R	Dusttight, rainproof, sleet-resistant
4	Watertight: can withstand a direct spray of water from a hose; used on ships and in food processing plants where washdown is required
4X	Watertight, corrosion-resistant
7	Hazardous locations, class I: can operate in areas where flammable gases or vapors are present
9	Hazardous locations, class II: combustible dust areas
12	Industrial use: resistant to dust, lint, oil, and coolants
13	Oiltight, dusttight

FIGURE 21–26 Six-step method of variable-speed AC motor control (Rockwell Automation/Allen-Bradley, Mayfield Heights, OH)

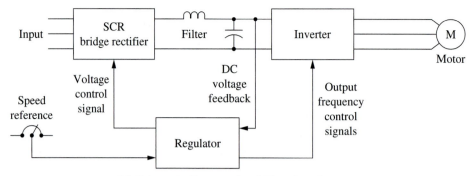

(a) Schematic diagram of variable voltage inverter

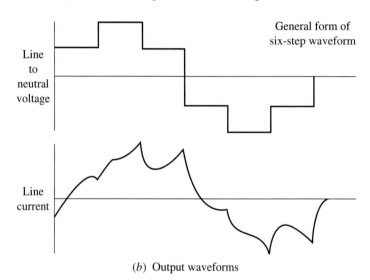

(b) Output waveforms

21–10
DC POWER

DC motors have several inherent advantages over AC motors, as discussed in the next section. A disadvantage of DC motors is that a source of DC power must be available. Most residential, commercial, and industrial locations have only AC power provided by the local utility. Three approaches are used to provide DC power:

1. *Batteries:* Typically batteries are available in voltages of 1.5, 6.0, 12.0, and 24.0 volts (V). They are used for portable devices or for mobile applications. The power is pure DC, but voltage varies with time as the battery discharges. The bulkiness, weight, and finite life are disadvantages.

2. *Generators:* Powered by AC electric motors, internal combustion engines, turbine engines, wind devices, water turbines, and so on; DC generators produce pure DC. The usual voltages are 115 and 230 V. Some industries maintain such generators to provide DC power throughout the plant.

3. *Rectifiers: Rectification* is the process of converting AC power with its sinusoidal variation of voltage with time to DC power, which ideally is nonvarying. A readily available device is the *silicon-controlled rectifier (SCR)*. One difficulty with rectification of AC power to produce DC power is that there is always some amount of "ripple," a small variation of voltage as a function of time. Excessive ripple can cause overheating of the DC motor. Most commercially available SCR devices produce DC power with an acceptably low ripple. Table 21–7 lists the

FIGURE 21–27
Pulse-width modulation method of variable-speed AC motor control (Rockwell Automation/Allen-Bradley, Mayfield Heights, OH)

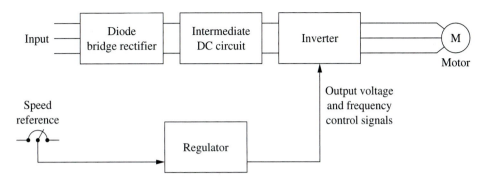

(a) Schematic diagram of pulse-width modulation (PWM) controller

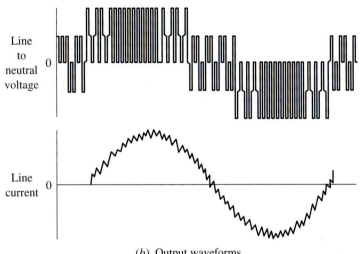

(b) Output waveforms

TABLE 21–7   DC motor voltage ratings

Input AC voltage	DC motor rating	NEMA code
115 V AC, one-phase	90 V DC	K
230 V AC, one-phase	180 V DC	K
230 V AC, three-phase	240 V DC	C or D
460 V AC, three-phase	500 V DC or	C or D
	550 V DC	
460 V AC, three-phase	240 V DC	E

commonly used DC voltage ratings for motors powered by rectified AC power as defined by NEMA.

21–11
DC MOTORS

The advantages of direct current motors are summarized below:

- The speed is adjustable by use of a simple rheostat to adjust the voltage applied to the motor.
- The direction of rotation is reversible by switching the polarity of the voltage applied to the motor.

- Automatic control of speed is simple to provide for matching of the speeds of two or more motors or to program a variation of speed as a function of time.

- Acceleration and deceleration can be controlled to provide the desired response time or to decrease jerk.

- Torque can be controlled by varying the current applied to the motor. This is desirable in tension control applications, such as the winding of film on a spool.

- Dynamic braking can be obtained by reversing the polarity of the power while the motor is rotating. The reversed effective torque slows the motor without the need for mechanical braking.

- DC motors typically have quick response, accelerating quickly when voltage is changed, because they have a small rotor diameter, giving them a high ratio of torque to inertia.

DC motors have electric windings in the rotor, and each coil has two connections to the commutator on the shaft. The commutator is a series of copper segments through which the electric power is transferred to the rotor. The current path from the stationary part of the motor to the commutator is through a pair of brushes, usually made of carbon, which are held against the commutator by light coil or leaf springs. Maintenance of the brushes is one of the disadvantages of DC motors.

DC Motor Types

Four commonly used DC motor types are the *shunt-wound, series-wound, compound-wound,* and *permanent magnet* motors. They are described in terms of their speed/torque curves in a manner similar to that used for AC motors. One difference here is that the speed axis is expressed in percentage of *full-load rated speed,* rather than a percentage of synchronous speed, as that term does not apply to DC motors.

Shunt-wound DC Motor.   The electromagnetic field is connected in parallel with the rotating armature, as sketched in Figure 21–28. The speed/torque curve shows relatively good speed regulation up to approximately two times full-load torque, with a rapid drop in speed after that point. The speed at no load is only slightly higher than the full-load speed. Contrast this with the series-wound motor described next. Shunt-wound motors are used mainly for small fans and blowers.

Series-wound DC Motor.   The electromagnetic field is connected in series with the rotating armature, as shown in Figure 21–29. The speed/torque curve is very steep, giving the motor a soft performance that is desirable in cranes, hoists, and traction drives for vehicles. The starting torque is very high, as much as 800% of full-load rated torque. A major difficulty, however, with series-wound motors is that the no-load speed is theoretically unlimited. The motor could reach a dangerous speed if the load were to be accidentally disconnected. Safety devices, such as overspeed detectors that shut off the power, should be used.

Compound-wound DC Motors.   The compound-wound DC motor employs both a series field and a shunt field, as sketched in Figure 21–30. It has a performance somewhat between that of the series-wound and the shunt-wound motors. It has fairly high starting torque and a soft speed characteristic, but it has an inherently controlled no-load speed. This makes it good for cranes, which may suddenly lose their loads. The motor would run slowly when heavily loaded for safety and control, and fast when lightly loaded to improve productivity.

FIGURE 21–28
Shunt-wound DC
motor performance
curve

FIGURE 21–29
Series-wound DC
motor performance
curve

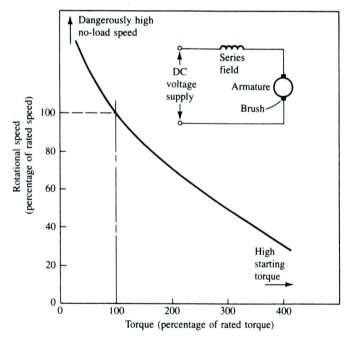

Permanent Magnet DC Motors.   Instead of using electromagnets, the permanent
magnet DC motor uses permanent magnets to provide the field for the armature. The direct
current passes through the armature, as shown in Figure 21–31. The field is nearly constant
at all times and results in a linear speed/torque curve. Current draw also varies linearly with
torque. Applications include fans and blowers to cool electronics packages in aircraft, small
actuators for control in aircraft, automotive power assists for windows and seats, and fans

FIGURE 21–30
Compound-wound DC
motor performance
curve

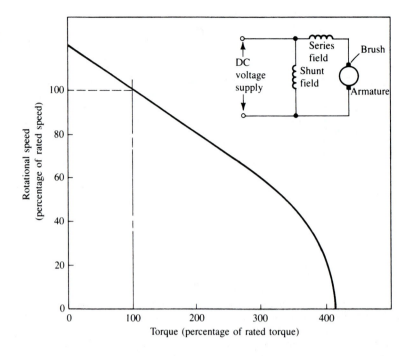

FIGURE 21–31
Permanent magnet DC
motor performance
curve

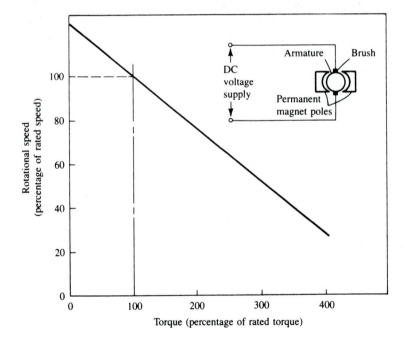

in automobiles for heating and air conditioning. These motors frequently have built-in gear-type speed reducers to produce a low-speed, high-torque output.

21–12
DC MOTOR
CONTROL

Starting DC motors presents essentially the same problems as discussed for AC motors in terms of limiting the starting current and the provision of switching devices and holding relays of sufficient capacity to handle the operating loads. The situation is made somewhat

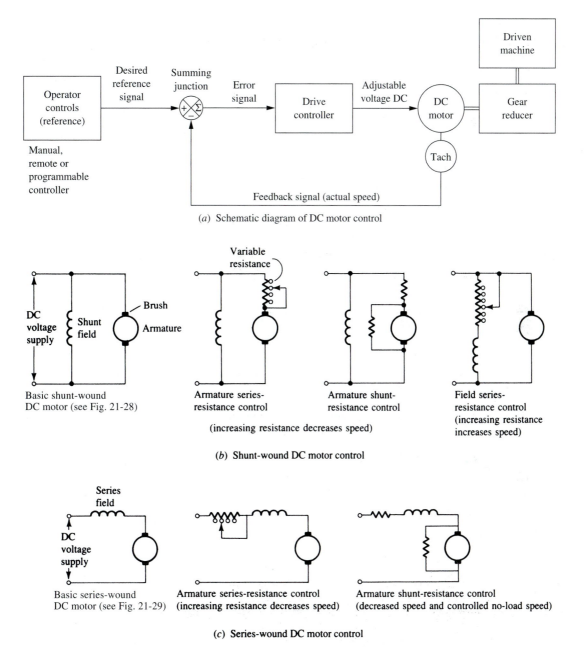

(a) Schematic diagram of DC motor control

(b) Shunt-wound DC motor control

(c) Series-wound DC motor control

FIGURE 21–32   DC motor control [Source for (a), Rockwell Automation/Allen-Bradley, Mayfield Heights, OH]

more severe, however, by the presence of the commutators in the rotor circuit which are more sensitive to overcurrent.

Speed control is provided by variation of the resistance in the lines containing the armature or the field of the motor. The details depend on whether the motor is a series, shunt, or compound type. The variable-resistance device, sometimes called a *rheostat,* can provide either stepwise variation in resistance or continuously varying resistance. Figure 21–32 shows the schematic diagrams for several types of DC motor speed controls.

<table>
<tr><td>

21–13
OTHER TYPES
OF MOTORS

</td></tr>
</table>

Torque Motors

As the name implies, *torque motors* are selected for their ability to exert a certain torque rather than for a rated power. Frequently this type of motor is operated at a stalled condition to maintain a set tension on a load. The continuous operation at slow speed or at zero speed causes heat generation to be a potential problem. In severe cases, external cooling fans may be required.

By special design, several of the AC and DC motor types discussed elsewhere in this chapter can be used as torque motors.

Servomotors

Either AC or DC *servomotors* are available to provide automatic control of position or speed of a mechanism in response to a control signal. Such motors are used in aircraft actuators, instruments, computer printers, and machine tools. Most have rapid response characteristics because of the low inertia of the rotating components and the relatively high torque exerted by the motor.

Figure 21–33 shows a schematic diagram of a servomotor controller system. Three control loops are shown: (1) the position loop, (2) the velocity loop, and (3) the current loop. Speed control is effected by sensing the motor speed with a tachometer and feeding the signal back through the velocity loop to the controller. Position is sensed by an optical encoder or a similar device on the driven load, with the signal fed back through the position loop to the controller. The controller sums the inputs, compares them with the desired value set by the control program, and generates a signal to control the motor. Thus, the system is a closed-loop servo-control. Typical uses for this type of control would be for numerical control machine tools, special-purpose assembly machines, and aircraft control surface actuators.

Stepper Motors

A stream of electronic pulses is delivered to a *stepper motor,* which then responds with a fixed rotation (step) for each pulse. Thus, a very precise angular position can be obtained by counting and controlling the number of pulses delivered to the motor. Several step an-

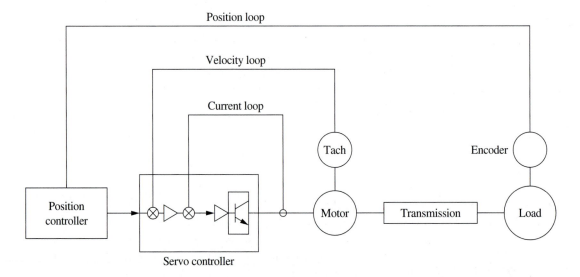

FIGURE 21–33   Servomotor controller system (Rockwell Automation/Allen-Bradley, Mayfield Heights, OH)

gles are available in commercially provided motors, such as 1.8°, 3.6°, 7.5°, 15°, 30°, 45°, and 90°. When the pulses are stopped, the motor stops automatically and is held in position. Because many of these motors are connected through a gear-type speed reducer to the load, very precise positioning is possible to a small fraction of a step. Also, the reducer provides a torque increase.

Brushless Motors

The typical DC motor requires brushes to make contact with the rotating commutator on the shaft of the motor. This is a major failure mode of such motors. In the *brushless DC motor,* the switching of the rotor coils is accomplished by solid-state electronic devices, resulting in very long life. The emission of electromagnetic interference is likewise reduced compared with brush-type DC motors.

Printed Circuit Motors

The rotor of the *printed circuit motor* is a flat disc that operates between two permanent magnets. The resulting design has a relatively large diameter and small axial length; sometimes it is called a *pancake motor.* The rotor has a very low inertia, so high acceleration rates are possible.

Linear Motors

Linear motors are electrically similar to rotary motors, except the components, stator and rotor, are laid flat instead of being formed into a cylindrical shape. Types include brush-type DC motors, brushless DC motors, stepping motors, and the single-phase type of motor. Capacity is measured in terms of the force that the motor can exert, ranging typically from a few pounds to 2 500 lb. Speeds range from about 40 to 100 in/s.

REFERENCES

1. Anderson, Leonard R., and Jack MacNeill. *Electric Machines and Transformers.* Englewood Cliffs, NJ: Prentice-Hall, 1988.

2. Avallone, Eugene P., and Theodore Baumeister III. *Marks' Handbook for Mechanical Engineers.* 10th ed. New York: McGraw-Hill, 1996.

3. Martin, Merrill. "How to Select a Variable-Speed Drive." *Machine Design Magazine* (October 25, 1990).

4. Penton Publishing, Inc. "Electrical/Electronic Reference Issue." *Machine Design Magazine* (May 19, 1988).

5. Rockwell Automation/Allen-Bradley. *Drives and Motion Controllers Catalog.* Costa Mesa, CA: Rockwell International Corporation.

6. Thorpe, James F. *Mechanical System Components.* Boston: Allyn & Bacon, 1989.

PROBLEMS

1. List six items that must be specified for electric motors.

2. List eight factors to be considered in the selection of an electric motor.

3. Define *duty cycle.*

4. How much variation in voltage will most AC motors tolerate?

5. State the relationship among torque, power, and speed.

6. What does the abbreviation "AC" stand for?

7. Describe and sketch the form of single-phase AC power.

8. Describe and sketch the form of three-phase AC power.

9. What is the standard frequency of AC power in the United States?

10. What is the standard frequency of AC power in Europe?

11. What type of electrical power is available in a typical American residence?

12. How many conductors are required to carry single-phase power? How many for three-phase power?

13. Assume that you are selecting an electric motor for a machine to be used in an industrial plant. The following types of AC power are available: 120-V, single-phase; 240-V, single-phase; 240-V, three-phase; 480-V, three-phase. In general, for which type of power would you specify your motor?

14. Define *synchronous speed* for an AC motor.

15. Define *full-load speed* for an AC motor.

16. What is the synchronous speed of a four-pole AC motor when running in the United States? In France?

17. A motor nameplate lists the full-load speed to be 3 450 rpm. How many poles does the motor have? What would its approximate speed be at zero load?

18. If a four-pole motor operates on 400-Hz AC power, what will its synchronous speed be?

19. If an AC motor is listed as a normal-slip, four-pole/six-pole motor, what will its approximate full-load speeds be?

20. What type of control would you use to make an AC motor run at variable speeds?

21. Describe a C-face motor.

22. Describe a D-flange motor.

23. What does the abbreviation "NEMA" stand for?

24. Describe a protected motor.

25. Describe a TEFC motor.

26. Describe a TENV motor.

27. What type of motor enclosure would you specify to be used in a plant that manufactures baking flour?

28. What type of motor would you specify for a meat grinder if the motor is to be exposed?

29. Figure 21–34 shows a machine that is to be driven by a 5-hp, protected, foot-mounted AC motor having a 184T frame. The motor is to be aligned with the shaft of the driven machine. Make a complete dimensioned drawing, showing standard side and top views of the machine and the motor. Design a suitable mounting base for the motor showing the motor mounting holes.

30. Define *locked-rotor torque*. What is another term used for this parameter?

31. What is meant if one motor has a poorer speed regulation than another?

32. Define *breakdown torque*.

33. Name the four most common types of single-phase AC motors.

34. Refer to the AC motor performance curve in Figure 21–35.
    (a) What type of motor is the curve likely to represent?
    (b) If the motor is a six-pole type, rated at 0.75 hp, how much torque can it exert at the rated load?
    (c) How much torque can the motor develop to start a load?
    (d) What is the breakdown torque for the motor?

35. Repeat Parts (b), (c), and (d) of Problem 34 if the motor is a two-pole type, rated at 1.50 kW.

36. A cooling fan for a computer is to operate at 1 725 rpm, direct-driven by an electric motor. The speed/torque curve for the fan is shown in Figure 21–36. Specify a suitable motor, giving the type of motor, horsepower, and number of poles.

37. Figure 21–37 shows the speed/torque curve for a household refrigeration compressor, designed to operate at 3 450 rpm. Specify a suitable motor, giving the type, power rating in watts, and number of poles.

38. How is speed adjusted for a wound-rotor motor?

39. What is the full-load speed of a 10-pole synchronous motor?

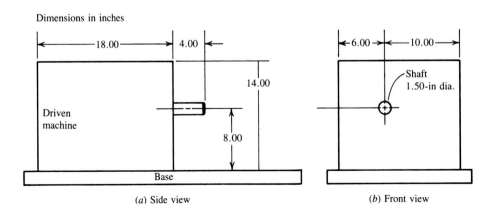

Dimensions in inches

(a) Side view (b) Front view

FIGURE 21–34 (Problem 29)

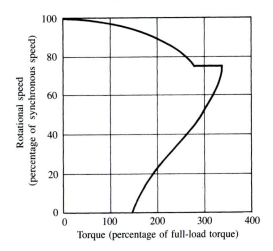

FIGURE 21–35 (Problems 34 and 35)

FIGURE 21–36 (Problem 36)

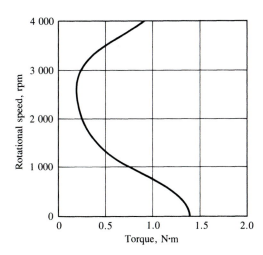

FIGURE 21–37 (Problem 37)

40. What is meant by the term *pull-out torque,* as applied to a synchronous motor?

41. Discuss the reasons that universal motors are frequently used for hand-held tools and small appliances.

42. Why is the adjective *universal* used to describe a universal motor?

43. Name three ways to produce DC power.

44. List 12 common DC voltages.

45. What is an SCR control? For what is it used?

46. If a DC motor drive advertises that it produces *low ripple,* what does the term mean?

47. If you want to use a DC motor in your home, and your home has only standard 115 VAC, single-phase power, what will you need? What type of motor should you get?

48. List seven advantages of DC motors with respect to AC motors.

49. Discuss two disadvantages of DC motors.

50. Name four types of DC motors.

51. What happens to a series-wound DC motor if the load on the motor falls to nearly zero?

52. Assume that a permanent magnet DC motor can exert a torque of 15.0 N·m when operating at 3 000 rpm. What torque could it exert at 2 200 rpm?

53. List 10 functions of a motor control.

54. What size motor starter is required for a 10-hp, three-phase motor, operating on 220 V?

55. A single-phase, 110-V AC motor has a nameplate rating of 1.00 kW. What size motor starter is required?

56. What does the term *plug stopping* mean, and how is it accomplished?

57. Why is a fuse an inadequate protection device for an industrial motor?

58. What type of motor control enclosure would you specify for use in a car wash?

59. What could you do to the control circuit for a standard series DC motor to give it a controlled no-load speed?

60. What happens if you connect a resistance in series with the armature of a shunt-wound DC motor?

61. What happens if you connect a resistance in series with the shunt field of a shunt-wound DC motor?

22

Motion Control: Clutches and Brakes

Motion Control: Clutches and Brakes

Discussion Map

- ☐ A *brake* is a device used to bring a moving system to rest, to slow its speed, or to control its speed to a certain value under varying conditions.
- ☐ A *clutch* is a device used to connect or disconnect a driven component from the prime mover of the system.

Discover

Where do you use brakes?

Consider the brakes for a car or a bicycle. Describe the components and the actuation cycle in as much detail as you can. Discuss your findings with your colleagues.

What other kinds of equipment besides vehicles use clutches or brakes? Describe some scenarios.

Describe the physics of the operation of a clutch and a brake, considering energy and inertia effects.

This chapter helps you explore all of these issues, and many convenient design and analysis equations are developed. Many kinds of commercially available clutches and brakes are pictured and described. Look through the chapter now.

Machine systems require control whenever the speed or the direction of the motion of one or more components is to be changed. When a device is started initially, it must be accelerated from rest to the operating speed. As a function is completed, the system must frequently be brought to rest. In continuously operating systems, changing speeds to adjust to different operating conditions is often necessary. Safety sometimes dictates motion control, as with a load being lowered by a hoist or an elevator.

In this chapter we will be concerned mostly with control of rotary motion in systems driven by electric motors, engines, turbines, and the like. Ultimately, linear motion may be involved through linkages, conveyors, or other mechanisms.

The machine elements most frequently used for motion control are the clutch and the brake, defined as follows:

- A *clutch* is a device used to connect or disconnect a driven component from the prime mover of the system. For example, in a machine that must cycle frequently, the driving motor is left running continuously, and a clutch is interposed between the motor and the driven machine. Then the clutch is cycled on and off to connect and disconnect the load. This permits the motor to operate at an efficient speed, and it also permits the system to cycle more rapidly because there is no need to accelerate the heavy motor rotor with each cycle.

- A *brake* is a device used to bring a moving system to rest, to slow its speed, or to control its speed to a certain value under varying conditions.

Where do you use brakes? An obvious answer is in a car or a bicycle where safe operation requires fast, smooth stopping when emergency conditions occur or simply when you need to stop at a stop sign or in your driveway. And you do not always need to bring the car or bicycle to a full stop. Slowing down to conform to prevailing traffic flow or to round a curve requires braking to decrease the speed from a higher to a lower speed.

What really happens when you apply the brakes of a bicycle? Can you describe the essential elements of the braking system? With hand brakes, your actuation of the brake lever pulls on a cable that, in turn, pulls on a linkage at the brake assembly mounted above the rim of a wheel. The linkage causes the friction pad to press against the rim. When you pull harder on the lever, a higher force is developed between the pad and the rim. This is called the *normal force*. Recall from physics and statics that a frictional force is created between surfaces moving relative to each other when a normal force presses them together. The frictional force acts in a direction opposite to the relative motion, and thus it tends to slow the motion. With a sufficiently large frictional force applied for a sufficient time, the wheel stops. Notice also that the frictional force acts at a fairly large radius from the center of the wheel. Thus, the force causes a frictional *torque* to be developed, and what is really happening is the deceleration of the angular velocity of the wheel. But since this is directly proportional to the linear speed of the bicycle, you experience the stopping action as a linear deceleration.

But that is not all! Have you ever felt the brake pad after a hard stop? The fact that it gets warm is an indication that the brake absorbs energy during the stopping action. Where does the energy come from? Can you compute the amount of energy that must be absorbed? What are the parameters involved in this calculation? Compare the amount of energy that must be absorbed in stopping a bicycle that you are riding with that involved in stopping a large airliner landing at 120 mph with a full load of people and baggage in addition to the huge weight of the airplane itself. Imagine what those brakes look like in comparison with the bicycle brake!

What other kinds of equipment besides transportation vehicles require brakes? Consider elevators, escalators, hoists, and winches which must stop and hold a load after lifting it. Machine tools, conveyors, and other manufacturing equipment must often be brought to a safe, quick stop.

But that equipment must also be *accelerated* to start a new cycle of operation. How can that be achieved? One way is to start and stop the motor or the engine that drives the equipment. However, that is inconvenient and time-consuming, and it may lead to early failure of the system.

What if you had to stop the engine of your car each time you came to a stop sign and then had to restart it to move on? How does the automotive system allow you to stop and then continue driving without shutting off the engine? Have you ever driven a standard shift car? A "stick shift"? A *clutch* is used to engage and disengage the drive train from the engine. In cars with automatic transmissions clutches are also built into the transmission.

What other kinds of equipment use clutches? Describe some examples from your personal experience, or create a scenario in which it would be desirable to employ a clutch.

Now, consider what the task of a clutch is. Some parts of a machine are rotating continuously, while others are temporarily stationary. Then you engage the clutch. What happens? Consider the physics of the situation. The stationary parts must be accelerated from zero speed to the desired speed according to the design of the drive system. Inertia must be overcome—sometimes rotational inertia; sometimes linear inertia; and sometimes both kinds in the same machine. How long do you want it to take to accelerate the load to its operating speed? You must realize that a shorter time requires a higher value of accelerating torque to be developed by the clutch, and it increases the technical demands on the system in terms of strength of its components, smoothness, and wear life of the frictional materials that actually accomplish the engagement.

This chapter helps you explore all of these issues, and many convenient design and analysis equations are developed. Several different styles of brakes and clutches are discussed, and photographs or detailed drawings of several commercially available designs are shown. Look through the chapter now to get a feel for the scope of the coverage of this chapter. It might be desirable to keep this book handy for future projects so that you can look up the details of different clutch and brake designs.

You Are the Designer

Your company manufactures conveyor systems for warehouses and truck terminals. The conveyors deliver cartons of materials to any of several stations where trucks are to be loaded. To save energy and to decrease the wear on operating parts of the conveyor system, it is decided to operate only the parts of the system that have a demand to deliver a carton. The system is to be operated automatically through a series of sensors, switches, programmable controllers, and an overall supervisory computer control. Your assignment is to recommend the type and the size of clutch and brake units to start and stop the various conveyors.

Some of the design decisions you will have to make are as follows:

1. How much time should be allowed to get the conveyors up to speed after the initial command to start has been given?
2. How fast should the conveyors come to a stop?
3. How many cycles per hour are expected?

4. How much room is available to install the clutch and brake units?
5. What means are available to actuate the clutch and brake units: electric power, compressed air, hydraulic pressure, or some other means?
6. What general type of clutch and brake should be used?
7. What size and model of units should be specified?

Along with these decisions, you will need information about the conveyor system itself, such as the following:

1. How much load will be on the conveyors when they are started and stopped?
2. What is the design of the conveyor system itself, and what are the weights, shapes, and dimensions of its components?
3. How is the conveyor driven: electric motor, hydraulic motor, or another means?
4. Are the products moved all on one level, or are there elevation changes in the system?

The information in this chapter will help you design such a system.

22–1
OBJECTIVES OF
THIS CHAPTER

After completing this chapter, you will be able to:

1. Define the terms *clutch* and *brake*.
2. Differentiate between a clutch and a *clutch coupling*.
3. Describe a fail-safe brake.
4. Describe a clutch-brake module.
5. Specify the required capacity of a clutch or a brake to drive a given system reliably.
6. Compute the time required to accelerate a system or to bring it to a stop with the application of a given torque.
7. Define the inertia of a system in terms of its Wk^2 value.
8. Compute the energy-dissipation requirements of a clutch or a brake.
9. Determine the response time for a clutch-brake system.
10. Describe at least five types of clutches and brakes.
11. Name six actuation means used for clutches and brakes.
12. Perform the design and analysis of plate-type, caliper disc, cone, drum and shoe, and band brakes and clutches.
13. Name nine other types of clutches and brakes.

22–2
DESCRIPTIONS
OF CLUTCHES
AND BRAKES

Several arrangements of clutches and brakes are shown in Figure 22–1. By convention, the term *clutch* is reserved for the application in which the connection is made to a shaft parallel to the motor shaft, as illustrated in Figure 22–1(a). If the connection is to a shaft in-line with the motor, the term *clutch coupling* is used, as shown in Figure 22–1(b).

(a) Clutch: Transmits rotary motion to a parallel shaft only when coil is energized, by using sheaves, sprockets, gears, or timing pulleys.

(b) Clutch coupling: Transmits rotary motion to an in-line shaft only when coil is energized. Split shaft applications.

(c) Brake: Stops (brakes) load when coil is energized. Panel shows clutch imparting rotary motion to the output (load) shaft while brake is de-energized. Conversely, de-energizing clutch coil and energizing brake coil causes load to stop.

(d) Fail-safe brake: Stops load by de-energization of coil; power off—brake on.

(e) Three types of mountings for clutch-brake modules: *Clutch brake* combines functions of clutch and brake in a complete preassembled package with input and output shafts. *C-flange clutch brake* performs same function but for use between NEMA "C" flange motor and speed reducer. *Motor clutch brake* is a preassembled module for mounting to a NEMA "C" flange motor and presenting an output shaft for connection to load.

FIGURE 22–1   Typical applications of clutches and brakes (Electroid Corp., Springfield, NJ)

Also by convention, a brake [Figure 22–1(c)] is actuated by some overt action: the application of a fluid pressure, the switching on of an electric current, or the moving of a lever by hand. A brake that is spring-applied automatically in the absence of an overt action is called a *fail-safe brake* [Figure 22–1(d)]: When the power goes off, the brake goes on.

When the functions of both a clutch and a brake are required in a system, they are frequently arranged in the same unit, the *clutch-brake module.* When the clutch is activated, the brake is deactivated, and vice versa [see Figure 22–1(e)].

A *slip clutch* is a clutch that, by design, transmits only a limited torque; thus, it will slip at any higher torque. It is used to provide a controlled acceleration to a load that is smooth and that requires a smaller motor power. It is also used as a safety device, protecting expensive or sensitive components in the event that a system is jammed.

Most of the discussion in this chapter will concern clutches and brakes that transmit motion through friction at the interface between two rotating parts moving at different speeds. Other types are discussed briefly in the last section.

References 1–9 provide additional information on the design, selection, and application of motion control systems, clutches, and brakes.

22–3 TYPES OF FRICTION CLUTCHES AND BRAKES

Clutches and brakes that use friction surfaces as the means of transmitting the torque to start or stop a mechanism can be classified according to the general geometry of the friction surfaces and by the means used to actuate them. In some cases, the same basic geometry can be used as either a clutch or a brake by selectively attaching the friction elements to the driver, the driven machine, or the stationary frame of the machine.

The following types of clutches and brakes are sketched in Figure 22–2:

1. *Plate clutch or brake:* Each friction surface is in the shape of an annulus on a flat plate. One or more friction plates move axially to contact a mating smooth plate, usually made of steel, to which the friction torque is transmitted.

2. *Caliper disc brake:* A disc-shaped rotor is attached to the machine to be controlled. Friction pads covering only a small portion of the disc are contained in a fixed assembly called a *caliper* and are forced against the disc by air pressure or hydraulic pressure.

3. *Cone clutch or brake:* A cone clutch or brake is similar to a plate clutch or brake except that the mating surfaces are on a portion of a cone instead of on a flat plate.

4. *Band brake:* Used only as a brake, the friction material is on a flexible band that nearly surrounds a cylindrical drum attached to the machine to be controlled. When braking is desired, the band is tightened on the drum, exerting a tangential force to stop the load.

5. *Block or shoe brake:* Curved, rigid pads faced with the friction material are forced against the surface of a drum, from either the outside or the inside, exerting a tangential force to stop the load.

Actuation

The following are the means used to actuate clutches or brakes. Each may be applied to several of the types described. Figures 22–3 through 22–9 show a variety of commercially available designs.

Manual.   The operator provides the force, usually through a lever arrangement to achieve force multiplication.

FIGURE 22–2 Types of friction clutches and brakes [(b), Tol-O-Matic, Minneapolis, MN]

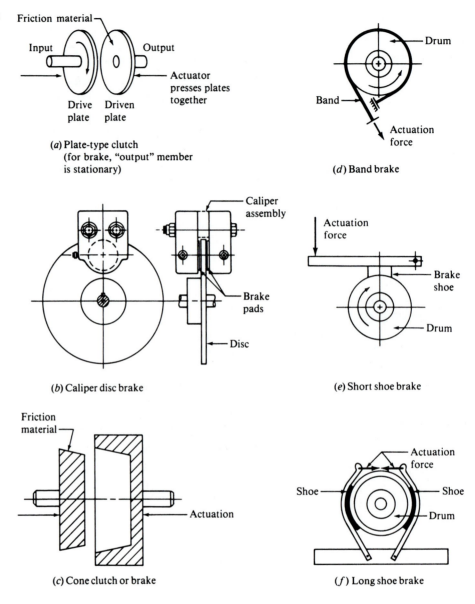

(a) Plate-type clutch (for brake, "output" member is stationary)

(b) Caliper disc brake

(c) Cone clutch or brake

(d) Band brake

(e) Short shoe brake

(f) Long shoe brake

FIGURE 22–3 Manually operated clutch (Rockford Division, Borg-Warner Corp., Rockford, IL)

FIGURE 22–4
Spring-applied, electrically released, long shoe brake (Eaton Corp., Cutler-Hammer Products, Milwaukee, WI)

Drum

Spring for brake application

Pivots

Solenoid for brake release

Shoes with friction pads

FIGURE 22–5 Slip clutch. Springs apply normal pressure on the friction plates. The spring force is adjustable to vary the torque level at which clutch will slip. (The Hilliard Corp., Elmira, NY)

Spring

Friction surfaces

Adjusting nut

Sleeve (output)

Hub (input)

Spring-applied.   Sometimes called a *fail-safe* design when applied to a brake, the brake is applied automatically by springs unless there is some opposing force present. Thus, if power fails, or if air pressure or hydraulic pressure is lost, or if the operator is unable to perform the function, the springs apply the brake and stop the load. The concept may also be used to engage or to disengage a clutch.

Centrifugal.   A centrifugal clutch is sometimes employed to permit the driving system to accelerate without a connected load. Then, at a preselected speed, centrifugal force moves the clutch elements into contact to connect the load. As the system slows, the load will be automatically disconnected.

Pneumatic.   Compressed air is introduced into a cylinder or some other chamber. The force produced by the pressure on a piston or a diaphragm brings the friction surfaces into contact with the members connected to the load.

Hydraulic.   Hydraulic brakes are similar to the pneumatic type, except that they use oil hydraulic fluids instead of air. The hydraulic actuator is usually applied where high actuation forces are required.

FIGURE 22–6 Air-actuated clutch or brake (Eaton Corp., Airflex Division, Cleveland, OH)

(*a*) Details of clutch design

Flexible tube
Air line
Friction pad
Drum

Air off:
Clutch disengaged

Air on:
Clutch engaged, friction pads forced against drum

(*b*) Clutch activation cycle

FIGURE 22–7 Hydraulically actuated disc brake assembly (Tol-O-Matic, Minneapolis, MN)

Hydraulic fittings

Caliper assembly

Brake pads

Caliper assembly

Brake pads

Disc

FIGURE 22-8
Electrically actuated
plate-type clutch or
brake (Warner Electric
Brake & Clutch Co.,
South Beloit, IL)

(a) Components of electromagnet

(b) Cutaway view of complete assembly

FIGURE 22-9
Electrically operated
clutch-brake module
(Electroid Corp.,
Springfield, NJ)

(a) Cutaway showing
internal components

Electromagnetic. An electric current is applied to a coil, creating an electromagnetic flux. The magnetic force then attracts an armature attached to the machine that is to be controlled. The armature is usually of the plate type.

**22-4
PERFORMANCE
PARAMETERS**

The principles of physics tell us that whenever the speed or the direction of the motion of a body is changed, there must be a force exerted on the body. If the body is in rotation, a torque must be applied to the system to speed it up or to slow it down. When a change in speed occurs, a corresponding change in the kinetic energy of the system occurs. Thus, motion control inherently involves the control of energy, either adding energy to accelerate a system or absorbing energy to decelerate it.

TABLE 22–2   Coefficients of friction

Friction material	Dynamic friction coefficient		Pressure range	
	Dry	In oil	(psi)	(kPa)
Molded compounds	0.25–0.45	0.06–0.10	150–300	1 035–2 070
Woven materials	0.25–0.45	0.08–0.10	50–100	345–690
Sintered metal	0.15–0.45	0.05–0.08	150–300	1 035–2 070
Cork	0.30–0.50	0.15–0.25	8–15	55–100
Wood	0.20–0.45	0.12–0.16	50–90	345–620
Cast iron	0.15–0.25	0.03–0.06	100–250	690–1 725
Paper-based		0.10–0.15		
Graphite/resin		0.10–0.14		

TABLE 22–3   Coefficient of friction classification codes of the Society of Automotive Engineers

Code letter	Coefficient of friction
C	Not over 0.15
D	Over 0.15 but not over 0.25
E	Over 0.25 but not over 0.35
F	Over 0.35 but not over 0.45
G	Over 0.45 but not over 0.55
H	Over 0.55
Z	Unclassified

For automotive applications, standards are set by the Society of Automotive Engineers. In SAE J866a, *Recommended Practice for Friction Identification System for Brake Linings and Brake Blocks for Motor Vehicles,* a set of codes is defined to classify friction materials according to the coefficient of friction, regardless of the material used. Table 22–3 lists these codes.

In problems in this book that require a coefficient of friction, we will use a value of 0.25 unless otherwise stated. Based on the values reported here, this is a relatively low value that should yield conservative designs.

**22–11
PLATE-TYPE
CLUTCH OR
BRAKE**

Figure 22–2(a) shows a simple sketch of a plate-type clutch, and Figure 22–8 shows a cutaway view of a commercially available, electrically actuated clutch or brake. In Figure 22–8, an electromagnet exerts an axial force that draws the friction surfaces together. When two bodies are brought into contact with a normal force between them, a friction force that tends to resist relative motion is created. This is the principle on which the plate-type clutch or brake is based.

As the friction plate rotates in relation to the mating plate with an axial force pressing them together, the friction force acts in a tangential direction, producing the brake or clutch torque. At any point, the local pressure times the differential area at the point is the *normal force.* The normal force times the coefficient of friction is the *friction force.* The friction force times the radius of the point is the *torque* produced at this point. The *total torque* is the sum of all of the torques over the entire area of the plate. We can find the sum by integrating over the area.

There is usually some variation of pressure over the surface of the friction plate, and some assumption about the nature of the variation must be made before the total torque can be computed. A conservative assumption that yields a useful result is that the friction surface will wear uniformly over the entire area as the brake or clutch operates. This assumption implies that the product of the local pressure, p, times the linear relative speed, v, between the plates is constant. Wear has been found to be approximately proportional to the product of p times v.

Taking all these factors into account and completing the analysis produces the following result for friction torque:

$$T_f = fN(R_o + R_i)/2$$

But the last part of this relation is the mean radius, R_m, of the annular plate. Then

Friction Torque on Annular Plate

$$T_f = fNR_m \qquad (22\text{–}8)$$

As stated before, this is a conservative result, meaning that the actual torque produced would be slightly larger than predicted.

Note that the torque is proportional to the mean radius, but that no area relationship is involved in Equation (22–8). Thus, the completion of the design for final dimensions requires some other parameter. The missing factor in Equation (22–8) is the expected wear rate of the friction material. It should be obvious that even with the same mean radius, a brake with a larger area would wear less than one with a smaller area.

Manufacturers of friction materials can assist in the final determination of the relationship between wear and the area of the friction surface. However, the following guidelines allow the estimation of the physical size of brakes and will be used for problem solutions in this book.

The wear rating, WR, will be based on the frictional power, P_f, absorbed by the brake per unit area, A, where

Frictional Power

$$P_f = T_f \omega \qquad (22\text{–}9)$$

and ω is the angular velocity of the disc. In SI units, with torque in N·m and ω in rad/s, the frictional power is in N·m/s or watts. In the U.S. Customary System, with torque in lb·in and angular velocity expressed as n rpm, the frictional power is in hp, computed from

$$P_f = \frac{T_f n}{63\ 000}\ \text{hp} \qquad (22\text{–}10)$$

For industrial applications, we will use

Wear Rating

$$WR = P_f/A \qquad (22\text{–}11)$$

where $WR = 0.04$ hp/in^2 for frequent applications, a conservative rating
$WR = 0.10$ hp/in^2 for average service
$WR = 0.40$ hp/in^2 for infrequently used brakes allowed to cool somewhat between applications

---

Example Problem 22–6   Compute the dimensions of an annular plate-type brake to produce a braking torque of 300 lb·in. Springs will provide a normal force of 320 lb between the friction surfaces. The coefficient of friction is 0.25. The brake will be used in average industrial service, stopping a load from 750 rpm.

Solution   ***Step 1.*** Compute the required mean radius. From Equation (22–8),

$$R_m = \frac{T_f}{fN} = \frac{300 \text{ lb·in}}{(0.25)(320 \text{ lb})} = 3.75 \text{ in}$$

Step 2. Specify a desired ratio of R_o/R_i, and solve for the dimensions. A reasonable value for the ratio is approximately 1.50. The range can be from 1.2 to about 2.5, at the designer's choice. Using 1.50, $R_o = 1.50R_i$, and

$$R_m = (R_o + R_i)/2 = (1.5R_i + R_i)/2 = 1.25 \, R_i$$

Then

$$R_i = R_m/1.25 = (3.75 \text{ in})/1.25 = 3.00 \text{ in}$$
$$R_o = 1.50R_i = 1.50(3.00) = 4.50 \text{ in}$$

Step 3. Compute the area of the friction surface:

$$A = \pi(R_o^2 - R_i^2) = \pi[(4.50)^2 - (3.00)^2] = 35.3 \text{ in}^2$$

Step 4. Compute the frictional power absorbed:

$$P_f = \frac{T_f n}{63\ 000} = \frac{(300)(750)}{63\ 000} = 3.57 \text{ hp}$$

Step 5. Compute the wear ratio:

$$WR = \frac{P_f}{A} = \frac{3.57 \text{ hp}}{35.3 \text{ in}^2} = 0.101 \text{ hp/in}^2$$

Step 6. Judge the suitability of *WR*. If *WR* is too high, return to Step 2 and increase the ratio. If *WR* is too low, decrease the ratio. In this example, *WR* is acceptable.

A more compact unit can be designed if more than one friction plate is used. We multiply the friction torque for one plate by the number of plates to determine the total friction torque. One disadvantage of this approach is that the heat dissipation is relatively poorer than for the single plate.

**22–12
CALIPER DISC
BRAKES**

The disc brake pads are brought into contact with the rotating disc by fluid pressure acting on a piston in the caliper. The pads are either round or in a short crescent shape to cover more of the disc surface [see Figure 22–8(b) and Figure 22–13]. However, one advantage of the disc brake is that the disc is exposed to the atmosphere, promoting the dissipation of heat. Because the disc rotates with the machine to be controlled, heat dissipation is further enhanced. The cooling effect improves the fade resistance of this type of brake relative to the shoe-type brake.

Designs for the friction torque and for the wear rating are similar to those explained for plate-type brakes.

**22–13
CONE CLUTCH
OR BRAKE**

The angle of inclination of the conical surface of the cone clutch or brake is typically 12°. A lower angle could be used with care, but there is a tendency for the friction surfaces to engage suddenly with a jerk. As the angle increases, the amount of axial force required to produce a given friction torque increases. Thus, 12° is a reasonable compromise.

Referring to Figure 22–16, we see that as an axial force F_a is applied by a spring, manually or by fluid pressure, a normal force *N* is created between the mating friction surfaces, all around the periphery of the cone. The desired friction force F_f is produced in the

FIGURE 22–16

Cone clutch or brake

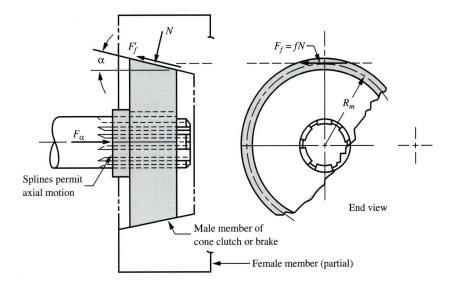

tangential direction, where $F_f = fN$. It is assumed that the friction force acts at the mean radius of the cone so that the friction torque is

$$T_f = F_f R_m = fN R_m \tag{22–8}$$

In addition to the tangentially directed friction force, a friction force develops along the surface of the cone and opposes the tendency for the member having the internal cone surface to move axially away from the external cone. We will call this force F'_f and compute it also from

$$F'_f = fN$$

For the equilibrium condition of the external cone, the sum of the horizontal forces must be zero. Then

$$F_a = N \sin \alpha + F'_f \cos \alpha = N \sin \alpha + fN \cos \alpha = N(\sin \alpha + f \cos \alpha)$$

or

$$N = \frac{F_a}{\sin \alpha + f \cos \alpha} \tag{22–12}$$

Substituting this into Equation (22–8) gives

⇨ **Frictional Torque on a Cone Clutch or Brake**

$$T_f = \frac{fR_m F_a}{\sin \alpha + f \cos \alpha} \tag{22–13}$$

Example Problem 22–7

Compute the axial force required for a cone brake if it is to exert a braking torque of 50 lb·ft. The mean radius of the cone is 5.0 in. Use $f = 0.25$. Try cone angles of 10°, 12°, and 15°.

Solution

We can solve Equation (22–13) for the axial force F_a:

$$F_a = \frac{T_f(\sin \alpha + f \cos \alpha)}{fR_m} = \frac{(50 \text{ lb·ft})(\sin \alpha + 0.25 \cos \alpha)}{(0.25)(5.0/12) \text{ ft}}$$

$$F_a = 480(\sin \alpha + 0.25 \cos \alpha) \text{ lb}$$

Then the values of F_a as a function of the cone angle are as follows:

For $\alpha = 10°$

$F_a = 202$ lb

For $\alpha = 12°$

$F_a = 217$ lb

For $\alpha = 15°$

$F_a = 240$ lb

22–14
DRUM BRAKES

Short Shoe Drum Brakes

Figure 22–17 shows a sketch of a drum brake in which the actuating force W acts on the lever that pivots on pin A. This creates a normal force between the shoe and the rotating drum. The resulting friction force is assumed to act tangential to the drum if the shoe is short. The friction force times the radius of the drum is the friction torque that slows the drum.

The objectives of the analysis are to determine the relationship between the applied load and the friction force and to be able to evaluate the effect of design decisions such as

FIGURE 22–17
Short shoe drum brake

(*a*) Lever design 1

(*b*) Lever design 2

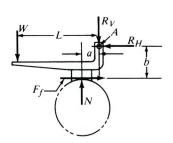

(*c*) Lever design 3

the size of the drum, the lever dimensions, and the placement of the pivot A. The free-body diagrams in Figure 22–17(a) support this analysis. For the lever, we can sum moments about the pivot A:

$$\Sigma M_a = 0 = WL - Na + F_f b \qquad (22\text{–}14)$$

But note that $F_f = fN$ or $N = F_f/f$. Then

$$0 = WL - F_f a/f + F_f b = WL - F_f(a/f - b)$$

Solving for W gives

$$W = \frac{F_f(a/f - b)}{L} \qquad (22\text{–}15)$$

Solving for F_f gives

Friction Force on Drum Brake

$$F_f = \frac{WL}{(a/f - b)} \qquad (22\text{–}16)$$

We can use these equations for the friction torque by noting

▷ **Friction Torque**

$$T_f = F_f D_d/2 \qquad (22\text{–}17)$$

where D_d = diameter of the drum

Note the alternate positions of the pivot in Parts (b) and (c) of Figure 22–17. In (b), the dimension $b = 0$.

Example Problem 22–8 Compute the actuation force required for the short shoe drum brake of Figure 22–17 to produce a friction torque of 50 lb·ft. Use a drum diameter of 10.0 in, $a = 3.0$ in, and $L = 15.0$ in. Consider values of f of 0.25, 0.50, and 0.75, and different points of location of pivot A such that b ranges from 0 to 6.0 in.

Solution The required friction force can be found from Equation (22–17):

$$F_f = 2T_f/D_d = (2)(50 \text{ lb·ft})/(10/12 \text{ ft}) = 120 \text{ lb}$$

In Equation (22–15), we can substitute for a, L, and F_f:

$$W = \frac{F_f(a/f - b)}{L} = \frac{120 \text{ lb}[(3.0 \text{ in})/f - b]}{15.0 \text{ in}} = 8(3/f - b) \text{ lb}$$

We can substitute the varying values of f and b into this last equation to compute the data for the curves of Figure 22–18, showing the actuating force versus the distance b for different values of f. Note that for some combinations, the value of W is *negative*. This means that the brake is *self-actuating* and that an upward force on the lever would be required to release the brake.

FIGURE 22–18
Results: actuating load
force vs. distance b

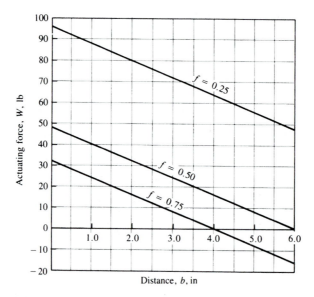

Long Shoe Drum Brakes

The assumption used for short shoe brakes, that the resultant friction force acted at the middle of the shoe, cannot be used in the case of shoes covering more than about 45° of the drum. In such cases, the pressure between the friction lining and the drum is very nonuniform, as is the moment of the friction force and of the normal force with respect to the pivot of the shoe.

The following equations govern the performance of a long shoe brake, using the terminology from Figure 22–19. (See Reference 3.)

1. **Friction torque on drum:**

$$T_f = r^2 f w p_{max}(\cos \theta_1 - \cos \theta_2) \tag{22–18}$$

2. **Actuation force:**

$$W = (M_N + M_f)/L \tag{22–19}$$

where M_N = moment of normal force with respect to the hinge pin
$$M_N = 0.25 p_{max} w r C[2(\theta_2 - \theta_1) - \sin 2\theta_2 + \sin 2\theta_1] \tag{22–20}$$
M_f = moment of friction force with respect to the hinge pin
$$M_f = f p_{max} w r[r(\cos \theta_1 - \cos \theta_2) + 0.25C(\cos 2\theta_2 - \cos 2\theta_1)] \tag{22–21}$$

The sign of M_f is negative ($-$) if the drum surface is moving away from the pivot and positive ($+$) if it is moving toward the pivot.

3. **Friction power:**

$$P_f = T_f n/63\ 000 \text{ hp} \tag{22–22}$$

where n = rotational speed in rpm

FIGURE 22–19

Terminology for long shoe drum brake

4. **Brake shoe area** (*Note:* Projected area is used):

$$A = L_s w = 2wr \sin[(\theta_2 - \theta_1)/2] \tag{22–23}$$

5. **Wear ratio:**

$$WR = P_f/A \tag{22–24}$$

The use of these relationships in the design and analysis of a long shoe brake is shown in Example Problem 22–9.

Example Problem 22–9   Design a long shoe drum brake to produce a friction torque of 750 lb·in to stop a drum from 120 rpm.

Solution   **Step 1.** Select a brake friction material, and specify the desired maximum pressure and the design value for the coefficient of friction. Table 22–2 lists some general properties for friction materials. Actual test values or specific manufacturer's data should be used where possible. The design value for p_{max} should be far less than the permissible pressure listed in Table 22–2 in order to improve wear life.

For this problem, let's choose a molded polymer compound and design for approximately 75 psi maximum pressure. Note, as shown in Figure 22–19, that the maximum pressure occurs at the section 90° from the pivot. If the shoe does not extend at least 90°, the equations used here are not valid. (See Reference 3.) Also, let's use $f = 0.25$ for the design.

Step 2. Propose trial values for the geometry of the brake drum and the brake pad. Several design decisions must be made here. The general arrangement shown in Figure 22–19 can be used as a guide. But your specific application and creativity may lead you to modify the arrangement.

Trial values are $r = 4.0$ in; $C = 8.0$ in; $L = 15$ in; $\theta_1 = 30°$; and $\theta_2 = 150°$.

Step 3. Solve for the required width of the shoe from Equation (22–18):

$$w = \frac{T_f}{r^2 f p_{max}(\cos \theta_1 - \cos \theta_2)}$$

For this problem,

$$w = \frac{750 \text{ lb·in}}{(4.0 \text{ in})^2(0.25)(75 \text{ lb/in}^2)(\cos 30° - \cos 150°)} = 1.44 \text{ in}$$

For convenience, let $w = 1.50$ in. Because the maximum pressure is proportional to the width, the actual maximum pressure will be

$$p_{max} = 75 \text{ psi}(1.44/1.50) = 72 \text{ psi}$$

Step 4. Compute M_N from Equation (22–20). The value of $\theta_2 - \theta_1$ must be in radians, with π radians $= 180°$. Then

$$\theta_2 - \theta_1 = 120°(\pi \text{ rad}/180°) = 2.09 \text{ rad}$$

The moment of the normal force on the shoe is

$$M_N = 0.25(72 \text{ lb/in}^2)(1.50 \text{ in})(4.0 \text{ in})(8.0 \text{ in})$$
$$[2(2.09) - \sin(300°) + \sin(60°)]$$

$$M_N = 5 \ 108 \text{ lb·in}$$

Step 5. Compute the moment of the friction force on the shoe, M_f, from Equation (22–21):

$$M_f = 0.25(72 \text{ lb/in}^2)(1.50 \text{ in})(4.0 \text{ in})$$
$$[(4.0 \text{ in})(\cos 30° - \cos 150°)$$
$$+ 0.25(8.0 \text{ in})(\cos 300° - \cos 60°)]$$

$$M_f = 748 \text{ lb·in}$$

Step 6. Compute the required actuation force, W, from Equation (22–19):

$$W = (M_N - M_f)/L = (5 \ 108 - 748)/(15) = 291 \text{ lb}$$

Note the minus sign because the drum surface is moving away from the pivot.

Step 7. Compute the frictional power from Equation (22–22):

$$P_f = T_f n/(63 \ 000) = (750)(120)/(63 \ 000) = 1.43 \text{ hp}$$

Step 8. Compute the projected area of the brake shoe from Equation (22–23).

$$A = L_s w = 2wr \sin[(\theta_2 - \theta_1)/2]$$
$$A = 2(1.50 \text{ in})(4.0 \text{ in})\sin(120°/2) = 10.4 \text{ in}^2$$

Step 9. Compute the wear ratio, WR:

$$WR = P_f/A = 1.43 \text{ hp}/10.4 \text{ in}^2 = 0.14 \text{ hp/in}^2$$

Step 10. Evaluate the suitability of the results. In this problem, we would need more information about the application to evaluate the results. However, the wear ratio seems reasonable for average service (see Section 22–11), and the geometry seems acceptable.

22–15
BAND
BRAKES
Figure 22–20 shows the typical configuration of a *band brake*. The flexible band, usually made of steel, is faced with a friction material that can conform to the curvature of the drum. The application of a force to the lever puts tension in the band and forces the friction material against

FIGURE 22–20

Band brake design

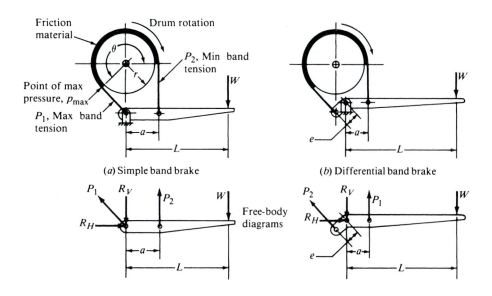

(*a*) Simple band brake   (*b*) Differential band brake

Free-body diagrams

the drum. The normal force, thus created, causes the friction force tangential to the drum surface to be created, retarding the drum.

   The tension in the band decreases from the value P_1 at the pivot side of the band to
P_2 at the lever side. The net torque on the drum is then

$$T_f = (P_1 - P_2)r \tag{22–25}$$

where r = radius of the drum

   The relationship between P_1 and P_2 can be shown (see Reference 3) to be the loga-
rithmic function

$$P_2 = P_1/e^{f\theta} \tag{22–26}$$

where θ = total angle of coverage of the band in radians

   The point of maximum pressure on the friction material occurs at the end nearest the
highest tension, P_1, where

$$P_1 = p_{max}rw \tag{22–27}$$

and w is the width of the band.

   For the two types of band brakes shown in Figure 22–20, the free-body diagrams of
the levers can be used to show the following relationships for the actuating force, W, as a
function of the tensions in the band. For the simple band brake of Figure 22–20(a),

$$W = P_2a/L \tag{22–28}$$

The style shown in Figure 22–20(b) is called a *differential band brake,* where the actuation
force is

$$W = (P_2a - P_1e)/L \tag{22–29}$$

   The design procedure is presented in Example Problem 22–10.

Example Problem
22–10

Design a band brake to exert a braking torque of 720 lb·in while slowing the drum from 120 rpm.

Solution

Step 1. Select a material and specify a design value for the maximum pressure. A woven friction material is desirable to facilitate the conformance to the cylindrical drum shape. Let's use p_{max} = 25 psi and a design value of f = 0.25. See Section 22–10.

Step 2. Specify a trial geometry: r, θ, w. For this problem, let's try r = 6.0 in, θ = 225°, and w = 2.0 in. Note that 225° = 3.93 rad.

Step 3. Compute the maximum band tension, P_1, from Equation (22–27):

$$P_1 = p_{max}rw = (25 \text{ lb/in}^2)(6.0 \text{ in})(2.0 \text{ in}) = 300 \text{ lb}$$

Step 4. Compute tension P_2 from Equation (22–26):

$$P_2 = \frac{P_1}{e^{f\theta}} = \frac{300 \text{ lb}}{e^{(0.25)(3.93)}} = 112 \text{ lb}$$

Step 5. Compute the friction torque, T_f:

$$T_f = (P_1 - P_2)r = (300 - 112)(6.0) = 1\,128 \text{ lb·in}$$

Note: Repeat Steps 2–5 until you achieve a satisfactory geometry and friction torque. Let's try a smaller design, say, r = 5.0 in:

$$P_1 = (25)(5.0)(2.0) = 250 \text{ lb}$$

$$P_2 = \frac{250 \text{ lb}}{e^{(0.25)(3.93)}} = 93.7 \text{ lb}$$

$$T_f = (250 - 93.7)(5.0) = 782 \text{ lb·in} \quad \text{(okay)}$$

Step 6. Specify the geometry of the lever, and compute the required actuation force. Let's use a = 5.0 in and L = 15.0 in. Then

$$W = P_2(a/L) = 93.7 \text{ lb}(5.0/15.0) = 31.2 \text{ lb}$$

Step 7. Compute the average wear ratio from $WR = P_f/A$:

$$A = 2\pi rw(\theta/360) = 2(\pi)(5.0 \text{ in})(2.0 \text{ in})(225/360) = 39.3 \text{ in}^2$$

$$P_f = T_f n/63\,000 = (782)(120)/(63\,000) = 1.50 \text{ hp}$$

$$WR = P_f/A = (1.50 \text{ hp})/(39.3 \text{ in}^2) = 0.038 \text{ hp/in}^2$$

This should be conservative for average service.

22–16
OTHER TYPES
OF CLUTCHES
AND BRAKES

The chapter thus far has concentrated on clutches and brakes using friction materials to transmit the torque between rotating members, but many other types are available. Brief descriptions are given below, but specific design information is not given. Most are unique to the manufacturer, and application data are available through catalogs.

Jaw Clutch

The teeth of the mating sets of jaws are brought into engagement by sliding one or both members axially. The teeth may be straight-sided or triangular, or they may incorporate some smooth curve to facilitate engagement. Once the teeth are engaged, there is a positive transmission of torque. The jaw clutch is normally engaged while the system is stopped or is running very slowly.

Ratchet

Although not strictly a clutch, the familiar ratchet and pawl permits alternate engagement and disengagement of moving members, and thus it can be used in similar applications. Typically, the ratchet moves only a small fraction of a revolution per cycle.

Sprag, Roller, and Cam Clutches

There are differences in the specific geometry of sprag, roller, and cam clutches, but they all perform similar functions. When the input shaft is rotating in the driving direction, the internal elements (sprags, rollers, or cams) are wedged between the driving and driven members and thus transmit torque. But when the input member rotates in the opposite direction, the internal elements move out of engagement, and no torque is transmitted. Thus, they can be used for applications similar to ratchets, but with much smoother operation and with a virtually infinite amount of incremental motion. Another application is *backstopping,* in which the clutch runs free when the machine is being driven in the intended direction. But if the drive is disengaged and the load starts to reverse direction, the clutch locks up and prevents motion. This type of clutch is also used for *overrunning:* a positive drive as long as the load rotates no faster than the driver. If the load tends to run faster (overrun) than the driver, the clutch elements disengage. This protects equipment that might be damaged by overspeeding.

Fiber Clutch

A fiber clutch operates in a manner similar to the overrunning clutches described previously. But instead of driving through solid elements, the torque is transmitted through stiff fibers that have a preferred orientation. When rotated in a direction opposite the preferred direction, the fibers "lie down," and no torque is transmitted.

Wrap-Spring Clutch

Again used in cases similar to the overrunning clutches, the *wrap-spring clutch* is made from a rectangular wire and normally has an inside diameter slightly larger than that of the shaft on which it is installed. Thus, no torque is transmitted. But when one end of the spring is restrained, the spring "wraps down" tightly on the surface of the shaft, and torque is transmitted positively through the spring.

Single-Revolution Clutches

It is frequently desired to have a machine cycle one complete revolution and then come to a stop. The *single-revolution clutch* provides this feature. After it is tripped, it drives the output shaft until reaching a positive stop at the end of one revolution. Some types can be engaged for more than one revolution, but they will return to a fixed position, for example, at the top of the stroke of a press.

Fluid Clutches

The *fluid clutch* is made up of two separate parts with no mechanical connection between them. A fluid fills a cavity between the parts, and as one member rotates, it tends to shear the fluid, causing torque to be transmitted to the mating element. The resulting drive is very smooth and soft because load peaks will simply cause one member to move relative to the other. In this situation, it is similar to the slip clutch described earlier.

Eddy Current Drive

When a conducting disc moves through a magnetic field, *eddy currents* are induced in the disc, causing a force to be exerted on the disc in a direction opposite to the direction of rotation. The force can be used to brake the disc or to transmit torque to a mating part, such as a clutch. An advantage of this type of unit is that there is no mechanical connection between the elements. The torque can be controlled by varying the current to the electromagnets.

Overload Clutches

The drive is positive, provided that the torque is below some set value. At higher torques, some element is disengaged automatically. One type uses a series of spherical balls positioned in detents and held under a spring force. When the tripping torque level is reached, the balls are forced out of the detents and disengage the drive.

REFERENCES

1. Avallone, Eugene A., and Theodore Baumeister III, eds. *Marks' Standard Handbook for Mechanical Engineers.* 10th ed. New York: McGraw-Hill, 1996.

2. Electroid Corporation. *Engineering Catalog No. 900E.* Springfield, NJ: Electroid Corporation.

3. Faires, V. M. *Design of Machine Elements.* 4th ed. New York: Macmillan, 1965.

4. Juvinall, Robert C., and Kurt M. Marshek. *Fundamentals of Machine Component Design.* 2d ed. New York: John Wiley & Sons, 1991.

5. *Machine Design Magazine, Power & Motion Control Reference Volume.* Cleveland, OH: Penton Publishing, Inc., June 1990.

6. Orthwein, William C. *Clutches and Brakes: Design and Selection.* New York: Marcel Dekker, 1986.

7. Shigley, J. E., and C. R. Mischke. *Mechanical Engineering Design.* 5th ed. New York: McGraw-Hill, 1989.

8. Society of Automotive Engineers. *Standard J866a, Recommended Practice for Friction Identification System for Brake Linings and Brake Blocks for Motor Vehicles.* Warrendale, PA: Society of Automotive Engineers, 1987.

9. Warner Electric Brake & Clutch Company. *Catalog P-137-15.* Beloit, WI: Warner Electric Brake & Clutch Company.

PROBLEMS

1. Specify the required torque rating for a clutch to be attached to a motor shaft running at 1 750 rpm. The motor is rated at 5.0 hp and is of the design B type.

2. Specify the required torque rating for a clutch to be attached to a diesel engine shaft running at 2 500 rpm. The engine is rated at 75.0 hp.

3. Specify the required torque rating for a clutch to be attached to an electric motor shaft running at 1 150 rpm. The motor is rated at 0.50 hp and drives a light fan.

4. An alternative design for the system described in Problem 1 is being considered. Instead of putting the clutch on the motor shaft, it is desired to place it on the output shaft of a speed reducer that rotates at 180 rpm. The power transmission is still approximately 5.0 hp. Specify the required torque rating for the clutch.

5. Specify the required brake torque rating for each of the conditions in Problems 1–4 for average industrial conditions.

6. Specify the required torque rating for a clutch in N·m if it is attached to a design B electric motor shaft rated at 20.0 kW and rotating at 3 450 rpm.

7. A clutch-brake module is to be connected between a design C electric motor and a speed reducer. The motor is rated at 50.0 kW at 900 rpm. Specify the required torque ratings for the clutch and the brake portions of the module for average industrial service. The drive is to a large conveyor.

8. Compute the torque required to accelerate a solid steel disc from rest to 550 rpm in 2.0 s. The disc has a diameter of 24.0 in and is 2.5 in thick.

9. The assembly shown in Figure 22–21 is to be stopped by a brake from 775 rpm to zero in 0.50 s or less. Compute the required brake torque.

FIGURE 22–21 (Problems 9, 14, and 16)

10. Compute the required clutch torque to accelerate the system shown in Figure 22–22 from rest to a motor speed of 1 750 rpm in 1.50 s. Neglect the inertia of the clutch.

FIGURE 22–22 (Problem 10)

11. A winch, sketched in Figure 22–23, is lowering a load at the speed of 50 ft/min. Compute the required torque rating for the brake on the winch shaft to stop the system in 0.25 s.

12. Figure 22–24 shows a tumbling barrel being driven through a wormgear reduction unit. Evaluate the torque rating required for a clutch to accelerate the barrel to 38.0 rpm from rest in 2.0 s (a) if the clutch is placed on the motor shaft and (b) if it is placed at the output of the reducer.

FIGURE 22–23 (Problem 11)

Neglect the inertia of the gear shafts, the bearing races, and the clutch. Consider the worm and wormgear to be solid cylinders.

13. Compute the dimensions of an annular plate-type brake to produce a braking torque of 75 lb·in. Air pressure will develop a normal force of 150 lb between the friction surfaces. Use a coefficient of friction of 0.25. The brake will be used in average industrial service, stopping a load from 1 150 rpm.

14. Design a plate-type brake for the application described in Problem 9. Specify the design coefficient of friction, the dimensions of the plate, and the axial force required.

15. Compute the axial force required for a cone clutch if it is to exert a driving torque of 15 lb·ft. The cone surface has a mean diameter of 6.0 in and an angle of 12°. Use $f = 0.25$.

16. Design a cone brake for the application described in Problem 9. Specify the design coefficient of friction, the mean diameter of the cone surface, and the axial force required.

17. Compute the actuation force required for the short shoe drum brake of Figure 22–17 to produce a friction torque of 150 lb·ft. Use a drum diameter of 12.0 in, $a = 4.0$ in, and $L = 24.0$ in. Use $f = 0.25$ and $b = 5.0$ in.

18. For all other data from Problem 17 being the same, determine the required dimension b for the brake to be self-actuating.

19. Design a short shoe drum brake to produce a torque of 100 lb·ft. Specify the drum diameter, the configuration of the actuation lever, and the actuation force.

20. Design a long shoe drum brake to produce a friction torque of 100 lb·ft to stop a load from 480 rpm. Specify the friction material, the drum size, the shoe configuration, pivot locations, and actuation force.

21. Design a band brake to exert a braking torque of 75 lb·ft while slowing a drum from 350 rpm to rest. Specify a material, the drum diameter, the band width, the angle of coverage of the friction material, the actuation lever configuration, and the actuation force.

FIGURE 22–24 (Problem 12)

23

Design Projects

23–1
OBJECTIVES OF
THIS CHAPTER

One of the primary focuses of this book has been to emphasize the integration of machine elements into complete mechanical designs. The interfaces between machine elements have been discussed for many examples. The forces exerted on one element by another have been computed. Commercially available components and complete devices have been shown in several figures throughout the book.

Although these discussions and examples are helpful, one of the better ways to learn mechanical design is to *do* mechanical design. You should decide the detailed functions and design requirements for the design. You should conceptualize several approaches to a design. You should decide which approach to complete. You should complete the design of each element in detail. You should make assembly and detail drawings to communicate your design to others who may use it or be responsible for its fabrication. You should specify completely the purchased components that are part of the design.

Following are several projects that call for you to do these operations. You or your instructor may modify or amplify the projects to suit individual needs or the available time or information, such as manufacturers' catalogs. As in most design projects, many solutions are possible. Different solutions from several members of a class could be compared and critiqued to enhance the learning achieved. It may be helpful at this time to review Sections 1–4 and 1–5 about the functions and design requirements for design and a philosophy of design. Also, the problems at the end of Chapter 1 asked that you write a set of functions and design requirements for several of the same devices. If you have already done so, they can be used as a part of these exercises.

23–2
DESIGN
PROJECTS

Automobile Hood Latch

Design a hood latch for an automobile. The latch must be able to hold the hood securely closed during operation of the vehicle. But it should be easy to open for servicing the contents of the engine compartment. Theft-proofing is an important design goal. Attachment of the latch to the frame of the car and to the hood should be defined. Mass production should be a requirement.

Hydraulic Lift

Design a hydraulic lift to be used for car repair. Obtain pertinent dimensions from representative cars for initial height, extended height, design of the pads that contact the car, and so on. The lift will raise the entire car.

Car Jack

Design a floor jack for a car to lift either the entire front end or the entire rear end. The jack may be powered by hand, using mechanical or hydraulic actuation. Or it may be powered by pneumatic pressure or electrical power.

Portable Crane

Design a portable crane to be used in homes, small industries, warehouses, and garages. It should have a capacity of at least 1 000 lb (4.45 kN). Typical uses would be to remove an engine from a car, lift machine components, or load trucks.

Can Crusher

Design a machine to crush soft-drink or beer cans. The crusher would be used in homes or restaurants as an aid to recycling efforts. It could be operated either by hand or electrically. It should crush the cans to approximately 20% of their original volume.

Transfer Device

Design an automatic transfer device for a production line. The parts to be handled are steel castings having the following characteristics:

>*Weight:* 42.0 lb (187 N)
>
>*Size:* Cylindrical; 6.75-in diameter and 10.0 in high. Exterior surface is free of projections or holes and has a reasonably smooth, as-cast finish.
>
>*Transfer rate:* Continuous flow, 2.00 s between parts.
>
>Parts enter at a 24.0-in elevation on a roller conveyor. They must be elevated to 48.0 inches in a space of 60.0 in horizontally. They leave on a separate conveyor.

Drum Dumper

Design a drum dumper. The machine is to raise a 55-gal drum of bulk material from floor level to a height of 60.0 in and dump the contents of the drum into a hopper.

Paper Feeder

Design a paper feed device for a copier. The paper must be fed at a rate of 120 sheets per min.

Gravel Conveyor

Design a conveyor to elevate gravel into a truck. The top edge of the truck bed is 8.0 ft (2.44 m) off the ground. The bed is 6.5 ft wide, 12.0 ft long, and 4.0 ft deep (1.98 m × 3.66 m × 1.22 m). It is desired to fill the truck in 5.0 min or less.

Construction Lift

Design a construction lift. The lift will raise building materials from ground level to any height up to 40.0 ft (12.2 m). The lift will be at the top of a rigid scaffold that is not a part of the design project. It will raise a load of up to 500 lb (2.22 kN) at the rate of 1.0 ft/s (0.30 m/s). The load will be on a pallet, 3.0 ft by 4.0 ft (0.91 m × 1.22 m). At the top of the lift, means must be provided to bring the load onto a platform that supports the lift.

Packaging Machine

Design a packaging machine. Toothpaste tubes are to be taken from a continuous belt and inserted into cartons. Any standard tube size may be chosen. The device may include the means to close the cartons after the tube is in place.

Carton Packer

Design a machine to insert 24 cartons of toothpaste into a shipping case.

Robot Gripper

Design a gripper for a robot to grasp a spare-tire assembly from a rack and insert it into the trunk of an automobile on an assembly line. Obtain dimensions from a particular car.

Weld Positioner

Design a weld positioner. A heavy frame is made of welded steel plate in the shape shown in Figure 23–1. The welding unit will be robot-guided, but it is essential that the weld line be horizontal as the weld proceeds. Design the device to hold the frame securely and move it to present the part to the robot. The plate has a thickness of 3/8 in (9.53 mm).

Garage Door Opener

Design a garage door opener.

Spur Gear Speed Reducer, Single-Reduction

Design a complete single-reduction, spur gear–type speed reducer. Specify the two gears, two shafts, four bearings, and a housing. Use any of the data from Chapter 9, Problems 59–69.

Spur Gear Speed Reducer, Double-Reduction

Design a complete double-reduction, spur gear–type speed reducer. Specify the four gears, three shafts, six bearings, and a housing. Use any of the data from Chapter 9, Problems 73–75.

Helical Gear Speed Reducer, Single-Reduction

Design a complete single-reduction, helical gear–type speed reducer. Use any of the data from Chapter 10, Problems 5–11.

Bevel Gear Speed Reducer, Single-Reduction

Design a complete single-reduction, bevel gear–type speed reducer. Use any of the data from Chapter 10, Problems 14–17.

Wormgear Speed Reducer, Single-Reduction

Design a complete single-reduction, wormgear–type speed reducer. Use any of the data from Chapter 10, Problems 18–24.

FIGURE 23–1
Frame to be handled by a weld positioner

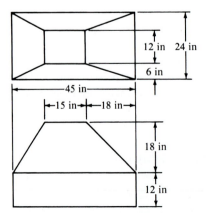

Weld all seams with weld line in horizontal plane

Lift Device Using Acme Screws

Design a device similar to that sketched in Figure 17–3. An electric motor drives the worm at a speed of 1 750 rpm. The two Acme screws rotate and lift the yoke, which in turn lifts the hatch. See Example Problem 17–1 for additional details. Complete the entire unit, including the wormgear set, the chain drive, the Acme screws, the bearings, and their mountings. The hatch is 60 in (1 524 mm) in diameter at its top surface. The screws should be nominally 30 in (762 mm) long. The total motion of the yoke will be 24 in, to be completed in 15.0 s or less.

Lift Device Using Ball Screws

Repeat the design of the lift device, using ball screws instead of Acme screws.

Brake for a Drive Shaft

Design a brake. A rotating load (as sketched in Figure 22–21) is to be stopped from 775 rpm in 0.50 s or less. Use any type of brake described in Chapter 22, and complete the design details, including the actuating means: springs, air pressure, manual lever, and so on. Show the brake attached to the shaft of Figure 22–21.

Brake for a Winch

Design a complete brake for the application shown in Figure 22–23 and described in Problem 11 in Chapter 22.

Indexing Drive

Design an indexing drive for an automatic assembly system. The items to be moved are mounted on a square steel fixture plate, 6.0 in (152 mm) on a side and 0.50 in (12.7 mm) thick. The total weight of each assembly is 10.0 lb (44.5 N). The center of each fixture (intersection of its diagonals) is to move 12.0 in (305 mm) with each index. The index is to be completed in 1.0 s or less, and the fixture must be held stationary at each station for a minimum of 2.0 s. Four assembly stations are required. The arrangement may be linear, rotary, or any other, provided that the fixtures move in a horizontal plane.

Child's Ferris Wheel

Design a child's Ferris wheel. It should be capable of holding one to four children weighing up to 80 lb (356 N) each. The rotational speed should be 1 rev in 6.0 s. It should be driven by an electric motor.

Merry-Go-Round

Design an amusement ride for small children, age 6 or younger, that they can sit in safely while enjoying an interesting motion pattern. At least two children at a time can ride the machine. The ride will be marketed to shopping centers and department stores to amuse customers' children.

Backyard Amusement Ride

Design a backyard amusement ride in which small coaster wagons are pulled along a circular path. The ride should be powered by an electric motor. Each wagon is 1.0 m long (39.4 in) and 0.50 m wide (19.7 in). The four wheels are 150 mm (6.0 in) in diameter. The wagon is to be attached to the drive rod at the point where the handle would normally be. The radial distance to the attachment point is to be 2.0 m (6.6 ft). The wagons are to make 1 rev in 8.0 s. Provide a means of starting and stopping the drive.

Transfer Device

Design a device to move automotive camshafts between processing stations. Each movement is to be 9.0 in (229 mm). The camshaft is to be supported on two unfinished bearing surfaces having a diameter of 3.80 in (96.5 mm) and an axial length of 0.75 in (19.0 mm). The spread between the bearing surfaces is 15.75 in (400.0 mm). Each camshaft weighs 16.3 lb (72.5 N). One motion cycle is to be completed each 2.50 s. Design the complete mechanism, including the drive from an electric motor.

Chain Conveyor

Design a powered, straight chain conveyor to move eight pallets along an assembly line. The pallets are 18 in long and 12 in wide. The maximum weight of each pallet and the product it carries is 125 lb. At the end of the conveyor, an external downward force of 500 lb is applied to the product which must be carried through the pallet to the conveyor structure. You may design the configuration of the sides and bottom of the pallet.

"You Are the Designer" Projects

At the beginning of each chapter in this book, a section called **You Are the Designer** appeared in which you were asked to imagine that you were the designer of some device or system. Choose any of those projects.

Appendixes

APPENDIX 1   PROPERTIES OF AREAS

(*a*) Circle

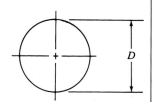

$$A = \pi D^2/4 \qquad r = D/4$$
$$I = \pi D^4/64 \qquad J = \pi D^4/32$$
$$S = \pi D^3/32 \qquad Z_p = \pi D^3/16$$

(*b*) Hollow circle (tube)

$$A = \pi(D^2 - d^2)/4 \qquad r = \sqrt{D^2 + d^2}/4$$
$$I = \pi(D^4 - d^4)/64 \qquad J = \pi(D^4 - d^4)/32$$
$$S = \pi(D^4 - d^4)/32D \qquad Z_p = \pi(D^4 - d^4)/16D$$

(*c*) Square

$$A = S^2 \qquad r = S/\sqrt{12}$$
$$I = S^4/12$$
$$S = S^3/6$$

(*d*) Rectangle

$$A = BH \qquad r_x = H/\sqrt{12}$$
$$I_x = BH^3/12 \qquad r_y = B/\sqrt{12}$$
$$S_x = BH^2/6$$

(*e*) Triangle

$$A = BH/2 \qquad r = H/\sqrt{18}$$
$$I = BH^3/36$$
$$S = BH^2/24$$

(*f*) Semicircle

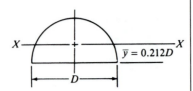

$A = \pi D^2/8 \qquad r = 0.132D$

$I = 0.007D^4$

$S = 0.024D^3$

(*g*) Regular hexagon

$A = 0.866D^2 \qquad r = 0.264D$

$I = 0.06D^4$

$S = 0.12D^3$

A = area

I = moment of inertia

S = section modulus

r = radius of gyration = $\sqrt{I/A}$

J = polar moment of inertia

Z_p = polar section modulus

APPENDIX 2   PREFERRED BASIC SIZES AND SCREW THREADS

TABLE A2–1   Preferred basic sizes

Fractional (in)				Decimal (in)			Metric (mm) First	Second	First	Second	First	Second
1/64	0.015 625	5	5.000	0.010	2.00	8.50	1		10		100	
1/32	0.031 25	$5\frac{1}{4}$	5.250	0.012	2.20	9.00		1.1		11		110
1/16	0.062 5	$5\frac{1}{2}$	5.500	0.016	2.40	9.50	1.2		12		120	
3/32	0.093 75	$5\frac{3}{4}$	5.750	0.020	2.60	10.00		1.4		14		140
1/8	0.125 0	6	6.000	0.025	2.80	10.50	1.6		16		160	
5/32	0.156 25	$6\frac{1}{2}$	6.500	0.032	3.00	11.00		1.8		18		180
3/16	0.187 5	7	7.000	0.040	3.20	11.50	2		20		200	
1/4	0.250 0	$7\frac{1}{2}$	7.500	0.05	3.40	12.00		2.2		22		220
5/16	0.312 5	8	8.000	0.06	3.60	12.50	2.5		25		250	
3/8	0.375 0	$8\frac{1}{2}$	8.500	0.08	3.80	13.00		2.8		28		280
7/16	0.437 5	9	9.000	0.10	4.00	13.50	3		30		300	
1/2	0.500 0	$9\frac{1}{2}$	9.500	0.12	4.20	14.00		3.5		35		350
9/16	0.562 5	10	10.000	0.16	4.40	14.50	4		40		400	
5/8	0.625 0	$10\frac{1}{2}$	10.500	0.20	4.60	15.00		4.5		45		450
11/16	0.687 5	11	11.000	0.24	4.80	15.50	5		50		500	
3/4	0.750 0	$11\frac{1}{2}$	11.500	0.30	5.00	16.00		5.5		55		550
7/8	0.875 0	12	12.000	0.40	5.20	16.50	6		60		600	
1	1.000	$12\frac{1}{2}$	12.500	0.50	5.40	17.00		7		70		700
$1\frac{1}{4}$	1.250	13	13.000	0.60	5.60	17.50	8		80		800	
$1\frac{1}{2}$	1.500	$13\frac{1}{2}$	13.500	0.80	5.80	18.00		9		90		900
$1\frac{3}{4}$	1.750	14	14.000	1.00	6.00	18.50					1 000	
2	2.000	$14\frac{1}{2}$	14.500	1.20	6.50	19.00						
$2\frac{1}{4}$	2.250	15	15.000	1.40	7.00	19.50						
$2\frac{1}{2}$	2.500	$15\frac{1}{2}$	15.500	1.60	7.50	20.00						
$2\frac{3}{4}$	2.750	16	16.000	1.80	8.00							
3	3.000	$16\frac{1}{2}$	16.500									
$3\frac{1}{4}$	3.250	17	17.000									
$3\frac{1}{2}$	3.500	$17\frac{1}{2}$	17.500									
$3\frac{3}{4}$	3.750	18	18.000									
4	4.000	$18\frac{1}{2}$	18.500									
$4\frac{1}{4}$	4.250	19	19.000									
$4\frac{1}{2}$	4.500	$19\frac{1}{2}$	19.500									
$4\frac{3}{4}$	4.750	20	20.000									

TABLE A2–2   American Standard screw threads

A. American Standard thread dimensions, numbered sizes

Size	Basic major diameter, D (in)	Coarse threads: UNC		Fine threads: UNF	
		Threads per inch, n	Tensile stress area (in²)	Threads per inch, n	Tensile stress area (in²)
0	0.060 0			80	0.001 80
1	0.073 0	64	0.002 63	72	0.002 78
2	0.086 0	56	0.003 70	64	0.003 94
3	0.099 0	48	0.004 87	56	0.005 23
4	0.112 0	40	0.006 04	48	0.006 61
5	0.125 0	40	0.007 96	44	0.008 30
6	0.138 0	32	0.009 09	40	0.010 15
8	0.164 0	32	0.014 0	36	0.014 74
10	0.190 0	24	0.017 5	32	0.020 0
12	0.216 0	24	0.024 2	28	0.025 8

B. American Standard thread dimensions, fractional sizes

Size	Basic major diameter, D (in)	Coarse threads: UNC		Fine threads: UNF	
		Threads per inch, n	Tensile stress area (in²)	Threads per inch, n	Tensile stress area (in²)
1/4	0.250 0	20	0.031 8	28	0.036 4
5/16	0.312 5	18	0.052 4	24	0.058 0
3/8	0.375 0	16	0.077 5	24	0.087 8
7/16	0.437 5	14	0.106 3	20	0.118 7
1/2	0.500 0	13	0.141 9	20	0.159 9
9/16	0.562 5	12	0.182	18	0.203
5/8	0.625 0	11	0.226	18	0.256
3/4	0.750 0	10	0.334	16	0.373
7/8	0.875 0	9	0.462	14	0.509
1	1.000	8	0.606	12	0.663
$1\frac{1}{8}$	1.125	7	0.763	12	0.856
$1\frac{1}{4}$	1.250	7	0.969	12	1.073
$1\frac{3}{8}$	1.375	6	1.155	12	1.315
$1\frac{1}{2}$	1.500	6	1.405	12	1.581
$1\frac{3}{4}$	1.750	5	1.90		
2	2.000	$4\frac{1}{2}$	2.50		

TABLE A2–3   Metric sizes of screw threads

Basic major diameter, D (mm)	Coarse threads		Fine threads	
	Pitch (mm)	Tensile stress area (mm^2)	Pitch (mm)	Tensile stress area (mm^2)
1	0.25	0.460		
1.6	0.35	1.27	0.20	1.57
2	0.4	2.07	0.25	2.45
2.5	0.45	3.39	0.35	3.70
3	0.5	5.03	0.35	5.61
4	0.7	8.78	0.5	9.79
5	0.8	14.2	0.5	16.1
6	1	20.1	0.75	22.0
8	1.25	36.6	1	39.2
10	1.5	58.0	1.25	61.2
12	1.75	84.3	1.25	92.1
16	2	157	1.5	167
20	2.5	245	1.5	272
24	3	353	2	384
30	3.5	561	2	621
36	4	817	3	865
42	4.5	1 121		
48	5	1 473		

APPENDIX 3 DESIGN PROPERTIES OF CARBON AND ALLOY STEELS

Material designation (AISI number)	Condition	Tensile strength (Ksi)	(MPa)	Yield strength (Ksi)	(MPa)	Ductility (percent elongation in 2 inches)	Brinell hardness (HB)
1020	Hot-rolled	55	379	30	207	25	111
1020	Cold-drawn	61	420	51	352	15	122
1020	Annealed	60	414	43	296	38	121
1040	Hot-rolled	72	496	42	290	18	144
1040	Cold-drawn	80	552	71	490	12	160
1040	OQT 1300	88	607	61	421	33	183
1040	OQT 400	113	779	87	600	19	262
1050	Hot-rolled	90	620	49	338	15	180
1050	Cold-drawn	100	690	84	579	10	200
1050	OQT 1300	96	662	61	421	30	192
1050	OQT 400	143	986	110	758	10	321
1117	Hot-rolled	62	427	34	234	33	124
1117	Cold-drawn	69	476	51	352	20	138
1117	WQT 350	89	614	50	345	22	178
1137	Hot-rolled	88	607	48	331	15	176
1137	Cold-drawn	98	676	82	565	10	196
1137	OQT 1300	87	600	60	414	28	174
1137	OQT 400	157	1 083	136	938	5	352
1144	Hot-rolled	94	648	51	352	15	188
1144	Cold-drawn	100	690	90	621	10	200
1144	OQT 1300	96	662	68	469	25	200
1144	OQT 400	127	876	91	627	16	277
1213	Hot-rolled	55	379	33	228	25	110
1213	Cold-drawn	75	517	58	340	10	150
12L13	Hot-rolled	57	393	34	234	22	114
12L13	Cold-drawn	70	483	60	414	10	140
1340	Annealed	102	703	63	434	26	207
1340	OQT 1300	100	690	75	517	25	235
1340	OQT 1000	144	993	132	910	17	363
1340	OQT 700	221	1 520	197	1 360	10	444
1340	OQT 400	285	1 960	234	1 610	8	578
3140	Annealed	95	655	67	462	25	187
3140	OQT 1300	115	792	94	648	23	233
3140	OQT 1000	152	1 050	133	920	17	311
3140	OQT 700	220	1 520	200	1 380	13	461
3140	OQT 400	280	1 930	248	1 710	11	555
4130	Annealed	81	558	52	359	28	156
4130	WQT 1300	98	676	89	614	28	202
4130	WQT 1000	143	986	132	910	16	302
4130	WQT 700	208	1 430	180	1 240	13	415
4130	WQT 400	234	1 610	197	1 360	12	461
4140	Annealed	95	655	60	414	26	197
4140	OQT 1300	117	807	100	690	23	235
4140	OQT 1000	168	1 160	152	1 050	17	341
4140	OQT 700	231	1 590	212	1 460	13	461
4140	OQT 400	290	2 000	251	1 730	11	578

Material designation (AISI number)	Condition	Tensile strength		Yield strength		Ductility (percent elongation in 2 inches)	Brinell hardness (HB)
		(Ksi)	(MPa)	(Ksi)	(MPa)		
4150	Annealed	106	731	55	379	20	197
4150	OQT 1300	127	880	116	800	20	262
4150	OQT 1000	197	1 360	181	1 250	11	401
4150	OQT 700	247	1 700	229	1 580	10	495
4150	OQT 400	300	2 070	248	1 710	10	578
4340	Annealed	108	745	68	469	22	217
4340	OQT 1300	140	965	120	827	23	280
4340	OQT 1000	171	1 180	158	1 090	16	363
4340	OQT 700	230	1 590	206	1 420	12	461
4340	OQT 400	283	1 950	228	1 570	11	555
5140	Annealed	83	572	42	290	29	167
5140	OQT 1300	104	717	83	572	27	207
5140	OQT 1000	145	1 000	130	896	18	302
5140	OQT 700	220	1 520	200	1 380	11	429
5140	OQT 400	276	1 900	226	1 560	7	534
5150	Annealed	98	676	52	359	22	197
5150	OQT 1300	116	800	102	700	22	241
5150	OQT 1000	160	1 100	149	1 030	15	321
5150	OQT 700	240	1 650	220	1 520	10	461
5150	OQT 400	312	2 150	250	1 720	8	601
5160	Annealed	105	724	40	276	17	197
5160	OQT 1300	115	793	100	690	23	229
5160	OQT 1000	170	1 170	151	1 040	14	341
5160	OQT 700	263	1 810	237	1 630	9	514
5160	OQT 400	322	2 220	260	1 790	4	627
6150	Annealed	96	662	59	407	23	197
6150	OQT 1300	118	814	107	738	21	241
6150	OQT 1000	183	1 260	173	1 190	12	375
6150	OQT 700	247	1 700	223	1 540	10	495
6150	OQT 400	315	2 170	270	1 860	7	601
8650	Annealed	104	717	56	386	22	212
8650	OQT 1300	122	841	113	779	21	255
8650	OQT 1000	176	1 210	155	1 070	14	363
8650	OQT 700	240	1 650	222	1 530	12	495
8650	OQT 400	282	1 940	250	1 720	11	555
8740	Annealed	100	690	60	414	22	201
8740	OQT 1300	119	820	100	690	25	241
8740	OQT 1000	175	1 210	167	1 150	15	363
8740	OQT 700	228	1 570	212	1 460	12	461
8740	OQT 400	290	2 000	240	1 650	10	578
9255	Annealed	113	780	71	490	22	229
9255	Q&T 1300	130	896	102	703	21	262
9255	Q&T 1000	181	1 250	160	1 100	14	352
9255	Q&T 700	260	1 790	240	1 650	5	534
9255	Q&T 400	310	2 140	287	1 980	2	601

NOTE: Properties common to all carbon and alloy steels:
  Poisson's ratio: 0.27
  Shear modulus: 11.5×10^6 psi; 80 GPa
  Coefficient of thermal expansion: 6.5×10^{-6} °F^{-1}
  Density: 0.283 lb/in^3; 7 680 kg/m^3
  Modulus of elasticity: 30×10^6 psi; 207 GPa

APPENDIX 4   PROPERTIES OF HEAT-TREATED STEELS

FIGURE A4–1
Properties of heat-treated AISI 1040, water-quenched and tempered (*Modern Steels and Their Properties,* Bethlehem Steel Co., Bethlehem, PA)

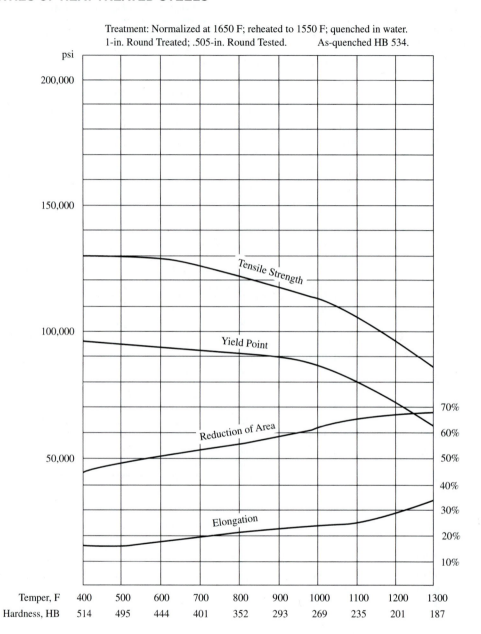

Treatment: Normalized at 1650 F; reheated to 1550 F; quenched in water.
1-in. Round Treated; .505-in. Round Tested.   As-quenched HB 534.

Temper, F	400	500	600	700	800	900	1000	1100	1200	1300
Hardness, HB	514	495	444	401	352	293	269	235	201	187

FIGURE A4–2
Properties of
heat-treated AISI 1144,
oil-quenched and
tempered (*Modern
Steels and Their
Properties,* Bethlehem
Steel Co., Bethlehem,
PA)

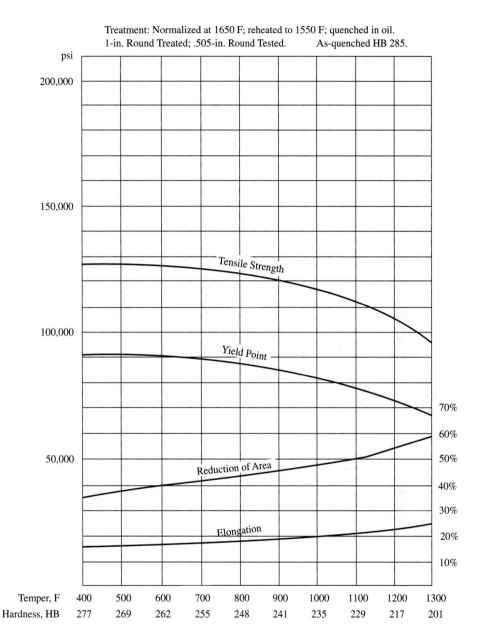

Treatment: Normalized at 1650 F; reheated to 1550 F; quenched in oil.
1-in. Round Treated; .505-in. Round Tested.   As-quenched HB 285.

Temper, F	400	500	600	700	800	900	1000	1100	1200	1300
Hardness, HB	277	269	262	255	248	241	235	229	217	201

FIGURE A4–3

Properties of
heat-treated AISI 1340,
oil-quenched and
tempered (*Modern
Steels and Their
Properties,* Bethlehem
Steel Co., Bethlehem,
PA)

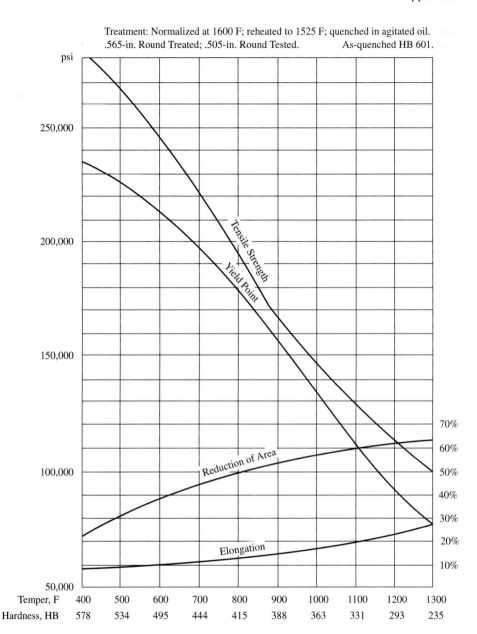

Treatment: Normalized at 1600 F; reheated to 1525 F; quenched in agitated oil.
.565-in. Round Treated; .505-in. Round Tested.   As-quenched HB 601.

Temper, F	400	500	600	700	800	900	1000	1100	1200	1300
Hardness, HB	578	534	495	444	415	388	363	331	293	235

FIGURE A4–4
Properties of
heat-treated AISI 4140,
oil-quenched and
tempered (*Modern
Steels and Their
Properties,* Bethlehem
Steel Co., Bethlehem,
PA)

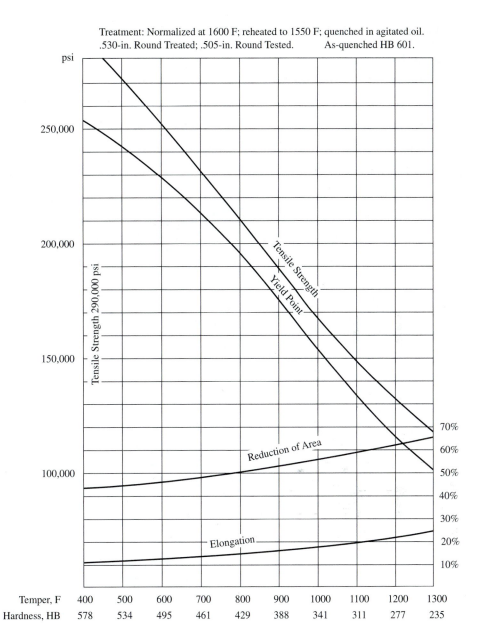

Treatment: Normalized at 1600 F; reheated to 1550 F; quenched in agitated oil.
.530-in. Round Treated; .505-in. Round Tested.   As-quenched HB 601.

Temper, F	400	500	600	700	800	900	1000	1100	1200	1300
Hardness, HB	578	534	495	461	429	388	341	311	277	235

FIGURE A4–5
Properties of
heat-treated AISI 4340,
oil-quenched and
tempered (*Modern
Steels and Their
Properties,* Bethlehem
Steel Co., Bethlehem,
PA)

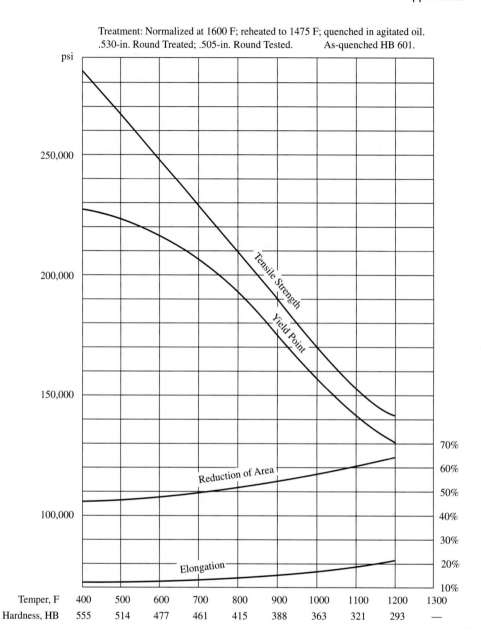

Treatment: Normalized at 1600 F; reheated to 1475 F; quenched in agitated oil.
.530-in. Round Treated; .505-in. Round Tested.   As-quenched HB 601.

Temper, F	400	500	600	700	800	900	1000	1100	1200	1300
Hardness, HB	555	514	477	461	415	388	363	321	293	—

FIGURE A4–6
Properties of
heat-treated AISI 6150,
oil-quenched and
tempered (*Modern
Steels and Their
Properties,* Bethlehem
Steel Co., Bethlehem,
PA)

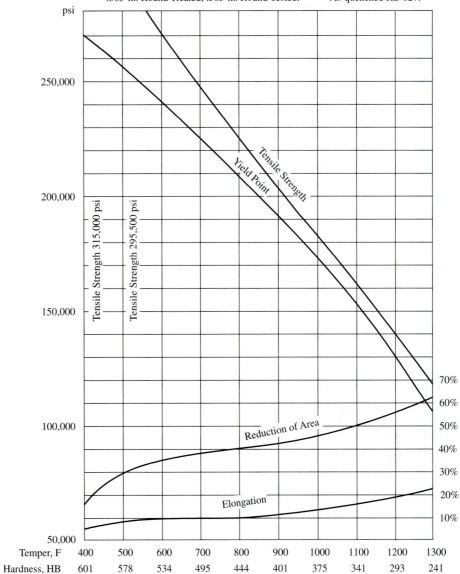

Treatment: Normalized at 1600 F; reheated to 1550 F; quenched in agitated oil.
.565-in. Round Treated; .505-in. Round Tested.   As-quenched HB 627.

A–14

APPENI

Materia
designati
(AISI num

1015
1020
1022
1117
1118

4118
4118
4118
4118

4320
4320
4320
4320

4620
4620
4620
4620

4820
4820
4820
4820

8620
8620
8620
8620

E931(
E931(
E931(
E931(

NOTES: Pro
SWQT: sin
SOQT: sing
DOQT: dou
300 and 45

A–16

Appendixes

APPENDIX 7 PROPERTIES OF STRUCTURAL STEELS

Material designation (ASTM number)	Grade or thickness	Tensile strength (Ksi)	Tensile strength (MPa)	Yield strength (Ksi)	Yield strength (MPa)	Ductility (percent elongation in 8 inches)
A36	$t \leq 8$ in	58	400	36	248	20
A242	$t \leq 3/4$ in	70	485	50	345	18
A242	$t \leq 1\frac{1}{2}$ in	67	460	46	315	
A242	$t \leq 4$ in	63	435	42	290	
A441	$t \leq 4$ in	63	435	42	290	18
A500	Cold-formed structural tubing, round or shaped					
	Round, Grade A	45	310	33	228	25
	Round, Grade B	58	400	42	290	23
	Round, Grade C	62	427	46	317	21
	Shaped, Grade A	45	310	39	269	25
	Shaped, Grade B	58	400	46	317	23
	Shaped, Grade C	62	427	50	345	21
A501	Hot-formed structural tubing, round or shaped	58	400	36	248	23
A514	Quenched and tempered, $t \leq 2\frac{1}{2}$ in	115	800	100	700	18% (in 2 inches)
A572	42, $t \leq 6$ in	60	414	42	290	
A572	50, $t \leq 4$ in	65	448	50	345	
A572	60, $t \leq 1\frac{1}{4}$ in	75	517	60	414	
A572	65, $t \leq 1\frac{1}{4}$ in	80	552	65	448	
A588	$t \leq 4$ in	70	485	50	345	18

NOTE: ASTM A572 is one of the high-strength, low-alloy steels (HSLA) and has properties similar to those of the SAE J410b steel specified by the SAE.

APPENDIX 8   DESIGN PROPERTIES OF CAST IRON

Material designation (ASTM number)	Grade	Tensile strength		Yield strength		Ductility (percent elongation in 2 inches)	Modulus of elasticity	
		(Ksi)	(MPa)	(Ksi)	(MPa)		(10^6 psi)	(GPa)
Gray iron								
A48-83	20	20	138			<1	12	83
	25	25	172			<1	13	90
	30	30	207			<1	15	103
	40	40	276			<1	17	117
	50	50	345			<1	19	131
	60	60	414			<1	20	138
Ductile iron								
A536-84	60-40-18	60	414	40	276	18	22	152
	80-55-06	80	552	55	379	6	22	152
	100-70-03	100	689	70	483	3	22	152
	120-90-02	120	827	90	621	2	22	152
Malleable iron								
A47-84(R1989)	32510	50	345	32	221	10	25	172
	35018	53	365	35	241	18	25	172
A220-88	40010	60	414	40	276	10	26	179
	45006	65	448	45	310	6	26	179
	50005	70	483	50	345	5	26	179
	70003	85	586	70	483	3	26	179
	90001	105	724	90	621	1	26	179

NOTES: Strength values are typical. Casting variables and section size affect final values. Modulus of elasticity may also vary. Density of cast irons ranges from 0.25 to 0.27 lb/in³ (6 920 to 7 480 kg/m³). Compressive strength ranges 3 to 5 times higher than tensile strength.

APPENDIX 9   TYPICAL PROPERTIES OF ALUMINUM

Alloy and temper	Tensile strength		Yield strength		Ductility (percent elongation in 2 inches)	Shearing strength		Endurance strength	
	(Ksi)	(MPa)	(Ksi)	(MPa)		(Ksi)	(MPa)	(Ksi)	(MPa)
1060-O	10	69	4	28	43	7	48	3	21
1060-H14	14	97	11	76	12	9	62	5	34
1060-H18	19	131	18	124	6	11	121	6	41
1350-O	12	83	4	28	28	8	55		
1350-H14	16	110	14	97		10	69		
1350-H19	27	186	24	165		15	103	7	48
2014-O	27	186	14	97	18	18	124	13	90
2014-T4	62	427	42	290	20	38	262	20	138
2014-T6	70	483	60	414	13	42	290	18	124
2024-O	27	186	11	76	22	18	124	13	90
2024-T4	68	469	47	324	19	41	283	20	138
2024-T361	72	496	57	393	12	42	290	18	124
2219-O	25	172	11	76	18				
2219-T62	60	414	42	290	10			15	103
2219-T87	69	476	57	393	10			15	103
3003-O	16	110	6	41	40	11	121	7	48
3003-H14	22	152	21	145	16	14	97	9	62
3003-H18	29	200	27	186	10	16	110	10	69
5052-O	28	193	13	90	30	18	124	16	110
5052-H34	38	262	31	214	14	21	145	18	124
5052-H38	42	290	37	255	8	24	165	20	138
6061-O	18	124	8	55	30	12	83	9	62
6061-T4	35	241	21	145	25	24	165	14	97
6061-T6	45	310	40	276	17	30	207	14	97
6063-O	13	90	7	48		10	69	8	55
6063-T4	25	172	13	90	22				
6063-T6	35	241	31	214	12	22	152	10	69
7001-O	37	255	22	152	14				
7001-T6	98	676	91	627	9			22	152
7075-O	33	228	15	103	16	22	152		
7075-T6	83	572	73	503	11	48	331	23	159

NOTE: Common properties:
   Density: 0.095 to 0.102 lb/in^3 (2 635 to 2 829 kg/m^3)
   Modulus of elasticity: 10 to 10.6 $\times$ 10^6 psi (69 to 73 GPa)
   Endurance strength at 5 $\times$ 10^8 cycles

APPENDIX 10   TYPICAL PROPERTIES OF ZINC CASTING ALLOYS

	Alloy				
	Zamak 3	Zamak 5	ZA-8	ZA-12	ZA-27
Composition (% by weight)					
Aluminum content	4	4	8.4	11	27
Magnesium content	0.035	0.055	0.023	0.023	0.015
Copper content		1.0	1.0	0.88	2.25
Properties (die-cast form)					
Tensile strength [Ksi (MPa)]	41 (283)	48 (331)	54 (374)	59 (404)	62 (426)
Yield strength [Ksi (MPa)]	32 (221)	33 (228)	42 (290)	46 (320)	54 (371)
% elongation in 2 inches	10	7	8	5	2.5
Modulus of elasticity [10^6 psi (GPa)]	12.4 (85)	12.4 (85)	12.4 (85)	12.0 (83)	11.3 (78)

APPENDIX 11   PROPERTIES OF TITANIUM ALLOYS

Material designation	Condition	Tensile strength		Yield strength		Ductility (percent elongation in 2 inches)	Modulus of elasticity	
		(Ksi)	(MPa)	(Ksi)	(MPa)		(10^6 psi)	(GPa)
Commercially pure alpha titanium (density = 0.163 lb/in^3; 4 515 kg/m^3)								
Ti-35A	Wrought	35	241	25	172	24	15.0	103
Ti-50A	Wrought	50	345	40	276	20	15.0	103
Ti-65A	Wrought	65	448	55	379	18	15.0	103
Alpha alloy (density = 0.163 lb/in^3; 4 515 kg/m^3)								
Ti-0.2Pd	Wrought	50	345	40	276	20	14.9	103
Beta alloy (density = 0.176 lb/in^3; 4 875 kg/m^3)								
Ti-3Al-13V-11Cr	Air-cooled from 1 400°F	135	931	130	896	16	14.7	101
Ti-3Al-13V-11Cr	Air-cooled from 1 400°F and aged	185	1 280	175	1 210	6	16.0	110
Alpha-beta alloy (density = 0.160 lb/in^3; 4 432 kg/m^3)								
Ti-6Al-4V	Annealed	130	896	120	827	10	16.5	114
Ti-6Al-4V	Quenched and aged at 1 000°F	160	1 100	150	1 030	7	16.5	114

APPENDIX 12   PROPERTIES OF BRONZES

Material	UNS number designation	Tensile strength		Yield strength		Ductility (percent elongation in 2 inches)	Modulus of elasticity	
		(Ksi)	(MPa)	(Ksi)	(MPa)		(10^6 psi)	(MPa)
Leaded phosphor bronze	C54400	68	469	57	393	20	15	103
Silicon bronze	C65500	58	400	22	152	60	15	103
Manganese bronze	C67500	65	448	30	207	33	15	103
	C86200	95	655	48	331	20	15	103
Bearing bronze	C93200	35	241	18	124	20	14.5	100
Aluminum bronze	C95400	85	586	35	241	18	15.5	107
Copper-nickel alloy	C96200	45	310	25	172	20	18	124
Copper-nickel-zinc alloy (also called nickel silver)	C97300	35	241	17	117	20	16	110

APPENDIX 13 TYPICAL PROPERTIES OF SELECTED PLASTICS

Material	Type	Tensile strength (Ksi)	Tensile strength (MPa)	Tensile modulus (Ksi)	Tensile modulus (MPa)	Flexural strength (Ksi)	Flexural strength (MPa)	Flexural modulus (Ksi)	Flexural modulus (MPa)	Impact strength IZOD (ft·lb/in of notch)
Nylon	6/6	12.0	83	420	2 900			410	2 830	1.0
	11	8.5	59	180	1 240			150	1 030	3.3
ABS	Medium-impact	6.0	41	360	2 480	11.5	79	310	2 140	4.0
	High-impact	5.0	34	250	1 720	8.0	55	260	1 790	7.0
Polycarbonate	General-purpose	9.0	62	340	2 340	11.0	76	300	2 070	12.0
Acrylic	Standard	10.5	72	430	2 960	16.0	110	460	3 170	0.4
	High-impact	5.4	37	220	1 520	7.0	48	230	1 590	1.2
PVC	Rigid	6.0	41	350	2 410			300	2 070	0.4–20.0 (varies widely)
Polyimide	25% graphite powder filler	5.7	39			12.8	88	900	6 210	0.25
	Glass-fiber filler	27.0	186			50.0	345	3 250	22 400	17.0
	Laminate	50.0	345			70.0	483	4 000	27 580	13.0
Acetal	Copolymer	8.0	55	410	2 830	13.0	90	375	2 590	1.3
Polyurethane	Elastomer	5.0	34	100	690	0.6	4			No break
Phenolic	General	6.5	45	1 100	7 580	9.0	62	1 100	7 580	0.3
Polyester with glass-fiber mat reinforcement (approx. 30% glass by weight)	Lay-up, contact mold	9.0	62			16.0	110	800	5 520	
	Cold press molded	12.0	83			22.0	152	1 300	8 960	
	Compression molded	25.0	172			10.0	69	1 300	8 960	

TABLE A14–1   Beam-deflection formulas for simply supported beams

(a)

$$y_B = y_{max} = \frac{-PL^3}{48EI} \quad \text{at center}$$

Between A and B:

$$y = \frac{-Px}{48EI}(3L^2 - 4x^2)$$

$a > b$

(b)

$$y_{max} = \frac{-Pab(L + b)\sqrt{3a(L + b)}}{27EIL}$$
$$\text{at } x_1 = \sqrt{a(L + b)/3}$$
$$y_B = \frac{-Pa^2b^2}{3EIL} \quad \text{at load}$$

Between A and B (the longer segment):

$$y = \frac{-Pbx}{6EIL}(L^2 - b^2 - x^2)$$

Between B and C (the shorter segment):

$$y = \frac{-Pav}{6EIL}(L^2 - v^2 - a^2)$$

At end of overhang at D:

$$y_D = \frac{Pabc}{6EIL}(L + a)$$

(c)

$$y_E = y_{max} = \frac{-Pa}{24EI}(3L^2 - 4a^2) \quad \text{at center}$$
$$y_B = y_C = \frac{-Pa^2}{6EI}(3L - 4a) \quad \text{at loads}$$

Between A and B:

$$y = \frac{-Px}{6EI}(3aL - 3a^2 - x^2)$$

Between B and C:

$$y = \frac{-Pa}{6EI}(3Lx - 3x^2 - a^2)$$

TABLE A14–1   *(continued)*

Total load = W = wL

w = uniformly distributed load

(d)

$$y_B = y_{max} = \frac{-5wL^4}{384EI} = \frac{-5WL^3}{384EI} \quad \text{at center}$$

Between A and B:

$$y = \frac{-wx}{24EI}(L^3 - 2Lx^2 + x^3)$$

At D at end:

$$y_D = \frac{wL^3 a}{24EI}$$

w = uniformly distributed load

(e)

Between A and B:

$$y = \frac{-wx}{24EIL}[a^2(2L - a)^2 - 2ax^2(2L - a) + Lx^3]$$

Between B and C:

$$y = \frac{-wa^2(L - x)}{24EIL}(4Lx - 2x^2 - a^2)$$

(f)

M_B = concentrated moment at B

Between A and B:

$$y = \frac{-M_B}{6EI}\left[\left(6a - \frac{3a^2}{L} - 2L\right)x - \frac{x^3}{L}\right]$$

Between B and C:

$$y = \frac{M_B}{6EI}\left[3a^2 + 3x^2 - \frac{x^3}{L} - \left(2L + \frac{3a^2}{L}\right)x\right]$$

(g)

At C at end of overhang:

$$y_C = \frac{-Pa^2}{3EI}(L + a)$$

At D, maximum upward deflection:

$$y_D = 0.06415\frac{PaL^2}{EI}$$

TABLE A14–1   *(continued)*

  (h)	At *C* at center: $$y = \frac{-W(L-2a)^3}{384EI}\left[\frac{5}{L}(L-2a) - \frac{24}{L}\left(\frac{a^2}{L-2a}\right)\right]$$ At *A* and *E* at ends: $$y = \frac{-W(L-2a)^3 a}{24EIL}\left[-1 + 6\left(\frac{a}{L-2a}\right)^2 + 3\left(\frac{a}{L-2a}\right)^3\right]$$
  (i)	At *C* at center: $$y = \frac{PL^2 a}{8EI}$$ At *A* and *E* at ends at loads: $$y = \frac{-Pa^2}{3EI}\left(a + \frac{3}{2}L\right)$$
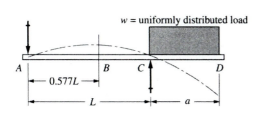 *(j)*	At *B*: $$y = 0.03208 \frac{wa^2 L^2}{EI}$$ At *D* at end: $$y = \frac{-wa^3}{24EI}(4L + 3a)$$

SOURCE: *Engineering Data for Aluminum Structures* (Washington, DC: The Aluminum Association, 1986), pp. 63–77.

TABLE A14–2 Beam-deflection formulas for cantilevers

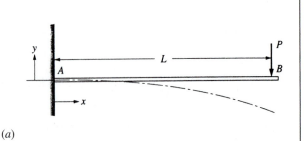

(a)

At B at end:

$$y_B = y_{max} = \frac{-PL^3}{3EI}$$

Between A and B:

$$y = \frac{-Px^2}{6EI}(3L - x)$$

(b)

At B at load:

$$y_B = \frac{-Pa^3}{3EI}$$

At C at end:

$$y_C = y_{max} = \frac{-Pa^2}{6EI}(3L - a)$$

Between A and B:

$$y = \frac{-Px^2}{6EI}(3a - x)$$

Between B and C:

$$y = \frac{-Pa^2}{6EI}(3x - a)$$

(c)

w = uniformly distributed load

W = total load = wL

At B at end:

$$y_B = y_{max} = \frac{-WL^3}{8EI}$$

Between A and B:

$$y = \frac{-Wx^2}{24EIL}[2L^2 + (2L - x)^2]$$

(d)

M_B = concentrated moment at end

At B at end:

$$y_B = y_{max} = \frac{-M_B L^2}{2EI}$$

Between A and B:

$$y = \frac{-M_B x^2}{2EI}$$

SOURCE: *Engineering Data for Aluminum Structures* (Washington, DC: The Aluminum Association, 1986), pp. 63–77.

TABLE A14–3   Beam diagrams and beam-deflection formulas for statically indeterminate beams

Deflections
At B at load:

$$y_B = \frac{-7}{768} \frac{PL^3}{EI}$$

y_{max} is at $v = 0.447L$ at D:

$$y_D = y_{max} = \frac{-PL^3}{107EI}$$

Between A and B:

$$y = \frac{-Px^2}{96EI}(9L - 11x)$$

Between B and C:

$$y = \frac{-Pv}{96EI}(3L^2 - 5v^2)$$

(a)

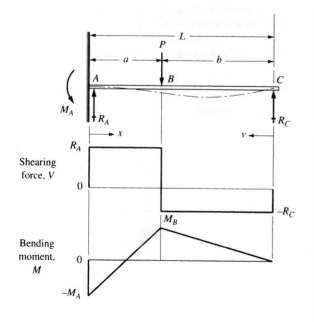

Reactions

$$R_A = \frac{Pb}{2L^3}(3L^2 - b^2)$$

$$R_C = \frac{Pa^2}{2L^3}(b + 2L)$$

Moments

$$M_A = \frac{-Pab}{2L^2}(b + L)$$

$$M_B = \frac{Pa^2b}{2L^3}(b + 2L)$$

Deflections
At B at load:

$$y_B = \frac{-Pa^3b^2}{12EIL^3}(3L + b)$$

Between A and B:

$$y = \frac{-Px^2b}{12EIL^3}(3C_1 - C_2x)$$

$$C_1 = aL(L + b); C_2 = (L + a)(L + b) + aL$$

Between B and C:

$$y = \frac{-Pa^2v}{12EIL^3}[3L^2b - v^2(3L - a)]$$

(b)

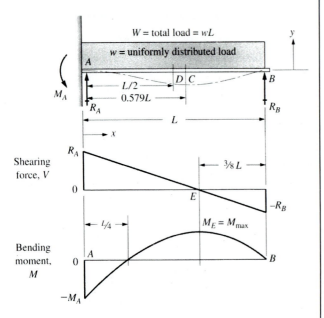

Reactions

$$R_A = \frac{5}{8}W$$

$$R_B = \frac{3}{8}W$$

Moments

$$M_A = -0.125WL$$
$$M_E = 0.0703WL$$

Deflections
At C at x = 0.579L:

$$y_C = y_{max} = \frac{-WL^3}{185EI}$$

At D at center:

$$y_D = \frac{-WL^3}{192EI}$$

Between A and B:

$$y = \frac{-Wx^2(L-x)}{48EIL}(3L-2x)$$

(c)

Reactions

$$R_A = \frac{-3Pa}{2L}$$

$$R_B = P\left(1 + \frac{3a}{2L}\right)$$

Moments

$$M_A = \frac{Pa}{2}$$

$$M_B = -Pa$$

Deflection
At C at end:

$$y_C = \frac{-PL^3}{EI}\left(\frac{a^2}{4L^2} + \frac{a^3}{3L^3}\right)$$

(d)

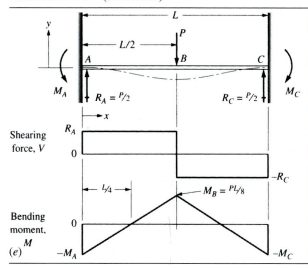

Moments

$$M_A = M_B = M_C = \frac{PL}{8}$$

Deflections
At *B* at center:

$$y_B = y_{max} = \frac{-PL^3}{192EI}$$

Between *A* and *B*:

$$y = \frac{-Px^2}{48EI}(3L - 4x)$$

Reactions

$$R_A = \frac{Pb^2}{L^3}(3a + b)$$

$$R_C = \frac{Pa^2}{L^3}(3b + a)$$

Moments

$$M_A = \frac{-Pab^2}{L^2}$$

$$M_B = \frac{2Pa^2b^2}{L^3}$$

$$M_C = \frac{-Pa^2b}{L^2}$$

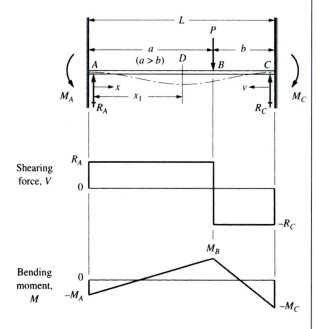

Deflections
At *B* at load:

$$y_B = \frac{-Pa^3b^3}{3EIL^3}$$

At *D* at $x_1 = \frac{2aL}{3a + b}$

$$y_D = y_{max} = \frac{-2Pa^3b^2}{3EI(3a + b)^2}$$

Between *A* and *B* (longer segment):

$$y = \frac{-Px^2b^2}{6EIL^3}[2a(L - x) + L(a - x)]$$

Between *B* and *C* (shorter segment):

$$y = \frac{-Pv^2a^2}{6EIL^3}[2b(L - v) + L(b - v)]$$

(*f*)

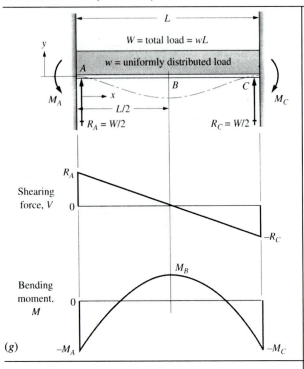

(g)

Moments

$$M_A = M_C = \frac{-WL}{12}$$

$$M_B = \frac{WL}{24}$$

Deflections
At *B* at center:

$$y_B = y_{max} = \frac{-WL^3}{384EI}$$

Between *A* and *C*:

$$y = \frac{-wx^2}{24EI}(L-x)^2$$

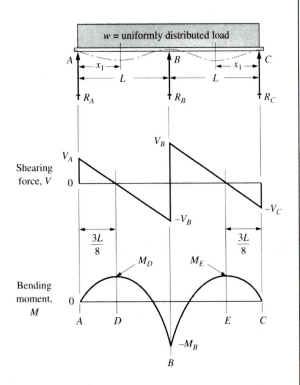

(h)

Reactions

$$R_A = R_C = \frac{3wL}{8}$$

$$R_B = 1.25wL$$

Shearing forces

$$V_A = V_C = R_A = R_C = \frac{3wL}{8}$$

$$V_B = \frac{5wL}{8}$$

Moments

$$M_D = M_E = 0.0703wL^2$$
$$M_B = -0.125wL^2$$

Deflections
At $x_1 = 0.4215L$ from *A* or *C*:

$$y_{max} = \frac{-wL^4}{185EI}$$

Between *A* and *B*:

$$y = \frac{-w}{48EI}(L^3x - 3Lx^3 + 2x^4)$$

TABLE A14–3 (*continued*)

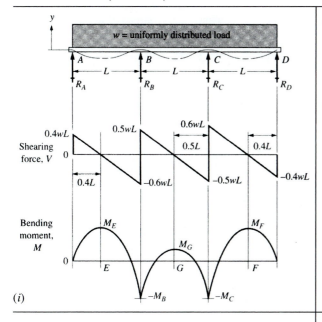

Reactions

$$R_A = R_D = 0.4wL$$
$$R_B = R_C = 1.10wL$$

Moments

$$M_E = M_F = 0.08wL^2$$
$$M_B = M_C = -0.10wL^2 = M_{max}$$
$$M_G = 0.025wL^2$$

(*i*)

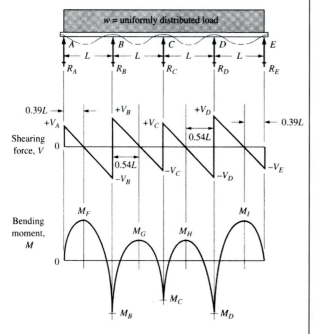

Reactions

$$R_A = R_E = 0.393wL$$
$$R_B = R_D = 1.143wL$$
$$R_C = 0.928wL$$

Shearing forces

$$V_A = +0.393wL$$
$$-V_B = -0.607wL$$
$$+V_B = +0.536wL$$
$$-V_C = -0.464wL$$
$$+V_C = +0.464wL$$
$$-V_D = -0.536wL$$
$$+V_D = +0.607wL$$
$$-V_E = -0.393wL$$

Moments

$$M_B = M_D = -0.1071wL^2 = M_{max}$$
$$M_F = M_I = 0.0772wL^2$$
$$M_C = -0.0714wL^2$$
$$M_G = M_H = 0.0364wL^2$$

(*j*)

SOURCE: *Engineering Data for Aluminum Structures* (Washington, DC: The Aluminum Association, 1986), pp. 63–77.

APPENDIX 15   STRESS CONCENTRATION FACTORS

FIGURE A15–1
Stepped round shaft

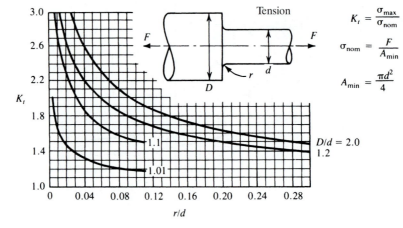

$$K_t = \frac{\sigma_{max}}{\sigma_{nom}}$$

$$\sigma_{nom} = \frac{F}{A_{min}}$$

$$A_{min} = \frac{\pi d^2}{4}$$

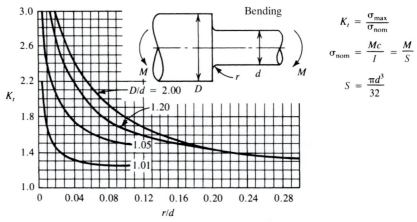

$$K_t = \frac{\sigma_{max}}{\sigma_{nom}}$$

$$\sigma_{nom} = \frac{Mc}{I} = \frac{M}{S}$$

$$S = \frac{\pi d^3}{32}$$

$$K_t = \frac{\tau_{max}}{\tau_{nom}}$$

$$\tau_{nom} = \frac{T}{Z_p}$$

$$Z_p = \frac{\pi d^3}{16}$$

FIGURE A15–2
Stepped flat plate with fillets

Curve A
Direct tension
on plate

$$\sigma_{nom} = \frac{F}{A_{net}} = \frac{F}{(w - d)t}$$

Curve B
Tension-load
applied through
a pin in the hole

$$\sigma_{nom} = \frac{F}{A_{net}} = \frac{F}{(w - d)t}$$

F = total load

Curve C
Bending in
the plane of
the plate

$$\sigma_{nom} = \frac{Mc}{I_{net}} = \frac{M}{S_{net}} = \frac{6Mw}{(w^3 - d^3)t}$$

Note : K_t = 1.0 for $d/w < 0.5$

FIGURE A15–3   Flat plate with a central hole

Note: K_{t_g} is based on the nominal stress in a round bar without a hole (gross section).

FIGURE A15–4 Round bar with a transverse hole

APPENDIX 16   STEEL STRUCTURAL SHAPES

TABLE A16–1   Properties of steel angles, equal legs and unequal legs, L-shapes*

Designation	Area (in²)	Weight per foot (lb)	Axis X–X			Axis Y–Y			Axis Z–Z	
			I (in⁴)	S (in³)	y (in)	I (in⁴)	S (in³)	x (in)	r (in)	α (deg)
L8 × 8 × 1	15.0	51.0	89.0	15.8	2.37	89.0	15.8	2.37	1.56	45.0
L8 × 8 × 1/2	7.75	26.4	48.6	8.36	2.19	48.6	8.36	2.19	1.59	45.0
L8 × 4 × 1	11.0	37.4	69.6	14.1	3.05	11.6	3.94	1.05	0.846	13.9
L8 × 4 × 1/2	5.75	19.6	38.5	7.49	2.86	6.74	2.15	0.859	0.865	14.9
L6 × 6 × 3/4	8.44	28.7	28.2	6.66	1.78	28.2	6.66	1.78	1.17	45.0
L6 × 6 × 3/8	4.36	14.9	15.4	3.53	1.64	15.4	3.53	1.64	1.19	45.0
L6 × 4 × 3/4	6.94	23.6	24.5	6.25	2.08	8.68	2.97	1.08	0.860	23.2
L6 × 4 × 3/8	3.61	12.3	13.5	3.32	1.94	4.90	1.60	0.941	0.877	24.0
L4 × 4 × 1/2	3.75	12.8	5.56	1.97	1.18	5.56	1.97	1.18	0.782	45.0
L4 × 4 × 1/4	1.94	6.6	3.04	1.05	1.09	3.04	1.05	1.09	0.795	45.0
L4 × 3 × 1/2	3.25	11.1	5.05	1.89	1.33	2.42	1.12	0.827	0.639	28.5
L4 × 3 × 1/4	1.69	5.8	2.77	1.00	1.24	1.36	0.599	0.896	0.651	29.2
L3 × 3 × 1/2	2.75	9.4	2.22	1.07	0.932	2.22	1.07	0.932	0.584	45.0
L3 × 3 × 1/4	1.44	4.9	1.24	0.577	0.842	1.24	0.577	0.842	0.592	45.0
L2 × 2 × 3/8	1.36	4.7	0.479	0.351	0.636	0.479	0.351	0.636	0.389	45.0
L2 × 2 × 1/4	0.938	3.19	0.348	0.247	0.592	0.348	0.247	0.592	0.391	45.0
L2 × 2 × 1/8	0.484	1.65	0.190	0.131	0.546	0.190	0.131	0.546	0.398	45.0

*Data are taken from a variety of sources. Sizes listed represent a small sample of the sizes available.

NOTES: Example designation: L4 × 3 × 1/2
4 = length of longer leg (in); 3 = length of shorter leg (in); 1/2 = thickness of legs (in)
Z–Z is axis of minimum moment of inertia (I) and radius of gyration (r).
I = moment of inertia; S = section modulus; r = radius of gyration.

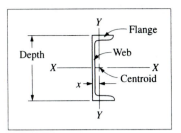

TABLE A16–2 Properties of American Standard steel channels, C-shapes*

Designation	Area (in^2)	Depth (in)	Web thickness (in)	Flange Width (in)	Flange Average thickness (in)	Axis X–X I (in^4)	Axis X–X S (in^3)	Axis Y–Y I (in^4)	Axis Y–Y S (in^3)	Axis Y–Y x (in)
C15 × 50	14.7	15.00	0.716	3.716	0.650	404	53.8	11.0	3.78	0.798
C15 × 40	11.8	15.00	0.520	3.520	0.650	349	46.5	9.23	3.37	0.777
C12 × 30	8.82	12.00	0.510	3.170	0.501	162	27.0	5.14	2.06	0.674
C12 × 25	7.35	12.00	0.387	3.047	0.501	144	24.1	4.47	1.88	0.674
C10 × 30	8.82	10.00	0.673	3.033	0.436	103	20.7	3.94	1.65	0.649
C10 × 20	5.88	10.00	0.379	2.739	0.436	78.9	15.8	2.81	1.32	0.606
C9 × 20	5.88	9.00	0.448	2.648	0.413	60.9	13.5	2.42	1.17	0.583
C9 × 15	4.41	9.00	0.285	2.485	0.413	51.0	11.3	1.93	1.01	0.586
C8 × 18.75	5.51	8.00	0.487	2.527	0.390	44.0	11.0	1.98	1.01	0.565
C8 × 11.5	3.38	8.00	0.220	2.260	0.390	32.6	8.14	1.32	0.781	0.571
C6 × 13	3.83	6.00	0.437	2.157	0.343	17.4	5.80	1.05	0.642	0.514
C6 × 8.2	2.40	6.00	0.200	1.920	0.343	13.1	4.38	0.693	0.492	0.511
C5 × 9	2.64	5.00	0.325	1.885	0.320	8.90	3.56	0.632	0.450	0.478
C5 × 6.7	1.97	5.00	0.190	1.750	0.320	7.49	3.00	0.479	0.378	0.484
C4 × 7.25	2.13	4.00	0.321	1.721	0.296	4.59	2.29	0.433	0.343	0.459
C4 × 5.4	1.59	4.00	0.184	1.584	0.296	3.85	1.93	0.319	0.283	0.457
C3 × 6	1.76	3.00	0.356	1.596	0.273	2.07	1.38	0.305	0.268	0.455
C3 × 4.1	1.21	3.00	0.170	1.410	0.273	1.66	1.10	0.197	0.202	0.436

*Data are taken from a variety of sources. Sizes listed represent a small sample of the sizes available.

NOTES: Example designation: C15 × 50
15 = depth (in); 50 = weight per unit length (lb/ft)
I = moment of inertia; S = section modulus.

TABLE A16–3 Properties of steel wide-flange shapes, W-shapes*

Designation	Area (in²)	Depth (in)	Web thickness (in)	Flange Width (in)	Flange Thickness (in)	Axis X–X I (in⁴)	Axis X–X S (in³)	Axis Y–Y I (in⁴)	Axis Y–Y S (in³)
W24 × 76	22.4	23.92	0.440	8.990	0.680	2100	176	82.5	18.4
W24 × 68	20.1	23.73	0.415	8.965	0.585	1830	154	70.4	15.7
W21 × 73	21.5	21.24	0.455	8.295	0.740	1600	151	70.6	17.0
W21 × 57	16.7	21.06	0.405	6.555	0.650	1170	111	30.6	9.35
W18 × 55	16.2	18.11	0.390	7.530	0.630	890	98.3	44.9	11.9
W18 × 40	11.8	17.90	0.315	6.015	0.525	612	68.4	19.1	6.35
W14 × 43	12.6	13.66	0.305	7.995	0.530	428	62.7	45.2	11.3
W14 × 26	7.69	13.91	0.255	5.025	0.420	245	35.3	8.91	3.54
W12 × 30	8.79	12.34	0.260	6.520	0.440	238	38.6	20.3	6.24
W12 × 16	4.71	11.99	0.220	3.990	0.265	103	17.1	2.82	1.41
W10 × 15	4.41	9.99	0.230	4.000	0.270	69.8	13.8	2.89	1.45
W10 × 12	3.54	9.87	0.190	3.960	0.210	53.8	10.9	2.18	1.10
W8 × 15	4.44	8.11	0.245	4.015	0.315	48.0	11.8	3.41	1.70
W8 × 10	2.96	7.89	0.170	3.940	0.205	30.8	7.81	2.09	1.06
W6 × 15	4.43	5.99	0.230	5.990	0.260	29.1	9.72	9.32	3.11
W6 × 12	3.55	6.03	0.230	4.000	0.280	22.1	7.31	2.99	1.50
W5 × 19	5.54	5.15	0.270	5.030	0.430	26.2	10.2	9.13	3.63
W5 × 16	4.68	5.01	0.240	5.000	0.360	21.3	8.51	7.51	3.00
W4 × 13	3.83	4.16	0.280	4.060	0.345	11.3	5.46	3.86	1.90

*Data are taken from a variety of sources. Sizes listed represent a small sample of the sizes available.

NOTES: Example designation: W14 × 43

14 = nominal depth (in); 43 = weight per unit length (lb/ft)

I = moment of inertia; S = section modulus.

TABLE A16–4   Properties of American Standard steel beams, S-shapes*

Designation	Area (in^2)	Depth (in)	Web thickness (in)	Flange Width (in)	Flange Average thickness (in)	Axis X–X I (in^4)	Axis X–X S (in^3)	Axis Y–Y I (in^4)	Axis Y–Y S (in^3)
S24 × 90	26.5	24.00	0.625	7.125	0.870	2 250	187	44.9	12.6
S20 × 96	28.2	20.30	0.800	7.200	0.920	1 670	165	50.2	13.9
S20 × 75	22.0	20.00	0.635	6.385	0.795	1 280	128	29.8	9.32
S20 × 66	19.4	20.00	0.505	6.255	0.795	1 190	119	27.7	8.85
S18 × 70	20.6	18.00	0.711	6.251	0.691	926	103	24.1	7.72
S15 × 50	14.7	15.00	0.550	5.640	0.622	486	64.8	15.7	5.57
S12 × 50	14.7	12.00	0.687	5.477	0.659	305	50.8	15.7	5.74
S12 × 35	10.3	12.00	0.428	5.078	0.544	229	38.2	9.87	3.89
S10 × 35	10.3	10.00	0.594	4.944	0.491	147	29.4	8.36	3.38
S10 × 25.4	7.46	10.00	0.311	4.661	0.491	124	24.7	6.79	2.91
S8 × 23	6.77	8.00	0.441	4.171	0.426	64.9	16.2	4.31	2.07
S8 × 18.4	5.41	8.00	0.271	4.001	0.426	57.6	14.4	3.73	1.86
S7 × 20	5.88	7.00	0.450	3.860	0.392	42.4	12.1	3.17	1.64
S6 × 12.5	3.67	6.00	0.232	3.332	0.359	22.1	7.37	1.82	1.09
S5 × 10	2.94	5.00	0.214	3.004	0.326	12.3	4.92	1.22	0.809
S4 × 7.7	2.26	4.00	0.193	2.663	0.293	6.08	3.04	0.764	0.574
S3 × 5.7	1.67	3.00	0.170	2.330	0.260	2.52	1.68	0.455	0.390

*Data are taken from a variety of sources. Sizes listed represent a small sample of the sizes available.

NOTES: Example designation: S10 × 35
10 = nominal depth (in); 35 = weight per unit length (lb/ft)
I = moment of inertia; S = section modulus.

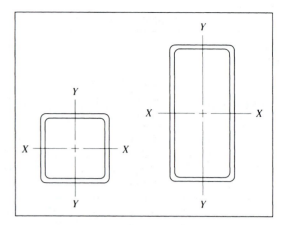

TABLE A16–5 Properties of steel structural tubing, square and rectangular*

Size	Area (in^2)	Weight per foot (lb)	Axis X–X			Axis Y–Y		
			I (in^4)	S (in^3)	r (in)	I (in^4)	S (in^3)	r (in)
8 × 8 × 1/2	14.4	48.9	131	32.9	3.03	131	32.9	3.03
8 × 8 × 1/4	7.59	25.8	75.1	18.8	3.15	75.1	18.8	3.15
8 × 4 × 1/2	10.4	35.2	75.1	18.8	2.69	24.6	12.3	1.54
8 × 4 × 1/4	5.59	19.0	45.1	11.3	2.84	15.3	7.63	1.65
8 × 2 × 1/4	4.59	15.6	30.1	7.52	2.56	3.08	3.08	0.819
6 × 6 × 1/2	10.4	35.2	50.5	16.8	2.21	50.5	16.8	2.21
6 × 6 × 1/4	5.59	19.0	30.3	10.1	2.33	30.3	10.1	2.33
6 × 4 × 1/4	4.59	15.6	22.1	7.36	2.19	11.7	5.87	1.60
6 × 2 × 1/4	3.59	12.2	13.8	4.60	1.96	2.31	2.31	0.802
4 × 4 × 1/2	6.36	21.6	12.3	6.13	1.39	12.3	6.13	1.39
4 × 4 × 1/4	3.59	12.2	8.22	4.11	1.51	8.22	4.11	1.51
4 × 2 × 1/4	2.59	8.81	4.69	2.35	1.35	1.54	1.54	0.770
3 × 3 × 1/4	2.59	8.81	3.16	2.10	1.10	3.16	2.10	1.10
3 × 2 × 1/4	2.09	7.11	2.21	1.47	1.03	1.15	1.15	0.742
2 × 2 × 1/4	1.59	5.41	0.766	0.766	0.694	0.766	0.766	0.694

*Data are taken from a variety of sources. Sizes listed represent a small sample of the sizes available.

NOTES: Example size: 6 × 4 × 1/4

6 = vertical depth (in); 4 = width (in); 1/4 = wall thickness (in).

I = moment of inertia; S = section modulus; r = radius of gyration.

TABLE A16–6 Properties of American National Standard Schedule 40 welded and seamless wrought steel pipe

Diameter (in)			Wall thickness (in)	Cross-sectional area of metal (in^2)	Properties of sections			
Nominal	Actual inside	Actual outside			Moment of inertia, I (in^4)	Radius of gyration (in)	Section modulus, S (in^3)	Polar section modulus, Z_p (in^3)
1/8	0.269	0.405	0.068	0.072	0.001 06	0.122	0.005 25	0.010 50
1/4	0.364	0.540	0.088	0.125	0.003 31	0.163	0.012 27	0.024 54
3/8	0.493	0.675	0.091	0.167	0.007 29	0.209	0.021 60	0.043 20
1/2	0.622	0.840	0.109	0.250	0.017 09	0.261	0.040 70	0.081 40
3/4	0.824	1.050	0.113	0.333	0.037 04	0.334	0.070 55	0.141 1
1	1.049	1.315	0.133	0.494	0.087 34	0.421	0.132 8	0.265 6
$1\frac{1}{4}$	1.380	1.660	0.140	0.669	0.194 7	0.539	0.234 6	0.469 2
$1\frac{1}{2}$	1.610	1.900	0.145	0.799	0.309 9	0.623	0.326 2	0.652 4
2	2.067	2.375	0.154	1.075	0.665 8	0.787	0.560 7	1.121
$2\frac{1}{2}$	2.469	2.875	0.203	1.704	1.530	0.947	1.064	2.128
3	3.068	3.500	0.216	2.228	3.017	1.163	1.724	3.448
$3\frac{1}{2}$	3.548	4.000	0.226	2.680	4.788	1.337	2.394	4.788
4	4.026	4.500	0.237	3.174	7.233	1.510	3.215	6.430
5	5.047	5.563	0.258	4.300	15.16	1.878	5.451	10.90
6	6.065	6.625	0.280	5.581	28.14	2.245	8.496	16.99
8	7.981	8.625	0.322	8.399	72.49	2.938	16.81	33.62
10	10.020	10.750	0.365	11.91	160.7	3.674	29.91	59.82
12	11.938	12.750	0.406	15.74	300.2	4.364	47.09	94.18
16	15.000	16.000	0.500	24.35	732.0	5.484	91.50	183.0
18	16.876	18.000	0.562	30.79	1 172	6.168	130.2	260.4

APPENDIX 17   ALUMINUM STRUCTURAL SHAPES

TABLE A17–1   Aluminum Association standard channels: dimensions, areas, weights, and section properties

Size Depth, A (in)	Size Width, B (in)	Area* (in²)	Weight† (lb/ft)	Flange thickness, t_1 (in)	Web thickness, t (in)	Fillet radius, R (in)	Axis X–X I (in⁴)	Axis X–X S (in³)	Axis X–X r (in)	Axis Y–Y I (in⁴)	Axis Y–Y S (in³)	Axis Y–Y r (in)	Axis Y–Y x (in)
2.00	1.00	0.491	0.577	0.13	0.13	0.10	0.288	0.288	0.766	0.045	0.064	0.303	0.298
2.00	1.25	0.911	1.071	0.26	0.17	0.15	0.546	0.546	0.774	0.139	0.178	0.391	0.471
3.00	1.50	0.965	1.135	0.20	0.13	0.25	1.41	0.94	1.21	0.22	0.22	0.47	0.49
3.00	1.75	1.358	1.597	0.26	0.17	0.25	1.97	1.31	1.20	0.42	0.37	0.55	0.62
4.00	2.00	1.478	1.738	0.23	0.15	0.25	3.91	1.95	1.63	0.60	0.45	0.64	0.65
4.00	2.25	1.982	2.331	0.29	0.19	0.25	5.21	2.60	1.62	1.02	0.69	0.72	0.78
5.00	2.25	1.881	2.212	0.26	0.15	0.30	7.88	3.15	2.05	0.98	0.64	0.72	0.73
5.00	2.75	2.627	3.089	0.32	0.19	0.30	11.14	4.45	2.06	2.05	1.14	0.88	0.95
6.00	2.50	2.410	2.834	0.29	0.17	0.30	14.35	4.78	2.44	1.53	0.90	0.80	0.79
6.00	3.25	3.427	4.030	0.35	0.21	0.30	21.04	7.01	2.48	3.76	1.76	1.05	1.12
7.00	2.75	2.725	3.205	0.29	0.17	0.30	22.09	6.31	2.85	2.10	1.10	0.88	0.84
7.00	3.50	4.009	4.715	0.38	0.21	0.30	33.79	9.65	2.90	5.13	2.23	1.13	1.20
8.00	3.00	3.526	4.147	0.35	0.19	0.30	37.40	9.35	3.26	3.25	1.57	0.96	0.93
8.00	3.75	4.923	5.789	0.41	0.25	0.35	52.69	13.17	3.27	7.13	2.82	1.20	1.22
9.00	3.25	4.237	4.983	0.35	0.23	0.35	54.41	12.09	3.58	4.40	1.89	1.02	0.93
9.00	4.00	5.927	6.970	0.44	0.29	0.35	78.31	17.40	3.63	9.61	3.49	1.27	1.25
10.00	3.50	5.218	6.136	0.41	0.25	0.35	83.22	16.64	3.99	6.33	2.56	1.10	1.02
10.00	4.25	7.109	8.360	0.50	0.31	0.40	116.15	23.23	4.04	13.02	4.47	1.35	1.34
12.00	4.00	7.036	8.274	0.47	0.29	0.40	159.76	26.63	4.77	11.03	3.86	1.25	1.14
12.00	5.00	10.053	11.822	0.62	0.35	0.45	239.69	39.95	4.88	25.74	7.60	1.60	1.61

SOURCE: Aluminum Association, *Aluminum Standards and Data,* 11th ed., Washington, DC, © 1993, p. 187.

*Areas listed are based on nominal dimensions.

†Weights per foot are based on nominal dimensions and a density of 0.098 lb/in³, which is the density of alloy 6061.

‡I = moment of inertia; S = section modulus; r = radius of gyration.

TABLE A17–2 Aluminum Association standard I-beams: dimensions, areas, weights, and section properties

| Size | | | | | | | Section properties‡ | | | | | |
Depth, A (in)	Width, B (in)	Area* (in^2)	Weight† (lb/ft)	Flange thickness, t_1 (in)	Web thickness, t (in)	Fillet radius, R (in)	Axis X–X			Axis Y–Y		
							I (in^4)	S (in^3)	r (in)	I (in^4)	S (in^3)	r (in)
3.00	2.50	1.392	1.637	0.20	0.13	0.25	2.24	1.49	1.27	0.52	0.42	0.61
3.00	2.50	1.726	2.030	0.26	0.15	0.25	2.71	1.81	1.25	0.68	0.54	0.63
4.00	3.00	1.965	2.311	0.23	0.15	0.25	5.62	2.81	1.69	1.04	0.69	0.73
4.00	3.00	2.375	2.793	0.29	0.17	0.25	6.71	3.36	1.68	1.31	0.87	0.74
5.00	3.50	3.146	3.700	0.32	0.19	0.30	13.94	5.58	2.11	2.29	1.31	0.85
6.00	4.00	3.427	4.030	0.29	0.19	0.30	21.99	7.33	2.53	3.10	1.55	0.95
6.00	4.00	3.990	4.692	0.35	0.21	0.30	25.50	8.50	2.53	3.74	1.87	0.97
7.00	4.50	4.932	5.800	0.38	0.23	0.30	42.89	12.25	2.95	5.78	2.57	1.08
8.00	5.00	5.256	6.181	0.35	0.23	0.30	59.69	14.92	3.37	7.30	2.92	1.18
8.00	5.00	5.972	7.023	0.41	0.25	0.30	67.78	16.94	3.37	8.55	3.42	1.20
9.00	5.50	7.110	8.361	0.44	0.27	0.30	102.02	22.67	3.79	12.22	4.44	1.31
10.00	6.00	7.352	8.646	0.41	0.25	0.40	132.09	26.42	4.24	14.78	4.93	1.42
10.00	6.00	8.747	10.286	0.50	0.29	0.40	155.79	31.16	4.22	18.03	6.01	1.44
12.00	7.00	9.925	11.672	0.47	0.29	0.40	255.57	42.60	5.07	26.90	7.69	1.65
12.00	7.00	12.153	14.292	0.62	0.31	0.40	317.33	52.89	5.11	35.48	10.14	1.71

SOURCE: Aluminum Association, *Aluminum Standards and Data,* 11th ed., Washington, DC, © 1993, p. 187.

*Areas listed are based on nominal dimensions.

†Weights per foot are based on nominal dimensions and a density of 0.098 lb/in^3, which is the density of alloy 6061.

‡I = moment of inertia; S = section modulus; r = radius of gyration.

APPENDIX 18   CONVERSION FACTORS

TABLE A18–1   Conversion of U.S. Customary units to SI units: basic quantities

Quantity	U.S. Customary unit		SI unit	Symbol	Equivalent units
Length	1 foot (ft)	= 0.304 8	meter	m	
Mass	1 slug	= 14.59	kilogram	kg	
Time	1 second	= 1.0	second	s	
Force	1 pound (lb)	= 4.448	newton	N	$kg \cdot m/s^2$
Pressure	1 lb/in^2	= 6 895	pascal	Pa	N/m^2 or $kg/m \cdot s^2$
Energy	1 ft-lb	= 1.356	joule	J	$N \cdot m$ or $kg \cdot m^2/s^2$
Power	1 ft-lb/s	= 1.356	watt	W	J/s

TABLE A18–2   Other convenient conversion factors

Length

1 ft	=	0.304 8	m
1 in	=	25.4	mm
1 mi	= 5 280		ft
1 mi	=	1.609	km
1 km	= 1 000		m
1 cm	=	10	mm
1 m	= 1 000		mm

Area

1 ft^2	=	0.092 9	m^2
1 in^2	=	645.2	mm^2
1 m^2	=	10.76	ft^2
1 m^2	=	10^6	mm^2

Volume

1 ft^3	=	7.48	gal
1 ft^3	= 1 728		in^3
1 ft^3	=	0.028 3	m^3
1 gal	=	0.003 79	m^3
1 gal	=	3.785	L
1 m^3	= 1 000		L

Volume flow rate

1 ft^3/s	=	449	gal/min
1 ft^3/s	=	0.028 3	m^3/s
1 gal/min	=	6.309×10^{-5}	m^3/s
1 gal/min	=	3.785	L/min
1 L/min	=	16.67×10^{-6}	m^3/s

Stress, pressure, or unit loading

1 lb/in^2	=	6.895	kPa
1 lb/ft^2	=	0.047 9	kPa
1 kip/in^2	=	6.895	MPa

Section modulus

1 in^3	=	1.639×10^4 mm^3

Moment of inertia

1 in^4	=	4.162×10^5 mm^4

Density

1 slug/ft^3	= 515.4	kg/m^3

Specific weight

1 lb/ft^3	= 157.1	N/m^3

Energy

1 ft·lb	=	1.356	J
1 Btu	=	1.055	kJ
1 W·h	=	3.600	kJ

Torque or moment

1 lb·in	=	0.113 0	N·m

Power

1 hp	= 550		ft·lb/s
1 hp	= 745.7		W
1 ft·lb/s	=	1.356	W
1 Btu/h	=	0.293	W

Temperature

$$T(°C) = [T(°F) - 32]5/9$$
$$T(°F) = \tfrac{9}{5}[T(°C)] + 32$$

APPENDIX 19   HARDNESS CONVERSION TABLE

Brinell		Rockwell		Steel: tensile strength, (1 000 psi approx.)	Brinell		Rockwell		Steel: tensile strength, (1 000 psi approx.)
Indent. dia. (mm)	No.*	B	C		Indent. dia. (mm)	No.*	B	C	
2.25	745		65.3		3.75	262	(103.0)	26.6	127
2.30	712				3.80	255	(102.0)	25.4	123
2.35	682		61.7		3.85	248	(101.0)	24.2	120
2.40	653		60.0		3.90	241	100.0	22.8	116
2.45	627		58.7		3.95	235	99.0	21.7	114
2.50	601		57.3		4.00	229	98.2	20.5	111
2.55	578		56.0		4.05	223	97.3	(18.8)	
2.60	555		54.7	298	4.10	217	96.4	(17.5)	105
2.65	534		53.5	288	4.15	212	95.5	(16.0)	102
2.70	514		52.1	274	4.20	207	94.6	(15.2)	100
2.75	495		51.6	269	4.25	201	93.8	(13.8)	98
2.80	477		50.3	258	4.30	197	92.8	(12.7)	95
2.85	461		48.8	244	4.35	192	91.9	(11.5)	93
2.90	444		47.2	231	4.40	187	90.7	(10.0)	90
2.95	429		45.7	219	4.45	183	90.0	(9.0)	89
3.00	415		44.5	212	4.50	179	89.0	(8.0)	87
3.05	401		43.1	202	4.55	174	87.8	(6.4)	85
3.10	388		41.8	193	4.60	170	86.8	(5.4)	83
3.15	375		40.4	184	4.65	167	86.0	(4.4)	81
3.20	363		39.1	177	4.70	163	85.0	(3.3)	79
3.25	352	(110.0)	37.9	171	4.80	156	82.9	(0.9)	76
3.30	341	(109.0)	36.6	164	4.90	149	80.8		73
3.35	331	(108.5)	35.5	159	5.00	143	78.7		71
3.40	321	(108.0)	34.3	154	5.10	137	76.4		67
3.45	311	(107.5)	33.1	149	5.20	131	74.0		65
3.50	302	(107.0)	32.1	146	5.30	126	72.0		63
3.55	293	(106.0)	30.9	141	5.40	121	69.8		60
3.60	285	(105.5)	29.9	138	5.50	116	67.6		58
3.65	277	(104.5)	28.8	134	5.60	111	65.7		56
3.70	269	(104.0)	27.6	130					

SOURCE: *Modern Steels and Their Properties,* Bethlehem Steel Co., Bethlehem, PA.

NOTE: This is a condensation of Table 2, *Report J417b, SAE 1971 Handbook.* Values in ( ) are beyond normal range and are presented for information only.

*Values above 500 are for tungsten carbide ball; below 500, for standard ball.

Answers to Selected Problems

Given here are the answers to problems for which there are unique solutions. Many of the problems for solution in this book are true design problems, and individual design decisions are required to arrive at the solutions. Others are of the review question form for which the answers are in the text of the associated chapter. It should also be noted that some of the problems require the selection of design factors and the use of data from charts and graphs. Because of the judgment and interpolation required, some of the answers may be slightly different from your solutions.

CHAPTER 1   The Nature of Mechanical Design

15. $D = 44.5$ mm

16. $L = 14.0$ m

17. $T = 1\,418$ N·m

18. $A = 2\,658$ mm^2

19. $Z = 2.43 \times 10^5$ mm^3

20. $I = 3.66 \times 10^7$ mm^4

21. $L\, 2\frac{1}{2} \times 2\frac{1}{2} \times 1/4$

22. $P = 5.59$ kW

23. $s_u = 876$ MPa

24. Weight $= 48.9$ N

25. $T = 20.3$ N·m
    $\theta = 0.611$ rad
    Scale $= 5.14$ lb·in/deg
    Scale $= 33.3$ N·m/rad

26. Energy $= 1.03 \times 10^{11}$ lb·ft/yr
    Energy $= 3.88 \times 10^7$ W·h/yr

27. $\mu = 540$ lb·s/ft^2
    $\mu = 25.9 \times 10^3$ N·s/m^2

28. 4.60×10^9 rev

CHAPTER 2   Materials in Mechanical Design

9. No. The percent elongation must be greater than 5.0 percent to be ductile.

11. $G = 42.9$ GPa

12. Hardness $= 52.8$ HRC

13. Tensile strength $= 235$ Ksi (Approximately)

Questions 14–17 ask what is wrong with the given statements.

14. Annealed steels typically have hardness values in the range from 120 HB to 200 HB. A hardness of 750 HB is extremely hard and characteristic of as-quenched high alloy steels.

15. The HRB scale is normally limited to HRB 100.

16. The HRC hardness is normally no lower than HRC 20.

17. The given relationship between hardness and tensile strength is only valid for steels.

18. Charpy and Izod

19. Iron and carbon. Manganese and other elements are often present.

20. Iron, carbon, manganese, nickel, chromium, molybdenum.

21. Approximately 0.40 percent.

22. Low carbon: Less than 0.30 percent
    Medium carbon: 0.30 to 0.50 percent
    High carbon: 0.50 to 0.95 percent

23. Nominally 1.0 percent.

24. AISI 12L13 steel has lead added to improve machinability.

25. AISI 1045, 4140, 4640, 5150, 6150, 8650.

26. AISI 1045, 4140, 4340, 4640, 5150, 6150, 8650.

27. Wear resistance, strength, ductility. AISI 1080.

28. AISI 5160 OQT 1000 is a chromium steel, having nominally 0.80 percent chromium and 0.60 percent carbon, a high carbon alloy steel. It has fairly high strength and good ductility. It was through-hardened, quenched in oil, and tempered at 1000°F.

29. Yes, with careful specification of the quenching medium. A hardness of HRC 40 is equivalent to HB 375. Appendix 3 indicates that oil quenching would not produce an adequate hardness. However, Appendix 4–1 shows that a hardness of HB 400 could be obtained by quenching in water and tempering in 700°F while still having 20% elongation for good ductility.

CHAPTER 3   Stress Analysis

1. $\sigma = 31.8$ MPa; $\delta = 0.12$ mm

2. $\sigma = 44.6$ MPa

3. $\sigma = 66.7$ MPa

4. $\sigma = 5375$ psi

5. $\sigma = 17\,200$ psi

6. For all materials, $\sigma = 34.7$ MPa
   Deflection:
   a. $\delta = 0.277$ mm   b. $\delta = 0.277$ mm
   c. $\delta = 0.377$ mm   d. $\delta = 0.830$ mm
   e. $\delta = 0.503$ mm   f. $\delta = 23.8$ mm
   g. $\delta = 7.56$ mm.

   Note: The stress is close to the ultimate strength for (f) and (g).

7. Force $= 2\,556$ lb; $\sigma = 2\,506$ psi

8. $\sigma = 595$ psi

9. Force $= 1\,061$ lb

10. $D = 0.274$ in

13. $\sigma_{AD} = \sigma_{DE} = 6\,198$ psi
    $\sigma_{EF} = 7\,748$ psi
    $\sigma_{BD} = 0$ psi
    $\sigma_{BE} = 5\,165$ psi
    $\sigma_{AB} = \sigma_{CE} = -4\,132$ psi
    $\sigma_{BC} = -3\,099$ psi
    $\sigma_{CF} = -5\,165$ psi

15. $\sigma = 144$ MPa

16. At pins A and C: $\tau_A = \tau_C = 7\,958$ psi
    At pin B: $\tau_B = 10\,190$ psi

19. $\tau = 98.8$ MPa

21. $\tau = 547$ MPa

22. $\tau = 32.6$ MPa

23. $\theta = 0.79°$

25. $\tau = 32\,270$ psi

28. $\tau = 70.8$ MPa; $\theta = 1.94°$

30. $T = 9\,624$ lb·in; $\theta = 1.83°$

31. Required section modulus $= 2.40$ in³. Standard nominal sizes given for each shape:
    a. Each side $= 2.50$ in
    c. Width $= 5.00$ in; height $= 1.75$ in
    e. S4 × 7.7
    g. 4-in schedule 40 pipe

32. Weights:
    a. 212 lb   c. 297 lb
    e. 77.0 lb   g. 107.9 lb

33. Maximum deflection   Deflection at loads
    a. 0.701 in   0.572 in
    c. 1.021 in   0.836 in
    e. 0.375 in   0.307 in
    g. 0.315 in   0.258 in

34. $M_A = 330$ N·m; $M_B = 294$ N·m; $M_C = -40$ N·m

36. a. $y_A = 0.238$ in; $y_B = 0.688$ in
    b. $y_A = 0.047$ in; $y_B = 0.042$ in

38. $\sigma = 3\,480$ psi; $\tau = 172$ psi

For Problems 25 through 36, the complete solutions require drawings. Listed below are the maximum bending moments only:

39. 480 lb·in

41. 120 lb·in

43. 93 750 N·mm

45. 8 640 lb·in

47. $-11\,250$ N·mm

49. -1.49 kN·m

51. $\sigma = 62.07$ MPa

53. a. $\sigma = 20.94$ MPa tension on top of lever
    b. At section B, $h = 35.1$ mm; at C, $h = 18$ mm

55. $\sigma = 84.6$ MPa tension

57. Sides $= 0.50$ in

59. Maximum $\sigma = -1.42$ MPa compression on the top surface between A and C

61. $\sigma = 86.2$ MPa

63. Left: $\sigma = 39\,750$ psi
    Middle: $\sigma = 29\,760$ psi
    Right: $\sigma = 31\,700$ psi

65. $\sigma = 96.4$ MPa

67. $\sigma = 32\,850$ psi

CHAPTER 4   Combined Stresses and Mohr's Circle

Answers to Problems 1 to 15:

	Maximum Principal Stress	*Minimum Principal Stress*	*Maximum Shear Stress*
1.	24.14 Ksi	−4.14 Ksi	14.14 Ksi
3.	50.0 Ksi	−50.0 Ksi	50.0 Ksi
5.	124.7 Ksi	0.0 Ksi	62.4 Ksi
7.	20.0 Ksi	−40.0 Ksi	30.0 Ksi
9.	144.3 MPa	−44.3 MPa	94.3 MPa
11.	61.3 MPa	−91.3 MPa	76.3 MPa
13.	168.2 MPa	0.0 MPa	84.1 MPa
15.	250.0 MPa	−80.0 MPa	165.0 MPa

CHAPTER 5   Design for Different Types of Loading

Stress Ratio

1. $\sigma_{max} = 44.6$ MPa
   $\sigma_{min} = 6.37$ MPa
   $\sigma_m = 25.5$ MPa
   $\sigma_a = 19.1$ MPa
   $R = 0.143$

3. $\sigma_{max} = 5375$ psi
   $\sigma_{min} = -750$ psi
   $\sigma_m = 2313$ psi
   $\sigma_a = 3063$ psi
   $R = -0.140$

5. $\sigma_{max} = 110.3$ MPa
   $\sigma_{min} = 50.9$ MPa
   $\sigma_m = 80.6$ MPa
   $\sigma_a = 29.7$ MPa
   $R = 0.462$

7. $\sigma_{max} = 9868$ psi
   $\sigma_{min} = 1645$ psi
   $\sigma_m = 5757$ psi
   $\sigma_a = 4112$ psi
   $R = 0.167$

9. $\sigma_{max} = 950$ MPa
   $\sigma_{min} = 594$ MPa
   $\sigma_m = 772$ MPa
   $\sigma_a = 178$ MPa
   $R = 0.625$

Endurance Strength

11. $s'_n = 230$ MPa

13. $S'_n = 31.2$ Ksi

Design and Analysis

19. $N = 1.38$ (low)

23. $N = 2.94$ at right load. OK

25. $N = 11.6$

27. $N = 9.93$

29. $D = 40$ mm; $a = 10.6$ mm

31. $2\frac{1}{2}$ in pipe

33. $b = 2.00$ in

35. $N = 9.11$

36. $\sigma = 34.7$ MPa

    a. $N = 5.96$ c. $N = 8.93$

    e. $N = 23.8$ g. $N = 1.30$ (low)

37. $N = 3.37$

39. Force $= 1061$ lb; $D = 5/16$ in

41. $N = 1.52$ (low)

43. $N = 4.79$

45. $D = 1.75$ in

49. $N = 7.95$

51. $N = 2.94$

57. $N = 1.72$ (low)

61. Design a

64. $N = 0.24$ at pin holes (Failure)

67. $N = 1.64$ at B (low)

69. $N = 1.44$ (low)

73. $N = 1.23$ (low)

75. $r_{\min} = 0.152$ in

CHAPTER 6   Columns

1. $P_{cr} = 4\,473$ lb

2. $P_{cr} = 14\,373$ lb

5. $P_{cr} = 4\,473$ lb

6. $P_{cr} = 32.8$ lb

8. a. Pinned ends: $P_{cr} = 7\,498$ lb
    b. Fixed ends: $P_{cr} = 12\,000$ lb
    c. Fixed-pinned ends: $P_{cr} = 10\,300$ lb
    d. Fixed-free ends: $P_{cr} = 1\,700$ lb

10. $D = 1.45$ in required. Use $D = 1.50$ in.

14. $S = 1.423$ in required. Use $S = 1.500$ in.

16. Use $D = 1.50$ in.

23. $P = 1\,189$ lb

25. $P = 1\,877$ lb

27. $\sigma = 212$ MPa; $y = 25.7$ mm

28. $\sigma = 6\,685$ psi; $y = 0.045$ in

30. $P = 37\,500$ psi

31. $P_a = 22\,600$ lb

33. $4 \times 4 \times 1/2$ is smallest outside dimension. $I = 12.3$ in^4. Weight $= 21.6$ lb/ft.
    $6 \times 4 \times 1/4$ is the lightest from those listed. $I_{\min} = I_y = 11.7$ in^4. Weight $= 15.6$ lb/ft.

35. $P_a = 11\,750$ lb

37. $P_a = 18\,300$ lb

39. Piston rod is safe.

CHAPTER 7   Belts and Chains

V-belts

1. 3V Belt, 75 inches long

2. $C = 22.00$ in

3. $\theta_1 = 157°$; $\theta_2 = 203°$

10. $v_b = 2\,405$ ft/min

13. $P = 6.05$ hp

Roller Chain

25. No. 80 chain

28. Design power rating $= 12.0$ hp; Type II lubrication (bath)

29. Design power rating $= 30.0$ hp

34. $L = 96$ in, 128 pitches

35. $C = 35.57$ in

CHAPTER 8   Kinematics of Gears

Spur Gears

1. $N = 44$; $P_d = 12$

    a. $D = 3.667$ in   b. $p = 0.261\,8$ in

    c. $m = 2.117$ mm   d. $m = 2.00$ mm

    e. $a = 0.083\,3$ in   f. $b = 0.104\,2$ in

    g. $c = 0.020\,8$ in   h. $h_t = 0.187\,5$ in

    i. $h_k = 0.166\,7$ in   j. $t = 0.131$ in

    k. $D_o = 3.833$ in

3. $N = 45$; $P_d = 2$

    a. $D = 22.500$ in   b. $p = 1.571$ in

    c. $m = 12.70$ mm   d. $m = 12.0$ mm

    e. $a = 0.500\,0$ in   f. $b = 0.625\,0$ in

    g. $c = 0.125\,0$ in   h. $h_t = 1.125\,0$ in

    i. $h_k = 1.000\,0$ in   j. $t = 0.785\,4$ in

    k. $D_o = 23.500$ in

5. $N = 22$; $P_d = 1.75$

    a. $D = 12.571$ in   b. $p = 1.795$ in

    c. $m = 14.514$ mm   d. $m = 16.0$ mm

    e. $a = 0.571\,4$ in   f. $b = 0.714\,3$ in

    g. $c = 0.142\,9$ in   h. $h_t = 1.285\,7$ in

    i. $h_k = 1.142\,9$ in   j. $t = 0.897\,6$ in

    k. $D_o = 13.714$ in

7. $N = 180$; $P_d = 80$

    a. $D = 2.250\,0$ in   b. $p = 0.039\,3$ in

    c. $m = 0.318$ mm   d. $m = 0.30$ mm

    e. $a = 0.012\,5$ in   f. $b = 0.017\,0$ in

    g. $c = 0.004\,5$ in   h. $h_t = 0.029\,5$ in

Data fr
length

CHAPTE

1. $T_B = 34$
   $F_{Bx} = $
   $F_{By} = $

3. $T_B = 6$
   $F_{Bx} = $
   $F_{By} = $

5. $T_D = 3$
   $W_{tD} = $
   $W_{rD} = $
   $F_{Dx} = $
   $F_{Dy} = $

7. $T_C = 6$
   $F_{Cx} = $
   $F_{Cy} = $

9. $T_C = 1$
   $F_{Cx} = $
   $F_{Cy} = $

11. $T_F = 1$
    $W_{tF} = $
    $W_{rF} = $
    $F_{Fx} = $
    $F_{Fy} = $

13. $T_A = $
    $F_{Ax} = $
    $F_{Ay} = $

15. $T_C = $
    $F_C = $
    $F_{Cx} = $
    $F_{Cy} = $

17. $T_C = $
    $F_{Cx} = $
    $F_{Cy} = $

19. $T_C = $
    $F_{Cx} = $
    $F_{Cy} = $

21. $T_E = $
    $F_E = $
    $F_{Ex} = $
    $F_{Ey} = $

22. $T_B = $
    $F_{Bx} = $
    $F_{By} = $
    W_{xB}
    194.$
    W_{xB}
    beari

23. $T_A = $
    F_{Ax}
    F_{Ay}

---

i. $h_k = 0.025\ 0$ in   j. $t = 0.019\ 6$ in
k. $D_o = 2.275\ 0$ in

9. $N = 28$; $P_d = 20$
   a. $D = 1.400\ 0$ in   b. $p = 0.157\ 1$ in
   c. $m = 1.270$ mm   d. $m = 1.25$ mm
   e. $a = 0.050\ 0$ in   f. $b = 0.062\ 0$ in
   g. $c = 0.012\ 0$ in   h. $h_t = 0.112\ 0$ in
   i. $h_k = 0.100\ 0$ in   j. $t = 0.078\ 5$ in
   k. $D_o = 1.500\ 0$ in

11. $N = 45$; $m = 1.25$
    a. $D = 56.250$ mm   b. $p = 3.927$ mm
    c. $P_d = 20.3$ d. $P_d = 20$
    e. $a = 1.25$ mm   f. $b = 1.563$ mm
    g. $c = 0.313$ mm   h. $h_t = 2.813$ mm
    i. $h_k = 2.500$ mm   j. $t = 1.963$ mm
    k. $D_o = 58.750$ mm

13. $N = 22$; $m = 20$
    a. $D = 440.00$ mm   b. $p = 62.83$ mm
    c. $P_d = 1.270$ d. $P_d = 1.25$
    e. $a = 20.0$ mm   f. $b = 25.00$ mm
    g. $c = 5.000$ mm   h. $h_t = 45.00$ mm
    i. $h_k = 40.00$ mm   j. $t = 31.42$ mm
    k. $D_o = 480.00$ mm

15. $N = 180$; $m = 0.4$
    a. $D = 72.00$ mm   b. $p = 1.26$ mm
    c. $P_d = 63.5$ d. $P_d = 64$
    e. $a = 0.40$ mm   f. $b = 0.500$ mm
    g. $c = 0.100$ mm   h. $h_t = 0.90$ mm
    i. $h_k = 0.80$ mm   j. $t = 0.628$ mm
    k. $D_o = 72.80$ mm

17. $N = 28$; $m = 0.8$
    a. $D = 22.40$ mm   b. $p = 2.51$ mm
    c. $P_d = 31.75$ d. $P_d = 32$
    e. $a = 0.80$ mm   f. $b = 1.000$ mm
    g. $c = 0.200$ mm   h. $h_t = 1.800$ mm
    i. $h_k = 1.60$ mm   j. $t = 1.257$ mm
    k. $D_o = 24.00$ mm

19. Problem 1: $P_d = 12$; backlash $= 0.006$ to 0.009 in
    Problem 12: $m = 12$; backlash $= 0.52$ to 0.82 mm

21. a. $C = 14.000$ in   b. $VR = 4.600$
    c. $n_G = 48.9$ rpm   d. $v_t = 294.5$ ft/min

23. a. $C = 2.266$ in   b. $VR = 6.25$
    c. $n_G = 552$ rpm   d. $v_t = 565$ ft/min

25. a. $C = 90.00$ mm   b. $VR = 3.091$
    c. $n_G = 566$ rpm   d. $v_t = 4.03$ m/s

27. a. $C = 162.0$ mm   b. $VR = 1.250$
    c. $n_G = 120$ rpm   d. $v_t = 1.13$ m/s

For problems 29 through 32, the following are errors in the given statements:

29. The pinion and the gear cannot have different pitches.

30. The actual center distance should be 8.333 in.

---

31. There are too few teeth in the pinion; interference is to be expected.

32. The actual center distance should be 2.156 in. Apparently, the outside diameters were used instead of the pitch diameters to compute C.

33. $Y = 8.45$ in; $X = 10.70$ in

35. $Y = 44.00$ mm; $X = 58.40$ mm

37. Output speed $= 111$ rpm CCW

39. Output speed $= 144$ rpm CW

Helical Gears

41. $p = 0.392\ 7$ in $p_n = 0.340\ 1$ in
    $P_{nd} = 9.238$ $P_x = 0.680$ in
    $D = 5.625$ in $\phi_n = 12.62°$
    $F/P_x = 2.94$ axial pitches in the face width

42. $P_d = 8.485$ $p = 0.370$ in
    $p_c = 0.261\ 8$ in $P_x = 0.370$ in
    $\phi_t = 27.2°$ $D = 5.657$ in
    $F/P_x = 4.05$ axial pitches in the face width

Bevel Gears

45. $\gamma = 18.43°$ $\Gamma = 71.57°$
    $h_t = 0.367$ in $h_k = 0.333$ in
    $c = 0.033\ 3$ in $a_G = 0.098\ 5$ in
    $a_p = 0.234\ 8$ in $D_o = 7.562$ in
    $d_o = 2.945\ 5$ in $A_o = 3.953$ in

47. $\gamma = 26.57°$ $\Gamma = 63.43°$
    $h_t = 0.2208$ in $h_k = 0.2000$ in
    $c = 0.0208$ in $a_G = 0.0655$ in
    $a_p = 0.1345$ in $D_o = 5.0586$ in
    $d_o = 2.7406$ in $A_o = 2.7951$ in

Wormgearing

52. $L = 0.314\ 2$ in $\lambda = 4.57°$
    $a = 0.100$ in $b = 0.1157$ in
    $D_{oW} = 1.450$ in $D_{RW} = 1.018\ 6$ in
    $D_G = 4.000$ in $C = 2.625$ in
    $VR = 40$

59. 0.4067 rpm

61. 0.5074 rpm

63. $N_P = 22$, $N_G = 38$

65. $N_P = 19$, $N_G = 141$

67. One possible solution: Double reduction; equal ratio;
    $N_P = 13$, $N_G = 119$, $n_{out} = 21.48$ rpm

69. One possible solution: Triple reduction with an idler
    $N_{P1} = 18$, $N_{G1} = 126$, $N_{P2} = 18$, $N_{G2} = 108$, $N_{P3} = 18$,
    $N_{G3} = 135$, $N_{idler} = 18$
    $n_{out} = 13.33$ rpm

CHAPTER 9 Spur Gear Design

1.
   a) $n_G = 486.1$ rpm
   b) $VR = m_G = 3.600$

c)
d)
e)
f)
g)
h)
i)

3.

a)
b)
c)
d)
e)
f)
g)
h)
i)

5.

a
b
c
d
e
f
g
h
i

8.

9.

11.

15.

26.

27.

33.

34.

35.

The
prob

37.

19. Life = 48 900 hours

21. Life = 25 300 hours

23. Life = 47 300 hours

CHAPTER 15   No practice problems

CHAPTER 16   Plain Surface Bearings

All problems in this chapter are design problems for which no unique solutions exist.

CHAPTER 17   Power Screws and Ball Screws

5. 2 1/2–3 Acme thread

6. $L > 1.23$ in

7. $T = 6\,974$ lb·in

8. $T = 3\,712$ lb·in

11. Lead angle = 4.72°; self-locking

12. Efficiency = 35%

13. $n = 180$ rpm; $P = 0.866$ hp

17. 24.7 years

CHAPTER 18   Fasteners

4. Grade 2 bolts: 5/16–18; $T = 70.3$ lb·in

5. $F = 1\,190$ lb

6. $F = 4.23$ kN

7. Nearest metric thread is M24 × 2. Metric thread is 1.8 mm larger (8% larger).

8. Closest standard thread is M5 × 0.8. (#10–32 is also close.)

9. 3.61 kN

10. a. 1 177 lb   c. 2 385 lb
    e. 2 862 lb   g. 1 081 lb
    i. 2 067 lb   k. 143 lb

CHAPTER 19   Springs

1. $k = 13.3$ lb/in

2. $L_f = 1.497$ in

3. $F_s = 47.8$ lb; $L_f = 1.25$ in

7. $ID = 0.93$ in; $D_m = 1.015$ in; $C = 11.94$; $N = 6.6$ coils

8. $C = 8.49$; $p = 0.241$ in; pitch angle = 8.70°; $L_s = 1.12$ in

9. $F_o = 10.25$ lb; stress = 74 500 psi

11. $OD = 0.583$ in when at solid length

12. $F_s = 26.05$ lb; stress = 189 300 psi (High)

31. Bending stress = 114 000 psi; torsion stress = 62 600 psi. Stresses are safe.

35. Torque = 0.91 lb·in to rotate spring 180°. Stress = 184 800 psi; OK for severe service

CHAPTER 20   Machine Frames, Bolted Connections and Welded Joints

Problems 1–16 are design problems for which there are no unique solutions.

17.

Material	Diameter (in)	Weight (lb per inch of length)
a. 1020 HR steel	0.638	0.0906
c. Aluminum 2014–T6	0.451	0.0160
e. Ti–6A1–4V (Aged)	0.285	0.0102

CHAPTER 21   Electric Motors

13. 480V, 3-Phase because the current would be lower and the motor size smaller

16. $n_s = 1\,800$ rpm in the United States
    $n_s = 1\,500$ rpm in France

17. 2-Pole motor; $n = 3\,600$ rpm at zero load (approximate)

18. $n_s = 12\,000$ rpm

19. 1 725 rpm and 1 140 rpm

20. Variable frequency control

34. a. Single phase, split phase AC motor
    b. $T = 41.4$ lb·in   c. $T = 61.7$ lb·in
    d. $T = 144$ lb·in

35. b. $T = 12.6$ N·m   c. $T = 18.9$ N·m
    d. $T = 44.1$ N·m

39. Full load speed = synchronous speed = 720 rpm

47. Use a NEMA Type K SCR control to convert 115 VAC to 90 VDC; use a 90 VDC motor.

51. Speed theoretically increases to infinity

52. $T = 20.5$ N·m

54. NEMA 2 starter

55. NEMA 1 starter

CHAPTER 22   Motion Control: Clutches and Brakes

1. $T = 495$ lb·in

3. $T = 41$ lb·in

5. Data from Problem 1: $T = 180$ lb·in
   Data from Problem 3: $T = 27.4$ lb·in

7. Clutch: $T = 2\,122$ N·m
   Brake: $T = 531$ N·m

8. $T = 143$ lb·ft

9. $T = 63.6$ lb·ft

11. $T = 223.6$ lb·ft

15. $F_a = 109$ lb

17. $W = 138$ lb

18. $b > 16.0$ in

Index